Chemistry and the Living Organism

Table of Atomic Masses and Numbers

Based on the 1989 Report of the Commission on Atomic Weights and Isotopic Abundances of the International Union of Pure and Applied Chemistry and for the elements as they exist naturally on earth. Scaled to the relative atomic mass of carbon-12. The estimated uncertainties in values, between ±1 and ±9 units in the last digit of an atomic mass, are in parentheses after the atomic mass. (From *Journal of Physical and Chemical Reference Data,* Vol. 20 (1991), pp. 1313–1325. Copyright © 1991 IUPAC.)

Element	*Symbol*	*Atomic number*	*Atomic mass*	
Actinium	Ac	89	227.0278	(L)
Aluminum	Al	13	26.981539(5)	
Americium	Am	95	243.0614	(L)
Antimony	Sb	51	121.757(3)	
Argon	Ar	18	39.948(1)	(g, r)
Arsenic	As	33	74.92159(2)	
Astatine	At	85	209.9871	(L)
Barium	Ba	56	137.327(7)	
Berkelium	Bk	97	247.0703	(L)
Beryllium	Be	4	9.012182(3)	
Bismuth	Bi	83	208.98037(3)	
Boron	B	5	10.811(5)	(g, m, r)
Bromine	Br	35	79.904(1)	
Cadmium	Cd	48	112.411(8)	(g)
Calcium	Ca	20	40.078(4)	(g)
Californium	Cf	98	251.0796	(L)
Carbon	C	6	12.011(1)	(r)
Cerium	Ce	58	140.115(4)	(g)
Cesium	Cs	55	132.90543(5)	
Chlorine	Cl	17	35.4527(9)	(m)
Chromium	Cr	24	51.9961(6)	
Cobalt	Co	27	58.93320(1)	
Copper	Cu	29	63.546(3)	(r)
Curium	Cm	96	247.0703	(L)
Dysprosium	Dy	66	162.50(3)	(g)
Einsteinium	Es	99	252.083	(L)
Erbium	Er	68	167.26(3)	(g)
Europium	Eu	63	151.965(9)	(g)
Fermium	Fm	100	257.0951	(L)
Fluorine	F	9	18.9984032(9)	
Francium	Fr	87	223.0197	(L)
Gadolinium	Gd	64	157.25(3)	(g)
Gallium	Ga	31	69.723(4)	
Germanium	Ge	32	72.61(3)	
Gold	Au	79	196.96654(3)	
Hafnium	Hf	72	178.49(2)	
Helium	He	2	4.002602(2)	(g, r)
Holmium	Ho	67	164.93032(3)	
Hydrogen	H	1	1.00794(7)	(g, m, r)
Indium	In	49	114.82(1)	
Iodine	I	53	126.90447(3)	
Iridium	Ir	77	192.22(3)	
Iron	Fe	26	55.847(3)	
Krypton	Kr	36	83.80(1)	(g, m)
Lanthanum	La	57	138.9055(2)	(g)
Lawrencium	Lr	103	262.11	(L)
Lead	Pb	82	207.2(1)	(g, r)
Lithium	Li	3	6.941(2)	(g, m, r)
Lutetium	Lu	71	174.967(1)	(g)
Magnesium	Mg	12	24.3050(6)	
Manganese	Mn	25	54.93805(1)	
Mendelevium	Md	101	258.10	(L)
Mercury	Hg	80	200.59(2)	
Molybdenum	Mo	42	95.94(1)	(g)
Neodymium	Nd	60	144.24(3)	(g)
Neon	Ne	10	20.1797(6)	(g, m)
Neptunium	Np	93	237.0482	(L)
Nickel	Ni	28	58.6934(2)	
Niobium	Nb	41	92.90638(2)	
Nitrogen	N	7	14.00674(7)	(g, r)
Nobelium	No	102	259.1009	(L)
Osmium	Os	76	190.2(1)	(g)
Oxygen	O	8	15.9994(3)	(g, r)
Palladium	Pd	46	106.42(1)	(g)
Phosphorus	P	15	30.973762(4)	
Platinum	Pt	78	195.08(3)	
Plutonium	Pu	94	244.0642	(L)
Polonium	Po	84	208.9824	(L)
Potassium	K	19	39.0983(1)	
Praseodymium	Pr	59	140.90765(3)	
Promethium	Pm	61	144.9127	(L)
Protactinium	Pa	91	231.03588(2)	
Radium	Ra	88	226.0254	(L)
Radon	Rn	86	222.0176	(L)
Rhenium	Re	75	186.207(1)	
Rhodium	Rh	45	102.90550(3)	
Rubidium	Rb	37	85.4678(3)	(g)
Ruthenium	Ru	44	101.07(2)	(g)
Samarium	Sm	62	150.36(3)	(g)
Scandium	Sc	21	44.955910(9)	
Selenium	Se	34	78.96(3)	
Silicon	Si	14	28.0855(3)	(r)
Silver	Ag	47	107.8682(2)	(g)
Sodium	Na	11	22.989768(6)	
Strontium	Sr	38	87.62(1)	(g, r)
Sulfur	S	16	32.066(6)	(g, r)
Tantalum	Ta	73	180.9479(1)	
Technetium	Tc	43	98.9072	(L)
Tellurium	Te	52	127.60(3)	(g)
Terbium	Tb	65	158.92534(3)	
Thallium	Tl	81	204.3833(2)	
Thorium	Th	90	232.0381(1)	(g)
Thulium	Tm	69	168.93421(3)	
Tin	Sn	50	118.7610(7)	(g)
Titanium	Ti	22	47.88(3)	
Tungsten	W	74	183.85(3)	
Unnilennium	Une	109	(266)	(L, n, s)
Unnilhexium	Unh	106	263.118	(L, n)
Unniloctium	Uno	108	(265)	(L, n, s)
Unnilpentium	Unp	105	262.114	(L, n)
Unnilquadium	Unq	104	261.11	(L, n)
Unnilseptium	Uns	107	262.12	(L, n)
Uranium	U	92	238.0289(1)	(g, m)
Vanadium	V	23	50.9415(1)	
Xenon	Xe	54	131.29(2)	(g, m)
Ytterbium	Yb	70	173.04(3)	(g)
Yttrium	Y	39	88.90585(2)	
Zinc	Zn	30	65.39(2)	
Zirconium	Zr	40	91.224(2)	(g)

(g) Geologically exceptional specimens of this element are known that have different isotopic compositions. For such samples, the atomic mass given here may not apply as precisely as indicated.

(L) The atomic mass is the relative mass of the isotope of longest half-life. The element has no stable isotopes.

(m) Modified isotopic compositions can occur in commercially available materials that have been processed in undisclosed ways, and the atomic mass given here might be quite different for such samples.

(n) Name and symbol are assigned according to systematic rules developed by the IUPAC.

(r) Ranges in isotopic compositions of normal samples obtained on earth do not permit a more precise atomic mass for the element, but the tabulated value should apply to any normal sample of the element.

(s) Element was not listed in the 1989 report but has been added here.

Chemistry and the Living Organism

Sixth Edition

Molly M. Bloomfield *Oregon State University*

Lawrence J. Stephens *Elmira College*

JOHN WILEY & SONS, INC.

New York • Chichester • Brisbane • Toronto • Singapore

ACQUISITIONS EDITOR Nedah Rose
DEVELOPMENTAL EDITOR Joan Kalkut
MARKETING MANAGER Catherine Faduska
OUTSIDE PRODUCTION York Production Services
COVER AND TEXT DESIGNER Karin Gerdes Kincheloe
MANUFACTURING MANAGER Mark Cirillo
PHOTO EDITOR Lisa Passmore
ILLUSTRATION COORDINATOR Jaime Perea
COVER ILLUSTRATION Carlyn Iverson
COVER PHOTOS (clockwise from top):
Scott Camazine/Photo Researchers; Michael Watson;
Leonard Lessin/Peter Arnold, Inc.

This book was set in Times New Roman by York Graphic Services and printed and bound by von Hoffman Press. The cover was printed by Phoenix Color.

Recognizing the importance of preserving what has been written, it is a policy of John Wiley & Sons, Inc. to have books of enduring value published in the United States printed on acid-free paper, and we exert our best efforts to that end.

The paper in this book was manufactured by a mill whose forest management programs include sustained yield harvesting of its timberlands. Sustained yield harvesting principles ensure that the number of trees cut each year does not exceed the amount of new growth.

Library of Congress Cataloging-in-Publication Data
Bloomfield, Molly M., 1944–
Chemistry and the living organism / by Molly Bloomfield and Lawrence Stephens.—6th ed.
p. cm.
Includes index.
ISBN 0-471-10777-8 (cloth : alk. paper)
1. Chemistry. 2. Biochemistry. I. Stephens, Lawrence. II. Title.
QD33.B672 1996
540—dc20 95-32172
CIP

ISBN 0-471-10777-8

Printed in the United States of America

10 9 8 7 6 5 4 3 2

Preface to the Sixth Edition

Chemistry and the Living Organism is written for a one- or two-term survey course for students in the allied health sciences and related fields. Our goal has always been to introduce students to general, organic, and biochemistry in a manner that is easy to understand and enjoyable to read. The fundamental concepts of chemistry, therefore, are presented through examples relevant to the students' own lives.

This sixth edition updates the content and format of the previous editions to make the textbook even more interesting to students and more useful to instructors. For example, in addition to its section on *Why Study Chemistry?,* the expanded introduction *To the Student* now also includes a discussion on how to study chemistry and how most effectively to use this textbook.

To further motivate students in their study of chemistry, several of the successful features of past editions have been revised and expanded. In addition, new features have been added to provide students and instructors with increased teaching and learning opportunities.

New Features

Continuing Theme The main theme that ties together all sections of the textbook is the relationship between basic chemical concepts and the chemistry of living organisms. To emphasize this relationship, the book is divided into three sections, with introductory passages motivating the study of each section. The first section starts with a discussion of the question, ''What makes something alive?'' The second opens by discussing the molecular structure of living organisms. The final section opens with a description of the cell as the simplest structure containing all the components of life.

Focus Story As always, each chapter opens with a story vividly demonstrating the application of that chapter's concepts to issues that directly concern students. Among the topical issues highlighted in the focus stories are alcoholism, radiation therapy, emergency room medicine, and the effects of food chemicals on the brain. Throughout each chapter we continue this strategy of illustrating fundamental chemical concepts through examples relevant to the students' own lives.

Perspectives Each chapter contains a number of *Perspectives,* which are short essays emphasizing the relevance of chemical principles to the students' personal and professional lives. Many new *Perspectives* have been added to this new edition. To further guide the student, the focus on each *Perspective* is now identified as historical, environmental, chemical, or health-related.

Your Perspective In today's increasingly complex world, the necessity and value of communication skills is becoming obvious. For those colleges pursuing programs of "writing across the curriculum," and for instructors who want to give their students practice in communicating the concepts they have learned, this edition of the textbook offers a new feature called *Your Perspective.* Appearing in the form of a question following selected *Perspectives,* this feature is designed to give students practice in verbal skills related to chemistry. Although intended to facilitate short written assignments, these questions may alternatively be used by instructors as topics for classroom or small-group discussion. The questions challenge students to organize their thoughts and to present them in some of the standard written formats they will commonly encounter: correspondence, memos, letters to the editor, persuasive essays, scripts, advertising copy, and so on.

Learning Objectives To help students understand why various topics are being presented, *Learning Objectives* now appear in the page margins accompanying discussion of the relevant concept. This is intended to aid the student as the chapter is being studied and to serve as a guideline for review prior to examinations.

STEP Problem Solving Strategy This important new feature provides students with a consistent method for solving all problems in the textbook. The four-step problem-solving strategy is introduced to students in the opening section of the book, and is then used consistently throughout the text to help students solve even the most challenging problems.

Check Your Understanding Each time the STEP problem-solving strategy is used, students are then asked to apply the STEP method to additional problems in a section called *Check Your Understanding.* The intent of these problems is to immediately reinforce the use of the STEP method and to demonstrate to students that they are now able to solve these problems on their own. To give students immediate feedback on their progress, we have included the answers to all of the *Check Your Understanding* questions in Appendix 3.

Full-color Art Work To clarify many of the technical diagrams and to provide greater reading enjoyment for students, this edition features full-color art work. We have worked closely with photo researchers and illustrators to develop an art program that complements the text and enhances student learning.

Chapter Summaries Each chapter concludes with a summary of *Key Concepts* introduced in the chapter. When applicable, we also present a review of the *Important Equations* covered in the chapter.

Problem Sets In this edition we have expanded the problem sets at the end of each chapter. They begin with *Review Problems* organized by section. These are drill problems that test a student's basic knowledge of the concepts presented in that section. The *Review Problems* are followed by *Applied Problems* that relate the concepts to specific situations. Finally, selected chapters also contain a set of *Integrated Problems* that challenge the student to relate and apply multiple concepts from previous chapters.

Other Learning Aids The first two *Appendices* of the textbook provide assistance to students who may lack some of the required mathematical skills. The third *Appendix* contains the answers to the *Check Your Understanding* problems and the odd-numbered problems at the end of each chapter. Finally, this edition features an extensive *Glossary* and *Index.*

Organization of the Book

The sixth edition combines some previous chapters and eliminates other subject matter, but it continues to include more topics than can be covered in one term. This strategy provides instructors with flexibility to choose those topics that best meet the needs of their students. In this edition, an explanation of the three states of matter is found in Chapter 3 before the discussion of atomic structure (Chapter 4). The material dealing with radioactivity has been shortened from the fifth edition and is now contained in Chapter 7. The organic chemistry section of the text has been shortened and completely rewritten as Chapters 13, 14, and 15. New in this edition is Chapter 21, discussing the body fluids and related topics.

Supplemental Materials

The following supplemental materials are available for use with the sixth edition:

Study Guide Prepared by Lawrence Stephens, Elmira College. Each chapter of the *Study Guide* begins with a summary and discussion of the important concepts covered in the chapter. A list of important terms is included for each section, as are numerous sample problems together with their solutions. Students will find many self-tests that enable them to assess their comprehension of the concepts presented. The *Study Guide* includes the answers to all of the even-numbered problems in the textbook.

Laboratory Experiments Prepared by David Macaulay, William Rainey Harper College. The popular laboratory manual by Joseph Bauer and Molly Bloomfield has been revised in this new edition by David Macaulay. The experiments continue to emphasize the application of chemical principles to biological systems. New to this edition is the use throughout of the National Fire Protection Association's (NFPA) classifications for health, flammability, and reactivity hazards. Each experiment also features recommendations for the disposal of wastes.

Teacher's Manual Prepared by David Macaulay. The *Teacher's Manual* contains answers to the laboratory exercises in the *Laboratory Experiments.* In addition, it contains lists of chemicals and equipment needed for the laboratory experiments.

Transparencies Over 100 four-color illustrations from the text are provided in a form suitable for projection in the classroom.

Test Bank Prepared by Catherine MacGowan, Armstrong State College. The Test Bank contains approximately 1,000 text items consisting of multiple-choice, short answer, and critical thinking questions.

Computerized Test Bank IBM and Macintosh versions of the entire Test Bank are available with full editing features to help you customize tests.

Acknowledgments

For the sixth edition of *Chemistry and the Living Organism,* Lawrence Stephens of Elmira College has joined with Molly Bloomfield in the revision of this popular textbook. Dr. Stephens has taught chemistry for 26 years, using each of the first five editions of the book for his courses. This new collaboration has been further strengthened by the contributions and suggestions of many professionals in both academic and applied fields. In particular, we would like to thank:

Professors Brian Dodd, Michael Mix, Dan Selivonchick, and Rosemary Wander of Oregon State University;

Professor William Lindsay, John Janikas, and Mary Draht, Elmira College;

Dr. Michael Huntington, Dr. William Lloyd, and Jim Randall of Good Samaritan Hospital, Corvallis, Oregon;

Allen Lefohn, A.S.L. & Associates, Helena, Montana;

Dr. Charles Kuttner, Albany, Oregon

We are pleased to give special thanks to Dr. Warren Sparks, whose scientific and medical knowledge is truly amazing. Dr. Sparks was extraordinarily generous in sharing his insights with us during the precious free time he was able to take from his medical practice.

The continual improvement and effectiveness that has marked the five editions of this textbook reflect the active participation of many colleagues throughout the country. Their careful review and innovative ideas have played a central role in shaping the content and format of this sixth edition. In particular we would like to thank:

Betty Klapper
Columbus State Community College
Gerald Weatherby
Oklahoma City University
Loretta Dorn
Fort Hayes State University
Robert Kolodny
Armstrong State College
John Jefferson
Luther College
Todd Tippets
College of Mt. St. Vincent
Mark Benvenuto
University of Detroit-Mercy
Catherine MacGowen
Armstrong State College
Mary Rekow
Jackson Community College
Steve Samual
SUNY-Old Westbury
Patricia Lorenz
Penn Valley Community College

At John Wiley, we would like especially to thank our editor, Joan Kalkut, Stella Kupferberg, Photo Research Director, and Karin Gerdes Kinchloe, designer, all of whom were willing to commit the planning time necessary for the smooth production of an outstanding product. We would also like to thank Lisa Passmore for locating the excellent color photographs that appear in this edition. Lori Stambaugh of York Production Services did an excellent job of coordinating the diverse contributors to a project as complex as this. Her diligence to detail kept the project on schedule.

The successful completion of any textbook can occur only with the support and patience of our families. Special thanks go to Molly's husband, Stefan, whose computer expertise and editing skills were indispensable to this project; to her college-age son, Jon, for his student-oriented contributions to the *Your Perspective* questions; to her daughter, Rebecca, for her unfailing support; and to Larry's wife, Terry, for her valuable proofreading assistance.

Molly Bloomfield
Oregon State University
Corvallis, Oregon

Lawrence Stephens
Elmira College
Elmira, New York

To the Student

Why Study Chemistry?

Chemistry is the study of the composition and interaction of substances. It tells us what substances are made of and helps us understand how they behave.

This may sound like a pretty broad definition for such a specialized field of study, but it is a good indication of just how thoroughly chemistry is involved in our lives. For example, you drink water from your faucet at home without a second thought because chemicals have been added to the water to make it safe. You seldom need to use an iron, thanks to the development of chemicals that give your clothes a permanent press. Just picture your daily routine: you wake up in the morning under sheets made of synthetic fibers that were chemically produced in a factory, or sheets made of cotton fibers that were created through chemical reactions in the blossoms of a cotton plant. You put on clothes made largely of synthetic materials, brush your teeth with toothpaste containing fluoride, and eat a breakfast cereal fortified with minerals and synthetic vitamins. You may drive to school in a car, which is powered by energy released through chemical reactions in the engine. Or perhaps you pedal a bicycle, which is powered by energy released through chemical reactions in your muscles. And now you're reading this textbook, whose paper was created through a chemical process and whose ink is a blend of chemicals. Chemistry, whether the synthetic chemistry of the test tube or the chemistry that makes up all nature, truly involves every aspect of your life.

Chemistry affects each of our lives also in very personal ways. For example, chemicals control your physical appearance. Chemical substances called hormones help determine your height, your weight, your build, and your sexual characteristics. Your good health depends on chemicals that preserve the food you eat, chemicals that protect you from disease, and chemicals (in the form of food) that supply your body with the nutrients it needs to function properly. Chemicals influence your behavior and your emotions. Much of your memory is chemical; your thoughts and experiences may be stored in your brain in the form of chemical compounds. That is why a basic knowledge of chemistry can help you become more aware of your total self and the way in which you interact with your environment.

This textbook will help you acquire that basic knowledge of the principles of chemistry. Each chapter begins with a focus story illustrating how the principles of chemistry in that chapter relate to your personal life or professional career. We carefully define new vocabulary terms whenever they are introduced and provide examples in each chapter to take you step by step through the mathematical and chemical skills you need to learn. When you have finished this course, you will have studied the fields of inorganic chemistry, organic chemistry, and biochemistry. You will learn that chemistry is an exciting science in which new discoveries are constantly being made. The chemistry that you learn in this course will help you better understand much that is happening around you every day.

How to Study Chemistry

Learning is an **active** pursuit. Research shows that student achievement increases greatly with the active participation of the learner. You can't just read this textbook, or passively listen to your professor, and expect to master the concepts of chemistry. So how can you become an active learner?

One of the keys to success in studying chemistry (or any other subject) is good time management. Putting off your chemistry reading or your assigned problems to the end of the day can leave you too tired to understand the material. Not keeping up with the weekly assignments can leave you with a huge amount of reading and cramming to do before an exam. In either case you will have a hard time getting the grades that you would like. Careful time management may seem restrictive, but it actually gives you a chance to spend your most valuable resource (your own time) in the way that you choose. You will not only be able to finish the things you *have* to do but will also have more time for the things you *want* to do. Buying a daily planner and using it consistently can help you reach your goal of successfully mastering this subject.

You will need to schedule at least two hours of study time for every hour you spend in your chemistry lecture. Try to work during the part of the day when you feel the sharpest. Procrastination can lead to disaster when you repeatedly fall asleep reading a critical assignment. It also helps to use a regular study area. The library is an ideal place because it is designed for learning and is away from the distractions of your home or your dorm room.

Many habits can help you in class. Read the assignments before you go to class so that you are familiar with the topics being covered. Write down what the instructor puts on the board or the overhead projector because these will be key concepts, terms, and examples. Actively ask questions. One helpful note-taking technique is to write only on the right side of your notebook during the lecture. You can then use the left side for information from your text reading, topics for your study group, or questions to ask your instructor. Read actively. If you can, highlight and underline important points in the text. Make notes in the margins to mark key concepts, things to review, or questions you want to ask your study partners or your instructor.

Form a study group. You can learn so much more when you study with other people. Work with your study group on the *Applied Problems* and *Integrated Problems* in this textbook. The very best way to learn a concept is to teach it to another person. In your study group, compare your class notes to help you better understand the ideas and principles presented in lecture. Review for exams by testing each other and making up questions that you think might appear on the exam. The most effective study groups meet regularly throughout the term, not just the night before an exam.

Ideas about how to become a successful student are found in many different books on study techniques and are often presented in classes at your college study center. One particularly helpful book we recommend is *Becoming a Master Student* by Dave Ellis, Houghton Mifflin Company, 1994.

Special Features of this Textbook

We have included many special features in this sixth edition to help you become a successful student of chemistry. One hint in *Becoming a Master Student* is to do a textbook "reconnaissance" before you begin a course and before you start each reading assignment. This gives you the big picture of the course or of that week's assignment. Our brains work best going from the general to the specific. Getting the overall picture helps you better to recall and understand the details later on. You might start by reading the introduction and the preface of this textbook. Look at the photos, the figures, and tables. Read the chapter summaries and the focus stories that begin each chapter. These give you the big picture of this course. You will gain an understanding of where you'll be going on your chemical journey and maybe even become excited about the things you'll be learning.

To help you in your reconnaissance of this textbook, let's look at some of its special features:

Focus Stories These stories, which appear at the beginning of each chapter, are based on real-life occurrences. They show you ways in which the chemical principles in the chapter relate to your personal life or professional career. Reading them all now will show you the different areas you will be studying in this course. Read them again as you begin an in-depth study of each chapter.

Learning Objectives It's often difficult to pick out the key concepts in a chapter filled with lots of information and new terms. In the margins of each chapter you will find *Learning Objectives* that tell you what you should be able to do by the time you finish your study of that chapter.

Boldfaced Terms Studying chemistry is much like studying a foreign language. First you have to master the vocabulary. We help you do this by highlighting important terms in boldface. Each new term is defined when it is first used. The textbook also contains a complete glossary to help you review these definitions.

Problem-Solving Strategy One thing that can make students nervous about studying chemistry is the math and problem solving that are part of the course. To overcome this anxiety, it helps to have a strategy or framework for solving mathematical problems. One such strategy is the four-step problem-solving method we use in this textbook. The strategy is called *STEP* because it asks you to

See the Question: Analyze the problem to determine exactly what you are being asked.

Think It Through: Using the information given in the problem and other facts that you know, choose a method for solving the problem.

Execute the Math: Do the mathematical calculations necessary to arrive at an answer. (For problems that don't involve mathematical calculations, this step is often combined with "Think It Through.")

Prepare the Answer: Check your answer to be sure it has the required units of measure and the correct number of significant figures.

After each set of examples are problems called *Check Your Understanding.* Use these problems to see if you understand the solutions presented in the examples. The answers to these problems are in Appendix 3.

Key Concepts The *Key Concepts* listed at the end of each chapter summarize the most important topics in the chapter. You might want to read the *Key Concepts* before beginning your study of a chapter to get an overview of what will be covered. The *Key Concepts* also provide a good review for exams and a framework to develop questions for your study group.

Chapter Problems Problems at the end of the chapters have been divided into three types.

1. Review Problems are arranged by chapter sections. You should try them by yourself to check your understanding of each topic. You will find the answers to the odd-numbered *Review Problems* in Appendix 3.

2. Applied Problems might also be called story problems. For each chapter, these problems ask you to apply the concepts you have learned to specific real-world situations. These problems are excellent for use in your study groups. The answers to the odd-numbered *Applied Problems* are found in Appendix 3.

3. Integrated Problems are sets of problems that cover groups of chapters. These problems are often rather complicated. They ask you to integrate concepts from several chapters and usually require more work than the Applied Problems. Your instructor may assign some of these problems for extra credit.

Perspectives These are short discussions of historical, chemical, environmental, or health topics that will help you understand chemistry in the real world. You might want to read these before you begin your study of the chapter, and again as you are studying the chapter. Some *Perspectives* end with a question that can serve as a topic for class discussion or as the prompt for a short reaction paper.

Appendices The first two *Appendices* contain a math review to help you solve the problems in Chapters 1, 3, and 11. The first appendix deals with numbers in exponential form, and the second with significant figures. If you need more help with these types of problems, you might want to purchase a copy of the *Student Study Guide.* The last Appendix contains answers to problems in the text. Appendix 3 provides the answers to the *Check Your Understanding* problems and the odd-numbered *Review, Applied* and *Integrated Problems.*

Glossary The *Glossary* contains definitions of all the terms printed in boldface in the text and all the medical terms that are used. In reading a chapter you might come across a term that was used in an earlier chapter. If you can't remember what the term means, just turn to the *Glossary* to refresh your memory.

Index The *Index* at the back of the textbook can be a very helpful tool when reviewing for exams or working on the *Integrated Problems.* By looking up a term or concept in the index you can find all of the places that it was mentioned in the text. This might allow you to find just the answer you needed to a perplexing question or problem.

Student Study Guide A student study guide is available to accompany the sixth edition. The study guide contains a brief summary of each section of the textbook and a list of important terms appearing in the section. The study guide features many worked-out examples and self-tests (with answers) to help you master these new concepts and to review for exams. It also contains the answers to the even-numbered problems in the textbook.

Brief Table of Contents

Table of Contents

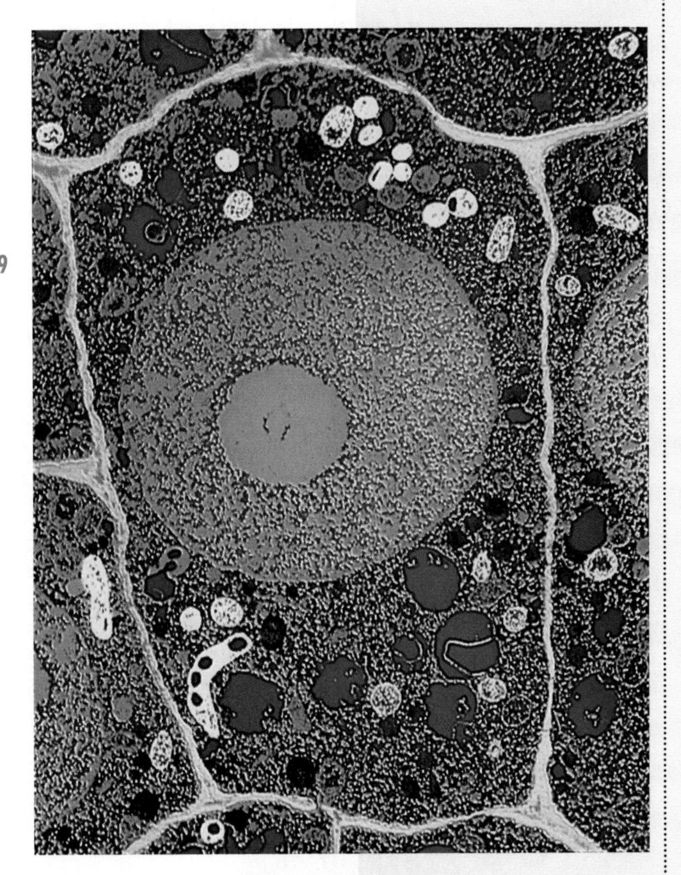

USA

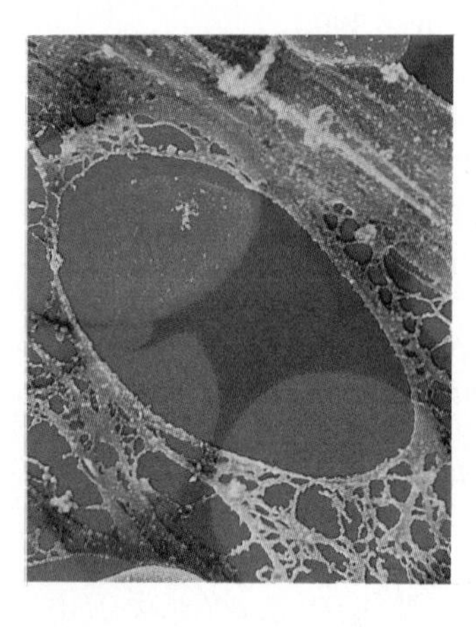

Section 1

The Living Organism: A Chemical Background

What makes something alive? You can certainly tell that a barking puppy is alive and a rock is not. But what about the orange patch of color you might see on a rock? It could come from iron oxide in the rock or could be a complex living organism called a lichen. In fact, no one characteristic determines whether something is living or nonliving. Many nonliving objects share some of the same characteristics as living organisms. For example, a snowflake has a very complicated, precise structure, and a river moves, but neither of these are living. So what group of properties shared by all living organisms allows us to say for sure that they are "alive"?

- **Living organisms all have a complex organization that they are able to maintain throughout their lives. Living organisms, even microscopic single-celled plants, have highly complex structures that allow them to carry out a great variety of activities.**
- **To maintain their complex organization, living organisms must take in energy and use it to do various types of work. Living organisms require a constant supply of energy to build and maintain their tissues and carry on other activities, such as movement or digestion. Plants obtain energy from sunlight that is absorbed by their leaves and converted into energy the plants can use. Animals obtain their energy from the food they eat and digest.**
- **A living organism uses energy and raw materials from the environment to grow in size and to change as it grows. An apple seed will grow into a large apple tree, and a fertilized human egg will develop into a complex person.**

- **Living organisms respond to the surrounding environment. Often such responses are necessary for the organism to survive. Trees grow to reach the sunlight; snakes sense changes in the surrounding temperature and air to capture their prey; and male peacocks display bright plumage to attract a nearby female.**
- **Some living organisms maintain a relatively constant internal environment. The chemical reactions that keep these organisms alive occur under very specific conditions. Even when external conditions change dramatically, such as with large swings in temperature or acidity, living organisms are able to maintain a fairly constant internal environment. This ability is called homeostasis.**
- **All of the above characteristics are important. But if living organisms could not reproduce, life would not continue. Living organisms produce offspring that have attributes similar to the parents. Later we will see that offspring inherit a genetic "blueprint" for development that is carried in genes passed from parents to their offspring during reproduction. For example, a fertilized salmon egg contains all the genetic information necessary to produce a young fish that will swim to the ocean, spend several years growing in saltwater, and then return up the same freshwater river to the same tiny stream in which the egg was laid.**

Organisms possessing these characteristics are recognized as living. All living organisms are composed of similar chemical compounds. The atoms and molecules that make up these compounds also are found in the nonliving world. The chemical and physical reactions that take place in living organisms obey the same physical and chemical laws that govern all matter. To understand the chemistry of living organisms, we start by exploring the chemical principles that govern all matter. The basic concepts of chemistry that we discuss in Section 1 are the key to understanding the processes of life.

Chapter 1

The Structure and Properties of Matter

Barry hurried home with the Jell-o and ginger ale. Kate, his 16-year-old daughter, had come down with the flu the first week of school and had been up all night with vomiting and diarrhea. Because Kate had seemed better this morning, Barry had made a quick trip to the grocery store to pick up some things for her. Calling her name as he entered the kitchen, Barry heard no response. He raced up the stairs to find Kate sitting on the bathroom floor, looking dazed and holding a towel to her bloody forehead. Kate explained that she had had some cramps and had hurried to the bathroom. There she must have passed out and hit her head on the edge of the sink. Concerned about these symptoms, Barry drove Kate to the hospital's emergency room.

After asking Kate about her symptoms and examining her dry mouth, the emergency room nurse took Kate's blood pressure and pulse. When lying down, Kate's

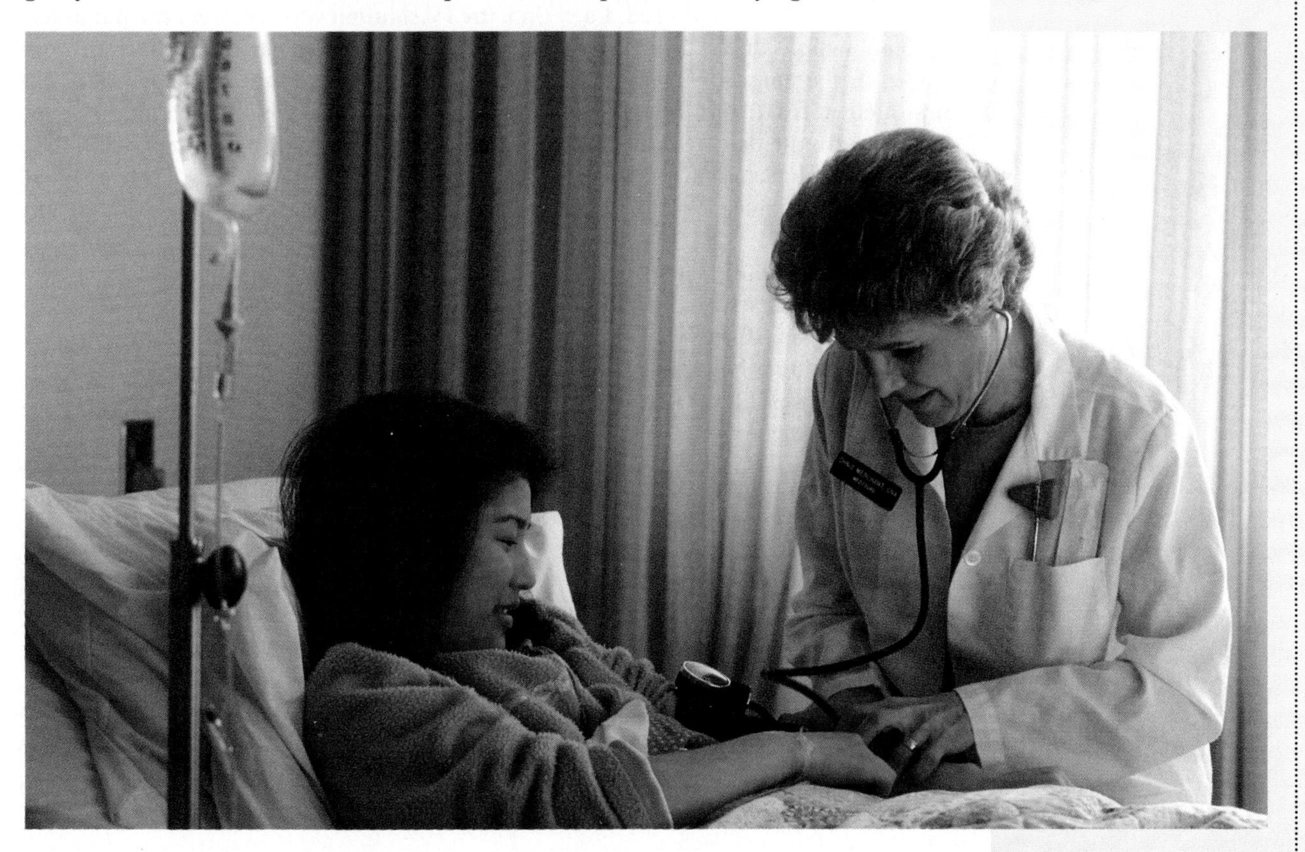

blood pressure was 110/60 and her pulse 80. When sitting up, her blood pressure lowered to 90/50, and her pulse increased to 120. When Kate stood, the nurse could not hear a heartbeat to take a blood pressure reading. Kate's pulse raced to 160, and she began perspiring heavily and felt faint. These readings indicated that Kate was suffering from dehydration. The nurse immediately started Kate on an intravenous (IV) solution of Ringer's lactate with 5% dextrose to replenish the body fluids and salts she had lost through vomiting and diarrhea.

Careful measurements of body functions are essential to medical practice. The nurse had measured Kate's blood pressure lying down, sitting, and then standing. This procedure, called taking orthostatic vital signs, (or the "tilt" test), is used to diagnose a patient's loss of blood volume. Loss of blood fluids can be caused by diarrhea and vomiting, miscarriages, or internal bleeding from wounds or injuries. Without immediate replacement of such lost blood fluids, a patient can go into shock and die.

But how much fluid should be given to a patient in these circumstances? In Kate's case the nurse infused the first liter of fluids very quickly (1 liter in 10 minutes) and then began a second bag at a slower rate. Kate's blood pressure was taken lying down, sitting, and standing every 5 minutes for the first 20 minutes and then every 15 minutes over the next hour. The blood pressure readings showed steady improvement. Soon the lying down and sitting readings became essentially normal (110/60), and Kate was no longer dizzy when she stood.

Blood pressure is not the only indicator that was carefully monitored in Kate's case. In addition to water, the IV solution contained salts and sugars to replace those lost in diarrhea and vomiting. This meant that Kate's blood sugar and ion levels also had to be carefully monitored. Each time the IV solution was replaced, the nurse sent a blood sample to the lab to make sure that the blood sugar and important ions stayed in the normal range. In Kate's case, the readings were all normal after the first hour of treatment. But if Kate's vomiting and diarrhea had been caused by the onset of juvenile diabetes rather than by the flu, the sugar in the IV drip could have been very dangerous. If the initial laboratory tests had shown high levels of blood sugar, the IV solution would have been changed immediately to one without any sugar.

As soon as Kate's blood pressure and blood ions were stabilized, she was sent home. She was instructed to eat popsicles and drink fluids to continue replacing her lost body fluids. If she were to again become faint, or if the vomiting and diarrhea were to continue, her father was to bring her back to the emergency room.

Careful observations and accurate measurements repeated over time are not only crucial to medicine but are the key to all of science. Just as a physician's diagnosis depends on the accuracy of data obtained from instruments and the clinical laboratory, the hypothesis of a scientist must be supported by data from careful and accurate measurements. In this chapter we introduce the vocabulary needed to describe and measure accurately the world around us.

1.1 What Is Matter?

Matter makes up the physical world in which we live. Chemistry is the science of matter: its structure, properties, and interactions. We define **matter** as anything that has mass and occupies space. Of course, that definition doesn't do you much good unless you know

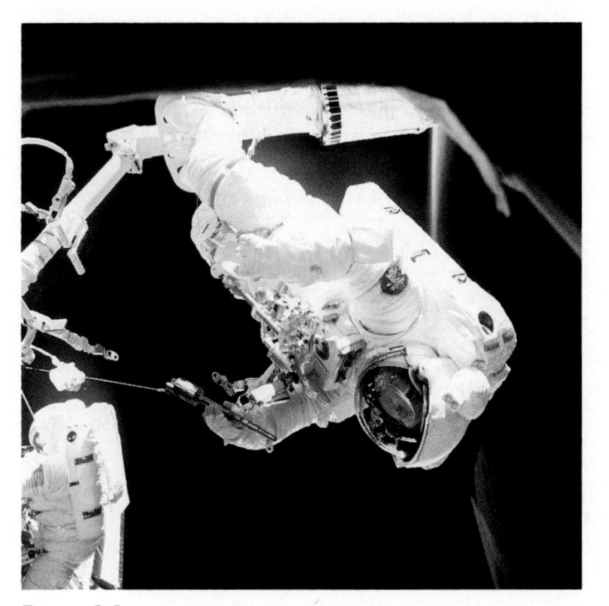

Figure 1.1 Although astronaut Kathryn Thornton is weightless in outer space, she has the same mass as on earth.

what mass is. You probably have a general feeling about the concept of mass and could tell which has greater mass—a brick or a feather. The **mass** of an object is a measure of how hard it is to start the object moving, or how hard it is to change its speed or direction once it is moving. For example, a bowling ball is harder to push than a balloon because the bowling ball has greater mass. The mass of an object is constant no matter where in the universe it is found (Fig. 1.1). Imagine a bowling ball and a balloon both floating weightlessly in the cabin of an orbiting spacecraft. The balloon would bounce harmlessly off an instrument panel, but the bowling ball could badly damage the delicate equipment.

◆ **Define mass and identify which of two given objects has the greater mass.**

The term weight is probably more familiar to you than mass, but you might not be certain of the difference between these terms. **Weight** is a measure of the force or attraction of gravity on an object; the mass of an object, however, does not depend on gravitational attraction, so it never changes. For example, because the moon's gravitational pull is roughly one-sixth that of the earth, an astronaut who weighs 180 pounds on the earth would weigh only 30 pounds on the moon. But the astronaut would have the same mass in either location. It is common and accepted practice, however, to use the terms *mass* and *weight* interchangeably, and we shall do so in this book.

1.2 Composition of Matter

Matter is composed of extremely small particles called **atoms.** The diameter of an atom is about eight-billionths of an inch (0.00000002 cm, or 2×10^{-8} cm—see Appendix 1 if this notation confuses you). It is very difficult to imagine anything so small. For example, a single page of this textbook is about 500,000 atoms thick.

As early as 400 B.C., Greeks pictured the atom as indivisible. The work of many scientists over the last 100 years has shown, however, that the atom is made up of smaller particles. Dozens of subatomic particles have now been identified, but only three are important for our discussions: the **proton,** the **neutron,** and the **electron.** It is the number

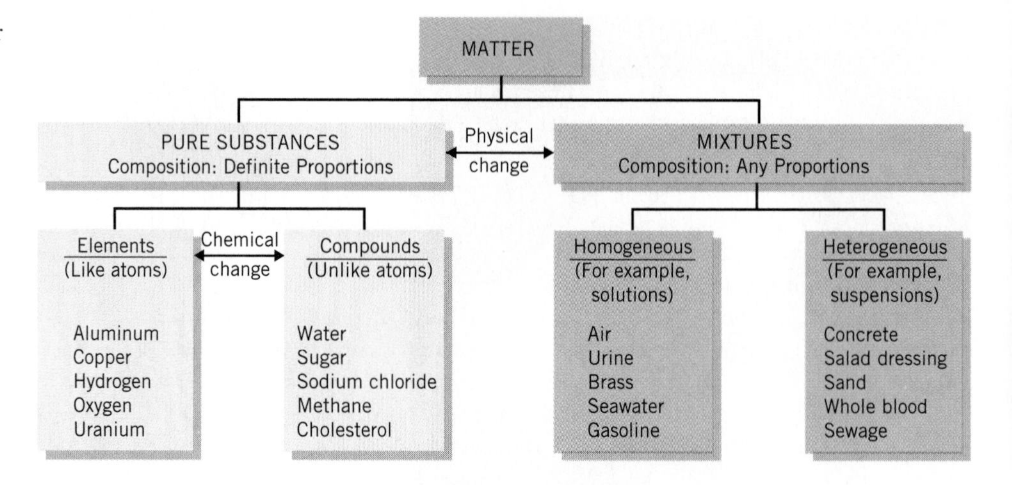

Figure 1.2 A sample of matter can be classified as either a pure substance or a mixture.

of these particles and the way in which they are arranged that give each atom its particular chemical properties. We discuss these three important subatomic particles in greater detail in Chapter 4.

1.3 Classes of Matter: Elements, Compounds, and Mixtures

Define element, compound, atom, and molecule.

Matter can be found in the form of pure substances—either elements or compounds—or as mixtures of elements or compounds (Fig. 1.2). An **element** is a substance that cannot be changed into simpler substances by ordinary chemical processes, such as heating, crushing, or exposure to acid. There are currently 110 known elements, the heaviest of which are synthetically produced.

An **atom** is the smallest unit of an element having the properties of that element. A **molecule** is a chemical unit containing two or more atoms joined together. The atoms making up the molecule can be of the same element, or they can be different elements. For example, atmospheric oxygen is found as a molecule containing two atoms of oxygen; a molecule of water has two atoms of hydrogen and one atom of oxygen (Fig. 1.3).

When two or more different elements combine chemically, a **compound** is formed. The properties of a compound are totally different from those of the elements that make it up. For example, we just mentioned that two atoms of the element hydrogen, which is a

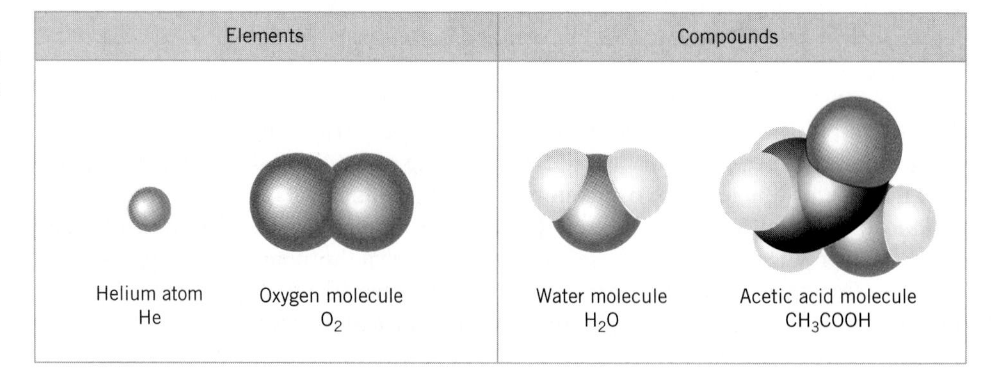

Figure 1.3 Many elements (such as helium) exist as single atoms, but other elements (such as oxygen) exist in molecular form. Compounds (such as water and acetic acid) exist as molecules containing atoms of more than one element.

gas that can burn, can combine with one atom of the element oxygen, which is a gas we breathe, to form the compound called water. Water won't burn, however, and no living organism can stay alive by breathing water. (You might think that fish breathe water, but they only filter water through their gills to remove the oxygen gas that is dissolved in the water.)

Compounds can be broken down into simpler substances by chemical means. Joseph Proust (1754–1826) observed that samples of copper carbonate, whether occurring naturally or synthesized in the laboratory, always had the same percentage of copper by mass. Proust's discovery formed the basis of a generalization known as the **law of definite proportions.** This law states that regardless of how a compound is formed or broken down, one fact is always true: *A compound is composed of specific elements in a definite proportion by weight.*

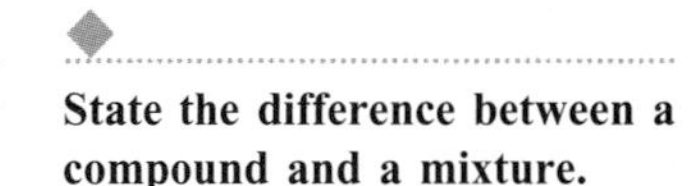

State the difference between a compound and a mixture.

A **mixture** differs from a compound in that it can be made up of two or more substances (elements or compounds) mixed in any proportion. When two substances chemically combine to form a compound, they form a new substance having different properties. But in a mixture, all the substances keep their individual properties.

State the difference between homogeneous and heterogeneous mixtures, and give three examples of each.

Sugared coffee, for example, is a mixture of sugar and coffee; the proportion of sugar to coffee in your cup depends on how sweet you like your coffee. The air that we breathe, the water that we drink, the ground on which we walk, and the gasoline that we put into our cars—even human beings—are each mixtures of various substances. Mixtures may be **homogeneous,** meaning they are so uniform that you can't tell one part from another, or **heterogeneous,** meaning that one part may be different from another (see Fig. 1.2). For example, a well-mixed cup of coffee with sugar in it is a homogeneous mixture. The sugar molecules are evenly distributed throughout the coffee, and each sip tastes the same. On the other hand, coffee with sugar in it that has not been totally dissolved and stirred is an example of a heterogeneous mixture. You could certainly tell the difference between the first few sips of coffee and the last.

1.4 Names and Symbols of the Elements

The **symbol** of an element is an abbreviation, or shorthand representation, of an element or an atom of that element. (The inside front cover of this book lists all known elements and their symbols.) Ordinarily, the symbol given to an element is the first letter in the name of that element: for example, H for hydrogen, C for carbon, and O for oxygen. It sometimes happens that more than one element has a name starting with the same letter. In such cases, the first two letters of the name are used for the symbol, with the first letter capitalized and the second lowercase: for example, Co for cobalt and Ca for calcium. Note that because chlorine and chromium share the same first two letters, these elements have the symbols Cl and Cr, respectively.

Figure 1.4 Candle wax melting is a physical change involving a change in form, but burning candle wax is an example of a chemical change involving a change in the basic nature of the wax.

1.5 The Three States of Matter

All matter can exist in three states: solid, liquid, or gas. These three states differ in the distance between the particles making up the substance, the attraction between the particles, and their amount of movement. Matter can be converted from one state to another, such as when a solid melts to form a liquid or a liquid evaporates to form a gas.

Describe the three states of matter.

In this book we study two types of change that matter can undergo: physical change and chemical change (Fig. 1.4). Common examples of substances undergoing physical change include water boiling to form steam, sugar dissolving in a cup of coffee, ham-

Historical Perspective / Names and Symbols

You may have noticed that the chemical symbols used for many of the elements have no apparent relation to their English names. The reason is that many elements, or the compounds from which they are derived, have been known for centuries and have names that originally came from Latin, Greek, Arabic, German, or other languages. Similarly, the symbols for many elements are derived from original names in other languages. Two of the elements that fall into this category are sodium and potassium, both of which are essential for human life.

The name *sodium* has its derivation from the Arabic *suwwad,* the name of a plant with a high concentration of soda (sodium carbonate). Mixed with water, this plant makes an alkaline solution. In medieval times, the Latin word for a headache remedy was *sodanum.* The symbol, Na, that we associate with sodium is thought to have come from both Hebrew and Latin. *Neter* (Hebrew) and *nitrum* (Latin) are names used in ancient times for alkali substances. In 15th-century Europe these alkali substances were often called *natron.* The metal in natron was later named *natrium* (Na). The suffix -ium denotes a metal.

Potassium (K) was named for the ashes where it was found. When plants are heated, a wood ash remains. When this ash is extracted with water and the resulting solution evaporated in iron pots, a solid is produced. This solid is called potash. The symbol we associate with this metal comes from the German and Scandinavian word for potassium, *kalium.*

burger grease solidifying on a plate, and wood being split by an ax. A **physical change** is one in which a substance changes form but still keeps its chemical identity. For example, water still keeps all of its chemical properties whether it is in the form of ice, water, or steam. When water boils and forms steam, the water has simply undergone a physical change and can be returned to its previous state by collecting and cooling the steam. Similarly, wax turning from a solid to a liquid as it drips down a candle undergoes another physical change, turning from a liquid to a solid as it cools. But the wax keeps the same chemical identity whether it is a liquid or a solid.

State the difference between a chemical change and a physical change.

Chemical changes are occurring when you leave your bicycle outside to rust in the rain, burn candlewax or wood in a campfire, accelerate your car from a stoplight, or digest your dinner. In a **chemical change,** the starting materials (reactants) are "used up," and different substances (products) are formed in their place. Obviously, the exhaust from your tail pipe has a different look and smell than the gasoline you put in the engine. If you were to collect smoke from a burning candle, the substances in the smoke would have none of the properties of the candlewax that was burning. The difference to remember is that a *physical change involves only a change of form,* whereas a *chemical change involves changes in the basic chemical composition of the substances involved.*

1.6 Scientific Method

Chemistry is the study of the composition and interaction of matter. To answer specific questions about the nature of matter, chemists observe the behavior of matter. Such observations are most often made under carefully controlled laboratory conditions. This allows other scientists to repeat the experiment to determine if the results are reproducible—that

Health Perspective / Public Health and the Scientific Method

Maria Rodriguez loved her job with the health department. She was responsible for finding and eliminating the causes of diseases that broke out in her community. Recently, she tracked down the cause of an outbreak of diarrhea caused by the *Escherichia coli* bacteria. In that case, she found that a chef at a local restaurant had used the same large knife first to bone raw chicken and then to cut the watermelon for the salad bar. But Maria was very puzzled by a new case she was working on.

A young boy named Miguel had been rushed to the emergency room from school, where he had become dizzy and had started sweating and vomiting. At the hospital, he seemed almost unconscious; his pupils became very small, he was twitching, and had severe cramps. The doctors were puzzled by the symptoms. They could have been caused by acute rheumatic fever or, because of the big industrial farms in the area, perhaps they were caused by insecticides. Data from special blood tests led the doctors to a diagnosis of poisoning by organophosphate insecticides. They suspected that Miguel might have inhaled some toxic insecticides from a crop duster or local farm sprayer.

To check this theory, Maria had called the local sprayers and was surprised to find that they were not using organophosphates on the crops. Meanwhile, Miguel had recovered. After six days in the hospital, he was being driven home when he suddenly became even more violently ill. His mother rushed him back to the hospital, where he was saved thanks to very quick action by the doctors. But why had Miguel become so violently sick the second time? He could not have been breathing spray from a crop duster. Perhaps the answer lay in Miguel's car or garage at home. Maria and her team checked Miguel's car, house, and garage for insecticides or insecticide residues and found none.

(continued on page 10)

is, to see if the same results occur each time the experiment is run. Chemists carefully record as data the observations made during experiments. If scientists notice patterns in the data they collect, they may propose a hypothesis (a suggested explanation) of why matter behaves as it does. This hypothesis must then be tested by further experiments. If data from additional experiments show the hypothesis to be incorrect, the hypothesis must be discarded and another one proposed. If over time, however, the data continue to support the hypothesis, the hypothesis then is treated as a scientific theory or law—that is, a unifying statement or mathematical relationship that explains the behavior of matter. Such theories may be accepted by scientists for years, even centuries, only to be shown to be incorrect as new data become available.

This careful approach to studying nature—through observation, development of a hypothesis, testing of the hypothesis, and then formulation of a scientific theory—is called the **scientific method.** Each of us deals with our surroundings in a similar way, even if we don't go about it as precisely as a scientist might.

Doctors use the scientific method in their diagnosis and treatment of an illness. They collect data about the disease, using blood tests, measurements of temperature and blood pressure, and data from diagnostic equipment such as electrocardiographs or magnetic resonance imagers. They then make a diagnosis (their hypothesis) and go on to test their hypothesis by prescribing a treatment. If the patient does not respond to the treatment, doctors must discard their hypothesis and collect more data—by further testing or perhaps exploratory surgery. They then propose a new or revised diagnosis and treatment.

While she was trying to piece together all of the data on this case, Maria got a call from Valley Children's Hospital. They had just treated a second boy, Jimmy, with symptoms very similar to Miguel's. Jimmy had been sent home from school with vomiting and had recovered slowly. But the day he returned to school he again became violently ill and was rushed to the hospital. Like Miguel, Jimmy was also found to be suffering from poisoning by organophosphate insecticides. What could be common to both boys' situations?

Maria questioned each of their mothers about the clothes the boys were wearing when they had gotten sick. Miguel had worn a new pair of jeans on the day he got sick and had put the same jeans on to go home from the hospital. Jimmy had also worn new jeans to school on both days he became sick. Jimmy's mother recalled buying the jeans not at a store, but at a salvage sale. The jeans had looked so new and were so cheap she had bought six pairs. Miguel's mother also bought his jeans at the salvage sale. To see if the jeans held the answer, the health investigators needed to do a controlled test. At a state health department laboratory, the jeans were put in a cage containing a colony of mosquitos. Similar jeans bought from a store were put in a second cage of mosquitos, and no jeans at all were put in a third cage. Within 15 minutes every mosquito in the cage with the contaminated jeans had died, but mosquitos in the two control cages remained healthy. The jeans the boys wore were contaminated with organophosphate insecticide.

Finding out how the jeans had become contaminated took a lengthy investigation. Maria and her team eventually learned that the jeans had been shipped in a trailer truck along with machinery and chemicals. One of the drums had leaked, soaking some of the jeans. When the stains eventually disappeared, the contaminated jeans were sold at the salvage sale.

1.7 Accuracy and Precision

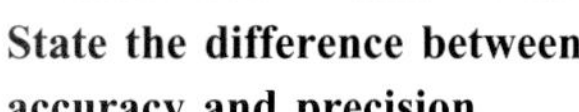

State the difference between accuracy and precision.

The testing of a hypothesis often requires careful measurement of the behavior of matter. The usefulness of experimental data collected in the laboratory depends on their accuracy and precision. An **accurate** measurement is one that is correct; the closer a measurement comes to the real value, the more accurate it is. You can see that an accurate measurement depends on the measuring device; it must be carefully calibrated and in good working order. A speedometer that registers 65 mph when you are going 75 mph is not very accurate, and it may result in an incorrect hypothesis that you are driving within the speed limit.

A **precise** measurement is one that is reproducible, so that repeating the measurement produces values that are very close to one another. For example, repeated readings of 155.5, 156.0, and 155.0 pounds on a bathroom scale are fairly precise measurements that may convince a person to begin a diet or to skip desserts. Precise measurements, however, are not always accurate. This person, for example, might find that the doctor's scale gives a reading of 147.3 pounds. The measurements on the bathroom scale were fairly precise but not very accurate (Fig. 1.5).

1.8 Significant Figures

The numbers you use when writing down experimental data indicate the precision with which you make the measurements. When making a measurement, a scientist records all

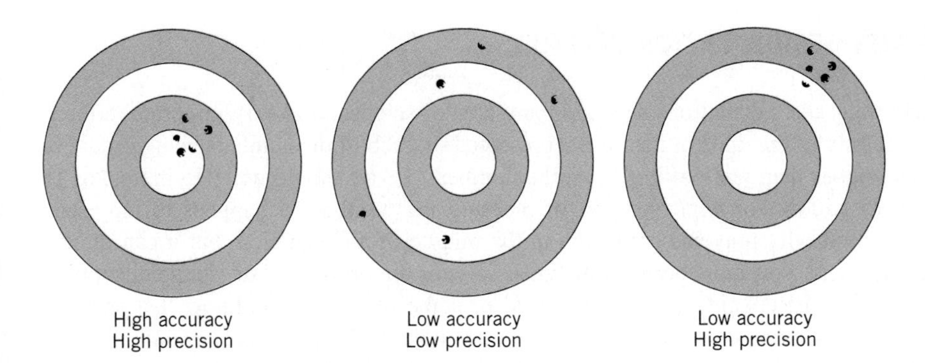

Figure 1.5 Examples of precision and accuracy.

the digits that are certain and then adds an additional digit that is uncertain. The digits in such a measurement are called **significant figures,** or **significant digits.** For example, suppose we want to measure the length of a block of wood with the two rulers shown here. Using ruler A, we know that the block is over 4 centimeters long, and we can estimate the total length to the nearest tenth of a centimeter: 4.8 centimeters. Other persons using ruler A to measure the block also would have to estimate how much the length exceeded 4 centimeters, and might record a value of 4.7 or 4.9 centimeters. Therefore, our recorded value of 4.8 centimeters is uncertain by 0.1 centimeter. The number of significant figures in this measurement is two (one digit that we know for certain and one that is uncertain).

◆ **Perform calculations with experimental data, maintaining the correct number of significant figures.**

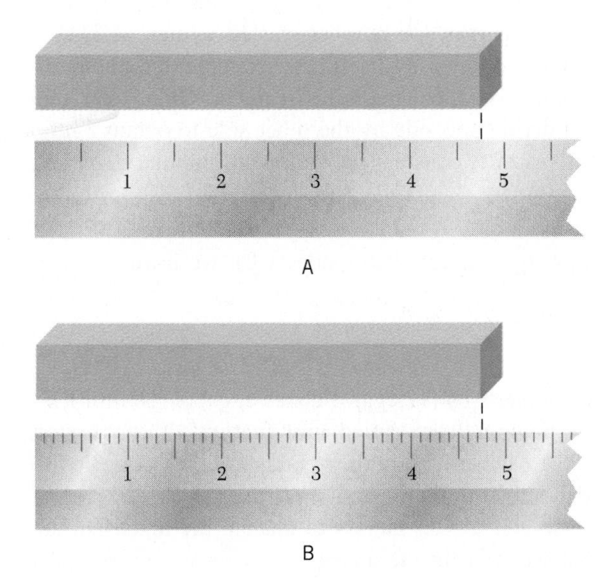

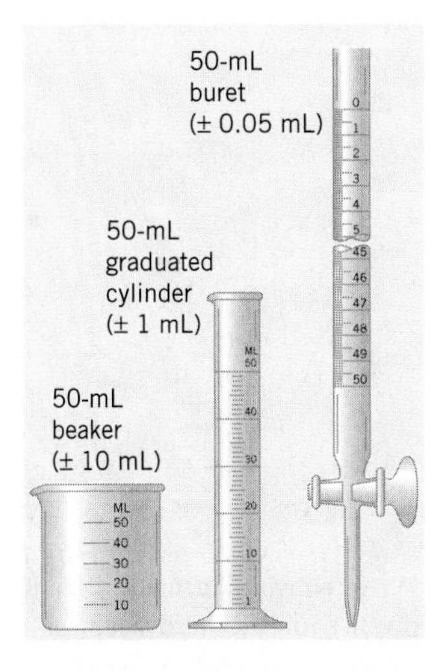

The error in a measurement depends on the equipment used.

Using ruler B, we know that the block is at least 4.7 centimeters long and can estimate the length to the nearest hundredth of a centimeter: 4.74 centimeters. This measurement is uncertain by about 0.01 centimeter, so the measurement of 4.74 centimeters has three significant figures.

Once data are collected in an experiment, precision can neither be gained nor lost during arithmetic operations on the data. Keeping track of significant figures is therefore very important in chemical calculations, especially if you use a calculator. Appendix 2 explains how to keep track of significant figures in mathematical calculations.

1.9 Conversion Factor Method

In performing the calculations needed in any area of science, it is very important to keep track carefully of the units of measure represented by each of the numbers being used. For example, rather than just writing down the number 15, you should write 15 grams, or 15 feet, or 15 gallons, or whatever unit of measure the number 15 happens to represent. Doing this not only prevents confusion in the middle of a calculation, but it can remind you of the steps you must perform to finish solving the problem. For example, suppose you must do a calculation to find the speed of some object. If you know that speed is commonly measured in such units as miles per hour (mi/h), feet per second (ft/s), or meters per second (m/s), you know that you will eventually need to divide some measurement of distance (in the appropriate units) by a measurement of time (in the appropriate units). For this reason, you will find the calculations required in this textbook much easier to perform if you make a point of keeping your numbers properly labeled with their unit of measure.

To perform chemical calculations, we make use of two properties of the number "one." The first property is that any number, when multiplied by 1, remains the same. More generally, any kind of quantity remains the same when multiplied by the number 1. For example,

$$36 \times 1 = 36$$

$$7 \text{ apples} \times 1 = 7 \text{ apples}$$

$$92 \text{ miles/hour} \times 1 = 92 \text{ miles/hour}$$

The second property of the number 1 is that this number can be written as 2/2, or 156/156, or as the quotient of any number divided by itself. For example, the number 1 can be represented by 5 apples/5 apples, or 156 camels/156 camels. Going one step further, we can take any equation and divide one side by the other side to obtain a ratio equal to the number 1. To see why this is true, consider the following familiar equation:

$$12 \text{ inches} = 1 \text{ foot}$$

If we divide both sides of the equation by the right-hand side (1 ft), we have

$$\frac{12 \text{ in.}}{1 \text{ ft}} = \frac{1 \text{ ft}}{1 \text{ ft}} \qquad \frac{12 \text{ in.}}{1 \text{ ft}} = 1$$

Write conversion factors from given equalities and use conversion factors to solve mathematical problems.

In the same way we can represent the number 1 by (12 eggs/1 dozen eggs) or (60 min/1 h). Such ratios that are equivalent to the number 1 are called **unit factors,** or **conversion factors.** They are the key to most of the calculations in this textbook. What you need to remember is that the numerator and denominator of conversion factors must always represent equivalent quantities. We shall use conversion factors to change the units that are given to us in a problem to units that we want in the answer.

To illustrate how conversion factors are used in a calculation, let's look at a problem you could easily solve: How many inches are in 2 feet? You would immediately say 24 inches, which you would have calculated by multiplying $2 \times 12 = 24$. But what did you do when you started with a distance that you called "2" and said that it is the same as a distance that you called "24"? What you actually did was to use the conversion factor that we derived above. To set up a problem using conversion factors, start with what is given in the problem (2 ft) and multiply it by a conversion factor that gives you the answer in the units required by the problem.

$$2\text{ ft} \times \frac{12\text{ in.}}{1\text{ ft}} = \frac{2 \times 12}{1}\ \frac{(\cancel{\text{ft}})(\text{in.})}{(\cancel{\text{ft}})} = 24\text{ in.}$$

You feel confident that the distance you started out with (2 ft) is the same distance that you ended with (24 in.). All you really did was to multiply your initial distance by a conversion factor—that is, by the number 1. When you make a point of keeping close track of the units of measure for each number in the problem (as we just did), you see that similar units in the numerator and denominator "cancel," leaving you with an answer in the desired unit of measure.

Check Your Understanding

Answers to *Check Your Understanding* problems are found in Appendix 3.

1. Write the two possible conversion factors from each of the following equalities:

   (a) 1 kilogram = 2.2 pounds

   (b) 1 mile = 5280 feet

   (c) 1 liter = 1000 milliliters

   (d) 1 μm = 10^{-6} meters

2. Which of the conversion factors that you wrote in Problem 1 can you use to solve the following problems?

   (a) How many kilograms are in 6 pounds?

   (b) How many feet are in 6.6 miles?

   (c) How many liters are in 65 milliliters?

   (d) If a white blood cell measures 10^{-5} meters across, how many μm is that?

1.10 The SI System of Units

Scientists throughout the world have long used the metric system for the measurement of matter. In 1960, however, at an international meeting of scientists in Paris, a revised set of units was proposed. This revised set, based on natural universal constants, is called the **SI** (*Système Internationale*) **system of units** and has been accepted by scientists as a means of easily exchanging data throughout the world. Table 1.1 lists the basic units of the SI system. In this book we use the SI units as well as the more common metric units in many of our discussions.

Table 1.1 Base Units in the SI System

Quantity	*Name*	*Symbol*
Length	meter	m
Mass	kilogram	kg
Time	second	s
Temperature	kelvin	K
Amount of substance	mole	mol
Electric current	ampere	A

Table 1.2 Common Prefixes Used in the Metric and SI Systems

Prefix	*Symbol*	*Multiple*	*Example*[a]
kilo-	k	$1000 = 10^3$	kilometer, km = 1000 meters
hecto-	h	$100 = 10^2$	hectometer, hm = 100 meters
deka-	da	$10 = 10^1$	dekameter, dam = 10 meters
			meter, m (basic unit)
deci-	d	$0.1 = 10^{-1}$	decimeter, dm = 0.1 meter
centi-	c	$0.01 = 10^{-2}$	centimeter, cm = 0.01 meter
milli-	m	$0.001 = 10^{-3}$	millimeter, mm = 0.001 meter
micro-	μ	$0.000001 = 10^{-6}$	micrometer, μm = 0.000001 meter
nano-	n	$0.000000001 = 10^{-9}$	nanometer, nm = 0.000000001 meter

[a]Conversions between numbers in the SI and metric systems are exact numbers and may be written with as many significant figures as necessary.

Using the familiar English system of measurements, we all have learned that 4 quarts makes a gallon, 12 inches make a foot, and 16 ounces make a pound. This is a complicated system of measurements because the relationship between the number of smaller units needed to make up a larger unit of measure is not consistent. By contrast, the great advantage of both the metric and SI systems is that all units of measure are related to their subunits by multiples of 10. Standard prefixes are used to show the number of multiplications or divisions by 10 that are required. The idea of a numerical prefix is not new to you; we all know that the prefix tri-, as in the words tricycle, tripod, or trio, indicates that the object being named has three of something. In the same way, the prefix kilo- used in the metric and SI systems means 1000. One kilometer, then, is 1000 meters, and one kilogram is 1000 grams. The use of decimals in the metric system makes calculations much easier to perform than in the English system. Table 1.2 lists some common prefixes used in the metric and SI systems. Figure 1.6 shows the relative sizes in biological terms of some of these units.

Convert measurements between the metric and English systems.

The back overleaf of this textbook contains common conversion factors between the metric and English systems. In Sections 1.11 to 1.14 we look at several examples of conversions between units in the metric system and the English system.

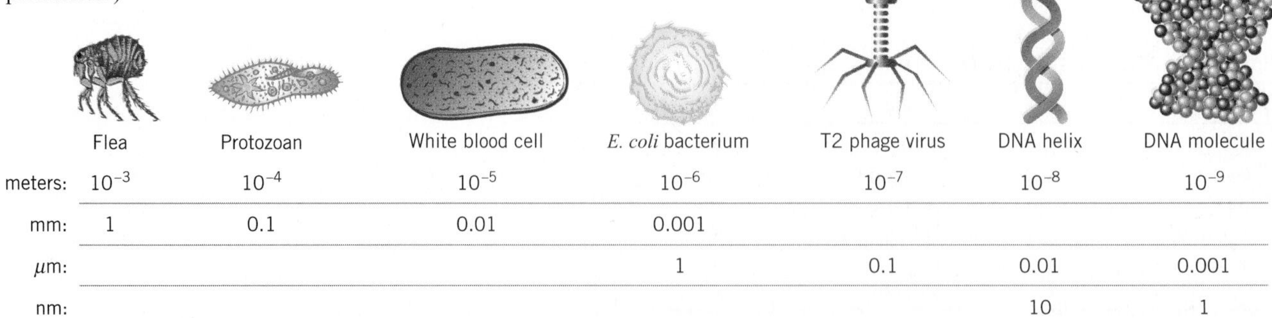

Figure 1.6 Using the metric system to compare sizes. A flea is six powers of 10 larger than a DNA molecule. The sizes listed are approximate. (Adapted from Figure 2-2 in Arthur Kornberg, *For the Love of Enzymes,* copyright 1989, Harvard University Press, Cambridge, MA. Used with permission.)

Historical Perspective / Body English

King Henry I

Have you ever wondered where we got the units used to measure length in the English system? It is interesting that many of these units have a relationship to the human body. When early humans needed to convey a specific measurement, they often used themselves as the measuring standard. The earliest recorded unit of measurement is found in Egyptian papyrus texts, where a symbol for the cubit is the figure of a forearm. The cubit was defined as the distance from an elbow to the tip of the middle finger. In the 10th century, the inch was defined as the distance between the knuckles on King Edgar's thumb. The emperor Charlemagne decreed that a foot was equal to the length of his own foot, and King Henry I established a yard as the distance from the tip of his nose to his fingertips. The mile was originally the distance covered by a Roman soldier when taking 1000 double steps; this equaled 5000 feet. In the 16th century, Queen Elizabeth I of England added 280 feet, so that a mile then measured 5280 feet and was equivalent to eight furlongs. (Don't worry, we won't get into furlongs!)

As you can see, such ''standards'' are really not sensible. A standard needs to be defined in terms of a reference point that can be checked to determine the exact size of the unit in question. (Even while King Edgar was alive, dragging him around from place to place to get an exact measure in inches was out of the question!) Early attempts to solve this problem resulted in the redefinition of the foot based on barleycorns. Specifically, the foot was standardized as the length of 36 grains of barleycorn laid end to end. The inch was then redefined as the width of three grains of barleycorn placed side by side. Of course not all barleycorns are the same size, so this standardization left much to be desired.

Today's standard inch, along with the other units of English measure, is much more reproducible. These measures are defined by items housed at the U.S. Bureau of Standards, but the conversion factors that must be used to convert one unit to another still create much confusion. Because the metric system has only one fundamental unit for each type of measurement, and all of its conversion factors are multiples of 10, it is unquestionably the superior system of measurement.

Your Perspective: Write a formal memorandum to your teacher proposing a new unit for measuring *time* based on something having to do with your body. Explain why this is a predictable unit of measure, and describe exactly how you would measure it.

1.11 Length

State the units of length used in the metric and SI systems.

The unit measure of length in the SI system is the **meter.** (Metre is the preferred international spelling for this unit, but meter is the spelling used in the United States and the spelling we use in this book.) This unit was defined in 1790 as one ten-millionth of the distance from the north pole to the equator. Although this standard has been redefined three times since then, its length has remained pretty much unchanged. One meter is equal to 3.28 feet and is, therefore, slightly longer than a yard (Fig. 1.7).

A kilometer is the unit that is used to measure large distances, such as the distance between cities (1 kilometer = 0.621 mile). There are 1000 meters in a kilometer. Commonly used units that are smaller than a meter are the millimeter and the centimeter.

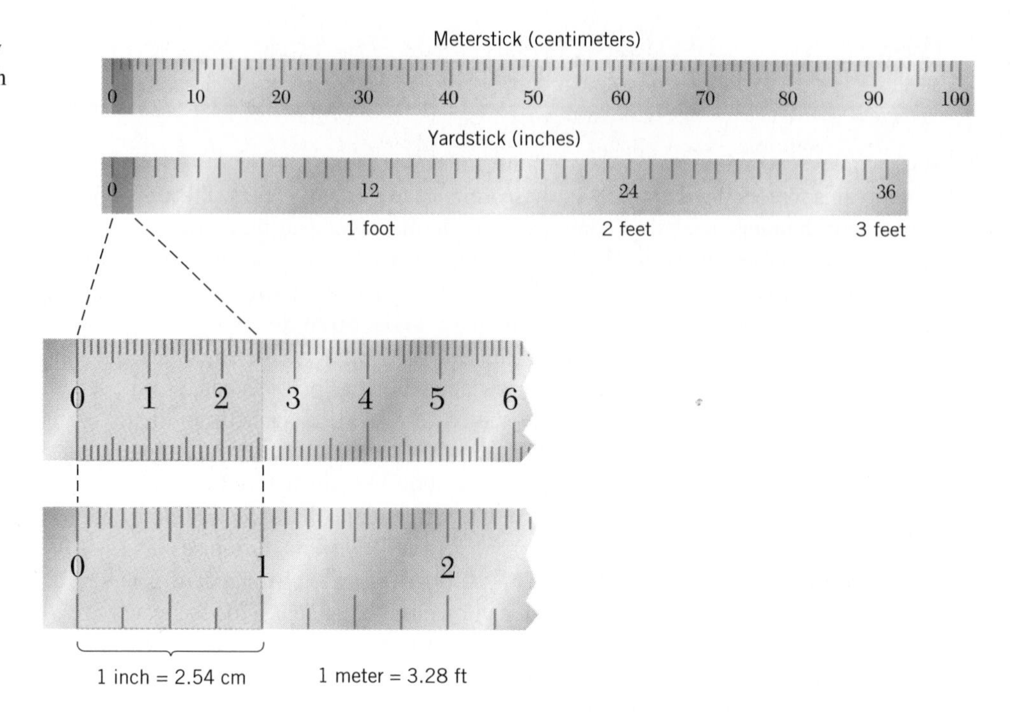

Figure 1.7 A meter is slightly longer than a yard, and an inch is a little bigger than 2.5 centimeters.

Example 1-1

Before beginning this exercise, reread the description of the *STEP* problem-solving strategy on page xi. This strategy for thinking through a problem will help you solve even the most complicated chemical and mathematical problems.

1. How many centimeters are in 4.35 meters?

***S*ee the Question** → The question in this problem is 4.35 meters = (?) centimeters

***T*hink It Through** → Because the problem requires us to change meters into centimeters, we must look for a conversion factor that shows the relationship between these two units of measure. We know that 1 meter = 100 centimeters. Using this equality, we can write two conversion factors that might be useful:

$$\frac{1\text{ m}}{100\text{ cm}} \quad \text{and} \quad \frac{100\text{ cm}}{1\text{ m}}$$

The second conversion factor is the correct one to choose because it allows units of measure to cancel, thereby giving us an answer in centimeters.

***E*xecute the Math** →

$$4.35\ \cancel{\text{m}} \times \frac{100\text{ cm}}{1\ \cancel{\text{m}}} = \frac{4.35 \times 100}{1}\text{ cm} = 435\text{ cm}$$

***P*repare the Answer** → Because the measurement ''4.35 meters'' has three significant figures, our answer 435 must also have three significant figures. The unit for our answer is centimeters (cm), so the answer to this problem is written as 435 cm.

2. A newborn baby measures 18 inches in length, but this measurement must be written in centimeters on the hospital chart. How many centimeters long is the baby?

S → This problem asks: 18 inches = (?) centimeters

The information that is needed to solve this problem is the relationship between centimeters and inches. We find that equality in the table on the back overleaf of the textbook. (Note that the "1 in." in the equality is an exact number (see Appendix 2), so we can write it to any number of significant figures. The 2.54 is not exact and is given to three significant figures.) ◄ *T*

$$1.00 \text{ in.} = 2.54 \text{ cm}$$

Use this equality to write the conversion factor we need to solve the problem. Because our answer must be in centimeters, we choose the conversion factor with centimeters in the numerator.

$$18 \text{ in.} \times \frac{2.54 \text{ cm}}{1.00 \text{ in.}} = 18 \times 2.54 \text{ cm} = 45.72 \text{ cm (calculator answer)}$$ ◄ *E*

The measurement given in the problem has two significant figures, so the answer should have two significant figures and be in centimeters. The baby is 46 cm long. (See Appendix 2 for the rules for rounding off numbers.) ◄ *P*

3. A white blood cell measures 42 μm (micrometers) in diameter. What is the diameter of this cell in inches?

In this question we need to convert a measurement in micrometers to one in inches: ◄ *S*

$$42 \text{ micrometers} = (?) \text{ inches}$$

To perform this conversion, we need several conversion factors: one that relates micrometers (μm) to meters (m), one for meters to centimeters (cm), and one for centimeters to inches (in.). ◄ *T*

$$1 \ \mu\text{m} = 10^{-6} \text{ m}, \qquad 1 \text{ m} = 100 \text{ cm}, \qquad 2.54 \text{ cm} = 1 \text{ in.}$$

To solve this problem, we can use conversion factors formed from each of these equalities in sequence, choosing the conversion factor that allows units of measure to cancel.

$$42 \ \mu\text{m} \times \frac{10^{-6} \text{ m}}{1 \ \mu\text{m}} \times \frac{100 \text{ cm}}{1 \text{ m}} \times \frac{1 \text{ in.}}{2.54 \text{ cm}} = 1653.5433 \times 10^{-6} \text{ in. (calculator answer)}$$ ◄ *E*

The result we get from the calculator is not our final answer. The measurement given in the problem (42 μm) has two significant figures, so our answer must have two significant figures. The answer to the problem can be expressed in exponential form or as a decimal fraction. If you have trouble working with these numbers, review the explanations and problems in Appendices 1 and 2. ◄ *P*

$$1.7 \times 10^{-3} \text{ in.} \qquad \text{or} \qquad 0.0017 \text{ in.}$$

Check Your Understanding

The route for the Boston Marathon measures 26.2 miles. Alberto Salazar won the race in 1982 with a time of 2 hours, 8 minutes, and 51 seconds.

(a) How many kilometers long is this race?

(b) On average, how many minutes did it take Salazar to run 1 mile?

(c) On average, how many minutes did it take Salazar to run 1 kilometer?

Health Perspective / Apothecary Measurements

If all we had to worry about was the conversion of English units to their metric equivalents, things wouldn't be so bad. But for many of you in your professions, it will be necessary to interpret drug orders and solve problems of dosages and solutions. To do this will require a knowledge of the other systems of measurement used in medical practice. Although doctors, pharmacists, and hospitals have joined with the pharmaceutical industry in trying to replace all other systems with the metric system, remnants of the past still linger. The most frequently encountered of these systems is the apothecaries' system of weights. Originating centuries ago in England, the apothecaries' system was used for weight and volume measurements involving the preparation, preservation, and dispensation of drugs. Familiarity with these units and their metric counterparts is important because they are still used by some doctors and hospitals. Some common apothecary units and their metric equivalents are shown here.

Dry Measures			*Liquid Measures*		
Metric		*Apothecary*	*Metric*		*Apothecary*
60 milligrams	=	1 grain	1 milliliter	=	15 minims
4 grams	=	60 grains or 1 dram	4 milliliters	=	1 fluid dram

1.12 Mass

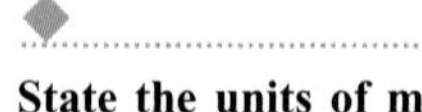

State the units of mass used in the metric and SI systems.

The unit of measure for mass in the metric system is the **gram.** The SI standard is a block of platinum metal weighing exactly one kilogram (1000 grams) that the International Bureau of Weights and Measures keeps in a vault in France. Units of mass commonly used are the kilogram, gram, and milligram (Fig. 1.8a).

$$1 \text{ kg} = 1000 \text{ g}$$

$$1 \text{ g} = 1000 \text{ mg}$$

$$1 \text{ kg} = 2.2 \text{ lb}$$

The mass of a sample is determined by comparing the weight of the sample to the weight of standard masses on a balance.

Example 1-2

1. How many milligrams are in 0.024 gram?

See the Question The problem asks: 0.024 g = (?) mg

Think It Through We know that 1 gram = 1000 milligrams, so we can write two conversion factors that show the relationship between grams and milligrams:

$$\frac{1 \text{ g}}{1000 \text{ mg}} \quad \text{or} \quad \frac{1000 \text{ mg}}{1 \text{ g}}$$

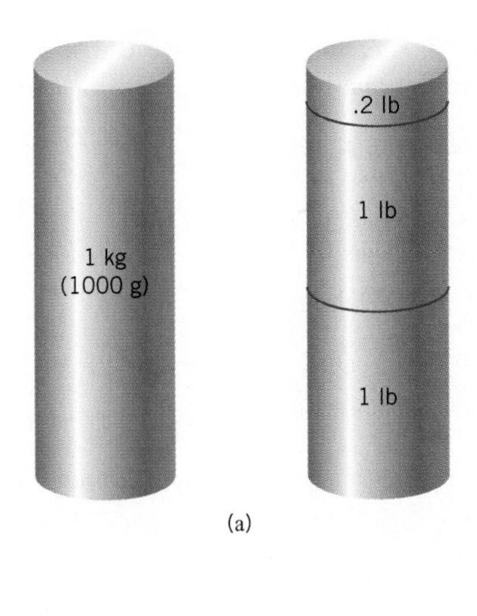

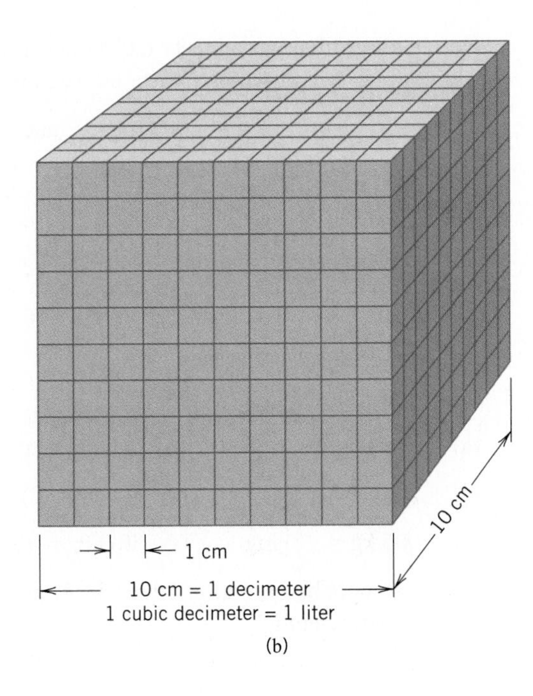

Figure 1.8 (a) A cylinder that weighs 1 kilogram in the metric system weighs 2.2 pounds in the English system. (b) The box has a volume of 1 liter. (Volume = 10 cm × 10 cm × 10 cm = 1000 cm^3 = 1000 mL = 1 L)

The solution to the problem must be in milligrams. For the units of grams to cancel, we choose the conversion factor with milligrams in the numerator.

$$0.024\,\cancel{g} \times \frac{1000\text{ mg}}{1\,\cancel{g}} = 0.024 \times 1000\text{ mg} = 24\text{ mg}$$

◄ ***E**xecute the Math*

The answer should have two significant figures and be in milligrams. The answer to this problem is 24 mg.

◄ ***P**repare the Answer*

2. Labels on drugs given to infants often list the dosage per kilogram of body weight. What is the weight in kilograms of a 16.5-lb baby?

In this problem, we need to find the number of kilograms in 16.5 lb.

◄ ***S***

$$16.5\text{ lb} = (?)\text{ kg}$$

The equality we need for writing conversion factors is the one between pounds and kilograms: 2.20 lb = 1.00 kg. Because our answer must be in kilograms, the appropriate conversion factor is the one with kilograms in the numerator.

◄ ***T***

$$16.5\,\cancel{lb} \times \frac{1.00\text{ kg}}{2.20\,\cancel{lb}} = \frac{16.5}{2.20}\text{ kg} = 7.5\text{ kg}$$

◄ ***E***

The calculator result is 7.5, but is this the answer we should record? The weight we are given in the problem (16.5 lb) has three significant figures, so our answer should also contain three significant figures. The correct answer to this problem is 7.50 kg.

◄ ***P***

3. A doctor prescribes a 0.1-g dose of medication. How many 25-mg tablets are required to fill the prescription?

This problem asks how many tablets are necessary to make 0.1 g of medication.

◄ ***S***

$$0.1\text{ g} = (?)\text{ tablets}$$

T No equality directly connects tablets and grams of medication. Instead, we use two equalities from which we can write conversion factors to solve the problem.

$$1 \text{ g} = 1000 \text{ mg} \quad \text{and} \quad 1 \text{ tablet} = 25 \text{ mg}$$

E

$$0.1 \text{ g} \times \frac{1000 \text{ mg}}{1 \text{ g}} \times \frac{1 \text{ tablet}}{25 \text{ mg}} = 4 \text{ tablets}$$

P The answer should have one significant figure, so it is 4 tablets.

Check Your Understanding

1. Make the following conversions:

   (a) 253 μg = (?) mg

   (b) 3.2 kg = (?) g

   (c) 0.005 kg = (?) mg

   (d) 0.34 kg = (?) oz

   (e) 681 g = (?) lb

   (f) 30.0 oz = (?) g

2. A premature baby born after 26 weeks of gestation weighed 832 g. How many pounds did the baby weigh?

1.13 Volume

State the units of volume used in the metric and SI systems.

You will often find the volume of an object stated in terms of some unit of length. For example, to calculate the volume of a box we might multiply the length by the width by the height. The SI unit of volume is the cubic meter, abbreviated m^3. This is a large unit of measure—a box with a volume of 1 m^3 could hold about 1060 quarts of milk. For most practical purposes, therefore, we use the metric unit of volume, the **liter (L),** which is just slightly larger than a quart (see Fig. 1.8b). (The international spelling is litre, but in this book we use the American spelling.)

$$1 \text{ m}^3 = 1000 \text{ liters (L)}$$

$$1 \text{ L} = 1000 \text{ milliliters (mL)}$$

$$1 \text{ mL} = 1 \text{ cubic centimeter (cm}^3 \text{ or cc)}$$

$$1 \text{ L} = 1.06 \text{ qt}$$

You will often find laboratory measuring devices, such as syringes or pipets, labeled in either cubic centimeters or milliliters (Fig. 1.9).

Example 1-3

1. A patient excretes 2.5 quarts of urine. How many liters of urine is this?

See the Question This problem asks: 2.5 qt = (?) liters

Think It Through To solve the problem we need to use the relationship between quarts and liters.

$$1.06 \text{ qt} = 1.00 \text{ L}$$

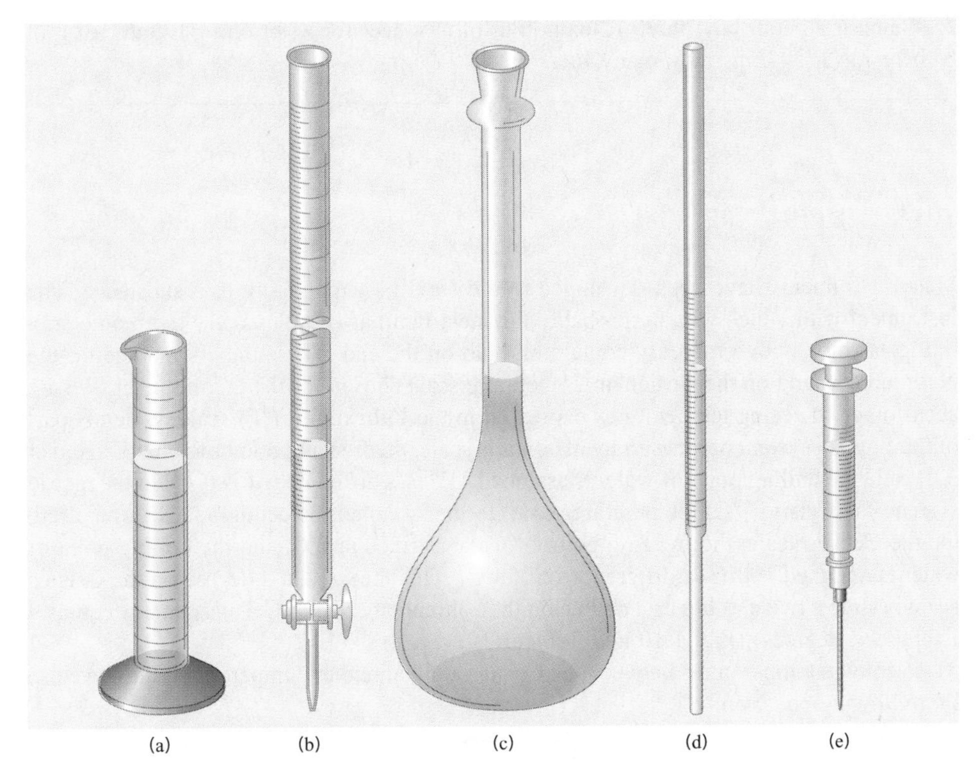

Figure 1.9 Laboratory equipment commonly used to measure volume: (a) graduated cylinder, (b) buret, (c) volumetric flask, (d) pipet, and (e) syringe.

Note that the 1 L in the equality is an exact number. For the answer to be in liters, we choose the conversion factor with liters in the numerator.

$$2.5 \text{ qt} \times \frac{1.00 \text{ L}}{1.06 \text{ qt}} = \frac{2.5 \times 1.00}{1.06} \text{ L} = 2.3584905 \text{ (calculator answer)}$$

***E**xecute the Math*

The calculator gives us a result with eight digits, but the volume given in the problem requires two significant figures. The correct answer is 2.4 L.

***P**repare the Answer*

2. A bottle of antibiotic contains 0.075 L. How many 5.0-mL doses does it contain?

This problem asks the following question: 0.075 L = (?) doses

S

Solving this problem requires two equalities that give us two conversion factors.

T

$$1 \text{ L} = 1000 \text{ mL} \qquad \text{and} \qquad 1 \text{ dose} = 5 \text{ mL}$$

$$0.075 \text{ liter} \times \frac{1000 \text{ mL}}{1 \text{ liter}} \times \frac{1 \text{ dose}}{5.0 \text{ mL}} = 15 \text{ doses}$$

E

The measurement given in the problem (0.075 L) has two significant figures. If you don't understand why, refer to the explanation and problems in Appendix 2. The answer to this problem is the same as the calculated result: 15 doses.

P

Check Your Understanding

1. Do the following conversions:

(a) 2.5 L = (?) mL **(d)** 329 gal = (?) L

(b) 345 mL = (?) L **(e)** 5.3 L = (?) qt

(c) 25 mL = (?) cc **(f)** 945 mL = (?) pint

2. Which is a better buy: three 12-oz bottles of root beer for $1 or one 1-L bottle of root beer for $1? (*Hint:* how many liters are in 12 oz?)

1.14 Temperature

State the units of temperature used in the metric and SI systems.

Many instruments have been developed to measure the temperature of a substance. The instrument with which you are probably the most familiar is the mercury thermometer, a thin glass tube with a mercury-containing bulb on the end. The numbers marked on the glass tube depend on the particular temperature scale chosen for the thermometer. We are accustomed to seeing temperatures expressed in the **Fahrenheit (°F)** scale, which is part of the English system of measurement. On this scale, the freezing point of water is fixed at 32°F, and the boiling point of water is assigned 212°F (a difference of 180°F). In the metric system, the **Celsius (°C)** temperature scale (formerly called the centigrade scale) is used. On the Celsius scale, the freezing point of water is fixed at 0°C, and the boiling point of water is assigned 100°C (a difference of 100°C). This means that a degree on the Celsius scale is almost twice as big as a degree on the Fahrenheit scale (5 Celsius degrees equals 9 Fahrenheit degrees) (Fig. 1.10 and Table 1.3).

To convert temperatures between the Celsius and Fahrenheit temperature scales, we use the following equation:

$$^{\circ}\mathrm{C} = \frac{5}{9}(^{\circ}\mathrm{F} - 32) \qquad \text{or} \qquad 1.8 \times {}^{\circ}\mathrm{C} = {}^{\circ}\mathrm{F} - 32$$

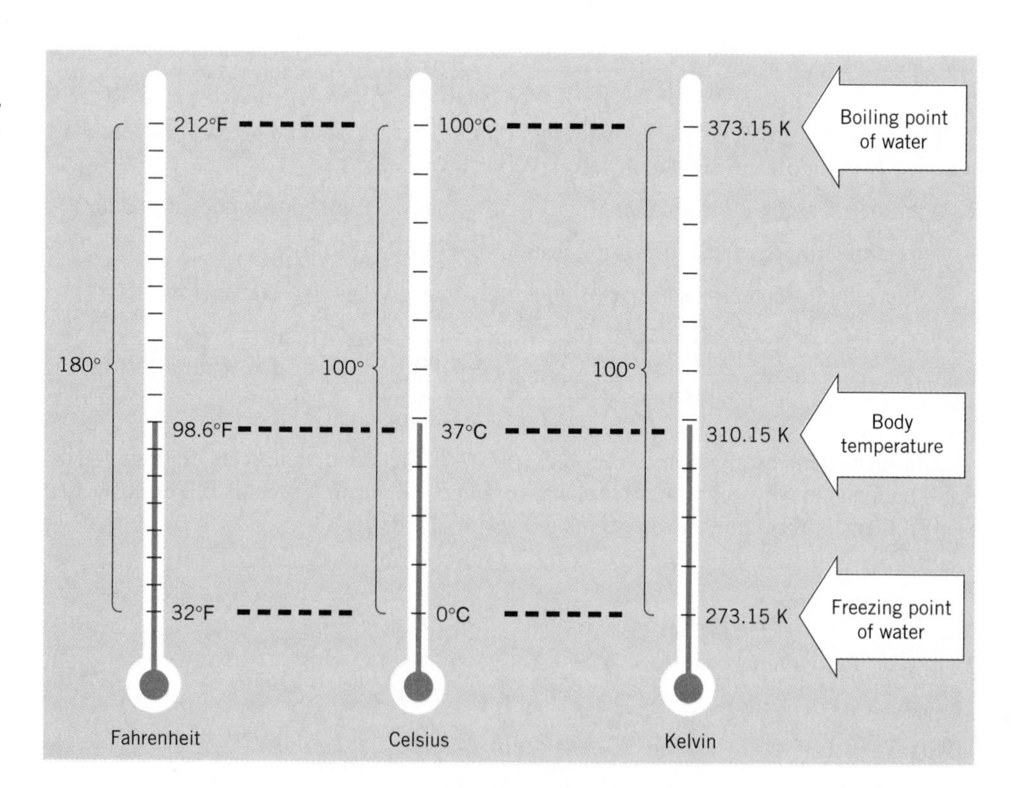

Figure 1.10 Relationships among the Fahrenheit, Celsius, and Kelvin temperature scales.

Health Perspective / Variations in Body Temperature

Have you ever noticed how some of your friends, especially the skinny ones, never seem affected by high temperatures on a hot summer day? They sit there, cool as cucumbers, while the rest of us swelter in the heat. You may have also noticed that friends who are considerably overweight suffer the most on hot humid days. The reason for this is that thin people are more efficient at radiating heat relative to their body weight than are overweight people. They can therefore maintain their body temperature at a lower level. On the other hand, thin people are often more affected by cold temperatures in the winter and need to dress more warmly.

Although no specific body temperature is "normal" for everyone, an average oral thermometer reading of 98.6°F (37.0°C) is used as a point of reference. This temperature varies depending on how much heat the body is producing and the rate at which this heat is lost from the body. When you first awake in the morning, your temperature may be 36°C or lower. As the day progresses and your activity level increases, so does your heat production. Hard physical labor or emotional stress can produce heat at a rate greater than the body can lose it, and body temperature can climb to 38°C or higher. A person involved in hard exercise may have a body temperature above 40°C!

The body loses approximately 80% of its excess heat through the skin. The remainder is lost in the evaporation of water vapor exhaled from the lungs. The flushed skin you see on a person involved in hard exercise is caused by excess blood being sent into skin capillaries, where it is cooled by exposure to the cooler outside air.

In very cold situations you can increase your body temperature by additional muscle work, such as shivering. If the temperature outside the body is extremely cold, the body holds back the blood from the skin to preserve heat. Unfortunately the skin cells can then freeze, causing frostbite in those areas.

Your Perspective: Write a short essay describing an actual situation in which your body became overheated (not from a fever). Explain what caused the overheating and what prevented your body from cooling itself down.

Table 1.3 Comparison of Temperatures on the Fahrenheit and Celsius Scales

	Fahrenheit (°F)	*Celsius (°C)*
A cold winter day	−5	−21
Freezing point of water	32	0
Room temperature	68	20
Body temperature	98.6	37
A hot summer day	100	38
Boiling point of water	212	100

A third temperature scale is the **Kelvin (K)** temperature scale, which belongs to the SI system. Kelvin temperatures have a degree unit called the kelvin, which is written as K, not °K. A kelvin is the same size as a degree Celsius. The freezing point of water is assigned 273.15 K on the Kelvin scale, and the boiling point 373.15 K. As with the

Celsius scale, this is a difference of 100 kelvins. You may wonder how a number such as 273.15 was chosen for the freezing point of water. This number was experimentally determined so that zero on the Kelvin scale is the lowest temperature theoretically possible. For this reason, 0 K is known as **absolute zero.**

To convert temperatures between the Kelvin and Celsius temperature scales we use the following equation:

$$K = {}^{\circ}C + 273.15$$

Example 1-4

1. When the thermometer reads 75°F, what is the temperature in degrees Celsius?

See the Question ► This problem is asking: 75°F = (?)°C

Think It Through ► We need to use the formula that converts temperatures between the Fahrenheit and Celsius temperature scales.

Execute the Math ►

$$1.8 \times {}^{\circ}C = {}^{\circ}F - 32$$

$$1.8 \times {}^{\circ}C = 75 - 32$$

$${}^{\circ}C = 23.8888\ldots \text{ (calculator answer)}$$

Prepare the Answer ► The answer should have two significant figures and be in degrees Celsius. The correct answer is 24°C.

2. A thermometer in Stockholm, Sweden, measures 11°C. What is the temperature in degrees Fahrenheit?

S ► The problem asks: 11°C = (?)°F

T ► We need to use the formula that converts temperatures between the Celsius and Fahrenheit temperature scales.

E ►

$$1.8 \times {}^{\circ}C = {}^{\circ}F - 32$$

$$1.8 \times 11 = {}^{\circ}F - 32$$

$${}^{\circ}F = 51.8$$

P ► The temperature given in the problem has two significant figures, so the answer to this problem is 52°F.

3. What does a thermometer with a Kelvin scale read if a thermometer in Celsius reads 56°C?

S ► The problem asks: 56°C = (?)K

T ► To calculate the answer to this problem we must use the formula that converts temperatures between the Celsius and Kelvin temperature scales.

E ►

$$K = {}^{\circ}C + 273.15$$

$$K = 56 + 273.15$$

$$K = 329.15$$

P

When adding numbers with significant digits, the number of decimal places in the answer equals the smallest number of decimal places in the numbers added. Therefore, the answer should be recorded as 329 K.

Check Your Understanding

1. Complete the following conversions:

   (a) $-85°F = (?)°C$ **(d)** $32.0°C = (?)°F$

   (b) $40.5°C = (?)°F$ **(e)** $142\ K = (?)°C$

   (c) $132°F = (?)°C$ **(f)** $37°C = (?)\ K$

2. Immediately after he won the 1982 Boston Marathon, Alberto Salazar collapsed. His body temperature at that moment was found to be 41.0°C. What was his body temperature in °F?

3. Lead melts at 621°F. What is its melting point in degrees Celsius? In kelvins?

1.15 Density and Specific Gravity

Define density and calculate the density, given the mass and volume of a given substance.

The **density** of a substance is defined as the mass of that substance per unit of volume (often expressed in grams per cubic centimeter: g/cm^3). Density depends on temperature because the volume of a substance can change with a change in temperature. Therefore, when comparing densities of several substances, the measurements are all taken at a constant temperature.

$$\text{Density} = \frac{\text{mass}}{\text{volume}}$$

The density of a substance is calculated from its measured mass and volume. If the object is irregular in shape, you can measure its volume by water displacement. Figure 1.11 and Table 1.4 compare the densities of some common substances.

The number used to express the density of a substance varies depending on the units used in measuring the density. The value for the density of water equals 1.00 if the mass is measured in grams and the volume in cubic centimeters; it equals 8.34 if the mass is in

Table 1.4 Densities of Some Common Substances

Substance	*Density (g/cm^3)*	*Substance*	*Density (g/cm^3)*	*Substance*	*Density (g/cm^3)*
Air (25°C)	0.041	Calcium carbonate	2.71	Copper	8.92
Water (4°C)	1.00	Sugar	1.59	Gold	19.3
Ice	0.917	Salt	2.18	Iron	7.86
Ethyl alcohol	0.789	Cement	2.7–3.0	Lead	11.3
Gasoline	0.660–0.69	Cork	0.22–0.26	Potassium	0.86
Olive oil	0.918	Bone	1.7	Sodium	0.97

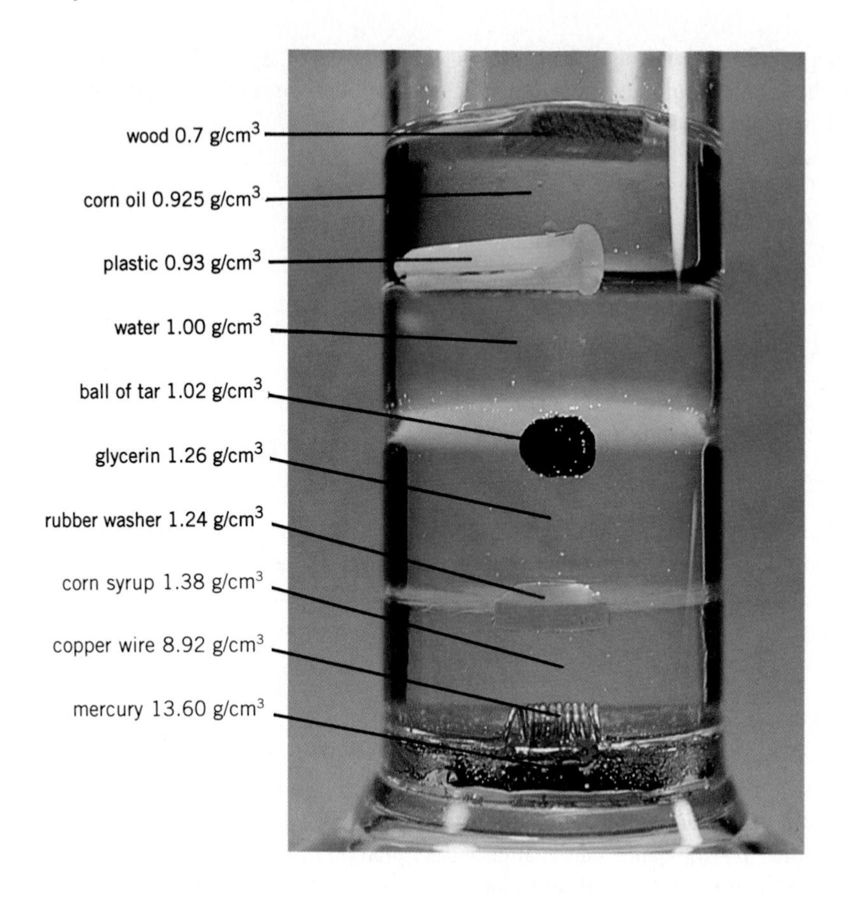

Figure 1.11 Because of their different densities, each substance will occupy a different level in this cylinder.

pounds and the volume in gallons. To avoid the problems that come from measuring densities using different units, we can instead calculate specific gravities. **Specific gravity** compares the density of a substance to the density of water (measured in the same units).

Define specific gravity and calculate the specific gravity, given the density of a substance.

$$\text{Specific gravity} = \frac{\text{density of sample}}{\text{density of water } (1.00 \text{ g/cm}^3 \text{ at } 4°\text{C})}$$

The specific gravity of a substance tells you how much denser the substance is than water and is most commonly used to compare the density of other liquids to water. For example, ethyl alcohol has a specific gravity of 0.789, meaning that it is less dense than water. Carbon tetrachloride, with a specific gravity of 1.59, is denser than water and twice as dense as ethyl alcohol.

Example 1-5

1. Calculate the density of nickel if a 5.00-cm^3 sample weighs 44.5 g.

See the Question → We need to calculate the density of nickel in grams per cubic centimeter.

Think It Through → Density is calculated by dividing the weight of a sample by its volume. In this case, we divide 44.5 g by 5.00 cm^3.

Chemical Perspective / Measuring Specific Gravity

Have you ever wondered why some objects float in water and others sink? When an object is placed in water it moves aside, or displaces, a certain amount of water. The weight of the water displaced is equal to the force holding up the object. An object floats if the weight of this displaced water is greater than the object's own weight. If we add dissolvable solids such as sugar to water, the solution becomes denser, and its specific gravity rises above 1.00 (the value for pure water). This means that the weight of liquid displaced also increases, making the force holding up an object that much greater. The higher the specific gravity of the liquid, the higher the object floats in it.

We can use this fact to measure the specific gravity of a liquid. A weighted, bulb-shaped instrument called a hydrometer is placed in the sample liquid. The higher the hydrometer floats, the less liquid it is displacing to equal its own weight and the denser the liquid. By viewing a scale calibrated in specific gravity readings on the side of the hydrometer, we can obtain a reading of the specific gravity of the sample. Hydrometers are used to measure the specific gravity of battery acid in car batteries and to keep track of the fermentation of wine and beer.

For example, when bottling home-brewed beer, a small amount of sugar is added to each bottle just before it is capped (this process is called priming.) The sugar serves as a source of carbon dioxide to give the finished product its fizz. If there is still some fermentable sugar remaining in the beer when the priming sugar is added, excessive amounts of carbon dioxide are generated. This can cause some bottles to blow their caps during the aging process.

One method of verifying the completion of the initial fermentation is by monitoring the specific gravity of the beer. The initial reading for beer at the beginning of the fermentation process is normally 1.035 to 1.042. As the yeast ferments the dissolved sugars into alcohol and carbon dioxide, the specific gravity of the liquid drops because of the decrease of sugar in the solution. The beer becomes less dense, and the hydrometer sinks lower in the liquid. When fermentation has stopped, the specific gravity remains constant. This usually occurs at a specific gravity of 1.005 to 1.017. At this point, it is safe to go ahead with the bottling.

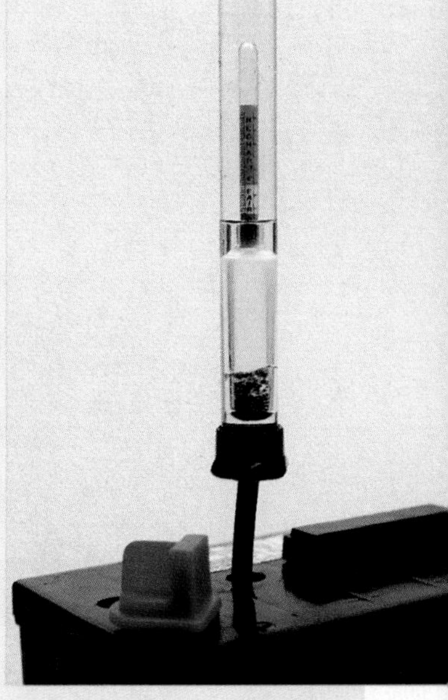

Hydrometer measuring the specific gravity of battery acid.

$$\text{Density of nickel} = \frac{44.5\text{ g}}{5.00\text{ cm}^3} = 8.9\text{ g/cm}^3$$

***E**xecute the Math*

***P**repare the Answer*

The calculator gives a result of 8.9, but each of the values in the problem has three significant digits. Therefore, the correct answer is 8.90 g/cm^3.

2. What volume does a 42.5-g sample of a liquid occupy if its density is 1.70 g/cm^3?

S

The problem asks us to find the volume of a specific sample of a liquid.

T

This question does not ask us to calculate the density of a substance. Instead, it gives us the density and asks for the volume occupied by a certain mass of the substance. We can use the density as a conversion factor to solve the problem. Because we want to solve for volume, we use the conversion factor that is the inverse of the density: 1.00 cm^3/1.70 g.

E

$$42.5\,\cancel{\text{g}} \times \frac{1.00\text{ cm}^3}{1.70\,\cancel{\text{g}}} = 25\text{ cm}^3$$

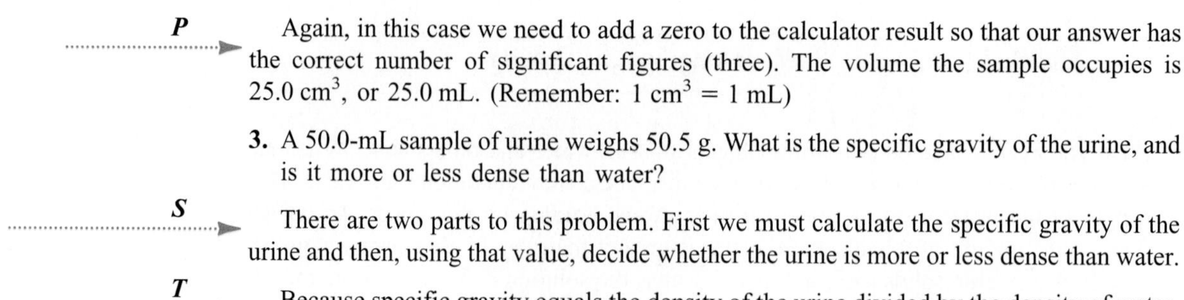

P Again, in this case we need to add a zero to the calculator result so that our answer has the correct number of significant figures (three). The volume the sample occupies is 25.0 cm^3, or 25.0 mL. (Remember: 1 cm^3 = 1 mL)

3. A 50.0-mL sample of urine weighs 50.5 g. What is the specific gravity of the urine, and is it more or less dense than water?

S There are two parts to this problem. First we must calculate the specific gravity of the urine and then, using that value, decide whether the urine is more or less dense than water.

T Because specific gravity equals the density of the urine divided by the density of water, we must first calculate the density of the urine. Density is mass, in grams, divided by volume, in cubic centimeters. We know that 1 cm^3 = 1 mL. The volume of the urine sample is therefore 50.0 cm^3.

E

$$\text{Density} = \frac{\text{mass}}{\text{volume}} = \frac{50.5\ \text{g}}{50.0\ \text{cm}^3} = 1.01\ \text{g/cm}^3$$

$$\text{Specific gravity} = \frac{\text{density of sample}}{\text{density of water}} = \frac{1.01\ \cancel{\text{g/cm}^3}}{1.00\ \cancel{\text{g/cm}^3}} = 1.01$$

P The answer for specific gravity is correct with three significant figures: 1.01. Because this value is slightly greater than 1, the urine in this sample is slightly denser than water.

Check Your Understanding

1. What is the density of lead if a 6.10 cm^3 sample weighs 68.9 g?

2. What volume does a 3.6-g sample of a liquid occupy if its density is 1.2 g/cm^3?

3. What does a bar of lead weigh if its dimensions are 3.00 cm by 5.50 cm by 2.00 cm? (Use the density value for lead from question 1.)

4. A 200-mL sample of ethylene glycol (antifreeze) weighs 222 g.

(a) What is the density of ethylene glycol?

(b) What is the specific gravity?

(c) Which is denser: water or ethylene glycol?

KEY CONCEPTS

Matter is anything that has mass and occupies space. Mass is a measure of how hard it is to change an object's speed or direction of movement. Weight is a measure of gravitational attraction on matter. The weight of an object can change, but the mass is always constant.

Pure substances are either elements or compounds. Elements can't be converted into simpler substances by ordinary chemical means. Compounds are composed of two or more elements chemically combined in definite proportions by weight. An atom is the smallest unit of an element having the properties of that element. A molecule contains two or more atoms chemically joined together.

Mixtures consist of two or more substances combined in any proportion. Mixtures can be homogeneous, meaning uniform throughout, or heterogeneous, meaning nonuniform.

Matter can exist in three states: solid, liquid, or gas. These three states differ in the closeness and the attraction between the particles of the substance.

Matter can undergo physical or chemical changes. Physical changes are changes in form; chemical changes are changes in the chemical composition of the substances involved.

The scientific method is the way in which scientists develop theories about the nature and behavior of matter. The data collected to support these theories should be both accurate and precise. The precision of a measurement is indicated by the number of significant figures or significant digits used to record the measurement.

The SI system of units is a modified metric system used by scientists to measure matter. All units in the SI system are related to their subunits by multiples of 10, and prefixes are used to show the power of 10 required.

The density of a substance indicates the mass of a designated unit of volume of that substance. The specific gravity of a substance compares the density of the substance to the density of water.

Important Equations

$$1.8 \times {}^\circ\text{C} = {}^\circ\text{F} - 32$$

$$\text{K} = {}^\circ\text{C} + 273.15$$

$$\text{Density} = \frac{\text{mass}}{\text{volume}}$$

$$\text{Specific gravity of a substance} = \frac{\text{density of the substance}}{\text{density of water}}$$

Review Problems

Section 1.1

1. What is the difference between the mass and the weight of an object?
2. If a person weighs 120 lb on the earth, what would her weight be on the moon? How much change is there in her mass?

Section 1.3

3. What is the difference between an atom and a molecule?
4. Give two examples of **(a)** an element, **(b)** a compound.

5. How does a mixture differ from a compound?
6. Give two examples of common foods in your kitchen that are **(a)** homogeneous, **(b)** heterogeneous.

Section 1.5

7. Label each of the following as either a physical or chemical change.
   (a) a log burning
   (b) milk souring
   (c) salt dissolving in water
   (d) ski goggles fogging
   (e) snow melting
   (f) a piece of chalk breaking
8. What is the difference between a physical change and a chemical change?

Section 1.6

9. What is the difference between a hypothesis and a scientific law?
10. Describe how the scientific method is used by a doctor in the diagnosis and treatment of a sickness.

Section 1.7

11. Two laboratory partners weighed the same sample of iron filings on two different balances. The weights they measured were as follows:

	Platform balance	*Analytical balance*
Iron filings:	10.4 g, 10.2 g	10.3507 g, 10.3509 g

Which balance is more precise? Which is more accurate? Give the reasons for your answers.

12. What is the difference between accuracy and precision?

Section 1.8

13. State the number of significant figures in each of the following measurements:
    (a) 0.0125 g
    (b) 150 mL
    (c) 12.060 km
    (d) 10.7 m
    (e) 4005.00 cm^3
    (f) 3.20×10^{-2} mg
    (g) 3000 mg
    (h) 0.00250 mm
    (i) 1604.732 kg
14. A meterstick divided into 1-cm units is used to measure the length of an object. To what degree of precision can we record our measurement of that length?

Section 1.9

15. Write two possible conversion factors for each of the following equalities:
    (a) 1000 g = 1 kg
    (b) 946 mL = 1 qt
    (c) 1 ton = 2000 lb

16. Which of the conversion factors in Problem 15 would you use to solve the following problems?

(a) How many kilograms are in 1500 g?

(b) How many milliliters are in 2.5 qt?

(c) How many tons are in 3300 lb?

Section 1.10–1.13

17. Set up the following problems using the correct conversion factor:

(a) 1.524 kg to g
(b) 2.55 qt to mL
(c) 1525 lb to tons
(d) 150 g to kg
(e) 550 mL to qt
(f) 2.1 tons to lb

18. Make the following conversions:

(a) 405 g to kg
(b) 1563 m to km
(c) 16 cm to mm
(d) 1.5 kg to g
(e) 15 mL to L
(f) 127 mm to m
(g) 0.07 g to mg
(h) 22.4 L to mL
(i) 0.3 m to cm
(j) 67 cm to m
(k) 125 mg to g
(l) 12 m to cm
(m) 456 μg to mg
(n) 1.3 m to mm
(o) 7.5 dL to L
(p) 5 mg to g
(q) 36 cL to dL
(r) 46 km to m
(s) 0.95 kg to g
(t) 0.02 L to mL
(u) 361 cm to m

19. Make the following conversions:

(a) 5.4 m to ft
(b) 1060 m to ft
(c) 76.2 cm to in.
(d) 85 km to mi
(e) 5.3 qt to mL
(f) 15.6 L to qt
(g) 2.0 gal to L
(h) 6.0 oz to g
(i) 105 g to lb
(j) 100 m to yd
(k) 14.5 oz to g
(l) 7.7 lb to kg
(m) 2.5 ft to cm
(n) 1.80 km to ft
(o) 227 mg to lb
(p) 1760 mL to gal
(q) 1.27 m to in.
(r) 0.55 lb to g
(s) 5.0 pints to L
(t) 4.7 L to pints
(u) 275 mL to pints

20. For each of the following, choose the answer that most closely applies:

(a) a pencil weighs 5 mg or 5 g or 5 kg?

(b) a man weighs 90 mg or 90 g or 90 kg?

(c) an aspirin tablet weighs 400 mg or 400 g or 4 kg?

(d) a coffee cup contains 25 mL or 250 mL or 2.5 L?

(e) a teaspoon contains 5 mL or 50 mL or 0.5 L?

(f) a quart of milk contains 10 mL or 0.1 L or 1 L?

(g) the length of a pencil is 15 mm or 15 cm or 15 m?

(h) the length of a football field is 92 cm or 92 m or 92 km?

Section 1.14

21. Make the following conversions:

(a) 71°F to °C

(b) 90.0°C to °F

(c) 151 K to °C

(d) −11°C to °F

(e) −49°F to °C

(f) 302°F to K

(g) 23°F to °C

22. The normal boiling point is shown for each of the following compounds. What is each boiling point in kelvins? In degrees Fahrenheit?

(a) silicon dioxide: 1610°C

(b) ethanol: 78.5°C

(c) hydrogen sulfide: −60.7°C

Section 1.15

23. For each of the following three liquids calculate (1) the density, (2) the specific gravity, (3) the weight of 250 mL of the liquid, (4) the volume in milliliters of 1.0 g of the liquid.

(a) liquid A: 100 mL weighs 79 g

(b) liquid B: 1 L weighs 1.1 kg

(c) liquid C: 500 mL weighs 490 g

24. What is the density of mercury if 15.0 mL weighs 204 g? What is the weight in kilograms of 250 mL of mercury?

25. A sample of water has a mass of 455 g at 25°C. The density of water at 25°C is 0.997 g/mL. What is the volume of this sample of water in milliliters? In liters?

Applied Problems

26. Your laboratory partner measures the length of a bar of iron (approximately 10 in.) as accurately as possible using a meterstick marked in millimeters. Which of the following measurements do you accept: 25.65 cm, 25.6 cm, 25 cm, 256.5 mm, 256 mm?

27. A sign on the freeway outside San Francisco reads "Los Angeles, 412 miles." How many kilometers is that?

28. How many square yards of carpet are needed to cover the floor of a rectangular bedroom measuring 12 ft by 10 ft? How many square meters of carpet?

29. Optical fibers 0.1 mm thick can be inserted into a patient's vein and maneuvered to the heart, where they deliver laser light that breaks up blood clots. What is the thickness of these optical fibers in inches?

30. The prescribed dose of an injected drug is 1.5 mL. If the syringe is graduated in cubic centimeters, to what mark do you draw the drug?
31. Wine is sold in bottles marked 750 mL. How many liters of wine are in 750 mL? How many quarts?
32. How many liters of water completely fill a swimming pool that measures 6.0 ft × 40 ft × 12 ft?
33. If you could put all of the toothpaste used in one day by Americans into one tube, it would fill a tube 68 ft long and would weigh 550,000 lb. **(a)** What would be the length of such a tube of toothpaste, in yards? **(b)** How many kilograms of toothpaste do Americans use every day?
34. Many people love cucumbers. In fact, Americans eat 2,800,000 lb per day. How many kilograms of cucumbers do Americans eat per week?
35. The sperm whale has the largest brain (about 9 kg) of any animal that has ever lived. This whale can dive to depths of more than 1.5 km while holding its breath for over 1.5 h.
    (a) What is the approximate weight of a sperm whale's brain in grams? In pounds?
    (b) What is the depth to which the whale can dive in meters? In miles?
36. In the treatment of ingrown toenails, the toe is sprayed with liquid nitrogen (N_2) at 77 K. The frozen skin dies, scales, and sloughs off, leaving the toenail unharmed. This eliminates the need for surgery and removal of the toenail. What is the temperature of the liquid nitrogen **(a)** in degrees Celsius, **(b)** in degrees Fahrenheit?
37. A weather forecaster on the television news states that it is 22°C in Paris and 32°C in Rome. What are the temperatures in Paris and Rome in degrees Fahrenheit?
38. Diamond has the highest melting point of any element: 3550°C. What is its melting point **(a)** in degrees Fahrenheit, **(b)** in kelvins?
39. A California couple was found dead in their hot tub, victims of hyperthermia (or heatstroke). Doctors and hot tub manufacturers recommend heating a tub to no more than 40°C; the water in the couple's tub was 46°C. What is the maximum recommended tub temperature and the couple's tub temperature in degrees Fahrenheit?
40. Do you expect gasoline to have a specific gravity greater than or less than that of water? Explain your answer.
41. One gallon of milk weighs 8.09 lb. Calculate the density of milk in grams per cubic centimeter.
42. The daily dosage of ampicillin for treating an ear infection is 100 mg per kilogram of body weight. What is the dosage for a 22-lb baby with an ear infection?
43. A nurse must administer the appropriate dosage of atropine sulfate to a 180-lb male patient. The dosage on the label of a bottle of atropine sulfate tablets reads 20 mg for every kilogram of body weight. If each tablet contains 5 grains of atropine sulfate, how many tablets are needed to give the correct dose to the patient?
44. The cells of *Clostridium botulinum* produce the toxin responsible for the vomiting and dizziness (and occasionally death) people experience who eat canned goods contaminated with this microorganism. A 200-lb person may suffer mild botulism poisoning from ingesting a dose of the toxin as small as 16 ng. Express this dose in milligrams per kilogram.
45. Two 16-oz bottles of a soft drink cost 62 cents. The same drink is sold in a 1-L bottle for 50 cents. Which is the better buy?

46. A nurse is to give a patient 3 L of fluids over a 24-hour period. How many cubic centimeters of fluid must he give each hour?

47. A circus fat lady, ''Dolly Dimples,'' was suffering from a life-threatening heart condition and had to lose weight. She went from 553 lb to 152 lb in a 14-month period. What was her average weight loss in kilograms per month?

48. A car is rated with a highway driving mileage of 35 miles per gallon. How many liters of gasoline are needed for a highway trip of 850 km?

49. A 12-oz can of a popular cola drink contains 12 tsp of sugar. If each teaspoon of sugar weighs 4.0 g, how many pounds of sugar are in a case of 12 cans? If a student drinks 2 cans of this cola a day, how many pounds of sugar will she consume over a year?

Chapter 2
Energy

*H*annah Segal, an energetic 81-year-old, lived alone in her Cleveland home. She volunteered time at the local library and never missed her Tuesday bridge game or Friday night bingo. One day in early December, however, things were quite different; she came into the library that morning confused, drowsy, and forgetful. A librarian, noticing how Hannah was stumbling as she walked among the shelves, became alarmed at this sudden change in behavior and decided to take her to the hospital. Hannah's condition continued to deteriorate even as they drove to the hospital, and she fell into a coma soon after arrival. The emergency room physician diagnosed a stroke and told Hannah's family that her chances of survival were slim. Even if she were to recover, he continued, the lasting effects of the stroke probably would require that she live in a nursing home. Luckily, an alert nurse noticed that Hannah's body was unusually cold to the touch. Using a special thermometer, she found that Hannah's temperature was 89°F (or 32°C). Hannah had not suffered a stroke but was instead suffering from hypothermia, a lowering of the body's inner temperature. This condition can be fatal unless the patient is quickly warmed and given intravenous fluids to prevent dehydration and restore the body's chemical balance. After 12 hours of treatment, Hannah's body temperature was back to normal. After several days, she was able to return to her home, eager to resume her regular activities.

Hypothermia is a condition that can be fatal when the body's temperature drops as few as 6°F below the normal of 98.6°F (a drop of 3°C from the normal 37°C). Surprisingly, a person can survive a tem-

perature drop of 40° to 50°F in the hands and feet, but only a small drop in the temperature of the body core can cause death. Hypothermia begins as soon as the body starts to lose energy faster than it can be produced. The hands and feet are affected first, as the body diverts warm blood to vital internal organs by constricting the blood vessels in all extremities. A drop of only 3°F in body temperature reduces manual dexterity to the point that the person is unable to perform the basic tasks necessary for survival. In addition, as the temperature of the body lowers, the central nervous system becomes depressed, resulting in progressive confusion, slurred speech, stumbling, seizures, and ultimately, coma and death.

Hypothermia has been diagnosed for centuries as the cause of death for such people as mountain climbers, skiers, and boaters who were inadequately protected against the cold. This type of hypothermia is called exposure hypothermia. Only recently has hypothermia been recognized as a common condition that affects elderly people who remain indoors, but who dress inadequately to protect themselves against house temperatures below 65°F. This form of hypothermia is called accidental hypothermia. Lack of activity of the thyroid gland or side effects of phenothiazines (drugs widely prescribed in the treatment of psychiatric conditions or nausea and vomiting) can also cause accidental hypothermia. Young people can counteract heat loss by regulating the amount of blood flow through the skin and by shivering. For unknown reasons, shivering, which increases heat production about five times, is often absent in elderly people. Also, older people seem to lose their perception of cold as hypothermia begins. Commonly, when elderly people have been found dead in their homes, their deaths have been attributed to natural causes or falls. Now it is suspected that hypothermia has often been the actual cause of death.

The need for a person to maintain a stable body temperature in a cold environment is only one example of the complex energy interactions between living systems and their environments. An understanding of these interactions requires an examination of the different types of energy that are involved and the energy changes that can occur.

2.1 What Is Energy?

Energy is a topic that seems to come up all the time. Sometimes it's in the context of the world's energy supply—referring to the petroleum and coal taken from the earth. Sometimes it's in discussions of nuclear energy, either peaceful energy from nuclear reactors or destructive energy from nuclear weapons. Sometimes it's in more personal terms, as in ''I just don't seem to have any energy today!'' Though you may use the word *energy* daily, you may have only a vague notion of precisely what energy is. Some days you may wake up feeling energetic, ready to get things accomplished—and the tasks you complete by the end of the day will have resulted from your expenditure of energy. It is precisely the ability to accomplish something—the capacity to do work—that defines **energy.** Energy is the ability to cause various kinds of change, such as moving an object from one place to another, breaking a chemical bond, or removing an electron from an atom. A rushing stream, a rock poised at the top of a hill, the gasoline in a car, the muscles in an arm, each has the capacity to cause change, to do work on an object. Each, therefore, possesses energy.

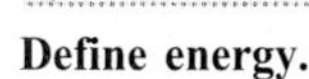

Define energy.

2.2 Kinetic Energy

Energy can take many forms, but it is often convenient to classify it as either energy of motion or energy of position. The name applied to energy of motion is **kinetic energy.** A car traveling at 20 mph, a rock hurtling through the air, and steam billowing from a water boiler all have energy by virtue of motion. That is, each possesses kinetic energy. The moving car could do a great deal of damage if it hit a parked car, the rock could break a window, and the steam could push the pistons of a steam engine. Although it is obvious that both the car and the rock have motion and, therefore, possess kinetic energy, you may wonder about the steam. The rock and the car are large objects whose motion we can see. If we had a ''super microscope'' that could let us see at the level of atoms and molecules, we would realize that all matter is in constant motion. The molecules of water in steam are moving extremely rapidly and, for this reason, possess a great deal of kinetic energy (Fig. 2.1).

Kinetic energy can be measured. Its numerical value depends on the mass of the particle and the particle's speed or velocity. Intuitively, you have a good general feeling for the kinetic energy possessed by an object. For example, you probably would rather have an eight-year-old throw a baseball at you than have a major league pitcher try it. In both cases, the baseballs would have the same mass, but they would be traveling at different speeds and therefore have different amounts of kinetic energy. The ball thrown by the child would be traveling at a much slower speed; it would have less kinetic energy and would do less damage to your hand. As a second example, would you rather have your parked car hit by a bicycle going 5 mph or a dump truck going 5 mph? In this example the speed of the two objects is the same; however, the much larger mass of the dump truck gives it a much larger kinetic energy and a much greater capacity to do work—that is, to crumple your fender.

A baseball, a car, and a dump truck are all objects that we can weigh, whose speed we can determine, and whose kinetic energy we can then calculate. But how can we determine the kinetic energy of particles that we can't see, such as the molecules of water in steam? The temperature of a substance allows us to measure the kinetic energy of its particles. Scientifically, **temperature** is a measure of average kinetic energy. The term *average kinetic energy* is used because all molecules, as with all people, are not alike. At any temperature, some molecules are moving very rapidly, some very slowly, and the majority somewhere in between. Just as we can talk about the average behavior or performance of

Figure 2.1 The kinetic energy possessed by steam molecules is used to power this steam engine.

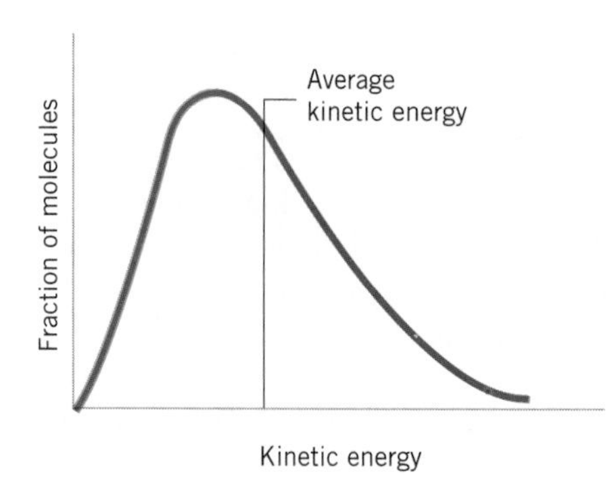

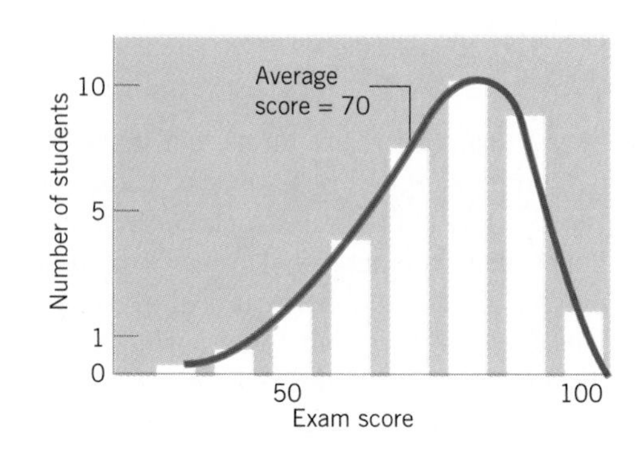

Figure 2.2 At any given temperature, each gas molecule has a specific kinetic energy, just as each student has a specific score on an exam. Just as there is an average score for the exam, however, there is an average kinetic energy for the gas molecules. Temperature is a measure of this average kinetic energy.

State which molecules have greater kinetic energy: those in a sample of water at 20°C, or those in steam at 100°C.

a group of students, however, so can we refer to the average behavior of a group of atoms or molecules (Fig. 2.2). As the average kinetic energy of a group of molecules increases, so does the temperature of that substance. From experience, you know that the hotter a pan of water, the more severely the water can burn your hand, or the faster it can cook vegetables. You now understand that the hotter water has a higher average kinetic energy and, therefore, a greater ability to do work, such as damaging tissues or cooking vegetables.

2.3 Potential Energy

State the difference between substances possessing kinetic energy and those with potential energy.

A rock hurtling through the air has kinetic energy; when it strikes a window, the rock can break it. A rock perched on top of a cliff also possesses energy; it has the capacity to do work. If it fell off the cliff onto a passing truck, it could do severe damage (Fig. 2.3). The rock on top of the cliff possesses **potential energy,** or energy of position. Such energy is not in use but is stored and has the capacity to do work when it is converted to other forms

Figure 2.3 The rocks (left) possess potential energy. These rocks could do significant damage if they slid off the hillside onto a passing truck (right).

of energy. For example, the water stored behind a dam possesses potential energy. When released in a controlled fashion, it can drive turbines to produce electric energy, or if released in an uncontrolled fashion by the collapse of the dam, the water could display its energy by destroying everything in its path.

The food we eat is another example of a substance possessing potential energy. The energy stored in food is called **chemical energy.** As the food molecules are broken down (or metabolized) in our bodies to simpler molecules with lower chemical energy, we use the released energy to contract muscles or maintain body heat.

2.4 Heat Energy

Energy can be transferred from one place to another in many forms, among which are electricity, sound, light, and heat. Heat is energy that is transferred from one place to another because of a difference in temperature. A **calorie (cal)** is the amount of energy required to raise the temperature of one gram of water one degree Celsius. The SI unit of heat energy is the **joule (J).**

Define calorie.

$$1 \text{ cal} = 4.1840 \text{ J}$$

The quantity of heat required to raise the temperature of one gram of a substance by one degree Celsius is called the **specific heat capacity,** or more commonly, the **specific heat** of the substance.

$$\text{Specific heat} = \frac{\text{cal}}{\text{g} \times {}^\circ\text{C}}$$

Liquid water has one of the highest specific heats known: 1.00 cal/g°C. This means that it takes a lot of heat energy to change the temperature of water. For example, one calorie of heat energy has about twice the effect on the temperature of a sample of ethyl alcohol (specific heat 0.586 cal/g°C) as it would have on the same-sized sample of water.

The amount of heat necessary to change the temperature of a substance depends on the mass of the substance, its specific heat, and the change in temperature (Δt = final temperature − initial temperature).

Use experimental data to calculate the number of calories in a food sample.

$$\text{Calories} = \text{grams} \times \text{temperature change } (\Delta t) \times \text{specific heat}$$

$$\text{cal} = \text{g} \times \Delta t \times \frac{1 \text{ cal}}{\text{g}^\circ\text{C}}$$

We can use this equation to calculate the number of calories required to raise the temperature of a cup of water (250 g) from 25° to 100°C to make a cup of coffee. In this case,

$$\Delta t = 100^\circ\text{C} - 25^\circ\text{C} = 75^\circ\text{C}$$

$$\text{cal} = 250 \cancel{\text{g}} \times 75 \cancel{^\circ\text{C}} \times \frac{1.00 \text{ cal}}{\cancel{\text{g}^\circ\text{C}}}$$

$$= 18{,}750 \text{ cal or } 19 \text{ kcal}$$

Because a calorie is a very small unit of energy, we often find it more convenient to talk in terms of 1000 cal, called a **kilocalorie (kcal).** The food Calorie (note the capital C) that

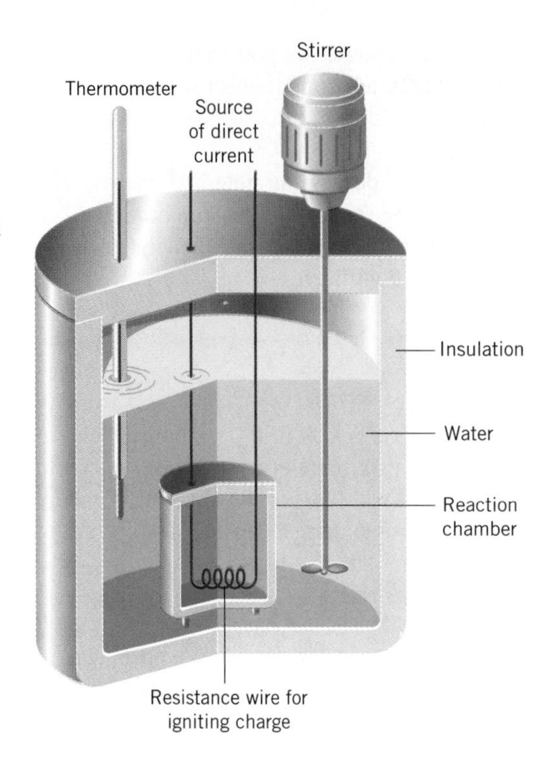

Figure 2.4 A calorimeter consists of a reaction chamber that is surrounded by water held in an insulated shell. As the sample is burned in the reaction chamber, the surrounding water is stirred and its temperature change measured.

you count when you are on a diet is actually a kilocalorie. That 1-oz bag of potato chips that you had for a snack contains 160 food Calories or 160 kcal. To avoid confusion, in this book we shall always refer to food energy content in terms of kilocalories.

The instrument that is used to find the calorie content of potato chips or other substances is called a **calorimeter** (Fig. 2.4). The sample to be tested is placed in the inner chamber and then is burned completely. The energy released from this burning warms the water in the surrounding container, allowing the number of calories transferred to be calculated from the rise in water temperature.

◆ **Define basal metabolism rate.**

The minimum amount of energy required daily to maintain the basic continuous processes of life in the body at rest is called the **basal metabolism rate (BMR).** This amount of energy varies for each individual: a woman weighing 121 lb (55 kg) requires about 1400 kcal/day, and a man weighing 143 lb (65 kg) about 1600 kcal/day. Calories that are consumed in excess of that amount are either used to supply the energy needed to do work or are deposited as fat. About 3500 excess kilocalories produce one pound of body fat.

Example 2-1

1. What is the average energy content of a peanut if the temperature of 1000 g of water in a calorimeter is increased by 50°C when 10 peanuts are burned?

See the Question →

This problem asks us to calculate the number of calories of food energy contained in each peanut.

$$1 \text{ peanut} = (?) \text{ calories}$$

Think It Through →

The problem gives us data from burning 10 peanuts. We can use these values to calculate the total number of calories, and then divide the answer by 10 to find the number of calories in one peanut.

Health Perspective / Calories and Exercise

Aren't you getting tired of all the radio, TV, and magazine ads for weight loss programs? Americans are obsessed with being thin. The problem is that most people live inactive lives and have eating habits that help put on extra pounds. Although crash diets can successfully take off a person's excess weight, such diets seldom keep the weight off. These diets are also often low in essential minerals and nutrients. The most sensible approach to weight loss is to combine a balanced low-calorie diet with a reasonable program of exercise.

It's a simple case of taking in less and putting out more. Each pound of fat contains approximately 3500 kcal. If you take in 1500 kcal a day and expend 2200 kcal during the same period, you can lose one pound of fat in five days. By exercising each day, your energy expenditure increases, and the weight loss occurs at a faster rate. Not only do you lose weight quicker, you also build muscle. Such an increase in muscle eventually leads to a higher metabolic rate, making it easier to maintain a lower body weight.

The average adult at rest metabolizes about 1 kcal/h for every kilogram (2.2 lb) of body weight. The caloric expenditures from daily activities vary considerably among individuals. The daily expenditure of energy for a typical student is 2300 to 3100 kcal. The table shows the relationship between certain activities and their energy expenditure.

Activity	*Energy Expended (kcal/h) by a 150-lb Adult*
Sleeping	80
Sitting	100
Walking (2.5 mph)	324
Cycling (5.5 mph)	330
Skiing (downhill)	486
Basketball	564
Swimming (fast crawl)	636

Your Perspective: Suppose that a medical clinic hired you to advertise their new weight-loss program, which is based on the above principles of good health. Write a script for a 30-second radio advertisement aimed at convincing people to join the program.

Execute the Math

$$\text{Calories} = \text{grams} \times \Delta t \times \text{specific heat}$$

$$= 1000\,\cancel{\text{g}} \times 50\,\cancel{°\text{C}} \times \frac{1.00\ \text{cal}}{\cancel{\text{g}}\,\cancel{°\text{C}}}$$

$$= 50{,}000\ \text{cal produced by 10 peanuts}$$

$$\frac{50{,}000\ \text{cal}}{10\ \text{peanuts}} = \frac{5000\ \cancel{\text{cal}}}{1\ \text{peanut}} \times \frac{1\ \text{kcal}}{1000\ \cancel{\text{cal}}} = \frac{5\ \text{kcal}}{1\ \text{peanut}}$$

Prepare the Answer

Each of our values has only one significant figure (the zeros are place markers), so our answer of 5 kcal per peanut is correct.

2. How many calories are required to raise the temperature of a 150-mL sample of water from 15.0°C to 20.0°C? (The density of water = 1.0 g/mL at 15°C)

In this question we are asked to calculate the number of calories needed to raise the temperature of a sample of water 5.0°C.

The formula for calculating calories requires us to know the change in temperature, the number of grams of water, and the specific heat. The problem tells us the volume of the water sample but not the number of grams. We can use the value of the density of water to calculate the weight of our sample.

$$150\ \text{mL} \times \frac{1.0\ \text{g}}{1\ \text{mL}} = 150\ \text{g}$$

$$\text{Calories} = 150\ \text{g} \times 5.0\text{°C} \times \frac{1.00\ \text{cal}}{\text{g°C}}$$

$$= 750\ \text{cal}$$

The smallest number of significant figures in the given data is two, so our answer should have two significant figures: 750 cal.

Check Your Understanding

1. How many kilocalories of energy are contained in 1.2 kg of body fat?
2. A glass of cold water is a refreshing drink on a hot day. When it enters your body the water is quickly warmed to body temperature. How many kilocalories of energy must be transferred within your body to warm 500 g of water from 4° to 37°C?

2.5 Changes in State

A **change in state** occurs when matter goes from one state to another, such as when a solid melts or a liquid evaporates. Changes in state are physical changes for which energy must either be added to make the process occur (as in melting or boiling) or be given off as the process occurs (as in condensation or freezing). A change that occurs only when energy is added is said to be **endothermic.** A change in which energy is given off as the process occurs is called **exothermic.** Melting is an endothermic process, whereas freezing is an exothermic process. For water to freeze, energy must be removed from the water. In a freezer, water loses energy and freezes. Because energy has been given up by the water, the process is exothermic.

Define exothermic and endothermic processes and give examples of each.

2.6 Electromagnetic Energy

Describe five types of electromagnetic radiation and list them in order of increasing energy and in order of increasing wavelength.

Light represents another form of energy that always surrounds us. The light we see is only a tiny part of an entire range of electromagnetic energy, from low-energy radio waves to very high-energy X rays and gamma rays. What we see as white light is actually made up of the whole visible spectrum that can be separated into its component colors, as seen in a rainbow. Figure 2.5 shows the location of this visible spectrum in the **electromagnetic spectrum.**

Health Perspective / Steam Burns

Anyone who has carelessly removed the lid from a pan of boiling water is painfully familiar with a steam burn. The burn caused by exposing your hand to an equivalent amount of boiling water, however, would not be nearly as severe.

Let's look at what happens when boiling water at 100°C touches the skin. Heat is immediately transferred to the skin until the temperature of the water equals the temperature of the skin. We assume the temperature of the skin to be 37°C. This means that the water changes its temperature by 63°C. The calories lost by the water are transferred to the skin, resulting in the burn.

If we assume the burn to be caused by 10 g of boiling water, the calories can be calculated as follows:

$$\text{Calories} = 10\,\cancel{\text{g}} \times 63\,\cancel{^\circ\text{C}} \times 1\ \text{cal}/\cancel{\text{g}}\,\cancel{^\circ\text{C}} = 630\ \text{cal}$$

Now look at the situation when the same amount of steam at 100°C touches the skin. Before the steam can lower its temperature, it must first change its phase from gas to liquid. This is called condensation. For water, a tremendous amount of energy is involved: 540 cal is released for every gram of steam that condenses to water. Therefore, 10 g of steam loses 5400 cal (10 g × 540 cal/g) just returning to the liquid phase. The resulting liquid water then loses energy until its temperature reaches that of the body, 37°C. The total heat lost by the steam is the sum of these two processes:

$$5400\ \text{cal} + 630\ \text{cal} = 6030\ \text{cal}$$

As you can see, because of the energy released, a steam burn is much more severe than one from boiling water.

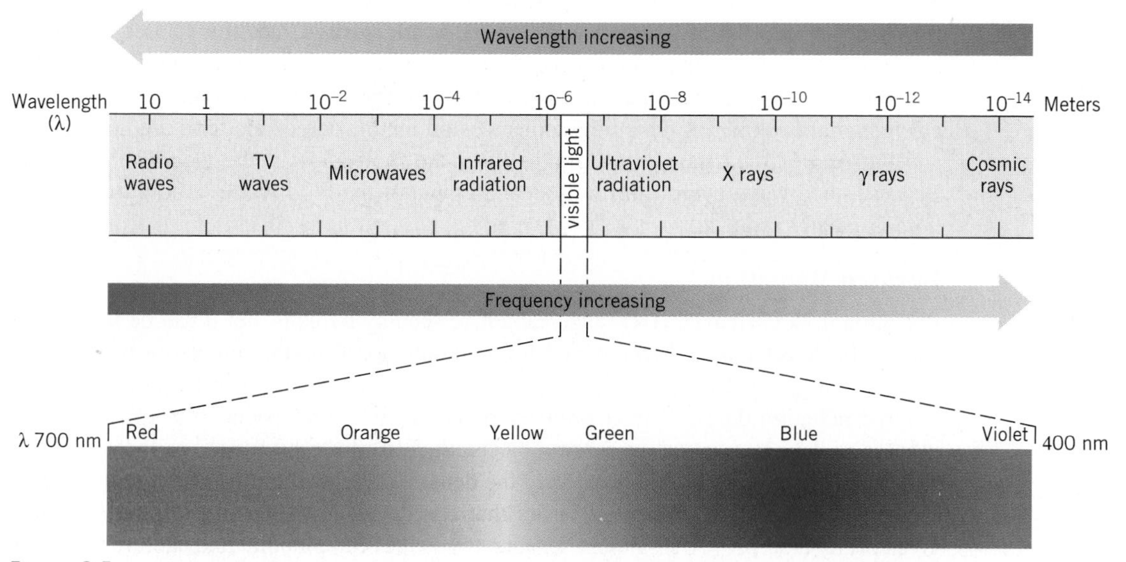

Figure 2.5 The electromagnetic spectrum, of which visible light makes up only a small part.

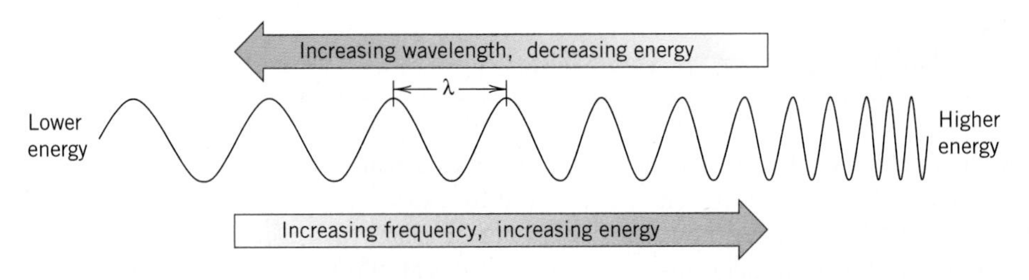

Figure 2.6 Light travels in waves. The wavelength of light can be defined as the distance from crest to crest. Each wavelength is associated with a particular energy: the longer the wavelength, the lower the energy.

Light waves may be compared to ocean waves, with their crests and troughs. The **wavelength** of light (represented by the Greek letter lambda, λ) is the distance from one crest to the next crest or from trough to trough, or from middle to middle (Fig. 2.6). The **frequency** of light is related to the time that it takes for one wavelength to pass a point or, stated another way, how many waves can pass a point in a given time. Because all light travels at a constant speed, the longer the wavelength of light the lower the frequency. Imagine a train and a car pass you, both traveling at 45 mph. It takes the train longer to pass you than it takes the car. In fact, many cars could pass during the time it takes one train to pass. In this analogy the train has the longer wavelength and, therefore, the lower frequency.

Each wavelength of light corresponds to a specific level of energy: the longer the wavelength, the lower the energy of the light. The relationship can be expressed mathematically by the equation

$$E = \frac{k}{\lambda}$$

where E is the energy of the electromagnetic radiation, and k is a constant related to the speed of light. Look at the visible spectrum shown in Figure 2.5. Visible light with the highest energy and shortest wavelength (highest frequency) is found in the purple region, whereas red light has the lowest energy and the longest wavelength (lowest frequency).

Microwave Radiation

You're familiar with the use of long-wavelength, low-energy radiation to send radio and television signals. Slightly shorter wavelength, or **microwave,** radiation is used as a popular means of rapid cooking. Microwaves interact with food by setting the molecules within the food into motion, causing the food to heat uniformly. This is quite different from conventional cooking, in which the molecules on the outside of the food are heated first, and the heat is then transferred gradually to the molecules in the center of the food. Because microwave ovens uniformly heat all parts of the food at the same time, cooking time is greatly reduced.

Infrared Radiation

Radiation in the **infrared (IR)** range cannot be seen by humans, but it can be felt as heat and can be detected by a thermometer. Infrared radiation from the sun warms us, and light fixtures that give off this radiation are used to keep food warm in restaurants. Although infrared radiation doesn't affect ordinary photographic film, special film can detect the infrared radiation given off by warm objects. Such film has been used to locate thermal pollution from power plants and to map the density of vegetation in certain regions. Two species of snakes have infrared sensors that can detect low-intensity infrared radiation given off by their prey. Using these sensors, the snakes can pinpoint and capture their prey in the dark and can locate hiding places having comfortable living temperatures.

Ultraviolet Radiation

Ultraviolet (UV) light has shorter wavelengths and, therefore, higher energy than visible light. It is penetrating radiation that can harm living tissue by causing changes in certain biological systems or by causing irreparable damage to cells. Fortunately, most of the ultraviolet radiation coming to the earth from the sun is absorbed by the ozone layer in the upper atmosphere and never reaches us. Because some wavelengths of ultraviolet light are very effective in killing bacteria, many sterilizing units use this type of radiation. It is perhaps surprising, then, that UV light is also important to our bodies. It causes the production within the skin of vitamin D, a compound necessary for the prevention of the bone disease rickets.

X Rays and Gamma Rays

X rays and **gamma rays** have even shorter wavelengths, and thus higher energies and greater penetrating power. On earth, X rays are produced by specially constructed X-ray tubes, whereas gamma rays come from natural sources. The shortest X rays can pass through steel walls, and gamma rays can pass through a 10-inch-thick lead plate. Because X rays do not pass as easily through bones and teeth as they do through tissue, they are a very useful diagnostic tool for the medical and dental professions. Penetration of cells by high-energy radiation such as X rays or gamma rays, however, can disrupt normal chemical processes, causing cells to grow abnormally or die. Because the effects of radiation build up over time, dosages and repeated exposure to X rays must be carefully controlled. Cancer cells are more sensitive to radiation than normal cells, so X rays and gamma rays can be used to treat certain forms of cancer. Medical radiation equipment is designed to control the exact degree of penetration of the radiation and to concentrate the radiation on the cancerous area to minimize the damage to normal tissue.

2.7 Law of Conservation of Energy

Energy can be converted from one form to another and ultimately ends up as heat energy. For example, electric energy is converted to heat energy in our furnaces, stoves, and toasters, and to light energy and heat energy in our lamps. Chemical energy is converted to heat energy when gas is burned in stoves or furnaces, or to mechanical energy and heat energy when gasoline is burned in car engines. Although energy can be converted from one form to another, the total amount of energy in the universe is constant. This means that energy can neither be created nor destroyed but only changed in form. In other words, the total amount of energy at the end of a reaction or process must equal the total amount of energy at the beginning. This is a statement of the **law of conservation of energy,** or the **first law of thermodynamics.**

State the first law of thermodynamics.

You know that gasoline is burned to provide energy for powering automobiles. But you might be surprised to learn that only about 20% of this energy is used in moving the car. Where does the rest of the energy go? (The law of conservation of energy says that it cannot just disappear.) In this case, the rest is lost as heat energy—*lost* in the sense that this energy does not do any useful work.

In the same way that gasoline provides energy for a car, the food we eat provides energy for our cells. In this case, too, the conversion of chemical energy in the food to energy that our cells can use is not 100% efficient. For example, only about 44% of the energy that can be transferred from a molecule of sugar is converted to energy that our cells can use to perform work. The rest ends up as heat energy that helps to maintain body

Describe how energy derived from the food we eat is used by the body.

Health Perspective / Why Sunscreens?

Sunlight can be both good news and bad news for our skin. The good news is that sunlight striking the skin stimulates the production of vitamin D, which is important for healthy bones and teeth. The bad news is that the ultraviolet radiation in sunlight can produce the familiar burn we call sunburn. Long exposure to ultraviolet radiation from the sun or a tanning booth can cause premature aging (wrinkling and "liver" spots), thick warty growths on the skin (a condition called keratosis), and even skin cancers. Just one blistering sunburn in childhood can double your chance of developing malignant melanoma, a particularly aggressive form of skin cancer. We are protected from most of the UV radiation coming to earth by a layer of ozone in the upper atmosphere. This ozone layer has been slowly thinning since the late 1970s, however. As a consequence, in the early 1990s large increases in the amount of UV radiation have been detected reaching some parts of the earth's surface.

Radiation in the ultraviolet part of the spectrum can be divided into two regions: UV-A and UV-B. UV-B radiation has shorter wavelengths and higher energy than UV-A. It causes damage quickly, producing the redness, blistering, and pain of sunburn. Such damage to the skin from sunburn is the same as that caused by intense heat. Long-term exposure to UV-B can produce skin cancers, especially in fair-skinned individuals. In contrast, UV-A radiation has longer wavelengths and is less energetic than UV-B. It acts more slowly on the skin, producing a slow tan rather than a sunburn. But UV-A radiation affects elastic tissues in the skin, converting them to more rigid collagens that produce premature aging and wrinkling of the skin. UV-A radiation may also play a role in the formation of melanomas, the most deadly form of skin cancer.

The cells of the skin do have some protection against ultraviolet light. First, the tough layer of dead cells on the surface of the skin absorbs some UV radiation. Second, exposure to the sun triggers cells in the skin to produce more of the skin pigment called

temperature. (Actually, a transformation of energy in any type of system results in the production of some waste heat.) Much to many people's unhappiness, our bodies must obey the law of conservation of energy. If we take in more food energy than our bodies need, this excess energy doesn't just disappear. Instead, it is stored within our tissues, mainly as unwanted fat. To get rid of this excess stored energy, we must take in less food energy than our bodies require so that our cells start using the fatty tissue as a source of energy to maintain body activities.

2.8 Entropy

You might be wondering why we need a constant supply of new energy from the sun to maintain life on earth if energy is conserved in all natural processes. Although the first law of thermodynamics tells us that the total quantity of energy remains unchanged, it tells us nothing about the quality of that energy. Only concentrated forms of energy can be used to do work, and work is essential to the maintenance of living organisms. The energy from the sun reaches us in the very concentrated form of light energy, but this energy is quickly converted to heat, a less concentrated form of energy.

melanin, which acts further to block the ultraviolet radiation. It is the appearance of this increased melanin in the skin that we call a suntan. But an effective tan takes three to five days to develop and even then reduces the penetration of UV radiation by only one-half.

Additional protection from UV radiation can be obtained by applying a sunscreen. Each lotion has a sunscreen protection factor (SPF), which is listed on the bottle. A sunscreen lotion with an SPF of 8 reduces the amount of ultraviolet radiation reaching the skin by a factor of 8. This means that if you apply this lotion and stay out in the sun for 8 hours, your skin is exposed to the same amount of UV radiation that would penetrate unprotected skin in 1 hour. Although you can find sunscreens with SPFs higher than 15, it is questionable how much additional protection they provide for normal activities in the sun.

Different sunscreen lotions contain chemicals that may block only UV-B, or that may block both types of UV radiation. The most effective sunscreens are those containing oxides of zinc or titanium. These compounds allow no sunlight to reach the skin. One popular sunscreen chemical, PABA (*para*-aminobenzoic acid), blocks only UV-B. It lets through the UV-A radiation that produces a tan. Too much exposure to UV-A, however, can cause a nasty skin rash called sun poisoning. This rash is often erroneously blamed on an allergy to PABA. Because of the long-term damaging effects of both UV-A and UV-B radiation, your best choice is a broad-spectrum sunscreen with an SPF of 15. The U. S. Environmental Protection Agency recommends that you protect yourself with a sunscreen whenever there is enough sun for you to see your shadow.

Your Perspective: Imagine you had a friend who was spending winter semester studying coral reefs in Tahiti, and suppose she has asked you for advice in safely acquiring a magnificent tan. Write a letter to your friend advising her how to proceed.

If you think about it for a moment, you will realize that all naturally occurring (or spontaneous) processes tend to go in one direction. Left alone, water always runs downhill, heat flows from a hot to a cold object, gases flow from regions of high pressure to regions of low pressure, and people grow old. We would be shaken if we were to see water flowing uphill on its own, but we still might ask what it is that prevents naturally occurring processes from reversing direction. The answer to this question is fairly obvious when the final state of the system is at a lower energy level than the initial state. Water flows downhill from higher to lower potential energy, and heat flows from a state of higher to lower kinetic energy. It is reasonable that there should be a general tendency for all substances to reach a state of lower energy.

But other spontaneous processes with which you are familiar don't seem to depend on the energy of the system. For example, if you put a drop of ink into a glass of water, the ink immediately begins to spread throughout the liquid, eventually giving the water a uniform tint (Fig. 2.7). This process certainly occurs in a single direction. You would be astonished if you ever saw a glass of tinted liquid suddenly change so that all the color moved through the liquid to form a single concentrated spot of dye.

This example leads us to conclude that another factor determines the direction of spontaneous processes. That factor is the amount of randomness, or disorder, of the sys-

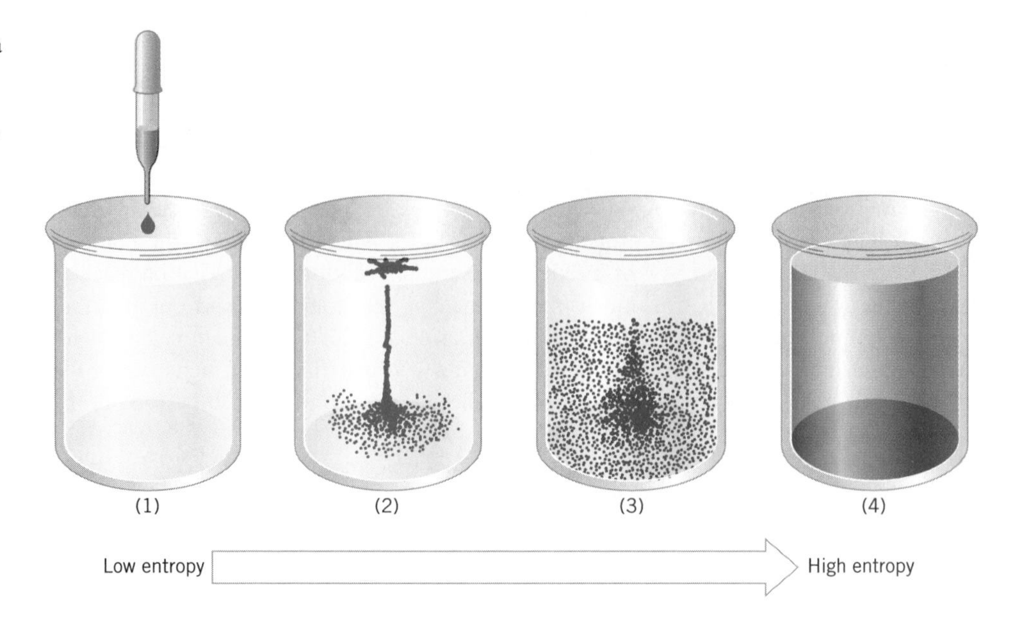

Figure 2.7 Ink dropped into a beaker of water quickly changes from a state of low entropy (the droplet) to a state of high entropy (spread throughout the water).

Define entropy and tell how the entropy changes in a spontaneous reaction.

tem. The term used to describe the disorder of a system is **entropy.** The more random and disordered a situation is, the greater its entropy. All spontaneous reactions go toward a condition of greater randomness, greater disorder, and greater entropy. We can now see that the drop of ink, which was originally in a small compact drop, spontaneously spreads throughout the liquid into a random and disordered arrangement of ink molecules, increasing the entropy of the system. The driving force behind the spreading of the drop of ink is the tendency for the entropy of the system to increase.

The **second law of thermodynamics** states that the entropy of the universe is increasing. This law is interwoven through all science and into each of our lives. Even the writers of nursery rhymes had an intuitive feeling for the second law when they wrote, ''All the king's horses and all the king's men couldn't put Humpty together again.'' You now see that the breaking of an egg is an irreversible process that leads to increased disorder of the egg parts and, therefore, greater entropy.

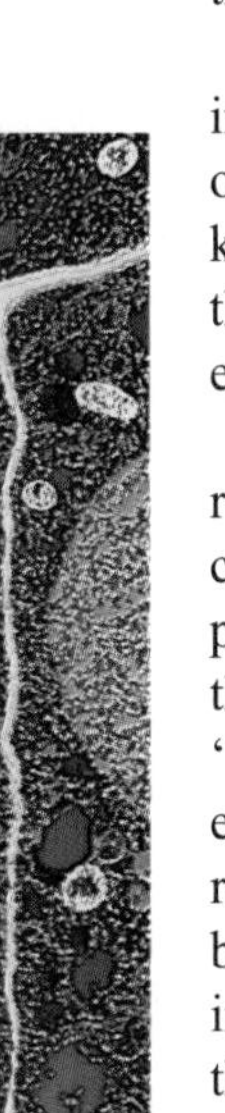

Figure 2.8 This micrograph of a root cell illustrates the highly organized, complex structures found in living organisms.

By now, it may have occurred to you that some naturally occurring processes can be reversed. Water can be pumped uphill to a storage reservoir, heat can be pumped from cold to hot areas in a refrigerator, and air can be pumped from low pressure to high pressure in a bicycle tire. The key word in each of these examples is *pump.* Each requires that energy be added; that is, work must be done. This means that each of these seemingly ''reversed'' processes must be coupled with, or related to, another reaction in which energy is produced. Electric energy is needed to power the water pump or to run the refrigerator, and chemical energy is needed to give our muscles the strength to push a bicycle tire pump. In the reactions required to produce this energy, however, entropy is increased. So, if we look at both of the reactions required to pump the water uphill (that is, the process of pumping the water uphill and the process of generating the energy necessary to do this pumping), we find that the total entropy of the combined processes has increased.

An awareness of entropy is very important in understanding the functioning of living organisms. Such organisms are composed of cells, which are highly complex structures (Fig. 2.8). The natural tendency is therefore for these cells to break down and increase the entropy of the system. To maintain the structure and functioning of the living system, energy must be added to counteract the natural drive toward increased entropy. We shall

Historical Perspective / Folk Culture and the Laws of Thermodynamics

Our folk culture—our songs, poems, nursery rhymes, and sayings—contains many examples of the laws of thermodynamics. Sometimes these examples are obvious, but more often their relationship to the laws of thermodynamics is subtle.

The first law of thermodynamics states that energy is conserved. In a cyclic process—one in which the conditions at the end of the process are the same as those at the beginning—there is no net gain or loss of energy. Sayings, such as "You can't get something for nothing" or "You can't have your cake and eat it too," express the idea of conservation.

Many people have tried to construct so-called perpetual-motion machines. Once started, such devices are supposed to run forever without adding any more energy. The attempt to make such a machine could be labeled "Mission Impossible" or "The Impossible Dream."

The second law of thermodynamics states that all spontaneous processes tend to go in one direction. They move toward increasing disorder, or entropy. Perhaps you learned the nursery rhymes "Humpty Dumpty," "Little Bo Peep," and "Little Boy Blue." Each of these expresses the idea of increasing disorder with time. "Jack and Jill" and "London Bridge" express the idea that things tend to go toward minimum energy.

You have probably heard many proverbs that relate to the second law. "There's no use crying over spilt milk" and "Water doesn't run uphill" are just two examples.

A poetic expression of the second law can be found in Ovid's *Metamorphoses:*

> *There's nothing constant in the universe,*
> *All ebb and flow, and every shape that's born*
> *Bears in its womb the seed of change.*

Your Perspective: Quote another nursery rhyme or folk song that can be interpreted in terms of the laws of thermodynamics. Describe in detail the connection between your quote and one of the laws.

see that this energy, which comes from the food that organisms eat, originates from the sun. If we want to consider the coupled reactions of this system, we see that the life processes that build up and maintain the complicated structures in living organisms depend on energy-producing reactions on the sun. It is amazing to realize that each of our bodies maintains its highly complex structures and functions only at the expense of huge entropy increases on the far-off sun.

KEY CONCEPTS

Energy is the capacity to cause a change or to do work on an object.

Potential energy is energy of position and kinetic energy is energy of motion. Temperature is a measure of the average kinetic energy of the particles of a substance.

There are many forms of energy. Heat is energy that is transferred from one place to another because of a difference in temperature. It can be measured in calories. Electromagnetic radiation is energy that travels in waves: the shorter the wavelength, the higher the energy of the radiation. Radiation of various wavelengths can have both beneficial and harmful effects on living organisms.

Important Equations

$$\text{Specific heat} = \frac{\text{cal}}{\text{g} \times {}^\circ\text{C}}$$

$$\text{calories} = \text{g} \times \Delta t \times \text{specific heat}$$

$$\text{Energy} = \frac{c}{\lambda}$$

Review Problems

Section 2.1

1. Define energy.
2. Give three examples of changes that can result from the use of energy.

Section 2.2

3. Which choice in each of the following pairs has the greater kinetic energy?
   (a) 100 g of water or 100 g of steam
   (b) a car moving at 25 mph or a car moving at 50 mph
   (c) a football linebacker or a sprinter, both running 100 yd in 12 seconds
4. What change in temperature occurs when the average kinetic energy of a group of molecules decreases?

Section 2.3

5. Which choice in each of the following pairs has the greater potential energy?
   (a) snow in the mountains or ocean water
   (b) a pendulum at the top or at the bottom of its swing
   (c) a resting bow and arrow or a drawn bow and arrow
6. Discuss the potential and kinetic energy of a skier as she:
   (a) stands in the lift line
   (b) travels to the top of the ski lift
   (c) stands at the top of the mountain
   (d) skis down the hill to the bottom

Section 2.4

7. How many calories are there in
   (a) 3.5 kcal **(b)** 0.25 kcal

(c) 125 food Calories (e) 146 J

(d) 0.01 kcal (f) 135 kJ

8. Calculate the number of calories and kilocalories transferred under each of the following sets of conditions:

	Grams of Water	*Initial Temp.*	*Final Temp.*
(a)	25.0 g	20.0°C	27.0°C
(b)	50.0 g	24.0°C	32.0°C
(c)	1000 g	25.0°C	76.0°C
(d)	500 g	45.0°C	23.0°C

9. When four cashews are burned in a calorimeter, the temperature of the 1000 g of water changes from 25° to 69°C. How many kilocalories of energy are transferred from one cashew? How many food Calories?

10. How many kilocalories of energy are transferred from 1 g of butter if burning 10 g of butter raises the temperature of 1000 g of water in a calorimeter from 20° to 90°C?

Section 2.5

11. Identify each of the following changes as endothermic or exothermic:

(a) water boiling (d) snow melting

(b) ice forming (e) water condensing

(c) water evaporating

12. Crystals of iodine change directly from the solid to the vapor state, in a process called sublimation. Is this an endothermic or exothermic process?

Section 2.6

13. Which contains more energy: orange light or yellow light?

14. List the following in order of **(a)** increasing energy and **(b)** increasing wavelength:

X rays, short waves, ultraviolet light, yellow light, microwaves

Section 2.7

15. In a nuclear power plant, only 40% of the energy released in the nuclear reaction is converted to electric energy. Why doesn't this violate the first law of thermodynamics?

16. In your own words explain how the build up of fat in people who eat more food than their bodies need is an application of the first law of thermodynamics.

Section 2.8

17. Which choice in each of the following pairs has the higher entropy?

(a) a glass of milk or spilled milk

(b) an assembled model airplane or an unassembled model airplane

(c) a virgin forest or a field of corn

(d) a log or the same log after it is converted to wood chips

18. The formation of cells, which are highly complex substances, from less complex substances involves a decrease in entropy. What is the energy source that makes this possible?

Applied Problems

19. A man is completely immersed in a tub containing 60 L of water. In 1 hour he raises the temperature of the water from 32.0° to 33.5°C. How many kilocalories of energy did the man give off in an hour?

20. A bowl of a popular oat cereal with $\frac{1}{2}$ cup of skimmed milk contains 150 kcal. How long does it take a 150-lb adult at rest to metabolize these kilocalories?

21. A pint of ice cream contains 600 kcal.

(a) How far must a person walk (at 2.5 mph) after eating a pint of ice cream to burn off all of the kilocalories?

(b) If you eat a pint of ice cream every week for a year (in addition to the food needed to maintain your body weight), how much weight will you have gained at the year's end, assuming you don't increase your exercise?

22. A 50.0-g sample of an unknown metal requires 300 kcal to change its temperature from 0° to 100°C. What is the specific heat of the metal in calories per gram per degree Celsius?

23. In steam boilers, fossil fuels are burned to convert liquid water into steam. How many kilocalories are required to convert 1 kg of liquid water at 20°C to steam at 100°C?

24. *Power* is a term that indicates the amount of energy used during a specific period of time. Electric power is often expressed in watts. By definition, one watt (W) is equal to one joule per second. If a 60-W light bulb burns for 6 hours, how many calories of energy are expended?

25. The British thermal unit (BTU) is often used to rate furnaces, air-conditioners, and other devices that consume large amounts of energy. The BTU is defined as the energy necessary to raise the temperature of one pound of water by one degree Fahrenheit. How many calories are in 1 BTU?

26. Each person in the United States uses directly or indirectly about 3.60×10^8 BTU of energy each year. If a pound of coal is equivalent to 1.33×10^4 BTU, how many tons of coal would have to be burned to meet the yearly need of one person?

27. A gas water heater is rated at 38,000 BTU/h. How long does it take to raise the temperature of 50 gal of water from 55° to 140°F?

28. A family uses 50 gal of hot water each day from a water heater set at 140°F. How many BTUs of energy would they save over a year's time if they lowered the setting of the water heater to 120°F?

29. In which place would you probably get a more severe sunburn on a clear day: summer skiing in the Rocky Mountains or swimming at a Delaware beach? Explain your answer.

30. Ultraviolet light can be classed into two groups: UV-A with longer wavelengths and UV-B with shorter wavelengths. Which type of ultraviolet light do you think is more likely to cause skin cancer? Give a reason for your answer.

31. When people put a cube of sugar into their morning coffee, the cube dissolves. Describe the entropy changes that occur as the cube dissolves in the coffee.

Chapter 3

The Three States of Matter

Cliff Brown was nearing retirement at the age of 65, but he was starting to wonder if he'd ever be able to enjoy the fishing and hunting he'd been looking forward to. Over the past five years he had been increasingly bothered by shortness of breath and continual coughing. He knew that his heavy cigarette smoking was to blame for many of these symptoms yet had never been able to cut down to less than two packs a day. Now his condition seemed to be getting worse—he was trying to recover from a bad cold, but he felt extremely tired all the time and was finding it more and more difficult to breathe.

After seeing his doctor, Cliff was sent to the hospital to have a series of tests done on his pulmonary system (the parts of his body involved in lung function). Chest X rays showed that his lungs were enlarged and his diaphragm flattened. Other tests revealed that his ability to exhale air was well below normal. The movement of oxygen into his blood and carbon dioxide out of the blood was not occurring at the proper rate. Blood tests verified the imbalance of oxygen and carbon dioxide in his blood. Oxygen in his blood was measured at an abnormally low tension, or pressure, of 40 mm Hg (normal is 80–100 mm Hg), and carbon dioxide was measured at a very high tension of 70 mm Hg (normal is 40 mm Hg).

The diagnosis was clear: Cliff was suffering from emphysema, one of the chronic obstructive pulmonary diseases (COPD). Emphysema involves the deteri-

oration of the tiny, bubblelike sacs, called alveoli, in the lungs. These sacs make possible the transport of oxygen into and carbon dioxide out of the blood. In emphysema the alveoli lose their ability to expand and contract. Eventually they break open, forming larger, inflexible sacs that cannot so easily exchange the oxygen and carbon dioxide. Emphysema is caused by an imbalance between two substances. One is the enzyme elastase, which cleans up foreign substances inside the lungs. In emphysema, however, this substance attacks the normal tissues of the alveoli. The other substance is a blood protein that normally keeps the action of elastase under control. This protein is in short supply in some people because of their genetic makeup or their heavy smoking.

Cliff was admitted to the hospital for treatment that would help ease the imbalance of oxygen and carbon dioxide in his blood and would reduce the amount of work required for him to breathe. This therapy provided Cliff with air to breathe that had a concentration of oxygen slightly higher than that found in normal room air. This extra oxygen was administered to Cliff at a very slow rate of 2 L/min because a much higher rate could have been quite dangerous for him. To understand this, you have to know that the brain keeps track of the concentration of carbon dioxide in the blood. High blood levels of carbon dioxide cause the brain to increase the rate of breathing. However, in an emphysema patient who has chronically high blood levels of carbon dioxide, the brain undergoes a change to become sensitive to blood oxygen levels instead of blood carbon dioxide levels. Low concentrations of oxygen in the blood of these persons triggers more rapid respiration. If oxygen is administered to an emphysema patient at a high rate, the blood oxygen level rapidly increases and the brain does not receive a signal for the need to breathe. The patient may therefore stop breathing and could easily die.

Cliff remained in the hosspital for a week. His upper respiratory infection was cleared up with antibiotics, and the oxygen tension in his blood was increased to an adequate level. At the end of the week Cliff was taken off the oxygen treatment, and his blood gases were monitored for 48 hours. This check revealed that Cliff's disease had progressed so far that he was unable to maintain enough oxygen in his blood by his own breathing efforts. To allow Cliff to return home, the hospital supplied him with a home oxygen concentration unit. This machine had long tubes that enabled Cliff to breathe oxygen as he moved around the house. Cliff also had a portable unit with liquid oxygen that he could carry over his shoulder when he wanted to leave the house for several hours.

The oxygen that Cliff needs to breathe can be conveniently stored as a liquid for his use. Liquid and gas are two of the three states in which matter can exist. The other state, of course, is solid. In this chapter we closely examine these different states of matter to see how they differ and how they can be converted from one to another. We then discuss some of the physical laws that describe the behavior of matter when it is in the gaseous state.

3.1 Solids

◆

State the difference between a solid, liquid, and gas.

As we just stated, matter can be found in the form of a gas, a liquid, or a solid (Fig. 3.1). These three states differ in the distance between, the attraction between, and the amount of movement of the particles making up the substance. In a solid, the attractive forces between the particles are relatively strong. The particles are packed close together in a rigid

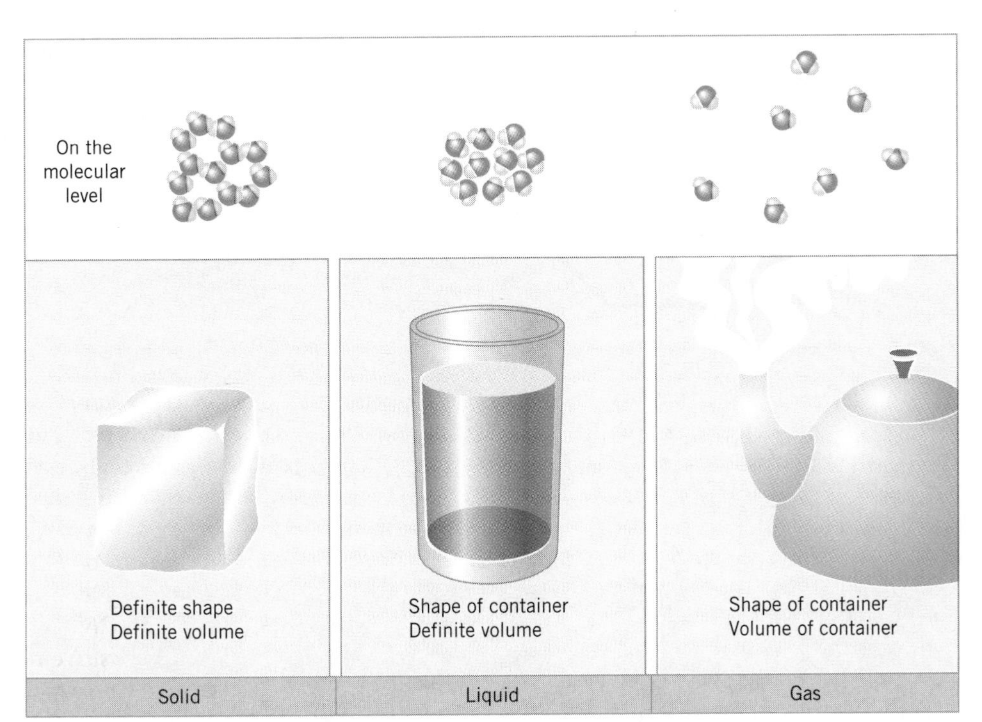

Figure 3.1 The three states of matter: water can exist as a solid, liquid, or gas.

structural arrangement, giving the solid a definite shape and volume. There are two types of solids, differing in the arrangement of particles they contain. The particles of **amorphous solids,** such as ordinary glass and most plastics, are ''frozen'' in a disordered state or randomly arranged pattern—much like the arrangement of the particles in the liquid state (Fig. 3.2(a)). The particles of **crystalline solids,** on the other hand, form a three-dimensional structure in which particles are arranged in a regular, repeating pattern. Examples of crystalline solids are quartz, table salt, sugar, and snowflakes (Figs. 3.2(b) and 5.1). Particles in a crystalline solid have specific positions and fixed neighboring particles, giving the solid a rigid structure. It is important to realize, however, that these particles are not totally motionless. They move back and forth and up and down, or vibrate around a fixed point.

When the temperature is low, these vibrations are small compared with the attraction between the particles of the solid. But if energy is added in the form of heat, the kinetic

(a)

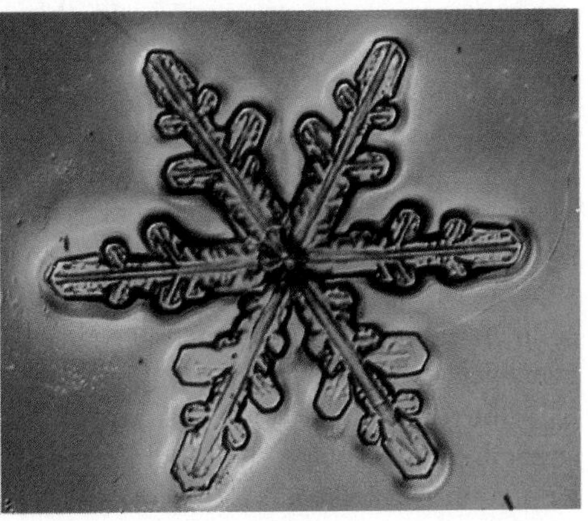

(b)

Figure 3.2 (a) Glass can be cooled to form any shape because it is an amorphous solid whose particles are randomly arranged. (b) The regular, intricate shapes of snowflakes result from the orderly arrangement of water molecules in ice crystals.

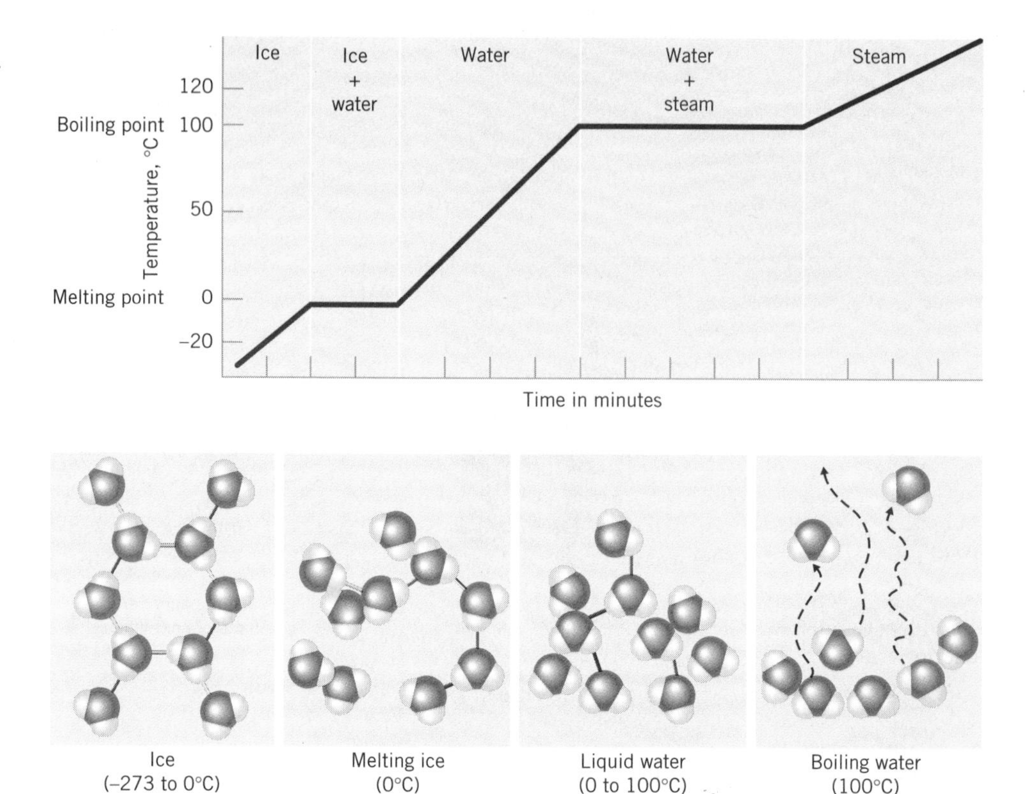

Figure 3.3 Heating curve showing the phase changes of water from solid to liquid to gas as heat is added at a constant rate.

Define melting point and heat of fusion.

energy of the particles increases. Their vibrations become more and more energetic until the solid finally begins to break apart, or melt. The specific temperature at which this occurs is called the **melting point** of the solid. The greater the attraction between the particles of a solid, the higher the melting point. When a crystalline solid has reached its melting point, additional heat energy does not go into increasing the average kinetic energy of the particles—that is, into increasing the temperature—but rather goes entirely into melting the solid. The amount of energy that must be added to change a solid to a liquid at the melting point is called the **heat of fusion** (expressed in calories per gram [cal/g]). The heat of fusion of water, for example, is 80 cal/g. (Note that the same amount of energy is given off when the liquid again becomes a solid.) Only after the solid is completely melted does the temperature of the substance begin to rise again as additional heat energy is added (Fig. 3.3).

3.2 Liquids

The particles in a liquid are not held together as tightly or as rigidly as they are in a solid. Although the particles are fairly close together, they can move from place to place by slipping past one another. This, for example, allows you to walk through a liquid but not a solid. A liquid can flow from one place to another and take on the shape of its new container, even though the volume of the liquid does not change.

Although a liquid flows, its particles are still strongly attracted to one another. The strength of this attraction determines many properties of a liquid, two of which are viscosity and surface tension. The **viscosity** of a liquid is a measure of its resistance to flow. The higher the attraction between the particles of the liquid, the greater the viscosity, or syrupy nature, of the liquid. Molasses is more viscous than water, which in turn is more viscous than gasoline. **Surface tension** is the resistance of particles on the surface of the liquid to the expansion of the liquid. It is the force that keeps the surface area of a liquid to a minimum (Fig. 3.4). Water has a high surface tension, whereas gasoline has a low surface tension. Both the viscosity and surface tension of a liquid decrease with an increase in temperature (that is, as the kinetic energy of the particles increases).

Very energetic particles near the surface of a liquid can break away from the surface. If these particles don't collide with air molecules and fall back to the surface of the liquid, they escape from the liquid completely. This is the process called **evaporation.** The rate of evaporation depends on the attraction between the liquid particles: the stronger the attraction, the slower the rate of evaporation. Because the most energetic particles are lost through evaporation, the average kinetic energy of the particles left behind is lower—that is, the liquid is cooler. This principle allows your body to rid itself of heat through the evaporation of sweat. Similarly, when alcohol is applied to the skin, the resulting cool feeling is caused by the rapid evaporation of alcohol molecules, leaving behind a cooler liquid on the skin. Note that as more heat energy is added to a liquid, the kinetic energy of the particles increases, which in turn increases the rate of evaporation.

◆ **Describe on the molecular level the processes of evaporation, condensation, and sublimation.**

Condensation is the opposite of evaporation. As gas particles are cooled, their kinetic energy decreases to a point at which the particles condense (or return to the liquid state). Good examples of condensation are the formation of dew on grass and water on the windshield of a car on a chilly morning.

A liquid evaporating and a gas condensing are physical changes that we call changes in state. Another change in state is the process called **sublimation:** the change of a solid directly into a gas. If the particles of a solid are held together very weakly, such solids tend to sublime—to "vanish into thin air." Examples of substances that sublime are moth balls (naphthalene) and Dry Ice (solid carbon dioxide).

Adding more heat energy to a liquid eventually increases the motion of the particles to the point that highly energetic collisions occur within the liquid, and bubbles of vapor (gas) are formed. This process is called **boiling.** The particular temperature at which boiling occurs is called the **boiling point** of a substance, and it depends on the atmo-

◆ **Define boiling point and heat of vaporization.**

Figure 3.4 The high surface tension of water is enough to support this insect and prevent it from sinking.

Describe the changes that occur on the molecular level as an ice cube is warmed from 5° to 110°C.

spheric pressure. (The boiling point of a substance under the pressure exerted by the atmosphere at sea level is called the *normal* boiling point of the substance.) At the boiling point, additional energy added to the substance goes into pulling the particles apart—that is, changing the liquid to a gas—and not raising the temperature (see Fig. 3.3).

The amount of energy required to change a substance from a liquid to a gas at constant temperature and pressure is called the **heat of vaporization** (expressed in calories per gram). The heat of vaporization of water at 100°C, for example, is 539 cal/g. (Note again that the same amount of energy is given off when the gas condenses to become a liquid.) Only when all the liquid has become a gas does the temperature of the substance again increase as more heat energy is added.

3.3 Gases

List four units used to measure pressure.

The particles of a gas are very far apart and travel at great speeds. A gas does not have a definite shape or volume but completely fills any container that it occupies. In doing so, the gas exerts pressure against the sides of the container. This pressure exerted by a gas results from the collisions of the gas particles against the sides of the container. **Pressure** is defined as the force exerted per unit of area. Gas pressure can be measured in several different units.

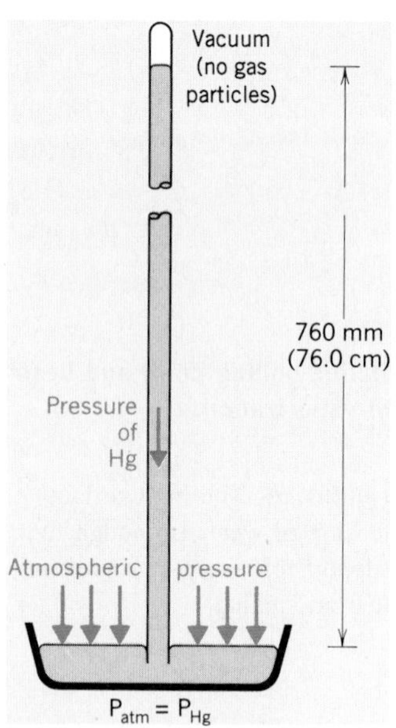

Figure 3.5 The mercury barometer is an instrument used to measure atmospheric pressure. The height of the column varies from place to place and from time to time as the atmospheric pressure changes. At sea level the atmosphere supports a column of mercury about 760 mm high, but at an altitude of 3.5 miles, the column is only about 380 mm high.

Millimeters of Mercury

One of the oldest units of pressure is **millimeters of mercury (mm Hg).** This unit comes from the use of a column of mercury to measure atmospheric pressure (Fig. 3.5). One millimeter of mercury (1 mm Hg) is the atmospheric pressure sufficient to support a column of mercury 1 mm in height. The height of the column of mercury in a barometer varies from place to place and from time to time as the atmospheric pressure changes. At sea level the atmosphere supports a column of mercury about 760 mm high, but at an altitude of 3.5 miles the column is only about 380 mm high.

Atmosphere

The height of the column of mercury that can be supported by atmospheric pressure depends on the prevailing weather conditions, as well as the geographic location of the measurement. **Standard atmospheric pressure** is the average pressure exerted by the earth's atmosphere at sea level when the temperature is 0°C. One standard **atmosphere (atm)** supports a column of mercury 760 mm high at 0°C.

$$1 \text{ atm} = 760 \text{ mm Hg}$$

The conditions called **standard temperature and pressure (STP)** are 0°C and 1 atm. In the English system, pressure is measured in pounds per square inch (psi). One atmosphere is equivalent to 14.7 psi.

$$1 \text{ atm} = 14.7 \text{ psi}$$

Torr

The **torr (torr)** is a small unit of pressure, named in honor of the 17th-century Italian physicist Evangelista Torricelli, the inventor of the mercury barometer. One torr equals 1/760 atm. Therefore,

$$1 \text{ atm} = 760 \text{ torr} \quad \text{and} \quad 1 \text{ torr} = 1 \text{ mm Hg}$$

Pascal

The SI unit of pressure is the **pascal (Pa),** named in honor of the 17th-century French scientist Blaise Pascal. The relationship between pascals and torr is

$$1 \text{ torr} = 133.3 \text{ Pa}$$

Example 3-1

1. How many torrs equal 2.75 atm?

This problem is a conversion problem asking: 2.75 atm = (?) torr ◄ *See the Question*

To solve the problem, we use the equality 1 atm = 760 torr to write the needed conversion factor: 760 torr/1 atm. ◄ *Think It Through*

$$2.75 \text{ atm} \times \frac{760 \text{ torr}}{1 \text{ atm}} = 2090 \text{ torr}$$ ◄ *Execute the Math*

The answer is in torrs and has three significant figures: 2090 or 2.09×10^3 torr. ◄ *Prepare the Answer*

2. Assume that the barometer reading in your chemistry laboratory is 743 mm Hg. What is this pressure in torrs and atmospheres?

We have two questions to answer: ◄ *S*

$$743 \text{ mm Hg} = (?) \text{ torr} \quad \text{and} \quad 743 \text{ mm Hg} = (?) \text{ atm}$$

We need to set up conversion factors from the following two equalities: ◄ *T*

$$1 \text{ torr} = 1 \text{ mm Hg} \quad \text{and} \quad 1 \text{ atm} = 760 \text{ mm Hg}$$

Solving for part (a) ◄ *E*

$$743 \text{ mm Hg} \times \frac{1 \text{ torr}}{1 \text{ mm Hg}} = 743 \text{ torr}$$

Solving for part (b)

$$743 \text{ mm Hg} \times \frac{1 \text{ atm}}{760 \text{ mm Hg}} = 0.9776315 \text{ atm}$$

Both answers should have three significant figures. The answers are then: ◄ *P*

(a) 743 torr **(b)** 0.978 atm

Check Your Understanding

A Canadian weather broadcast reported the air pressure inside a cyclone to be 660 mm Hg. What is this pressure in atmospheres and torrs?

3.4 Boyle's Law: The Relationship Between Pressure and Volume

◆ **State Boyle's law in your own words, and give an everyday example of its application.**

In the 17th century, the British chemist Robert Boyle discovered that the volume of a given mass of a gas varies inversely with applied pressure if the temperature is kept constant (Fig. 3.6). In other words, if you increase the pressure, the volume decreases.

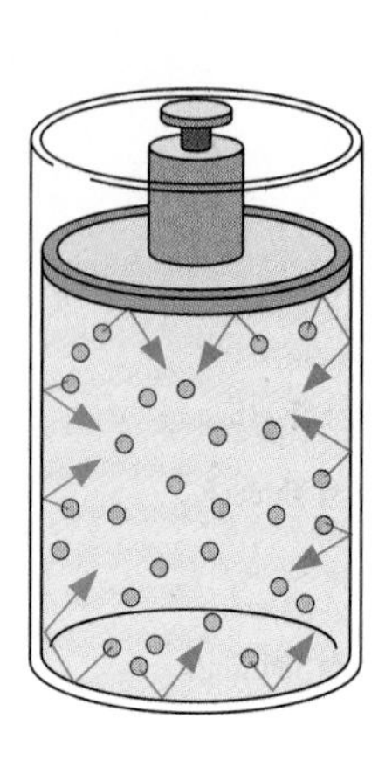

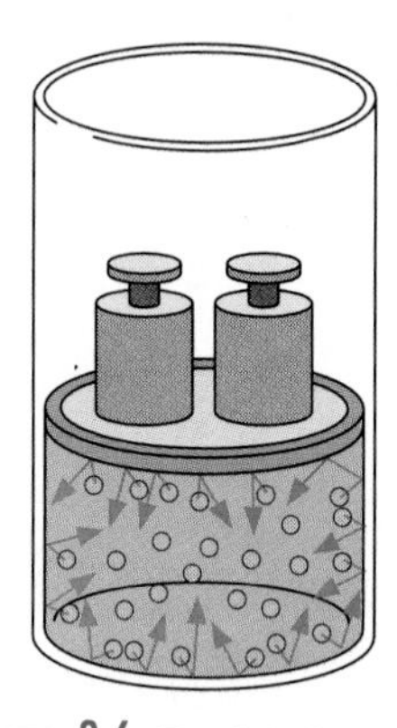

Figure 3.6 Boyle's law states that when the volume of a gas decreases, the pressure increases (if the temperature remains constant).

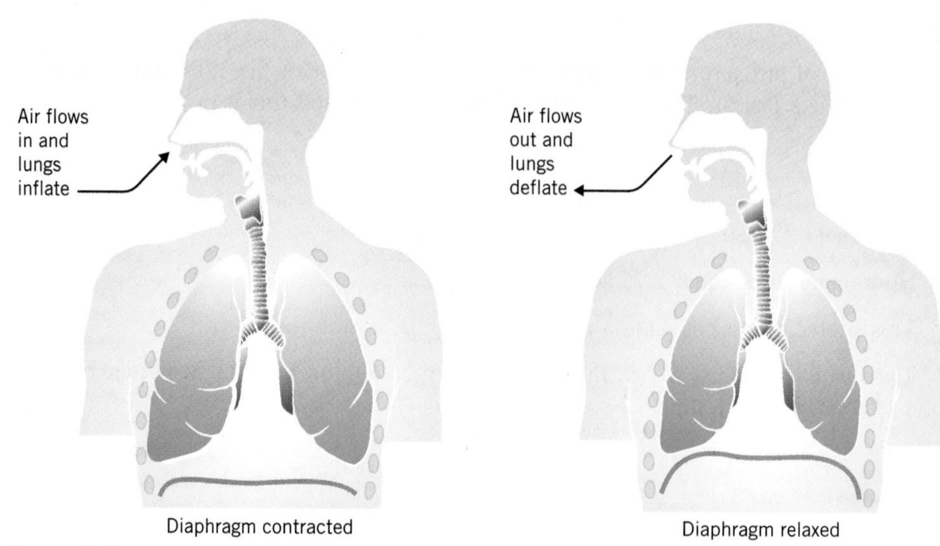

Figure 3.7 Breathing is an example of Boyle's law in action.

Decrease the pressure and the volume increases. When the volume occupied by a gas decreases, there is less space for the particles to move around in, so they collide with the sides of the container much more often. This appears as an increase in pressure.

Breathing is an illustration of Boyle's law at work in our bodies (Fig. 3.7). Your lungs are located in the thoracic cavity, surrounded by the ribs and a muscular membrane called the diaphragm. To inhale, the diaphragm contracts and flattens out, increasing the volume of the thoracic cavity. This lowers the air pressure in the cavity below that of the atmosphere, causing air to flow into the lungs. To exhale, the diaphragm relaxes and pushes up into the thoracic cavity, increasing the pressure above that of the atmosphere and causing air to flow out of the lungs.

We just described Boyle's law in words, but this inverse relationship can also be stated mathematically. We state Boyle's law mathematically as follows:

$$PV = \text{constant} \quad (\text{when } T \text{ is constant})$$

where P = pressure; V = volume; and T = Kelvin temperature.

And, when given two conditions of pressure and volume,

$$P_1V_1 = P_2V_2 \quad (\text{when } T \text{ is constant})$$

◆ **Perform calculations using Boyle's law.**

where P_1 and V_1 are the starting conditions of pressure and volume, and P_2 and V_2 are the final conditions. The following examples show ways to think through various types of problems, as well as how to use the equation for Boyle's law.

Example 3-2

1. Imagine that a gas is inside a cylinder with a movable piston. If the volume of the gas is 3.0 L when the pressure is 760 torr, what is the volume if the pressure is increased to 1140 torr while the temperature is held constant?

See the Question

In this problem we are asked to find the value of V_2: $V_2 = (?)\ \text{L}$

Think It Through

The first step is to identify the quantities that are stated in the problem.

$$P_1 = 760 \text{ torr} \qquad P_2 = 1140 \text{ torr} \qquad V_1 = 3.0 \text{ L}$$

Does the starting volume of 3.0 L increase or decrease when the pressure is increased from 760 to 1140 torr? Boyle's law states that the volume of a gas decreases if the pressure is increased. Our answer should be less than 3.0 L.

Execute the Math

Substituting the values in the Boyle's law equation, we have

$$P_1V_1 = P_2V_2$$

$$760 \text{ torr} \times 3.0 \text{ L} = 1140 \text{ torr} \times V_2$$

$$\frac{760 \cancel{\text{torr}} \times 3.0 \text{ L}}{1140 \cancel{\text{torr}}} = V_2$$

$$2 \text{ L} = V_2$$

Prepare the Answer

Our calculator result of 2 does not have enough significant figures. The correct number of significant figures is two. Happily, the calculated result agrees with our prediction and is less than the starting volume. The final answer to this problem is:

$$V_2 = 2.0 \text{ L}$$

2. Suppose that an 8.0-L tank contains oxygen at a pressure of 44 atm. What pressure does this amount of oxygen exert in a 55-L tank at the same temperature?

S

We need to find the correct value for P_2: $P_2 = (?)$ atm

T

First identify the given quantities:

$$V_1 = 8.0 \text{ L} \qquad P_1 = 44 \text{ atm} \qquad V_2 = 55 \text{ L}$$

Does the starting pressure increase or decrease when the volume is increased from 8.0 to 55 L? We know from Boyle's law that if the volume is increased, the pressure decreases. Because the increase in volume is large, we should expect to see a large decrease in pressure.

E

Substituting the given quantities in the Boyle's law equation, we have

$$44 \text{ atm} \times 8.0 \text{ L} = P_2 \times 55 \text{ L}$$

$$\frac{44 \text{ atm} \times 8.0 \cancel{\text{L}}}{55 \cancel{\text{L}}} = P_2$$

$$6.4 \text{ atm} = P_2$$

P

The answer of 6.4 atm has the correct number of significant figures and meets our prediction of a large decrease in pressure.

$$P_2 = 6.4 \text{ atm}$$

Check Your Understanding

1. Suppose that the volume of a balloon is 3.50 L in New York City when the atmospheric pressure measures 760 torr. What is the atmospheric pressure in Mexico City (which is about 7000 ft above sea level) if the balloon has a volume of 4.43 L there? Assume that the temperature is the same in both cities.

Health Perspective / Heimlich Maneuver

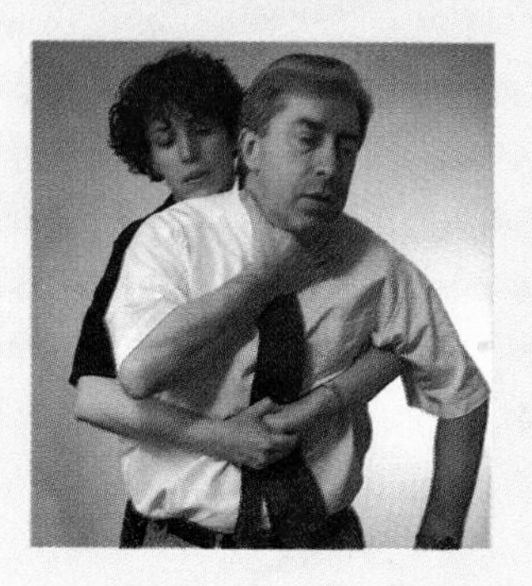

You might be surprised at the number of applications of Boyle's law in our everyday lives. One such application can be used to save a person who is choking from food caught in the windpipe: the Heimlich maneuver. To perform this maneuver you stand behind, and put your arms around, the choking person. Then you thrust upward with your fists on the abdomen. This causes a rapid decrease in volume of the chest, resulting in an increase of pressure inside the lungs—which can expel the obstruction out of the windpipe like a cork out of a champagne bottle.

A medical syringe also operates using Boyle's law. By drawing back the plunger you increase the volume inside the chamber. The increase in volume lowers the pressure within the syringe, drawing the liquid up into the chamber.

2. A sample of oxygen has a volume of 840 mL at a pressure of 800 torr. What is the volume of this sample if the pressure is reduced to standard atmospheric pressure while the temperature is held constant?

3.5 Charles's Law: The Relationship Between Volume and Temperature

State Charles's law in your own words, and give an everyday example of its application.

In the early 19th century, the French physicist Jacques Charles discovered that the volume occupied by a given mass of gas varies directly with Kelvin temperature when the pressure is constant. That is, when the temperature of a gas is increased, the volume increases (Fig. 3.8). On the molecular level, increasing the temperature means increasing the kinetic energy of the gas particles. Because these particles now move faster, they collide with the sides of the container much more often and with more force per collision. The only way that the pressure can remain constant under these circumstances is to increase the volume

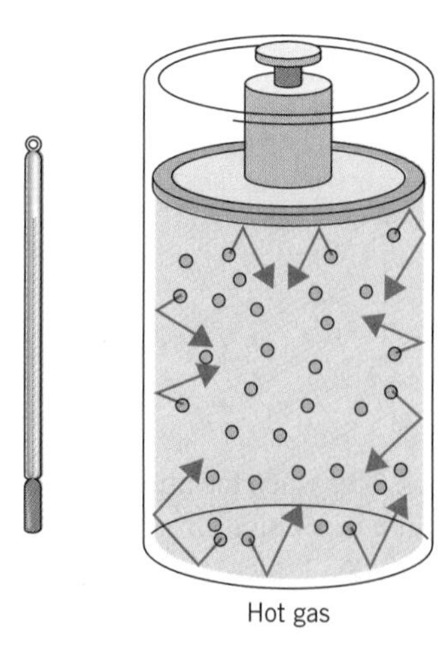

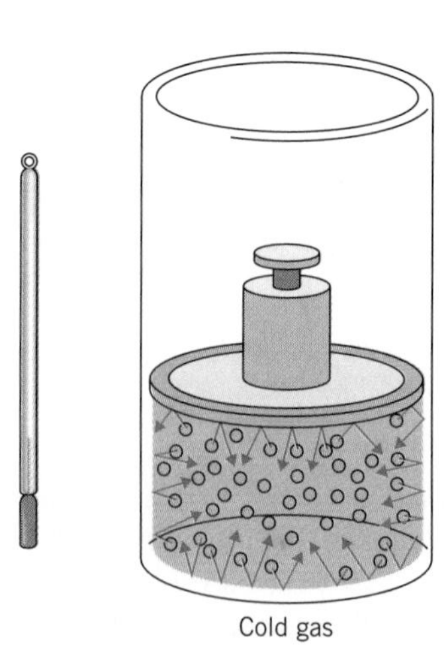

Figure 3.8 Charles's law states that when the temperature of a gas is increased, the volume of the gas increases (if the pressure remains constant).

of the container. Similarly, when gas particles are cooled, they slow down. If the pressure is to remain constant, the volume must decrease so that the gas particles exert the same force on the sides of the container as they did before.

Charles's law can be expressed mathematically by the following direct relationship:

$$\frac{V}{T} = \text{constant} \quad \text{(when } P \text{ is constant)}$$

And, when given two conditions of volume and temperature,

Perform calculations using Charles's law.

$$\frac{V_1}{T_1} = \frac{V_2}{T_2} \quad \text{(when } P \text{ is constant)}$$

where V_1 and T_1 are the starting conditions of volume and temperature (in kelvins), and V_2 and T_2 are the final conditions.

Example 3-3

1. Imagine that a gas occupies a volume of 2.0 L at a temperature of 27°C. To what temperature in degrees Celsius must the gas be cooled to reduce its volume to 1.5 L if the pressure is held constant?

   See the Question

   This problem is asking us to calculate T_2: $T_2 = (?)°\text{C}$

   Think It Through

   First we need to identify the known quantities from the problem statement.

   $$V_1 = 2.0\text{ L} \qquad T_1 = 27 + 273 = 300\text{ K} \qquad V_2 = 1.5\text{ L}$$

   (Remember that the temperature must be in kelvins, K = °C + 273.) In this problem we are working with changes in volume and temperature. From Charles's law we know that decreasing the volume of a gas means that the temperature must also decrease.

   Execute the Math

   Substituting the known quantities in the question for Charles's law, we have

   $$\frac{2.0\text{ L}}{300\text{ K}} = \frac{1.5\text{ L}}{T_2}$$

   When we cross multiply, we get the following equation:

   $$2.0\text{ L} \times T_2 = 1.5\text{ L} \times 300\text{ K}$$

   Dividing both sides of the equation by 2.0 L, we have

   $$T_2 = \frac{1.5\text{ \cancel{L}} \times 300\text{ K}}{2.0\text{ \cancel{L}}} = 225\text{ K}$$

   Prepare the Answer

   The question asks for the answer in degrees Celsius, so we need to convert 225 K to degrees Celsius:

   $$225\text{ K} = °\text{C} + 273$$

   $$-48 = °\text{C}$$

   The answer agrees with our reasoning, showing a decrease in temperature.

2. Imagine that you are skiing on a beautiful, clear day with a temperature outside of −3°C. The cold air that you breathe is warmed to a body temperature of 37°C as it

travels to your lungs. If you inhale 425 mL of air at −3°C, what volume does it occupy in your lungs? (Assume that the pressure is constant.)

S ➤ This question is asking you to calculate V_2: V_2 = (?) mL

T ➤ Identify the known quantities from the stated problem:

$$T_1 = -3 + 273 = 270\ \text{K} \qquad V_1 = 425\ \text{mL} \qquad T_2 = 37 + 273 = 310\ \text{K}$$

The temperature of the gas is increasing as it enters your lungs, so from Charles's law we know that the volume must also increase.

E ➤ Substituting the known quantities in the equation for Charles's law, we have

$$\frac{425\ \text{mL}}{270\ \text{K}} = \frac{V_2}{310\ \text{K}}$$

When we cross multiply, we get

$$425\ \text{mL} \times 310\ \text{K} = V_2 \times 270\ \text{K}$$

Dividing both sides of the equation by 270 K, we have

$$\frac{425\ \text{mL} \times 310\ \cancel{\text{K}}}{270\ \cancel{\text{K}}} = V_2$$

$$487.96296\ \text{mL} = V_2$$

P ➤ The calculator answer (487.96296) must be rounded to three significant figures: 488 mL. As we predicted, the volume increases.

$$V_2 = 488\ \text{mL}$$

3. A cylinder contains 255 mL of carbon dioxide at 27°C and 812 torr. What volume does this amount of carbon dioxide occupy at standard temperature and pressure (1 atm or 760 torr, and 0°C)?

S ➤ This problem is asking us to determine the value of V_2: V_2 = (?) mL

T ➤ Identify the known quantities from the stated problem:

$$P_1 = 812\ \text{torr} \qquad T_1 = 27 + 273 = 300\ \text{K} \qquad V_1 = 255\ \text{mL}$$

$$P_2 = 760\ \text{torr} \qquad T_2 = 0 + 273 = 273\ \text{K} \qquad V_2 = ?$$

We can solve this problem by reasoning in two steps. First, notice that the pressure is decreasing. From Boyle's law, we know that this should make the volume increase. So we multiply the starting volume by a ratio of pressures that increases the volume. Second, notice that the temperature is decreasing. From Charles's law, we know that a decrease in temperature causes the volume to decrease. So the ratio of temperatures to use is the one that decreases the volume.

E ➤ Multiplying the initial volume by these ratios we have

$$255\ \text{mL} \times \frac{812\ \cancel{\text{torr}}}{760\ \cancel{\text{torr}}} \times \frac{273\ \cancel{\text{K}}}{300\ \cancel{\text{K}}} = 247.9271\ \text{mL}$$

We can combine the equations for Boyle's and Charles's laws into a general equation that predicts the behavior of a gas under various conditions of volume, temperature, and pressure.

$$\frac{P_1V_1}{T_1} = \frac{P_2V_2}{T_2}$$

This general equation gives us another way to solve the same problem. By substituting known quantities into the general equation, we get

$$\frac{812 \text{ torr} \times 255 \text{ mL}}{300 \text{ K}} = \frac{760 \text{ torr} \times V_2}{273 \text{ K}}$$

$$\frac{812 \text{ torr} \times 255 \text{ mL} \times 273 \text{ K}}{300 \text{ K} \times 760 \text{ torr}} = V_2$$

$$247.9271 \text{ mL} = V_2$$

Be certain to check the units you use in these gas law problems carefully, labeling all quantities and making certain that the units cancel. In this problem the units cancel to give us an answer in units of volume, milliliters. The calculator gives us an answer with four decimal places. Checking the number of significant figures in the given data, we see that the answer must have only three significant figures:

P

$$V_2 = 248 \text{ mL}$$

Check Your Understanding

1. Suppose that a balloon having a volume of 2.0 L at room temperature (20°C) is taken outside on a warm summer day. What is the temperature (in °C) outside if the volume of the balloon outside is 2.1 L? (Assume that the pressure remains constant.)
2. An anesthesiologist administers a gas at 20°C to a patient whose body temperature is 37°C. What is the change in volume in milliliters of a 1.20-L sample of gas as it goes from room temperature to body temperature? (Assume that the pressure remains constant.)

3.6 Henry's Law

When you open a bottle of carbonated beverage and listen to the hiss of the escaping carbon dioxide, you are experiencing Henry's law in action. At the beginning of the 19th century, the English chemist William Henry discovered that the solubility of a gas in a liquid at a given temperature is directly proportional to the pressure of that gas on the liquid. In other words, the higher the pressure, the more gas that dissolves in a liquid when the temperature does not change. Carbonated beverages are bottled under high pressure. When you open the cap you reduce the pressure in the bottle, thus lowering the solubility of the carbon dioxide and allowing the gas to escape from the beverage.

State Henry's law in your own words, and give an example of its application.

At the beginning of this chapter we described a person suffering from emphysema. Such patients can benefit from an application of Henry's law. The amount of oxygen dissolved in their blood increases when emphysema patients breathe air containing higher than normal amounts of oxygen (and, therefore, having a higher oxygen pressure.)

Health Perspective / Scuba Diving and the Gas Laws

Before leading a group of scuba divers into the water, the dive master tells them about the dive site. She describes interesting things to look for, such as a cave, a particularly pretty coral formation, or perhaps a large moray eel. She also stresses the use of safe diving practices, such as staying close to a dive buddy or using a safety line if entering a wreck. She also tells the group the maximum depth for the dive and what time to return to the dive boat.

If the dive is as deep as 60 to 120 ft, the divers are given special safety instructions. They are warned to ascend to the surface slowly, at no more than 60 ft/min. They also must make a safety stop 15 ft below the surface, the length of the safety stop depending on the depth and duration of the dive. Divers who do not follow these instructions risk suffering decompression sickness, better known as the ''bends.'' A person suffering from the bends may experience severe pains in the joints and muscles, deafness, paralysis and sometimes even death.

Scuba tanks contain compressed air, which is mainly nitrogen and oxygen. As we know from Henry's law, increasing the pressure of a gas increases its solubility. As a diver descends, pressure increases to as much as three or four times that of atmospheric pressure. This increases the solubility of gases in the diver's blood. If the diver ascends too fast, the solubility of these gases decreases so rapidly that they form bubbles in the blood vessels, a phenomenon called cavitation. Such tiny bubbles (air emboli) block the passage of blood into the capillaries, causing the symptoms of the bends. If the diver ascends more slowly, the change in solubility is more gradual. The nitrogen coming out of the blood is carried to the lungs and exhaled, whereas the oxygen is carried to the tissues and used in metabolism.

Deep-sea divers who must work for long periods at great pressure breathe a mixture of helium and oxygen. Like nitrogen, helium is a very inert gas. Unlike nitrogen, however, helium is much less soluble in the blood. This lower solubility means that less helium has to escape from the blood when the diver ascends to the surface. Because helium is lighter than nitrogen, it also diffuses from the bloodstream into the lungs more rapidly than nitrogen. As a result, the use of helium rather than nitrogen reduces the risk of the bends.

At a depth greater than 60 ft, scuba divers can suffer from nitrogen narcosis. This condition, caused by the increased concentration of nitrogen in the blood, can have symptoms very similar to alcohol intoxication. Such narcosis is particularly dangerous because divers begin to develop a sense of euphoria that, in some cases, has led them to remove their respirators and other equipment. Nitrogen narcosis disappears quickly on ascent.

Your Perspective: Suppose that you have been asked to write a short article for *Diving Illustrated* magazine describing various ways that a diver can pass the time while waiting at decompression stops when coming up from a deep dive. Use your imagination to write such a short article.

3.7 Dalton's Law of Partial Pressures

We have seen that the pressure of a gas comes from the number and the force of the collisions of gas particles against the walls of the container. There are two ways we can increase the frequency of collisions and, therefore, increase the pressure. One way is to increase the temperature of the gas. This increases the kinetic energy of the gas particles, which increases both the number and force of the collisions that occur. The second way is to increase the number of particles of gas in the container.

The pressure exerted by a gas at constant temperature does not depend on the type of gas particles, but instead depends only on the number of particles of gas that are present. Gas particles behave independently of one another: each exerts pressure as if it were by itself. For example, we can double the pressure in a container by adding an equal number of particles of the same gas or of a different gas (Fig. 3.9). In a mixture of gases, the pressure exerted by each gas is called the **partial pressure (P)** of that gas. This partial pressure depends only on the number of particles of that gas present. **Dalton's law** states that the total pressure of a mixture of gases is equal to the sum of the partial pressures of each gas in the mixture. For a mixture of four gases, A, B, C, and D,

State Dalton's law of partial pressures, and give an example of its application.

$$P_{\text{total}} = P_{\text{A}} + P_{\text{B}} + P_{\text{C}} + P_{\text{D}}$$

The earth's atmosphere is a mixture of nitrogen (N_2), oxygen (O_2), argon (Ar), and other gases found in small amounts. By Dalton's law, then,

$$P_{\text{atmosphere}} = P_{N_2} + P_{O_2} + P_{\text{Ar}} + P_{\text{other}} \tag{3.1}$$

When gases are prepared in a laboratory, they are often collected over water in an apparatus such as the one shown in Figure 3.10. The bottle contains both the gas being collected and water vapor, which together are called a wet gas. Water vapor exerts a pressure just like any other gas; the partial pressure of the water vapor in the bottle is called the **vapor pressure.** The amount of water vapor that is present—and therefore the vapor pressure—depends only on the temperature of the water. As you can see from Table 3.1, the vapor pressure of water increases as the temperature increases. The total pressure of the wet gas in the jar (which equals atmospheric pressure if the water levels are equal) is the sum of the partial pressures of the gas collected and the vapor pressure of water.

$$P_{\text{wet gas}} = P_{\text{dry gas}} + P_{\text{water}}$$

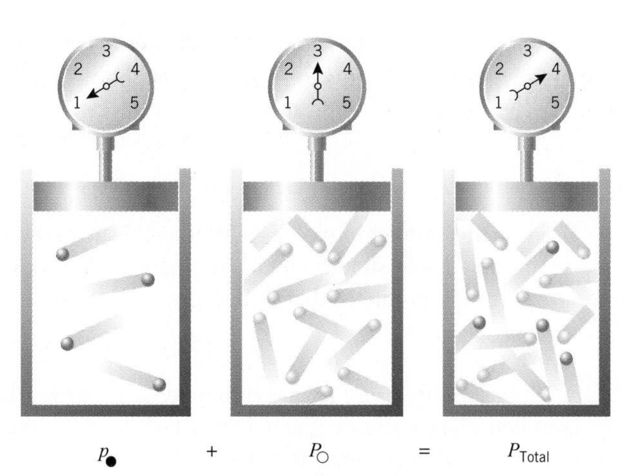

Figure 3.9 Dalton's law states that the total pressure in a container equals the sum of the partial pressures of the gases in that container.

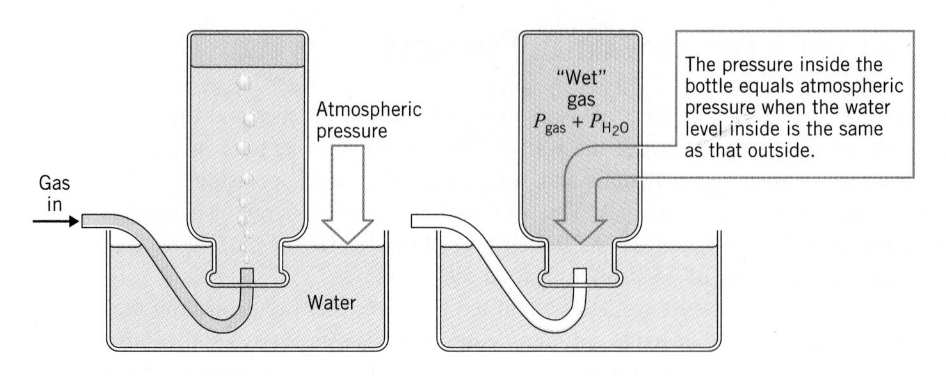

Figure 3.10 When a gas is collected over water, the bottle contains both gas and water vapor. The total pressure of the wet gas in the bottle is the sum of the pressure of the dry gas and the pressure of the water vapor.

Table 3.1 **Vapor Pressure of Water**

Temperature (°C)	*Vapor Pressure (torr)*	*Temperature (°C)*	*Vapor Pressure (torr)*
0	4.6	40	55.3
10	9.2	50	92.5
15	12.8	60	149.4
20	17.5	70	233.7
25	23.8	80	355.1
30	31.8	90	525.8
35	42.2	100	760.0

Example 3-4

1. What is the partial pressure of oxygen in the air if $P_{N_2} = 593.0$ torr, $P_{Ar} = 7.0$ torr, and $P_{other} = 0.2$ torr, when the atmospheric pressure is measured to be 760.0 torr?

See the Question The question asks us to determine the partial pressure of oxygen in air.

$$P_{O_2} = (?) \text{ torr}$$

Think It Through To solve this problem, we need to substitute the known values in equation (3.1).

Execute the Math

$$760.0 \text{ torr} = 593.0 \text{ torr} + P_{O_2} + 7.0 \text{ torr} + 0.2 \text{ torr}$$

So that,

$$P_{O_2} = 760.0 \text{ torr} - 600.2 \text{ torr}$$

$$P_{O_2} = 159.8 \text{ torr}$$

Prepare the Answer Because this problem involves addition and subtraction of data, the number of decimal places in the answer should equal the smallest number of decimal places among all the numbers. In this problem, that is one decimal place.

$$P_{O_2} = 159.8 \text{ torr}$$

2. A cylinder contains a mixture of oxygen and nitrous oxide (N_2O), which is used as an anesthetic. The pressure gauge on the tank reads 1.20 atm. If the partial pressure of the oxygen is 137 torr, what is the partial pressure of the nitrous oxide?

The total pressure in torrs can be determined using the relationship

$$1 \text{ atm} = 760 \text{ torr}$$

The problem asks, What is the partial pressure of the nitrous oxide? ◄ *S*

$$P_{N_2O} = (?) \text{ torr}$$

This problem can be solved using Dalton's law of partial pressures. But in the problem statement the total pressure is given in atmospheres, whereas the partial pressure of the oxygen is in torr. To solve the problem we must have both pressures expressed in the same units. The total pressure can be converted to torrs using a conversion factor from the equality 1 atm = 760 torr. ◄ *T*

$$1.20 \text{ atm} \times \frac{760 \text{ torr}}{1 \text{ atm}} = 912 \text{ torr}$$ ◄ *E*

Using Dalton's law, we obtain

$$P_{N_2O} + P_{O_2} = P_{\text{total}}$$

$$P_{N_2O} + 137 \text{ torr} = 912 \text{ torr}$$

$$P_{N_2O} = 775 \text{ torr}$$

After checking the units and significant figures: ◄ *P*

$$P_{N_2O} = 775 \text{ torr}$$

Check Your Understanding

1. A student collects oxygen gas at 25°C by displacement of water until the levels of the water inside and outside the container are equal (See Table 3.1). If we assume that the barometer reads 755 torr, what is the partial pressure of the oxygen gas in the container?
2. Identify the gas law that describes each of the following situations.
   (a) A balloon expands when heated.
   (b) When you push down on the handle of a bicycle pump, the plunger goes down in the air cylinder.
   (c) A hissing sound is heard when you open a bottle of carbonated beverage.
   (d) A tire blows out when a car is driven at high speeds on a very hot summer day.

3.8 Diffusion of Respiratory Gases

The diffusion of respiratory gases (oxygen and carbon dioxide) within our bodies is directly related to the partial pressures of these gases (Fig. 3.11). All gases diffuse from a region of higher partial pressure to a region of lower partial pressure. For example, venous blood entering the lungs from the tissues has been depleted of its oxygen supply and is carrying the waste product carbon dioxide from the cells. Oxygen diffuses from the lungs

Health Perspective / Loss of Airplane Cabin Pressure

"If the cabin should lose air pressure, an oxygen mask will drop down from the overhead compartment. . . . " The flight attendant speaking these familiar words is warning passengers about an important aspect of Dalton's law of partial pressures. Ordinarily, cabin pressure in an airplane is maintained at about 650 mm Hg. A sudden decrease in total pressure would cause a corresponding decrease in the partial pressure of oxygen. Since 21% of the molecules in air are oxygen molecules, oxygen must account for 21% of the total air pressure. At sea level, the partial pressure of O_2 is 21% of 760 mm Hg, or 160 mm Hg. If an airplane is descending to an elevation of 13,000 ft, the outside total atmospheric pressure is only 490 mm Hg. The partial pressure of the oxygen is 21% of 490 mm Hg, or 103 mm Hg.

It is the pressure difference, or pressure gradient, between the alveoli in the lungs and the surrounding capillaries that allows O_2 to diffuse into red blood cells. When the partial pressure of the oxygen decreases, as can happen on an airplane, the pressure gradient decreases and so does the diffusion of O_2. The resulting lack of oxygen in the blood and tissues has many side effects, including difficulty seeing and thinking clearly. If you should ever find yourself in such a situation, immediately put on your oxygen mask. Don't stop to help others until you have done this, especially if you are with small children. It is important that you be able to think clearly. Only then will you be capable of acting to save others.

Your Perspective: Write a script that can be used by airplane cabin attendants to reassure passengers when cabin pressure is lost and the oxygen masks drop down. The script should tell passengers what to do, while helping them calm down and avoid panic.

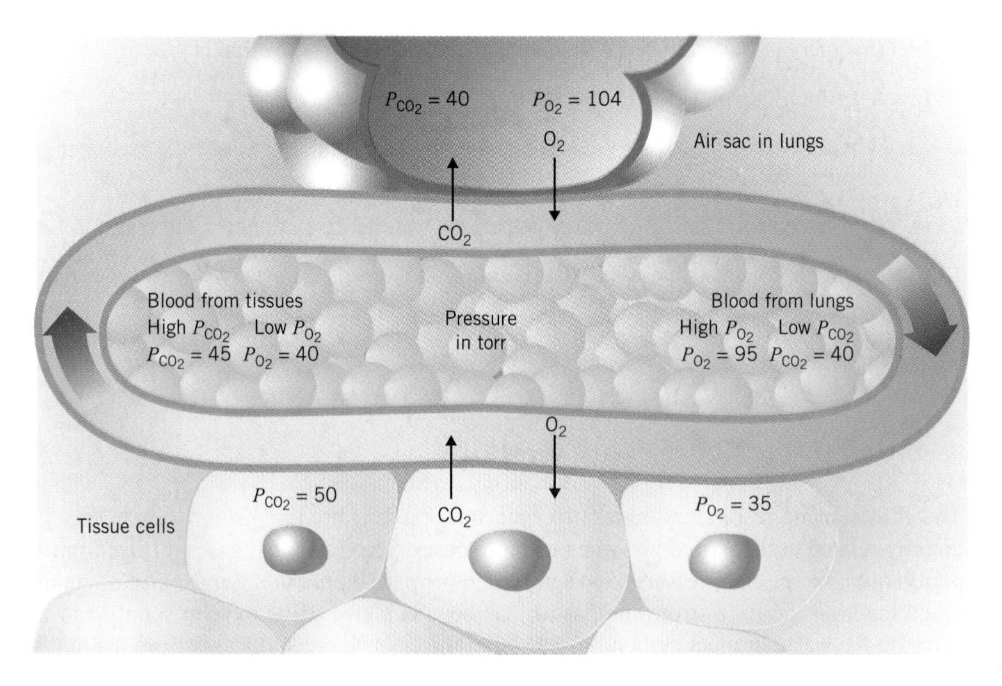

Figure 3.11 Movement of oxygen and carbon dioxide in blood and tissues depends on the partial pressure of each gas.

Health Perspective / Blood Pressure

Elevated blood pressure (called hypertension) affects over 58 million Americans. The small blood vessels of the brain, eyes, heart, and kidneys are particularly susceptible to the damaging effects of high blood pressure. It can lead to death through strokes, heart attacks, and kidney failure. The cause of hypertension is largely unknown.

To better understand hypertension, picture the heart as a pump with two chambers. First the heart muscle contracts, which shrinks the heart chambers and pushes the blood out through the arteries (this is called systole). Then the heart muscle relaxes, making the heart chambers larger, which lowers the pressure and allows the blood in the veins to return to the heart (called diastole). This cycle of high pressure followed by low pressure repeats on the average 72 times a minute. The high-pressure, or systolic, blood pressure averages 120 mm Hg, and the low-pressure, or diastolic, blood pressure averages 80 mm Hg. We record this average blood pressure as 120/80. If the blood pressure repeatedly measures over 140/90, the person is suffering from high blood pressure (or hypertension). The higher the blood pressure, the more dangerous the condition and the sooner treatment is required. Diet, exercise, stress reduction, and drugs can all be used to lower blood pressure. Age, sex, and race play a part in determining which treatments should be used to reduce hypertension.

Your Perspective: Suppose you are helping put together a book on health intended for 10-year-olds, and you must write a short section on the subject of blood pressure. In words your young audience would understand, explain the meaning of a blood pressure reading of 120/70.

(P_{O_2} = 104 torr) to the blood (P_{O_2} = 40 torr), and carbon dioxide diffuses from the blood (P_{CO_2} = 45) to the lungs (P_{CO_2} = 40 torr) to be exhaled. The oxygen, most of which is held by the carrier molecule hemoglobin, is then transported by the arterial blood to the tissues. Tissue cells are constantly using oxygen, so the partial pressure of oxygen (or the oxygen tension) in the cells is low. Because the arterial blood has a higher oxygen tension, the oxygen diffuses from the blood (P_{O_2} = 95 torr) to the tissues (P_{O_2} = 35 torr). Carbon dioxide, which is produced in the tissues by the cells, diffuses from the cells (P_{CO_2} = 50 torr) to the bloodstream (P_{CO_2} = 45 torr) to be carried back to the lungs.

3.9 Kinetic Theory of Gases

The gas laws were developed by 19th-century scientists from their observations of the predictable behavior of gases. Trying to find reasons for this behavior, they developed a theory of what gases must be like. The **kinetic theory of gases** is based on five key assumptions and describes the properties of an ideal gas.

Describe the properties of an ideal gas in the kinetic theory of gases.

- *Gases are made up of very small particles, with an enormous amount of space between them.* A gas consists mostly of empty space. Gases act as though their particles have no volume. This explains why it is so much easier to run through air than through water.

- *Gas particles travel at high speeds in straight lines.* This explains why gases fill the container in which they are stored. A gas that fills a small flask also fills a large room if the flask is opened. Each particle of gas moves in a straight line until it strikes another particle or the walls of the container. If we could track the movement of one particle, we would see it move in an apparently random zig-zag pattern, as it bounced off other particles.
- *The collisions of gas particles with one another and with the container are perfectly elastic.* This means that no energy is lost when gas molecules collide. If energy were lost with each collision, the particles would slow down slightly every time they collided. But a slowing down of the gas particles would mean a decrease in kinetic energy, and thus a decrease in temperature. This does not occur in an ideal gas.
- *There is no attraction or repulsion between gas particles.* This means that each particle acts independently. Dalton's law of partial pressures can hold only if each gas particle acts independently.
- *The kinetic energy of gas particles increases with temperature.* When gases are heated, the particles move faster and faster. When gases are cooled, the particles move slower and slower until, theoretically, a temperature is reached at which all motion ceases. This temperature, called absolute zero, is equal to −273.15°C or 0 K.

Define absolute zero and state its value in degrees Celsius and in kelvins.

The kinetic theory describes the behavior of an *ideal* gas. Truly ideal gases, of course, do not exist. The particles in real gases actually have volume, and small attractive forces exist between them. Because of this, real gases do not always behave exactly as described by the kinetic theory. This is especially true under conditions of very high pressure or very low temperature. Under less extreme conditions, real gases behave much like ideal gases. The kinetic theory of gases works under these more typical conditions, such as we find in living organisms.

KEY CONCEPTS

Matter can exist in three states: solid, liquid, and gas. Solids have a rigid structure, with definite shape and volume. Liquids have a fixed volume, but take on the shape of their container. Gases have no fixed volume or shape.

Adding energy to a solid increases the kinetic energy of the particles until the solid melts. The temperature at which the solid melts to form a liquid is called the melting point. The amount of energy that must be added to convert 1 g of a substance from a solid to liquid (at the melting point) is called the heat of fusion.

Adding energy to a liquid increases the kinetic energy of the liquid particles until the liquid boils. The temperature at which this occurs under conditions of normal atmospheric pressure is called the boiling point. The amount of energy needed to convert 1 g of liquid to gas (at the boiling point) is called the heat of vaporization.

The kinetic theory of gases describes the physical properties and behavior of an ideal gas. Real gases behave as ideal gases except under conditions of very high pressures or very low temperatures.

The pressure of a gas depends on the number of particles of gas in the container, not on the nature of the particles. The partial pressure of a gas is the pressure that the gas would exert if it were the only gas in the container.

Dalton's law of partial pressures states that the total pressure of a mixture of gases in a container equals the sum of the partial pressure of each gas.

Boyle's law states that if the pressure of a gas is increased, the volume decreases when the temperature is constant.

Charles's law states that the volume of a gas increases when the temperature increases, if the pressure remains constant.

Henry's law says that the higher the pressure of a gas over a liquid, the greater the solubility of the gas in the liquid.

Important Equations

Boyle's law

$$P_1V_1 = P_2V_2 \quad \text{(when } T \text{ is constant)}$$

Charles's law

$$\frac{V_1}{T_1} = \frac{V_2}{T_2} \quad \text{(when } P \text{ is constant)}$$

Combined gas law

$$\frac{P_1V_1}{T_1} = \frac{P_2V_2}{T_2}$$

Dalton's law

$$P_{\text{total}} = P_A + P_B + P_C + P_D$$

Review Problems

Section 3.1

1. Describe the difference between an amorphous solid and a crystalline solid, and give an example of each.
2. How many kilocalories of energy are given off when 50 g (about one ice cube) of water is frozen at 0°C?

Section 3.2

3. Which substance in each of the following pairs is more viscous?
   (a) motor oil or gasoline
   (b) water or honey
   (c) water at 10°C or water at 70°C
4. Ethanol, or ethyl alcohol, has a melting point of −117°C and a boiling point of 78.5°C. Draw a heating curve for ethanol (similar to Figure 3.4), and label the following:

(a) melting point

(b) boiling point

(c) region A—ethanol is a gas

(d) region B—ethanol is a liquid

(e) region C—ethanol is a solid

(f) region D—ethanol exists as solid and liquid

(g) region E—ethanol exists as liquid and gas

5. How many kilocalories of energy are needed to convert 50 g of water from a liquid to a gas at 100°C.

6. How many kilocalories of energy are released when 250 g of water condenses from a gas to a liquid at 100°C?

7. At room temperature ammonia is a gas, water is a liquid, and sugar is a solid. Which substance has the strongest attraction between its particles? Which has the weakest?

8. Describe in your own words what happens at the molecular level when water is cooled from 110° to −5°C.

Section 3.3

9. Set up the following conversions using the conversion factor method:

(a) 5.0 atm = (?) mm Hg **(b)** 63 torr = (?) Pa

10. Complete the following conversions:

(a) 5.0 atm = (?) mm Hg

(b) 190 mm Hg = (?) atm

(c) 70 torr = (?) mm Hg

(d) 2.7 atm = (?) torr

(e) 63 torr = (?) Pa

(f) 0.9 atm = (?) kPa

(g) 737 torr = (?) psi

(h) 890 torr = (?) atm

Section 3.4

11. When the barometer reads 720 torr, a sample of oxygen occupies a volume of 250 mL. What volume does the sample occupy when the barometer reads 750 torr and the temperature remains constant?

12. A sample of helium occupies a volume 325 mL at a pressure of 650 mm Hg. At what pressure does the sample occupy 250 mL?

13. If a sample of gas occupies 10 L at 1 atm pressure, what volume does it occupy at 5 atm pressure?

14. A sample of gas occupies 12 L at a pressure of 630 mm Hg. What pressure is required to compress the sample to 9 L?

Section 3.5

15. If a sample of nitrogen occupies 505 mL at −23°C, what volume does it occupy at 23°C if the pressure remains constant?

16. If a sample of ammonia occupies a volume of 1.2 L at 45°C, to what temperature, in degrees Celsius, must the gas be lowered to reduce its volume to 1.0 L when the pressure is held constant?

17. What volume does a gas occupy at 27°C and 760 mm Hg, if it occupies 2.8 L at 30°C and 909 mm Hg?

18. A 1.0-L sample of gas is collected under a pressure of 912 torr and 0°C. What is the volume of the gas at standard temperature and pressure?

Section 3.6

19. Assume that you have two flasks containing equal amounts of water into which equal amounts of carbon dioxide have been introduced. The pressure in one container is 1 atm, and in the other container the pressure is 2.5 atm. Which container has more carbon dioxide dissolved in the water? Give the reason for your answer.
20. State Henry's law in your own words, then explain why you hear a hiss of escaping gas when you open a container of carbonated beverage.

Section 3.7

21. The pressure in a bottle containing a mixture of oxygen and nitrogen is 1.00 atm. The partial pressure of the oxygen is 76 mm Hg. What is the partial pressure of the nitrogen?
22. A chemistry student collected a 250-mL sample of methane over water at 20.0°C. What is the partial pressure of the methane if the barometer reading in the classroom is 756.2 torr?

Section 3.8

23. Compare the partial pressure of carbon dioxide in the lungs with its partial pressure in the cells.
24. Compare the oxygen tension in arterial blood with that of venous blood.

Section 3.9

25. State in your own words the kinetic theory of gas behavior.
26. Use the kinetic theory of gases to explain Boyle's law.

APPLIED PROBLEMS

27. If there is danger of a frost in late spring, grape growers sprinkle their vineyards with water to protect their grapes from freezing. Suggest a reason why this is effective.
28. Everyone has observed that if the sun comes out immediately after a rain shower, the puddles dry up much faster than if it remains cloudy. Explain, at the molecular level, why this occurs.
29. Fans are often used to cool us off when the weather is hot and humid. But the air blown at us by the fan is the same temperature as the air in the room. Why does this same temperature air have a cooling effect?
30. Rooms that must be sterile are often kept at a pressure slightly higher than atmospheric pressure. Explain how this procedure helps to keep out dust and microorganisms.
31. Use Boyle's law to explain the following:
    (a) An inflated balloon, when released, propels itself across the room like a rocket.
    (b) Liquid is drawn up through a straw when we suck on it.

(c) A food particle stuck in the trachea of a choking person can be expelled using the Heimlich maneuver.

32. A diver collecting samples at a depth of 100 m exhales a bubble having a volume of 100 mL. The pressure at this depth is 11 atm. What is the volume of the bubble when it reaches the surface of the ocean (if we assume the water temperature is constant)?

33. Warm air rises and cold air falls. This is the reason that early morning frost often occurs in a valley while the surrounding hills remain well above the freezing point. Explain why warm air is less dense than cold air.

34. Although it is not recommended, people sometimes inflate their cold tires to the maximum pressure of 35 psi before heading off on a long trip. Driving at high speeds on hot pavement rapidly heats up the tires. If you filled your tires to 36 psi when they were cold (60°F), what would the pressure in the tires be after the tire had warmed to 100°F?

35. Exhaled breath is a mixture of nitrogen, oxygen, carbon dioxide, and water vapor. What is the vapor pressure of the water in exhaled breath at 37°C (body temperature) if the partial pressure of the oxygen is 116 torr, the nitrogen 569 torr, and the carbon dioxide 28 torr? (Assume that the atmospheric pressure is 1 atm.)

36. A cyclopropane–oxygen mixture can be used as an anesthetic. If the partial pressure of cyclopropane is 255 torr and the partial pressure of oxygen is 855 torr:

(a) What is the total pressure in the tank?

(b) What is the total pressure expressed in atmospheres?

37. We are accustomed to breathing air that is 21% oxygen and has an oxygen partial pressure of 159.6 torr. When we go to a high-altitude mountaintop or to a city at a higher elevation, the air is less dense and the total air pressure is lower. For this reason, breathing becomes more difficult. If oxygen still makes up 21% of the air, calculate the oxygen partial pressure at the top of Mt. Everest, where the total pressure is 240.16 torr.

Chapter 4

Atomic Structure

In 1918, Ruth Adams, then 16 years old, was excited to have finally found a job. She was hired by the Radium Luminous Materials Corporation of Orange, New Jersey, to apply luminous paint to watch dials and instrument dials. As you can imagine, this was painstaking work that required great precision. To help her get started, the other women showed Ruth how to keep a fine point on her brush by turning the bristles on her tongue and lips. The paint she used was invented by the president of the company and contained phosphorescent zinc sulfide with a small amount of radium and adhesive added.

At that time, neither Ruth nor anyone else realized the dangers that radioactive material posed to human tissues. It wasn't long, however, before the possible harm of internal exposure to radium was very clearly brought to the public's attention. In 1925, the *New York Times* reported that 5 watch dial painters had died, and 10 others had been stricken with "radium necrosis," a general breakdown of the bone tissues. But still, so little was known about the nature of radium and its effects on human tissues that the company, when brought to trial, defended itself with the claim that small quantities of radium were actually beneficial to health!

The watch dial painters who died in the 1920s suffered a variety of symptoms that included severe anemia, tumors of the sinuses, and inflammation of the bone marrow in the jaw and other bone structures of the body. Although some women who worked in this industry during that time are still living and in good health, many others died after developing bone cancer 20 to 30 years later. Ruth Adams worked in the watch dial factory for six years before leaving to get married. She showed no ill effects until some 25 years later, when she died of bone cancer which had spread throughout her body.

The disabilities or deaths of the women who were exposed to radium resulted from the destruction of their bone tissue, especially the tissue of the bone marrow, which produces the blood cells for the body. You might wonder what special property of radium caused it to collect in the bones of their bodies, as opposed to other tissues. We shall see that radium (Ra) has chemical properties very similar to those of calcium (Ca), which is the main component of bones and teeth. Unfortunately, the body is unable to tell the difference between the toxic radium and the essential calcium. Both elements are therefore deposited in the bones, where the radioactive radium does its damage. To understand why calcium and radium have similar chemical properties, we must study the structure of the atom.

4.1 The Parts of the Atom

List three subatomic particles, their relative mass, their charge, and their location.

Describe the structure of an atom.

In Chapter 1 we stated that atoms are composed of many types of particles, but that we would be focusing our attention on three such particles: **protons, neutrons,** and **electrons** (Table 4.1). Protons and electrons are electrically charged particles, whereas a neutron is neutral (that is, it has no charge). A proton is assigned the smallest unit of positive charge (1+) that will just cancel the negative charge on an electron (1−). Interestingly, Benjamin Franklin first used the terms *positive* and *negative* more than 50 years before the discovery of the electron and proton. A proton repels other protons (charges that are alike repel one another) and attracts electrons (unlike charges attract one another).

An atom has a small, dense **nucleus** that contains protons and neutrons. Because we have just said that protons repel one another, you might wonder what holds the nucleus together. Although nobody completely understands nuclear forces, the neutrons seem to

Table 4.1 **Subatomic Particles**

Name of Particle	*Location in the Atom*	*Charge*	*Symbol*	*Relative Mass (amu)*
Proton	Nucleus	1+	p, ${}^{1}_{1}p$, ${}^{1}_{1}\mathrm{H}$	1
Electron	Around the nucleus	1−	e, e^-, ${}^{0}_{-1}e$	1/1837
Neutron	Nucleus	0	n, ${}^{1}_{0}n$	1

Historical Perspective / John Dalton and His Atomic Theory

John Dalton (1766–1844)

John Dalton (1766–1844) was an English chemist and physicist who is considered the founder of modern atomic theory. His ideas about atomic theory came from many years of studying the properties of air and other gases. Dalton reasoned that if matter consisted of atoms, as proposed many centuries earlier by the Greeks, then these atoms must have properties that would explain the following discoveries made by other scientists of his time: (1) a number of different elements exist; (2) no mass is lost when substances combine chemically (the law of conservation of mass); and (3) a compound contains two or more elements in a definite, fixed proportion by mass (the law of definite proportions). In 1808, Dalton published *A New System of Chemical Philosophy* in which he listed the properties of atoms:

1. Matter consists of indivisible particles called atoms.
2. All the atoms of one particular element are identical in mass and other properties.
3. The atoms of different elements differ in mass and other properties.
4. Atoms are indestructible and merely rearrange themselves in chemical reactions. They do not break apart.
5. When atoms of different elements combine to form compounds, they form new, more complex particles. The particles of any compound always contain the same fixed ratio of atoms.

This list of the properties of atoms is known as Dalton's atomic theory. His theory has been modified over the years as new scientific discoveries have been made. For example, we now know that atoms can be broken into smaller pieces and that atoms of one element can have different masses. Dalton's atomic theory continues to describe how matter combines chemically: the law of definite proportions and the law of conservation of mass.

play an important role in binding together the positive protons. The electrons are found in the region surrounding the nucleus, but for the most part an atom is just empty space. To give you an idea of the relative positions of the subatomic particles in an atom, suppose you are in a large baseball stadium. If we let a flea on second base represent the nucleus of the atom, the nearest electron would be found somewhere in the top deck of the stands (Fig. 4.1).

4.2 Atomic Number and Mass Number

◆ **Define atomic number and mass number.**

The special characteristic that determines which element an atom represents is the number of protons in its nucleus. In Section 1.3 we defined an element as a substance that cannot be broken down into simpler substances by ordinary chemical processes. More specifically an **element** is a substance that consists of atoms, all of which have the same number of protons in their nuclei. The **atomic number (*Z*)** of an element represents the number of

Figure 4.1 An atom is mostly empty space. If we imagine a flea on second base to be the nucleus of an atom, the nearest electron would be a speck of dust somewhere in the top deck of the left-field stands.

protons in the nucleus of any atom of that element. (The atomic numbers of the elements are listed on the inside front cover of this book.)

Because the electric charge on a proton just cancels the electric charge on an electron, we can see that the atomic number of an element also tells us how many electrons are in a neutral atom of that element. For example, the element sodium has an atomic number of 11. Each neutral atom of sodium therefore has 11 protons in the nucleus and 11 electrons surrounding the nucleus. But, no matter how many electrons (or neutrons) there may be, the identity of an element is always determined by the number of protons in the nucleus—that is, by its atomic number.

Protons, electrons, and neutrons are extremely small particles. Protons and neutrons have a mass of 1.7×10^{-24} g; electrons are even lighter, having a mass only 1/1837 that of a proton. In fact, the mass of the electrons in an atom is so small in comparison with the mass of the protons and neutrons that it is ignored when calculating the mass of an atom. We define the **mass number** of an atom as equal to the number of protons plus the number of neutrons in the nucleus of the atom.

◆

Given the atomic number and mass number of any element, state the number of protons, neutrons, and electrons in an atom of the element.

$$\text{Mass number} = \text{protons} + \text{neutrons}$$

or

$$M = p + n$$

Chemists often use shorthand methods to express the mass number and the atomic number of an atom. For example, an atom of carbon (C) has six protons and six neutrons. The atomic number (Z) is 6, and the mass number (M) is 12. Chemists may refer to atoms of carbon having the mass number of 12 as carbon-12, ^{12}C, or $^{12}_{6}C$, where the upper number is the mass number and the lower number is the atomic number. In general, then, the symbol $^{M}_{Z}X$ gives you the symbol of the element (X), the mass number (M), and the atomic number (Z).

Example 4-1

1. State the number of protons, neutrons, and electrons in a neutral atom of

(a) $^{14}_{6}C$ **(b)** cobalt-60 **(c)** ^{235}U

See the Question

In each case, we need to determine the number of protons, neutrons, and electrons in a neutral atom of the element.

Think It Through

We can use the table on the inside front cover of this book to identify the element and find the atomic number. We use the equation for the mass number and the shorthand representation in the question to obtain the information needed to find each answer.

Execute the Math

(a) Using the table on the inside front cover of this book, we can identify this element as carbon. The symbol $^{14}_{6}C$ tells us that the mass number is 14 and the atomic number is 6. The atomic number gives both the number of protons and the number of electrons in a neutral atom.

$$p = 6 \qquad \text{and} \qquad e^- = 6$$

We know that

$$M = p + n$$

therefore

$$14 = 6 + n$$
$$8 = n$$

(b) Looking up cobalt, we find that the atomic number is 27. The mass number is 60. From the atomic number we then have

$$p = 27 \qquad \text{and} \qquad e^- = 27$$

$$60 = 27 + n$$

$$33 = n$$

(c) U is the symbol for the element uranium, which has an atomic number of 92. This atom of uranium has a mass number of 235. From the atomic number,

$$p = 92 \qquad \text{and} \qquad e^- = 92$$

From the mass number,

$$235 = 92 + n$$

$$143 = n$$

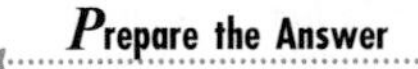

(a) $^{14}_{6}C$ contains 6 protons, 8 neutrons and 6 electrons.

(b) Cobalt-60 contains 27 protons, 33 neutrons, and 27 electrons.

(c) ^{235}U contains 92 protons, 143 neutrons, and 92 electrons.

2. State three ways of writing the symbol for an atom of iodine with 78 neutrons.

S

The question asks us to write three shorthand symbols for an atom of iodine that contains 78 neutrons.

T

We need to determine the mass number and the atomic number for iodine to write the answers. From the inside front cover we learn that the symbol for iodine is I and its atomic number is 53.

E The mass number for such an atom is therefore

$$M = 78 + 53 = 131$$

P The three different notations are

$$\text{iodine-131} \qquad {}^{131}\text{I} \qquad \text{and} \qquad {}^{131}_{53}\text{I}$$

Check Your Understanding

1. State the number of protons, neutrons, and electrons in a neutral atom of each of the following:

(a) ^{222}Ra **(b)** chromium-51 **(c)** $^{203}_{80}$Hg

2. What is the atomic number, mass number, and symbol for an atom having 15 protons and 16 neutrons?

4.3 Isotopes

In the early 19th century, John Dalton developed an atomic theory based on the idea that each atom of a given element is exactly alike. It was not until 100 years later that Frederick Soddy proved part of this theory to be wrong by showing that the element neon consisted of not one, but two types of atoms. Some of these neon atoms had a mass number of 20, and some had a mass number of 22. (A third type of neon atom with the mass number of 21 also exists.) Neon has an atomic number of 10, so each atom of neon must have 10 protons in the nucleus. Therefore, these two different types of neon atoms must have different numbers of neutrons in their nuclei. Soddy chose the term **isotopes** to describe atoms of an element containing different numbers of neutrons in the nucleus. For neon, then, the isotopes are

Describe how isotopes of an element differ.

	Atomic Number	*Mass Number*	*Number of p*	*Number of n*
Neon-20	10	20	10	10
Neon-21	10	21	10	11
Neon-22	10	22	10	12

A few of the elements have only one type of atom, but most elements have two or more naturally occurring isotopes; in fact, tin (Sn) has 10!

4.4 Atomic Weight

Because atoms are so small and light, it is impractical to measure the mass of a small number of atoms. It is possible, however, to measure the relative masses of atoms of different elements. A scale based on these relative measurements has been developed, with the most common isotope of carbon being arbitrarily assigned a value of exactly 12 **atomic mass units (amu).** It has been determined by indirect measurement that the mass of one atom of carbon-12 is 1.992×10^{-23} g (a *very* small mass!). Therefore, one atomic mass unit equals one-twelfth of this mass: 1 amu = 1.660×10^{-24} g.* If an isotope of

*Some scientists use the term *dalton* (after John Dalton) for one atomic mass unit: 1 dalton = 1 amu.

Table 4.2 **Relative Abundance of the Isotopes of Several Elements**

Isotope	*Natural Abundance (%)*	*Isotope*	*Natural Abundance (%)*
Hydrogen-1	99.99	Silicon-28	92.21
Hydrogen-2	0.01	Silicon-29	4.70
Carbon-12	98.89	Silicon-30	3.09
Carbon-13	1.11	Chlorine-35	75.53
Nitrogen-14	99.63	Chlorine-37	24.47
Nitrogen-15	0.37	Zinc-64	48.89
Oxygen-16	99.76	Zinc-66	27.81
Oxygen-17	0.04	Zinc-67	4.11
Oxygen-18	0.20	Zinc-68	18.57
Fluorine-19	100.00	Zinc-70	0.62
		Bromine-79	50.54
		Bromine-81	49.46

another element were one-half as heavy as the carbon-12 isotope, then its mass in atomic mass units would be $1/2 \times 12$, or 6; if an isotope were 2.84 times as heavy as the carbon-12 isotope, its mass would be $2.84 \times 12 = 34.1$ amu.

Most elements have at least two naturally occurring isotopes, so a sample of any of these elements contains a mixture of the different isotopes. Table 4.2 lists the share of the sample, or percentage abundance, of the naturally occurring isotopes of several elements. Suppose, for example, you are asked for the weight of a single atom of chlorine. According to Table 4.2, sometimes the mass of a single chlorine atom is 35 amu and sometimes 37 amu, depending on the atom that happens to be chosen. Because any sample of chlorine contains both kinds of chlorine atoms, we can approach this question by speaking instead of the *average* weight of a chlorine atom. If 75.53% of the atoms in the sample have a mass of 35.0 amu, and 24.47% have a mass of 37.0 amu, then the average mass of a chlorine atom taken from this sample is

$$(35.0 \text{ amu} \times 0.7553) + (37.0 \text{ amu} \times 0.2447) = 35.5 \text{ amu}$$

For questions dealing with the total mass of this sample, therefore, we could act as if each chlorine atom had a mass of 35.5 amu. This number is called the atomic weight of chlorine. More generally, the **atomic weight** of an element is the weighted average of the masses of the naturally occurring isotopes of the element, expressed in atomic mass units. (Note that such a number really ought to be called the ''atomic mass'' of the element, but this term is seldom used.) The atomic weight of each element is listed on the inside front cover of this book.

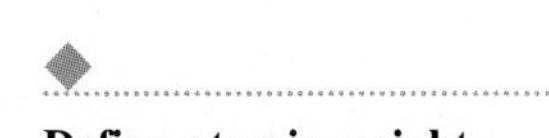

Define atomic weight.

Example 4-2

1. Teachers often use a weighted average in assigning final grades in a course. Suppose a teacher tells you that two midterm exams will each determine 25% of your final grade, and the final exam will be worth 50% of the final grade. If you received 76 on your first midterm, 64 on your second midterm, and 90 on your final exam, what is your grade in the course?

See the Question You need to determine the final grade using weighted averages.

Think It Through To determine your final grade, multiply each grade by its percent weight in the final grade, and then add up these products.

Execute the Math

$$\begin{aligned} 76 \times 25\% &= 76 \times 0.25 = 19 \\ 64 \times 25\% &= 64 \times 0.25 = 16 \\ 90 \times 50\% &= 90 \times 0.50 = \underline{45} \\ & \qquad\qquad\qquad\quad 80 \end{aligned}$$

Prepare the Answer Your grade in the course is 80.

2. Boron has two isotopes (boron-10 and boron-11) whose percentage abundances are 19.6% and 80.4%, respectively. What is the atomic weight of boron?

S The problem asks: Atomic weight of boron = (?) amu

T To calculate the atomic weight of boron, we use a procedure similar to that shown in Problem 1. The mass of each isotope is multiplied by its percentage abundance, and the products are then added up to give the atomic weight.

E

$$\begin{aligned} 10.0 \text{ amu} \times 19.6\% &= 10.0 \times 0.196 = 1.96 \\ 11.0 \text{ amu} \times 80.4\% &= 11.0 \times 0.804 = \underline{8.84} \\ & \qquad\qquad\qquad\quad 10.80 \text{ amu} \end{aligned}$$

P In the first mathematical steps we are multiplying data with three significant figures. Each of the products has three significant figures. Adding these two numbers results in the uncertainty being in the hundredth decimal place.

Atomic weight of boron = 10.80 amu

Check Your Understanding

Calculate the atomic weight of silicon using the data listed in Table 4.2.

4.5 The Quantum Mechanical Model of the Atom

In the beginning of this chapter, we said an atom of any element contains a small, dense, positive nucleus surrounded by negative electrons. Remember that a neutral atom contains the same number of electrons as there are protons in the nucleus. The number of electrons in a neutral atom of an element is therefore equal to the atomic number of that element. Because protons and electrons are too small to be seen with even the most powerful microscopes, we must depend on various theoretical models to describe the arrangement of the electrons around the nucleus.

◆ **Describe the Bohr model of the atom.**

In the early 1900s, Niels Bohr developed one of the first models of the atom, called the planetary model. He based his theory on the experimental data of Ernest Rutherford. Over the years, this theory has been modified, but it is helpful for us to understand Bohr's simple model. He pictured the atom as consisting of a small, dense nucleus surrounded by electrons traveling in orbits, similar to planets traveling around the sun. In his model the electrons could occupy only certain positions around the nucleus. Each of these positions

Table 4.3 **Energy Levels in the Atom**

Energy level number	1	2	3	4	5	6	7
Maximum number of electrons allowed in theory	2	8	18	32	50	72	98
Maximum number of electrons actually found in nature	2	8	18	32	32	18	8

corresponds to a certain energy value, so these energy positions were called **energy levels,** or **energy shells.** Electrons closest to the nucleus are in the lowest possible energy state. As electrons are found farther from the nucleus, they have higher and higher energy states.

The energy levels within an atom are labeled with numbers, with the first energy level being the one closest to the nucleus. As the number of the energy level increases, it is found farther from the nucleus and contains electrons having greater and greater energy. Each energy level can contain only a certain maximum number of electrons, but in nature this maximum number is, in most cases, never realized (Table 4.3).

Indicate the maximum number of electrons possible in energy levels 1 through 7.

The simple planetary model of the atom developed by Bohr did not explain some of the experimental data that were later collected. In the late 1920s Erwin Schrödinger, P. A. M. Dirac, Werner Heisenberg, and others developed a new model of the atom called the **quantum mechanical model.** The model is based on complicated mathematical concepts, and we shall look at only some of the ways this theory changed the idea of the atom developed by Bohr.

Although the Bohr atom described electrons orbiting around the nucleus, scientists found it impossible to pinpoint accurately the position and speed of an electron. The electron's position therefore could not be described as following a definite path. Instead, scientists could only specify regions in which there is a large probability of finding one or two electrons. These probability regions are known as atomic **orbitals.** Each energy level has energy sublevels consisting of one or more atomic orbitals. An orbital can contain zero, one, or two electrons (Fig. 4.2).

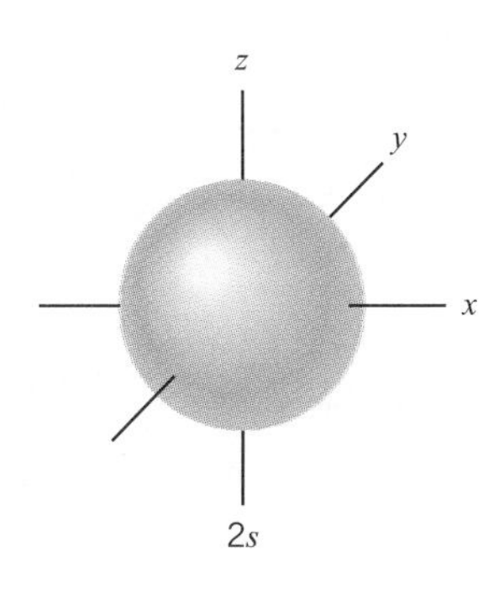

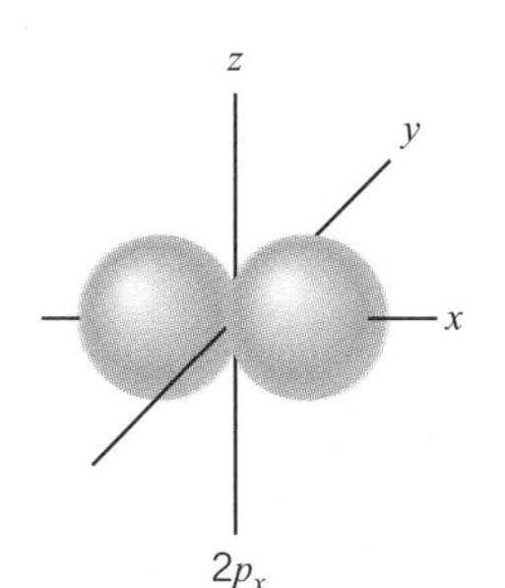

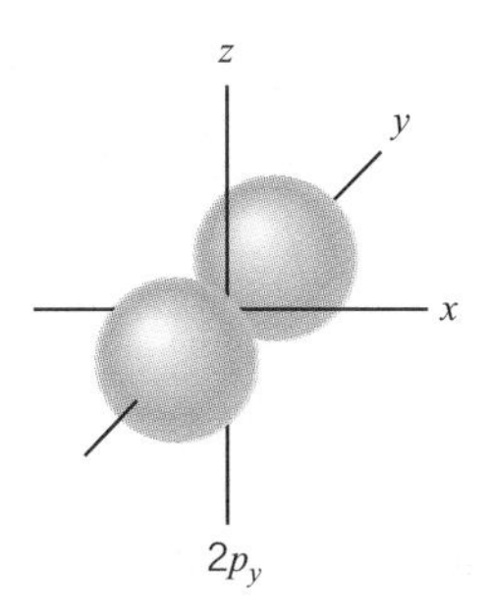

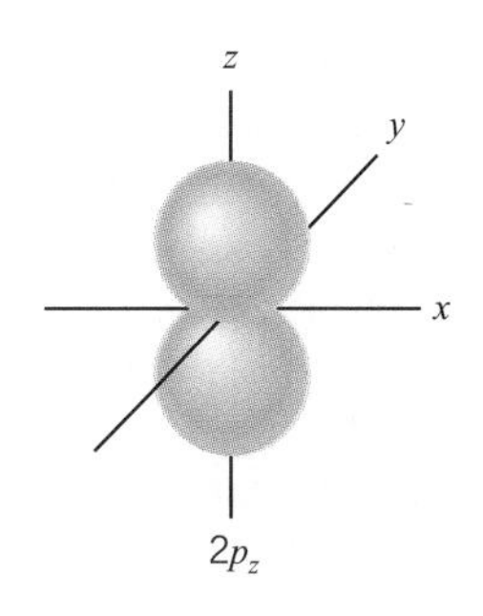

Figure 4.2 Representations of the probability regions (areas in which electrons are most likely to be found) for the 1*s*, 2*s*, and 2*p* orbitals.

The first energy level contains one atomic orbital, labeled the 1*s* orbital. The probability region described by this orbital is a sphere with its center at the nucleus. The second energy level contains four atomic orbitals. One of the orbitals is spherical in shape and is called the 2*s* orbital. The probability regions of the other three orbitals are shaped like dumbbells that are at right angles to each other about the nucleus (see Fig. 4.2). These orbitals, called the 2*p* orbitals, have slightly more energy than the 2*s* orbital.

Each higher energy level contains one *s* and three *p* orbitals. In addition to these are two other types of orbitals: the *d* orbitals, which appear in groups of five starting from energy level 3, and the *f* orbitals, which appear in groups of seven starting from energy level 4. We shall not discuss the shapes of the *d* and the *f* orbitals in this textbook.

To determine the arrangement of electrons in the energy levels and atomic orbitals of an atom, we can use the following rules:

1. Electrons fill the lowest possible energy levels first.
2. An orbital can hold no more than two electrons. To hold two electrons, the electrons must be spinning in opposite directions. We can indicate the direction of the spin with arrows: "↑" for clockwise and "↓" for counterclockwise.
3. Electrons do not pair in an orbital if another orbital is available at the same energy. That is, electrons pair up only when all orbitals of the same energy contain at least one electron. For example, each of the three 2*p* orbitals must contain one electron before electrons pair up in one of the 2*p* orbitals.
4. Electrons are found in orbitals in order of increasing energy as shown in Figure 4.3. The orbitals in the higher energy levels overlap. For example, the 4*s* orbital has less energy than the 3*d* orbitals. Figure 4.4 gives you a way to remember the order of the orbitals. Just list the orbitals in order of increasing energy level and then follow the ordering shown by the diagonal arrows beginning at the bottom.

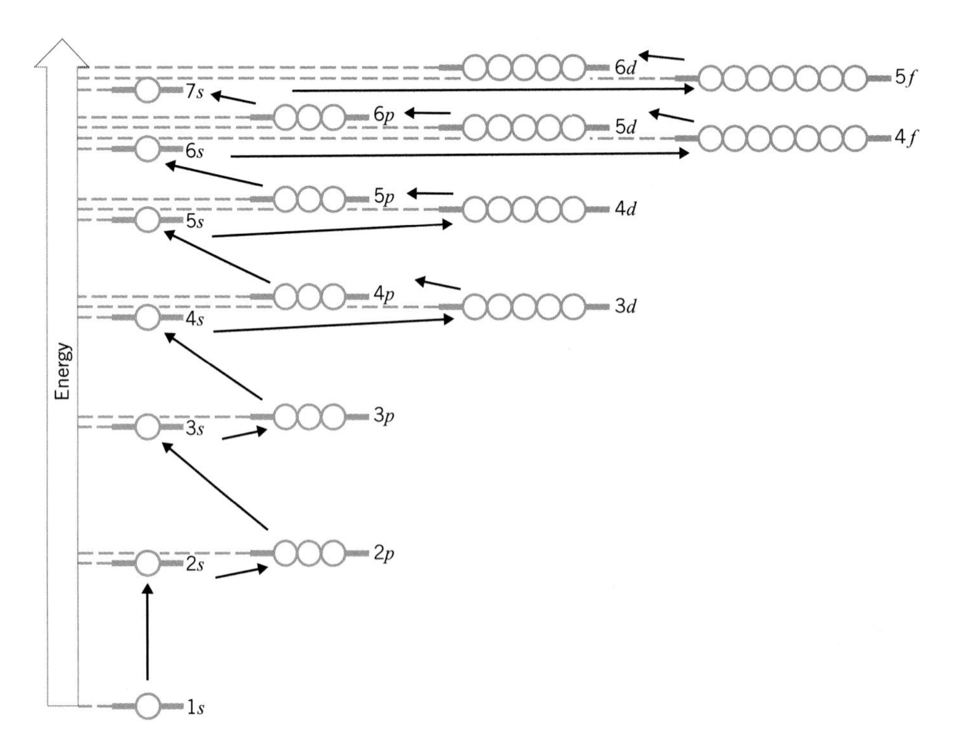

Figure 4.3 The order in which electrons fill orbitals is given by the arrows, beginning with the 1*s* orbital.

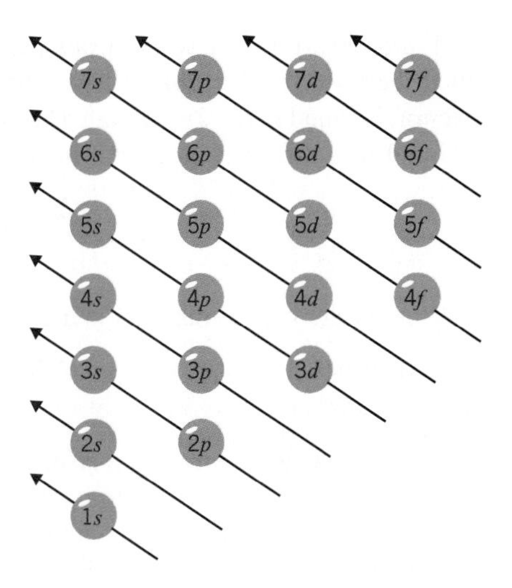

Figure 4.4 To remember the order in which electrons fill the orbitals, follow the diagonal arrows beginning at the bottom.

4.6 Electron Configurations

Write the electron configuration and orbital diagram of a given element.

There are several ways to indicate the arrangement of electrons, or the **electron configuration,** in an atom of an element. Let's use boron (atomic number 5) as an example. The atomic number tells us that a neutral atom of boron has five electrons. Using Figure 4.3, we see that two of the electrons fill the 1*s* orbital, two more the 2*s* orbital, and the remaining electron is found in one of the 2*p* orbitals. We can write this electron configuration as follows:

$$\mathrm{B} \quad 1s^2 2s^2 2p^1$$

A second way to show the electron configuration of boron is to use an **orbital diagram,** in which each orbital is represented by a circle and each electron by an arrow.

	$1s$	$2s$	$2p_x$	$2p_y$	$2p_z$
B	(↑↓)	(↑↓)	(↑)	( )	( )

Example 4-3

In the following problems the *STEP* strategy is used, but the *Execute the Math* step is not necessary. This will be the case in a number of problems that do not require mathematical calculations. For clarity, the *T* and *E* symbols are combined for these problems. As you *Think It Through* for each problem, decide whether mathematical calculations are required for the solution or not.

1. Show the orbital diagram and the electron configuration of carbon, C.

See the Question

The question asks us to draw the orbital diagram and the electron configuration of the element carbon.

Think It Through (Execute the Math)

From the inside front cover, we find that carbon has an atomic number of 6. Using Figure 4.3 as a guide, we can fill the orbitals with the six electrons. The 1*s* orbital is filled with two electrons, leaving four electrons still to be assigned. Next, the 2*s* orbital is filled with two electrons. These electrons pair up in the $2s$ orbital before the $2p$ orbitals are filled, because the $2p$ orbitals have slightly more energy than the $2s$ orbital. The next electron is found in the $2p_x$ orbital. The last electron does not pair up in the $2p_x$ but instead is found in a different $2p$ orbital having the same energy.

Prepare the Answer

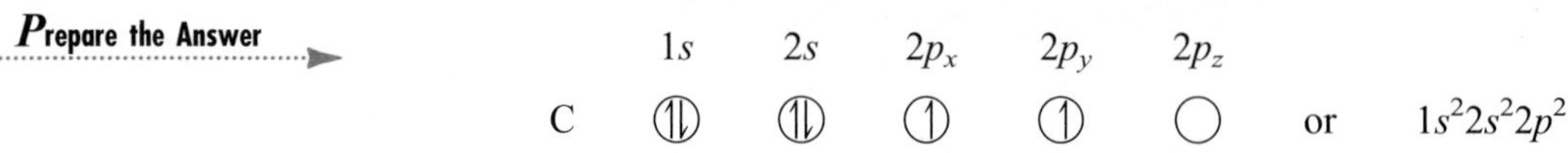

2. What is the orbital diagram and electron configuration of calcium, Ca?

S

This question asks us to draw the orbital diagram and electron configuration of the element calcium.

T • E

Calcium has an atomic number of 20, so a neutral atom of calcium has 20 electrons. Again, using Figure 4.3 as a guide, we can fill the orbitals. The $1s$ orbital is filled with two electrons, the $2s$ orbital is filled with two electrons, and the three $2p$ orbitals are filled with six electrons. That makes a total of 10 electrons that we have so far placed, with 10 left to go. The $3s$ orbital is filled with two electrons, and the three $3p$ orbitals with six electrons. We now have only two electrons remaining. Notice from Figure 4.3 that the $4s$ orbital has less energy than the $3d$ orbital. The $4s$ orbital therefore is filled first with two electrons, completing the electron configuration for calcium. (Note that the $3d$ orbitals are filled in elements having atomic numbers 21 through 30, and the $4p$ orbitals are then filled in the elements having atomic numbers 31 through 36.)

P

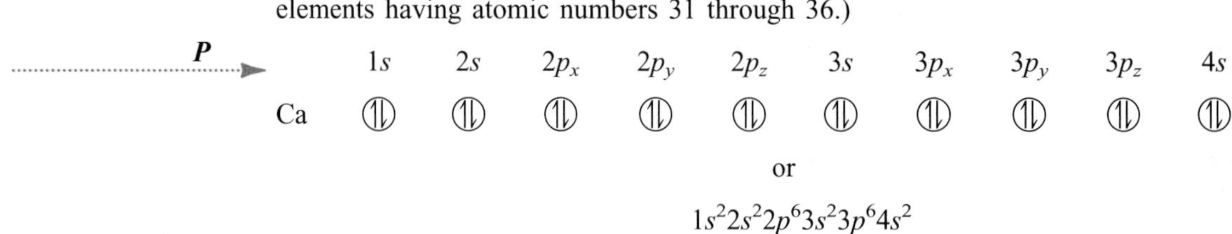

Check Your Understanding

Write the orbital diagram and the electron configuration for a neutral atom of the following elements:

(a) sodium **(b)** phosphorus **(c)** chlorine

4.7 Formation of Ions

State the difference between an atom in the ground state and an excited atom.

In a normal atom, all the electrons are found in the lowest possible energy levels. Such an atom is in the **ground state.** If the atom absorbs energy (such as heat or light) from an external source, some electrons might jump to a higher energy level. When this occurs, each such electron absorbs an amount of energy exactly equal to the energy difference between the two orbitals. An atom having one or more electrons in a higher than normal energy state is called an **excited atom.** Excited atoms are unstable, and in such atoms electrons drop back to lower energy levels by giving off energy, usually in the form of radiant energy (UV, IR, or visible light waves). The light given off by an atom as an electron drops from one energy level to another has a specific energy or wavelength corresponding to the energy difference between the two levels (Fig. 4.5).

Chemical Perspective / Chemiluminescence

When an element or compound becomes excited by absorbing energy from an outside source and then gives off that energy as visible light, we call it *luminescent.* There are two kinds of luminescence: fluorescence, in which the element or compound stops giving off light when the energy being supplied to the compound is stopped; and phosphorescence, in which the element or compound continues to give off light for a short time after the incoming energy has stopped.

You are familiar with many examples of luminescence. Laundry whiteners and brighteners make use of fluorescent dyes that attach to clothing in the laundering process. These dyes absorb energy from the ultraviolet light in sunlight and then give off light in the visible region. The clothing, through a combination of reflection and fluorescence, gives off more visible light than falls on it—making the clothing appear ''brighter.'' Red reflective tape, which you might see attached to bicycles or cars to increase their visibility at night, contains compounds that absorb nonred light and give off the red-orange light we see. The luminous dial you may have on your watch contains tiny amounts of radioactive material mixed with a powdered phosphor. The radioactive material releases a small but steady amount of energy, which is absorbed by the atoms in the phosphor. When the electrons in these atoms return to the ground state, they release the visible light that you see as a faint glow on the watch dial.

Living organisms make use of luminescence for sexual attraction, protection, and hunting. The firefly uses chemicals that luminesce to attract a mate. These chemicals produce the firefly's blink as their electrons return to the ground state. Green plants use the reverse of this process to produce food: energy from sunlight excites electrons in molecules of the plant's green pigment, called chlorophyll. As these electrons return to the ground state, they do not release their energy as light. Instead the energy is trapped as chemical energy in a molecule of sugar. This process of using the sun's energy to produce a sugar molecule from carbon dioxide and water is called photosynthesis.

Your Perspective: Write a letter to the Consumer Products Protection Agency requesting a new law that would require all bicycles to be painted in luminous paint. Justify your request, making up any supporting data you might need.

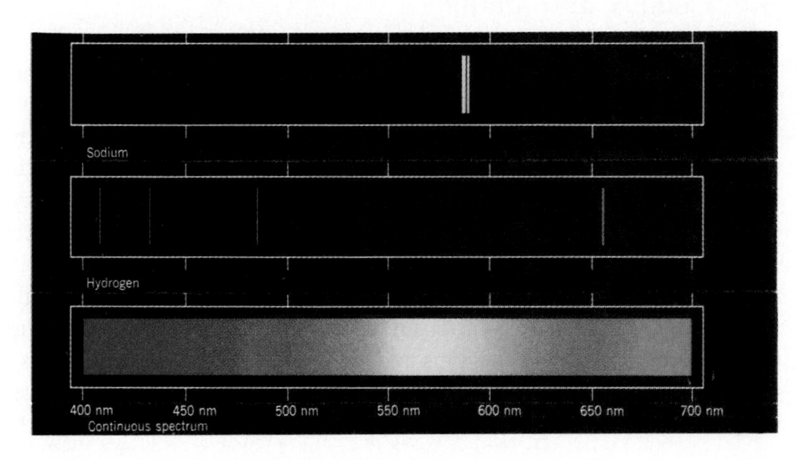

Figure 4.5 When white light produced by a light bulb is passed through a prism, it produces a continuous spectrum. When atoms of elements such as hydrogen and sodium are excited and allowed to return to the ground state, they also emit light. But this light, when passed through a prism, produces a series of lines called the atomic spectrum of that element.

If enough energy is supplied to an atom, one of its electrons may jump completely away from the atom, leaving the atom with one less electron than it has protons. This new particle is then no longer a neutral atom but a positively charged particle called a positive ion. An **ion** is a positively or negatively charged particle that is formed when an atom loses or gains electrons. Positive ions are called **cations,** and negative ions are called **anions.** A calcium atom that loses two electrons forms the cation Ca^{2+}. A chlorine atom that gains one electron forms the anion Cl^{-}.

Define cation and anion and give an example of each.

4.8 The Periodic Table

When discussing radium, we mentioned that the chemical properties of each element depend on that element's electron configuration. It would certainly be quite a chore to memorize the electron configuration of all the chemical elements. But take heart; there is a way of writing down the chemical elements so that their similarities are easily seen. This method makes use of the **periodicity,** or repeating nature, of properties of the elements. The periodicity results from the way in which electrons fill in the *s*, *p*, *d*, and *f* orbitals. The way of arranging and writing down elements to show this periodicity is known as the **periodic table of the elements.** Since 1871 more than 700 versions of the periodic table have been published, showing large amounts of information about each element. One preferred version of the periodic table is shown on the inside front cover of this book.

Define periodicity.

Let's closely examine this periodic table. Notice that each element appears in a box containing specific information about that element. The box lists the symbol of the element, the atomic number, and the atomic weight.

Use the periodic table to state the symbol, atomic number, and atomic weight of an element.

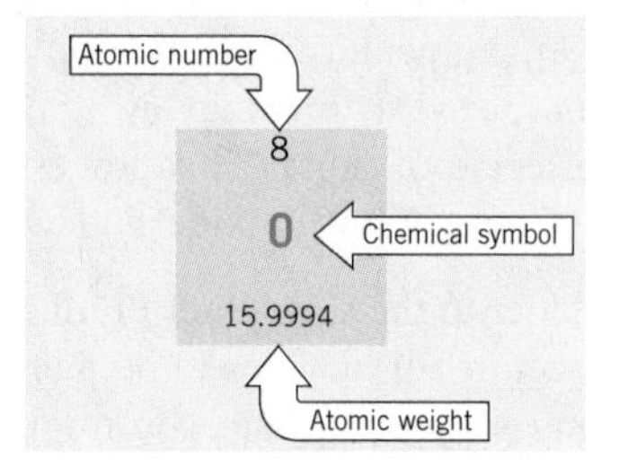

4.9 Periods and Groups

Each horizontal row of the periodic table is called a **period.** The periods are numbered from one to seven, corresponding to the seven energy levels of an atom that can contain electrons. This means that any element in, say, period 4 has its outermost electrons assigned to the fourth energy level. For example, find sodium (symbol Na, atomic number 11) on the table. Sodium has only one electron in its outermost energy level. Because sodium appears in period 3 of the periodic table, it contains that one electron in the third energy level.

Use the periodic table to locate members of a given period or group.

Each column of the periodic table is called a **group,** or **chemical family.** Dmitri Mendeleev noticed that the various members of a group show similar chemical behavior. Chemical reactions between atoms involve interactions between electrons in the outermost energy levels. The number of electrons in this outermost energy level determine the chemical characteristics of the element. These outermost electrons are given the name

Historical Perspective / The Periodic Table

The periodic table of the elements shown on the inside front cover of this text represents the efforts of many 19th-century researchers. In the 1800s, chemists were trying to find some logical order among the known chemical elements, in much the same way that biologists classified animals and plants according to their similarities.

The first chemist to recognize patterns among the elements was Johann Döbereiner. He found that some of the elements could be arranged into groups of three, which he called triads. Döbereiner recognized similarities in the properties of lithium, sodium, and potassium as well as among chlorine, bromine, and iodine. However, his attempts to expand this model of triads to the other elements was unsuccessful. In 1864, the English chemist John Newlands arranged the elements in groups of eight, which he called octaves after the eight notes in an octave of music. In this arrangement, the eighth element in a sequence had chemical properties very similar to the starting one. Beyond the element calcium, however, the arrangement did not work well at all. His fellow members of the Chemical Society of London were so unimpressed that they refused to publish his paper on the subject.

In 1869, the Russian chemist Dmitri Mendeleev and the German chemist J. L. Meyer independently developed a much more successful proposal, which is the basis of the present-day periodic table. Because Mendeleev also showed how the existence and properties of yet-unknown elements could be predicted, most of the credit for developing the periodic table is given to him. By arranging the known elements in order of increasing atomic weight, Mendeleev made a chart that contained eight "families" of elements. Unlike Newlands, Mendeleev realized that the elements beyond calcium would fit properly only if empty spaces were left in the periodic table. He correctly believed that some of the gaps in his chart were due to elements that should exist but had not yet been discovered. He was able to predict the existence of gallium, scandium, and germanium from these gaps in his periodic table. About 45 years later, H. G. J. Moseley found that if Mendeleev's table were slightly rearranged so that the elements were listed by atomic number rather than by atomic weight, some of the irregularities left in Mendeleev's table would be eliminated.

Dmitri Mendeleev (1834–1907)

valence electrons. Each member of a chemical family has the same number of valence electrons (Table 4.4).

There are several different systems in use today for numbering the groups or columns on the periodic table. One system proposed by the International Union of Pure and Applied Chemistry (IUPAC) is to number each column from left to right using the numbers 1 through 18. (See the periodic table on the inside front cover.) The system we shall use in this book uses the Roman numerals I through VIII and the letters A and B (Fig. 4.6). The A groups are the **representative,** or **main-group elements.** The group numbers I through VIII at the top of these columns indicate the number of valence electrons for each member of the group. For example, the members of group IIA are beryllium (Be), magnesium (Mg), calcium (Ca), strontium (Sr), barium (Ba), and radium (Ra). Each neutral atom of these elements contains two valence electrons. These two electrons are located on the fourth energy level in an atom of calcium, and on the seventh energy level in an atom of radium.

◆ **State the number of valence electrons in an atom of any representative element.**

Table 4.4 Electron Configurations of Three Chemical Families

Group Number	*Chemical Family*	*Element*	*Atomic Number*	*Electron Configuration by Energy Level*						
IA	Alkali metals	Lithium	3	2	1					
		Sodium	11	2	8	1				
		Potassium	19	2	8	8	1			
		Rubidium	37	2	8	18	8	1		
		Cesium	55	2	8	18	18	8	1	
		Francium	87	2	8	18	32	18	8	1
IIA	Alkaline earth metals	Beryllium	4	2	2					
		Magnesium	12	2	8	2				
		Calcium	20	2	8	8	2			
		Strontium	38	2	8	18	8	2		
		Barium	56	2	8	18	18	8	2	
		Radium	88	2	8	18	32	18	8	2
VIIA	Halogens	Fluorine	9	2	7					
		Chlorine	17	2	8	7				
		Bromine	35	2	8	18	7			
		Iodine	53	2	8	18	18	7		
		Astatine	85	2	8	18	32	18	7	

Locate the metals, nonmetals, metalloids, representative elements, transition elements, and inner transition elements on the periodic table.

The B group elements consist of the three rows of 10 elements in the middle of the table and are called the **transition elements,** or **transition metals** (see Fig. 4.6). They often show similar chemical properties not only within groups, but also along periods. This similarity in chemical behavior results from the fact that across the transition elements, electrons are being added to the next-to-outermost level rather than the outermost energy level. (The 3*d* orbitals are being filled in the transition elements of period 4, the 4*d* orbitals in the transition elements of period 5, etc.) The outermost level of each transition element contains only one or two electrons. For example, iron (Fe), cobalt (Co), and nickel (Ni), located on the fourth row of the periodic table, all have two electrons on the same outermost energy level (the fourth). Although they contain different total numbers of electrons, they are very similar in chemical behavior and are often grouped together for this reason.

We have seen that the same number of electrons in the outermost energy level of the transition elements masks the difference in the number of electrons in the next-to-outermost level. This is doubly true for the **inner transition elements** (also called the lanthanide and actinide series), the two rows of 14 elements at the bottom of the table (see Fig. 4.6). Each of these rows is named after the element it follows in the main body of the periodic table. The differences in the electron arrangements of the inner transition elements are hidden beneath two identical outer shell electron configurations. (Electrons in these elements are filling the *f* orbitals—the 4*f* in the lanthanide series and the 5*f* in the actinide series.) As a result, these elements have very similar properties.

Some groups on the periodic table are known by common names. The elements in group IA are called the **alkali metals,** those in group IIA the **alkaline earth metals,** and the elements in group VIIA the **halogens.** The elements of group VIIIA are called the **noble gases.** Some of these gases can be forced to form compounds with oxygen or with members of the halogen family, but otherwise they are extremely unreactive. Each noble gas, except for helium (He), has an electron configuration of eight electrons in the outer-

Representative Elements

Transition Elements

Noble Gases

Inner Transition Elements

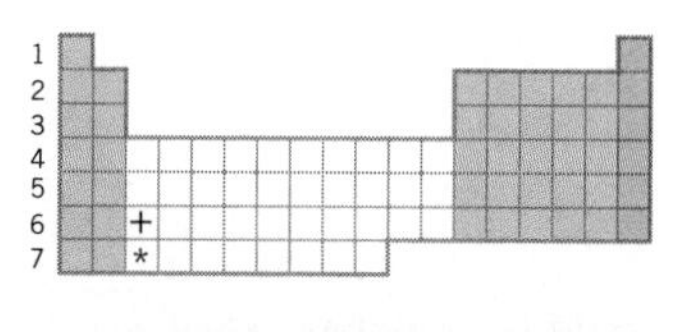

Figure 4.6 Locations of the representative elements, the transition elements, and the inner transition elements.

The neon lights used to make this bicycle contain mixtures of the noble gases.

most energy level. The chemical stability of these elements comes from this stable arrangement of electrons. Although helium contains only two electrons, these two valence electrons completely fill the first energy level of the atom, giving helium its stability.

We have said that members of the same chemical family show similar chemical behavior. For example, chlorine (Cl_2) and iodine (I_2), both members of the halogen family, are highly effective in killing bacteria and are used as disinfectants and antiseptics, respectively. Germanium (Ge) and silicon (Si), members of group IVA, are both widely used as semiconductors in transistors. The ''hardness'' in water can be caused by either calcium (Ca^{2+}) or magnesium (Mg^{2+}) ions because both of these group IIA ions undergo similar chemical reactions with soap. Two radioactive wastes of special concern are strontium-90 and cesium-137. Strontium, like radium, is in group IIA, so it too has chemical properties similar to calcium. Strontium-90 in the atmosphere falls onto the grass, is eaten by cows, and is consumed by humans when they drink milk. As with radium, the human body mistakes strontium-90 for calcium and deposits it in bones and teeth. Cesium is in group IA, and cesium-137 is mistakenly used in the body in much the same way as sodium, an element that is essential for many body functions.

4.10 Metals, Nonmetals, and Metalloids

Various physical and chemical properties of elements allow us to classify the elements of the periodic table as **metals** or **nonmetals** (Fig. 4.7). You are probably familiar with many properties of metals. Metals are generally shiny, dense, malleable (able to be hammered or rolled into sheets), and ductile (able to be drawn out into wires). Metals have high melting points: all except mercury are solids at room temperature. Metals are also excellent conductors of electricity and heat.

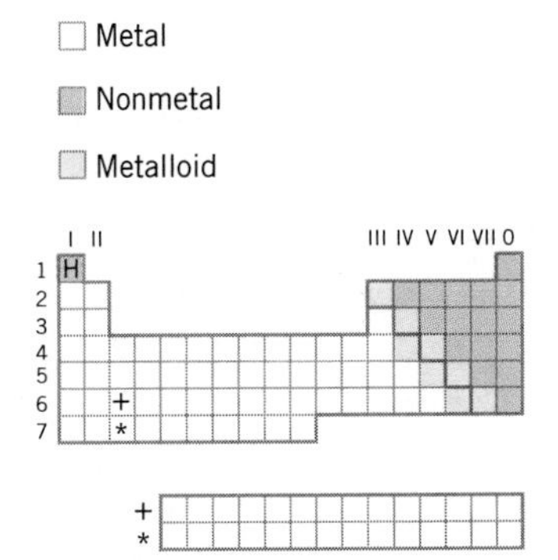

Figure 4.7 Locations of the metals, nonmetals, and metalloids.

Nonmetals, on the other hand, tend to have low density and to be brittle when solid. They are poor conductors of heat and, except for the graphite form of carbon, are poor

Samples of the nonmetal bromine (right), the metal copper (left), and the metalloid silicon.

conductors of electricity. Most nonmetals have low melting points, and many are gases at room temperature. All the group VIIIA elements are gases; the other gaseous nonmetals are hydrogen, oxygen, nitrogen, fluorine, and chlorine.

The heavy jagged line on the periodic table divides those elements that are metals from the nonmetals. The metals are found to the left of the line and clearly make up the majority of elements. There is no sharp distinction, however, between the elements having metallic properties and those having nonmetallic properties; the elements next to the jagged line may exhibit properties of both metals and nonmetals. Such elements are known as **metalloids,** or **semimetals.** For example, arsenic (As, atomic number 33) is a brittle gray solid. When freshly cut, however, it has a bright metallic shine. Arsenic acts as a metal in forming compounds with oxygen and chlorine but acts as a nonmetal in other chemical reactions. The most important physical property of metalloids is their ability to conduct electricity. Because they do not do this as well as metals, they are known as **semiconductors.** This behavior, especially in silicon and germanium, has led to the development of integrated circuits on silicon chips, making possible such electronic devices as personal computers.

Notice that hydrogen, at the top of group IA, does not fit into the division of metals and nonmetals. Hydrogen displays nonmetallic properties under normal conditions, although its valence electron configuration is the same as that of the group IA metals. It has been found, however, that hydrogen shows metallic properties similar to the alkali metals under extreme high pressures.

4.11 The Periodic Law

Several physical and chemical properties of atoms vary periodically with their atomic number. This fact is often referred to as the **periodic law.** By examining three of these periodic properties, we can better understand the physical and chemical properties of the elements—especially the different chemical properties of metals and nonmetals.

Predict the trends in atomic size, ionization energy, and electron affinity going down a group or across a period on the periodic table.

Atomic Size

One property of the elements that shows a periodic relationship is the size of their atoms. It is impossible to measure the exact size of an atom because we cannot pinpoint the exact position of electrons around the nucleus. The atomic radius is therefore determined by taking half the distance between nuclei of adjacent atoms in a compound. With only a few exceptions, the atomic radius decreases as you move across a period of the periodic table (for example, from Li to F). As you move down a group or chemical family, the atomic radius increases as the atomic number increases. One result of this pattern is that the metallic elements at the left-hand side of the periodic table have large atomic radii, and the nonmetals at the upper right-hand part of the table have small radii.

Ionization Energy

A second important property of an element is its ionization energy. The **ionization energy** of an element is the amount of energy that must be added to a gaseous atom to remove its outermost (or most loosely held) electron. Put another way, the ionization energy shows how strongly the nucleus attracts its valence electrons. The stronger the attraction for these valence electrons, the greater is the ionization energy. It is easy to see that the size of an atom affects how strongly the nucleus attracts the valence electrons. In small atoms the nucleus and the valence electrons are close together, and the electric attraction is strong. In larger atoms the outermost energy level is farther from the nucleus and, therefore, the attraction for the valence electrons is weaker. So, in general, the trend is for the ionization energy to increase across a period as the radius decreases. On the other hand, as we move down any group, the atomic radii are increasing. So the attraction of the nucleus becomes weaker and the ionization energy decreases down a chemical family.

Metals generally have lower ionization energies than do the nonmetals, so that it is easier to remove a valence electron from a metal than from a nonmetal. The noble gases, with their very stable configuration of electrons, all have very high ionization energies. This electron configuration somehow makes it extremely difficult to remove an electron from the outermost energy level of the atom.

Electron Affinity

A third property showing a periodic relationship among the elements is electron affinity. The **electron affinity** of an element is the amount of energy released when an electron is added to a neutral gaseous atom of that element. Some elements, such as the noble gases, have no affinity for extra electrons. In general, however, the trend in electron affinities is to increase from left to right across a period as the atomic size decreases and to decrease from top to bottom within a group as atomic size increases. Metals have low electron affinities and nonmetals have high electron affinities.

4.12 The Elements Necessary for Life

Living organisms, like all other matter on earth, are composed of atoms of the naturally occurring elements. Not all 90 of these elements are found in such organisms, however. Table 4.5 shows the 25 elements that have so far been shown to be essential to life. Hydrogen, carbon, nitrogen, and oxygen are the most plentiful, or abundant, elements in living organisms. They make up 99.3% of all the atoms in your body, whereas the remaining 21 elements account for only 0.7% (Fig. 4.8). Although these 21 elements are a very small proportion of the atoms in the human body, they perform a wide variety of functions critical to life. Depriving a living organism of any of these elements results in disease and possibly death.

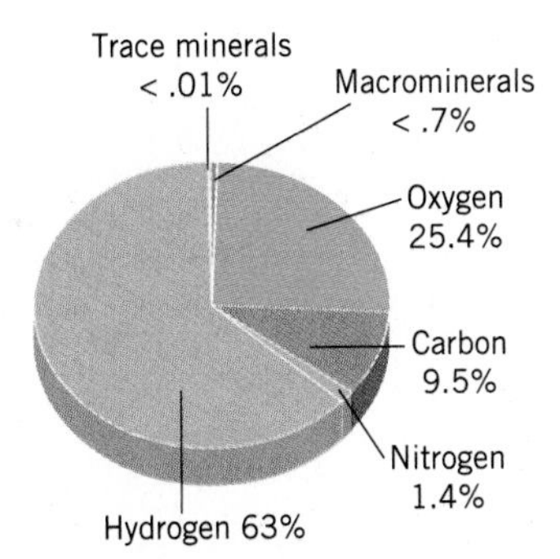

Figure 4.8 Percentage abundance of atoms in the human body.

The remaining 21 elements can be divided into two groups. One group contains seven elements, called the **macrominerals,** which are found in greater concentrations in the body than are the remainder. This group of elements contains potassium, magnesium, sodium, calcium, phosphorus, sulfur, and chlorine. The remaining elements are found in such very small amounts that they are called the **trace elements** (Table 4.6). In the past the necessity of these trace elements to good nutrition was largely overlooked, with most

Table 4.5 **Elements Essential to Life**

Arsenic	As	Hydrogen	H	Oxygen	O
Boron	B	Iodine	I	Phosphorus	P
Calcium	Ca	Iron	Fe	Potassium	K
Carbon	C	Magnesium	Mg	Selenium	Se
Chlorine	Cl	Manganese	Mn	Silicon	Si
Chromium	Cr	Molybdenum	Mo	Sodium	Na
Cobalt	Co	Nickel	Ni	Sulfur	S
Copper	Cu	Nitrogen	N	Zinc	Zn
Fluorine	F				

Table 4.6 **The Trace Elements**

Element	*Daily Requirement*[a]	*Function*
Nonmetals		
Fluorine	1.5–4.0 mg[b]	Found in bones and teeth; important in prevention of dental caries
Iodine	0.15 mg	Required for normal thyroid function; found in thyroid hormones
Selenium	0.070 μg for men 0.055 μg for women	Required for the prevention of white muscle and liver disease in some animals and Keshan disease in humans; part of the enzyme glutathione peroxidase
Silicon	Unknown[c]	Required for bone growth and connective tissue development in animals
Arsenic	Unknown[c]	Required for adequate growth and reproduction in animals
Boron	Unknown[c]	Enhances parathormone action and the metabolism of Ca^{2+}, P, and Mg^{2+}
Metals		
Iron	10 mg for men 15 mg for women	Found in hemoglobin and many enzymes; important in prevention of anemia
Copper	1.5–3.0 mg[b]	Part of enzymes essential for the formation of hemoglobin, blood vessels, bones, tendons, and the myelin sheath
Zinc	15 mg for men 12 mg for women	Essential for many enzymes, normal liver function, and DNA synthesis
Cobalt	Unknown[c]	Part of vitamin B_{12} molecule
Manganese	2.0–5.0 mg[b]	Essential for several enzymes, bone and cartilage growth, and brain and thyroid function
Chromium	0.05–0.2 mg[b]	Lowers blood sugar level by increasing the effectiveness of insulin
Molybdenum	0.075–0.25 mg[b]	Required for the function of several enzymes
Nickel	Unknown[c]	Aids iron absorption, needed for optimal growth and reproduction in animals

[a]Recommended daily allowances set by the Food and Nutrition Board, National Academy of Sciences—National Research Council, in 1989.
[b]Estimated safe and adequate daily dietary intake for adults.
[c]Requirements in humans have not yet been established.
Abbreviation: DNA = deoxyribonucleic acid.

emphasis focused on the importance of vitamins. Today it is recognized that the trace elements play an equally important nutritional role. Their functions are varied and may depend on their chemical form or their location in the body's tissues or fluids. Many trace elements form important parts of enzymes, which are critical biological molecules that increase the rate of chemical reactions in cells. Because enzymes can be used over and over again, the trace elements can be effective even when present in only very low concentrations in the cells of the body. The transition metals, in particular, have chemical bonding properties that make them especially important in enzyme molecules.

Because the trace elements are found in the body in such minute amounts, it is very difficult to determine exactly which elements are essential to life. Recent research, for

Health Perspective / Selenium—Essential and Toxic

In 1857, a disease that was causing horses to lose hair and hooves was reported in North Dakota. Called "blind staggers," this disease affected other grazing animals as well. Symptoms included impaired vision, muscle weakness, and sometimes necrosis of the liver and death from respiratory failure. Not until 1930 did scientists identify selenium, which was present in large amounts in the region's soils, as the element causing the disease. The animals suffering from blind staggers were grazing on plants that accumulated the selenium from the soils. By way of contrast, in the late 1950s, scientists confirmed that animals living in regions with soils poor in selenium were suffering from other diseases. By supplementing their animals' feed with selenium, ranchers in these areas could prevent white muscle disease in cattle, horses, and sheep, liver disease in pigs, and pancreatic disease in chickens.

Similarly, in a region of China where food was grown on selenium-deficient soils, as many as 22,000 women and children suffered from a disease of the heart muscle called Keshan disease. This heart disease largely disappeared when the people in the area were treated with selenium salts. Likewise, studies have shown that the death rate from heart disease is higher in regions of the United States where soils are deficient in selenium. Scientists in China have found that the minimum daily requirement for selenium in adult men is 0.070 mg, but that an intake of 1 mg daily can cause toxic effects. So you can see that the daily intake of selenium must fall in a narrow range to avoid toxic effects.

Why is selenium essential to life? Selenium is part of the enzyme glutathione peroxidase, which protects cellular membranes. The enzyme prevents the accumulation of hydrogen peroxide and organic peroxides in cells. Such organic peroxides are thought to play a role in cancer development. Selenium also can protect the body against the toxicity of heavy metals, such as cadmium and mercury. Why is selenium toxic? No one knows for sure, but the toxicity of this and other trace elements is an area of active research.

Your Perspective: Imagine you are a rancher with a herd that appears to be suffering from selenium-related symptoms. Write a letter to the Department of Soil Science at your state university requesting that someone test your soil for selenium content. Explain the circumstances behind this request and the importance of such testing to you.

example, has revealed that essential roles are played by chromium, selenium, and arsenic—elements that had previously been considered toxic to the body. Safe and adequate daily amounts for some trace elements are fairly well established, but for others there are not yet enough data to determine a recommended daily allowance (see Table 4.6).

It is important to understand that for each trace element within the body a specific range of concentration allows the body to function normally. If the minimum concentration is not present, signs of a deficiency begin to appear. As the concentration of a trace element continues to decrease, a deficiency disease develops and death can result. Each of the trace elements is also potentially toxic if the range of safe and adequate concentration for that element is exceeded. Recently, megadoses (very large doses) of dietary mineral supplements, especially zinc and selenium, have been promoted as cures or preventive medicine for a variety of medical problems. What is not mentioned in the advertising is that large doses of trace elements, like large doses of certain vitamins, can be very dangerous to your health. For example, we must have zinc in our diets to promote growth, taste foods, heal wounds, and improve night vision. Zinc supplements are now being advertised to help cure prostate problems and impotence. Not only is there no evidence to support these claims, but excess zinc can be harmful to the body. It can interfere with the body's absorption of copper, hinder blood cell formation and immune function, and reduce levels of "good," high-density lipoprotein (HDL) cholesterol. Adequate levels of zinc and the other trace elements can easily be maintained by eating a balanced diet. Mineral supplements are unnecessary.

KEY CONCEPTS

The atom consists of a small, dense nucleus containing positive protons and neutral neutrons, and a region surrounding the nucleus containing negative electrons. Atoms of the same element that differ in the number of neutrons in their nuclei are called isotopes.

The **atomic number** of an element indicates the number of protons in the nucleus of each atom of that element. The **mass number** is the sum of the number of protons and neutrons in the nucleus of an atom. The **atomic weight** of an element is a weighted average of the masses of the naturally occurring isotopes of that element, expressed in atomic mass units.

The electron configuration of an element is the arrangement of electrons around the nucleus of its atoms. Electrons occupy only certain energy states, called energy levels. Each energy level can hold no more than a certain number of electrons. Within each energy level are regions called orbitals where electrons are most likely to be found. An orbital can contain zero, one, or two electrons. Each type of orbital (*s*, *p*, *d*, *f*) has a specific shape. Electrons occupy orbitals according to a specific set of rules.

Electrons are normally found in the lowest possible energy state. The energy level closest to the nucleus contains electrons with the least amount of energy. For an electron to move from one energy level to another, an atom must absorb or give off energy exactly equal to the energy difference between the two orbitals. If an atom absorbs energy from an outside source, an electron in that atom can move from a lower energy orbital to a higher energy orbital. Such an atom is said to be excited. If enough energy is absorbed by an atom, one of its electrons may jump completely away from the atom, forming a positively charged ion.

Chemical and physical properties of elements show periodicity. These properties repeat according to the location of the elements on the periodic table. Three such properties are atomic size, ionization energy, and electron affinity.

When elements are arranged according to their atomic numbers on a periodic table, members of each vertical column, called a group or family, display similar properties. Each member of a group has the same number of valence, or outermost energy level, electrons.

The periodic table arranges elements into classes according to specific properties. Elements are divided into metals and nonmetals. They can be further classified as representative elements, transition elements, and inner transition elements.

Important Equations

$$\text{Atomic number } (Z) = \text{protons } (p)$$

$$\text{Mass number } (M) = \text{protons } (p) + \text{neutrons } (n)$$

Review Problems

Section 4.1

1. Complete the following chart:

	Subatomic Particle	*Location*	*Charge*	*Relative Mass*
(a)			1+	
(b)				1
(c)				1/1837

2. Compare, in your own words, the relative positions of the subatomic particles.

Section 4.2

3. Complete the following chart (assume the atom is neutral):

				Number of		
	Atomic Number	*Symbol*	*Mass Number*	*p*	e^-	*n*
(a)	3		7			
(b)		S				16
(c)				11		12
(d)	26					30
(e)	35		80			
(f)		Sr	88			
(g)		Sn				66
(h)		Hg	200			
(i)	88					138

4. Calculate the atomic number and the mass number, and use two different shorthand methods to denote an atom of each of the following elements.

	Name	*Number of p*	*Number of n*
(a)	chlorine (Cl)	17	18
(b)	cobalt (Co)	27	33
(c)	hydrogen (H)	1	2

5. State the number of protons, neutrons and electrons in a neutral atom of each of the following:
   (a) $^{55}_{25}Mn$ (b) $^{122}_{51}Sb$ (c) $^{39}_{19}K$
6. What is the atomic number, mass number, and symbol for the following atoms?
   (a) an atom with 8 protons and 8 neutrons
   (b) an atom with 90 protons and 142 neutrons
   (c) an atom with 47 protons and 60 neutrons

Section 4.3

7. Boron has two isotopes, boron-10 and boron-11. State the number of protons, neutrons, and electrons in each isotope.
8. State the number of protons, neutrons, and electrons in a neutral atom of each of the five isotopes of zinc (see Table 4.2).
9. Using the information from problem 8, describe how isotopes differ.
10. Atoms of cobalt-60, iron-59, and copper-62 are not isotopes, but what do they have in common?

Section 4.4

11. Using Table 4.2, calculate the atomic weights of bromine and zinc to three significant figures.
12. Calculate the atomic masses to three significant figures for copper and magnesium, given the following data:

Isotope	*Natural Abundance (%)*
copper-63	69.77
copper-65	30.23
magnesium-24	78.99
magnesium-25	10.00
magnesium-26	11.01

Section 4.5

13. Describe the quantum mechanical model of the atom.
14. What is the maximum number of electrons possible in the fourth energy level? In the sixth?

Section 4.6

15. Fill in the following table:

	Symbol	*Atomic Number*	*Atomic Weight*	*Electron Configuration*
(a) lithium				
(b) nitrogen				
(c) neon				
(d) magnesium				
(e) aluminum				
(f) chlorine				

16. Write the electron configuration for a neutral atom of each of the following elements (include the orbital diagram for parts a, d, and f):

(a) beryllium	**(f)** copper
(b) iodine	**(g)** manganese
(c) silicon	**(h)** calcium
(d) sulfur	**(i)** radium
(e) selenium	**(j)** potassium

17. Which atoms have the following electron configurations?

(a) $1s^22s^22p^63s^1$

(b) $1s^22s^22p^5$

(c) $1s^22s^22p^63s^23p^64s^23d^3$

(d) $1s^22s^22p^63s^23p^64s^23d^6$

(e) $1s^22s^22p^63s^23p^64s^23d^{10}4p^3$

18. What is the electron configuration of a ground state atom containing 16 electrons?

Section 4.7

19. Identify each of the following as a cation or anion:

(a) Li^+ **(b)** Cl^- **(c)** Br^- **(d)** Na^+

20. Look at the following electron configuration for a certain magnesium atom, atomic number 12:

energy level	1	2	3	4	5
electron configuration	2	8	1	1	

Is this atom in the ground state? Why or why not?

Section 4.8

21. What is meant by the term *periodicity*?

22. List three properties of elements that exhibit periodicity.

Section 4.9

23. State the number of valence electrons in a neutral atom of each of the following elements:

(a) rubidium	**(e)** beryllium
(b) indium	**(f)** silicon
(c) phosphorus	**(g)** bromine
(d) krypton	**(h)** selenium

24. Identify the following as either a representative element or a transition element:

(a) strontium	**(e)** copper
(b) selenium	**(f)** fluorine
(c) iron	**(g)** silicon
(d) germanium	**(h)** potassium

25. Identify the following elements:

(a) group IIA, period 5	**(c)** group VIIA, period 2
(b) group IVB, period 4	**(d)** group IIA, period 2

Which of these four elements have similar chemical properties?

26. Data for the nuclei of three neutral atoms are shown here. Which two atoms have similar chemical properties? Give a reason for your choice.

    atom A contains 4 protons and 5 neutrons
    atom B contains 8 protons and 8 neutrons
    atom C contains 12 protons and 12 neutrons

Section 4.10

27. Identify the elements listed in problem 23 as metal, metalloid, or nonmetal.
28. Arrange the following elements in order from the most metallic to the least metallic:
    (a) sulfur, chlorine, silicon, phosphorus
    (b) tin, rubidium, silver, palladium

Section 4.11

29. Arrange each of the following groups of atoms in order of increasing atomic size:
    (a) B, Al, Ga
    (b) Sn, Sb, Te
    (c) Cd, Si, Ga
    (d) As, P, Cl
    (e) O, F, Cl
30. Arrange the following groups of atoms in order of increasing ionization energy:
    (a) Be, Mg, Ca **(b)** Te, I, Xe **(c)** S, Cl, F
31. For each of the following pairs, predict which element has (1) the larger radius and (2) the larger ionization energy.
    (a) Na and Cl **(e)** Ne and Xe
    (b) C and O **(f)** N and Sb
    (c) Li and Rb **(g)** Sr and Si
    (d) As and F **(h)** Fe and Br
32. For each of the following pairs, predict which element has the higher electron affinity:
    (a) Na and Cl **(b)** Cl and I **(c)** Cs and Cl

Section 4.12

33. What are the four most abundant elements in the body?
34. What is meant by the term *trace elements?*

Applied Problems

35. It is hard to imagine something that has a mass as small as that of an electron. Calculate how many electrons are needed to equal 1 g.
36. Dmitri Mendeleev predicted the properties of elements that had not yet been discovered at the time he developed his periodic table of the elements. We can do the same thing. Element number 116 has not yet been discovered or synthesized. To what chemical group would element 116 belong?

37. Lithium ions (Li^+) are used in treating the manic stage of manic-depressive psychosis. In nature, lithium consists of two isotopes: 92.58% of the atoms have a mass number of 7, and 7.42% have a mass of 6.

(a) For each of isotope of lithium, calculate the following:

(1) atomic number (3) number of protons

(2) mass number (4) number of neutrons

(b) Calculate the atomic weight of lithium.

38. The atomic mass unit is based on the most common isotope of carbon being arbitrarily assigned a value of exactly 12 amu. Other values could have been used; in fact, at one time another standard was used. Imagine that the **average** isotopic mass of carbon was assigned a value of 12. What masses of phosphorous (P) and selenium (Se) would appear on the periodic table?

39. The quantum mechanical model of the atom permits the prediction of electronic configurations for elements that do not exist and that, in some cases, probably never will exist. Predict the electronic configuration for the next alkaline earth metal, as it would be listed below radium in Table 4.4.

40. Sometimes what occurs in nature is not exactly what is predicted by the quantum mechanical model. As shown in Table 4.3, the maximum number of electrons allowed in theory for the fifth, sixth, and seventh energy levels are not actually found in nature. Predict the number of electrons allowed, in theory, for the eighth energy level.

41. Why is the 19th electron of a potassium atom located in the fourth energy level rather than the third energy level?

Chapter 5

Combinations of Atoms

A light wind was blowing as the afternoon freight train rumbled through a small town in southeastern Louisiana. The train was just starting to gather speed about 2 miles outside of town when a freight car suddenly jumped the tracks, pulling 18 cars with it. One of those 18, a tank car containing 30 tons of liquid chlorine (Cl_2), lay on its side with a huge gash ripped open. The chlorine, which had immediately vaporized to a greenish yellow gas, now poured out of the tank car and was carried by the breeze back toward the town.

The Harrison family lived in a nearby farmhouse, unaware of the approaching danger. Within minutes of the train accident the irritating odor of chlorine began to fill the house. Mrs. Harrison suddenly had trouble breathing, and her two small children began retching and vomiting. Her husband came running from the barn with his eyes streaming and quickly loaded his family into the car. Their desperate drive to the hospital was accompanied by the sound of the volunteer fire depart-

ment's siren screaming out the emergency signal, and the sight of the sheriff's car helping to evacuate people from nearby farms.

The deadly cloud of chlorine gas forced nearly 1000 people to flee from their homes, offices, and schools. Hundreds of farm animals died from exposure to the gas, and no one could live in several square miles of the countryside for several days after the spill. Fifty people were treated at the hospital for severe irritation caused by the chlorine. Ten, including the Harrison family, were hospitalized with critical poisoning.

Chlorine, at room temperature, is a greenish yellow gas that has a characteristic irritating and suffocating odor. In low concentrations it irritates mucous membranes and the respiratory system, and in high concentration it causes difficulty breathing—leading in extreme cases to death from suffocation.

Another dangerous substance is the element sodium (Na). This element is an alkali metal which is so highly reactive that it is never found in a pure state in nature. (When isolated in the pure form, it is a soft, silvery metal that can easily be cut with a knife.) Great care must be taken in handling sodium to be sure it is kept away from water. If it contacts water, sodium reacts extremely vigorously, releasing hydrogen gas (H_2) that can be ignited by the heat from this reaction.

Vigorous reaction of sodium with chlorine gas.

Chlorine and sodium, then, are both highly reactive elements that are potentially dangerous to living tissue. Suppose, however, you were to drop a piece of freshly cut sodium into a container of chlorine gas and warm the container. You would soon see a white powder start to form—a chemical reaction is taking place. The product of this reaction is sodium chloride (NaCl), the substance more commonly known as table salt. But table salt has none of the properties of the reactants, sodium and chlorine. In fact, sodium chloride is an essential part of our diets. It plays an important role in maintaining the proper amount of water in our cells and tissues and is needed for the contraction of muscles and the transport of nerve impulses.

Sodium chloride is a chemical compound, a homogeneous substance produced by the reaction of different elements. This chemical reaction, like many other spontaneous processes in nature, results in the formation of a more stable substance. The elements sodium and chlorine are both relatively unstable and react spontaneously (the addition of heat only speeds up the reaction) to form sodium chloride, which is very stable.

5.1 The Octet Rule

In the previous chapter, we saw that the elements of group VIII are called the noble gases. They have also been known as the rare gases and as the inert gases. The term *noble* here means that these elements are unreactive. The noble metals, such as platinum, are also unreactive.

It was not until 1962 that the first compound of a noble gas, xenon hexafluoride (XeF_6), was synthesized. Only a few other compounds of krypton, xenon and radon with fluorine have been synthesized since then, and no compounds of helium or neon exist. This unreactivity of the noble gases has led to their use when an inert atmosphere is needed. For example, argon is used in light bulbs, rather than air, because the oxygen in air would react with the hot metal filament.

Each of the noble gases, except helium, has eight valence electrons in its outermost energy level. Helium has two electrons in its first energy level, which is the maximum possible. This arrangement of outermost energy level electrons seems to be responsible for the great stability of the noble gases.

This idea became quite important when scientists observed that the representative elements in groups IA through VIIA enter into chemical combinations that involve the loss, gain, or sharing of electrons. Such sharing, or transferring, of electrons is not haphazard, however. Rather, the representative elements share, or transfer, exactly the number of electrons needed to attain the electron arrangement of a noble gas. That is, they attain eight electrons in their outermost energy level when they form chemical bonds.

By the early 1900s, these observations led Richard Abegg, J. J. Thomson, G. N. Lewis, and Irving Langmuir to develop a new model of atomic structure. Their model is based on the tendency of elements to react so as to attain an outermost energy level arrangement of eight electrons. The result is an octet of valence electrons. This tendency to attain eight electrons is called the **octet rule.** Although there are many exceptions to the octet rule among the heavier elements, the rule is useful for predicting the results of reactions between lighter elements (those with atomic numbers 1 to 22).

State the octet rule.

5.2 The Ionic Bond

There are two ways in which chemical elements can attain a stable octet of electrons: by gaining or losing electrons, or by sharing electrons. The first of these ways, the transfer of electrons from one atom to another, causes electrically neutral atoms to become ions. The force of attraction between ions of opposite charge is called an **ionic bond.** A transfer of electrons occurs when an element such as chlorine, which has a very strong attraction for additional electrons (that is, high electron affinity), reacts with an element such as sodium, which has a weak attraction for its valence electron (that is, low ionization energy). If an atom of sodium loses its single valence electron, it forms the cation Na^+. Notice that a positively charged sodium ion has 10 electrons (eight in the outermost energy level)—the same number as an atom of neon (Ne), the rightmost entry on the previous period on the periodic table. Similarly, if an atom of chlorine gains an electron (such as might be lost from a sodium atom), it forms the anion Cl^-. Such a negatively charged chloride ion has 18 electrons (eight in the outermost energy level), the same number as the closest noble gas, argon (Ar) (Table 5.1).

Describe an ionic bond.

$$Na \longrightarrow Na^+ + e^-$$
$$e^- + Cl \longrightarrow Cl^-$$

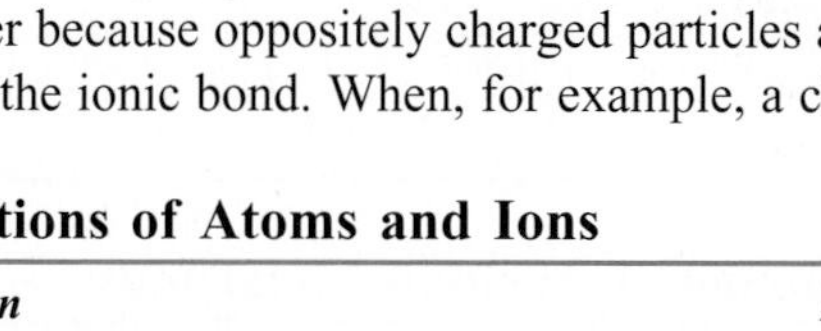

Na + Cl ⟶ Na^+ Cl^-

Na• + •$\ddot{\underset{\cdot\cdot}{Cl}}$: ⟶ Na^+ :$\ddot{\underset{\cdot\cdot}{Cl}}$:

In general, we can expect that elements with few valence electrons (the metals in groups IA, IIA, and IIIA) lose electrons when they react with elements that have almost eight valence electrons (the nonmetals in groups VIA and VIIA). The ions formed by such a transfer are attracted to each other because oppositely charged particles attract. It is this attraction between ions that forms the ionic bond. When, for example, a chunk of freshly

Table 5.1 Electron Configurations of Atoms and Ions

	Atomic Number	*Electron Configuration by Energy Level*				*Atomic Number*	*Electron Configuration by Energy Level*		
Sodium, Na	11	2	8	1	Chlorine, Cl	17	2	8	7
Sodium ion, Na^+	11	2	8		Chloride ion, Cl^-	17	2	8	8
Neon, Ne	10	2	8		Argon, Ar	18	2	8	8

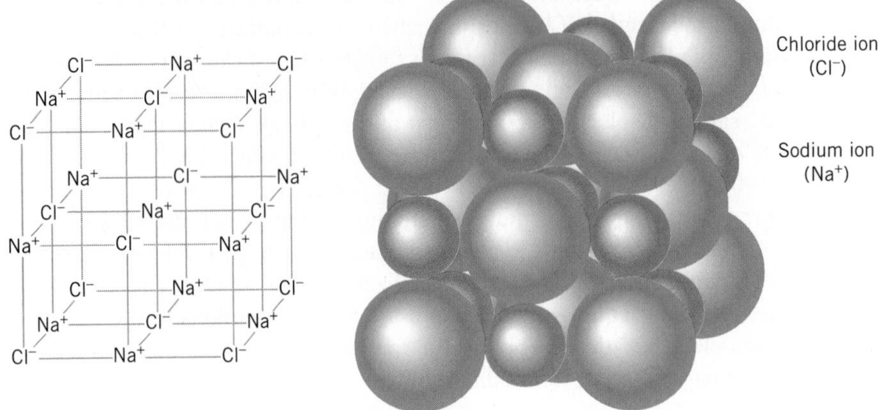

Figure 5.1 A sodium chloride crystal consists of sodium and chloride ions in a closely packed arrangement.

Describe an ionic compound.

cut sodium is added to a container of chlorine gas, many billions of atoms are involved in the reaction (that is, in the transfer of electrons). The attraction between the positive and negative ions causes these ions to group together in an orderly three-dimensional pattern called a **crystal lattice.** This entire large grouping of ions is then called an **ionic compound** (Fig. 5.1). In a crystal of sodium chloride, six chloride ions surround each sodium ion, and six sodium ions surround each chloride ion. Because there is one sodium ion, Na^+, for every chloride ion, Cl^-, a crystal of sodium chloride is electrically neutral. This is always true—the ratio of ions in the crystal lattice of an ionic compound always results in an electrically neutral compound.

An ionic compound contains no specific molecules. No ion is attracted exclusively to another ion; rather, each ion is attracted to all the oppositely charged ions surrounding it. Again, the ionic bond is not a thing or a substance but is simply the force of attraction between oppositely charged ions.

In general, ionic compounds are formed between metals and nonmetals. For a representative element, knowing its group on the periodic table allows you to predict the number of electrons its atoms must gain or lose and, therefore, the charge on the ions that form (Table 5.2). The transition metals are harder to predict, because they can often form more than one ion. For example, copper can form either Cu^+ or Cu^{2+}. Table 5.3 lists some common ions formed by the transition metals and other metals that form more than one ion.

Table 5.2 **Common Ions of the Representative Elements**

Group IA	*Group IIA*	*Group IIIA*	*Group VA*	*Group VIA*	*Group VIIA*
Li^+	Be^{2+}	Al^{3+}	N^{3-}	O^{2-}	F^-
Na^+	Mg^{2+}		P^{3-}	S^{2-}	Cl^-
K^+	Ca^{2+}			Se^{2-}	Br^-
Rb^+	Sr^{2+}			Te^{2-}	I^-
Cs^+	Ba^{2+}				

Table 5.3 **Common Transition Metal Ions**

Metal	*Ion*	*Metal*	*Ion*	*Metal*	*Ion*
Cadmium	Cd^{2+}	Iron	Fe^{2+}, Fe^{3+}	Nickel	Ni^{2+}
Chromium	Cr^{2+}, Cr^{3+}	Lead	Pb^{2+}, Pb^{4+}	Silver	Ag^+
Cobalt	Co^{2+}, Co^{3+}	Manganese	Mn^{2+}, Mn^{3+}	Tin	Sn^{2+}, Sn^{4+}
Copper	Cu^+, Cu^{2+}	Mercury	Hg_2^{2+}, Hg^{2+}	Zinc	Zn^{2+}

Health Perspective / Essential Ions in the Body

In the body, many elements necessary for life are found in the form of their ions. These ions act as control agents in a number of processes involving energy production and cell building and maintenance. The table shows the function of four such ions, the effects of a deficiency of the ion, and the sources of these ions in our diets.

Ion	*Function*	*Results of Deficiency*	*Dietary Sources*
Calcium, Ca^{2+}	Bone formation, tooth formation, blood clotting, muscle contraction and relaxation, heart function	Rickets, porous bones, poor tooth formation, delayed blood clotting	Milk and milk products, cheese, green leafy vegetables, whole grains, egg yolk, legumes, nuts
Sodium Na^{+}	Water and acid–base balance in extracellular fluids	Imbalance in water shifts and controls, imbalance in buffer system	Table salt, milk, meat, eggs, carrots, beets, spinach, celery
Potassium, K^{+}	Water balance in cells, muscle and nerve action, protein synthesis	Water imbalance, irregular heart beat, cardiac arrest, tissue breakdown	Whole grains, meat, legumes, fruits, vegetables
Magnesium, Mg^{2+}	Transmission of nerve impulses, activation of certain enzymes, muscle contraction	Muscle tremor, heart spasms, convulsions, seizures and delirium	Dairy products, flour, cereal products, dry beans, nuts, peas, green leafy vegetables

Your Perspective: You feel your school's food service needs to offer an assortment of foods that supply all essential nutrients. Write a letter to them suggesting some menus you would like them to offer.

5.3 Lewis Electron Dot Diagrams

In Chapter 4 we learned how to write the complete electron configuration of an atom, but now we shall focus our attention primarily on valence electrons. We can represent an atom's configuration of valence electrons using **Lewis electron dot diagrams.** In these diagrams a dot is used to represent each valence electron, and these dots are placed around the symbol of the element. On each side of the element's symbol there is room for two dots. Beginning on any side, the dots are first placed singly, then paired for elements having more than four valence electrons. Table 5.4 shows the electron dot diagrams for the elements in period 2. Remember that the group number of a representative element tells you the number of valence electrons for each member of that group or family. For example, sodium, which is in group I, has one valence electron (Na ·).

Draw the electron dot diagram for an ionic compound formed from representative elements.

Table 5.4 Electron Dot Diagrams for the Elements of Period 2

Element	*Symbol*	*Atomic Number*	*Group*	*Valence Electrons*	*Electron Dot Diagram*
Lithium	Li	3	IA	1	Li·
Beryllium	Be	4	IIA	2	Be· (with a dot below)
Boron	B	5	IIIA	3	·B· (with a dot below)
Carbon	C	6	IVA	4	·C· (with dots above and below)
Nitrogen	N	7	VA	5	·N: (with dots above and below)
Oxygen	O	8	VIA	6	·O: (with a dot above and a pair below)
Fluorine	F	9	VIIA	7	:F: (with a dot above and a pair below)
Neon	Ne	10	VIIA	8	:Ne: (with pairs above and below)

Example 5-1

1. Write the electron dot diagram for phosphorus.

See the Question We are asked for the electron dot diagram for phosphorus: P = (?)

Think It Through From the inside front cover, we see that the symbol for phosphorus is P and that it is found in group VA on the periodic table.

Execute the Math Phosphorus therefore has five valence electrons.

Prepare the Answer Place the first four dots on each side of the symbol P, then pair up the fifth dot with one of the first four. (*Note:* Any of the following dot diagrams are equally correct.)

$$\cdot\overset{..}{\underset{\cdot}{P}}\cdot \quad \text{or} \quad :\overset{\cdot}{\underset{\cdot}{P}}\cdot \quad \text{or} \quad \cdot\overset{\cdot}{\underset{..}{P}}\cdot \quad \text{or} \quad \cdot\overset{\cdot}{\underset{\cdot}{P}}:$$

2. Use a Lewis diagram to show the reaction between sodium and chlorine to form sodium chloride.

S The answer to this problem requires drawing electron dot diagrams for the elements sodium and chlorine and the ionic compound sodium chloride.

T • E When sodium reacts with chlorine it loses its one valence electron to chlorine, forming the sodium cation (Na^+) and the chloride anion (Cl^-). The Lewis diagram for sodium with its one valence electron is $\cdot Na$ and that for chlorine with its seven valence electrons is $\cdot\overset{..}{\underset{..}{Cl}}:$.

P

$$Na^{\odot} + \cdot\overset{..}{\underset{..}{Cl}}: \longrightarrow Na^+:\overset{..}{\underset{..}{Cl}}:^-$$

3. Write the electron dot diagram for the reaction between magnesium and chlorine.

S The answer to this problem requires writing the dot diagrams for magnesium, chlorine, and magnesium chloride.

T • E From the inside front cover, we see that the symbol for magnesium is Mg. It is in group IIA on the periodic chart, so it has two valence electrons. Chlorine, in group VIIA, has seven valence electrons. To attain an octet of electrons, magnesium must lose two electrons and chlorine must gain one electron. Two chlorines must therefore react with each magnesium to form an electrically neutral ionic compound.

P

$$:\overset{..}{\underset{..}{Cl}}\cdot \;{}^{\odot}Mg^{\odot} + \cdot\overset{..}{\underset{..}{Cl}}: \longrightarrow Mg^{2+} + 2[:\overset{..}{\underset{..}{Cl}}:^-]$$

(The 2 written before the brackets shows that there are two chloride ions.)

Check Your Understanding

1. Write the Lewis electron dot diagrams for the following elements:

   (a) Rb **(b)** Si **(c)** I

2. Use Lewis electron dot diagrams to show the reaction between the following elements:

   (a) potassium and iodine **(b)** magnesium and iodine

5.4 The Covalent Bond

Ionic bonds cannot form between elements that are alike in their attraction for electrons. To complete their octet and become stable they must share electrons, thereby forming a **covalent bond.** A covalent bond results when two positive nuclei attract the same electrons, thus holding the two nuclei close together. When two or more atoms share electrons through covalent bonds, a single (electrically neutral) unit called a **molecule** is formed. **Covalent compounds** are composed of molecules, which in turn are composed of atoms held together by covalent bonds. Nonmetallic elements normally form covalent bonds (remember that nonmetals are elements having high ionization energies and, therefore, strong attraction for their valence electrons).

Describe a covalent bond.

Describe a covalent compound.

Some nonmetallic elements exist in nature not as individual atoms, but as two atoms of the element covalently bonded together. The resulting molecules are called **diatomic.** Elements that exist as diatomic molecules are hydrogen (H_2), nitrogen (N_2), oxygen (O_2), fluorine (F_2), chlorine (Cl_2), bromine (Br_2), and iodine (I_2). Let's look at a diatomic molecule of chlorine. Each chlorine atom has seven valence electrons and needs one more electron to reach the stable electron configuration of argon. By sharing a pair of electrons (thus forming a covalent bond), each chlorine atom can achieve this stable octet. A shorthand way of representing a covalent bond is to draw a dash between the chemical symbols of the elements involved. Such a shorthand representation is called a **bond diagram** (Fig. 5.2).

Draw the electron dot diagram and bond diagram for a covalent compound formed from representative elements.

Hydrogen is another element that is found as a diatomic molecule, H_2. Each hydrogen atom has one valence electron and needs one more electron to reach the stable electron configuration of the closest noble gas, helium. Two hydrogen atoms, then, share a pair of electrons to become stable (Fig. 5.3).

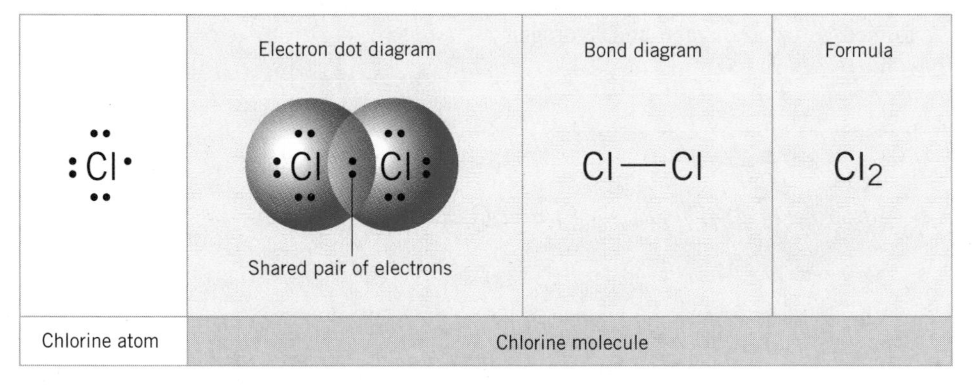

Figure 5.2 The diatomic chlorine molecule contains a single covalent bond, resulting in an octet of electrons around each chlorine atom.

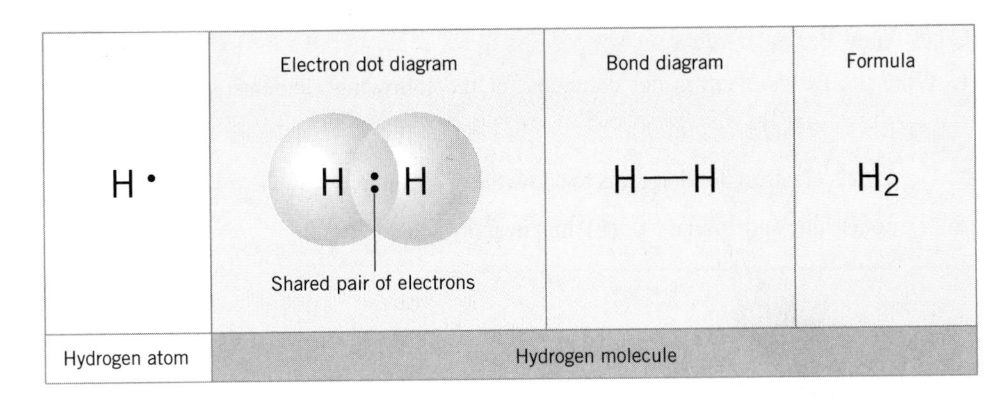

Figure 5.3 The diatomic hydrogen molecule contains a single covalent bond.

5.5 Multiple Bonds

Often a stable octet of electrons can be attained only if more than one pair of electrons is shared between two nuclei. A single shared pair of electrons results in a **single covalent bond,** which we have represented by a dash between the symbols of the elements. A **double bond** is formed when two pairs of electrons are shared between two nuclei, and is represented by two dashes, =. Carbon dioxide (CO_2) is an example of a compound containing double covalent bonds. The carbon atom must share two pairs of electrons with each oxygen atom to reach a stable octet of electrons (Fig. 5.4).

Nitrogen is found in the atmosphere as the diatomic molecule N_2. Each nitrogen atom has five valence electrons and needs to share three more electrons to reach a stable octet. This happens when each of two nitrogen atoms shares three electrons with the other atom. These three shared pairs of electrons form a **triple bond,** represented by three dashes, ≡ (Fig. 5.5). Nitrogen, which accounts for 80% of the gases in the atmosphere, is stable and relatively unreactive. Because of its unreactivity, nitrogen as N_2 is useless to most forms of life. Only one class of organism can force atmospheric nitrogen to form chemical combinations with other atoms, thus making it usable by other forms of life. These microorganisms, known as nitrogen-fixing bacteria, live in the soil or in the roots of plants such as peas and alfalfa. The bacteria convert gaseous nitrogen to compounds that can be used by plants. Animals then obtain the nitrogen they need by eating the plants.

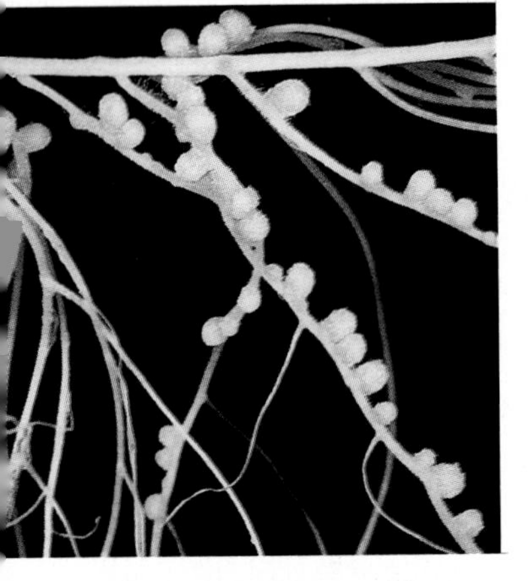

The root nodules on this pea plant contain nitrogen-fixing bacteria that allow the plant to utilize atmospheric nitrogen.

Example 5-2

1. Write the electron dot and bond diagrams for water, which is the covalent compound formed between oxygen and hydrogen.

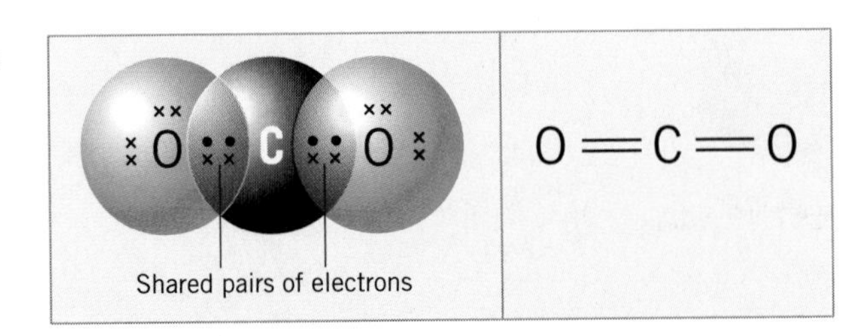

Figure 5.4 The carbon dioxide molecule contains two double covalent bonds, resulting in an octet of electrons around each atom.

See the Question

We are asked to draw the electron dot diagram and bond diagram for the compound water.

Think It Through (Execute the Math)

Oxygen is in group VIA, so it has six valence electrons. This means that oxygen needs to share two more electrons to become stable. Hydrogen is in group IA. It has one valence electron and needs to share one more electron to have the electron configuration of helium, the nearest noble gas. This can happen if the oxygen atom shares one electron with each hydrogen atom.

Prepare the Answer

Using electron dot diagrams to represent this, we have

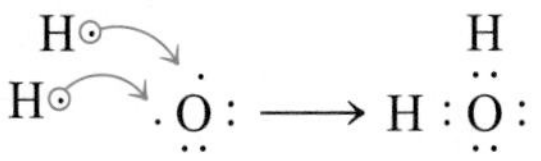

$$\mathrm{H}\cdot + \mathrm{H}\cdot + \cdot\ddot{\underset{\cdot\cdot}{\mathrm{O}}}: \longrightarrow \mathrm{H}:\ddot{\underset{\cdot\cdot}{\mathrm{O}}}: \text{ (with H above O)}$$

Therefore, the bond diagram is

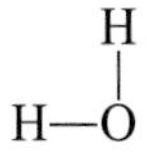

$$\mathrm{H{-}O{-}H} \text{ (one H above O, one H to the left)}$$

(Note that these diagrams are not meant to show the actual shape of the molecule.)

2. Write the electron dot and bond diagrams for the covalent compound formed between carbon and bromine.

S

We are asked to draw the electron dot diagram and the bond diagram for the compound formed between carbon and bromine.

T • E

Carbon is in group IVA. It has four valence electrons and needs to share four additional electrons to reach a stable octet. Bromine is in group VIIA, has seven valence electrons, and needs to share one more electron to attain a stable octet. Carbon must therefore share one electron with each of four bromines.

P

The electron dot diagram and bond diagram are

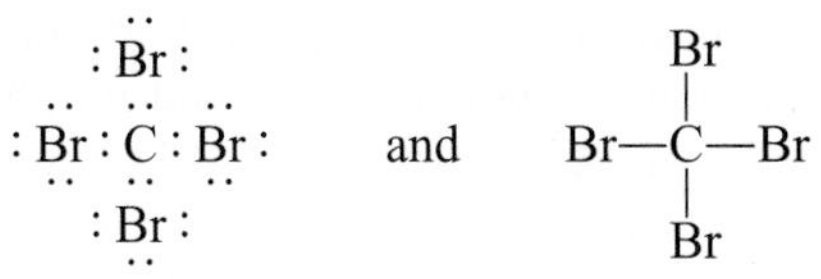

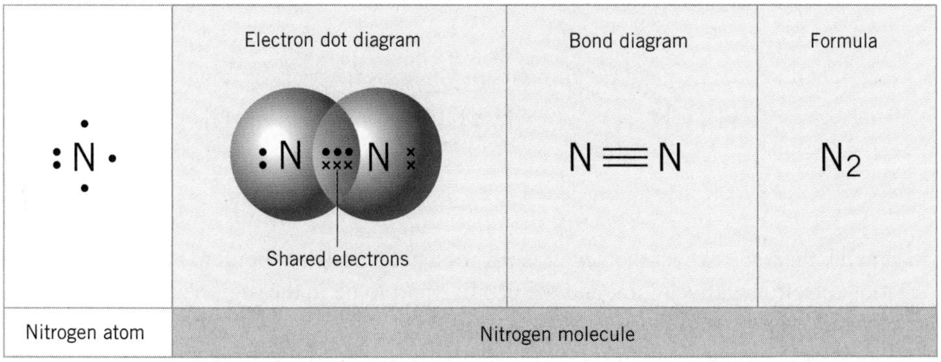

$$\begin{array}{ccc} & :\ddot{\mathrm{Br}}: & \\ :\ddot{\mathrm{Br}}: & \ddot{\mathrm{C}} & :\ddot{\mathrm{Br}}: \\ & :\underset{\cdot\cdot}{\mathrm{Br}}: & \end{array} \quad \text{and} \quad \begin{array}{ccc} & \mathrm{Br} & \\ & | & \\ \mathrm{Br}- & \mathrm{C} & -\mathrm{Br} \\ & | & \\ & \mathrm{Br} & \end{array}$$

	Electron dot diagram	Bond diagram	Formula
:N· (with dots above and below)	:N ⁝⁝⁝ N (Shared electrons)	N≡N	N_2
Nitrogen atom	Nitrogen molecule		

Figure 5.5 The diatomic nitrogen molecule contains a triple covalent bond, resulting in an octet of electrons around each atom.

Check Your Understanding

Write the electron dot diagrams and bond diagrams for the covalent compounds formed between the following elements.

(a) hydrogen and chlorine

(b) hydrogen and nitrogen

(c) carbon and sulfur

5.6 Electronegativity

Define electronegativity, and describe the electronegativity trends on the periodic table.

The relative attraction of an atom for the shared electrons in a chemical bond is called the **electronegativity** of the atom. Linus Pauling devised a scale of relative electronegativities based on bond energies. In each period of the periodic table the element with the smallest radius is the most electronegative; among all the elements, fluorine is the most electronegative. The electronegativity of the elements increases from left to right along a period and decreases from top to bottom within a group (Fig. 5.6). Imagine two identical atoms sharing a pair of electrons in a covalent bond—for example, Cl_2. The two atoms have the same electronegativity, so we expect the electrons to be shared equally between the two nuclei. This results in the center of positive charge and the center of negative charge of the molecule occurring midway along the bond between the two nuclei. Such a bond is called a **nonpolar covalent bond** (Fig. 5.7a).

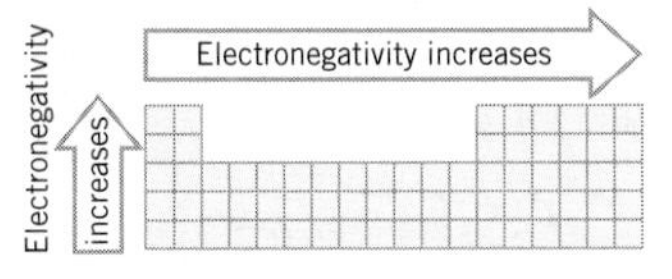

Figure 5.6 Metals have the lowest and nonmetals have the highest electronegativities.

Now imagine that two different atoms are sharing electrons. We expect that one atom might attract the electrons more strongly than the other, so the electrons are to be found closer to the more electronegative (or electron-attracting) atom. In such a bond the center of positive charge is not found in the same place as the center of negative charge, producing an unequal distribution of charge (Fig. 5.7b). Such an unequal distribution of charge results in the formation of an electric dipole—the molecule has two electric poles, just as a magnet has two magnetic poles. This kind of bond is called a **polar covalent bond.**

Describe the difference between a polar and a nonpolar covalent bond.

For example, consider hydrogen chloride, HCl. Chlorine is the more electronegative element. Thus, the shared electrons are pulled closer to the chlorine nucleus in the hydrogen–chlorine bond. This gives the chlorine end of the bond a partial negative charge and the hydrogen end a partial positive charge.

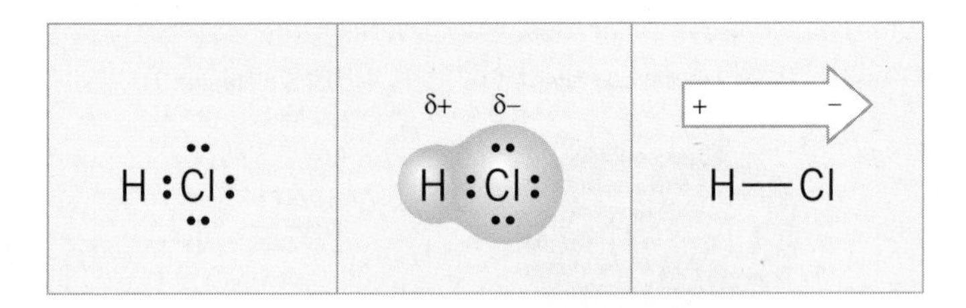

In general, then, we expect the negative part of a polar covalent bond to be in the region of the more electronegative atom, and the positive part to be in the region of the less electronegative atom.

Atoms of the same element form nonpolar covalent bonds, with centers of positive and negative charges coinciding. Atoms of unlike elements that have different electronegativities form bonds with the centers of electric charge separated. When the difference in

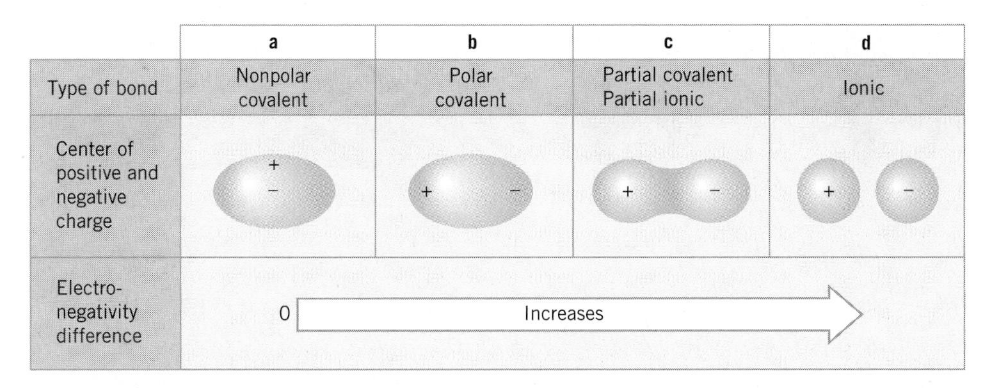

a
b
c
d
Type of bond
Nonpolar covalent
Polar covalent
Partial covalent Partial ionic
Ionic
Center of positive and negative charge
+
−
+
−
+
−
+
−
Electro-negativity difference
0
Increases

Figure 5.7 There is no absolute break between ionic and covalent bonding, but rather a continuous range from nonpolar covalent bonding to ionic bonding.

electronegativity becomes very large, the electric charges are completely separated to the two opposite ends of the bond, and an ionic bond forms. We can imagine a continuous range from nonpolar covalent bonding to complete ionic bonding, with different compounds appearing at specific places along this continuum (see Fig. 5.7). In general, we can say that bonds formed between metals and nonmetals are ionic, and bonds formed between nonmetals are covalent. Bonds are nonpolar covalent when the electronegativity of the atoms is the same, and polar covalent when one atom has greater electronegativity than the other. As we shall see, the type of bonding in a compound determines many of the compound's important properties, such as boiling point and solubility.

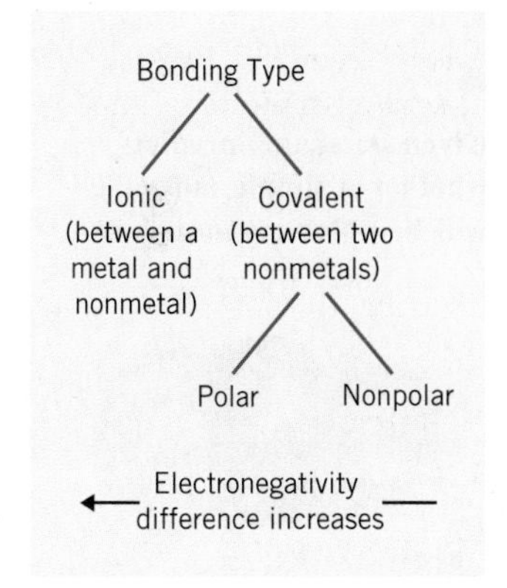

5.7 Polar and Nonpolar Molecules

The properties and behavior of a molecular compound depend not only on the type of bonding between the atoms in the molecule but also on the shape or arrangement of these atoms in the molecule. The properties affected by the shape of the molecule range from the odor of the compound to the role that the molecule plays in regulating the chemical reactions in living organisms. For example, all hallucinogenic drugs seem to affect one site in the brain. It may be that this site is sensitive to, or able to ''recognize,'' a specific three-dimensional shape. According to one theory, the molecules of hallucinogenic drugs each have particular regions that closely resemble this three-dimensional shape and, therefore, each trigger hallucinations.

In Section 4.9, we discussed the fact that elements within the same group (chemical family) have similar chemical properties. For nonmetals in groups IVA through VIIA, covalent compounds in which the central element is from the same group have similar shapes. Specifically, when all bonds in the molecule are single bonds, the arrangement of atoms around the elements in group IVA are tetrahedral in shape, group VA are triangular-pyramidal, group VIA bent, and group VIIA linear (Fig. 5.8).

A molecule can have polar bonds and still be nonpolar. It is the shape of a molecule that determines whether the molecule is polar or nonpolar. The polarity or nonpolarity of a molecule plays a large part in its behavior and role in living organisms. We shall see that in living systems polar molecules are found with other polar molecules, and nonpolar molecules with other nonpolar molecules.

As with covalent bonds, the centers of positive and negative charges coincide in a nonpolar molecule and do not coincide in a polar molecule. It is important to understand that a polar molecule is still electrically neutral, even though a separation of charge occurs in the molecule, resulting in regions of positive and negative charge.

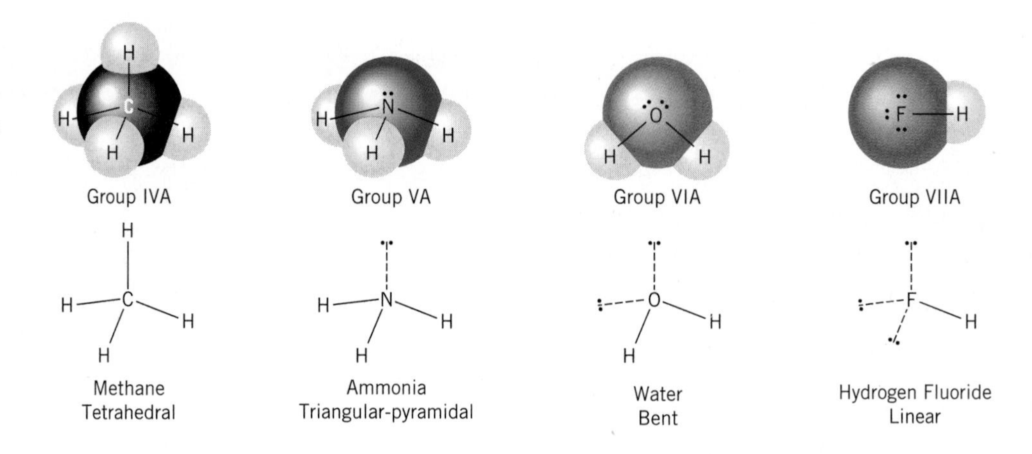

Figure 5.8 Shapes of covalently bonded molecules formed when elements in groups IVA through VIIA are singly bonded to another element.

Given its shape, predict whether a simple molecule will be polar or nonpolar.

Several examples illustrate how the shape of a molecule can determine whether the molecule is polar or nonpolar (Table 5.5). Note that in each of our examples, however, all of the bonds that are formed are polar. Hydrogen chloride (HCl), as we have seen, is a linear molecule containing one polar bond. The molecule is polar, with the hydrogen at the positive end and the chlorine at the negative end. Carbon dioxide (CO_2) is also a linear molecule, containing two polar double bonds. The linear arrangement of atoms in the molecule, however, makes the centers of positive and negative charge coincide, so the entire carbon dioxide molecule is nonpolar. A molecule of water contains three atoms and two polar bonds, just as we found in carbon dioxide. Because the oxygen atom at the center of the water molecule has two sets of unpaired electrons, the water molecule has a bent (rather than linear) shape. This bent shape produces a separation between the centers of positive and negative charge. A water molecule is therefore polar. You can now see that for molecules with more than two atoms, we must know the molecular geometry or shape to predict whether the molecule is polar or nonpolar.

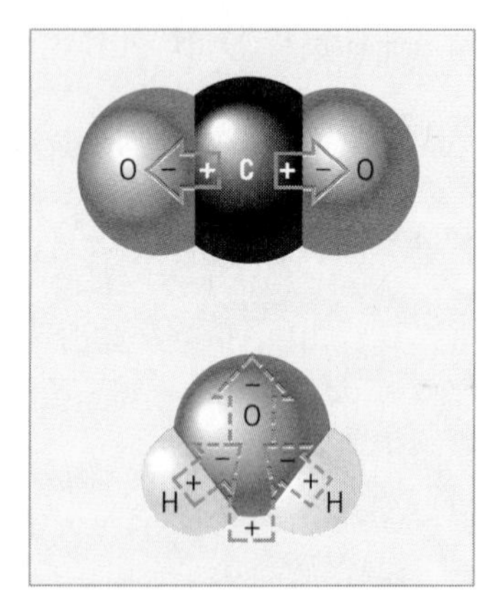

Both CO_2 and H_2O contain two polar covalent bonds, but the difference in shape of the molecules causes a molecule of CO_2 to be nonpolar and a molecule of H_2O to be polar.

5.8 Hydrogen Bonding

Molecules containing hydrogen attached to a highly electronegative atom, such as fluorine, oxygen, or nitrogen, have an intermolecular (between molecules) type of bonding called **hydrogen bonding.** The partially positive hydrogen region on one molecule is attracted to the partially negative fluorine, oxygen, or nitrogen region on a neighboring

Table 5.5 **Molecular Shape and the Polarity of Molecules**

Compound	*Formula*	*Shape*	*Bond Diagram*	*Center of Charge*	*Type of Molecule*
Hydrogen chloride	HCl	Linear	$\overset{\delta+}{H}—\overset{\delta-}{Cl}$	(+ −)	Polar
Carbon dioxide	CO_2	Linear	$\overset{\delta-}{O}=\overset{\delta+}{C}=\overset{\delta-}{O}$	(±)	Nonpolar
Water	H_2O	Bent	$\delta-$ O, $\delta+$ H, $\delta+$ H	− / +	Polar

Health Perspective / Taste and Molecular Shapes

How often have you heard someone complain about lack of self-control when it comes to sweets? We often refer to this weakness as our "sweet tooth." The substances most often associated with the sweet taste we crave are sugars. A typical American consumes, on the average, at least 170 g of sugar each day. Even if we avoid adding sugar to our food at home, we receive an alarmingly high amount in the processed foods we eat. In fact, we eat more sugar indirectly through processed foods than directly by deliberate consumption.

The problem is that sugar provides "empty calories." In other words, it lacks any nutritional value outside its role as an energy source. Diets rich in sugar, as with diets rich in fats, are a major cause of obesity. Because we consume too much sugar, artificial sweeteners have gained widespread use as food additives. It is their molecular shape that makes these substances taste similar to the sugars that they replace.

In biological systems, we find that the three-dimensional molecular structure of substances is very important. The reason is that within the fluids that make up a biological system, the appropriate molecules somehow must find and react with each other. To recognize each other among the hundreds of different molecules present, the reacting partners depend on structure. It is their specific structures that allow only the appropriate molecules to come together in a chemical reaction. In this way, molecules can convey messages by fitting into the appropriate receptor sites on "receiver" molecules. An example of this process is the stimulation of taste buds by natural sugars. These sugars fit into the sites on the taste buds that produce a sweet response in the brain. Artificial sweeteners are molecules whose shapes are designed to fit these same sites. The difference is that they are not broken down in the digestive process like the natural sugars. For this reason, artificial sweeteners don't provide additional calories.

Two noncaloric artificial sweeteners in current use are saccharin and aspartame. Saccharin is 450 times sweeter than table sugar but has almost none of its calories. There is some question about the safety of saccharin because in some tests rats have developed cancer when fed this compound; other tests, however, have not supported these results. The other artificial sweetener, aspartame, is 150 times sweeter than table sugar. So far tests have shown aspartame to be safe. As chemists become more familiar with the parts of molecules that are responsible for taste, new molecules may be designed that better fit the sweet receptor sites on the taste buds. So, our list of artificial sweeteners will continue to grow.

Foods containing Aspartame

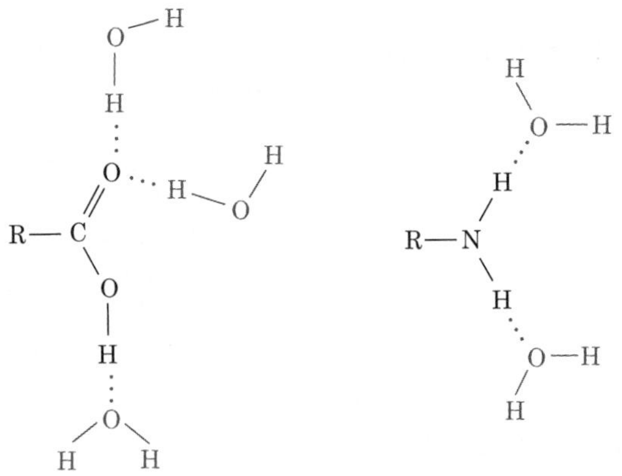

Figure 5.9 Examples of hydrogen bonding that occur in biological molecules: (a) a carboxylic acid and (b) an amine.

molecule (Fig. 5.9). This means that the hydrogen is actually attracted to two other atoms: one by means of the covalent bond, and the other through a special type of attraction known as a hydrogen bond. Such hydrogen bonds have a strength about one-tenth that of ordinary covalent bonds. In some very large biological molecules, hydrogen bonding may occur between different parts of the same molecule, causing the molecule to bend back on itself (called intramolecular hydrogen bonding).

Describe a hydrogen bond.

Hydrogen bonding plays an important role in nature. Hydrogen bonds are responsible for the unusual properties of water—such as its high melting and boiling points—that make this fluid so important to all living organisms. These properties are discussed in detail in Chapter 9. Hydrogen bonds also determine the shapes of large biological molecules in living organisms and, on a more familiar level, explain why the lanolin in skin creams causes hands to become soft, why hard candy gets sticky, and why cotton fabrics take longer to dry than synthetic fabrics.

5.9 Polyatomic Ions

Define polyatomic ion, and give several examples of polyatomic ions.

A **polyatomic ion** is a group of covalently bonded atoms that carries an electric charge and that is so stable that the entire group participates in most chemical reactions as a unit—without breaking apart. Table 5.6 lists some polyatomic ions and their formulas.

Table 5.6 **Common Polyatomic Ions**

Name of Ion (Common Name)	*Formula*	*Name of Ion (Common Name)*	*Formula*
Ammonium ion	NH_4^+	Oxalate ion	$C_2O_4^{2-}$
Acetate ion	$C_2H_3O_2^-$	Permanganate ion	MnO_4^-
Carbonate ion	CO_3^{2-}	Peroxide ion	O_2^{2-}
Hydrogen carbonate ion (bicarbonate)	HCO_3^-	Phosphate ion	PO_4^{3-}
		Monohydrogen phosphate ion	HPO_4^{2-}
Chlorate ion	ClO_3^-	Dihydrogen phosphate ion	$H_2PO_4^-$
Chromate ion	CrO_4^{2-}	Sulfate ion	SO_4^{2-}
Dichromate ion	$Cr_2O_7^{2-}$	Hydrogen sulfate ion (bisulfate)	HSO_4^-
Cyanide ion	CN^-		
Hydroxide ion	OH^-	Sulfite ion	SO_3^{2-}
Hypochlorite ion	ClO^-	Hydrogen sulfite ion (bisulfite)	HSO_3^-
Nitrate ion	NO_3^-		
Nitrite ion	NO_2^-	Thiosulfate ion	$S_2O_3^{2-}$

Health Perspective / Ionic Compounds with Medicinal Uses

Learning how to name ionic compounds and how to recognize their formulas are skills that can be helpful in your everyday life. In your home life and in your chosen careers you will encounter many compounds that are ionic. Some of these are listed in the table with an example of their medicinal use.

Formula	*Name*	*Medicinal Use*
$Al(OH)_3$	Aluminum hydroxide	Antacid
$CaSO_4$	Calcium sulfate	Casts for bone fractures
$FePO_4$	Iron(III) phosphate	Iron supplement in enriched breads
$FeSO_4$	Iron(II) sulfate	Used to treat iron deficiency
KI	Potassium iodide	Used to treat asthma
$K_2C_2O_4$	Potassium oxalate	Used in storage of whole blood
$HgCl_2$	Mercury(II) chloride	Surgical hand wash
$MgSO_4$	Magnesium sulfate	Used for sitz baths
$NaHCO_3$	Sodium bicarbonate	Antacid
NH_4NO_3	Ammonium nitrate	Used to acidify urine
$(NH_4)_2CO_3$	Ammonium carbonate	Expectorant
ZnO	Zinc oxide	Astringent
ZrO_2	Zirconium oxide	Poison ivy ointment

Compounds containing polyatomic ions are so common and play such a central role in our daily lives that you need to become familiar with their names and formulas. Examples of such familiar compounds are sodium bicarbonate (sodium hydrogen carbonate) $NaHCO_3$, which is baking soda, and magnesium hydroxide, $Mg(OH)_2$, which is a common antacid.

5.10 Chemical Formulas

Write the formula for an ionic or covalent compound when given its name.

Chemical formulas are shorthand ways of representing chemical compounds. The formula contains two pieces of information: it identifies the atoms or ions that are present, and indicates the number of atoms in the molecule or the ratio of ions in the compound. The type of atom is indicated by the symbol of the element, and the number of atoms of each element is shown by subscripts following the symbol. For ionic compounds, the symbol of the positive ion (or cation) is always placed first in the formula.

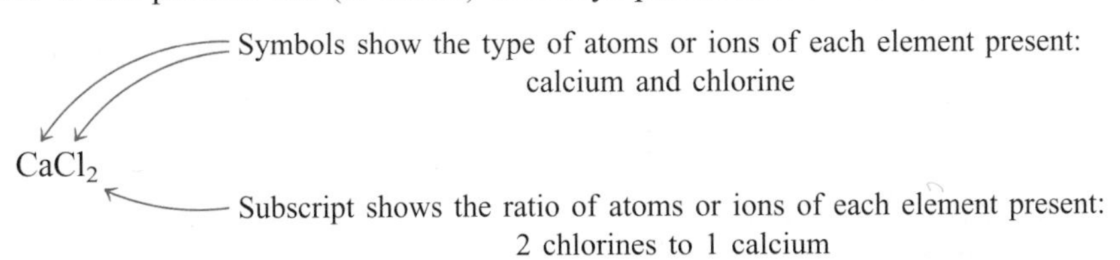
Symbols show the type of atoms or ions of each element present: calcium and chlorine
$CaCl_2$
Subscript shows the ratio of atoms or ions of each element present: 2 chlorines to 1 calcium

Both ionic and covalently bonded compounds are electrically neutral. The ratio of ions used in a chemical formula is the lowest set of whole numbers that produces such electrical neutrality. This means that the formula for calcium chloride ($CaCl_2$) could not be expressed as $Ca_{1/2}Cl$ or Ca_4Cl_8. Note also that the charges on the ions are not written as part of the chemical formula. When a compound contains more than one of the same polyatomic ion, the symbol of the polyatomic ion is put in parentheses with the subscript following the parentheses. For example, the two polyatomic hydroxide ions (OH^-) in calcium hydroxide are shown as $Ca(OH)_2$.

5.11 Naming Chemical Compounds

Write the name of a compound when given its formula.

Binary ionic compounds (those containing only two elements) are named for the ions from which they are formed. Positive ions carry the same name as their parent element: for example, Na^+ is the sodium ion and Ca^{2+} is the calcium ion. If a metal ion forms more than one positive ion, the charge on the ion is indicated by Roman numerals after the name of the element. For example, Fe^{3+} is the iron(III) ion and Cu^+ is the copper(I) ion. Negative ions are named by using a prefix taken from the name of the nonmetal element and then adding the suffix "-ide." For example, Br^- is the bromide ion and O^{2-} is the oxide ion. When the name of a binary ionic compound is written, the name of the positive ion appears first. For example, NaBr is sodium bromide and $CaCl_2$ is calcium chloride.

For metals that form more than one cation, we must know the charge on the ion before we can name a compound. For example, what is the name of the compound CuO? The oxygen needs two electrons to become stable, so the copper must donate two electrons—forming the Cu^{2+} ion. The name of this compound is copper(II) oxide.

Binary covalent compounds, such as SO_2, are named in much the same way. The name of the element with the lower electronegativity is placed first. The second element is named by adding the suffix -ide to the name of the parent element. Prefixes are also added to the name of each element to show the number of atoms of that element in the molecule, as follows:

Mono—1	Tetra—4
Di—2	Penta—5
Tri—3	Hexa—6

The prefix "mono-" is usually left out, unless it helps distinguish between two different compounds. For example, CO is carbon monoxide and CO_2 is carbon dioxide.

Many ionic compounds consist of a metal and a polyatomic ion. These compounds are named by placing the name of the metal first, followed by the name of the polyatomic ion. For example, Na_2CO_3 is sodium carbonate, and $KHSO_4$ is potassium hydrogen sulfate (or potassium bisulfate).

Example 5-3

Think It Through

(These problems require only the ***T*** of our *STEP* problem-solving strategy to arrive at the answer.)

1. Name the following compounds:

(a) *KBr* KBr is an ionic compound formed between the metal potassium and the nonmetal bromine. (Use the inside front cover to find the names and symbols of the elements. Then use the periodic table to find their group number and whether they are metals or nonmetals.) Placing the name of the metal ion first, we have potassium bromide.

(b) *Cu_2O* From Table 5.3, we learn that copper forms two ions. We must determine which of the two ions is involved in this compound. Each oxygen atom requires two electrons to become stable, and there are two copper ions. This means that each copper ion must donate one electron, which results in the copper(I) ion, Cu^+. The name is therefore copper(I) oxide.

(c) *HBr* HBr is a binary covalent compound. The hydrogen has the lower electronegativity, so it is named first. There is just one atom of each element in the molecule, so the name is hydrogen bromide.

(d) *CCl_4* CCl_4 is a binary covalent compound in which carbon has the lower electronegativity. There are four chlorine atoms, so the prefix "tetra" must be added to the chloride. The name is carbon tetrachloride.

(e) *$CaCO_3$* $CaCO_3$ is an ionic compound that contains the positive calcium ion and the polyatomic carbonate ion (see Table 5.6). The name is calcium carbonate.

2. Write the chemical formulas for the following compounds:

(a) *Sodium fluoride* From Table 5.2, we learn that sodium, a metal in group IA, forms the cation Na^+. Fluorine, a group VIIA halogen, forms the anion F^-. We therefore need one sodium for every fluorine: The ratio is 1:1. The correct formula is NaF (notice that the subscript "1" is not written).

(b) *Iron(III) oxide* The name iron(III) tells us that we are dealing with the Fe^{3+} ion. The oxide ion is the anion O^{2-}. To have an electrically neutral compound, we need two iron ions and three oxide ions.
Hint: An easy method to use when writing formulas involving two ions is to "crisscross" the numbers of their charges, as follows:

$$Fe^{③+} \quad O^{②-} \longrightarrow Fe_2O_3$$

(c) *Calcium oxide* From Table 5.2, we see that calcium is a group IIA metal that forms the positive ion Ca^{2+}. The oxide ion is O^{2-}. Using the crisscross method:

$$Ca^{②+} \quad O^{②-} \longrightarrow Ca_2O_2$$

But we know that chemical formulas for ionic compounds are written using the lowest whole-number subscripts, so the correct formula is CaO.

(d) *Magnesium hydroxide* From Table 5.2, we see that magnesium is a group IIA metal that forms the ion Mg^{2+}. Hydroxide is the polyatomic ion OH^-. We therefore need one magnesium ion for every two hydroxide ions. The correct formula is $Mg(OH)_2$. (Remember: when there is more than one polyatomic ion in a formula, the polyatomic ion is enclosed in parentheses.) We also could have used the crisscross method to answer this problem.

$$Mg^{②+} \quad OH^{①-} \longrightarrow Mg(OH)_2$$

(e) *Diphosphorus pentoxide* The prefixes in the name of this covalent compound indicate that there are two atoms of phosphorus and five atoms of oxygen in the molecule. The formula is therefore P_2O_5.

....

Health Perspective / Common Names

We often refer to many chemicals used in the allied health fields by their common names. Listed in the table are just a few of these compounds along with their chemical name and formula.

Common Name	*Chemical Name*	*Formula*
Baking soda	Sodium bicarbonate	$NaHCO_3$
Borax	Sodium tetraborate decahydrate	$Na_2B_4O_7 \cdot 10H_2O$
Caustic soda, lye	Sodium hydroxide	$NaOH$
Epsom salts	Magnesium sulfate heptahydrate	$MgSO_4 \cdot 7H_2O$
Gypsum	Calcium sulfate dihydrate	$CaSO_4 \cdot 2H_2O$
Laughing gas	Nitrous oxide	N_2O
Milk of magnesia	Magnesium hydroxide	$Mg(OH)_2$
Muriatic acid	Hydrochloric acid	HCl
Plaster of paris	Hydrated calcium sulfate	$CaSO_4 \cdot \frac{1}{2}H_2O$
Washing soda	Sodium carbonate decahydrate	$Na_2CO_3 \cdot 10H_2O$

Check Your Understanding

1. Name the following compounds:

(a) MgI_2 **(d)** H_2S

(b) FeO **(e)** $NaHSO_4$

(c) SO_2 **(f)** $K_2Cr_2O_7$

2. Write the formula for each of the following compounds (referring, when necessary, to Tables 5.2, 5.3, and 5.6):

(a) magnesium oxide **(d)** sulfur trioxide **(g)** tin(IV) chloride

(b) nickel chloride **(e)** silicon tetrafluoride **(h)** copper(II) bisulfate

(c) potassium sulfide **(f)** dinitrogen pentoxide **(i)** calcium phosphate

KEY CONCEPTS

The **octet rule** states that most elements become more stable by entering into reactions in which they lose, gain, or share electrons to attain eight valence electrons.

An ionic bond is formed when electrons are transferred from one atom to another, forming positive and negative ions. The force of attraction between these ions forms an ionic bond. An ionic compound is a group of ions combined in an orderly fashion called a crystal lattice. In general, ionic bonds form between metals and nonmetals.

A covalent bond is formed when electrons are shared between two atoms, creating electrically neutral units called molecules. Two atoms can share two electrons in a single covalent bond, four electrons in a double covalent bond, or six electrons in a triple covalent bond.

Hydrogen bonding is a type of bonding that is extremely important in living organisms. A hydrogen bond is formed between a hydrogen covalently bonded to a highly electronegative atom, and a fluorine, oxygen, or nitrogen atom on a neighboring molecule or on another part of the same molecule.

Electronegativity is the tendency for an atom of an element to attract the electrons in a covalent bond. When electrons are shared equally between two atoms, a nonpolar bond forms. If there is a difference in the electronegativity of the two atoms, a polar bond forms. Molecules can be polar or nonpolar depending on the shape of the molecule and the types of bonds in the molecule.

Polyatomic ions are groups of covalently bonded atoms that, as a group, carry an electric charge and stay together as a unit through most chemical reactions.

Review Problems

Section 5.1

1. Why are the elements of group VIIIA called the noble gases?
2. How are the electronic configurations of all of the noble gases (except helium) similar?

Section 5.2

3. Which groups of elements tend to form ionic bonds?
4. Why is it unlikely that (a) calcium would form an ion with a 1+ charge, (b) potassium would form an ion with a 2+ charge?

Section 5.3

5. Draw the electron dot diagram for each of the following atoms:

   (a) cesium **(e)** arsenic
   (b) germanium **(f)** aluminum
   (c) calcium **(g)** sulfur
   (d) neon **(h)** iodine

6. Draw the electron dot diagrams for NaF and MgO. Show that the ions formed in these compounds have the same electron configurations.

Section 5.4

7. Which groups of elements tend to form covalent bonds?
8. What is the difference between the units that make up an ionic compound and the units that make up a covalent compound?

Section 5.5

9. State the difference between a single, double, and triple covalent bond. Give an example of each.

10. Draw the electron dot and bond diagrams for each of the following compounds, all of which contain a multiple bond:

(a) C_2H_4 **(c)** C_2H_2
(b) H_2CO_2 **(d)** SiO_2

11. Draw the electron dot diagrams for the following molecules:

(a) HI **(d)** HCN **(g)** CH_2I_2
(b) F_2 **(e)** OF_2 **(h)** SO_2
(c) CH_3Cl **(f)** PCl_3 **(i)** $CHBr_3$

12. Draw the bond diagrams for each of the following compounds:

(a) NH_3 **(d)** C_2Br_2
(b) CCl_4 **(e)** SF_2
(c) HCl **(f)** I_2

Section 5.6

13. Arrange each of the following sets of atoms in order of increasing electronegativity:

(a) Cl, H, Br, C
(b) B, F, P, O
(c) Al, K, N, C

14. Predict whether the bond formed between the following pairs of atoms is polar or nonpolar. If the bond is polar, indicate which is the more electronegative atom.

(a) Cl and Cl **(c)** C and N **(e)** N and O **(g)** F and Br
(b) H and Br **(d)** S and Cl **(f)** P and O **(h)** O and O

15. In general, what conditions cause two atoms to combine to form **(a)** an ionic bond, **(b)** a covalent bond?

16. Indicate which element is the more positive and which is the more negative in the following compounds:

(a) MgO **(d)** NH_3 **(g)** HBr
(b) NO **(e)** CH_4 **(h)** CCl_4
(c) OF_2 **(f)** IBr **(i)** PbS

Section 5.7

17. Is it possible for a molecule to be nonpolar even though it contains polar bonds? Explain.

18. Predict whether the molecules in problem 12 are polar or nonpolar.

19. Predict whether the following molecules are polar or nonpolar:

(a) oxygen difluoride, OF_2 **(d)** methane, CH_4
(b) fluorine, F_2 **(e)** chloromethane, CH_3Cl
(c) hydrogen iodide, HI **(f)** trichloromethane, $CHCl_3$

20. Classify each of the compounds in problem 16 as ionic, polar covalent, or nonpolar covalent.

Section 5.8

21. Do you expect hydrogen bonds to form between molecules of hydrogen fluoride, HF? Why or why not?

22. Which of the following molecules exhibits hydrogen bonding with another molecule of the same compound?

(a)
$$\begin{array}{ccccccc} & & \text{H} & & & & \\ & & | & & & & \\ \text{H} & — & \text{C} & — & \text{O} & — & \text{H} \\ & & | & & & & \\ & & \text{H} & & & & \end{array}$$

(b)
$$\begin{array}{ccccccccccccc} & & \text{H} & & \text{H} & & & & \text{H} & & \text{H} & & \\ & & | & & | & & & & | & & | & & \\ \text{H} & — & \text{C} & — & \text{C} & — & \text{O} & — & \text{C} & — & \text{C} & — & \text{H} \\ & & | & & | & & & & | & & | & & \\ & & \text{H} & & \text{H} & & & & \text{H} & & \text{H} & & \end{array}$$

(c)
$$\begin{array}{ccccccc} & & \text{H} & & \text{H} & & \\ & & | & & | & & \\ \text{H} & — & \text{C} & — & \text{C} & = & \text{O} \\ & & | & & & & \\ & & \text{H} & & & & \end{array}$$

(d)
$$\begin{array}{ccccccc} & & \text{H} & & \text{H} & & \\ & & | & & | & & \\ \text{H} & — & \text{N} & — & \text{C} & — & \text{H} \\ & & & & | & & \\ & & & & \text{H} & & \end{array}$$

(e)
$$\begin{array}{ccccccccc} & & \text{H} & & \text{O} & & & & \\ & & | & & \| & & & & \\ \text{H} & — & \text{C} & — & \text{C} & — & \text{O} & — & \text{H} \\ & & | & & & & & & \\ & & \text{H} & & & & & & \end{array}$$

(f)
$$\begin{array}{ccccccccc} & & & & \text{H} & & & & \\ & & & & | & & & & \\ & & & \text{H} — & \text{C} & — \text{H} & & & \\ & & & & | & & & & \\ & & \text{H} & & | & & \text{H} & & \\ & & | & & | & & | & & \\ \text{H} & — & \text{C} & — & \text{N} & — & \text{C} & — & \text{H} \\ & & | & & & & | & & \\ & & \text{H} & & & & \text{H} & & \end{array}$$

Sections 5.9–5.10

23. Write the correct formula for the compounds composed of the following pairs of ions:

(a) K^+, CO_3^{2-} **(g)** Cu^{2+}, $C_2H_3O_2^-$

(b) Na^+, S^{2-} **(h)** Ba^{2+}, SO_4^{2-}

(c) Ca^{2+}, NO_3^- **(i)** Al^{3+}, SO_3^{2-}

(d) Sr^{2+}, S^{2-} **(j)** Sn^{4+}, NO_3^-

(e) Cr^{3+}, Cl^- **(k)** Be^{2+}, F^-

(f) Fe^{3+}, HPO_4^{2-} **(l)** Cs^+, Br^-

24. Write the correct formula for the compounds composed of the following pairs of ions:

(a) Li^+, $C_2H_3O_2^-$ **(f)** Sn^{2+}, PO_4^{3-}

(b) Mg^{2+}, HCO_3^- **(g)** Ag^+, SO_3^{2-}

(c) Al^{3+}, CrO_4^{2-} **(h)** Mn^{3+}, O^{2-}

(d) Sr^{2+}, P^{3-} **(i)** Pb^{4+}, $H_2PO_4^-$

(e) Co^{3+}, $C_2O_4^{2-}$ **(j)** NH_4^+, N^{3-}

25. How many atoms of each element are found in one unit of each of the compounds in problem 24?

26. How many atoms of each element are found in one unit of each of the following?

(a) $(NH_4)_3PO_4$ **(c)** $Ca(C_2H_3O_2)_2$

(b) $Al_2(HPO_4)_3$ **(d)** $Fe_4[Fe(CN)_6]_3$

Section 5.11

27. Name each of the compounds for which you wrote the formula in problem 23.

28. Write the correct formulas for each of the following sets of compounds:

Set 1	*Set 2*	*Set 3*
(a) lithium fluoride	ammonium carbonate	iodine pentafluoride
(b) potassium sulfide	calcium bisulfate	hydrogen selenide
(c) magnesium bromide	magnesium bicarbonate	dinitrogen tetroxide
(d) silver chloride	lithium phosphate	hydrogen bromide
(e) iron(II) sulfide	barium nitrite	boron trichloride
(f) barium iodide	magnesium dihydrogen phosphate	sulfur dioxide
(g) aluminum oxide	iron(III) sulfate	carbon disulfide
(h) copper(I) oxide	potassium cyanide	sulfur dichloride
(i) mercury(II) nitride	strontium chromate	disilicon hexabromide
(j) lead(IV) chloride	beryllium nitrate	phosphorus trifluoride
(k) cesium iodide	chromium(III) sulfite	oxygen fluoride
(l) gallium nitride	nickel acetate	selenium trioxide

29. Name the following compounds:

(a) NH_4I **(g)** O_2F_2 **(l)** $NaMnO_4$

(b) PCl_3 **(h)** $Zn(NO_2)_2$ **(m)** Li_2S

(c) $Ca(OH)_2$ **(i)** P_2O_5 **(n)** K_2SO_3

(d) CBr_4 **(j)** NaH **(o)** CaC_2

(e) $FeCl_2$ **(k)** HI **(p)** $Al_2(SO_4)_3$

(f) $(NH_4)_2Cr_2O_7$

30. Name each of the compounds for which you wrote the formula in problem 24.

APPLIED PROBLEMS

31. Consider the imaginary compound XY_2. This compound is not polar, yet the electronegativity of X is 2.5, whereas the electronegativity of Y is 3.6. What does this information tell us about the shape of the molecule?

32. The compounds ethyl ether and 1-butanol contain the same number of atoms of C, H, and O, yet 1-butanol has a much higher boiling point than ethyl ether. Examine the structures of these two compounds and see if you can offer any explanation for this difference in boiling points:

```
    H  H  H  H                H  H     H  H
    |  |  |  |                |  |     |  |
H—C—C—C—C—O—H          H—C—C—O—C—C—H
    |  |  |  |                |  |     |  |
    H  H  H  H                H  H     H  H
     1-butanol                 ethyl ether
```

33. Calcium chloride is often used instead of rock salt for thawing icy streets and sidewalks. Write the formula for this compound. Is this compound ionic or covalent? What particles make up a crystal of this substance?

34. In addition to water, hydrogen and oxygen can combine to form hydrogen peroxide, H_2O_2. This substance is used as a bleaching agent and as an antiseptic. When you purchase a bottle of hydrogen peroxide, approximately 97% of the solution is water and the other 3% is actually hydrogen peroxide. Draw the electron dot and bond diagrams for this substance. What does the fact that this compound is polar tell you about the shape of a molecule of hydrogen peroxide?

35. Although hydrogen cyanide (HCN) is a deadly gas, compounds containing the cyanide ion (CN^-) are used in metal-plating baths. Draw the electron dot diagram for hydrogen cyanide. Then draw the electron dot diagram of the cyanide ion.

36. Although the octet rule is a powerful device for predicting bonding, it doesn't work all of the time. In some cases, atoms violate the rule. In the compound SF_4, sulfur violates the rule, but the fluorines do not. Draw the electron dot structure for this compound.

37. Potassium metal reacts vigorously with the red liquid bromine to produce potassium bromide, which is a white solid ionic compound. Explain this reaction in terms of the octet rule. Write the electron configuration of the two ions formed in this reaction.

Chapter 6

Chemical Equations and the Mole

At 45 years of age, Betty Johns felt that she was in perfect health. It had been 8 years since her last physical examination, however, and she thought now was a good time to schedule another checkup. At Betty's exam the doctor measured her blood pressure at the very high level of 180/120. Although she had not suspected it, this meant that Betty was suffering from hypertension (the medical term for high blood pressure).

Initial treatment with a diuretic did not lower Betty's blood pressure to normal levels, so the doctor next prescribed the drug α-methyldopa (sold as Aldomet). This muscle relaxant helps reduce blood pressure by dilating the small arteries. When Betty returned to the doctor for her appointment a month later, she complained of a constant feeling of fatigue and lack of energy. This time the doctor found that Betty's blood pressure was within the normal range at 140/90, but her hemoglobin level had fallen from a normal level of 13.5 to a very low level of 10. Further tests confirmed the diagnosis that Betty was now suffering from hemolytic anemia, a destruction of red blood cells caused by the drug α-methyldopa. The doctor immediately stopped the use of this drug and prescribed a different smooth-muscle relaxant for Betty. Within three weeks her hemoglobin level had returned to normal.

It is an unfortunate fact of modern medicine that the treatment given for one medical problem may very well cause an-

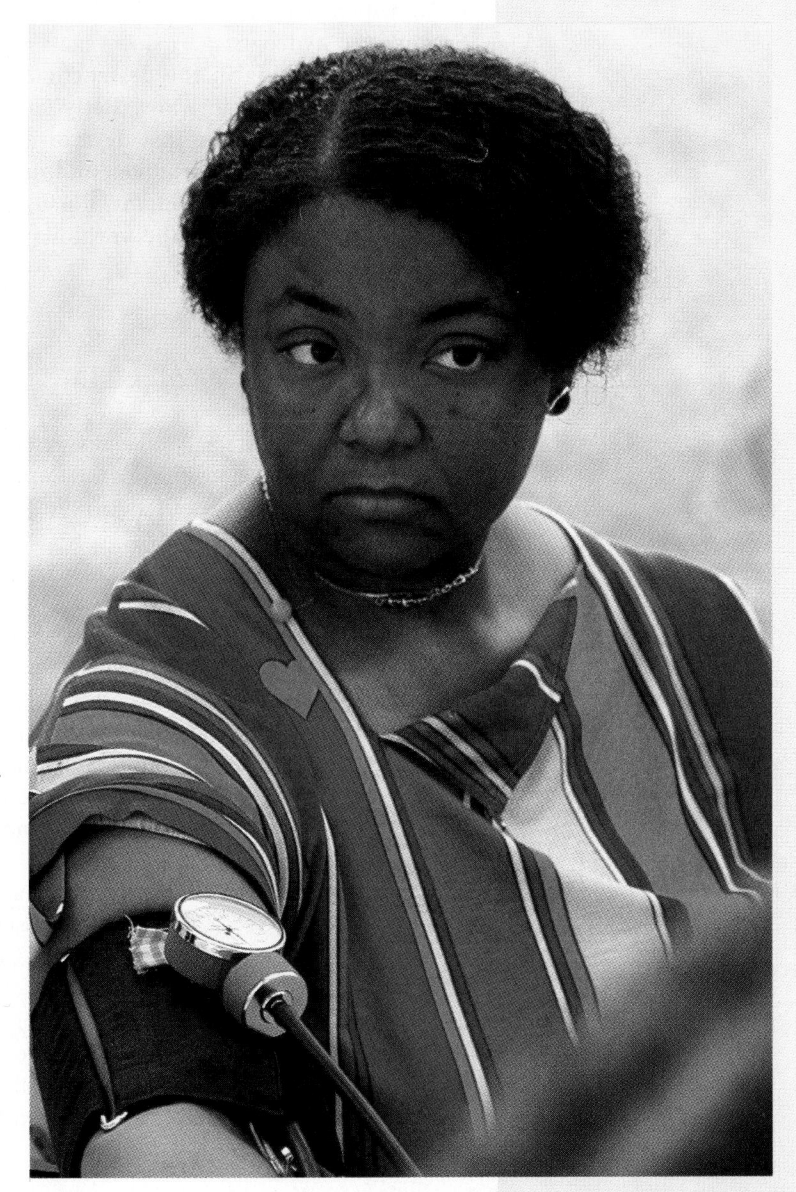

other. The important question for us at this moment, however, is what exactly caused the destruction of Betty's red blood cells? The α-methyldopa that Betty was taking for her high blood pressure reacts with oxygen in the body to form the compound hydrogen peroxide, H_2O_2. This is not usually a problem because most people have two enzymes in their body that protect against the buildup of hydrogen peroxide by breaking down this compound to form water. Enzymes are special molecules that control the rate of chemical reactions in the body. But some people are missing one or both of these particular enzymes, leaving their cells without protection against the harmful effects of hydrogen peroxide.

Betty's red blood cells lacked the selenium-containing enzyme called glutathione peroxidase, which is the main enzyme that destroys hydrogen peroxide in red blood cells. This resulted in a buildup of hydrogen peroxide, which can damage red blood cells in two ways. First, H_2O_2 changes the iron found in hemoglobin from Fe^{2+} to Fe^{3+}, which makes the hemoglobin lose its ability to carry oxygen in the blood. Second, the hydrogen peroxide reacts with molecules in the membrane of the red blood cell, resulting in the destruction (hemolysis) of these cells. This caused the hemolytic anemia from which Betty suffered.

Each process that we just described—the breakdown of hydrogen peroxide to water, the change in the iron ion, and the destruction of red blood cells—takes place as a result of chemical reactions. For us to study and understand such processes we need to be familiar with the vocabulary and symbols used to represent chemical reactions.

6.1 Writing Chemical Equations

Chemical equations are a shorthand way of representing what occurs in a chemical reaction. A chemical equation contains the formulas of the starting materials, called **reactants,** separated by an arrow from the resulting materials, called **products.**

$$\underset{\text{Reactants}}{A + B} \longrightarrow \underset{\text{Products}}{C + D}$$

The reactants and products may be made up of atoms, ions, molecules, or ion groups. In this chapter we are concerned mainly with reactants between atoms and molecules. We study reactions between ions in greater detail in Chapter 9.

Atoms are neither created nor destroyed in a chemical reaction, so the chemical equation representing the reaction must be **balanced.** That is, for each element involved in the reaction the equation must show the same number of atoms on the product side as are found on the reactant side. The **law of conservation of mass** summarizes this fact by stating that the total mass of the products equals the total mass of the reactants.

State the law of conservation of mass.

For example, both hydrogen peroxide and water can be formed from hydrogen and oxygen. As we have seen, however, our tissues are greatly affected by which one of these two compounds is formed. Hydrogen peroxide can badly damage cells, whereas water is an essential part of all cells. In writing a chemical equation for the formation of either hydrogen peroxide or water, the reactants or starting materials are oxygen (O_2) and hydrogen (H_2). Although the hydrogen needed for this reaction in our tissues does not actually come from the elemental hydrogen molecule H_2 (it comes from other molecules containing hydrogen), we use elemental hydrogen to simplify this introductory discussion:

$$H_2 + O_2 \longrightarrow$$

To complete the equation, we must write the correct chemical formula for the product: water or hydrogen peroxide. We know that water is H_2O. From Table 5.6 we can find that the peroxide ion is O_2^{2-}. To maintain electric neutrality we therefore need two hydrogen ions, H^+. This gives us the correct formula for hydrogen peroxide, H_2O_2. Now we can write the reactants and products of the two equations we have discussed:

$$\text{(Hydrogen peroxide)} \qquad H_2 + O_2 \longrightarrow H_2O_2$$

$$\text{(Water)} \qquad H_2 + O_2 \longrightarrow H_2O$$

Let's check to see if these two equations are balanced. Starting with the equation for the formation of hydrogen peroxide, we must check the number of atoms of each element on both sides of the equation. If there is no coefficient in front of a formula in an equation, it is understood that the coefficient is 1. You can see that there are two hydrogen atoms on the reactant side and two on the product side. There are two oxygen atoms on the reactant side and two on the product side. This equation is therefore balanced.

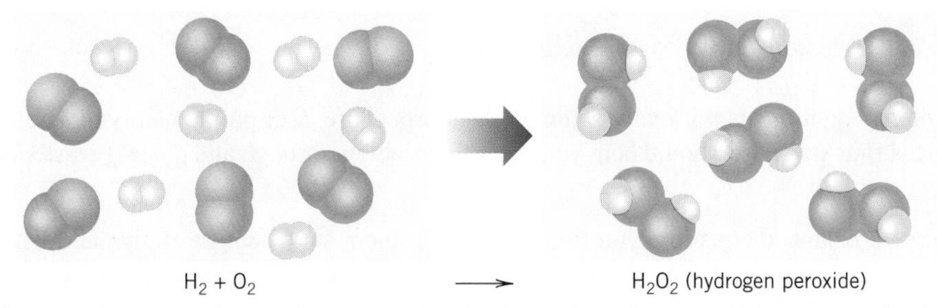

$H_2 + O_2 \longrightarrow H_2O_2$ (hydrogen peroxide)

Now look at the chemical equation for the formation of water. There are two hydrogen atoms on each side of the arrow. There are two oxygen atoms on the reactant side, but only one on the product side. To balance an equation the *only* numbers that can be changed (after the correct chemical formulas have been written) are the numbers, or coefficients, that appear in front of the formulas of the substances involved. So, to obtain two oxygen atoms on the product side of the equation we must place a coefficient of 2 in front of the formula for water.

$$H_2 + O_2 \longrightarrow 2H_2O$$

Now the oxygen atoms balance, but the hydrogen atoms are unbalanced—there are four hydrogen atoms on the product side and only two on the reactant side. Placing a coefficient of 2 in front of the formula for hydrogen on the reactant side balances the hydrogens. Now you can check to see that the equation is balanced (Fig. 6.1).

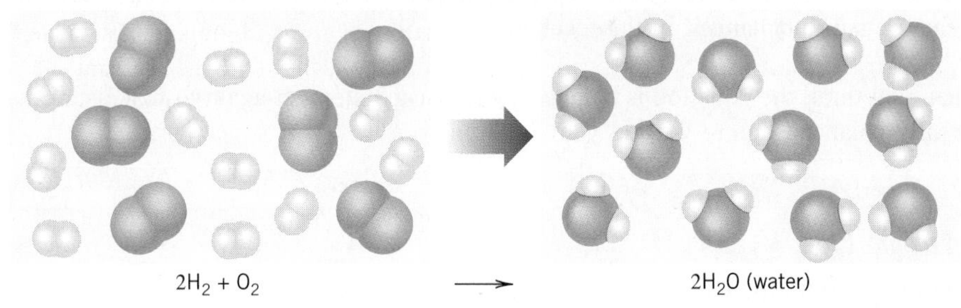

$2H_2 + O_2 \longrightarrow 2H_2O$ (water)

By comparing these two balanced equations, you can now see that the reactions for the formation of water and hydrogen peroxide are actually different. In one reaction, one molecule of hydrogen combines with one molecule of oxygen to form one molecule of

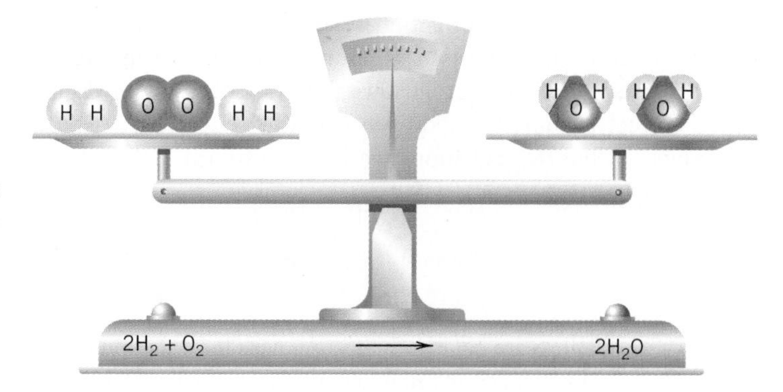

Figure 6.1 When an equation is balanced, the number of atoms of each element on the reactant side equals the number of atoms of each element on the product side. The total mass of the reactants therefore equals the total mass of the products.

hydrogen peroxide. In the other reaction, two molecules of hydrogen combine with one molecule of oxygen to form two molecules of water.

◆

Write a balanced chemical equation given the names of the reactants and the products.

6.2 Balancing Chemical Equations

Balancing equations is not really a complicated procedure. Our problem-solving strategy involves four steps that should help you think the problem through and quickly master this skill:

See the Question

To write a balanced equation, you first read the problem statement carefully and identify those substances that are the reactants and those that are the products. Then, you must write the correct chemical formula for each reactant and each product. Once this is done, you should never change the formula of a substance; only the coefficients can be changed. Any change of subscript would change the compound. For example, CO and 2CO represent one and two molecules of the compound carbon monoxide. But CO and CO_2 are entirely different substances, and you wouldn't want to confuse the two. Carbon monoxide is deadly, but carbon dioxide is produced in, and exhaled from, our bodies.

Think It Through (Execute the Math)

First, because balancing equations involves juggling coefficients (not subscripts!), it is often easiest to begin by giving the coefficient 1 to the compound having the most complicated formula.

Second, start balancing each element. *Hint:* You will often find it easiest if you balance oxygen last.

Third, treat polyatomic ions as one unit (just as though they were a single element) if they remain unchanged in the reaction.

Prepare the Answer

When you think the equation is balanced, check each element again to make sure that it is really balanced before writing your final answer.

Example 6-1

1. Balance the following equation: $Na + Cl_2 \longrightarrow NaCl$

See the Question

The formulas of the reactants and products are already written.

This is not a balanced equation: there are two chlorine atoms on the reactant side of the equation and only one on the product side. Placing a coefficient of 2 in front of the formula for sodium chloride balances the chlorine atoms in this equation.

Think It Through (Execute the Math)

$$Na + Cl_2 \longrightarrow 2NaCl$$

But now there are two sodium atoms on the product side, and only one on the reactant side. Placing a coefficient of 2 in front of the sodium on the reactant side balances the equation.

$$2Na + Cl_2 \longrightarrow 2NaCl$$

We have two atoms of sodium and two atoms of chlorine on each side of the equation. It is balanced.

Prepare the Answer

2. A small amount of the pollutant sulfur trioxide is released into the air when sulfur-containing coal or petroleum is burned. Write the balanced chemical equation for this reaction.

When a substance burns, it reacts with oxygen in the air. The reactants for this chemical reaction are elemental sulfur and oxygen (remember that elemental oxygen exists as a diatomic molecule). The product is sulfur trioxide.

S

$$S + O_2 \longrightarrow SO_3$$

When the equation is written as above, the sulfur atoms balance but not the oxygen atoms. For the oxygen to balance, we must have the same number of atoms on both sides, and the lowest possible number is six.

T • E

$$S + 3O_2 \longrightarrow 2SO_3$$

Now to balance the sulfur,

$$2S + 3O_2 \longrightarrow 2SO_3$$

We have two sulfur atoms on each side and six oxygen atoms on each side.

P

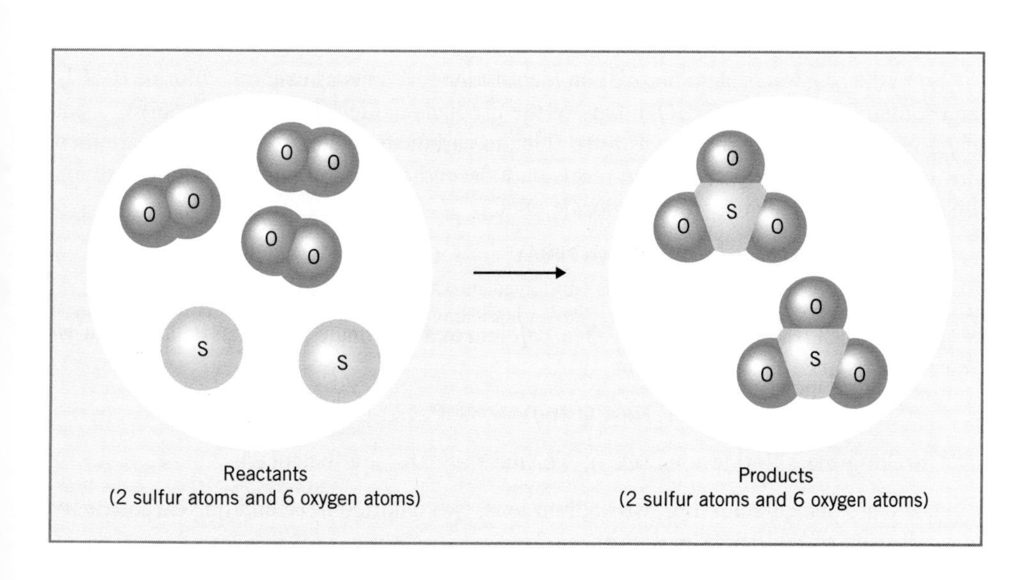

3. When aluminum reacts with sulfuric acid (H_2SO_4), hydrogen gas and aluminum sulfate are produced. Write the balanced chemical equation for this reaction.

S .... The reactants in this equation are aluminum and sulfuric acid, and the products are hydrogen gas and aluminum sulfate. Using Tables 5.2 and 5.6, we find that aluminum forms an Al^{3+} ion and that the polyatomic ion sulfate has a charge of 2−. The formula for aluminum sulfate, therefore, is $Al_2(SO_4)_3$.

$$Al + H_2SO_4 \longrightarrow H_2(g) + Al_2(SO_4)_3 \qquad (g) = \text{gas}$$

T • E .... Assign $Al_2(SO_4)_3$ a coefficient of 1. Then, balance the aluminum.

$$2Al + H_2SO_4 \longrightarrow H_2(g) + Al_2(SO_4)_3$$

Next balance the sulfate, SO_4^{2-}, by adding the coefficient 3 in front of H_2SO_4.

$$2Al + 3H_2SO_4 \longrightarrow H_2(g) + Al_2(SO_4)_3$$

Finally, balance the hydrogens.

$$2Al + 3H_2SO_4 \longrightarrow 3H_2(g) + Al_2(SO_4)_3$$

P .... There are now two atoms of aluminum on each side, six atoms of hydrogen on each side, and three sulfate ions on each side. Another way to check our equation is to set up the following table:

Symbol of Atom	*Number of Atoms on the Reactant Side*	*Number of Atoms on the Product Side*
Al	2	2
H	6	6
S	3	3
O	12	12

4. Hard water contains dissolved calcium chloride ($CaCl_2$) that forms a scum with soap. You can ''soften'' the water by adding sodium carbonate. The calcium reacts with the carbonate to form calcium carbonate, which is insoluble in water. Write a balanced equation for this reaction.

S .... From the problem statement, we can identify the reactants as calcium chloride ($CaCl_2$) and sodium carbonate (Na_2CO_3). One of the products is identified in the problem: ''calcium reacts with the carbonate to form calcium carbonate ($CaCO_3$)''; however, are there any other products? When calcium reacts with the carbonate, sodium and chloride remain. So the other product is sodium chloride, NaCl.

$$CaCl_2(aq) + Na_2CO_3(aq) \longrightarrow CaCO_3(s) + NaCl(aq)$$

aq = aqueous (dissolved in water), s = solid

T • E .... Assign $CaCO_3$ a coefficient of 1. The calcium and carbonate ions then balance, but we must still balance the sodium ions.

$$CaCl_2(aq) + Na_2CO_3(aq) \longrightarrow CaCO_3(s) + 2NaCl(aq)$$

Checking the chloride ions (Cl^-), we find they are now balanced.

P .... There are one calcium ion, two sodium ions, two chloride ions, and one carbonate ion on each side of the equation.

....

Check Your Understanding

Use the *STEP* problem-solving strategy discussed in Section 6.2 to balance the following equations:

1. $P + O_2 \rightarrow P_4O_{10}$
2. $NOCl \rightarrow NO + Cl_2$
3. $CH_4 + O_2 \rightarrow CO_2 + H_2O$
4. $Ca(OH)_2 + HCl \rightarrow CaCl_2 + H_2O$
   (*Note:* H_2O can also be written as HOH.)
5. Magnesium reacts with oxygen gas to form magnesium oxide.
6. Lead(II) sulfide reacts with oxygen gas to form lead(II) oxide and sulfur dioxide gas.
7. Sodium carbonate reacts with magnesium nitrate to form magnesium carbonate and sodium nitrate.

6.3 Oxidation–Reduction Reactions

Many chemical reactions involve the transfer of electrons from one atom to another. These reactions are called **oxidation–reduction,** or **redox,** reactions. The battery in your car uses redox reactions to produce the energy that starts the engine, and the cells in your body use redox reactions to produce the energy that keeps you alive. In a reaction involving electron transfer, **oxidation** refers to the loss of electrons by an atom, and **reduction** to the gain of electrons by another atom. Oxidation and reduction reactions are complementary processes. They always occur simultaneously and in equal amounts: the number of electrons lost by a substance must equal the number of electrons gained by another substance. In the formation of sodium chloride, for example, the sodium atom loses an electron while the chlorine atom gains an electron. The sodium atom is oxidized, and the chlorine atom is reduced.

Give an example of an oxidation reaction and a reduction reaction. Identify the oxidizing agent and the reducing agent.

$Na \longrightarrow Na^+ + e^-$ oxidation reaction (loss of electron)

$Cl + e^- \longrightarrow Cl^-$ reduction reaction (gain of electron)

We call the element that gains electrons the **oxidizing agent** and the element that loses the electrons the **reducing agent.** For example, in the reaction between calcium and chlorine we have:

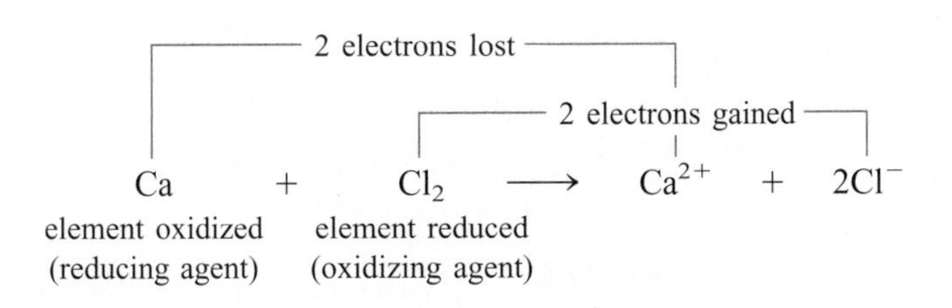

Chemical Perspective / Photography

An important example of a reduction reaction is found in the chemistry of photography. The photographic process is based on the reaction that occurs when energy from light decomposes silver halides like AgBr and AgCl into their original elements. Photographic film is made up of cellulose acetate that has been coated with a gelatin emulsion containing millions of tiny AgBr crystals. Each AgBr crystal contains equal numbers of Ag^+ and Br^- ions. When photons of light hit such a crystal, some electrons are knocked out of the Br^- ions, changing them to neutral bromine atoms. These bromine atoms pair up to form Br_2 molecules. The electrons that have been knocked off the bromine ions pair up with the Ag^+ ions to form neutral silver atoms. This reduction of silver ions results in tiny black specks of silver metal in the areas of the photographic film that have been struck by light. Crystals of AgBr that have not been hit by light remain a pale, yellowish white.

The fraction of a second that the film is exposed to light allows only a few silver ions to be reduced to the metallic silver, but each microscopic crystal contains billions of Ag^+ ions. However, the crystals containing these reduced silver atoms are very sensitive to reducing agents, which can cause the remaining silver ions in these crystals to be reduced to silver metal thousands of times faster than the silver ions in the unsensitized crystals. When the exposed film is placed into a reducing solution called a developer, the Ag^+ ions in these sensitized crystals are rapidly and completely changed to Ag atoms. Meanwhile, the silver ions in the unsensitized crystals are left virtually unchanged. This process produces a complete negative of the scene photographed, but our negative isn't quite finished yet. If we were to leave the darkroom with the negative at this time, all the unexposed silver bromide would immediately be exposed to light, and the entire negative would turn black. To prevent this, the negative is washed in a solution that removes any unexposed AgBr crystals, leaving behind just the blackened image. This solution is called the fixer because it fixes the image to the film.

To create the final photograph from the finished negative, we need to repeat the entire reduction process. This time, the negative is placed above a paper that has been coated with AgBr, and light is shone through the negative. The light can pass through the negative only where it hasn't already been blackened, so the bottom paper gets blackened in those areas but remains unblackened in the others. The final result, after developing and fixing, is a positive photograph of the original scene.

President Abraham Lincoln and General McClellan at Antietam, photographed by William Brady in 1862.

6.4 The Mole

When sodium (Na) is placed in a container of chlorine gas (Cl_2), sodium atoms and chlorine atoms undergo a chemical reaction to produce the compound sodium chloride (NaCl). One sodium atom reacts with each chlorine atom—more precisely, two sodium atoms react with each chlorine molecule to produce two sodium chloride ion units. (Remember that ionic compounds do not exist as single molecules.) If a chemist wanted to produce a certain exact amount of sodium chloride, however, it would be impossible to measure out the sodium and chlorine atoms individually. Atoms are too small to count out one at a time, or even millions at a time. A unit of measurement therefore had to be invented that would allow us to measure out equal numbers of sodium and chlorine atoms: this unit is the mole (Fig. 6.2).

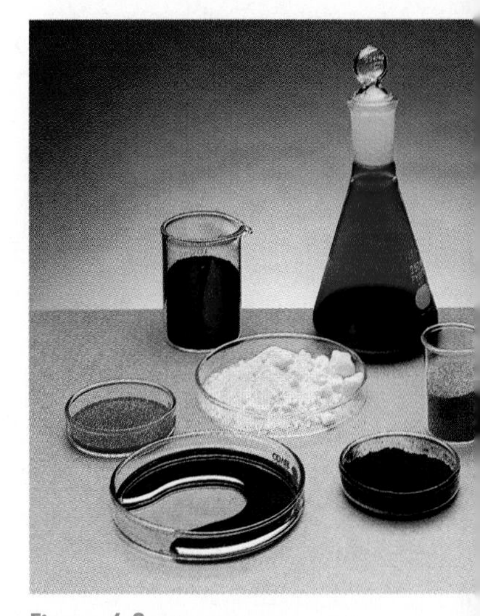

Figure 6.2 There is one mole of an element in each of these containers. Top: carbon (12.0 g), bromine (79.9 g). Middle: copper (63.6 g), sulfur (32.1 g), aluminum (27.0 g). Bottom: Mercury (200.6 g), boron (10.8 g).

In 1896, Wilhelm Ostwald first introduced the term "mole," which comes from the Latin word *moles,* meaning heap or pile. A **mole (mol)** is defined to be the number of atoms in 12.0000 g of carbon-12. Because atoms are so small, this number of atoms is very large: 6.02×10^{23}. It is hard to imagine just how large this number is, but another way to write 6.02×10^{23} is $602{,}000 \times 1$ million $\times$ 1 million $\times$ 1 million. This number is called **Avogadro's number,** named for the 19th-century scientist Amedeo Avogadro, who contributed a great deal to our knowledge of atomic weights. A mole, then, contains 6.02×10^{23} units. These units can be atoms, molecules, ions, electrons, or anything else. A mole of sodium atoms contains 6.02×10^{23} atoms; a mole of chlorine molecules contains 6.02×10^{23} chlorine molecules; a mole of marshmallows contains 6.02×10^{23} marshmallows, enough to cover all 50 states of the United States with a blanket of marshmallows 60 miles deep!

Define a mole.

How much does one mole weigh? It depends on the nature of the particles, just as a dozen lemons, grapefruit, or pumpkins each has a different total weight. The weight of one mole of atoms of any monatomic element is exactly equal to the *atomic weight in grams* of that element. For example, aluminum has an atomic weight of 27.0 amu, so one mole of aluminum weighs 27.0 g and contains 6.02×10^{23} atoms of aluminum. Likewise, one mole of uranium weighs 238.0 g. (For calculations involving atomic weights, we round off the atomic weights to the nearest tenths.)

Example 6-2

1. Determine the weight in grams of each of the following:

(a) one mole of magnesium atoms **(b)** 3.50 mol of iron

See the Question

In both (a) and (b), we need to find the atomic weight of each element to determine the weight of the given number of moles.

Think It Through

From the inside front cover of this book, we find that the atomic weight of magnesium is 24.3 amu and iron is 55.8 amu. From the definition of a mole, the weight of one mole of an element is equal to the atomic weight in grams.

$$1 \text{ mol Mg} = 24.3 \text{ g} \qquad 1 \text{ mol Fe} = 55.8 \text{ g}$$

We have the answer to (a), but (b) requires us to set up conversion factors from the equality for iron.

$$\frac{1 \text{ mol Fe}}{55.8 \text{ g}} \quad \text{and} \quad \frac{55.8 \text{ g}}{1 \text{ mol Fe}}$$

***E**xecute the Math* Because the answer needs to be in grams, the second conversion factor is the one to use.

$$3.50 \text{ mol Fe} \times \frac{55.8 \text{ g}}{1 \text{ mol Fe}} = 195.3 \text{ g} \quad \text{(calculator answer)}$$

***P**repare the Answer* **(a)** 1 mol Mg = 24.3 g

(b) Each of the numbers that we multiplied to find the answer has three significant figures, so the answer should have three significant figures.

$$3.50 \text{ mol Fe} = 195 \text{ g}$$

2. Determine the number of moles in 1.80×10^{24} atoms of lead.

S The problem asks: 1.80×10^{24} atoms of lead = (?) mol lead

T We know that 1 mol Pb = 6.02×10^{23} atoms, so we can construct a conversion factor that allows us to convert atoms of Pb into moles of Pb.

E

$$1.80 \times 10^{24} \text{ atoms Pb} \times \frac{1 \text{ mol Pb}}{6.02 \times 10^{23} \text{ atoms Pb}} = \frac{1.80}{6.02} \times \frac{10^{24}}{10^{23}} \text{ mol Pb}$$
$$= 0.2990033 \times 10 \text{ mol Pb} \quad \text{(calculator answer)}$$

(If you are not sure how this calculation was done, see Appendix 1.)

P The calculator answer has too many digits. Each given value in the problem has three significant figures, so the answer should have three significant figures.

$$1.80 \times 10^{24} \text{ atoms Pb} = 2.99 \text{ mol Pb}$$

3. How many moles are in 275 g of boron?

S The question asks: 275 g of boron = (?) mol boron

T The atomic weight of boron is 10.8 amu, so 1 mol B = 10.8 g.

E Using the conversion factor that gives the answer in moles, we have

$$275 \text{ g} \times \frac{1 \text{ mol B}}{10.8 \text{ g}} = 25.462962 \text{ mol B}$$

P Rounding off the calculator answer to the correct number of significant figures, the answer is

$$275 \text{ g B} = 25.5 \text{ mol B}$$

Checking Your Understanding

1. Determine the weight of each of the following:

(a) 0.0500 mol gold **(c)** 0.100 mol sulfur

(b) 2.00 mol zinc **(d)** 2.50×10^{20} atoms of magnesium

2. Determine the number of moles in each of the following:

(a) 32.4 g silver **(c)** 0.0202 g neon

(b) 980 g silicon **(d)** 1.20×10^{25}atoms of uranium

Historical Perspective / Amedeo Avogadro

Despite the jokes on late-night talk shows, not all lawyers are bad. In fact, some of them have even contributed to society in the realm of science. Consider the example of one, Amedeo Avogadro. Born in 1776 in Turin, Italy, he was a lawyer-turned-scientist. Avogadro was intrigued by a discovery made by Joseph Louis Gay-Lussac that all gases expand by the same amount when their temperatures increase. Avogadro believed this meant all gases must have an identical number of particles per unit volume when they were at the same temperature. He published this hypothesis in 1811, suggesting that the particles could be both individual atoms and combinations of atoms. He named these combinations of atoms *molecules,* a word he originated. Although his hypothesis was later universally accepted by the scientific community, this acceptance did not come until after his death in 1856.

Today, the name of Avogadro is familiar to every beginning chemistry student. It is used to indicate the number of atoms or molecules present in a mole. This number, 6.02×10^{23}, is called Avogadro's number. You might be interested to know that Avogadro's full name was on a scale with this number itself: Lorenzo Romano Amedeo Carlo Avogadro di Quaregna e di Cerreto.

Amedeo Avogadro (1776–1856)

6.5 Formula Weight

When atoms react to form compounds, there is no net gain or loss of weight. The particle that forms, whether a molecule or ion group, has a **formula weight** that is equal to the sum of the atomic weights of all the atoms appearing in its chemical formula. For example, the formula weight for sodium chloride, NaCl, is 58.5 amu—the atomic weight of one sodium atom (23.0 amu) plus the atomic weight of one chlorine atom (35.5 amu). The formula weight of carbon tetrachloride, CCl_4, equals the sum of the atomic weight of one carbon atom plus the atomic weight of four chlorine atoms.

Calculate the formula weight of a given compound.

$$\begin{aligned} C &= 12.0 \text{ amu} \\ 4 \times Cl = 4 \times 35.5 &= \underline{142.0 \text{ amu}} \\ \text{Formula weight } CCl_4 &= 154.0 \text{ amu} \end{aligned}$$

One mole of any substance has a mass equal to the *formula weight* of that substance expressed in grams: 1 mol = formula weight (g). For example, one mole of chlorine molecules (Cl_2) weighs 2×35.5, or 71.0 g. One mole of the compound sodium carbonate (Na_2CO_3) weighs $2 \times 23.0 + 12.0 + 3 \times 16.0$, or 106 g. If you were to weigh out 106 g of sodium carbonate, your sample would contain 6.02×10^{23} Na_2CO_3 ion groups.

Example 6-3

1. Calculate the formula weight in grams for (a) HF, (b) $Ca(OH)_2$, and (c) $Mg_3(PO_4)_2$.

See the Question

For each part, the question asks: Formula weight = (?) g

Think It Through

The formula weight is the sum of the atomic weights of each of the elements in the formula.

Execute the Math

(a) From the periodic table, we find that the atomic weight of hydrogen is 1.0 amu and that of fluorine is 19.0 amu.

$$HF = 1.0 + 19.0 = 20.0 \text{ amu}$$

(b) From the periodic table, we see that Ca = 40.1 amu and OH = 16.0 + 1.0 = 17.0 amu. Therefore,

$$\begin{aligned} Ca &= 40.1 \text{ amu} \\ 2 \times OH = 2 \times 17.0 &= \underline{34.0 \text{ amu}} \\ Ca(OH)_2 &= 74.1 \text{ amu} \end{aligned}$$

(c) The formula weight of $Mg_3(PO_4)_2$ can be calculated as follows:

$$\begin{aligned} 3 \times Mg = 3 \times 24.3 &= 72.9 \text{ amu} \\ 2 \times P = 2 \times 31.0 &= 62.0 \text{ amu} \\ 2 \times 4 \times O = 8 \times 16.0 &= \underline{128.0 \text{ amu}} \\ Mg_3(PO_4)_2 &= 262.9 \text{ amu} \end{aligned}$$

Prepare the Answer

In each of the additions, the uncertainty is in the tenth's place. The answers should therefore be written to the tenth's place.

(a) 20.0 amu **(b)** 74.1 amu **(c)** 262.9 amu

2. What is the weight of one mole of $Ca(OH)_2$?

S The problem asks: 1 mol $Ca(OH)_2$ = (?) g

T One mole of $Ca(OH)_2$ has a mass equal to the formula weight in grams.

E The formula weight was calculated in 1(b).

P One mole of $Ca(OH)_2$ weighs 74.1 g.

3. How many HF molecules are in 20.0 g HF?

S The problem asks: 20.0 g HF = (?) molecules HF

T One mole contains 6.02×10^{23} molecules.

E In 1(a), we calculated the weight of one mole of HF as 20.0 g.

P Therefore, 20.0 g HF contains 6.02×10^{23} molecules.

Check Your Understanding

1. Calculate the formula weight (to three significant figures) for each of the following:

(a) I_2 **(d)** $KClO_3$

(b) NaBr **(e)** $Al_2(SO_4)_3$

(c) PbS **(f)** C_4H_{10}

2. What is the weight of one mole of each of the following?

(a) sodium bromide

(b) potassium chlorate

(c) aluminum sulfate

3. What is the weight in grams of 6.02×10^{23} molecules of I_2?

6.6 Molar Volume of a Gas

After comparing the volumes of reacting gases and their gaseous products, Amedeo Avogadro formulated a hypothesis that has become known as **Avogadro's law.** This law states that, under identical conditions of temperature and pressure, equal volumes of different gases contain an equal number of gas particles—that is, an equal number of moles of gas. Under the conditions of standard temperature and pressure (STP = 0°C and 760 mm Hg), one mole of any gas occupies a volume of 22.4 L.

◆ **State the volume of one mole of gas at STP.**

1 molar volume of any gas = 22.4 L/mol (at STP)

Example 6-4

What volume does 56 g of oxygen gas occupy at STP?

The question in this problem is: 56 g = (?) L at STP

See the Question

What equalities help us solve this problem? We know that

Think It Through

$$1 \text{ mol gas} = 22.4 \text{ L at STP} \qquad \text{and} \qquad 1 \text{ mol } O_2 = 2 \times 16.0 = 32.0 \text{ g}$$

Using these equalities, we need to set up conversion factors whose units cancel to give us an answer in liters.

$$56\,\cancel{\text{g}} \times \frac{1\,\cancel{\text{mol}}\ O_2}{32.0\,\cancel{\text{g}}} \times \frac{22.4 \text{ L}}{1\,\cancel{\text{mol}}} = 39.2 \text{ L}$$

Execute the Math

Checking significant figures, we see that our answer should contain only two significant figures, so the answer is

Prepare the Answer

56 g O_2 occupies 39 L at STP

Check Your Understanding

A sample of dinitrogen tetroxide occupies a volume of 10.3 L at STP. How much does this sample weigh?

6.7 Using the Mole in Problem Solving

As we saw in Section 6.4, problems involving mole calculation often require the use of conversion factors to convert from one unit of measure to another. Suppose, for example, a chemist needs 0.45 mol of potassium chlorate, $KClO_3$. How many grams must we weighed out? This problem requires that we convert from moles of $KClO_3$ to grams, so we must construct an appropriate conversion factor. We know how to calculate the weight of one mole of $KClO_3$.

◆ **Calculate the mass of a given number of moles of a substance.**

$$\begin{aligned} K &= 39.1 \\ Cl &= 35.5 \\ 3 \times O = 3 \times 16.0 &= 48.0 \\ \text{Formula weight of } KClO_3 &= 122.6 \end{aligned}$$

Therefore,

$$1 \text{ mol } KClO_3 = 122.6 \text{ g}$$

From this equality, we can write the conversion factor that allows us to convert from moles to grams,

$$0.45 \cancel{\text{mol } KClO_3} \times \frac{122.6 \text{ g}}{1 \cancel{\text{mol } KClO_3}} = 55 \text{ g}$$

Study Example 6-5 carefully. You will see that we use the following four *STEP*s in solving these problems:

See the Question ➤ What is the problem asking?

Think It Through ➤ Set up conversion factors necessary to make the conversions.

Execute the Math ➤ Make certain the units cancel.

Prepare the Answer ➤ Check that your answer is in the desired units and has the correct number of significant figures.

Example 6-5

1. If you have a flask containing 9.0 g of water, how many moles of water are in the flask?

See the Question ➤ The problem asks, 9.0 g = (?) mol of water

Think It Through ➤ We need to find the relationship between grams and moles of water.

$$\text{Formula weight of } H_2O = (2 \times 1.0) + 16.0 = 18.0$$

Therefore,

$$1 \text{ mol of } H_2O = 18.0 \text{ g}$$

Execute the Math ➤ From the previous equality, we can construct the conversion factor necessary to give us an answer in moles.

$$9.0 \cancel{\text{g}} \times \frac{1 \text{ mol } H_2O}{18.0 \cancel{\text{g}}} = 0.50 \text{ mol } H_2O$$

Prepare the Answer ➤ The answer should be in moles of H_2O and have two significant figures.

$$9.0 \text{ g} = 0.50 \text{ mol } H_2O$$

2. How many molecules of chlorine gas are in a tank containing 7.10 g of chlorine?

S ➤ The problem asks, 7.10 g = (?) molecules of Cl_2

T ➤ To solve the problem, we need to establish a relationship between grams of chlorine and molecules of chlorine. We know that

$$1 \text{ mol } Cl_2 = 2 \times 35.5 = 71.0 \text{ g}$$

and

$$1 \text{ mol } Cl_2 = 6.02 \times 10^{23} \text{ molecules}$$

Chemical Perspective / Avogadro's Number and the Mole

How big is the number 6.02×10^{23}? In our discussions of Avogadro's number and the mole, we often lose sight of the size of this number. Even our national debt is minuscule in comparison. Just to give you some idea of how large this number is, consider that a ''mole'' of pennies laid side by side would stretch for more than 7.1×10^{18} miles. This is more than 1.2 million light-years, a distance that would take you far beyond the limits of our galaxy. If this ''mole'' of pennies were evenly distributed to all of the world's 5 billion people, each man, woman, and child would receive more than $1.2 trillion.

Conversely, the extremely small size of atoms can be illustrated by the size of 1 mol of carbon, which weighs only 12 g. It's hard to believe, but in 1 teaspoonful of carbon there are 6.02×10^{23} atoms. Taking this one step further, the number of atoms in the period at the end of this sentence is about 5×10^{18}. If you don't believe it, just go ahead and count them.

Your Perspective: Use your imagination to describe what a day would be like if the planet earth had to accommodate a mole of marshmallows (see Section 6.4).

Andromeda Galaxy

Therefore,

$$71.0 \text{ g } Cl_2 = 6.02 \times 10^{23} \text{ molecules}$$

Using the previous equality, we can write a conversion factor that gives us an answer in molecules of chlorine. ◄ ***E***

$$7.10 \text{ g } \cancel{Cl_2} \times \frac{6.02 \times 10^{23} \text{ molecules}}{71.0 \text{ g } \cancel{Cl_2}} = 6.02 \times 10^{22} \text{ molecules}$$

Our answer should have three significant figures and be in molecules of chlorine. ◄ ***P***

$$7.10 \text{ g} = 6.02 \times 10^{22} \text{ molecules of } Cl_2$$

Check Your Understanding

Use the four *STEP*s to solve the following problems:

1. Calculate the weight in grams of each of the following (using the formula weights you calculated in the first Check Your Understanding problem following Example 6-3):

(a) 0.500 mol I_2 **(c)** 0.0350 mol $KClO_3$

(b) 2.82 mol PbS **(d)** 4.00 mol $Al_2(SO_4)_3$

2. How many moles are in each of the following?

(a) 500 g HF

(b) 17.4 g C_4H_{10}

(c) 1.76 g $Ca(NO_3)_2$

3. How many molecules are in 1.45 g C_4H_{10}?

6.8 Calculations Using Balanced Equations

Use a balanced chemical equation to determine the amount of reactants necessary to produce a given amount of product.

A balanced chemical equation contains a great deal of information for the chemist. Not only does it indicate the identity of the reactants and products, it also tells us the relative amounts of these substances involved in the reaction. Consider the following reaction:

$$2H_2 + O_2 \longrightarrow 2H_2O$$

This equation can be read: ''Two hydrogen molecules react with one oxygen molecule to produce two water molecules.'' Or, using quantities with which we can work, it says: ''Two moles of hydrogen react with one mole of oxygen to produce two moles of water.'' This, in turn, tells us that 4 g of hydrogen react with 32 g of oxygen to produce 36 g of water.

Example 6-6

The following equation gives three sets of information. Write out each in sentence form.

$$2Al + 3H_2SO_4 \longrightarrow 3H_2 + Al_2(SO_4)_3$$

1. Two atoms of aluminum react with three molecules of hydrogen sulfate (sulfuric acid) to produce three molecules of hydrogen and one aluminum sulfate ion group.
2. Two moles of aluminum react with three moles of sulfuric acid to produce three moles of hydrogen and one mole of aluminum sulfate.
3. Fifty-four grams of aluminum react with 294 g of sulfuric acid to produce 6 g of hydrogen and 342 g of aluminum sulfate.

Check Your Understanding

Write three sentences (in terms of atoms, moles, and then grams) that express the information given by the following equation:

$$\underset{\text{methane}}{CH_4} + 2O_2 \longrightarrow CO_2 + 2H_2O$$

As you go through the exercises and problems that follow, you can use the *STEP* method.

See the Question Decide what the problem is asking and, if it is not given in the problem, write a balanced equation for the reaction.

Think It Through Determine what conversions are necessary to solve the problem, and set up a conversion factor for each one. Remember especially that you cannot convert directly from grams of one substance to grams of another substance in analyzing a chemical reaction. You must first convert to moles. To get from grams of substance A to grams of substance B use the following relationship:

$$\text{grams A} \longrightarrow \text{moles A} \longrightarrow \text{moles B} \longrightarrow \text{grams B}$$

The conversion factors you need come from the balanced equation and from the weights of one mole of the substances involved.

Do the calculations, making certain that units cancel so that your answer is in the desired units. ◄ ***Execute the Math***

Check that the answer is in the desired units and has the correct number of significant figures. ◄ ***Prepare the Answer***

As an example, let's determine how many moles of hydrogen combine with 7.0 mol of oxygen to form water.

Earlier we wrote the balanced equation for the formation of water. This gives us the relationship between moles of reactants and products in the reaction. ◄ ***S***

$$2H_2 + O_2 \longrightarrow 2H_2O$$

The problem is asking for the number of moles of hydrogen that react with a given number of moles of oxygen. From the coefficients in the balanced equation we can write the ratio of moles that must hold among the reactants (oxygen and hydrogen). This allows us to construct conversion factors for solving the problem. ◄ ***T***

$$\frac{2 \text{ mol } H_2}{1 \text{ mol } O_2} \quad \text{and} \quad \frac{1 \text{ mol } O_2}{2 \text{ mol } H_2}$$

The question asks how many moles of H_2 react with 7.0 mol of O_2, so the first conversion factor is the one that allows our units to cancel properly. ◄ ***E***

$$7.0 \cancel{\text{mol } O_2} \times \frac{2 \text{ mol } H_2}{1 \cancel{\text{mol } O_2}} = 14 \text{ mol } H_2$$

The value for moles of oxygen given in the problem has two significant figures, and the problem asks for the answer in moles: 14 mol of H_2 is needed for the reaction. ◄ ***P***

Next, let's calculate the number of grams of water that are formed when 6.4 g of oxygen is available to react.

The problem refers to the same balanced equation that we used previously. ◄ ***S***

$$2H_2 + O_2 \longrightarrow 2H_2O$$

In this case, we are being asked to convert from grams of oxygen to grams of water. But the conversion factors that come from the balanced chemical equation give the relationship between moles of oxygen and moles of water. We must therefore use a series of conversion factors to solve the problem, going through the following conversions: ◄ ***T***

$$\text{Grams of } O_2 \xrightarrow{\textbf{(a)}} \text{moles of } O_2 \xrightarrow{\textbf{(b)}} \text{moles of } H_2O \xrightarrow{\textbf{(c)}} \text{grams of } H_2O$$

We can write the conversion factors for each step. ◄ ***E***

(a) 6.4 g O_2 = (?) mol O_2

We know that 1 mol O_2 = 32 g

This allows us to construct a conversion factor that gives us a result in moles.

$$6.4 \cancel{\text{g } O_2} \times \frac{1 \text{ mol } O_2}{32 \cancel{\text{g } O_2}} = 0.20 \text{ mol } O_2$$

(b) Now we must determine how many moles of water are formed from the reaction of 0.20 mol of oxygen. The conversion factors that we obtain from the ratio of coefficients in our balanced equation are

$$\frac{1 \text{ mol } O_2}{2 \text{ mol } H_2O} \quad \text{and} \quad \frac{2 \text{ mol } H_2O}{1 \text{ mol } O_2}$$

Therefore,

$$0.20 \text{ mol } O_2 \times \frac{2 \text{ mol } H_2O}{1 \text{ mol } O_2} = 0.40 \text{ mol } H_2O$$

(c) Now we need to convert our answer from moles of H_2O to grams of H_2O. We know that

$$1 \text{ mol } H_2O = (2 \times 1.0) + 16.0 = 18.0 \text{ g}$$

Constructing the appropriate conversion factor from this equality so that our units cancel properly, we can now write

$$0.40 \text{ mol } H_2O \times \frac{18.0 \text{ g}}{1 \text{ mol } H_2O} = 7.2 \text{ g}$$

To save time and extra calculations, this problem could have been solved in just one step by using a series of conversion factors. The trick is to carefully arrange the conversion factors in such a way that all the units cancel except the unit desired for the final answer. In this case, we would have done the following calculation:

$$6.4 \text{ g } O_2 \times \frac{1 \text{ mol } O_2}{32 \text{ g } O_2} \times \frac{2 \text{ mol } H_2O}{1 \text{ mol } O_2} \times \frac{18.0 \text{ g } H_2O}{1 \text{ mol } H_2O} = 7.2 \text{ g } H_2O$$

P ▸ The given value had two significant figures, and the stated problem asked for grams of water. The answer is 7.2 g H_2O.

Now use the *STEP* problem-solving method in the following examples and problems:

Example 6-7

1. Imagine that you are working in a laboratory and want to obtain 2.40 mol of magnesium chloride, a compound that is produced (along with water) when magnesium oxide is heated with hydrochloric acid (HCl). How many grams of magnesium oxide do you need for this reaction?

See the Question ▸ The problems asks how many grams of magnesium oxide are necessary to produce 2.40 mol of magnesium chloride. First we need to write a balanced chemical equation for the reaction. A balanced equation is necessary to solve any of these problems.

$$\text{Magnesium oxide} + \text{hydrochloric acid} \longrightarrow \text{magnesium chloride} + \text{water}$$

$$\text{Unbalanced:} \quad MgO + HCl \longrightarrow MgCl_2 + H_2O$$

$$\text{Balanced:} \quad MgO + 2HCl \longrightarrow MgCl_2 + H_2O$$

Think It Through ▸ We need to write the conversion factors necessary for the following conversions:

$$\text{Moles } MgCl_2 \xrightarrow{\textbf{(a)}} \text{moles MgO} \xrightarrow{\textbf{(b)}} \text{grams MgO}$$

(a) Looking at the ratio of coefficients in our balanced equation, we obtain the following conversion factors:

$$\frac{1 \text{ mol MgO}}{1 \text{ mol MgCl}_2} \quad \text{and} \quad \frac{1 \text{ mol MgCl}_2}{1 \text{ mol MgO}}$$

(b) To obtain the conversion factor for the second step, we must know the weight of one mole of MgO.

$$1 \text{ mol MgO} = 24.3 + 16.0 = 40.3 \text{ g}$$

We therefore have

$$\frac{1 \text{ mol MgO}}{40.3 \text{ g}} \quad \text{and} \quad \frac{40.3 \text{ g}}{1 \text{ mol MgO}}$$

Let's solve the problem in a one-step calculation, choosing the conversion factors so that all the units cancel except for grams of MgO. ◄ ***E**xecute the Math*

$$2.40 \cancel{\text{mol MgCl}_2} \times \frac{1 \cancel{\text{mol MgO}}}{1 \cancel{\text{mol MgCl}_2}} \times \frac{40.3 \text{ g MgO}}{1 \cancel{\text{mol MgO}}} = 96.72 \text{ g MgO}$$

The calculator answer has one too many digits. The correct answer is 96.7 g MgO ◄ ***P**repare the Answer*

2. Antacid tablets are taken by millions of persons to reduce the discomfort of an upset stomach. The active ingredient in some commercial antacid tablets is magnesium hydroxide, $Mg(OH)_2$, which reacts with stomach acid (HCl) to produce magnesium chloride ($MgCl_2$) and water. One popular tablet contains 0.10 g of $Mg(OH)_2$. How many grams of stomach acid does this tablet neutralize?

The problem asks how many grams of HCl react with 0.10 g of $Mg(OH)_2$. First, we must write the balanced equation for this reaction. ◄ ***S***

Unbalanced: $Mg(OH)_2 + HCl \longrightarrow MgCl_2 + H_2O$

Balanced: $Mg(OH)_2 + 2HCl \longrightarrow MgCl_2 + 2H_2O$

Write the conversion factors needed for the following conversions: ◄ ***T***

$$\text{Grams Mg(OH)}_2 \xrightarrow{\text{(a)}} \text{moles Mg(OH)}_2 \xrightarrow{\text{(b)}} \text{moles HCl} \xrightarrow{\text{(c)}} \text{grams HCl}$$

(a) $1 \text{ mol Mg(OH)}_2 = 24.3 + 2 \times (16.0 + 1.0) = 58.3 \text{ g}$

$$\frac{1 \text{ mol Mg(OH)}_2}{58.3 \text{ g}} \quad \text{and} \quad \frac{58.3 \text{ g}}{1 \text{ mol Mg(OH)}_2}$$

(b) From the coefficients in our balanced equation,

$$\frac{1 \text{ mol Mg(OH)}_2}{2 \text{ mol HCl}} \quad \text{and} \quad \frac{2 \text{ mol HCl}}{1 \text{ mol Mg(OH)}_2}$$

(c) $1 \text{ mol HCl} = 1.0 + 35.5 = 36.5 \text{ g}$

$$\frac{1 \text{ mol HCl}}{36.5 \text{ g}} \quad \text{and} \quad \frac{3.5 \text{ g}}{1 \text{ mol HCl}}$$

To solve the problem in one step, choose the appropriate conversion factors from (a), (b), and (c) shown previously. ◄ ***E***

$$0.10\ \text{g}\ \cancel{Mg(OH)_2} \times \frac{1\ \cancel{\text{mol}\ Mg(OH)_2}}{58.3\ \text{g}\ \cancel{Mg(OH)_2}} \times \frac{2\ \cancel{\text{mol}\ HCl}}{1\ \cancel{\text{mol}\ Mg(OH)_2}} \times \frac{36.5\ \text{g}\ HCl}{1\ \cancel{\text{mol}\ HCl}}$$

$$= 0.12521\ \text{g}\ HCl$$

P The given value 0.10 contains two significant digits, so our answer must also have two significant figures. Therefore, the correct answer to this problem is 0.13 g HCl.

Check Your Understanding

1. How many grams of MgO are needed to produce 418 g of $MgCl_2$ using the reaction described in question 1 of Example 6-7?
2. In question 2 of Example 6-7, how many grams of hydrochloric acid would have been neutralized if 0.700 mol of H_2O was produced? If 0.190 g of $MgCl_2$ was produced?

KEY CONCEPTS

A chemical equation is a shorthand way of showing what occurs in a chemical reaction. The formulas of the reactants (or starting substances) are separated from the formulas of the products (or resulting substances) by an arrow. A balanced chemical equation is one in which the number of atoms of each element on the product side of the equation equals the number of atoms of that same element on the reactant side.

A mole of a substance contains 6.02×10^{23} particles and has a mass equal to the formula weight of the substance expressed in grams. A mole is a measure that allows equal numbers of particles of different substances to be weighed out. The formula weight of a compound is equal to the sum of the atomic weights of each atom in the formula of that compound.

Avogadro's law states that, under identical conditions of temperature and pressure, equal volumes of different gases contain equal numbers of particles of gas. At standard temperature and pressure, one mole of any gas occupies 22.4 L.

Important Equations

$$\text{Number of moles} = \frac{\text{grams of substance}}{\text{formula weight of substance}}$$

$$1\ \text{Molar volume} = 22.4\ \text{L/mol} \quad \text{(at STP)}$$

REVIEW PROBLEMS

Sections 6.1–6.2

1. Write a balanced equation for each of the following reactions:
   (a) Hydrogen reacts with bromine to form hydrogen bromide

(b) Calcium bicarbonate, when heated, breaks apart to form calcium carbonate, water, and carbon dioxide.

(c) Silver nitrate reacts with copper to form copper(II) nitrate and silver.

(d) Hydrogen reacts with nitrogen to form ammonia (NH_3)

(e) Methane gas (CH_4) and chlorine gas react to form carbon tetrachloride and hydrogen chloride.

2. Balance the following equations:

(a) $Na + H_2O \rightarrow NaOH + H_2(g)$

(b) $KClO_3 \rightarrow KCl + O_2(g)$

(c) $MnO_2 + HCl \rightarrow Cl_2(g) + MnCl_2 + H_2O$

(d) $C_3H_8 + O_2 \rightarrow CO_2 + H_2O$

(e) $NH_3 + O_2 \rightarrow NO + H_2O$

(f) $CO_3^{2-} + H^+ \rightarrow CO_2 + H_2O$

3. Balance the following equations:

(a) $HCl + Cr \rightarrow CrCl_3 + H_2(g)$

(b) $FeCl_3 + Na_2CO_3 \rightarrow Fe_2(CO_3)_3 + NaCl$

(c) $PbS + H_2O_2 \rightarrow PbSO_4 + H_2O$

(d) $C_4H_{10} + O_2 \rightarrow CO_2 + H_2O$

(e) $CrO_4^{2-} + H^+ \rightarrow Cr_2O_7^{2-} + H_2O$

4. Balance the following equations:

(a) $AgNO_3 + CaCl_2 \rightarrow Ca(NO_3)_2 + AgCl$

(b) $Fe_2O_3 + H_2 \rightarrow Fe + H_2O$

(c) $Al(NO_3)_3 + H_2SO_4 \rightarrow Al_2(SO_4)_3 + HNO_3$

(d) $C_3H_8O + O_2 \rightarrow CO_2 + H_2O$

(e) $Al + H_2SO_4 \rightarrow Al_2(SO_4)_3 + H_2$

(f) $Fe_2O_3 + C \rightarrow Fe + CO_2$

Section 6.3

5. For each of the following reactions, identify the element that is oxidized and the element that is reduced:

(a) $Ca + Cl_2 \rightarrow CaCl_2$

(b) $2Mg + O_2 \rightarrow 2MgO$

6. For each of the reactions in problem 5, identify the oxidizing agent and the reducing agent.

Section 6.4

7. Determine the weight of each of the following:

(a) 1.25 mol sodium

(b) 2.36 mol manganese

(c) 0.115 mol barium

(d) 3.75×10^{22} atoms of selenium

8. Determine the number of moles in each of the following:

(a) 18.4 g bromine

(b) 200 g copper

(c) 25.6 g oxygen

(d) 1.51×10^{23} atoms carbon

Section 6.5

9. Calculate the formula weight to three significant figures for each of the following compounds:

Group 1	*Group 2*	*Group 3*
(a) sodium hydroxide	KCl	$Ca(OH)_2$
(b) calcium chloride	H_2SO_4	$(NH_4)_2SO_4$
(c) sulfur dioxide	$Mg_3(PO_4)_2$	CH_4
(d) sodium phosphate	CO_2	HCl

10. Calculate the formula weight to three significant figures for each of the following compounds:

Group 1	*Group 2*	*Group 3*
(a) barium sulfate	$NaHCO_3$	$NaNO_3$
(b) hydrogen bromide	$KMnO_4$	$AgNO_3$
(c) boron trifluoride	HNO_3	H_2CO_3
(d) water	$CuSO_4$	$CaCO_3$

Section 6.6

11. What volume does 1 mol of hydrogen gas occupy at STP?

12. What volume does 1.054×10^{23} molecules of hydrogen (H_2) occupy at STP?

13. What is the volume occupied by 98.0 g of nitrogen gas at 0.750 atm and 300 K?

14. A sample of chlorine gas weighs 1.31 g at STP.

(a) Calculate the volume this sample of chlorine occupies under the following new conditions:

(1) 3.20 atm and 0.00°C

(2) 760 torr and −23.0°C

(3) 400 torr and 100°C

(b) Calculate the pressure of the chlorine gas under the following new conditions:

(1) 1.25 L and 0.00°C

(2) 415 mL and 45.0°C

(3) 138 mL and −100°C

(c) Calculate the temperature of the chlorine gas under the following new conditions:

(1) 830 mL and 1 atm

(2) 0.415 L and 404 kPa

(3) 2.49 L and 507 torr

Section 6.7

15. What is the weight in grams of each of the following? (Use the formula weights calculated in questions 9 and 10.)

Group 1	Group 2	Group 3
(a) 0.15 mol of sodium hydroxide	4.20 mol of KCl	0.95 mol of $Ca(OH)_2$
(b) 2.50 mol of calcium chloride	1.50 mol of H_2SO_4	4.60 mol of $(NH_4)_2SO_4$
(c) 0.80 mol of sulfur dioxide	0.015 mol of $Mg_3(PO_4)_2$	12.5 mol of CH_4
(d) 0.50 mol of sodium phosphate	5.50 mol of CO_2	0.025 mol of HCl
(e) 3.60 mol of barium sulfate	0.32 mol of $NaHCO_3$	0.62 mol of $NaNO_3$
(f) 0.020 mol of hydrogen bromide	0.0750 mol of $KMnO_4$	0.50 mol of $AgNO_3$
(g) 0.125 mol of boron trifluoride	0.10 mol of HNO_3	6.50 mol of H_2CO_3
(h) 0.0010 mol of water	0.060 mol of $CuSO_4$	0.0001 mol of $CaCO_3$

16. Calculate the number of moles in each of the following examples. (Use the formula weights calculated in questions 9 and 10.)

Group 1	Group 2	Group 3
(a) 0.0120 g of sodium hydroxide	89.4 kg of KCl	407 g of $Ca(OH)_2$
(b) 33.3 g of calcium chloride	7.35 g of H_2SO_4	3.3 μg of $(NH_4)_2SO_4$
(c) 2.48 kg of sulfur dioxide	131 μg of $Mg_3(PO_4)_2$	148.8 g of CH_4
(d) 0.82 g of sodium phosphate	198 g of CO_2	2.19 mg of HCl
(e) 2.33 g of barium sulfate	67.2 mg of $NaHCO_3$	64.6 g of $NaNO_3$
(f) 52.65 mg of hydrogen bromide	498.8 g of $KMnO_4$	0.68 g of $AgNO_3$
(g) 122.4 g of boron trifluoride	94.5 μg of HNO_3	80.6 mg of H_2CO_3
(h) 64.8 μg of water	3.19 g of $CuSO_4$	1.25 kg of $CaCO_3$

17. Calculate the following:

(a) the number of moles in 3.6×10^{23} molecules of oxygen

(b) the weight in grams of 1.5×10^{23} molecules of chlorine

18. Determine the number of oxygen atoms present in each of the following quantities:

(a) 2.0 mol of $Al_2(CO_3)_3$

(b) 18 g of $C_6H_{12}O_6$

(c) 1.0 g of $CuSO_4 \cdot 5H_2O$

Section 6.8

19. Use the balanced equation to answer the following questions:

$$2Na + Cl_2 \longrightarrow 2NaCl$$

(a) How many moles of Cl_2 are needed to react with 3.5 mol of Na?

(b) How many grams of NaCl are produced from 1.75 mol of Cl_2?

(c) What weight of Na is required to produce 73.11 g NaCl?

20. Use the balanced equation to answer the following questions:

$$2\,Al + Fe_2O_3 \longrightarrow Al_2O_3 + 2Fe$$

(a) How many moles of Al are needed to react with 2.5 mol of Fe_2O_3?

(b) How many grams of Al_2O_3 are produced from 2.5 mol of Fe_2O_3?

(c) How many grams of Fe_2O_3 are needed to react with 13.5 g Al?

21. Use the balanced equation to answer the following questions:

$$2S + 3O_2 \longrightarrow 2SO_3$$

(a) How many moles of oxygen are needed to react with 4.80 g sulfur?

(b) How many moles of sulfur are needed to produce 8.80 g sulfur trioxide?

(c) What is the maximum number of moles of sulfur trioxide that can be produced from 4.0 moles of sulfur and 5.0 moles of oxygen?

22. Use the balanced equation to answer the following questions:

$$Fe_2O_3 + 3H_2 \longrightarrow 2Fe + 3H_2O$$

(a) What weight of Fe_2O_3 is needed to react with 150 mg H_2?

(b) What weight of Fe can be produced from 2.0×10^{-2} mol of Fe_2O_3?

(c) How many moles of water are produced from 160 mg Fe_2O_3?

23. For each of the following reactions, determine the reactant that is completely consumed in the reaction (called the limiting reactant) and the reactant that is still left over. The quantities available to be reacted are shown below each reactant.

(a) $\underset{25.0\ g}{Mg} + \underset{34.0\ g}{S} \longrightarrow MgS$

(b) $\underset{94.4\ g}{SO_3} + \underset{250\ g}{2HNO_3} \longrightarrow H_2SO_4 + N_2O_5$

(c) $\underset{16.8\ g}{3Fe} + \underset{12.0\ g}{4H_2O} \longrightarrow Fe_3O_4 + 4H_2$

24. Use the balanced equation to answer the following questions:

$$Mg_3N_2 + 6H_2O \longrightarrow 3Mg(OH)_2 + 2NH_3$$

(a) How many moles of $Mg(OH)_2$ are produced from the reaction of 0.10 mol of Mg_3N_2?

(b) How many moles of $Mg(OH)_2$ and of NH_3 are produced from the reaction of 500 g of Mg_3N_2?

(c) How many grams of Mg_3N_2 and H_2O must react to produce 0.060 mol of $Mg(OH)_2$?

(d) How many grams of Mg_3N_2 are needed to produce 52.2 g of $Mg(OH)_2$?

(e) What is the maximum number of grams of $Mg(OH)_2$ that can be produced by the reaction of 10.0 g of Mg_3N_2 and 14.4 g of H_2O?

Applied Problems

25. Wine is produced by the process of fermentation, in which the sugar in grapes is converted by yeast to ethyl alcohol and carbon dioxide.

$$\underset{\text{sugar}}{C_6H_{12}O_6} \xrightarrow{\text{yeast}} \underset{\text{ethyl alcohol}}{2C_2H_5OH} + 2CO_2(g)$$

How many kilograms of ethyl alcohol are produced if all of the 5.00 kg of sugar in a batch of grapes fermented?

26. Aspirin is produced by the reaction of salicylic acid with acetic anhydride.

$$\underset{\substack{\text{salicylic}\\ \text{acid}}}{C_7H_6O_3} + \underset{\substack{\text{acetic}\\ \text{anhydride}}}{C_4H_6O_3} \longrightarrow \underset{\text{aspirin}}{C_9H_8O_4} + C_2H_4O_2$$

How many grams of salicylic acid are required to produce an aspirin tablet that contains 0.33 g of aspirin?

27. Oxygen gas is produced by the decomposition of potassium chlorate, $KClO_3$.

$$2KClO_3 \longrightarrow 2KCl + 3O_2$$

If 1.00 g of $KClO_3$ is completely decomposed by this reaction, how many grams of oxygen are formed? How many grams of KCl?

28. Natural gas is used extensively as a clean-burning energy source. In addition to its use in furnaces and water heaters in homes, it is used to power some vehicles and is the fuel for some electric generating plants. The products from the combustion of any fossil fuel are water and carbon dioxide. How many grams of water are formed when 100 g of natural gas (CH_4) is burned completely?

29. Ammonia is used in the production of fertilizers. The fertilizer urea, $(NH_2)_2CO$, is formed by reacting ammonia with carbon dioxide.

(a) Write the balanced equation for this reaction. (Water is the other product of the reaction.)

(b) How many metric tons of urea are produced by the reaction of 6.00 metric tons (1 metric ton = 10^3 kg, or 2200 lb) of ammonia?

30. Refining aluminum metal from aluminum ore involves the reaction of aluminum oxide with carbon, producing aluminum and carbon dioxide.

(a) Balance the equation for the refining of aluminum.

(b) If 5.40 metric tons of aluminum are produced in the United States each year, how many metric tons of aluminum oxide are used?

(c) How many metric tons of carbon are consumed in the production of 1.00 metric ton of aluminum?

31. Sodium lauryl sulfate, a detergent, can be prepared by the following reactions:

$$\underset{\substack{\text{lauryl}\\ \text{alcohol}}}{C_{12}H_{25}OH} + H_2SO_4 \longrightarrow \underset{\substack{\text{lauryl sulfonic}\\ \text{acid}}}{C_{12}H_{25}OSO_3H} + H_2O$$

$$\underset{\substack{\text{lauryl}\\ \text{sulfonic acid}}}{C_{12}H_{25}OSO_3H} + NaOH \longrightarrow \underset{\text{detergent}}{C_{12}H_{25}OSO_3Na} + H_2O$$

If a day's production of detergent is 11 metric tons, how many metric tons of lauryl alcohol are required?

32. Butane, the gas often used in cigarette lighters, burns in the presence of oxygen according to the reaction:

$$2C_4H_{10}(g) + 13O_2(g) \longrightarrow 8CO_2(g) + 10H_2O(g)$$

(a) If 4.0 mol of C_4H_{10} and 4.0 mol of O_2 are reacted, how many moles of CO_2 are produced? (*Hint:* There is an excess of one of the reactants. The reaction continues until one of the reactants is used up.)
(b) If 120.0 g of C_4H_{10} and 120.0 g of O_2 are reacted, how many grams of CO_2 are produced?
(c) If 50.0 g of C_4H_{10} and 25.0 g of O_2 are reacted together in a closed container, what compounds are present in the container after the reaction is complete, and how many grams of each compound are present?

Integrated Problems

1. One of the most deadly of the poisonous mushrooms is the amanita, or death angel.
(a) If 50 g of amanita mushrooms is enough to kill a 180-lb man, what is the lethal dose of these mushrooms calculated in milligrams per kilogram of body weight?
(b) Using the lethal dose you have just calculated, how many grams of amanita mushrooms are sufficient to kill a 120-lb woman? A 90-lb teenager?

2. A sample of an unknown metal weighs 14.50 g. The sample is placed in an empty 25.00-mL graduated cylinder. If 15.65 g of methyl alcohol (density 0.794 g/mL) is needed to fill the cylinder to the 25-mL mark, what is the density of the unknown metal?

3. If you had one mole of typing paper, each piece with a thickness of 0.004 in., how high would this stack of typing paper be? Give your answer in miles.

4. A 200.0-g iron bar at 85.0°C is placed in 1.00 L of water at 20.0°C. As the heat lost by the iron is gained by the water, the iron and water eventually come to the same temperature. What is the final temperature, assuming that no heat is lost to the outside? (The specific heat of iron is 0.115 cal/g°C.)

5. Jogging may or may not be an efficient way of losing weight. One kilogram of body fat contains about 7700 kcal of energy. If you run at the moderate pace of 1 km every 5 minutes, you expend energy at the rate of 95 kcal/km. How far do you have to run to burn 1 kg of body fat? (*Note:* A marathon is about 42 km in length.)

6. Assume that your weight remains steady on an intake of 2400 kcal/day. How long would it take to lose a pound of body fat if you started on a diet in which you reduced your daily intake of food energy to 1400 kcal/day?

7. In 1982, the eruption of the El Chichon volcano in Mexico blasted away the upper 220 m of the mountain. Water collected in the crater, forming a hot acidic lake whose temperature reached 51.4°C.
(a) How many feet of the mountain top were blasted away by the eruption?
(b) What was the temperature of the lake in degrees Fahrenheit?

8. Sugars are the most widely used food additive in the United States. Ketchup is 29% sugar, Coffeemate creamer 65.4% sugar, and Ritz crackers 11.8%. On the average, we each consume 62,000 g of sugar a year.
(a) How many pounds of sugar do we each, on the average, consume in a year?
(b) When this sugar is metabolized by our bodies, 4.0 kcal is produced for each gram consumed. How many extra kilocalories are we consuming each week from this added sugar?

9. Low-density lipoprotein (LDL), the major cholesterol carrier in the blood, is a spherical particle with a mass of 3 million daltons and a diameter of 22 nm.
   (a) What is the mass of an LDL particle in micrograms?
   (b) What is the diameter of the particle in millimeters?

10. You are running a carefully controlled experiment that keeps track of the use of sodium ions by a certain one-celled organism. You discover that your culture water has been contaminated by trace amounts of the following ions: potassium ion, barium ion, and iodide ion. Do you think any of these ions would interfere with your data? Why or why not? Design an experiment that supports your conclusion.

11. Sulfur trioxide can be produced by the reaction of the air pollutant sulfur dioxide with oxygen in the air.
    (a) Write the balanced equation for this reaction.
    (b) How many grams of sulfur dioxide must be released into the air to produce 1.00 kg of sulfur trioxide?

12. Sulfur trioxide is removed from the atmosphere by rain, which reacts with the sulfur trioxide to form sulfuric acid (H_2SO_4). Along major freeways, rainwater has been found to be quite acidic and damaging to plant life. For every 1.00 kg of sulfur trioxide that reacts with rain, how many grams of sulfuric acid are formed?

13. Deep in the Pacific Ocean, at pressures of 265 atm, researchers have discovered vents that shoot out water at 350°C. What is the temperature of this water in degrees Fahrenheit? What is the pressure at this depth in millimeters of mercury?

Chapter 7

Radioactivity and the Living Organism

Nineteen-year-old Becca White was entering her sophomore year majoring in business administration. Proud of her good health and boundless energy, she assured her parents that she would have no problem handling 18 hours of classes and a part-time job at a local computer store. The only thing that worried her was the 30 lbs she had put on during her freshman year. Becca decided to see a nurse practitioner at the health center for a checkup and some advice on losing weight.

During the checkup, the nurse discovered a lump in the lower portion of Becca's neck, just above the right collar bone. Becca, knowing that her mother had had a benign thyroid tumor removed many years earlier, was not particularly alarmed by this discovery. The nurse referred Becca to an endocrinologist who determined, through a physical exam and an analysis of serum thyroxine levels, that Becca had normal thyroid activity.

To determine the type of cells in the lump (or nodule), the endocrinologist ordered a scan of Becca's thyroid gland. The thyroid produces a hormone called thyroxine, each molecule of which contains four atoms of iodine. To be able to synthesize this hormone, the cells of the thyroid collect any iodine ingested by the body. Twenty-four hours after Becca swallowed 10 millicuries (mCi) of radioactive iodine-131, a scan of her thyroid was taken with a special camera that detects gamma radia-

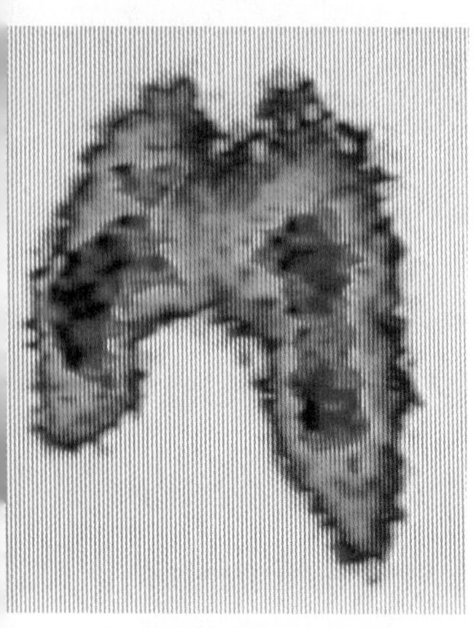
A thyroid scan

tion given off by the iodine-131. The scan showed a low level of iodine-131 uptake in the area of the nodule. Because thyroid cancer cells take up iodine much more slowly than normal thyroid cells, the scan indicated that the lump was cancerous. To verify the diagnosis, a surgeon took a sample of the nodule using a needle biopsy. The pathologist confirmed the results of the thyroid scan: Becca had a type of thyroid cancer, and surgery would be necessary.

The following week Becca went ahead with the surgery, which involved the total removal of the thyroid gland. Although she didn't like her "necklace" scar or taking thyroid hormone tablets each day, Becca didn't regret her decision. Within two weeks she was back in school. Two years later she graduated in business with a minor in computer science, and went on to become manager of the computer store. Everything was going great, or so Becca thought. But her fifth yearly scan of the thyroid area showed a disturbing pattern. There were areas of isotope uptake both where the thyroid gland had been removed and also lower on the left side of her neck. This pattern suggested that her cancer had returned. The surgeon explained that further surgery was not advisable because of the scar tissues in her neck from the first surgery.

Becca next met with a radiation oncologist who explained that the best treatment now would be a very high dose of iodine-131, so high a dose that it would kill all the thyroid cells that absorb it. Iodine-131 is an effective radioisotope for such use because of its short, 8-day half-life. This allows the iodine-131 to remain in the system long enough to kill the thyroid cells, but not long enough to damage other healthy tissues. The radiation oncologist explained that Becca probably would experience mild radiation sickness the week after receiving the iodine-131, and could lose some or all of her hair. There was also a 1% chance that the dose of iodine-131 would cause Becca to have leukemia later in life. Finally, the thyroid cancer might return even after the iodine-131 treatment.

Becca was crushed and overwhelmed by the news, but then she attacked the problem like everything else in her life. She talked to family and friends, read articles from medical journals, and consulted several physicians. In the end, she decided the benefits were worth the risks and chose to have the treatment. Becca entered the hospital and was placed in a room with a bed surrounded by lead shields. She drank 5 mL of a solution containing 150 mCi of iodine-131, which tasted a little like seawater. Becca remained in her room, separated from other people by lead shielding, until her radioactivity measured from 1 m away was less than 2 millirem per hour (mrem/h). After three days in the hospital, Becca returned home. She suffered some nausea and weakness, but it quickly went away. It took the doctors several months to regulate her dosage of thyroxine to the point that she felt her old self again. Now, three years later, Becca's physical examinations and thyroid scans show no evidence of recurrent thyroid cancer. She has resumed her career in the computer industry, lost 10 pounds, found romance through a computer dating service, and is reading up on investments for her retirement.

The use of radioactive materials is a standard procedure in many areas of medical diagnosis and treatment. In the case of Becca White, one radioactive isotope was used very effectively for both diagnosis and treatment of her cancer. Such materials are also used to sterilize medical products and foods, discover metabolic pathways, and track medicines through the body. Their use, however, is not without risk. In this chapter we discuss the different effects of radiation on living organisms and describe the use of radiation in the diagnosis and treatment of disease.

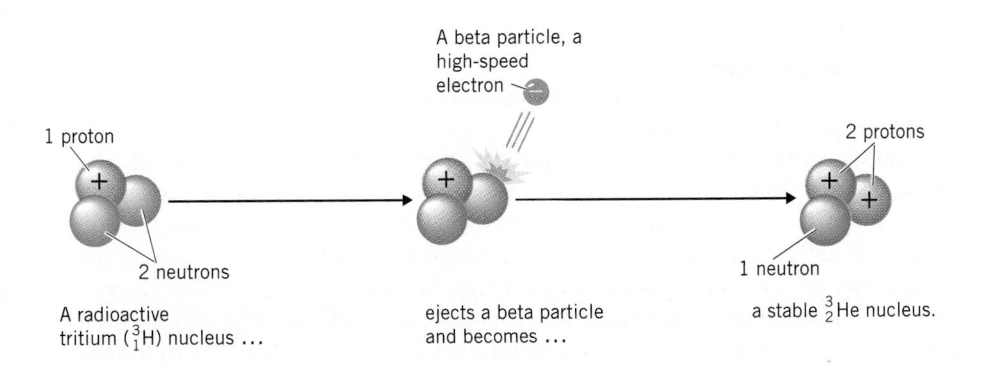
A beta particle, a
high-speed
electron
1 proton
2 neutrons
2 protons
1 neutron
A radioactive
tritium (^3_1H) nucleus ...
ejects a beta particle
and becomes ...
a stable ^3_2He nucleus.

Figure 7.1 Radioactivity is the term used to describe the emission of radiation from an unstable nucleus.

7.1 What Is Radioactivity?

Some atoms have nuclei that are unstable because of the ratio of protons and neutrons in their nucleus. Such an unstable nucleus decays (that is, it "spits out" particles), leaving behind a new nucleus called a **daughter nucleus.** An unstable daughter nucleus decays again, and this process continues until a stable daughter nucleus is formed (Fig. 7.1). Such a series of spontaneous decays is called a **transformation series.** (You may also hear this kind of process being referred to as a decay series or disintegration series.) The decay of a nucleus can result in the giving off, or emission, of several different kinds of radiation. The main types of such radiation are alpha, beta, and gamma radiation (Table 7.1). **Radioactivity** is the term used to describe the emission of radiation from isotopes of elements or their compounds. About 50 of the 350 naturally occurring isotopes are radioactive, as are all of the synthetically produced isotopes. These radioactive isotopes are called **radionuclides,** or **radioisotopes.**

Define transformation series.

Define radioactivity and radionuclide.

Alpha Radiation

One way in which nuclei can become more stable is by giving off alpha radiation. Alpha radiation consists of streams of alpha particles (α particles), each of which is made up of two protons and two neutrons—the nucleus of a helium atom, ^4_2He. By giving off an alpha particle, the atomic number of a nucleus is reduced by two and the mass number reduced by four. A well-known source of alpha radiation is the most abundant isotope of uranium, uranium-238. A uranium-238 atom decays by giving off an alpha particle to form an atom of thorium-234. We can write an equation, called a nuclear equation, for this reaction as follows:

Describe the three types of radiation given off by radioactive material.

Write a balanced nuclear equation for a radioactive decay.

$$^{238}_{92}U \longrightarrow ^{234}_{90}Th + ^4_2He$$

Table 7.1 **Particles and Radiation Emitted by Radionuclides**

Particle or Radiation	*Type*	*Charge*	*Symbol*
Alpha	Particle, helium nucleus	2+	α, ^4_2He
Beta	Particle, electron	1−	β, $^{0}_{-1}e$
Gamma	Electromagnetic radiation	0	γ
Neutron	Particle	0	^1_0n
Proton	Particle	1+	^1_1p, ^1_1H
Positron	Particle	1+	^0_1e

Environmental Perspective / Household Radon

In December, 1984, a worker at a nuclear power plant in Pennsylvania set off the building's contamination detectors. This mystified plant officials because the alarms were set off as he *entered* work that day, not as he left after working in the facility. Armed with radiation detection devices, the company's safety officers went to the worker's home to see if they could locate the source of the contamination. They discovered that the atmosphere of his home was contaminated with very high levels of the radioactive element radon. The U. S. Environmental Protection Agency (EPA) recommends that the level of radon in homes be no more than 4 picocuries (pCi) per liter, but the radiation level measured in the worker's living room was 3200 pCi/L. This meant that the family, which had lived in the house for a year, had been exposed to a level of radiation whose potential harm was the same as smoking more than 200 packs of cigarettes a day!

Radon is a naturally occurring radioactive gas that is produced from the decay of uranium. Uranium is found in virtually all rocks and soils but is often concentrated in certain areas that are rich in granite or shales. The worker's home was built on top of a geological formation that is particularly rich in uranium and its decay products. Radon gas released from the soil quickly disperses in the open air, but if it seeps into a house through holes or cracks in the foundation, it can accumulate to dangerous levels. Tests by the EPA in uranium-rich areas have shown that about 21% of the homes have radon levels above the recommended maximum. Radon gas is often drawn into houses by an air pressure difference (or gradient), which causes the interior air pressure to be less than that outdoors. Such a gradient can form from the suction caused when air is pumped outside by a clothes dryer, from the rising air in burning fireplaces, or from warm air rising inside a heated house.

Although radon is radioactive, it is not itself a health threat. The problem is caused by the decay products, or daughters, that radon produces. Radon is an inert gas. When it is breathed in, it is either exhaled or carried away from the lungs by the blood. This is different for radon's daughters (primarily isotopes of polonium, an alpha and gamma emitter). They are chemically reactive and attach to tiny airborne dust particles that, when inhaled, stick to the lung tissues and remain there. In this way, the high-energy alpha particles directly irradiate lung cells. Over time, the radioactive daughters can damage lung tissue and may eventually cause lung cancer. Spread over the whole population, a person's lifetime risk of dying from radon-related lung cancer may be much higher than the typical risk of dying from exposure to dangerous materials such as asbestos, pesticides such as ethylene bromide, or air pollutants such as benzene. This risk is further multiplied at least 10 times for smokers who live in a house with high radon levels because the radioactive daughters can adhere to particles in cigarette smoke. It is fairly easy to test for the radon levels in individual homes. If a health risk is found, a ventilation system can be installed to lower the radon level below 4 pCi/L.

Your Perspective: In a letter to the editor of your local newspaper, explain the need for the city's health department to test every house in the city for radon levels. Be sure to explain why this suggestion should be seriously considered.

The starting, unstable nucleus is shown on the left-hand side of the arrow, and the products resulting from the radioactive decay of this nucleus are shown on the right-hand side. To ensure this nuclear equation is written correctly, we must check that the number of protons and neutrons on one side of the arrow is equal to the number of protons and neutrons on the other side. That is, the sum of the mass numbers on each side of the arrow must be equal, and the sum of the atomic numbers on each side of the arrow also must be equal. For the decay of uranium-238,

	Reactants	$\longrightarrow$	Products
Mass number:	238	=	234 + 4
Atomic number:	92	=	90 + 2

Beta Radiation

Beta radiation, like alpha radiation, consists of streams of particles. In this case, the particle is an electron ($_{-1}^{0}e$) that is produced inside the nucleus. During beta decay, a neutron in the nucleus is changed into a proton and an electron. The electron (or β particle) is ejected from the nucleus, leaving behind a daughter nucleus having the same mass number as the original atom but a different atomic number. For example, the thorium atom produced by the alpha decay of uranium-238 is a beta emitter.

$$^{234}_{90}\text{Th} \longrightarrow {}^{234}_{91}\text{Pa} + {}_{-1}^{0}e$$

To check if this nuclear equation is written correctly, we have

	Reactants	$\longrightarrow$	Products
Mass number:	234	=	234 + 0
Atomic number:	90	=	91 + (−1)

Gamma Radiation

Gamma radiation (γ rays) does not consist of particles but is high-energy radiation similar to X rays. (Look again at the energy spectrum shown in Figure 2.5.) Often, the daughter nucleus produced by an alpha or beta emitter is in a high-energy, or excited, state. It can become more stable by releasing this energy in the form of gamma rays—short, intense bursts of electromagnetic radiation. The release of gamma rays usually occurs together with alpha or beta radiation. For example, radium-226 has a radioactive nucleus that releases alpha and gamma radiation when it decays.

$$^{226}_{88}\text{Ra} \longrightarrow {}^{222}_{86}\text{Rn} + {}^{4}_{2}\text{He} + \gamma$$

Example 7-1

1. Determine the symbol of the element that belongs in the following nuclear equation.

$$^{3}_{1}\text{H} \longrightarrow (?) + {}_{-1}^{0}e$$

See the Question

We must determine which element is produced in this nuclear equation.

Think It Through

To have a balanced nuclear equation, the sum of the mass numbers and the atomic numbers on both sides of the arrow must be equal. We can use this fact to calculate the atomic number and mass number of the unknown element.

Execute the Math

$$3 = M + 0 \qquad M = 3$$

$$1 = Z + (-1) \qquad Z = 2$$

Prepare the Answer From the front inside cover of this book, we find that the element having an atomic number of 2 is helium, He. The complete decay, therefore, is

$${}^{3}_{1}\text{H} \longrightarrow {}^{3}_{2}\text{He} + {}^{0}_{-1}e$$

2. Polonium-214 is an alpha and gamma emitter that decays to form an isotope of lead. Write the nuclear equation for this decay.

S We need to write a balanced nuclear equation for the decay of polonium-214.

T To begin, we use Table 7.1 and the inside front cover to write the necessary symbols.

$$\text{Polonium-214} = {}^{214}_{84}\text{Po}$$

$$\text{Alpha particle} = {}^{4}_{2}\text{He}$$

$$\text{Gamma ray} = \gamma$$

$$\text{Lead} = {}^{M}_{82}\text{Pb}$$

We do not know the mass number of the lead isotope. We can write the equation and use the fact that the mass numbers must be equal on both sides of the equation to determine the mass number of lead.

E

$${}^{214}_{84}\text{Po} \longrightarrow {}^{M}_{82}\text{Pb} + {}^{4}_{2}\text{He} + \gamma$$

$$214 = M + 4$$

$$M = 210$$

P The correct nuclear equation is

$${}^{214}_{84}\text{Po} \longrightarrow {}^{210}_{82}\text{Pb} + {}^{4}_{2}\text{He} + \gamma$$

Check Your Understanding

1. Write the symbol that belongs in the following nuclear equations:

(a) $(?) \rightarrow {}^{222}_{86}\text{Rn} + {}^{4}_{2}\text{He}$

(b) ${}^{28}_{13}\text{Al} \rightarrow (?) + {}^{0}_{-1}e$

(c) ${}^{227}_{89}\text{Ac} \rightarrow {}^{223}_{87}\text{Fr} + (?)$

2. Write the nuclear equations for

(a) The beta decay of ${}^{45}_{20}\text{Ca}$ (the decay of the ${}^{45}_{20}\text{Ca}$ isotope by emission of a β particle)

(b) The beta decay of ${}^{14}_{6}\text{C}$

(c) The alpha decay of samarium-149.

3. Cesium-137 is a radioactive waste produced by nuclear power plants. It decays by giving off beta and gamma radiation. Write the nuclear equation for this decay.

Historical Perspective / Marie Curie

Marie Curie (1867–1934)

Marie Skłodowska was born in Warsaw, Poland, in 1867. In those days Poland was not an independent nation, but a part of czarist Russia. In 1831, the Polish citizens had unsuccessfully revolted in an attempt to gain independence. This led the Russian government to impose many restrictions on the Polish people: the Polish language could not be used in churches, schools, or newspapers, and the Russian secret police were everywhere. In her late teens, Marie joined an underground group of young men and women intent on overthrowing the Russian occupation. During this time she wrote for an underground newspaper, calling for a revolution. The Russian police soon cracked down on this group and arrested many. Fortunately Marie escaped the police, but to avoid being used as a witness against one of her friends, she was forced to flee Warsaw. In 1891, Marie arrived in Paris, where she entered the Sorbonne to pursue a doctor's degree in physical chemistry. Even though she struggled to survive—fainting from hunger in the classroom on one occasion—Marie graduated at the top of her class. In 1895, she married Pierre Curie, a well-known French physicist.

Marie Curie began an investigation of emissions produced by uranium (a phenomenon she called radioactivity). She showed that it was the uranium atom itself that was the source of the radiations. Soon afterward Pierre Curie, realizing the genius of his wife, gave up his research and joined Marie in her work. Together they searched for other radioactive elements they believed to be present in uranium ore. By July 1898, they had separated from uranium ore an element many times more radioactive than the uranium atom. They named it ''polonium'' after Marie's native Poland. In December 1898, they discovered the even more radioactive element radium. It took them four years to isolate from tons of uranium ore just one gram of radium salts.

In 1903, Marie and Pierre Curie and Antoine Becquerel (the first person to discover radiation) were awarded the Nobel Prize in physics. Three years later Pierre was killed in a traffic accident, and Marie assumed his professorship at the Sorbonne—the first woman ever to hold such a position. She continued her work with radioactivity and in 1911 was awarded her second Nobel Prize (for chemistry), thereby becoming the first person to win two Nobel Prizes.

On July 4, 1934, Marie Curie died of leukemia, most likely brought on by long exposure to radiation from the materials she had studied. In 1935, her daughter, Irène Joliot-Curie, with Irene's husband Frédéric, also won a Nobel Prize in chemistry for their work with radioactive materials. Like her mother, Irène later died of leukemia, probably from the same type of long-term exposure to radiation.

7.2 Half-Life

Each radioactive substance has its own special rate of decay—that is, the number of emissions (or transformations) that occur each minute. This rate of decay is expressed by a number called the half-life of the isotope. The **half-life ($t_{1/2}$)** of a radionuclide is the length of time it takes for one-half of the atoms in a given sample to undergo radioactive decay. This means that, after one half-life, one-half of a sample of the radioactive isotope

◆ **Define half-life.**

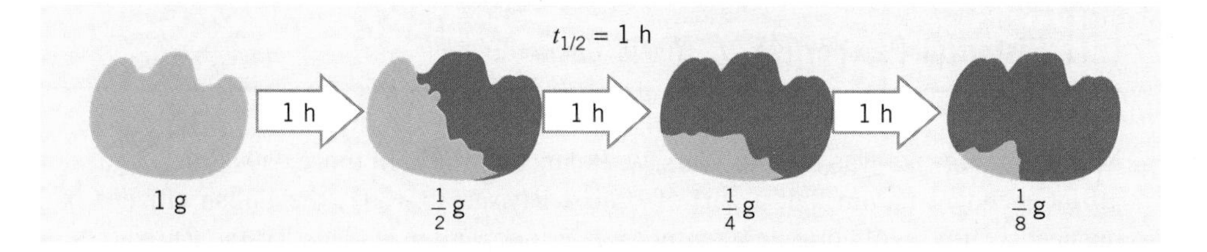

Figure 7.2 After one half-life, one-half of a sample of radionuclide has transformed into a new element and one-half remains. After two half-lives, one-fourth of the original sample remains among the transformation products.

has decayed to form a new substance, and one-half remains. After two half-lives, one-half of one-half (or one-fourth) remains. After three half-lives, one-half of one-fourth (or one-eighth) remains (Fig. 7.2). For example, nitrogen-13 has a half-life of 10 minutes. If you have 1 g of nitrogen-13, 10 minutes later 0.5 g of nitrogen-13 will have decayed to form carbon-13, and 0.5 g of nitrogen-13 will remain. Mathematically,

Calculate the amount of radionuclide remaining after a given number of half-lives.

$$\text{Quantity remaining after } n \text{ half-lives} = \text{initial quantity} \times (\tfrac{1}{2})^n$$

As another example, a radionuclide of technetium is widely used in medical diagnosis. This isotope has a half-life of 6 hours, which is favorable for purposes of diagnosis but which requires the laboratory to constantly renew its supply. If 1 g of this isotope is in the laboratory at 6:00 PM on Friday, only 0.25 g remains for use at 6:00 AM on Saturday morning (two half-lives later). Continuing in this way, only 1 mg would still be found among the decay products when the laboratory opened Monday morning (Fig. 7.3).

Half-lives of the different radionuclides can be as short as fractions of a second or as long as billions of years. Half-lives indicate how stable the isotope is. Most artificially produced radionuclides are highly unstable and have very short half-lives. Table 7.2 gives some examples of these isotopes and their half-lives.

The half-lives of radioactive elements give us a useful tool for determining the age of archaeological objects. Carbon-14, which has a half-life of 5730 years, is used to date once-living material. This carbon isotope is created in the earth's upper atmosphere when nitrogen atoms are bombarded by cosmic rays (which are streams of particles pouring into the atmosphere from the sun and outer space). The procedure of carbon-14 dating assumes

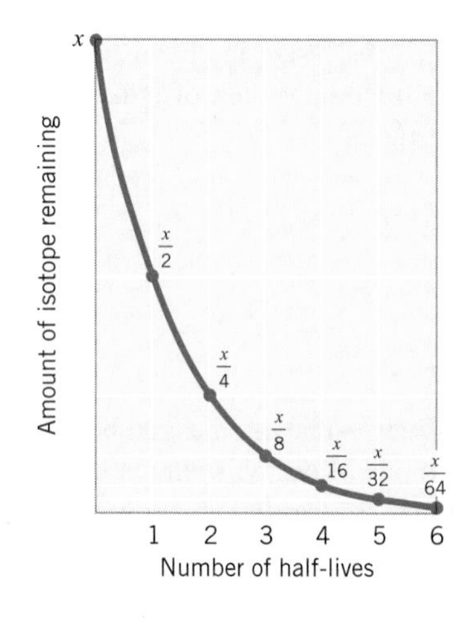

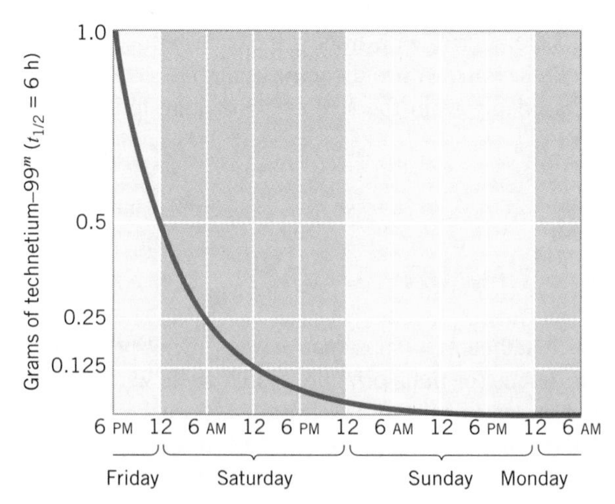

Figure 7.3 The half-life of technetium-99^m is 6 h. If you have a 1-g sample of technetium-99^m at 6 PM on Friday, you would have only 0.5 g at 12 midnight, and by 6 AM Monday only 1 mg would be left among the transformation products.

Table 7.2 **Some Radionuclides and Their Half-lives**

Element	*Isotope*	*Half-life*	*Radiation Given Off*
Hydrogen	$^{3}_{1}H$	12.3 years	Beta
Carbon	$^{14}_{6}C$	5730 years	Beta
Phosphorus	$^{32}_{15}P$	14.3 days	Beta
Potassium	$^{40}_{19}K$	1.28×10^{9} years	Beta and gamma
Cobalt	$^{60}_{27}Co$	5.27 years	Beta and gamma
Strontium	$^{90}_{38}Sr$	29.0 years	Beta
Technetium	$^{99m}_{43}Tc$	6.01 hours	Gamma
Iodine	$^{131}_{53}I$	8.04 days	Beta and gamma
Cesium	$^{137}_{55}Cs$	30.2 years	Beta
Polonium	$^{214}_{84}Po$	1.63×10^{-4} seconds	Alpha and gamma
Radium	$^{226}_{88}Ra$	1600 years	Alpha and gamma
Uranium	$^{235}_{92}U$	7.04×10^{8} years	Alpha and gamma
	$^{238}_{92}U$	4.46×10^{9}years	Alpha
Plutonium	$^{239}_{94}Pu$	2.41×10^{4} years	Alpha and gamma

that the ratio of carbon-14 to the stable carbon-12 isotope remains constant in living organisms. When an organism dies, the total amount of carbon-12 it has accumulated during its life becomes fixed and cannot change. However, one-half of the carbon-14 it has accumulated will be gone in 5730 years. Measuring the ratio of carbon-14 to carbon-12 in material that was once alive is therefore a fairly accurate way to date objects that died within the last 40,000 years.

Although carbon-14 is very useful for dating once-living things, radionuclides with much longer half-lives must be used to date geologic periods in the earth's history. Uranium-235 ($t_{1/2}$ = 704 million years) decays through a disintegration series to produce the stable lead-207 isotope. By measuring the ratio of uranium-235 to lead-207 in certain rock formations, meteorites, and moon rocks, scientists have estimated the age of the earth to be between 4.56 and 4.57 billion years.

Example 7-2

1. Phosphorus-32 is often used to study chemical reactions in biological research. The half-life of ^{32}P is 14 days. If a research laboratory received 525 mg of phosphorus-32, how many milligrams of ^{32}P would remain for use after 70 days?

   We want to know: ^{32}P = (?) g after 70 days

   See the Question

   Think It Through

   To solve this problem, we must first know how many half-lives will occur in 70 days. For ^{32}P, we can use the fact that 1 half-life = 14 days to set up the conversion factor to solve for the number of half-lives. Then, we can use the equation for calculating the quantity remaining after n half-lives to calculate the amount of phosphorus-32 remaining.

Execute the Math

$$70 \text{ days} \times \frac{1 \text{ half-life}}{14 \text{ days}} = 5 \text{ half-lives}$$

$$^{32}\text{P remaining} = 525 \text{ mg} \times \left(\frac{1}{2}\right)^5$$

$$= \frac{525 \text{ mg}}{32} = 16.40625 \text{ mg}$$

Prepare the Answer

The initial weight given in the problem has three significant figures, but the value for the half-life has only two significant figures. Therefore, the answer should have two significant figures.

$$^{32}P = 16 \text{ mg after 70 days}$$

2. You are in charge of reordering the phosphorus-32. If you start with 1.0 g and must reorder when the supply reaches 125 mg, how long will it be until you have to reorder?

S The problem is asking: After how many days would you have to reorder?

T To know the number of days, we need to determine how many half-lives it takes for 1.0 g (1000 mg) of ^{32}P to decay so that we have only 125 mg remaining. We can do this mathematically through step-by-step thinking or by substituting values in the equation from problem 1. Once we know the number of half-lives, we can calculate the number of days using a conversion factor from the equality: 1 half-life ^{32}P = 14 days.

E **(a)** To calculate the half-lives:

$$\text{Half-lives} \quad 1000 \text{ mg} \xrightarrow{1} 500 \text{ mg} \xrightarrow{2} 250 \text{ mg} \xrightarrow{3} 125 \text{ mg}$$

or

$$125 \text{ mg} = 1000 \text{ mg} \times \left(\frac{1}{2}\right)^n$$

$$\left(\frac{1}{2}\right)^n = \frac{125 \text{ mg}}{1000 \text{ mg}} = \frac{1}{8}$$

(b) To determine the number of days:

$$\text{Since, } \frac{1}{2} \times \frac{1}{2} \times \frac{1}{2} = \frac{1}{8}, \quad \text{then } n = 3$$

$$3 \text{ half-lives} \times \frac{14 \text{ days}}{1 \text{ half-life}} = 42 \text{ days}$$

P Our answer should have two significant figures: You should reorder after 42 days.

Check Your Understanding

Iodine-123 is better than iodine-131 for diagnosing thyroid function because it has a half-life of only 13.3 hours (compared with eight days for iodine-131).

1. If a dose of 0.24 mg of iodine-123 is used in a diagnostic test, how many milligrams remain after three days? (Round to the nearest half-life.)

2. How soon does a hospital need to reorder iodine-123 if it receives a shipment of 5.12 g and reorders when 90.0 mg remain?

Historical Perspective / The Shroud of Turin

Of all the applications of carbon-14 dating, perhaps the most fascinating have been in the field of biblical archaeology. One of the best examples is the use of carbon-14 dating to establish the authenticity of the Dead Sea scrolls.

A similar biblical puzzle is posed by the Shroud of Turin. A linen cloth measuring over 4 m in length, the shroud bears a faint image—like a photographic negative—of a crucified man. Many people believe that this shroud is the burial cloth of Jesus Christ. In 1978, officials of the Roman Catholic church agreed to allow nondestructive testing of the shroud to investigate its authenticity. Of special interest to the team of scientists who conducted the tests were the blood spots on the cloth and the detailed image on the shroud of the front and back of a human body.

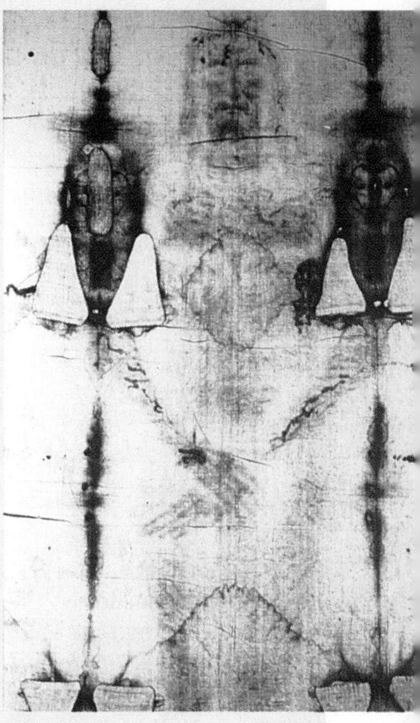

The Shroud of Turin

Doctors who studied the blood spots determined that they were located precisely where they should be for a person executed by crucifixion. In addition, there were blood spots around the head area as would be caused by many small wounds, and other blood spots on the back and shoulders that could have resulted from a beating with a scourge (a flesh-ripping whip). Extensive chemical testing showed these spots to be human blood. Further study showed that the blood had been deposited on the shroud before the image-forming process had taken place. If the shroud was simply the work of an artist, he had painted the blood spots first before applying the image around them. But how the negative image was produced presented an even greater challenge.

A dehydration–oxidation process seemed to have produced the image, but scientists were at a loss to explain how either a medieval artist or a natural process could have caused this to occur. When it was suggested that low-energy X rays could have caused the dehydration, chemist Alan Adler pointed out that "anyone who was *that* radioactive would have been dead long before he was crucified." To this day, no one has been able to explain satisfactorily how the image on the shroud was produced.

The one thing scientists believe they have determined is the approximate age of the shroud. In 1988, using advanced techniques of carbon-14 dating, three separate analyses of the flax from which the linen was made were carried out. These tests showed the linen to be about 660 years old. In effect, there was no chance that the shroud came from the time of Jesus. But the mystery of the imaging process still remains. Somehow, without the apparent use of paint, an unquestionably accurate, negative image of a crucified man was transferred to the shroud.

Your Perspective: For years, the Roman Catholic Church refused requests by scientists to test the Shroud of Turin for authenticity. Write a short essay explaining why the church might have resisted these requests, and why it might have changed its mind.

7.3 Nuclear Fission

Energy produced from atomic nuclei has an enormous influence on our modern world. It has peaceful uses, such as the generation of electricity in nuclear power reactors, and destructive uses, such as the explosive power in nuclear weapons. In all cases, this energy source has imposed new responsibilities on the citizens of our planet.

Describe the process of nuclear fission.

When bombarded with neutrons, the nuclei of several isotopes (^{235}U, ^{238}U, and ^{239}Pu) can break apart or undergo **fission** to form smaller, more stable nuclei. For example,

$$^{1}_{0}n + ^{235}_{92}U \longrightarrow ^{236}_{92}U \longrightarrow ^{140}_{56}Ba + ^{93}_{36}Kr + 3^{1}_{0}n + \text{Energy}$$

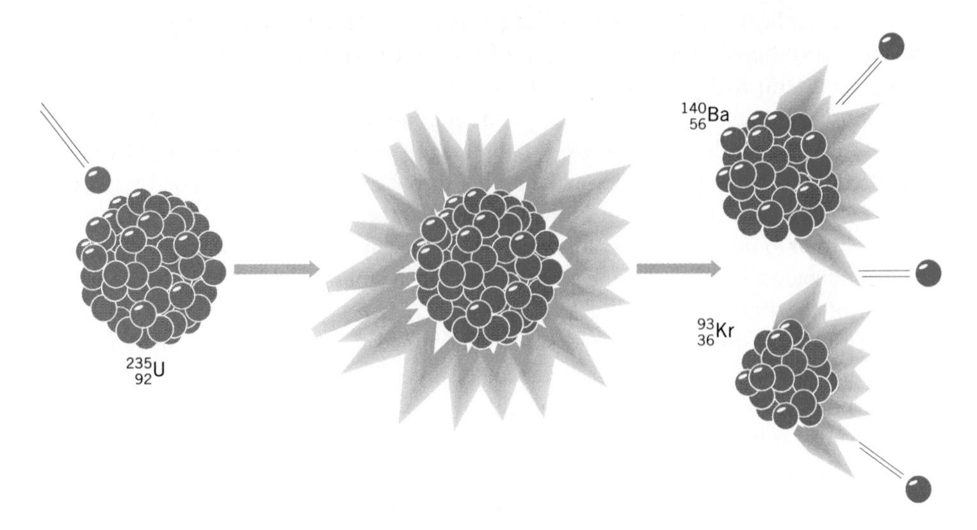

Describe how nuclear fission can produce tremendous amounts of energy.

This fission process yields great amounts of energy. The energy released by the complete fission of just 1 g of uranium-235 is enough to supply the energy needs of almost seven households for a full year. A 1000-megawatt (MW) nuclear power plant consumes about 150 g of uranium-235 an hour. Equivalent oil-burning and coal-burning power plants consume about 1900 barrels (79,800 gal) of oil an hour or 430 tons of coal an hour.

In nuclear fission, an atom of fuel (such as ^{235}U) is struck by a neutron and breaks into two small fragments that fly apart at high speeds. The kinetic energy of these fast-moving particles is converted to heat as they collide with surrounding molecules. Besides these two fragments, two or three neutrons are released, which can strike and react with other uranium-235 nuclei. If this process continues, a **chain reaction** occurs (Fig. 7.4). When the rate of fission in a chain reaction is uncontrolled and extremely rapid, all the energy is

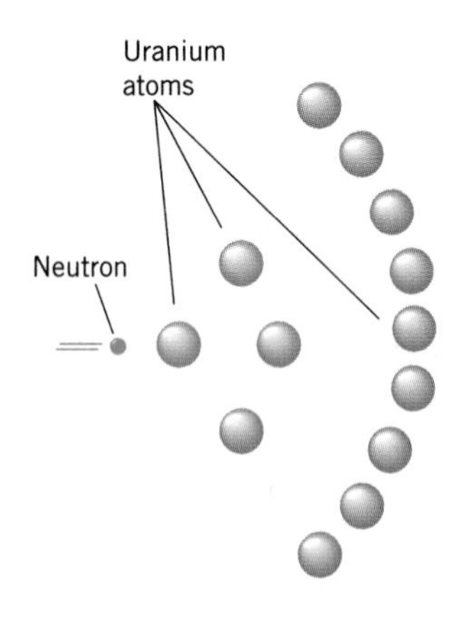

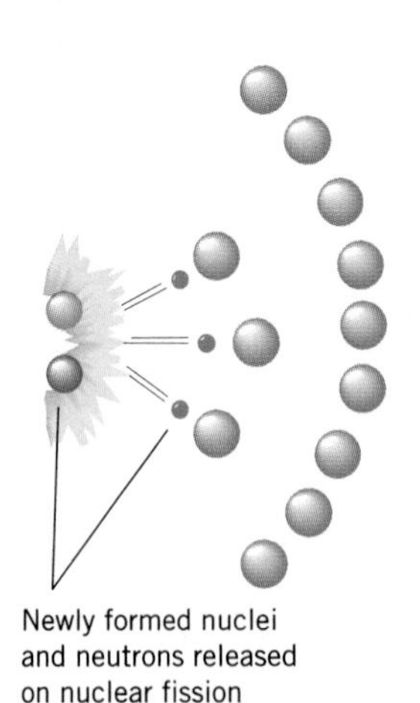

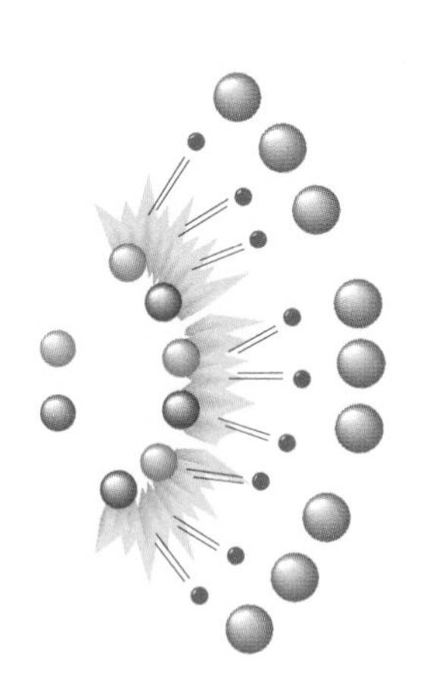

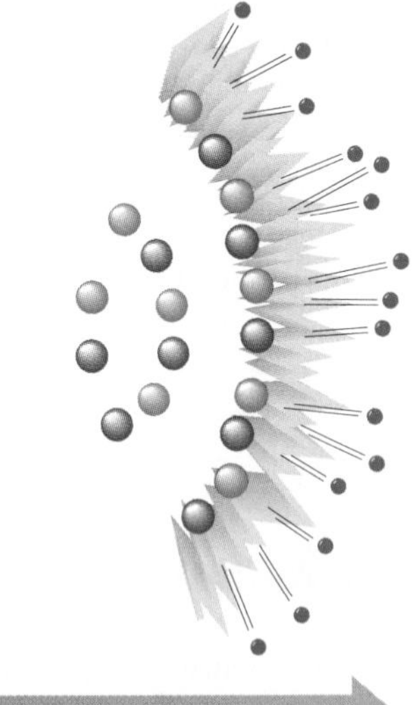

Figure 7.4 Nuclear chain reaction. When a nucleus undergoes fission, two or three neutrons are produced that are capable of reacting with other nuclei and causing a chain reaction.

Historical Perspective / Lise Meitner

Lise Meitner (1878–1968)

Lise Meitner was a brilliant Jewish physicist whose work in the 1920s brought worldwide recognition to the Kaiser Wilhelm Institute in Berlin. In 1934, intrigued by the neutron irradiation experiments carried out by Enrico Fermi, she established a research team at the institute to investigate neutron bombardment of uranium. Joined by the chemists Otto Hahn and Fritz Strassman, she spent the next four years leading the investigation into the synthesis and identification of several transuranium elements. Hahn and Strassman painstakingly separated one element after another, recording their chemical and radioactive properties, while Meitner tried to identify and interpret the nuclear processes involved.

Unfortunately, the political environment in Germany was steadily growing worse. The Nazis had come to power in 1933, and by 1938 the persecution of Jews had reached the point where Meitner was forced to flee the country. Arriving in Stockholm in September 1938, she took a position at the Nobel Institute for Physics. From there she kept in contact with Hahn by mail.

It was during this time that Hahn wrote to Meitner about some puzzling experimental results. It appeared that one product of their neutron irradiation of uranium was not an isotope of radium—a normal decay product of uranium—but rather an isotope of barium, a much smaller element. Hahn was unable to explain the presence of this element, and he urged Meitner to come up with some explanation. Could they conceivably have split the uranium atom? Lise wrote back, encouraging Hahn to publish his results and to suggest in the article that they had actually split the uranium nucleus. She then joined with her nephew, physicist Otto Frisch, to develop the first theory explaining nuclear fission. In calculating the energy associated with the fission of uranium, they found it to be several times greater than any previously known nuclear reaction. Meitner and Frisch also saw in this splitting something of even greater importance: the release of more neutrons. This meant that the neutrons so produced could split other uranium atoms, yielding phenomenal amounts of energy in a chain of reactions.

The same racial policies that had driven Meitner out of Germany made it dangerous for Hahn to acknowledge his continued cooperation with the Jewish physicist. (The fact that Hahn was regarded as anti-Nazi had already put his professional survival in question.) To secure his position and enhance his prestige, he cut his politically dangerous ties to Meitner. He not only ignored Meitner's role in the discovery, he also denied the value of nearly everything she had done during the four years they had worked together. In 1946, Otto Hahn alone was awarded the Nobel Prize in chemistry for his research in nuclear fission.

Your Perspective: In a letter to the Nobel Prize committee in Stockholm, urge them to award a Nobel Prize to Lise Meitner. Explain why this request should be granted after so many years have passed.

released in a very short time—resulting in the explosion of an atomic bomb. But when the rate of fission in the chain reaction is held to a slower, constant rate, the result is the controlled energy release of a nuclear reactor. A chain reaction cannot occur, however, unless at least a minimum number of fuel nuclei (called the **critical mass**) are present in one place.

7.4 Nuclear Wastes

Long-lived radioactive wastes are produced in the fission process, and they must be safely stored for long periods of time—at least 20 half-lives. There are four categories of radioactive wastes: mill tailings, low-level waste, transuranic waste, and high-level waste. Mill tailings are made up of naturally occurring radioactive rock that comes from mining operations. They contain small amounts of radium, which decays to radon. Mill tailings are disposed of in isolated places and covered with enough soil to protect the public and the environment from radon. Low-level waste contains small amounts of radioactivity distributed within large volumes of material. It is produced in many commercial, medical, and industrial processes. Low-level waste is buried in special containers in government-approved landfills. Transuranic wastes are produced when nuclear fuels are reprocessed. These wastes have long half-lives and require the same storage as high-level wastes.

High-level wastes are made up of spent fuel from nuclear power plants and waste from nuclear-related defense activities. This waste contains elements that decay slowly and may remain radioactive for thousands of years. In the United States in 1990, 110 nuclear power plants operated by utilities in 33 states generated approximately 20% of the nation's electricity. The nuclear fuel rods used in those plants operate efficiently for about 3 years before they are removed. The removed spent fuel is currently stored in deep, steel-lined pools within the reactor site. By the year 2000, the amount of this spent fuel stored at nuclear facilities is expected to total approximately 40,000 metric tons. The U.S. Department of Energy is responsible for establishing a permanent storage site for such high-level wastes in a deep underground repository. The site of this repository is a critical issue that is yet to be settled.

7.5 Nuclear Fusion

Enormous amounts of energy are given off when small nuclei (such as those of hydrogen, helium, or lithium) combine to form heavier, more stable nuclei. This process is called **nuclear fusion.** Nuclear fusion reactions are the source of the energy released by the sun and by the hydrogen bomb. The fusion reaction occurring on the sun involves the combining of four hydrogen atoms; it releases four times as much energy per gram of fuel as does a fission reaction. The following fusion reaction is the one that is most likely to be used here on earth because it occurs at the lowest temperature and because deuterium can be obtained from the sea at a very low cost.

$$\underset{\text{Deuterium}}{{}^{2}_{1}\text{H}} + \underset{\text{Tritium}}{{}^{3}_{1}\text{H}} \longrightarrow {}^{4}_{2}\text{He} + {}^{1}_{0}n + \text{Energy}$$

Deuterium + Tritium → Helium + Neutron + Energy

Compare and contrast nuclear fission and nuclear fusion.

The use of nuclear fusion is proving to be much more difficult than the development of nuclear fission because temperatures of 100 million degrees Celsius or higher are required for the fusion reaction to occur. Scientists have been able to create these conditions on a small scale using powerful lasers.

If fusion reactors can be developed at a cost comparable to fission reactors, the fusion process would have great advantages over fission as a source of the world's energy supply. It would be more efficient; the fuel would be cheap and almost inexhaustible; there is no possibility of runaway accidents from a reactor core melting (as could occur in fission reactions); and there is very little radioactive waste. Tritium, which is one of the least toxic radioactive substances, may be produced, but it can be returned to the system as a fuel.

7.6 Ionizing Radiation

Although there is controversy over the use of nuclear reactions for power generation, there are no doubts about the beneficial use of these reactions in diagnosing and treating disease. The use of nuclear reactions in the treatment of disease is based on the interaction of radiation with living tissue. Alpha, beta, and gamma radiation each have different penetrating power and interactions with living tissue. Alpha particles are the largest particles emitted by radioactive substances and have very little penetrating power. Even when traveling through air, alpha particles lose energy very quickly through collisions with air molecules and stop within a few inches. They can be stopped by a piece of paper and cannot penetrate even the layer of dead cells on the surface of your skin. Alpha particles can do a great deal of damage, however, if they are emitted inside the body, which might result from inhaling or swallowing an alpha emitter. Although they don't travel very far, their mass and electric charge can damage the atoms and molecules with which they collide. If the damaged molecule is essential to the life of the cell, the cell can die or become cancerous.

Beta particles are 7000 times lighter than alpha particles and have much more penetrating power. Beta particles can pass through a piece of paper but are stopped by a piece of wood. Beta radiation can penetrate the dead outer layer of skin. When it stops within the skin layer, it can damage the skin tissue in the same manner as a burn.

Because they have such high energy, gamma rays easily pass through paper and wood but can be reduced to background levels by lead blocks or thick concrete walls. Gamma rays completely penetrate the human body (Fig. 7.5). Gamma radiation does less damage than alpha or beta radiation over the same distances.

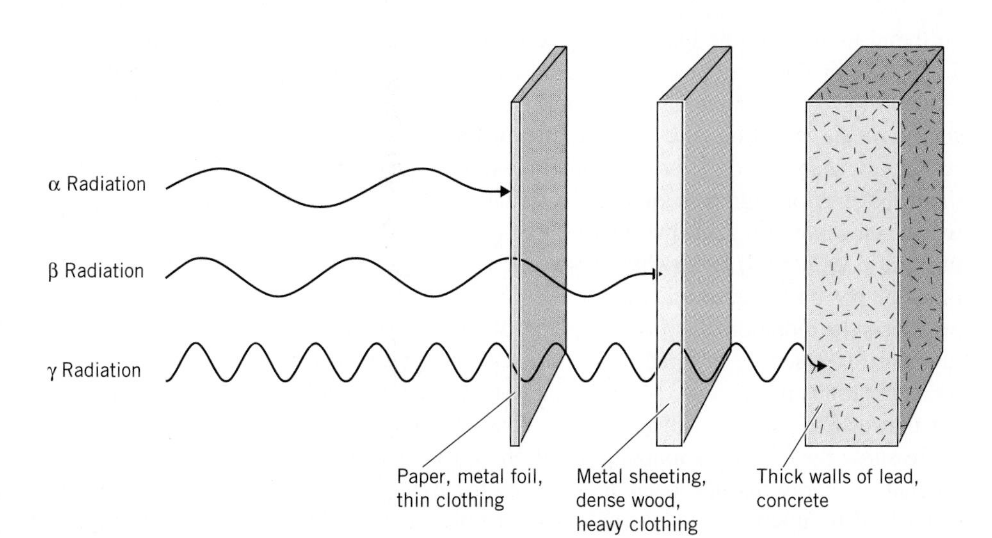

Figure 7.5 Each type of radiation (alpha, beta, or gamma) has a different penetrating power.

Define ionizing radiation.

Describe two ways in which ionizing radiation can affect living tissue.

How do these types of radiation affect living tissues? When they hit living tissue, alpha, beta, and gamma radiation can each produce unstable and highly reactive particles called ion pairs. For this reason, these radiations are called **ionizing radiation.** In addition, such radiation can transfer so much energy to the molecules in living tissue that the molecules vibrate apart, forming high-energy uncharged fragments called **free radicals.** Free radicals are even more reactive than ions; they can pull other molecules apart and completely disrupt cells. The most vital molecule in a living cell is the deoxyribonucleic acid (DNA) molecule, which carries all the information necessary for the cell to divide and reproduce. If DNA is destroyed by interaction with ionizing radiation, then the cell cannot divide and dies. When such cells die without replacement, the entire irradiated tissue eventually dies. Even if the DNA molecule is only damaged and not destroyed completely, it may cause the cell to divide abnormally into new cells with altered DNA. Such cells are called **mutant cells.** A mutant cell may have its DNA so altered that it is no longer under the body's control. The cell may then begin to grow and divide in an uncontrolled fashion, destroying the normal cells around it. Cells that behave in this manner are called **cancerous,** or **malignant.** It is important to understand that in spite of repeated exposure to ionizing radiation (naturally occurring or artificially made), the vast majority of damaged cells do not become malignant. This is because living cells are equipped to repair their DNA as soon as it becomes damaged.

Ions or free radicals produced in tissues by ionizing radiation may recombine (resulting in little damage), or they may combine with other molecules to form new substances foreign to the cell. The new substances produced in this way are often highly reactive chemically but are not themselves radioactive. One such substance produced when ionizing radiation interacts with water in the body is hydrogen peroxide (H_2O_2), which is harmful to living tissue. Luckily, within cells and throughout the body are antioxidant molecules that work to neutralize quickly any free radicals or peroxides that are formed. Such antioxidants as vitamins C and E help protect the cell from the adverse effects of exposure to ionizing radiation.

7.7 Radiation Dosage

Ionizing radiation is measured in terms of many different scientific units. The different units help answer specific questions about the nature of the radioactive source, its effects on living tissue, and the long-term biological risk. Some of these units are described in Table 7.3. In this textbook we concentrate on just two units: the curie and the rem.

Define curie and rem.

The first aspect of ionizing radiation we want to describe is the activity level of the source of the radiation: the number of transformations that occur per unit of time. A commonly used unit of activity is the **curie (Ci),** named after Marie Curie (1867–1934), the discoverer of radium. One curie is equal to 3.7×10^{10} transformations per second, which is the rate of transformation of 1 g of radium-236. It is important to understand that this unit does not tell us anything about the type of radiation produced or the effect this radiation has on tissue or other substances. It describes only the rate at which radiation is produced. In radiation therapy, hospitals often use cobalt-60 as the source of gamma radiation. If a source is rated at 140 Ci, it disintegrates at a rate of $140 \times 3.7 \times 10^{10}$ transformations per second. A 2-Ci cobalt-60 source delivers twice as many gamma rays per unit of time to a patient as a 1-Ci cobalt-60 source.

To study the effects of ionizing radiation on living tissue, we must try to measure the amount of energy that has been absorbed by the tissue. The dose absorbed by a tissue is influenced by many factors, including the nature of the radioactive source, the type of ionizing radiation, the energy of the radiation, the distance from the source to the tissue,

Table 7.3 **Units Used to Measure Ionizing Radiation**

Measurements	*Unit*	*Definition*
Rate of transformation	curie (Ci)	1 Ci = 3.7×10^{10} transformations per second
	becquerel (Bq)[a]	1 Bq = 1 transformation per second
Absorbed dose of radiation	rad	1 rad = absorbed radiation that liberates 100 ergs per gram of tissue
	gray (Gy)[a]	1 Gy = 100 rad
Dose equivalent	rem	1 rem = absorbed dose that produces same biological effect as 1 rad of X rays
	sievert (Sv)[a]	1 sievert = 100 rem

[a]SI unit.

the nature of the tissue itself, and the duration of exposure. Scientists have discovered that if living tissue is hit by different types of ionizing radiation, even though the energy absorbed by each type might be exactly the same, the effects on the biological system can be different. For example, it takes twice the dose of cobalt-60 gamma radiation as it does a specific energy of X rays to produce the same disruption in a certain type of pollen grain. To study the cumulative effects of different types of radiation on humans, a unit called the **rem** (**r**oentgen **e**quivalent for **m**an) was therefore devised. When a dose is stated in rems, there is no need to specify the type of radiation because the biological effects are the same for one rem of each type. For example, 450 rem of any type of radiation over the whole body is sufficient to kill 50% of all the irradiated people (Table 7.4). The government has set radiation exposure limits for workers. Workers currently have a dose limit of 5000 mrem/year (1 rem = 1000 mrem). Studies of the long-term risk from low levels of radiation exposure are also based on measurements of exposures in millirems. In your reading you might also see radiation exposure limits expressed in sieverts. One sievert (Sv) is equal to 100 rems.

7.8 Background Radiation

Radiation is all around us; there is no way of totally escaping exposure to ionizing radiation. In fact, the naturally occurring radionuclides within the average human body undergo several hundred thousand radioactive transformations per minute. Fifty of the 350 natu-

Table 7.4 **Radiation Doses Required for Biological Effects**

Effect	*Dose*[a] *(mrem)*
No measurable effect	25,000–50,000
Blood changes, no illness	100,000
Radiation sickness, no deaths	200,000
Death to 50% of irradiated people	450,000
Death to 100% of irradiated people	800,000–1,000,000

[a]The values apply only to whole-body radiation exposure over a short period of time.

Table 7.5 **Average Annual Radiation Exposure per Person in the United States**[a]

Radiation Source		*Dose (mrem/year)*
Environmental		
Natural		
Cosmic rays	27	
Terrestrial		
External	28	295
Internal	40	
Inhaled (radon)	200	
Global fallout		1
Nuclear fuel cycle		0.05
Medical		
Diagnostic X rays	39	53
Nuclear medicine	14	
Occupational		1
Miscellaneous Consumer Products		13
	Total	363

[a]Data from *Ionizing Radiation Exposure of the Population of the United States.* National Council on Radiation Protection and Measurements Report No. 93, 1987, Bethesda, MD.

rally occurring isotopes are radioactive, and small amounts of radioactive material are found in the soil we walk on, the food we eat, the water we drink, and the air we breathe.

List three sources of background radiation.

On average, we each receive a very small dose (363 millirem or about 0.4 rem) of background radiation per year. Eighty-one percent of this radiation is from outer space and natural materials in the soil, water, and air; 15% comes from medical radiation, such as medical and dental X rays; 0.3% is from radioactive fallout and pollution from nuclear power plants; and 4% is from occupational and other miscellaneous sources (Table 7.5). We cannot control the natural background radiation, but we can control radiation exposure from other sources. Very few people would argue that we should stop all uses of radioactive materials, for these materials play a large role in improving living conditions around the world. But when considering decisions that might increase the level of background radiation, it becomes important to weigh the benefits to humanity against the risk. Such decisions are especially difficult because so little is known about the long-term effects on health of very low levels of radiation.

7.9 Protection Against Radiation Exposure

Earlier in this chapter we saw that various kinds of radiation differ in their penetrating power. Alpha and beta radiation have low penetrating power and are easily stopped. X rays and gamma rays have high penetrating power, but their penetrations are greatly reduced by lead or thick concrete. Because lead is an effective shield, doctors and technicians use lead aprons to protect parts of the patient's body from radiation during medical or dental X rays.

Besides the use of shielding, another important safety factor is the distance from the source of the radiation. The farther you are from the source, the lower the intensity of the

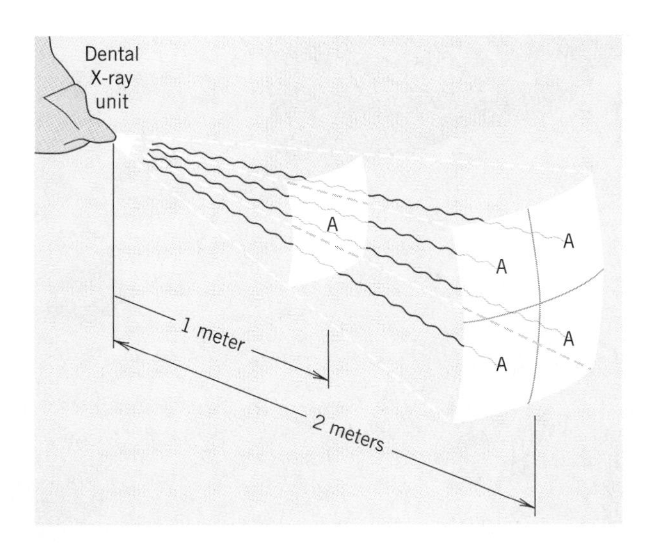

Figure 7.6 Illustration of the inverse square law. The same number of radiations hitting area *A* at 1 m from the radioactive source hit an area four times larger at 2 m from the source. The intensity of the radiation at 2 m is therefore one-fourth the intensity at 1 m.

exposure. Because the rays spread out in a cone shape as they move away from the source, the number of radiations striking any object decreases as the object moves away (Fig. 7.6). In more mathematical language, the intensity of radiation over a given surface area decreases with the square of the distance from the source. Scientists call such a relationship an **inverse square law.** X-ray technicians minimize their exposure both by being as far away from the radiation source as possible and by standing behind a shielded wall when taking X rays.

Even with good shielding, the most sensible safety precaution is to minimize your exposure to ionizing radiation. This is especially important because some harmful effects of ionizing radiation are known to build up over time. Studies on survivors of the atomic bombs dropped on Hiroshima and Nagasaki (as well as other cases) have shown that a one-time exposure to high doses of ionizing radiation increases the risk of cancer in humans. The effects of long-term, low-level exposure are not totally known or understood, however. Living cells have an amazing ability to repair much of the damage from ionizing radiation: Enzymes can repair DNA, and chemicals, such as vitamins C, E, and A, help prevent the damaging effects of free radicals.

We have seen that it is important to monitor carefully the exposure of researchers, X-ray technicians, nurses, and others who work with ionizing radiation. To make this as easy as possible, small portable devices have been designed to measure radiation dose accurately. Two commonly used devices are **pocket dosimeters** and **film badges** worn by individuals. The dosimeter has a built-in scale showing the radiation exposure of the person wearing it. The film in the badge must be developed to obtain a reading. The degree of darkening in the resulting negative indicates the total amount of radiation to which the wearer has been exposed (Fig. 7.7).

7.10 Medical Diagnosis

Describe two clinical procedures that use ionizing radiation to aid in diagnosis.

Major advances in medical diagnosis have been made possible by the ability to trace the paths taken by specific chemical compounds in living systems. **Radioactive tracers** are chemicals that contain radioactive atoms and that have the same chemical nature and behavior as the compounds they trace. A living system treats the tracer just as it would the

Figure 7.7 Film badges are analyzed after specific lengths of time. The exposure to radiation for each individual is carefully logged to avoid overexposure.

normal compound, but the radioactive atoms make it possible to follow the path of the compound through the system.

In 1923, Georg von Hevesy was the first to use a radioactive tracer in his study of the uptake of lead-212 in plants. Artificially produced radionuclides were first used in 1936, when Joseph G. Hamilton and Robert S. Stone of the University of California at Berkeley used radioactive sodium produced in a cyclotron to study the uptake and excretion of sodium. Since then, the use of natural and artificially produced radionuclides has greatly increased.

In early tracer diagnostic procedures, the movement of the radionuclides was followed with a Geiger counter. The 1950s saw the development of instruments called scanners. These devices have a sensitive measuring head that moves slowly back and forth across the patient's body to detect radioactivity, producing a picture that is built up line by line. In the 1970s, computer programs were developed to analyze the scanning data to produce much clearer and sharper pictures. The gamma camera now in use does not have to move across the patient; it records all the radiation coming from a specified region of the patient, producing images in several seconds to a few minutes.

In 1979, Allan Cormack and Godfrey Hounsfield were awarded the Nobel Prize in medicine for developing a sophisticated X-ray machine called a computerized axial tomography (or CAT) scanner—now called a CT scanner—which has revolutionized medical diagnosis. Ordinary X rays produce two-dimensional images whose details are often obscured by overlapping tissues and in which certain types of tissue do not sufficiently stand out. For some diagnostic procedures, this requires that a series of X rays be taken from various angles. For a CT scan, the patient lies in a doughnut-shaped detector while the X-ray tube sweeps thousands of beams through the patient's body. (The patient is exposed, however, to no more radiation than with a standard series of X rays.) The rays strike the scanner's sensitive detectors, producing signals that are analyzed by a computer (Fig. 7.8). With the computer manipulating the data, cross-sectional pictures of the patient's body can be made from any angle. These images have different shades or colors to represent specific types of tissue or other body materials ranging in density from fluid to bone. Using a CT scan, doctors can identify and locate to within a fraction of a millimeter cancerous tumors, blood clots, herniated discs in the spinal column, and obstructions in

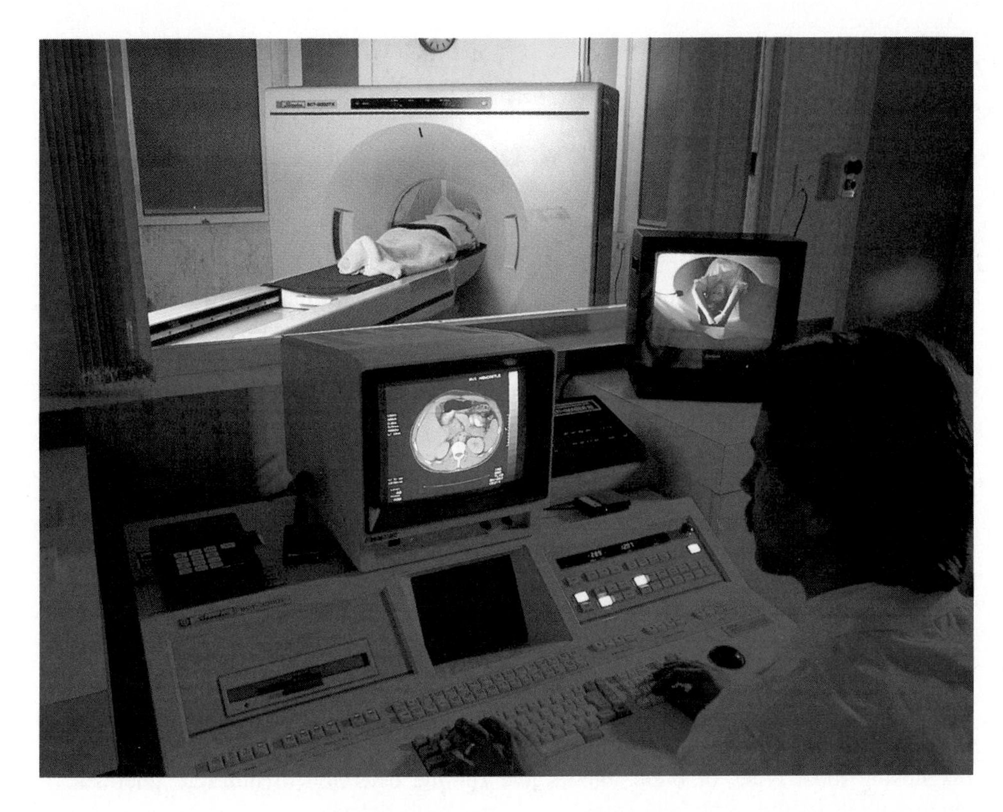

Figure 7.8 A CT scanner can complete a scan of the head or body in 5 seconds, producing a cross-sectional image showing all types of tissues.

the bile ducts (Fig. 7.9a). The CT scanner is eliminating the need for many invasive diagnostic tests and most exploratory surgery. Combining CT scans with virtual reality technology is allowing physicians to diagnose colon cancer, the second leading cause of cancer deaths, without invasive tubal examinations (Fig. 7.9b).

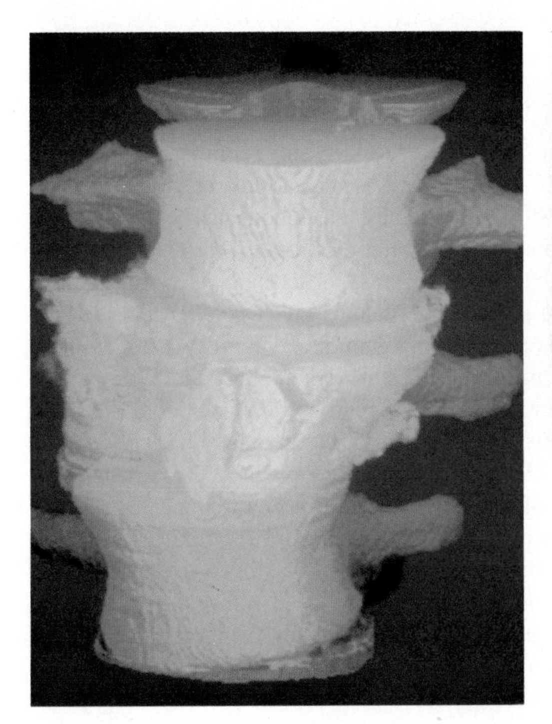

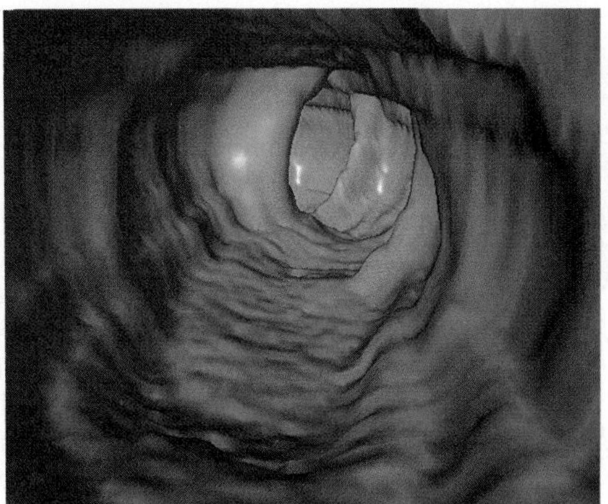

Figure 7.9 (a) A three-dimensional image of a damaged section of vertebrae produced on a computer from multiple CT scans. (b) A dynamic three-dimensional view of a normal rectum produced from multiple CT scans using virtual reality technology.

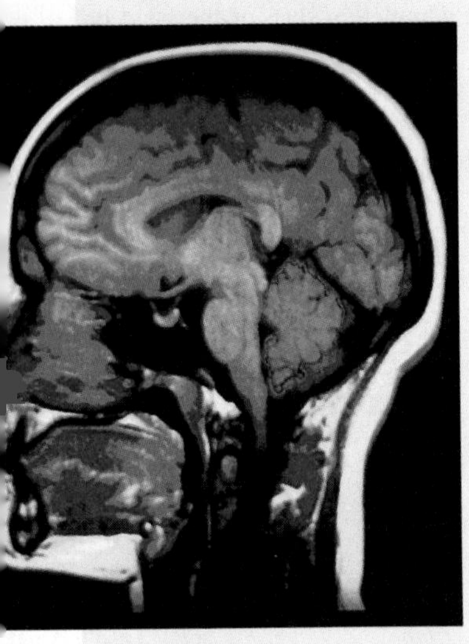

A MRI scan

Although body scanning techniques like CT and PET are very useful for medical diagnosis, they expose the patient to ionizing radiation. A scanning technique developed during the 1980s gets around this problem by applying nuclear magnetic resonance (NMR) to body scanning. This technique, called magnetic resonance imaging (MRI), uses the fact that the nuclei of atoms having an odd number of protons and neutrons have particular magnetic properties. These nuclei act like tiny magnets, spinning about their axes. In living systems, the nuclei of hydrogen atoms are the most abundant of these ''nuclear magnets'' because hydrogen is a component of water and therefore is present in most biological molecules.

When a patient is placed in a strong magnetic field and radio waves are passed through the body, these nuclear magnets ''flip'' their spins. Such a flipping of spins, called resonance, can be detected as low-energy electromagnetic radiation. This kind of radiation is biologically harmless. A computer analyzes the changes in this electromagnetic radiation and transforms them into an image much like those formed in CT and PET scans.

Magnetic resonance imaging is especially useful in the study of soft tissues, an area where X-ray imaging is not very effective. Soft tissues, even tumors, differ in the amounts of water or fat molecules they contain, and these differences show up clearly in MRI. Magnetic resonance imaging has rapidly become the preferred method for diagnosing most types of brain tumors and for the detection of multiple sclerosis, abnormalities of the joints and spinal cord, and birth defects.

Your Perspective: Magnetic resonance imaging, although an excellent diagnostic aid, is a very expensive one. This may prevent many patients without medical health coverage from affording this service. Write a short essay discussing whether this system is fair, and what you would suggest to improve it.

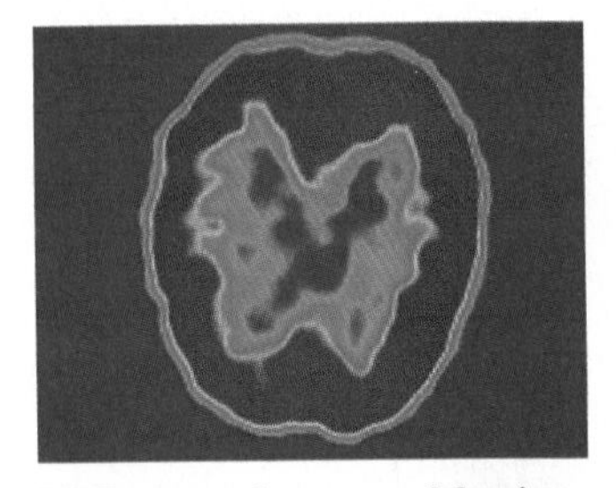

PET scan of a normal brain.

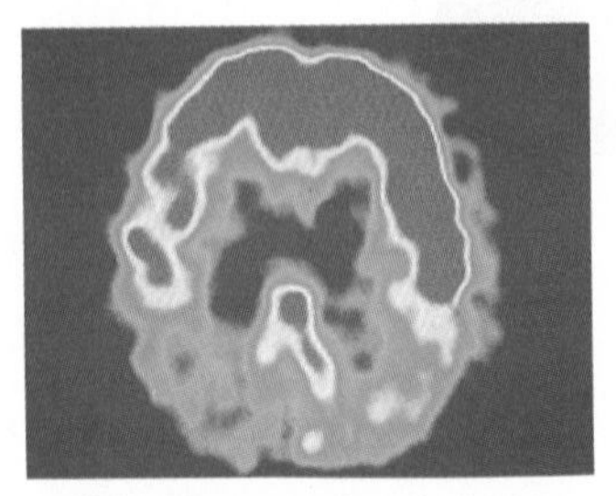

PET scan of the brain of an Alzheimer's patient.

One shortcoming of CT scans is inability to detect the early stages of disease in tissues. By the time a disease shows up on a CT scan, it may be too far advanced for effective treatment. To identify tissue breakdown in its early stages, a technique has been developed that combines computerized image reconstruction with radioactive tracer diagnosis. This technique uses synthetic radionuclides that emit positrons, particles that are like electrons but carry a positive charge. Positrons are converted to gamma rays when they collide with electrons within the body. Although it is the resulting gamma radiation that is measured, this diagnostic technique is still called positron emission tomography, or PET.

The first step in a PET scan involves synthesizing a specialized radioactive tracer. This is done by incorporating a positron-emitting radionuclide into a compound that concentrates in the tissue under investigation. This compound is then given to the patient either orally or by injection. After a period long enough for the tracer to concentrate in the desired tissue, a scanner detects the gamma radiation emitted from the tissue in a manner similar to a CT scan. The difference is that in a CT scan X rays are passed through the body, but in a PET scan the radiation being measured is produced within the tissue being analyzed. The PET scan has been particularly useful in detecting biochemical problems within the brain. Doctors use PET scans to analyze the brain function of patients with Alzheimer's disease, stroke, Parkinson's disease, and mental disorders like schizophrenia.

7.11 Radionuclides Used in Diagnosis

Over the past 20 years many different radionuclides have been used to produce images of various body organs. Technetium-99^m, however, because of its more suitable chemical and physical properties, is now replacing many of these other radionuclides in common diagnostic procedures. Technetium is one of four elements lighter than uranium that do not occur naturally and, therefore, must be artificially produced. It is obtained from the decay of molybdenum-99.

$$^{99}_{42}\text{Mo} \xrightarrow[t_{1/2}\ 67\ \text{h}]{} {}^{99m}_{43}\text{Tc} + {}^{0}_{-1}e + \gamma$$

The m after the mass number means that the nucleus of the isotope is **metastable,** or in an energy state higher than normal. When it decays, technetium-99^m releases this excess energy as gamma rays and forms the lower energy isotope technetium-99.

$$^{99m}_{43}\text{Tc} \xrightarrow[t_{1/2}\ 6.02\ \text{h}]{} {}^{99}_{43}\text{Tc} + \gamma$$

The advantages of technetium-99^m over other isotopes are (1) the energy of the gamma rays can easily be detected by the cameras now in use, (2) no alpha or beta radiations are given off (which would increase the absorbed dose to the patient), and (3) the 6-hour half-life allows enough time for the isotope to localize in the body after injection, and yet allows the isotope to be administered in amounts that limit the radiation exposure to that of a comparable X-ray procedure.

Various technetium compounds are used in diagnostic procedures. Technetium diphosphonate is used to detect whether cancer cells have spread, or **metastasized,** to the bones. Bone cells absorb the diphosphonate and incorporate it into the structure of the bone in those places where cancerous cells are destroying bone tissue. In this case, the highest radioactivity is found in areas of the metastasized cells (Fig. 7.10).

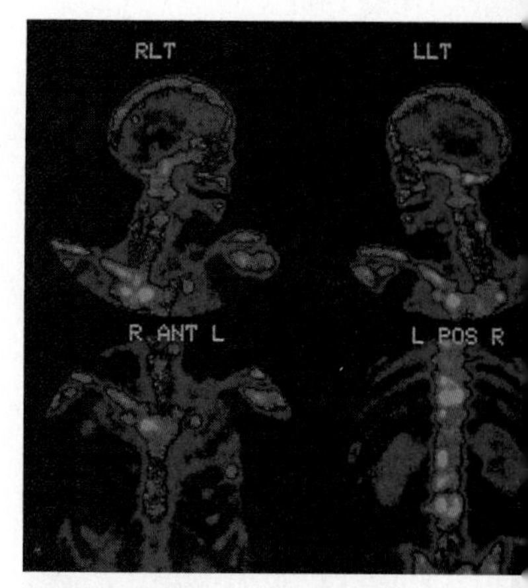

Figure 7.10 Bright spots in the backbone and ribs indicate the active uptake of technetium-99^m by metastasized cancer cells.

Technetium-99^m, in several forms, is also allowing doctors to determine the location and extent of both old and new damage to the heart. Such procedures can help a doctor decide what therapy the patient should or should not receive. Radioactive technetium (as with calcium) deposits in the mitochondria of heart cells that are irreversibly damaged. Thallium-201, which is produced in a cyclotron, is being used with technetium to diagnose heart damage. Thallium, unlike technetium, acts like potassium and concentrates in normal heart muscle. It can be used, for example, to determine whether a patient who enters the hospital with chest pains, but with few other symptoms of a heart attack, has a deficiency in the blood supply to the heart that might result in a future heart attack. Both radionuclides can be administered intravenously, and their use is safer, more comfortable, and less expensive than the angiographic procedure of inserting a catheter through an artery into the heart.

7.12 Radiation Therapy

Describe three different techniques for using radioactive materials in medical therapy.

Cells are especially sensitive to ionizing radiation when they are dividing. Because cancerous cells are often dividing much more rapidly than the surrounding tissue, they are much more vulnerable to damage from ionizing radiation. This is the key fact that allows doctors to use radiation in killing cancerous tissue while leaving normal tissue relatively unharmed.

Chemical Perspective / Radiochemical Cows

Although the short half-life of technetium-99^m makes it an excellent radionuclide for diagnosis, transporting and storing such a short-lived material is difficult. To get around this problem, an innovation called a "radiochemical cow" was developed. Like cows that produce milk that is removed by milking, radioisotopes in the radiochemical cows decay to produce other isotopes that can be removed by "milking." The radiochemical cow contains a parent isotope whose daughter isotope is the desired element. As decay occurs, the desired daughter isotope replaces the parent. By using a chemical solution that removes the daughter but leaves the parent, the desired radionuclide is "milked" from the radiochemical cow.

For technetium-99^m, the parent radioisotope is molybdenum-99. Molybdenum-99 has a half-life of 67 hours, so there is sufficient time to ship this material to the hospital or laboratory. When too little molybdenum-99 remains, the "cow" is shipped back to a reactor for refilling, somewhat in the same manner that a real cow is put out to pasture.

Your Perspective: In a technical memorandum to the board of directors of your local hospital, explain why it is difficult to transport and store the short-lived technetium-99^m isotope. Describe the kind of hospitals that would be most affected by this problem, and explain why.

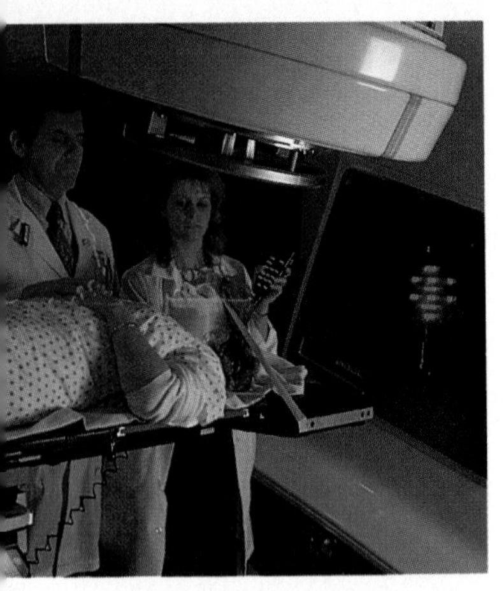

Figure 7.11 Physicians prepare a patient for radiation therapy with a linear accelerator treatment unit. Laser beams are used to precisely align the patient's head before treatment.

High-intensity radiation is used to destroy cancerous tissue that can't be removed by surgery or which, because of its sensitivity to radiation or its location, is better treated with irradiation (perhaps in combination with surgery or chemotherapy). In such therapy, radiation from an X-ray machine, a cobalt-60 source, or a particle accelerator is focused into a beam that is directed at the cancerous tissue. In the treatment of internal cancers, the patient is carefully positioned on a treatment table. The source of radiation is then rotated around the patient (or directed at the patient from different angles) so that the radiation produces minimal skin tissue damage but is constantly focused on the tumor cells (Fig. 7.11).

A second medical procedure uses a needle to insert a radionuclide (in the form of a seed) into the area to be treated. Seeds containing gold-198 or iridium-192 are sometimes implanted in the tumors of patients with advanced inoperable tumors to slow tumor growth. They are also implanted in tumors located near body surfaces, such as in the mouth, breast, prostate, and cervix, to destroy the cancer.

A third method of using ionizing radiation to kill cancerous tissue involves administering radionuclides in chemical forms designed so that they concentrate in specific areas of the body. For example, because radioiodine is concentrated in the thyroid, it can be used to destroy thyroid cells in the treatment of an overactive thyroid (hyperthyroidism). Phosphorus-32 is used to treat the disease polycythemia vera, which causes an abnormal increase in the number of red blood cells. There is no known cure for the disease, but the radiophosphorus destroys the cells that produce red blood cells. This slows down the formation of red blood cells and temporarily alleviates the symptoms of the disease.

KEY CONCEPTS

Radioactive decay occurs when the nuclei of certain isotopes are unstable and emit particles to form a new daughter nucleus. There are several types of radiation, with the three most common being alpha, beta, and gamma radiation. Radiation given off by radioactive substances is called ionizing radiation because it can produce ions and free radicals in material through which it passes.

The half-life of an isotope is the time it takes for half of a sample of radioactive material to decay. Each radionuclide has its own characteristic rate of decay.

Nuclear fission occurs when a nucleus is bombarded by neutrons and splits apart to form several smaller nuclei, more neutrons, and large amounts of energy.

Nuclear fusion results when several small nuclei are fused together to form a heavier, more stable nucleus. This reaction takes place at very high temperatures and generates large amounts of energy. It is the source of the energy released by the sun.

Ionizing radiation can be measured in several units. The curie measures the number of transformations per second. The rem measures the effects of different types of radiation on humans.

Background radiation is the amount of radiation occurring naturally in the environment. To protect yourself from the dangerous effects of ionizing radiation, you should try to minimize your exposure to such radiation by staying as far as possible from the source and using lead or other shielding when working with ionizing radiation.

Although ionizing radiation in large doses can be harmful to living tissue, it has proven extremely useful in the diagnosis and treatment of disease. A CT scan uses X rays to create very detailed images of the interior of the human body, thereby helping the physician locate tumors, blood clots, and broken bones. Radionuclides injected into the body are used to analyze heart and thyroid functions and to locate metastasized cancers. The ionizing radiation given off by radioactive elements can also be used to destroy cancerous cells and tumors.

Important Equations

$$\text{Quantity remaining after } n \text{ half-lives} = \text{initial quantity} \times \left(\frac{1}{2}\right)^n$$

REVIEW PROBLEMS

Section 7.1

1. What causes a substance to be radioactive?
2. What are several differences between the three most common radiations given off by radioactive substances?

3. Radon-222 is an alpha-particle emitter that decays to polonium-218. Write the balanced nuclear equation for this radioactive decay.
4. Bismuth-210 is a beta emitter that decays to form an isotope of polonium. Write the balanced nuclear equation for this radioactive decay.
5. When the thorium-230 isotope decays, it gives off an alpha particle and gamma radiation. Write the balanced nuclear equation for this radioactive decay.
6. When plutonium-239 decays, it emits a beta particle and gamma radiation. Write the balanced nuclear equation for this radioactive decay.
7. Write balanced equations for the following nuclear reactions:
   (a) the alpha decay of beryllium-8
   (b) the beta decay of iodine-135
   (c) the beta decay of oxygen-20
   (d) the alpha decay of berkelium-245
8. Neptunium-241 forms a stable isotope through a transformation series that involves the following successive particle emissions:

$$\alpha,\ \beta,\ \beta,\ \alpha,\ \alpha,\ \alpha,\ \alpha,\ \alpha,\ \beta,\ \beta,\ \alpha,\ \beta,\ \beta,\ \alpha$$

   Write the symbol (including mass and atomic number) for each nuclide formed in this series.

Section 7.2

9. Some scientists advise that radioactive wastes be stored for at least 20 half-lives. How long would that be for cesium-137 (a radioactive waste produced in nuclear power plants)?
10. How much of a 100-g sample of cesium-137 remains after 150 years?
11. If you had $10 to gamble in a slot machine in Las Vegas but could bet only half the first hour and half of the remainder in each succeeding hour, at the end of what hour would you have $1.25 left? (Ignore any earnings.) How is this similar to problems involving the half-life of a radioactive substance? How is it different?
12. Strontium-90 is a beta emitter. After how many years is a 5.00-g sample of this isotope reduced to 1.25 g of strontium-90?

Section 7.3

13. What are the bombarding particles in nuclear fission?
14. Why is the series of reactions that occur in nuclear fission called a ‘‘chain reaction’’?

Section 7.4

15. List four categories of nuclear waste. Explain which category poses the greatest threat to human health.
16. Radium-226 is found in mine tailings from the mining of uranium. How long will it take for a 10-g sample of this isotope to be reduced to 2.5 g?

Section 7.5

17. What is the difference between nuclear fission and nuclear fusion? Which process is presently being used in power plants?

18. Complete the following fission reaction by filling in the other fission product.

$$^{235}_{92}\text{U} + ^{1}_{0}n \longrightarrow ^{94}_{38}\text{Sr} + 2\,^{1}_{0}n + ?$$

Section 7.6

19. What is ionizing radiation? Name three types of ionizing radiation.

20. Describe how the three types of radiation in problem 19 damage living tissue.

21. Compare the penetrating power of these three types of radiation.

22. What are two ways in which the body protects itself from the harmful effects of ionizing radiation?

Section 7.7

23. A cobalt-60 source used for radiation therapy is rated at 170 Ci. How many transformations does it deliver in 15 seconds?

24. A radionuclide has an activity of 100 mCi/g. How many transformations does 5.0 g of this isotope deliver per minute?

Section 7.8

25. A patient with abdominal pain was given four X rays (exposure 10 rem each) and two radioactive barium enemas (exposure 300 mrem each). How have these tests affected the patient's annual exposure to background radiation compared with the average in the United States (Table 7.5)?

26. Should the patient in problem 25 have routine dental X-rays taken? Why or why not?

Section 7.9

27. Compare the intensity of radiation 3 m from a radioactive source to that 1 m from the same source.

28. You are setting up a dental office in a building that is being remodeled. The contractor asks you where you would like the operating controls for the X-ray machine to be located. Given the following three possibilities, what is your order of preference? Give the reasons for your order.

First possibility: On a wall 5 ft from the examining chair, with convenient access to the patient.

Second: On a wall 5 ft from the patient but separated from the patient by a concrete dividing wall.

Third: On a wall 11 ft from the patient.

Section 7.10

29. Why is the image produced by a CT scan more helpful than the image produced by ordinary X rays?

30. What shortcoming of CT scans do PET scans overcome?

Section 7.11

31. What are three advantages of the use of technetium-99^m for diagnostic purposes compared with other isotopes?

32. After giving her patient a complete examination, a doctor feels that the patient may be suffering from an abnormal thyroid. How might radioisotopes be used to confirm the diagnosis?

Section 7.12

33. Strontium-90 decays as follows:

$$^{90}_{38}\text{Sr} \longrightarrow {}^{90}_{39}\text{Y} + {}^{0}_{-1}e$$

Iridium-192 decays as follows:

$$^{192}_{77}\text{Ir} \longrightarrow {}^{192}_{78}\text{Pt} + {}^{0}_{-1}e + \gamma$$

Assume that a patient is suffering from an inoperable cancer in the nose. Which of the two above radioisotopes is better for treating this tumor? Give a reason for your choice.

34. Why can ionizing radiation be used to treat cancerous tissue while leaving normal tissue relatively unharmed?

Applied Problems

35. Iodine-131 is used to destroy thyroid tissue in the treatment of an overactive thyroid. If a hospital receives a shipment of 20.0 g of iodine-131, how much ^{131}I remains in the laboratory after 24 days?

36. In 1963, the United States, the Soviet Union, and several other countries signed a treaty that banned the above-ground testing of nuclear weapons. Strontium-90 that had been released in previous nuclear explosions still remains in the environment. Assuming that 100% of the strontium-90 released was present in 1963, what percentage remains today? (*Hint:* Use a plot like that shown in Fig. 7.3 but replace the number of half-lives on the x-axis with the actual years they represent.)

37. During an archaeological dig, scientists discovered a sandal made from the bark of a tree. If the sandal was found to be emitting only 25% as much carbon-14 radiation per gram of carbon as was the bark on newly cut trees, how old was it?

38. The question of when humans first appeared on earth is hotly debated. Some anthropologists believe that this may have occurred in Africa over 3 million years ago. They base this belief on bones that have been discovered on that continent. If there had been wood or charcoal found along with these bones, could carbon-14 dating be used to determine their age? Explain why.

39. During the neutron fissioning of 1 lb of uranium-235, the amount of mass converted to energy is approximately 0.37 g. If the conversion of 1 g of mass yields 9×10^{10} kJ of energy, what is the energy obtained from the fissioning of 30 tons of uranium-235 over the period of 1 year?

40. A patient is to receive 10.0 mCi of a radioactive isotope intravenously. On hand at the hospital is a vial containing 9.0 mL of a liquid form of this isotope with an activity of 0.90 Ci. How much of this liquid (in milliliters) should be used for the injection?

41. Many watch-dial painters such as those described in the opening to Chapter 4 suffered from bone cancer and damage to the bone marrow. Why was the damage concentrated in the bones? Was the damage to the tissues done mostly through direct or indirect action? Explain your answer.

42. A diagnostic test to measure the absorption of vitamin B_{12} uses the fact that each molecule of vitamin B_{12} contains an atom of cobalt. Molecules of vitamin B_{12} can be labeled with radioactive cobalt and the absorption of the vitamin measured. Two radioisotopes of cobalt are available: ^{57}Co with a half-life of 270 days and ^{60}Co with a half-life of 5.3 years. Which of the two isotopes would you use to label vitamin B_{12}? Explain your answer.

43. Boron-10 neutron capture therapy is one of the most imaginative methods of radiation therapy used in the treatment of cancer. First, the patient is given a dose of a boron-containing compound that is taken up by the tumor. Then the patient is irradiated externally with neutrons. The boron absorbs the neutrons, producing alpha particles and a nontoxic lithium nucleus.

$$^{10}_{5}B + ^{1}_{0}n \longrightarrow ^{7}_{3}Li + ^{4}_{2}He$$

Why do you think this might be an especially good way to treat cancer?

Chapter 8

Reaction Rates and Chemical Equilibrium

The day was cold and clear. Four-year-old Jimmy Tontlewicz had been spending an enjoyable winter afternoon with his father sledding on his favorite hill beside Lake Michigan. His father chuckled as he noticed Jimmy fall off his sled, but his chuckle turned to cries of horror when he saw Jimmy running after the sled as it continued out onto the frozen lake. Jimmy heard his father's screams just as he reached the sled, but by that time the ice was already breaking beneath him. It was only a second before he disappeared into the icy waters of the lake.

His father jumped into the water but was unable to find Jimmy under the ice. It wasn't until 20 minutes later that fire department scuba divers found Jimmy and brought his lifeless body to the surface. He was clinically dead: his skin was ashen, his pupils fixed and dilated, and there was no sign of a pulse or breathing.

How could it be possible, then, that Jimmy is alive today? When he fell into the water, Jimmy instinctively held his breath. No one, however, can hold his breath for much more than a minute under such conditions. Eventually Jimmy was forced to breathe in the cold lake water. On entering the lungs, this icy water quickly cooled the blood flowing through the lungs. The cool blood made its way through the heart and into the circulatory system, where it cooled (down to 85°F, or 29.4°C) all the tissues it contacted. This drastically lowered the metabolic rate of the tissues and their need for oxygen. Such a rapid lowering of the rate of reaction in tissue cells can delay tissue death for 45 minutes or more after breathing stops.

The special treatment that Jimmy received after being pulled from the water was critical to his recovery. Paramedics immediately began heart massage and other reviving techniques, but did not try to raise his body temperature. In the emergency room, Jimmy's heart was started with electric shocks, and he was put on a respirator. Heat lamps and warm intravenous fluids were used to very slowly warm his body to 91°F (32.8°C) to prevent further tissue damage. Meanwhile, doctors gave Jimmy massive doses of barbiturates to maintain his coma. Such drugs reduce the brain's need for oxygen and glucose and lessen the chance of lethal swelling. Only after three days did the doctors stop giving Jimmy barbiturates. His brain function slowly began to return over the next several weeks.

Under normal conditions the brain undergoes irreparable damage if it is cut off from its oxygen supply for as little as 3 minutes. Jimmy survived because his brain tissue was cooled by the icy water he had inhaled, slowing down all the chemical reactions in the brain. This allowed the brain tissue to survive without its normal supply of oxygen.

The rate of reactions can mean the difference between life and death to a young child such as Jimmy. Similarly, for many industrial processes it is critically important to control the rate of chemical reactions. If the rate is too slow, the process may not work. If the rate is too fast, a disastrous explosion could occur. To understand how scientists and engineers control reaction rates, it is important to study the various factors that influence the rate at which chemical reactions occur.

8.1 Activation Energy

Define activation energy.

Define activated complex.

For a reaction to occur between two particles, they must be brought close enough together for their outermost energy level electrons to interact. In fact, they must collide. Not only must they collide, but they must do so with enough energy to overcome the repelling forces between the electrons surrounding the two nuclei. The amount of energy necessary for a successful collision is called the **activation energy, E_{act}.** For a reaction to occur, molecules must collide with energy at least equal to the activation energy. When such a collision occurs, the molecules form a reactive group called the **activated complex,** or **transition state.** This complex of atoms is neither the reactants nor the products; rather, it is a highly unstable combination of atoms that represents a state of change between the reactants and the products.

It may be helpful to think of a chemical reaction in terms of driving a car over a mountain range. The activation energy of the reaction is like a mountain pass. Just as each mountain has a pass at a different height, each chemical reaction has a different activation energy. The travelers (the reactants) must get over the mountain pass (the activation energy) before they reach their destination (become products) on the other side of the mountain range (Fig. 8.1). Specifically, the reactants must gain enough energy, usually

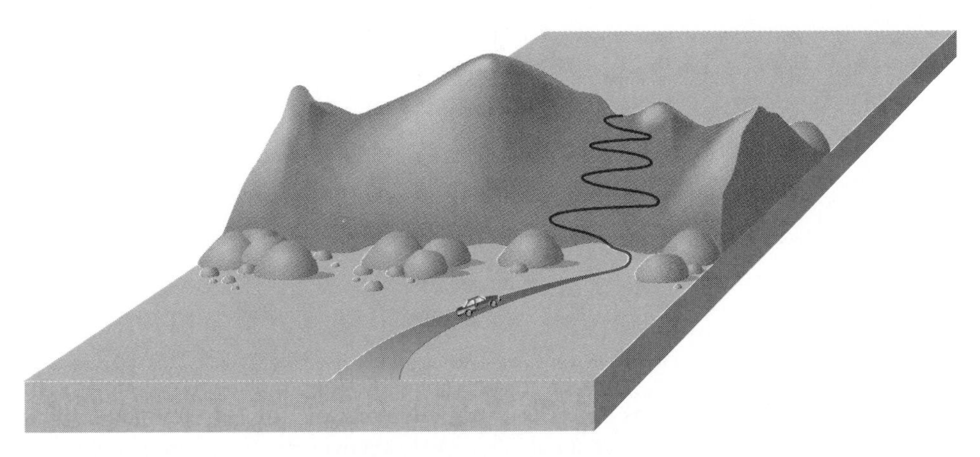

Figure 8.1 Just as a car can drive over a mountain pass only when it has been given enough energy, so a reaction occurs between two molecules only when they collide with sufficient energy to overcome the activation energy barrier.

heat, to reach the activated complex (the top of the mountain pass). This is why most collisions between reactant particles do not result in the formation of a product. For most reactions, energy must be added for the reaction to occur.

8.2 Exothermic and Endothermic Reactions

You know that a candle does not spontaneously burst into flame at room temperature, but you may wonder what allows it to keep burning once it is lit. The reaction between candle wax and oxygen has a high activation energy, and very few molecules have enough energy at room temperature to react. The heat of a match is required to cause a large number of molecules to overcome the activation energy barrier and make the wax burn. The reaction between candle wax and oxygen is **exothermic,** meaning that energy is released as the reaction occurs. Exothermic reactions result when the products of the reaction have less potential energy than the reactants; therefore, once the molecules begin to react, this reaction releases enough energy to boost additional molecules over the activation energy barrier. Hence, the candle continues to burn (Fig. 8.2).

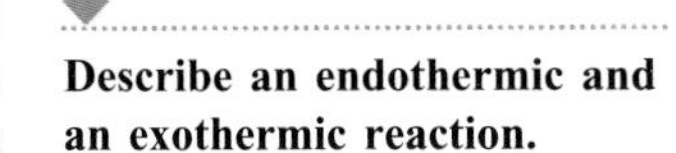

Describe an endothermic and an exothermic reaction.

Some highly exothermic reactions can be explosive. Dynamite, for example, must be set off with a percussion cap. The very small explosion of the percussion cap releases enough energy to cause a small number of dynamite molecules to react. The energy released by these molecules is enough to make the remainder of the molecules react all at once, releasing a tremendous amount of energy.

Other reactions, such as the decomposition of water into hydrogen and oxygen or the formation of sugar molecules during photosynthesis, require the addition of energy for the reaction to continue. Reactions requiring a continuous input of energy are called **endothermic.** In such reactions, the potential energy of the products is greater than that of the

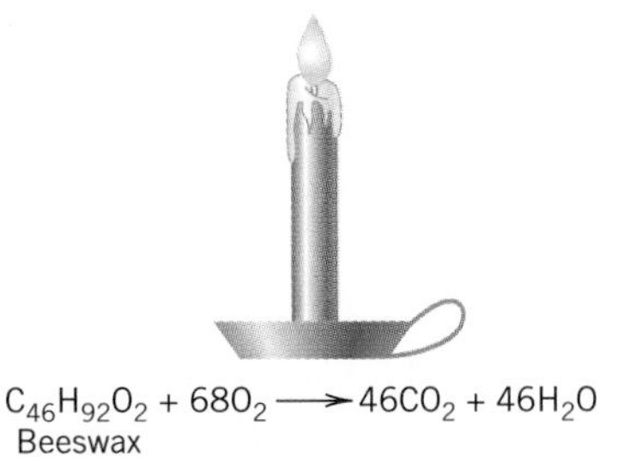

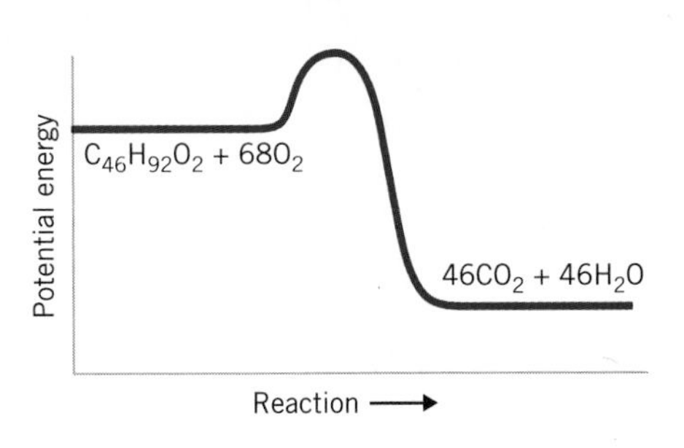

Figure 8.2 The burning of candle wax is an exothermic reaction because the products have less potential energy than the reactants.

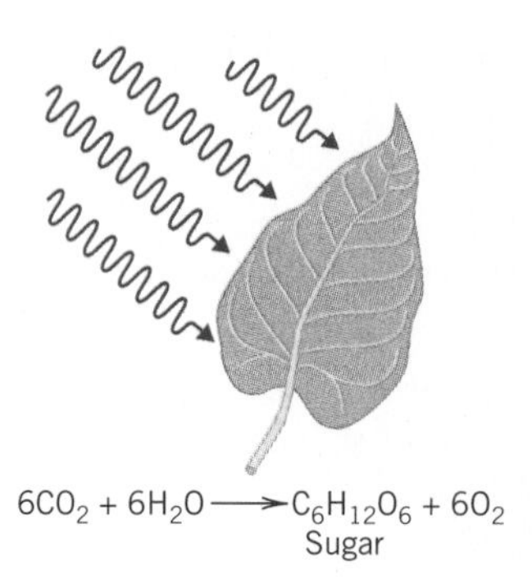

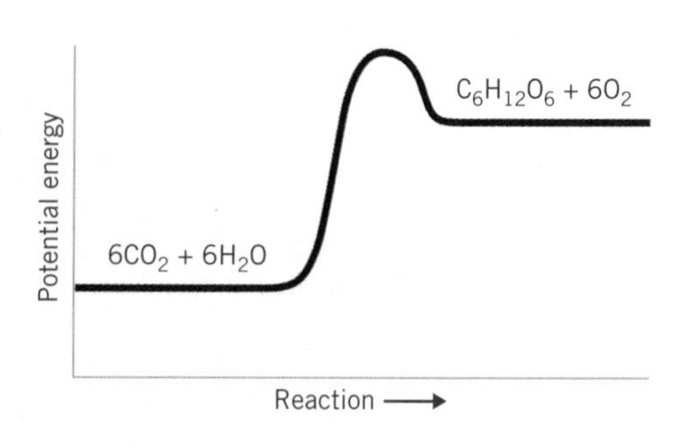

Figure 8.3 Photosynthesis is an endothermic process requiring a continuous supply of energy from the sun.

reactants. In effect, energy has been absorbed by the reaction. In endothermic reactions, reactants require not only an initial input of energy to get over the activation energy barrier but also a continuing supply of energy to keep the reaction going. Stop the supply of electricity, and the decomposition of water comes to a halt; keep a green plant in the dark, and photosynthesis stops, causing the plant to die (Fig. 8.3).

Draw potential energy diagrams for endothermic and exothermic reactions.

Figure 8.4 contains potential energy diagrams for an exothermic and an endothermic reaction. The energy required or released by a reaction is called the **heat of reaction, ΔH.** This change in heat content (also called enthalpy) is expressed in kilocalories or kilojoules per mole of reactant. The value of ΔH is the difference between the potential energy of the products and the potential energy of the reactants. By convention, the heat of reaction has a positive value for an endothermic reaction and a negative value for an exothermic reaction. Each reaction has its own characteristic activation energy and heat of reaction.

Example 8-1

These questions involve the ***T*** of our *STEP* problem-solving strategy.

1. Are the following reactions endothermic or exothermic?

(a) $2Na_2O_2 + 2H_2O \rightarrow 4NaOH + O_2 + 30.2\ \text{kcal}$

(b) $Si + 2H_2 + 8.2\ \text{kcal} \rightarrow SiH_4$

Think It Through

Reaction (a) is exothermic because energy is given off as a product of the reaction. Reaction (b) is endothermic because energy is required by the reaction; that is, energy is one of the reactants.

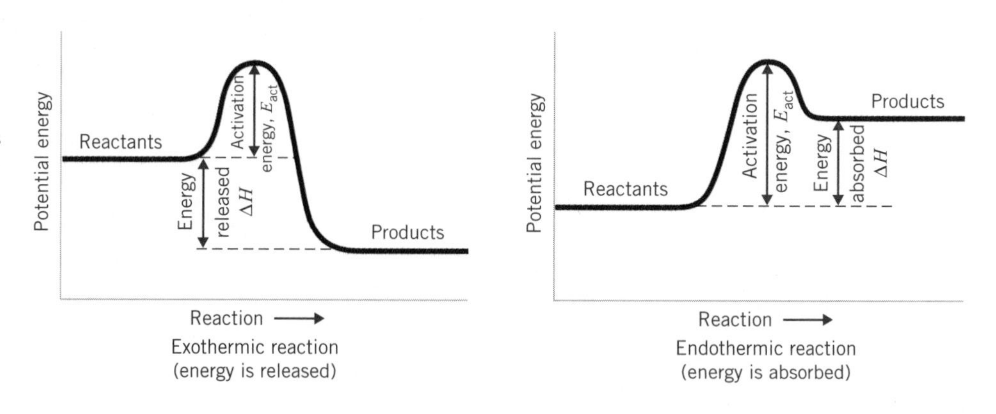

Figure 8.4 Potential energy diagrams for exothermic and endothermic reactions. E_{act} is the activation energy and ΔH is the heat of reaction (the energy released or absorbed).

2. What is the value of the heat of reaction ΔH for the reactions in the previous problem?

(a) $\Delta H = -30.2$ kcal. The heat of reaction for an exothermic reaction has a negative value. ◄ *T*

(b) $\Delta H = +8.2$ kcal. The heat of reaction for an endothermic reaction has a positive value.

3. Draw the potential energy diagram for the following reaction, whose activation energy E_{act} is 39 kcal:

$$2ICl + H_2 \longrightarrow I_2 + 2HCl + 53 \text{ kcal}$$

To draw the potential energy diagram, we must first decide if the reaction is endothermic or exothermic. Because energy is given off by the reaction, it is exothermic. This means that the potential energy of the reactants is greater than the potential energy of the products—in this case, 53 kcal greater. There are several ways to draw the diagram, but to simplify things we can arbitrarily assign the substances with the least potential energy (in this reaction, the products) a relative potential energy of 100 kcal. The reactants, therefore, have a potential energy of 153 kcal in relation to the products. The activation energy is 39 kcal, so the top of the curve is at 153 kcal (the potential energy of the reactants) + 39 kcal (the activation energy), or 192 kcal. ◄ *T*

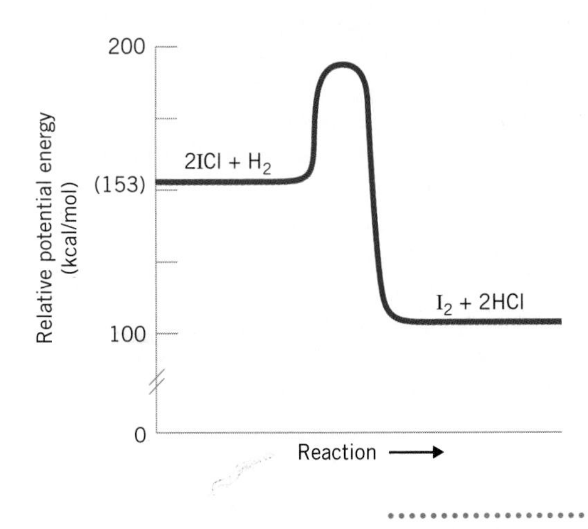

Check Your Understanding

1. The heat of reaction for the following reaction is +31.4 kcal.

$$C(s) + H_2O(g) \longrightarrow CO(g) + H_2(g)$$

(a) Is this an exothermic or endothermic reaction?

(b) Is energy absorbed or released by this reaction?

(c) Rewrite the equation so that the heat of reaction appears in the equation.

2. The following reaction has a heat of reaction ΔH of +3 kcal and an activation energy E_{act} of 43.8 kcal. Draw the potential energy diagram for this reaction, labeling the reactants, products, activated complex, activation energy, and heat of reaction.

$$2HI \longrightarrow H_2 + I_2$$

Health Perspective / Homeostasis

You are certainly familiar with your body's responses to large changes in temperature: You shiver when you get cold, and sweat when you get hot. These are just some of the ways in which your body works to keep its internal temperature constant. These mechanisms are critically important because biochemical processes within the body are temperature dependent. The enzymes that catalyze metabolic reactions function properly only within a very limited temperature range (97°–104°F).

Of the calories that we burn each day, only 20% is used to do cellular work. The rest is given off within our bodies as heat. A 150-lb person consuming a diet of 3000 kcal/day produces enough excess heat to raise the body temperature by more than 100°F. This is obviously a great deal more than the 7°F range the body can tolerate. The way the body prevents such a drastic change in temperature is by maintaining a balance between heat production and heat loss. The word used to describe this fine balance between heat production and heat loss, which allows us to maintain a relatively constant body temperature, is *homeostasis.*

The body loses most of its excess heat through the skin by means of radiation, conduction, and evaporation. Of these three processes, radiation accounts for the largest amount of heat loss. The process goes like this: As the body temperature increases, more blood enters the skin capillaries. From the skin, heat radiates to the cooler surroundings. In addition to radiation of heat from the skin, conduction also takes place. This occurs when warm air next to the body rises and is replaced by cooler air. Heat from the skin is conducted away by contact with this cooler air. The remainder of the excess heat lost through the skin leaves by the evaporation of sweat. Evaporating 1 g of water requires 540 cal of heat from the body. A 150-lb person loses, on the average, 0.75 L of water each day through the cooling process of evaporation.

Your Perspective: The 1994 World Cup soccer matches were played in very hot weather. After each game, selected players were tested for drug usage through urinalysis. In some cases, players required several hours to produce a urine specimen, even though they drank large quantities of water after the game. Write a column for the sports page of your local newspaper explaining why this occurred.

8.3 Factors Affecting Reaction Rates

Nature of the Reactants

◆ **List four factors that affect the rate of a chemical reaction.**

The nature of the reactants influences the rate of a chemical reaction. For example, when colorless nitric oxide escapes from a test tube into the air, reddish brown nitrogen dioxide forms very quickly. When carbon monoxide from automobile exhaust fumes is released into the air, however, the reaction with oxygen to form carbon dioxide is slow (unfortunately for the residents of large urban areas).

$$2NO + O_2 \longrightarrow 2NO_2 \quad \text{Fast reaction at 25°C}$$

$$2CO + O_2 \longrightarrow 2CO_2 \quad \text{Very slow reaction at 25°C}$$

These two balanced equations look similar, but the large difference in their rates of reaction comes from the nature of the carbon monoxide and nitric oxide molecules themselves.

Concentration of Reactants

The concentration of reactants can greatly influence the rate of a chemical reaction. We have seen that particles must collide to react, so it seems reasonable that the more reactant particles there are in a given space, the more collisions will occur. (We see the same thing on our highways: The more crowded the highways, the higher the probability of a collision. This is especially apparent when you consider the high death rates on holiday weekends.) Increasing the concentration of the reactants increases the number of collisions and, therefore, the reaction rate.

Physicians make use of this principle when prescribing medicines to treat specific diseases. The appropriate treatment for tonsillitis or pneumonia may involve a dose of 250 mg of penicillin, but meningitis requires two to five times that dosage. The higher concentration of penicillin increases the rate of absorption into the bloodstream and, therefore, the effective concentration in the blood.

Surface Area of a Solid

Increasing the surface area per unit of volume of a solid reactant increases the rate of reaction. An increase in the surface area increases the number of solid particles that are exposed, thereby increasing the number of collisions possible between the reactants. Increasing the surface area can sometimes increase the rate of reaction to explosive levels. For example, lumber mills seldom worry about log piles spontaneously catching fire, but sawdust piles must be kept wet because they can burst into flame. You probably don't think of flour as potentially dangerous, but finely divided flour dust can explode (Fig. 8.5). As another example, a crushed aspirin tablet, because of its increased surface area, relieves your headache faster than a whole tablet.

Figure 8.5 Grain elevators can be destroyed by the force of an explosion of grain dust.

Temperature of the Reaction

Increasing the temperature of the reactants increases the kinetic energy of the particles. This not only increases the frequency of collisions but also increases the likelihood that the colliding particles will have enough energy to overcome the activation energy barrier (Fig. 8.6). (This is similar to the idea that increasing speed limits on highways not only increases the number of collisions but also increases the likelihood that any collision will be a fatal one.)

Changes in temperature have important effects on living organisms. A fever increases the rate of chemical reactions in the body, as shown by the increased pulse rate, increased

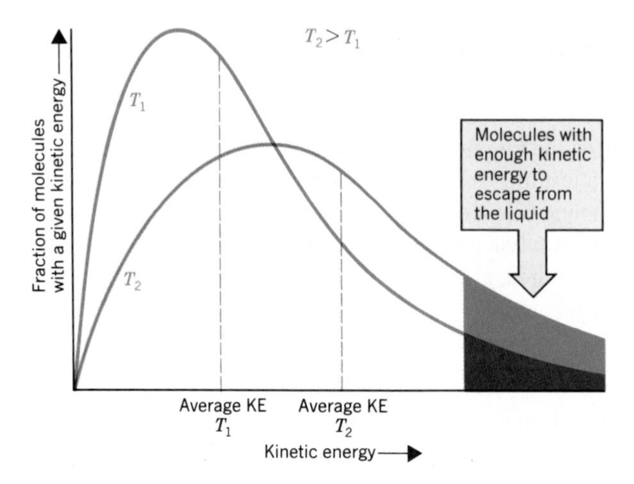

Figure 8.6 Increasing the temperature of a reaction increases the average kinetic energy of the molecules. This increases the number of molecules having enough energy to collide successfully and produce products.

breathing rate, and abnormalities in digestive and nervous systems. When you run a fever, your basal metabolism rate goes up by 5% for each degree rise in body temperature. The effects on living organisms of an increase in external temperatures are illustrated by what happens when warm water from power plants is released into streams and lakes, causing an increase in the metabolic rate in the aquatic life. To support their new higher rate of metabolism, the organisms require more oxygen. But the warming of the water also decreases the concentration of oxygen in the water, contributing to the death of fish in these streams and lakes. Another detrimental effect of these higher water temperatures is a resulting increase in the sensitivity of fish to pollutants in the water.

Decreasing the body's temperature slows down bodily reaction rates. During open heart surgery, a patient's body temperature is usually lowered 4° to 5°F to decrease the metabolism rate and oxygen requirements. At the beginning of this chapter we saw an extreme example of the decrease in bodily reactions from a lowering of body temperatures. As another example, an athletic trainer might spray a surface anesthetic called ethyl chloride on the injured knee of a basketball player to relieve the pain. The rapid evaporation of the ethyl chloride lowers the temperature of the tissue enough to greatly slow the reactions responsible for the transmission of nerve impulses.

Many animals hibernate during the cold winter months. Their body temperatures fall to a few degrees above freezing, and all chemical reactions in their bodies slow down. Their breathing and heart rates become very slow, and they require much less energy to sustain life. This allows the animals to live on stored body fat alone. To illustrate the extent of this slowdown, consider that a woodchuck's heart beats about 80 times a minute while it is active, but when hibernating, its heart beats only 4 times a minute. The interesting idea of keeping human beings in suspended animation by lowering their body temperatures has been the subject of many science fiction stories. In 1974, scientists from the Darwin Research Institute discovered that bacteria which had been frozen in the cold rock of Antarctica 10,000 to 1 million years ago could be revived in the warmth of the laboratory and, incredibly, could even reproduce.

Catalysts

◆ **Define catalyst, and explain how a catalyst affects the rate of a chemical reaction.**

Many reactions that normally proceed slowly can be made to occur at a more rapid rate by the introduction of substances called catalysts. A **catalyst** is a substance that increases the rate of a chemical reaction without being consumed in the reaction. The effect of the catalyst is to lower the activation energy required for the reaction. In our illustration of driving over the mountain pass, the action of the catalyst is like that of opening a tunnel

Health Perspective / Hypothermia

Hypothermia is the word used to describe the lowering of the body's temperature and the metabolic changes that accompany it. As we saw earlier in our discussion of homeostasis, heat produced from the core of the body is lost through the skin. If the body loses more heat than it produces, the overall body temperature drops.

When the body temperature falls to less than 95°F (35°C), both the mind and muscles become incapable of functioning adequately. If the body temperature drops low enough, the heart stops beating and death quickly follows.

The onset of hypothermia can be very fast. If a person is wet, it can happen almost immediately. Initial symptoms often go unnoticed because victims appear normal except for a seeming change of attitude: they may drop out of the conversation or may become mildly depressed and less enthusiastic about simple tasks. As the brain becomes more affected by the cold, small decisions become big problems, and forgetfulness increases.

Because heat flows from a warm region to a colder one, hypothermia can occur under ordinary conditions even when the air temperature is relatively mild. Being wet increases the flow of heat from the body, because water conducts heat about 240 times better than still air. The speed with which hypothermia develops varies considerably depending on the energy reserves of the individual and the nature of the situation. Still, anything that can be done to prevent heat loss is extremely important in conditions when hypothermia is possible. The most important thing you can do is to be careful about your clothing. Clothing should be worn in layers. Wool should be worn next to the body, because wool is one of the few materials that insulates when it is wet. Your outer layer of clothing should be windproof and water-repellent. Because much of your heat loss occurs from the head and neck, wearing a cap and scarf is always a good idea. Alcohol should be avoided because it dilates your capillaries, allowing your body to pump more heat from the body core to the skin—where it is lost. In fact, consumption of alcohol is one of the most common contributors to hypothermia.

Your Perspective: Suppose Columbia Sportswear asked you to help them design a parka that is comfortable to wear and is stylish, but is also especially helpful in preventing hypothermia. Write a memo to the company explaining what the parka should look like and how it should be constructed, explaining each of your statements.

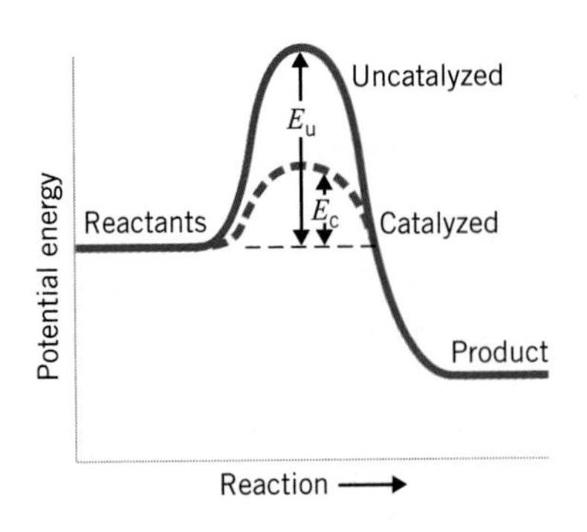

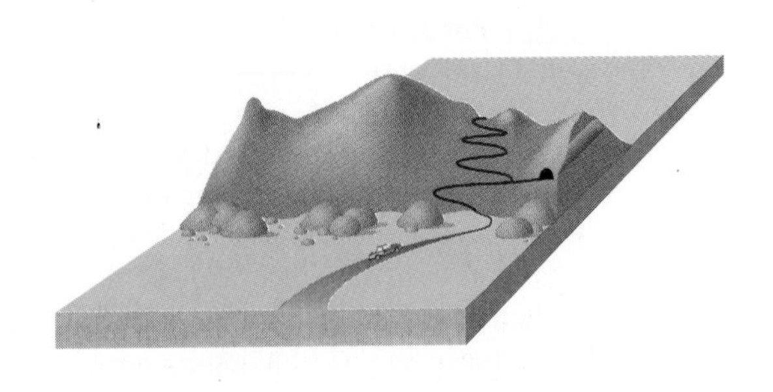

Figure 8.7 A catalyst lowers the activation energy, increasing the chances for a successful collision between reactant particles.

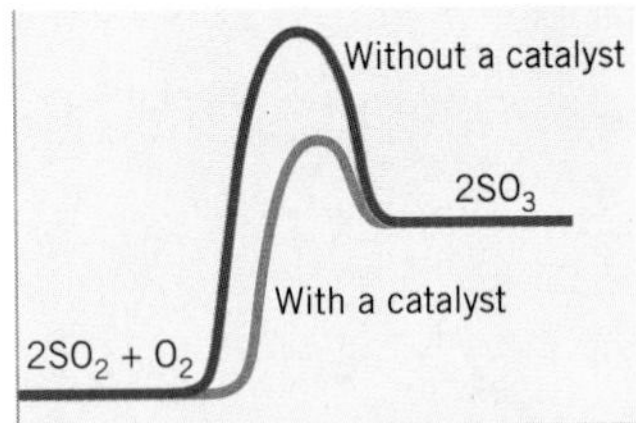

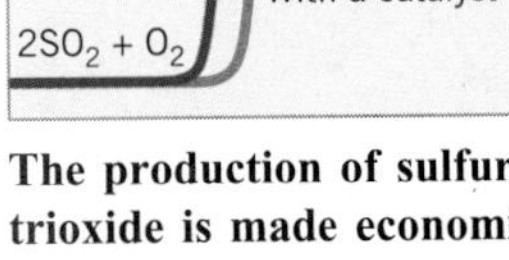

The production of sulfur trioxide is made economical by the addition of a platinum catalyst.

through the mountain. With a tunnel, you can expend a lot less energy (gasoline) and still reach the other side of the mountain range (Fig. 8.7).

Industry makes wide use of catalysts, especially catalysts that allow companies to save on energy costs by producing large amounts of a product at a lower temperature. An example of this occurs in the production of sulfuric acid, H_2SO_4, the so-called king of chemicals. Over 20 million tons of sulfuric acid is used each year in the United States, mostly by the steel, fertilizer, and petroleum industries. One step in the production of sulfuric acid involves the production of sulfur trioxide (SO_3) from sulfur dioxide (SO_2) and oxygen (O_2). This reaction has a very high activation energy and is quite slow even at high temperatures. The reaction has been made economically worthwhile, however, by the introduction of a catalyst, such as finely divided platinum, which greatly increases the reaction rate. Similarly, the cost of producing gasoline from much larger molecules of crude petroleum is kept low by using catalysts in the "cracking" process—the process by which large molecules are broken into smaller pieces.

Define enzyme, and give an example.

Most of the chemical reactions that occur in our bodies would not ordinarily occur at body temperature at a high enough rate to sustain life. The body, however, produces special compounds called **enzymes,** which act as biological catalysts. They permit essential reactions to occur readily at body temperature. For example, carbon dioxide, which is produced as a waste product by our tissues, is converted to carbonic acid in red blood cells before being transported to the lungs to be exhaled.

$$CO_2 + H_2O \longrightarrow H_2CO_3$$

This reaction is very slow at body temperature. But red blood cells contain an enzyme called carbonic anhydrase that increases the rate of the uncatalyzed reaction, making it 10 million times faster!

Damaging or removing body enzymes can have disastrous effects. For example, the fish poison rotenone, which is used to clear lakes and ponds of undesirable species, kills the fish by preventing enzymes from catalyzing reactions essential to the production of cellular energy. The digestive tracts of 90% of non-Caucasian adults lack the enzyme lactase, which catalyzes the breakdown of lactose (the sugar found in milk). Because these individuals cannot digest the lactose, drinking milk results in nausea, gas, and diarrhea.

8.4 What Is Chemical Equilibrium?

At one time or another you may have added sugar to a glass of iced tea, stirred, and watched the sugar dissolve. You may have noticed that additional sugar added to this solution continued to dissolve until you reached a certain point, after which any more

Chemical Perspective / Hydrogen Peroxide

One popular antiseptic used for the treatment of minor cuts and scratches is hydrogen peroxide. No home medicine cabinet is complete without the small brown bottle containing this amazing liquid. Like most of the antiseptics and disinfectants used in medicine, hydrogen peroxide is an oxidizing agent. Oxidizing agents are effective in inhibiting the growth of and in killing microbes. They interfere with normal cellular activities and prevent growth and reproduction. What makes hydrogen peroxide unique is the foaming action that takes place whenever it is placed on a cut. Some people used to believe this foaming action indicated the presence of an infection. In reality, hydrogen peroxide foams any time it comes in contact with blood. An enzyme in the blood catalyzes the decomposition of hydrogen peroxide, which produces water and oxygen gas. A benefit of the foaming action is that the hydrogen peroxide acts as a cleansing agent, bringing any embedded dirt up to the surface.

Hydrogen peroxide is such an extremely strong oxidizing agent that it is sold as a solution, containing just 3% hydrogen peroxide in water. Although this makes it much safer, hydrogen peroxide tends to decompose in water and lose its oxidizing power. Because its decomposition has such a low activation energy, simply exposing hydrogen peroxide to light increases this decomposition. This is why hydrogen peroxide solutions are stored in brown bottles.

Your Perspective: Suppose that the World Health Organization asked you to develop a 30-second radio advertisement, in the form of a dialog, urging people to use hydrogen peroxide as an antiseptic. Write a script for such an ad, using language that requires no scientific or medical knowledge on the part of the listener.

added sugar just settled to the bottom of the glass. Although you may have continued stirring or let the glass sit for a time, neither the amount of sugar on the bottom of the glass nor the amount of sugar dissolved in the iced tea changed.

Under these circumstances, the iced tea contained all the sugar molecules it could possibly hold. As we will see, the molecules of sugar dissolved in the iced tea and the molecules of sugar in the crystals on the bottom of the glass had come into equilibrium. A molecular equilibrium of this type is not a static, motionless condition such as the equilibrium reached when two children are equally balanced on a seesaw. Rather, a chemical equilibrium is a dynamic state in which events are constantly occurring but at exactly equal rates. In the glass of iced tea, for example, new sugar molecules are constantly dissolving and other sugar molecules are constantly crystallizing out of solution, but the rate of each of the processes is the same. There is therefore no overall change in the numbers of sugar molecules dissolved in the iced tea or lying on the bottom of the glass. In other words, for every sugar molecule that dissolves, one sugar molecule crystallizes out of solution (Fig. 8.8).

Let's examine the iced tea example in chemical terms. We said that when we first add sugar to the iced tea, it dissolves. We can represent this process by the following equation.

$$\text{Sugar}(s) \longrightarrow \text{sugar}(aq)$$

where *aq* stands for aqueous (dissolved in water), and *s* represents solid.

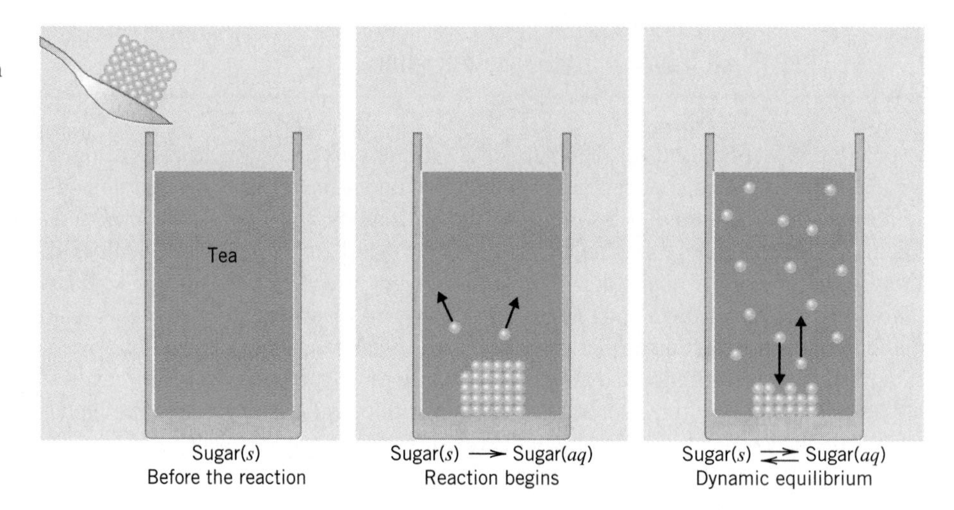

Figure 8.8 A dynamic equilibrium is established when the rate of the forward process equals the rate of the reverse process.

As more and more sugar molecules are added to the iced tea, the number dissolved in the tea increases to the point that the reverse of this process (that is, the crystallization of sugar from the tea) begins to occur.

$$\text{Sugar}(s) \rightleftharpoons \text{sugar}(aq)$$

Notice that in this equation we use two arrows pointing in different directions to indicate that the process is reversible. The lengths of the two arrows indicate the rate at which the two processes are occurring. As the number of sugar molecules in the tea continues to increase, the rate of the reverse process increases until it equals the rate of the forward process. If we express the rate of the forward process as rate_f and the rate of the reverse process as rate_r, we then have $\text{rate}_f = \text{rate}_r$.

$$\text{Sugar}(s) \underset{\text{rate}_r}{\overset{\text{rate}_f}{\rightleftharpoons}} \text{sugar}(aq)$$

At this point, a state of equilibrium has been reached, and no net change occurs in our glass of sugared iced tea.

The same concept of equilibrium holds for chemical reactions. For example, carbon dioxide is a waste product from the energy-producing reactions of the cell. It reacts with water in the blood to form carbonic acid, H_2CO_3.

$$CO_2(aq) + H_2O \longrightarrow H_2CO_3(aq)$$

As the concentration of carbonic acid increases, the rate of the reverse reaction increases until it equals the rate of the forward reaction. This establishes an equilibrium between the dissolved carbon dioxide and the carbonic acid in the blood.

$$CO_2(aq) + H_2O \underset{\text{rate}_r}{\overset{\text{rate}_f}{\rightleftharpoons}} H_2CO_3(aq)$$

It is again important to emphasize that in any system at equilibrium, the forward process and the reverse process continue to occur, but at equal rates. In particular, a **chemical equilibrium** is a dynamic state in which the rate of the forward reaction equals the rate of the reverse reaction. Note especially that the *only* things that are equal in a chemical equilibrium are the reaction rates.

Define chemical equilibrium, and give two examples.

8.5 Altering the Equilibrium

Altering a chemical equilibrium is an important technique in industrial or biological chemistry. It is especially useful in forcing a reaction to go further toward completion even though the equilibrium state is reached early in the reaction. Various factors affecting the rate of a chemical reaction can alter a chemical equilibrium. Let's review each of these factors.

◆ **State two ways a chemical equilibrium can be disrupted.**

Changes in Concentration

Changing the concentration of a reactant or product in a reaction that is at equilibrium can affect the rate of the forward or reverse reaction. Increasing the concentration of a reactant increases the number of collisions between reactant molecules, thereby increasing the rate of the forward reaction. When this occurs, the system is no longer at equilibrium.

$$\text{Rate}_f > \text{rate}_r$$

The forward reaction proceeds at a higher rate, producing more product molecules until the rate of the reverse reaction again increases enough to equal the rate of the forward reaction. At this point a new equilibrium is established.

One colorful example of the effect on a chemical equilibrium of changing reactant concentrations can be seen in the equilibrium between the yellow chromate ion, CrO_4^{2-}, and the orange dichromate ion, $Cr_2O_7^{2-}$.

$$\underset{\text{yellow}}{2CrO_4^{2-}(aq) + 2H^+(aq)} \underset{\text{rate}_r}{\overset{\text{rate}_f}{\rightleftharpoons}} \underset{\text{orange}}{Cr_2O_7^{2-}(aq) + H_2O}$$

Increasing the concentration of hydrogen ions on the reactant side of the equation increases the number of collisions between the chromate ions and the hydrogen ions, thus causing an increase in the rate of the forward reaction. As more dichromate forms, the rate of the reverse reaction also increases until it again equals the rate of the forward reaction. At that point a new equilibrium is established. This new equilibrium has a higher concentration of dichromate ions and a lower concentration of chromate ions than the previous equilibrium. The solution therefore appears more orange. It is common for chemists to add more reactant or to remove a product in a reaction in order to upset a chemical equilibrium and to drive a reaction toward completion.

As we have seen, an equilibrium exists in the blood between carbon dioxide and carbonic acid.

$$\underset{\text{carbon dioxide}}{CO_2(aq)} + H_2O \underset{\text{rate}_r}{\overset{\text{rate}_f}{\rightleftharpoons}} \underset{\text{carbonic acid}}{H_2CO_3(aq)}$$

In the tissues, carbon dioxide is a waste product that enters the blood. As the concentration of carbon dioxide in the blood increases, the reaction is driven to the right ($\text{rate}_f > \text{rate}_r$), and carbonic acid, which can be transported by the bloodstream, is formed. When the blood reaches the lungs, we exhale carbon dioxide. This lowers the concentration of carbon dioxide in the blood, driving the reaction to the left and allowing us to rid the bloodstream of the waste product carbonic acid (Fig. 8.9).

Figure 8.9 Changes in the concentration of carbon dioxide in the body affect the carbon dioxide/carbonic acid equilibrium.

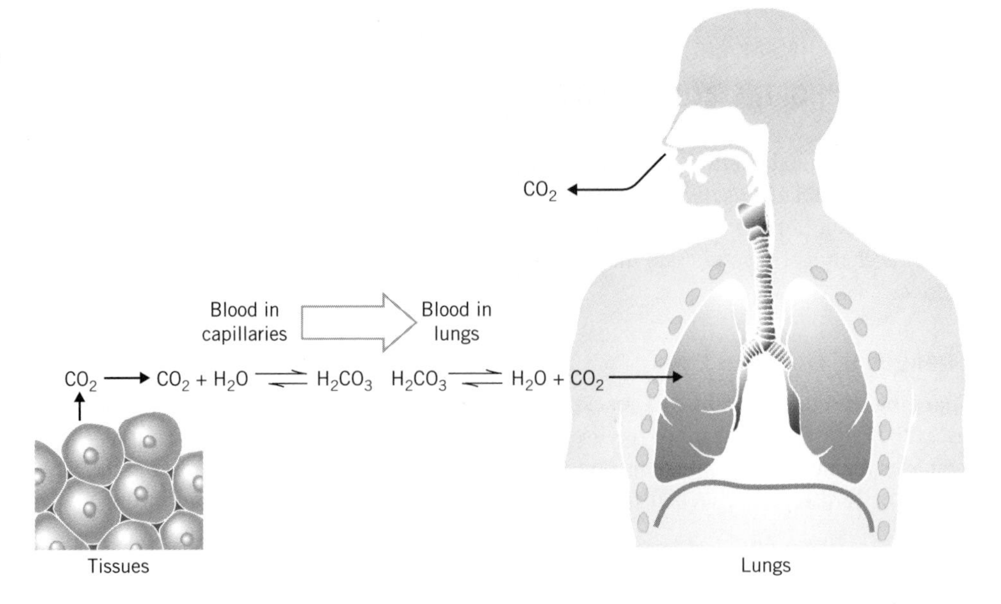

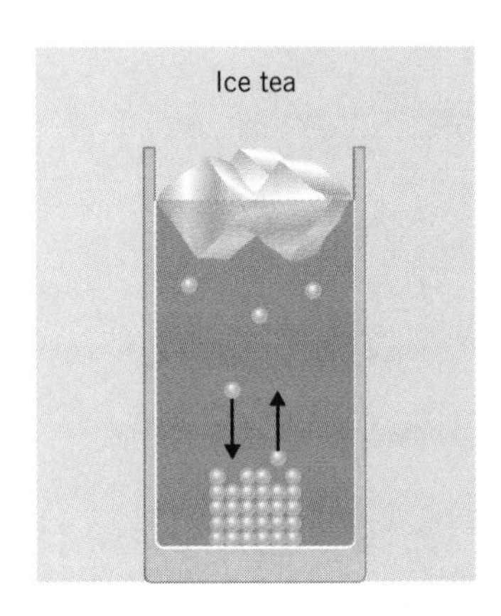

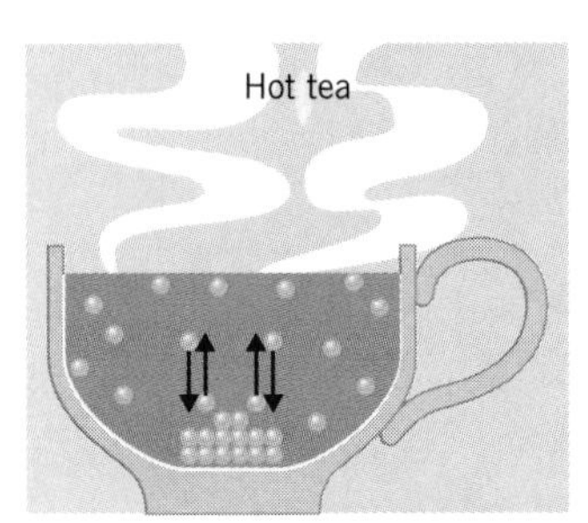

Changes in Temperature

If a reaction is endothermic in one direction, it must be exothermic in the other direction. Thus, changing the temperature of a system in equilibrium alters the equilibrium. Increasing the temperature favors the endothermic reaction (the reaction requiring energy), and lowering the temperature favors the exothermic reaction (the reaction in which energy is released). For example, let's return to our glass of iced tea. The process of dissolving sugar is an endothermic process. Increasing the temperature therefore increases the number of sugar molecules that dissolve.

$$\text{sugar}(s) + \text{energy} \underset{\text{rate}_r}{\overset{\text{rate}_f}{\rightleftharpoons}} \text{sugar}(aq)$$

Warming our tea increases rate_f, because the forward process is the endothermic process. We therefore have $\text{rate}_f > \text{rate}_r$, and more sugar dissolves than crystallizes. If we raise the temperature and hold it constant, rate_r increases until it equals rate_f, thus establishing a new equilibrium with a larger number of sugar molecules dissolved in the tea. Obviously, if you like your tea sweet you should drink it warm!

Catalysts

We have discussed the fact that a catalyst increases the rate of a chemical reaction, but you may be surprised to learn that a catalyst has no effect on the equilibrium concentrations of the reactants and products in an equilibrium system. Why is this? Remember that the way a catalyst increases the rate of a chemical reaction is by lowering the activation energy of the reaction (see Fig. 8.7). At the same time, however, the activation energy for the reverse reaction is lowered by an equal amount, so the rate of the reverse reaction also increases. Because the catalyst increases the rate of both the forward and the reverse reactions by the same amount, it has no effect on the equilibrium system.

8.6 Le Chatelier's Principle

We have seen that a chemical system is in equilibrium when the rates of the forward and reverse reactions are equal. Moreover, changes made in such systems can disrupt the equilibrium, which alters the equilibrium concentrations of the reactants and products. After studying many equilibrium systems, the French chemist Henri Louis Le Chatelier developed a principle that helps predict whether the reactants or products will be favored in response to a given change. **Le Chatelier's principle** states that a system at equilibrium resists attempts to change its temperature or pressure, or the concentration of a reactant or product. In other words, when upset by a change in temperature, pressure, or concentration, an equilibrium system shifts in the direction necessary to resist the change and to reestablish an equilibrium. Le Chatelier's principle tells us, for example, that if we increase the concentration of a reactant, the system moves in a direction to remove the increase; that is, $rate_f$ becomes greater than $rate_r$. If we increase the temperature, the reaction that uses up that increase in energy is favored; that is, the rate of the endothermic reaction is greater than the rate of the exothermic reaction, until a new equilibrium has been established.

State Le Chatelier's principle.

Let's consider the effect of various changes on the following reaction at equilibrium:

$$2CO(g) + O_2(g) \underset{rate_r}{\overset{rate_f}{\rightleftharpoons}} 2CO_2(g) + heat$$

1. *Increase the temperature:* If we increase the temperature, the system shifts in the direction that resists the change. This is the direction that removes heat. That means that the endothermic, or reverse, reaction is favored. $Rate_r$ is therefore greater than $rate_f$ until a new equilibrium point is reached.

$$2CO(g) + O_2(g) \underset{rate_r}{\overset{rate_f}{\longleftarrow\!\!\!\rightleftharpoons}} 2CO_2(g) + heat$$

2. *Increase the concentration of CO:* Increasing the concentration of a reactant increases the rate of the forward reaction. (Remember that increasing the concentration of a reactant increases the rate of the reaction). $Rate_f$ is therefore greater than $rate_r$ until a new equilibrium point is reached.

$$2CO(g) + O_2(g) \underset{rate_r}{\overset{rate_f}{\rightleftharpoons\!\!\!\longrightarrow}} 2CO_2(g) + heat$$

3. *Increase the pressure:* When the stress on the equilibrium is an increase in pressure, the favored reaction is the one that produces a system exerting less pressure. This system has fewer gas molecules. In this case, the forward reaction produces a system with two gas molecules, whereas the reverse reaction produces a system with three gas molecules. The forward reaction is therefore favored, and $rate_f$ is greater than $rate_r$ until a new equilibrium point is reached.

$$2CO(g) + O_2(g) \underset{rate_r}{\overset{rate_f}{\rightleftharpoons\!\!\!\longrightarrow}} 2CO_2(g) + heat$$

Le Chatelier's principle can be very helpful in explaining the results of stresses on equilibrium systems in living organisms. For example, *calculus* is a word used in medicine to describe an abnormal buildup of ionic solids (called mineral salts) in a framework or matrix of other substances, such as proteins. Kidney stones, bladder stones, and gallstones are examples of calculi. The mineral salts found in kidney and bladder stones are formed from ions circulating in body fluids: the positive ions are calcium (Ca^{2+}) and magnesium

Predict the changes that occur in an equilibrium when a given stress is applied.

Chemical Perspective / Sunglasses

Have you seen the sunglasses that turn dark in sunlight and get lighter when they are indoors? They are called photochromic glasses, and they provide a perfect example of Le Chatelier's principle at work.

Photochromic lenses have silver chloride (AgCl) crystals incorporated directly into the glass. When ultraviolet radiation in sunlight strikes the silver chloride crystals, they darken. This occurs when the silver ions (Ag^+) are reduced to metallic silver (Ag) by the chloride ions (Cl^-), which become elemental chlorine atoms:

$$\underset{\substack{\text{silver}\\ \text{chloride}\\ \text{(colorless)}}}{AgCl} + \text{light energy} \rightleftharpoons \underset{\substack{\text{silver}\\ \text{atom}\\ \text{(dark)}}}{Ag} + \underset{\substack{\text{chlorine}\\ \text{atom}}}{Cl}$$

The resulting millions of silver atoms give a dark color to the glass. In the reverse reaction, silver atoms recombine with chlorine atoms to form AgCl plus energy. The reverse reaction takes place because the silver and chlorine atoms are trapped in the glass and can't escape. The greater the intensity of light, the greater the number of silver atoms formed. Adding more light drives the reaction to the right, causing the glass to get darker. Removing some of the light by going into a dimly lit room, drives the reaction to the left, causing the glass to get lighter. If you wear such glasses, you might say that Le Chatelier's principle is being demonstrated right before your very eyes!

Your Perspective: Develop a magazine advertisement for a brand of photochromic lenses. In writing your advertisement, do your best to convince the readers of how glamorous, stylish, and useful these lenses would be.

(Mg^{2+}), and the negative ions are phosphate (PO_4^{3-}), monohydrogen phosphate (HPO_4^{2-}), and dihydrogen phosphate ($H_2PO_4^-$). One of the equilibria involved is

$$3Ca^{2+}(aq) + 2PO_4^{3-}(aq) \underset{\text{rate}_r}{\overset{\text{rate}_f}{\rightleftharpoons}} Ca_3(PO_4)_2(s)$$

A person can develop kidney stones when something disrupts the system controlling the amount of calcium and magnesium ions in body fluids. If the concentration of calcium ions goes up, this stress increases the rate of the forward reaction, namely, the formation of the solid calcium phosphate, $Ca_3(PO_4)_2$. The stone grows slowly as this solid, as well as other salts, is deposited in the matrix.

As another example, aspirin, which is taken by millions of people for relief of headaches, is a very safe drug. In some people, however, aspirin can disrupt the normal functioning of the stomach lining and can cause bleeding. The process by which aspirin enters the stomach lining involves various stresses on the equilibrium between the neutral aspirin molecule and the aspirin ion that is formed when the aspirin molecule is dissolved in water.

$$\underset{\text{(nonpolar)}}{\text{Aspirin}} \underset{\text{rate}_r}{\overset{\text{rate}_f}{\rightleftharpoons}} \underset{\text{(polar)}}{\text{aspirin}^- + H^+}$$

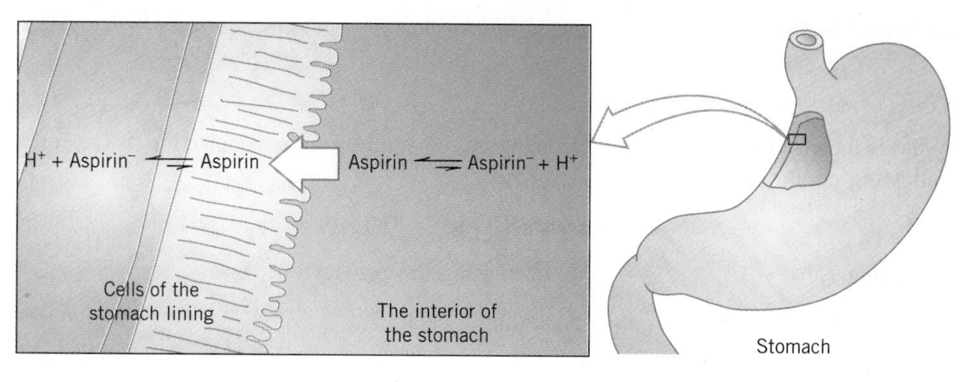

Figure 8.10 The passage of aspirin molecules through the stomach wall is accelerated by the stress applied to the aspirin/aspirin ion equilibrium by the hydrogen ions in stomach fluid.

When aspirin is dissolved in water, an equilibrium is formed between the polar and nonpolar forms of this compound. When you swallow an aspirin with some water, a significant amount of the aspirin is found in the polar ion form, which cannot pass through the protective lining of the stomach. The contents of the stomach, however, are very high in hydrogen ions (H^+). This puts a stress on the equilibrium, increasing the rate of the reverse reaction and forming more nonpolar aspirin molecules, which can pass through the protective lining. Once the aspirin molecules have passed into the cells of the stomach, the concentration of hydrogen ions becomes very low and the equilibrium shifts to the right. This again produces the polar aspirin ions, which cannot pass back into the stomach. Thus, these ions become trapped in the cells, where they produce the damage that causes the bleeding (Fig. 8.10).

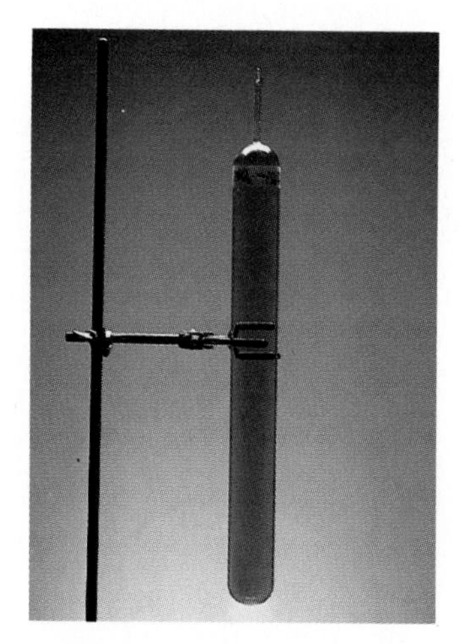

Example 8-2

Use the ***T*** of our *STEP* problem-solving strategy to reason through the following problems.

Suppose that a glass cylinder with a movable piston at one end contains the brown gas nitrogen dioxide, NO_2, in equilibrium with the colorless gas dinitrogen tetroxide, N_2O_4.

$$\underset{\text{brown}}{2NO_2(g)} \rightleftharpoons \underset{\text{colorless}}{N_2O_4(g)} + 13.6\text{ kcal}$$

What effect would the following changes have on the color of the gas in the cylinder (that is, on the equilibrium concentration of NO_2)?

Think It Through

1. *Removing N_2O_4 from the cylinder.* The brown color decreases. The equilibrium system resists the removal of the N_2O_4 by increasing the forward reaction, thereby reducing the number of NO_2 molecules in the flask and decreasing the color.

2. *Putting the cylinder in an ice-water bath.* The brown color decreases. Putting the flask in an ice-water bath decreases the temperature of the flask. The system resists this change by shifting in the direction of the exothermic reaction—the forward reaction—lowering the concentration of NO_2. (See bottom photo.)

3. *Putting the cylinder in a boiling water bath.* The brown color increases. Because the temperature is increased, the system shifts in the direction of the endothermic reaction.

4. *Decreasing the volume of the cylinder.* The brown color decreases. Decreasing the volume increases the pressure within the cylinder. The system, therefore, shifts in the direction of the forward reaction, which is the direction that produces fewer gas molecules and lowers the pressure. The concentration of NO_2 decreases.

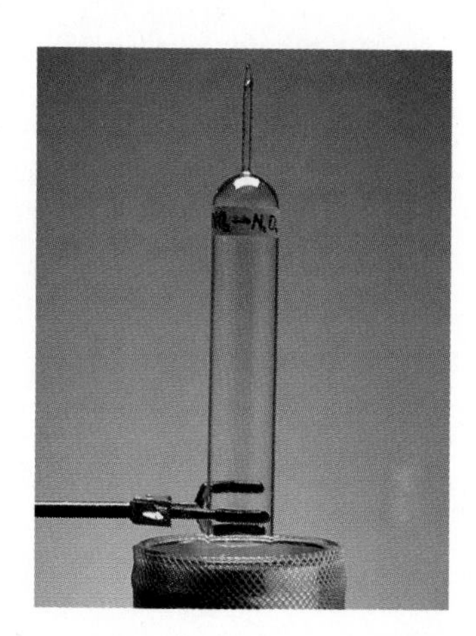

Check Your Understanding

What is the effect of each of the following on the equilibrium concentration of O_2 in the following equilibrium system?

$$2N_2O_5(g) \rightleftharpoons 4NO_2(g) + O_2(g) + \text{heat}$$

1. adding NO_2

2. removing N_2O_5

3. increasing the temperature

4. decreasing the pressure

5. adding a catalyst

KEY CONCEPTS

For a chemical reaction to occur, the reactants must collide with enough energy to overcome an energy barrier called the activation energy, E_{act}. When this occurs, the particles first form a reactive group called the activated complex (a transition state between reactants and products) and then continue on to form the products.

The amount of energy given off or absorbed by a reaction is called the heat of reaction, ΔH. A reaction that gives off energy is called exothermic, and one that absorbs energy is called endothermic.

A chemical equilibrium occurs when the rate of the forward reaction equals the rate of the reverse reaction. This results in no net change in the concentration of reactants and products. Factors that affect the rate of a chemical reaction are the nature of the reactants, the concentration of the reactants, the temperature at which the reaction takes place, and the addition of a catalyst.

Le Chatelier's principle states that when a system at equilibrium is put under stress (by changes in temperature or pressure, or in the concentration of reactants or products), the system changes in a direction that tends to remove the stress.

REVIEW PROBLEMS

Section 8.1

1. Using the concept of activation energy, explain why a dropped glass might remain intact on one occasion and shatter on another.

2. Why do most collisions between reactant particles not result in the formation of a product?

Section 8.2

3. Draw potential energy diagrams for a general endothermic reaction and an exothermic reaction. Label the reactants, the products, the activation energy E_{act}, and the heat of reaction ΔH.

4. Which of the following reactions are endothermic and which are exothermic?

(a) $2H_2(g) + O_2(g) \longrightarrow 2H_2O(g) + 115.6\text{ kcal}$

(b) $N_2(g) + 2O_2(g) + 16.2\text{ kcal} \longrightarrow 2NO_2(g)$

(c) $2NH_3(g) + 22\text{ kcal} \longrightarrow N_2(g) + 3H_2(g)$

(d) $3C(s) + 2Fe_2O_3(s) + 110.8\text{ kcal} \longrightarrow 4Fe(s) + 3CO_2(g)$

5. The following is the potential energy curve for the reaction

$$CO + NO_2 \longrightarrow CO_2 + NO$$

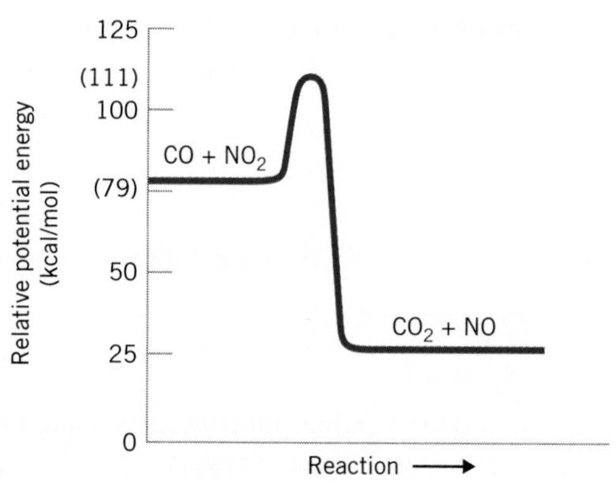

(a) Is this reaction endothermic or exothermic?

(b) What is the heat of reaction ΔH?

(c) What is the activation energy E_{act} for the forward reaction?

(d) What is the activation energy E_{act} for the reverse reaction,

$$CO_2 + NO \longrightarrow CO + NO_2$$

6. The heat of reaction for the conversion of graphite to diamond is small, yet this reaction takes place only under extremely high temperatures and pressure. Why is this? Draw a potential energy diagram for the reaction.

$$0.45\text{ kcal} + C_{(graphite)} \longrightarrow C_{(diamond)}$$

7. What are three things that can be done to speed up the rate of a reaction?

8. Explain why special industrial procedures must be designed for handling large amounts of finely divided, dry, combustible materials.

9. What is a catalyst?

10. Why do labels on certain antibiotic drugs say the drugs must be kept refrigerated?

11. The activation energy for the following reaction is 43.8 kcal. Adding platinum to the reaction mixture lowers the activation energy to 29 kcal.

$$2HI + 3\text{ kcal} \longrightarrow I_2 + H_2$$

(a) What effect does the platinum have on the rate of the reaction? Why?

(b) Draw potential energy diagrams for the catalyzed and uncatalyzed reactions.

(c) What is the activation energy of the reverse reaction before adding the platinum? After adding the platinum?

(d) What is the heat of reaction ΔH for the reverse reaction before adding the platinum? After adding the platinum?

Section 8.4

12. What is *equal* in a reaction at equilibrium?

13. Why is a chemical equilibrium described as dynamic rather than static?

Section 8.5

14. What is the effect of a catalyst on a reaction at equilibrium?
15. List two factors that can alter a system at equilibrium.
16. Dissolving sodium hydroxide in water is an exothermic process. Does sodium hydroxide dissolve more easily in hot water or in cold water?
17. Is an equilibrium established in a can of ether with the lid off or with the lid on? Give a reason for your answer.

Section 8.6

18. State three ways to increase the concentration of NOCl in the following reaction:

$$2NO(g) + Cl_2(g) \rightleftharpoons 2NOCl(g)$$

19. State Le Chatelier's principle in your own words.
20. What effect do the following changes have on the equilibrium concentration of $O_2(g)$ in the following reaction? Give a reason for each of your answers.

$$4HCl(g) + O_2(g) \rightleftharpoons 2H_2O(g) + 2Cl_2(g) + 27 \text{ kcal}$$

   (a) increasing the temperature of the reaction
   (b) increasing the pressure
   (c) decreasing the concentration of Cl_2
   (d) increasing the concentration of HCl
   (e) adding a catalyst

21. The following reaction is exothermic.

$$4HCl(g) + O_2(g) \rightleftharpoons 2H_2O(g) + 2Cl_2(g)$$

   What happens to the equilibrium if the reaction temperature is lowered from 300° to 290°C?

22. Consider the following reaction:

$$C(s) + CO_2(g) + 41.3 \text{ kcal} \rightleftharpoons 2CO(g)$$

   In which direction will the equilibrium shift when the reaction is subjected to each of the following conditions?

   (a) the amount of carbon is increased
   (b) the amount of carbon monoxide is increased
   (c) the reaction mixture is heated
   (d) the amount of carbon dioxide is increased
   (e) the pressure is increased
   (f) a catalyst is added to the mixture

23. What effect does each of the following changes have on the equilibrium concentration of PCl_3 in the system:

$$PCl_3(g) + Cl_2(g) \rightleftharpoons PCl_5(g) \qquad \Delta H = -21 \text{ kcal}$$

   (a) decreasing the concentration of PCl_5
   (b) increasing the concentration of Cl_2
   (c) increasing the temperature
   (d) decreasing the volume of the reaction container

Applied Problems

24. The compound nitrogen triiodide, NI_3, is a solid that is very sensitive to shock. Even brushing it with a feather can cause the NI_3 to decompose to $N_2(g)$ and $I_2(g)$ so rapidly that an explosion results. Draw a potential energy diagram for this reaction and explain how a feather can cause such a large change in the reaction rate.
25. Sliced peaches turn brown because molecules on the surface of the peach react with oxygen in the air. Explain on the molecular level why covering the sliced peaches with plastic wrap and placing them in the refrigerator slows the browning process.
26. Many animals, such as snakes, are cold-blooded; that is, their body temperature stays the same as the temperature of their surroundings. Using your knowledge of reaction rates, explain why snakes are very sluggish on cold mornings and often sun themselves on rocks before searching for food.
27. People exposed to extremely low temperatures face the very real danger of freezing to death. Staying awake and moving as much as possible is important, but staying awake is often difficult. People caught out in the extreme cold may eventually fall asleep and freeze to death. Why is it so hard to stay awake in such conditions?
28. Fever is a protective mechanism that uses an increase in body temperature to fight a disease. On the other hand, if the body temperature rises too much it can be quite harmful to the person. Explain this in terms of reaction rates.
29. An important industrial process for the production of ammonia uses the following reaction:

$$N_2(g) + 3H_2(g) \rightleftharpoons 2NH_3(g) + 22 \text{ kcal}$$

This process uses a catalyst.

(a) Draw the potential energy diagram for the uncatalyzed and the catalyzed reaction.

(b) Describe three steps that manufacturers could take to increase the yield of ammonia in the reaction.

30. Nitrogen dioxide, NO_2, causes the reddish brown haze seen in air pollution. It is formed by the reaction of nitric oxide, NO, with oxygen.

$$2NO(g) + O_2(g) \rightleftharpoons 2NO_2(g)$$

What is the effect of each of the following on this system at equilibrium?

(a) doubling the volume of the reaction flask

(b) adding more oxygen to the reaction flask

(c) increasing the pressure in the flask

31. An industrially important reaction for the production of hydrogen gas is

$$CO(g) + H_2O(g) \rightleftharpoons H_2(g) + CO_2(g) + 9.9 \text{ kcal}$$

Describe how each of the following changes affects the equilibrium concentration of H_2.

(a) addition of H_2O

(b) addition of CO_2

(c) removal of CO

(d) increasing the temperature of the reaction

(e) decreasing the volume of the reaction container

Integrated Problems

1. The aroma of vanillin can be detected by the human nose at the lowest concentration of any chemical: 2.0×10^{-11} g/L of air. The hangar used by the Goodyear blimp has a volume of 5.5×10^{7} ft^3.
   (a) How many grams of vanillin are enough to be detected anywhere in the hangar? How many milligrams?
   (b) If 500 g of vanillin cost $20.45, what is the cost of enough vanillin to be detected anywhere in the hangar?

2. On August 29, 1982, element 109 was produced by bombarding bismuth-209 with iron-58.
   (a) State the atomic number, mass number, number of protons, and number of neutrons in (1) an atom of bismuth-209, and (2) an atom of iron-58.
   (b) Write the balanced nuclear equation for this reaction.
   (c) Element 109 could be expected to have chemical properties similar to what other elements?

3. Strontium-90, a beta emitter, is a radioactive waste product of nuclear testing. When released into the atmosphere, it settles back to earth and is ingested by grazing animals such as cows. Why does strontium-90 in cows' milk pose a serious health risk?

4. In 1983, in the Mexican city of Juarez, an obsolete cancer therapy machine containing 6000 pellets of cobalt-60 was mistakenly broken up as scrap and refined into metal table legs. The six men who were breaking up the machine could have been exposed to as much as 500 rem of whole body radiation.
   (a) If the exposure was that high, how many of these men would be expected to be alive after one month?
   (b) If the activity of the cobalt-60 source was 400 Ci, what is the rate of decay of this source in transformations per second?
   (c) Several children played in the bed of the truck that had carried the machine. The truck was later measured to be giving off 50 rem/h of radiation. If you examine the chromosomes in the children's blood cells, what do you expect to find?

5. Iodine-131 is used in the treatment of thyroid cancer. A patient admitted to the hospital for treatment is placed in a special room in which the bed is surrounded by thick lead shields. The patient is given an oral dose of 150 mCi of iodine-131 and then remains in the room for about four days until the radiation given off by his body falls to a safe level.
   (a) Iodine-131 is produced in the laboratory by bombarding tellurium-130 with neutrons. Write the balanced nuclear equation for this reaction.
   (b) Iodine-131 decays by giving off beta and gamma radiation. Write the balanced equation for this transformation.
   (c) If a hospital receives a shipment of 20.0 g of iodine-131, how much iodine-131 remains in the laboratory after 24 days?
   (d) Why is it important that the nurses caring for the patient wear film badges and not go inside the shielding unless absolutely necessary?
   (e) Must extra care be taken in handling the urine and blood of this patient while he is under treatment? Why?
   (f) If the half-life of iodine-131 is eight days, why is it that the patient can go home in four days?

6. One of the arguments for shutting down nuclear reactors is the high cost of storing radioactive wastes and the lack of a permanent waste disposal site. Many of those who oppose nuclear power plants would replace them with coal-fired power plants. But persons living downwind from a coal-fired power plant can receive between 16 and 20 times the radiation of a person living downwind from a nuclear power plant! Why do you think this is the case?

7. The hemoglobin molecule in red blood cells carries oxygen from the lungs to the tissues. This results in the following equilibrium:

$$\underset{\text{hemoglobin}}{HHb^+} + O_2 \rightleftharpoons \underset{\text{oxyhemoglobin}}{HbO_2} + H^+$$

A second equilibrium system that is present in the blood involves carbon dioxide, carbonic acid, and the bicarbonate ion:

$$\underset{\text{carbon dioxide}}{CO_2} + H_2O \rightleftharpoons \underset{\text{carbonic acid}}{H_2CO_3} \rightleftharpoons \underset{\text{bicarbonate ion}}{HCO_3^-} + H^+$$

Use Le Chatelier's principle and Figure 3.11 to explain how the production of the waste product carbon dioxide by the tissues and the concentration of oxygen in the lungs cause the two equilibrium systems to shift in such a way as to aid in the transport of oxygen to the tissues and the exhalation of carbon dioxide from the lungs.

Chapter 9

Water, Solutions, and Colloids

Judy examined Amaresh and shook her head in wonder and frustration. It had taken Amaresh and her three children 30 days to make the terrifying journey from their Ethiopian village in Tigre to the refugee camp in the desert across the Sudanese border. They had walked the whole way, hiding during the day and traveling only at night to avoid the bombs and bullets of the Ethiopian military planes. They started their journey with only a small amount of food, which had been used up long before they reached the Sudanese border.

The trucks they met at the border took them directly to the refugee camp. Judy, one of the camp's volunteer health workers, had examined many such families as they arrived weak with starvation and fatigue. She assigned Amaresh's two youngest children to a special feeding program to combat their severe malnutrition and then showed the family their new home: a single 10-ft-by-10-ft tent that already housed 15 of their relatives.

The camp routine never varied. Each day, Amaresh went to collect her family's allotment of raw wheat, beans, and cooking oil. She mixed these together and boiled them in the blackish green water drawn from the camp's water tanks. Until recently, water had been taken directly from a nearby irrigation canal, and diarrhea had been rampant in the camp. Unfortunately, many refugees drinking this water were so weakened by malnutrition that even simple diarrhea was life-threatening, especially for children and the elderly. The health workers had put a high priority on constructing the water tank, which allowed the canal water to be chlorinated and to settle before being used in the camp.

The water was not the only source of worry for the health workers. The camp lacked sanitary facilities. All the camp's 12,000 refugees used the open field next to the camp as their toilet. The danger of this arrangement became obvious a week after Amaresh's arrival at the camp, when a fierce sandstorm destroyed most of the tents and the camp's water system. The only available water was that pumped directly from the canal, without any chlorination or settling. By the next morning, with the temperature rising to 100°F, everyone in the camp was suffering from terrible diarrhea.

Judy was not surprised when Amaresh came to her that evening carrying her youngest son. He was very pale and unresponsive and was obviously suffering from severe dehydration from his diarrhea. Because no antibiotics were available, she gave Amaresh a yellow plastic jug and several small sealed packets containing oral rehydration salts (ORS). Judy instructed Amaresh to pour the white powder from a packet into the jug and then to fill the jug with chlorinated water up to a clearly marked line. She was to feed as much of this liquid to her children as they could drink. Judy predicted that the children would be out playing with their friends in only a few days.

The oral rehydration salts that Judy distributed have become almost a miracle treatment for refugees suffering from dehydration caused by diarrhea. It was remarkable that in the six weeks Judy had worked at the camp not one child had died of dehydration from diarrhea. Of the 27.5 g of white powder in each packet of ORS, 7.5 g are sodium chloride, sodium bicarbonate, and potassium chloride. When added to water, these salts help restore the fluids and electrolytes lost from the body during diarrhea. The remaining 20 g is glucose, a sugar that gives the children the calories they need to regain their strength. The additional water also helps control the body's temperature, an extremely important concern in the sub-Saharan heat of the refugee camp. By restoring normal fluid and ion balance, ORS therapy allows the body's immune system to take over and fight the cause of the diarrhea.

Why is dehydration so potentially dangerous? Water is critical to the survival of all living organisms, and it is second only to oxygen in importance for human survival. We can survive for several weeks without food, but only for a few days without water (and only a few minutes without oxygen). The water content of living organisms varies from less than 50% in some bacterial cells to 97% in some marine invertebrates. It is by far the most abundant chemical compound in the human body, making up 60% to 70% of the adult body weight. Water is found inside the cells, in the extracellular (or interstitial) fluid that bathes the cells, and in the blood plasma (Fig. 9.1).

Water performs many biological functions. It is the fluid found throughout living organisms, transporting food and oxygen to the cells and carrying away wastes. It is the fluid in which digestion takes place, and it is a lubricant for the cells and tissues.

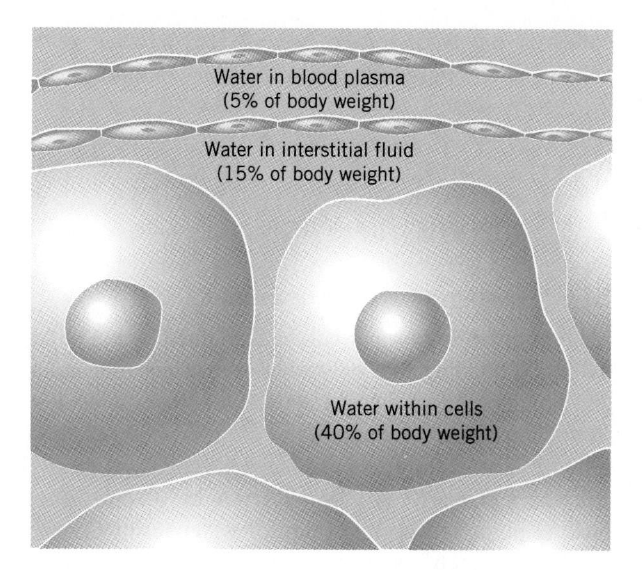

Figure 9.1 Water is the most abundant compound in the human body. It is distributed within cells, in the interstitial fluid that bathes the cells, and in the blood plasma.

Water also plays a major role in the regulation of body temperature and the acid–base balance of body fluids. Water is an important reactant and product in many chemical reactions that take place in living organisms.

Each day, you lose between 1500 and 3000 mL of water in the form of urine, perspiration, water vapor in your breath, and feces. This lost water must be replaced through the liquids you drink or the foods you eat. Dehydration can be caused by diarrhea, high fever, bleeding, burns, or ulcers, all of which can disrupt the normal water balance in the body. For adults, a 10% loss in total body fluids causes serious dehydration and disruption of normal body chemistry; a 20% loss can be fatal. To understand the chemistry of living organisms, we need to examine more closely the properties of water that make it so essential for life.

9.1 Molecular Shape of Water

Water is the most abundant liquid in the world, yet knowledge of its structure is far from complete. It is the special properties of water that make it so vital to life. Water, H_2O, is a molecule containing an oxygen atom covalently bonded to two hydrogen atoms. The shape of the molecule is bent.

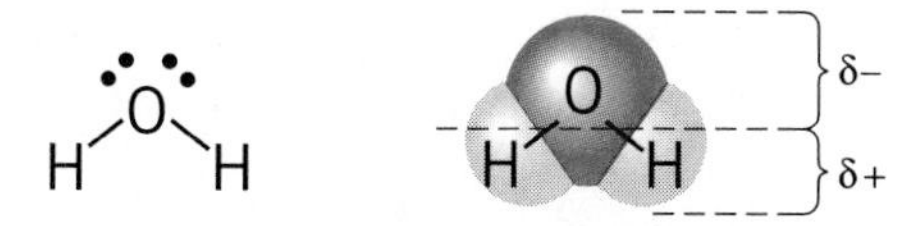

The oxygen atom is much more electronegative than the hydrogen atom, so each covalent bond is polar. In addition, the bent shape of the molecule makes the entire molecule polar, with the oxygen at the negative end and the hydrogens at the positive end. The polarity of water allows the formation of hydrogen bonds between water molecules. Both the polarity of the molecules and the possibility of hydrogen bonding give water its solvent properties. (A solvent is a substance in which other substances dissolve.) Water is called the *universal solvent* because it is a better solvent than most other liquids. It dissolves ionic compounds and molecular compounds that are polar or that contain polar groups.

9.2 Properties of Water

Describe three properties of water that result from its polarity.

High Melting and Boiling Points

Many properties of water can be explained by the polarity of water molecules and by the hydrogen bonding that exists between water molecules. Water has a high melting point and a high boiling point compared with other molecules of similar formula weight. In fact, most compounds of comparable formula weight are gases at the temperature that water is a liquid. For example, water has a formula weight of 18 and a boiling point of 100°C, whereas methane (CH_4, formula weight 16) has a boiling point of −164°C. The high melting point and boiling point of water result from the strong attraction and hydrogen bonding between water molecules, which require more energy to be pulled apart.

Density of Ice

Recall that the density of a substance is a measure of the mass of a substance per unit volume (density = mass/volume). When heated, nearly all substances increase in volume. Because the mass of the substance remains constant, its density decreases as the temperature increases (Fig. 9.2a). As water cools, it contracts until it reaches its maximum density at 3.98°C. As it continues to cool and then freeze at 0°C, however, it expands by 9% (Fig. 9.2b). It is the open lattice structure of ice, in which each molecule of water is hydrogen-bonded to four other water molecules, that makes ice less dense than the more compact liquid water (Fig. 9.3). This is why pipes break in winter when the water in them expands as it freezes. The fact that ice is less dense than water is critical to aquatic life. When the surface of a lake freezes, the ice floats and the denser water (at 4°C) sinks to the bottom, giving plants and animals a place to survive. If ice were denser than water, lakes and oceans would freeze from the bottom up and probably would never completely thaw in summer.

Surface Tension

We mentioned in Section 3.2 that water has a high surface tension. Water molecules beneath the surface are strongly attracted in all directions to other water molecules by hydrogen bonding. The molecules on the surface, however, are not attracted to the non-

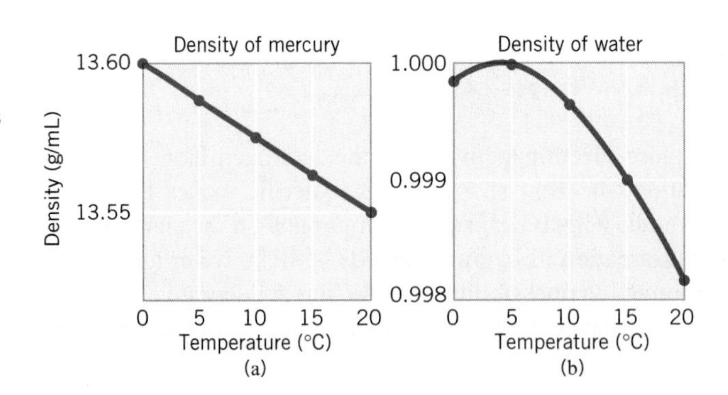

Figure 9.2 (a) The density of most substances, such as mercury, gradually increases as the temperature decreases. (b) The density of water, however, increases to a maximum at 4°C and then decreases as the temperature decreases.

Figure 9.3 This iceberg floats because ice is less dense than water.

polar air molecules. Rather than being attracted in all directions, these surface molecules are therefore attracted only downward and inward toward the water. This surface tension pulls the surface molecules together, creating the effect of a thin, elastic membrane on the surface. This is why, for example, dust particles or water spiders can remain on the water's surface even though they are denser than water. You may also have noticed that water, when dropped onto waxed paper, forms small beads of liquid. This occurs because the surface water molecules are attracted neither to the nonpolar wax nor to the surrounding air, and thus the attraction between water molecules draws the liquid up to form a bead. Glass, on the other hand, is polar like water. Water is attracted to glass and, when placed on a glass surface, spreads out rather than forming beads. Substances that reduce the surface tension of water are called **surface-active agents,** or **surfactants.** Surfactants, such as soap or detergents, allow water to penetrate the dirt and oil in clothes in the laundry, and surfactants such as bile aid in the digestion of fats.

Water forms spherical drops because of its high surface tension.

High Heat of Vaporization

Water has a high heat of vaporization (539 cal/g). The strong attractive forces between water molecules cause the liquid to boil away very slowly. It takes nearly seven times as much energy to boil away 1 L of water at 100°C as it does to heat it from 21°C (room temperature) to 100°C. It is important to realize that evaporation is essentially the same process as boiling; it involves changing liquid water to gas, only at a slower rate. As with boiling, evaporation uses up a great deal of energy in the form of heat. Animals make use of this principle to rid their bodies of excess heat through the evaporation of sweat.

High Heat of Fusion

Water also has a high heat of fusion (80 cal/g); that is, a great deal of heat must be released before ice can be formed. To convert 1 kg of liquid water at 0°C to 1 kg of ice requires almost four times as much refrigeration as is needed to cool this much water from 21°C to 0°C. This explains why ice forms slowly in the winter and melts slowly in the spring. Water stored as snow in the mountains runs off relatively slowly, helping to prevent disastrous floods in the spring.

Chemical Perspective / Heat of Fusion

Scott and Diane McIntire, with their five-month-old daughter Emily, were driving down a logging road in the Oregon Cascade Mountains when snow began falling from the suddenly dark sky. It wasn't long before Scott, unable to keep the car from sliding uncontrollably, skidded into a ditch and became stuck. Unable to get the car out of the ditch, they decided to spend the night in the car. Finding themselves completely snowed in the following morning, the McIntires made a fateful decision. Rather than waiting in their car to be found, they decided to hike to a ranger station that Scott believed to be only 5 miles away. Dressed only in light clothing they walked down the road in search of the ranger station but found themselves lost as evening drew near. Unable to find shelter, they settled down by a log. Diane had been eating snow throughout the day to keep her fluid level high enough to breast-feed the baby. As night descended, she grew increasingly cold. The next morning, cold and incoherent, Diane died. For the next two days, Scott remained by her body and managed to keep himself and his child alive until they were rescued.

Why had Diane McIntire died? In her attempt to keep her baby alive, she gave her own life by eating snow. Had she drunk water from a stream or even water from thawing snow or icicles, she would probably be alive today. Instead, however, Diane robbed her body of precious heat in order to melt the snow she had eaten. As you know, water has a very high heat of fusion. The 80 cal necessary to melt each gram of snow meant that Diane lost more than 36,000 cal for each pint of water she obtained. This tremendous heat loss led to Diane's death from hypothermia.

Your Perspective: You have probably seen cartoons of Saint Bernard dogs being used in the Alps to rescue mountain climbers lost in the snow. These dogs are often shown carrying a small barrel of brandy around their necks for the victims. Write a short essay explaining why this may or may not make sense, and suggest what the rescue dogs might carry instead.

High Specific Heat

The specific heat of a liquid is the amount of heat required to raise the temperature of one gram of that liquid by one degree Celsius. Water has a high specific heat compared with other liquids (Table 9.1). The higher the specific heat of a substance, the less its temperature changes when it absorbs a given amount of heat. The high specific heat of water enables this fluid to keep the temperature of an organism relatively constant in the face of fluctuating internal or external heat levels. On a larger scale, the water in lakes and oceans absorbs and stores large quantities of solar energy, explaining why such large bodies of water have a moderating effect on local climates.

Table 9.1 **Specific Heat of Several Compounds**

Compound	*Specific Heat (cal/g)*	*Compound*	*Specific Heat (cal/g)*
Water	1.0	Chloroform	0.23
Ethanol	0.58	Ethyl acetate	0.46
Methanol	0.60	Liquid ammonia	1.12
Acetone	0.53		

Table 9.2 **Comparison of Properties of Solutions, Colloids, and Suspensions**

Property	*Solutions*	*Colloids*	*Suspensions*
Particle size	<1 nm[a]	1–1000 nm	>1000 nm
Filtration	Passes through filters and membranes	Passes through filters but not membranes	Stopped by filters and membranes
Visibility	Invisible	Visible in an electron microscope	Visible to the eye or in a light microscope
Motion	Molecular motion	Brownian movement	Movement only by gravity
Passage of light	Transparent, no Tyndall effect	May be transparent or often translucent or opaque, Tyndall effect	Often opaque, may be translucent

[a]1 nanometer (nm) = 10^{-9} meter.

9.3 Three Important Mixtures

In Chapter 1, we learned that matter can be classified as elements, compounds, and mixtures. Mixtures make up most of the things we see around us. The soil, buildings, lakes, the cells of our body—all are made up of mixtures. Mixtures are categorized into three types depending on their particle size: solutions have very small particles, colloids have medium-sized particles, and suspensions have the largest particles. The size of the particles determines various properties as described in Table 9.2.

Describe and compare the three different types of mixtures.

9.4 Suspensions

Suspensions are heterogeneous (nonuniform) mixtures containing large particles that, in time, settle out. Medicines such as milk of magnesia that have labels saying ''Shake well before using'' are suspensions. Suspensions made up of solid particles suspended in a liquid can be separated using filter paper or a centrifuge. Whole blood is an example of a suspension; the red and white blood cells settle out in time, or they can be separated from the plasma using a centrifuge.

9.5 Colloids

Define colloid and give an example.

Colloidal dispersions, or **colloids,** are mixtures whose particles are larger than the molecules of ions forming solutions but smaller than particles forming suspensions. Colloids are usually homogeneous. The particles in a colloidal dispersion cannot be separated out by ordinary filter paper. Homogenized milk, for example, is a colloidal dispersion of butterfat in water; it passes right through filter paper. Colloids can be classified by the solvent (or dispersing medium) and by the colloidal matter (or the dispersed medium). The eight classes that result are shown in Table 9.3. Colloidal chemistry is important in the study of biological systems. Tissues and cells are colloidal dispersions, and reactions occurring within them take place through colloidal chemistry. Food digestion, for exam-

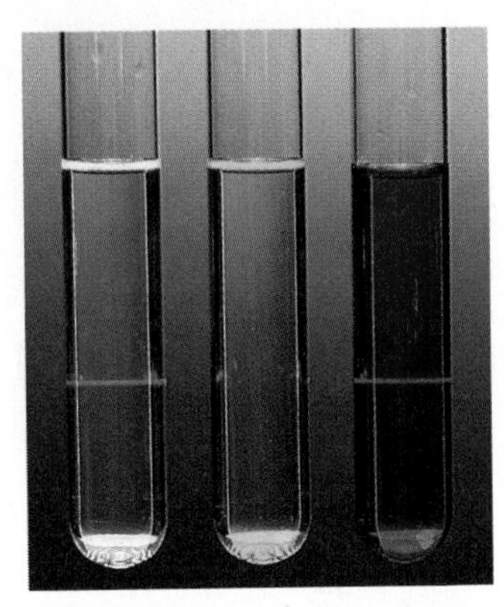

Figure 9.4 The Tyndall effect. A laser beam passes through three test tubes. The middle tube contains a salt solution, and the two other tubes contain a colloidal dispersion of starch.

Table 9.3 **Classes of Colloids**

Class	*Examples*
1. Solid in solid	Colored glass, certain alloys
2. Solid in liquid	Gelatin in water, protein in water, starch in water
3. Solid in gas	Aerosols, dust in air, smoke
4. Liquid in solid	Water in gems (opals and pearls), jellies, butter, cheese
5. Liquid in liquid	Emulsions, milk, mayonnaise, protoplasm
6. Liquid in gas	Aerosols, fog, mist, clouds
7. Gas in solid	Activated charcoal, styrofoam, marshmallows
8. Gas in liquid	Foam, whipped cream, suds
(Gas in gas)	(A solution, not a colloid)

ple, involves the formation of colloids before the food can be digested. The contraction of muscles can be explained by colloidal chemistry, and the body's proteins are of colloidal size.

State three properties that are unique to colloids.

Have you ever watched particles of dust dancing in a sunbeam or moving randomly about in the light from a movie projector? The random movement of particles in a colloid is caused by the bombardment of these particles by the solvent molecules. The resulting motion is called **Brownian movement** after the English botanist Robert Brown. He first observed this irregular movement of particles when looking through a microscope at pollen grains suspended in water. This constant bombardment by solvent molecules helps keep the colloidal particles from settling out.

When you watch dust dancing in a sunbeam, you are not seeing the actual dust particles—they are too small to be seen with the naked eye. What you are actually seeing is the scattering of light by the particles. Unlike the extremely small particles in a solution, colloidal particles scatter light. You can distinguish a dilute colloidal dispersion that may look like a solution from a true solution by passing a strong beam of light through the two mixtures. The path of the light beam is clearly visible in the colloidal dispersion because of the scattering of the light by the colloidal particles (Fig. 9.4). This property of colloids is known as the **Tyndall effect.** Light rays can be seen streaming through a forest, for example, as a result of the scattering of light by colloidal particles suspended in the forest air (Fig. 9.5).

Figure 9.5 Sunbeams decorate the forest as a result of the Tyndall effect.

Colloidal particles range in size from 1 to 1000 nanometers (1 nm $= 1 \times 10^{-9}$ m) in diameter. Particles of this size have very large surface areas compared with their volumes. This large surface area gives colloidal particles the ability to take up, or **adsorb,** substances on their surface. Powdered charcoal is an example of a substance that has many practical uses because of the colloidal size of its particles. It is put in gas masks to adsorb poisonous gases in the air, and it is used to remove gases and odors from city water supplies. It is also used to remove colored impurities from solutions in the laboratory and in industry and even serves as an antidote for swallowed poisons. A person who has swallowed poison or taken an overdose of drugs can often be saved by passing their blood through a machine containing a charcoal filter that removes the toxic substance from the blood.

9.6 Solutions

A **solution** is a homogeneous, or uniform, mixture of two or more substances (called crystalloids) whose particles are of atomic or molecular size. Club soda (carbon dioxide in water), air (gas in gas), rubbing alcohol (alcohol in water), and steel (carbon in iron) are all

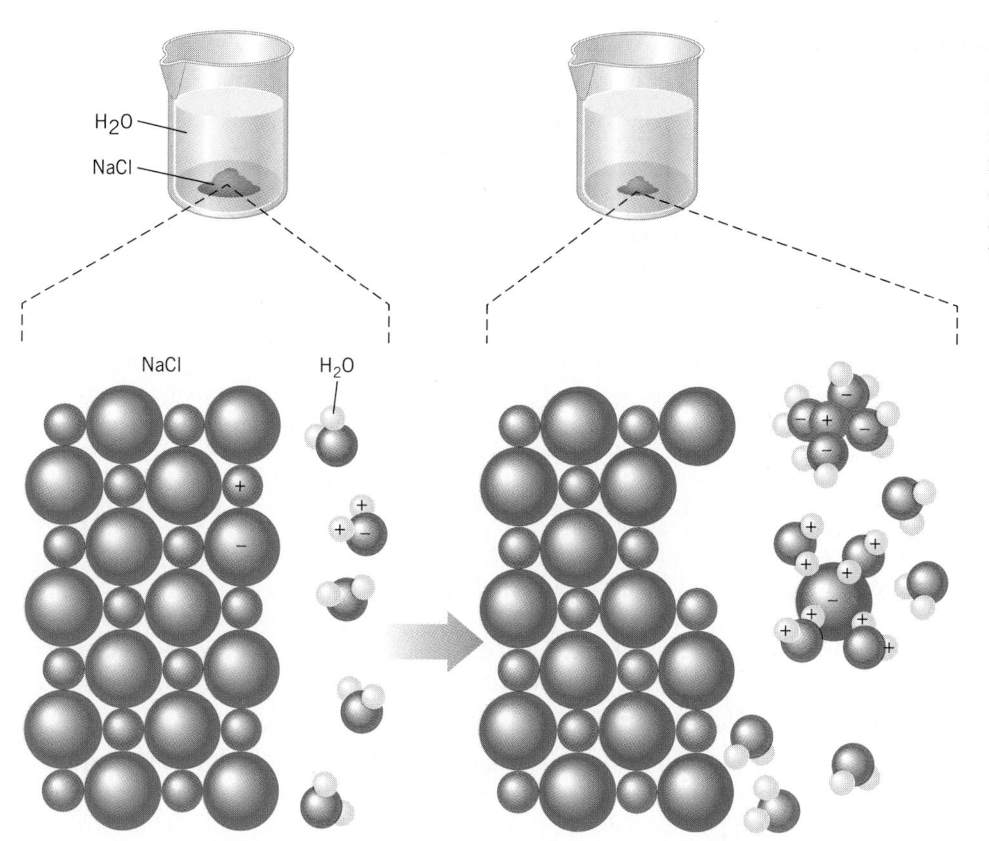

Figure 9.6 The ionic compound sodium chloride dissolves in water because the attraction of the water molecules for the sodium and chloride ions is greater than the attraction between the ions in the crystal.

examples of solutions. We call the substance being dissolved the **solute;** the **solvent** is the substance in which the solute is dissolved. Our major interest in this text is **aqueous** solutions, which are solutions in which water is the solvent.

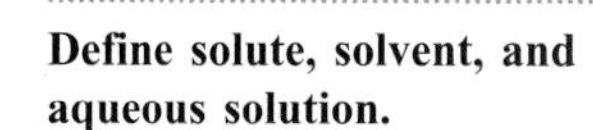

Define solute, solvent, and aqueous solution.

Water is often called the universal solvent because its highly polar nature allows a large number of substances to dissolve in it. The process of dissolving occurs as follows. When, for example, a sodium chloride crystal is placed in water, the surface of the crystal is bombarded by the water molecules. The polar ends of the water molecules exert an attractive force on the surface ions of the crystal and pull them loose (Fig. 9.6). The water molecules then surround each ion as it is pulled loose. This action prevents the sodium and chloride ions from rejoining. Such an ion surrounded by water molecules is said to be **hydrated.** Note, however, that not all ionic substances dissolve in water. For some crystals, the attraction of the water molecules may not be strong enough to overcome the attraction between the oppositely charged ions within the crystal.

Highly polar molecular substances, such as sugar, tend to be soluble in water. The hydration of sugar molecules occurs because of the attraction and hydrogen bonding between oppositely charged polar regions on the sugar and water molecules (Fig. 9.7).

9.7 Electrolytes and Nonelectrolytes

In the dissolving process, solute particles take two forms: they may be charged (ionic) particles or uncharged (molecular) particles. Substances such as sugar, which form uncharged molecular particles when dissolved in a solvent, are called **nonelectrolytes.**

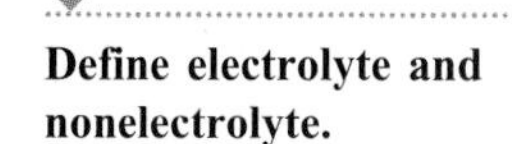

Define electrolyte and nonelectrolyte.

$$\text{Sugar}(s) + H_2O \longrightarrow \text{Sugar}(aq)$$

Chemical Perspective / Cold Packs and Hot Packs

An ionic solid dissolving in water to form an aqueous solution can be either an endothermic or exothermic process. Both types of dissolving processes, in the form of cold packs and hot packs, have been put to use for administering first aid.

Cold packs illustrate a relatively new application of heat of reaction. These single-use packets, commonly seen at athletic events and in hospital emergency rooms, are especially useful where ice is not available. A cold pack contains ammonium nitrate (NH_4NO_3) in a sealed compartment surrounded by water. When the seal separating the two is broken, the following endothermic reaction occurs:

$$NH_4NO_3(s) \longrightarrow NH_4^+(aq) + NO_3^-(aq)$$

Every mole of ammonium nitrate absorbs 6.72 kcal when it dissolves, causing a significant drop in the temperature of the surrounding water.

Hot packs serve as another illustration of a practical application of heat of reaction. The only difference between hot packs and cold packs is in the salt used to make the solution. Calcium chloride is the salt usually used for hot packs because, when it dissolves, it gives off 14 kcal of heat per mole.

Your Perspective: Imagine that you write a monthly health column for the newsletter of the local youth soccer league. Suppose that a parent coach writes to ask you whether she should carry a supply of both cold packs and hot packs. Write a column for the newsletter answering this question, giving examples of possible uses if you believe that either should be carried.

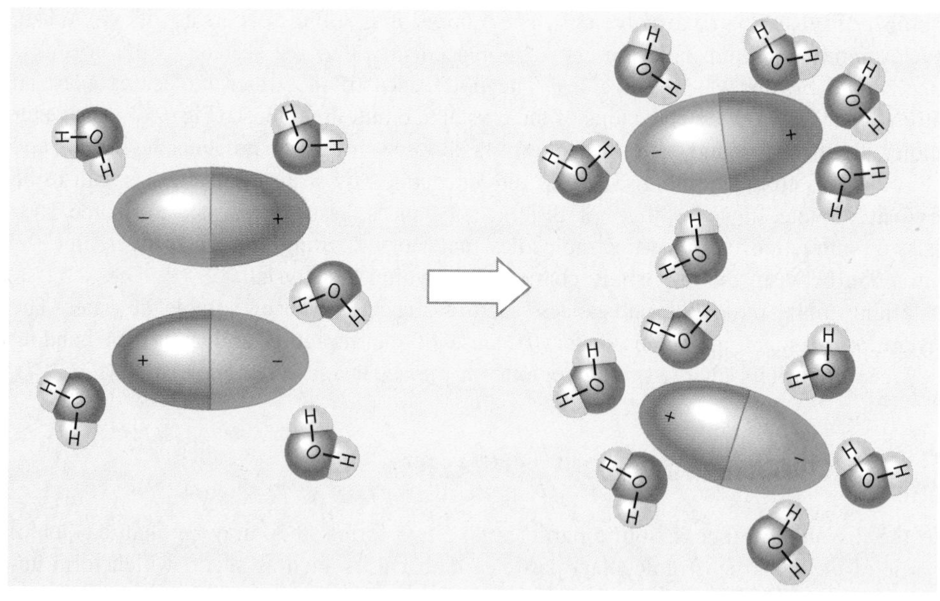

Figure 9.7 Highly polar molecular substances are soluble in water, especially if they contain atoms, such as oxygen or nitrogen, that can form hydrogen bonds with the water molecules.

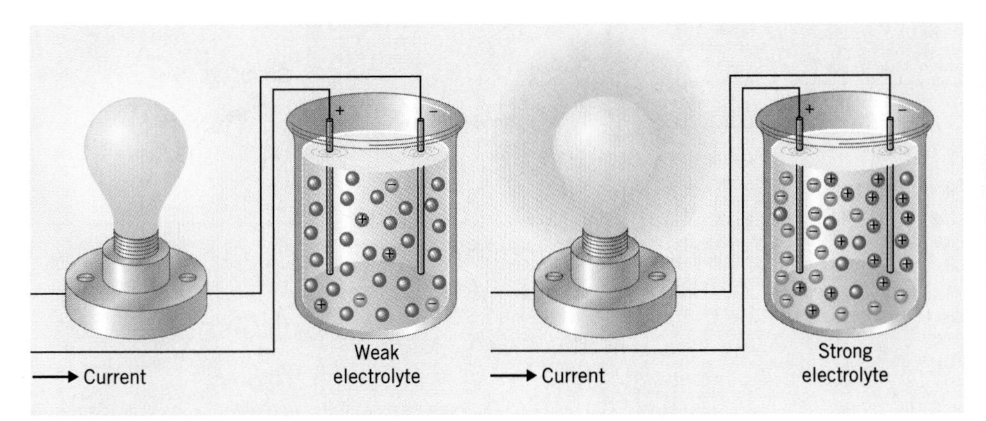

Figure 9.8 Solutions of electrolytes conduct electricity. The stronger the electrolyte, the better the solution conducts electricity.

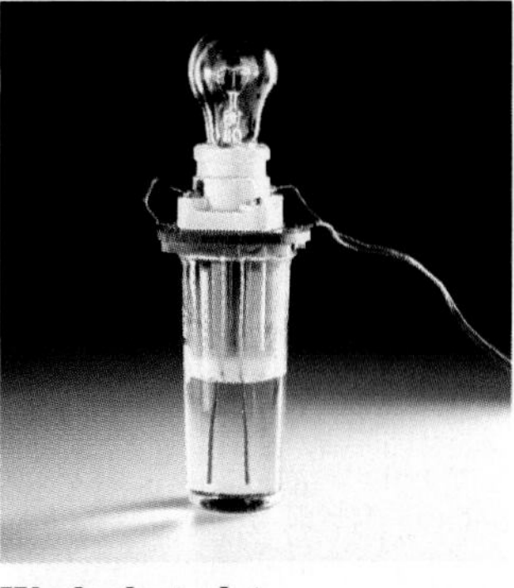

Weak electrolyte

Strong electrolyte

Ionic compounds such as sodium chloride, which dissolve (or dissociate) to form charged (ionic) particles in solution, are called **electrolytes.**

$$NaCl(s) + H_2O \longrightarrow Na^+(aq) + Cl^-(aq)$$

Some covalent compounds are so highly polar that the molecules are pulled apart, or ionized, by water molecules. For example, hydrogen chloride in water forms an aqueous solution of hydrochloric acid. These polar compounds also form electrolytes in solution.

$$HCl(g) + H_2O \longrightarrow H_3O^+(aq) + Cl^-(aq)$$

State the difference between strong and weak electrolytes.

The term electrolyte refers to the fact that solutions of electrolytes conduct electricity. Electrolytes may be classified as either strong or weak, depending on the number of charged (ionic) particles formed when the substance dissolves. A **strong electrolyte** is a substance that completely ionizes, or dissociates into ions, in solution. A **weak electrolyte** is a substance that only partially ionizes in solution (Fig. 9.8). Potassium fluoride and hydrogen chloride are examples of strong electrolytes, whereas acetic acid and carbonic acid are weak electrolytes.

Electrolytes perform many important regulatory roles in our bodies, and they are responsible for maintaining the acid–base and water balances. The most abundant positive ions, or cations, found in living tissue are Na^+, K^+, Ca^{2+}, and Mg^{2+}. The most abundant negative ions, or anions, are HCO_3^-, Cl^-, HPO_4^{2-}, SO_4^{2-}, organic acids, and proteins (Fig. 9.9).

9.8 Factors Affecting the Solubility of a Solute

Many factors affect the solubility of a solute. A major factor is the nature of the solvent and the solute. As a rule, like dissolves like: polar solvents dissolve polar solutes, and nonpolar solvents dissolve nonpolar solutes. For most solid substances, increasing the temperature increases the solubility of the solute. There are a few compounds, however, such as $CaCr_2O_7$, $CaSO_4$, and $Ca(OH)_2$, whose solubility decreases with an increase in temperature.

The solubility of a gas in a liquid decreases with an increase in temperature. For example, we have discussed how the warming of river water caused by heat discharged

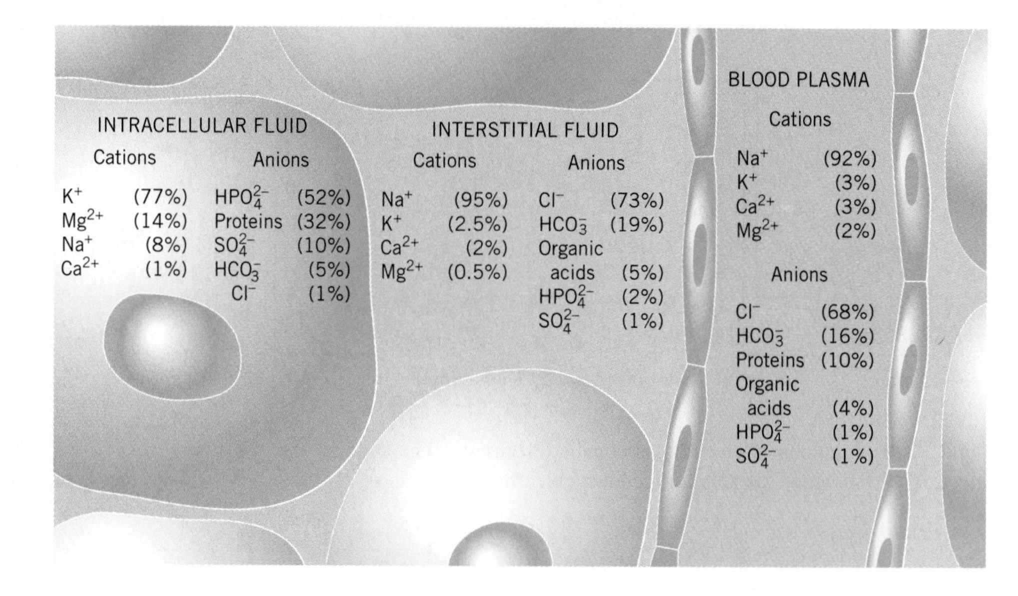

Figure 9.9 Principal electrolytes found in body fluids.

from a nuclear power plant decreases the supply of oxygen available for the river's fish. Also, think about how much more gas rushes out when you open a warm bottle of pop than a cold one. Finally, as mentioned in Chapter 3 the solubility of a gas in a liquid increases with an increase in pressure (Henry's law).

9.9 Solubility of Ionic Solids

We have said that ionic solids vary in their solubility in water. Although the exact solubility of each compound must be determined by experiment, it is possible to make a few general statements about the solubility or insolubility of various solids.

I	*II*	*III*
Compounds containing the following cations are generally soluble:	Compounds containing the following anions are generally soluble:	All ionic compounds formed from ions not listed in I or II are generally not soluble or, at most, only slightly soluble
Li^+	NO_3^-, ClO_3^-, CH_3COO^-	
Na^+	Cl^-, Br^-, and I^- (except with Ag^+, Pb^{2+}, and Hg_2^{2+})	
K^+	SO_4^{2-} (except with Pb^{2+}, Sr^{2+}, and Ba^{2+})	
NH_4^+		

Predict whether a precipitate will form when two ion-containing solutions are mixed.

Using the table shown here, we can predict whether an insoluble solid, called a **precipitate,** will form when two aqueous solutions of soluble ionic compounds are mixed. For example, will a precipitate form when aqueous solutions of sodium iodide (NaI) and silver nitrate ($AgNO_3$) are mixed? A solution of sodium iodide contains sodium ions (Na^+) and iodide ions (I^-), and a solution of silver nitrate contains silver ions (Ag^+) and nitrate ions (NO_3^-). The two new substances that can be formed from this combination are AgI and $NaNO_3$. We see from the table that $NaNO_3$ is soluble, but AgI is not. Thus, we expect a precipitate of AgI to form. We can write these results in equation form:

$$Na^+(aq) + I^-(aq) + Ag^+(aq) + NO_3^-(aq) \longrightarrow AgI(s) + Na^+(aq) + NO_3^-(aq)$$

Cancelling the ions that appear on both sides of the equation but were not involved in the reaction, we get

$$Ag^+(aq) + I^-(aq) \longrightarrow AgI(s)$$

This equation is called the **net-ionic equation,** because it shows only those ions that react.

Example 9-1

Kidney stones are formed from the precipitate of the reaction between sodium phosphate and calcium chloride found in solution in the kidneys. Write the net-ionic equation for the formation of kidney stones.

See the Question

This problem requires us to determine the chemical formula for the precipitate and then write the net-ionic equation for its formation.

Think It Through

We need to write the correct formulas for the reactants and then determine what ions are formed in solution. From these ions and the rules stated previously we can determine the identity of the precipitate.

Reactant	*Ions formed in solution*
Na_3PO_4	Na^+, PO_4^{3-}
$CaCl_2$	Ca^{2+}, Cl^-

The two new substances formed from this reaction are sodium chloride (NaCl) and calcium phosphate [$Ca_3(PO_4)_2$]. From the solubility rules we see that sodium chloride is soluble, but calcium phosphate is not. Calcium phosphate, therefore, forms the precipitate that results in kidney stones.

Execute the Math

First, we must write a complete, balanced equation for this reaction.

Unbalanced: $$Na_3PO_4(aq) + CaCl_2(aq) \longrightarrow Ca_3(PO_4)_2(s) + NaCl(aq)$$

Balanced: $$2Na_3PO_4(aq) + 3CaCl_2(aq) \longrightarrow Ca_3(PO_4)_2(s) + 6NaCl(aq)$$

Now we need to expand this equation to show all the ions that are free to react.

$$6Na^+(aq) + 2PO_4^{3-}(aq) + 3Ca^{2+}(aq) + 6Cl^-(aq) \longrightarrow Ca_3(PO_4)_2(s) + 6Na^+(aq) + 6Cl^-(aq)$$

Prepare the Answer

Cancelling the "spectator" ions on both sides of the equation, we get the following net-ionic equation:

$$2PO_4^{3-}(aq) + 3Ca^{2+}(aq) \longrightarrow Ca_3(PO_4)_2(s)$$

Check Your Understanding

1. Does a precipitate form when aqueous solutions of the following compounds are mixed?

   (a) NaOH and $MgCl_2$ **(c)** NaCl and $Pb(NO_3)_2$

   (b) NaCl and $(NH_4)_2SO_4$ **(d)** $Ba(NO_3)_2$ and Li_2SO_4

2. Using the steps discussed in Example 9-1, write the net-ionic equations for the reactions in Question 1 that produce a precipitate.
3. Write the net-ionic equation for the reaction between calcium chloride and sodium carbonate, forming the insoluble precipitate calcium carbonate.

KEY CONCEPTS

Water's polar structure gives it unique properties that make water critical to all living organisms. It is a good solvent for polar substances, has a high freezing and a high boiling point, and is less dense as a solid than as a liquid at 0°C. It also has a high surface tension, heat of vaporization, heat of fusion, and specific heat.

Solutions, colloids, and suspensions are three types of mixtures that differ according to the size of their particles. A suspension is a heterogeneous mixture containing relatively large particles that settle out in time. A colloid is a mixture in which particles of the dissolved substance are larger than the particles in a solution but smaller than those in a suspension. Brownian movement, the Tyndall effect, and the ability to adsorb large amounts of other substances onto the surface of the colloidal particles are all properties of colloids.

A solution is a homogeneous mixture of two or more substances whose particles are of atomic or small molecular size. The dissolving medium is called the solvent, and the substance that is dissolved is the solute. An aqueous solution is one in which the solvent is water. The solute may be an electrolyte, which forms ions in solution, or a nonelectrolyte, which forms uncharged particles in solution.

The solubility of a substance in a solvent depends on the nature of the solute and solvent, the temperature and, for a gas, the pressure. When two solutions containing electrolytes are mixed, a precipitate may form if any of the possible combinations of ions is insoluble in the solvent.

REVIEW PROBLEMS

Section 9.1

1. Why is water called the universal solvent?
2. Which do you expect to be more soluble in water: carbon tetrachloride or ammonia? Give a reason for your answer.

Section 9.2

3. Suppose an ice cube at 0°C weighs 15.2 g. How many calories of heat energy are needed to melt the ice cube, raise the temperature of the resulting water to 100°C, and convert it to steam at 100°C?
4. How many calories of heat energy are needed to raise the temperature of a 22-g sample of water from 15°C to 90°C?
5. How many calories of heat energy must be removed from a 34-g sample of water at 45°C to cool it to 0°C and then freeze it at 0°C?

6. How many calories of heat energy are required to heat a 30-g sample of methanol from 25°C to 52°C?

7. A 250-g sample of water was heated from 10° to 15°C. If the water had been replaced with an equal amount of chloroform at 10°C and the same amount of heat added, what would be the final temperature?

8. Explain the reason for each of the following properties of water in terms of the polarity of the water molecule and the hydrogen bonding that exists in water.
   (a) high boiling point
   (b) density of ice
   (c) high heat of vaporization
   (d) solubility of molecular substances

Section 9.3

9. Based on their particle sizes, what are the three classes of mixtures?

10. What is the most important difference in composition between elements, compounds, and mixtures?

Section 9.4–9.6

11. State what test can be used to distinguish between (a) a solution and a colloidal dispersion, and (b) a suspension and a colloidal dispersion.

12. What causes the Brownian movement observed in a colloidal dispersion?

13. Using the information in Table 9.3, classify the following colloids:
    (a) cheese
    (b) cement
    (c) hair spray
    (d) pudding
    (e) blood plasma
    (f) soap suds

14. What is the difference between a solvent, a solute, and a solution?

15. State in your own words the events that occur when sodium chloride is dissolved in water.

16. Why do highly polar molecules like sugar dissolve in water?

Section 9.7

17. When dissolved in water, which of the following compounds form solutions that conduct electricity? Identify each as an electrolyte or nonelectrolyte.
    (a) ethanol (C_2H_6O)
    (b) potassium iodide
    (c) carbonic acid

18. Which of the following, when dissolved in water, form solutions that conduct electricity?
    (a) acetone (C_3H_6O)
    (b) sodium sulfate
    (c) lithium nitrate

19. Which of the following are cations and which are anions?
    (a) PO_4^{3-} **(b)** H^+ **(c)** Br^- **(d)** Ba^{2+}

20. What is the difference between a weak electrolyte and a strong electrolyte?

Section 9.8

21. Explain why a soft drink goes ''flat'' faster at room temperature than in the refrigerator.

22. What effect does an increase in temperature have on the solubility of sodium chloride in water?

Section 9.9

23. State which of the following ionic solids are soluble in water:

(a) $MgCO_3$ **(c)** NH_4Br **(e)** KNO_3

(b) NaOH **(d)** $PbSO_4$ **(f)** AgCl

24. Does a precipitate form when aqueous solutions of the following compounds are mixed?

(a) $Pb(NO_3)_2$ and $Fe_2(SO_4)_3$ **(c)** $AgNO_3$ and NaCl

(b) NaBr and LiOH **(d)** NH_4Cl and $Pb(ClO_3)_2$

25. Hydrochloric acid, HCl(*aq*), reacts with sodium hydroxide, NaOH(*aq*), to produce water, H_2O, and sodium chloride, NaCl(*aq*). Write the net-ionic equation for this reaction.

26. Sodium carbonate, $Na_2CO_3(aq)$, reacts with sulfuric acid, $H_2SO_4(aq)$, to form sodium sulfate, $Na_2SO_4(aq)$, water, H_2O, and carbon dioxide, $CO_2(g)$. Write the net-ionic equation for this reaction.

Applied Problems

27. A student adds 250 g of ice cubes at 0°C to 3.0 L of tea at 50.0°C. If all the heat lost by the ice as it melts is absorbed by the tea, what is the temperature of the tea when all the ice has melted? (Assume tea has the same specific heat as water, and ignore the added volume of the melted ice cubes.)

28. Temperatures in Death Valley often climb well above 120°F during the day, but at night they sometimes drop to below freezing (32°F). At the same time, temperatures in nearby Los Angeles may reach a maximum of only 80°F during the day and drop only slightly into the high 60s at night. What is the reason for this drastic difference in temperature fluctuations?

29. Silver bromide, the light-sensitive coating on photographic film, is made in a darkroom by mixing aqueous solutions of silver nitrate and sodium bromide. Write the net-ionic equation for this reaction.

30. Barium ions, when swallowed, are highly toxic. If a person swallows some barium chloride, sodium sulfate is given as an antidote. Explain why this is an effective antidote, and write the net-ionic equation for the reaction that takes place.

31. Urea is a water-soluble product formed in the liver from the ammonia produced in the breakdown of protein. Urea is toxic in high concentrations and, therefore, is excreted by the body in the urine. Explain why people often feel thirsty after a meal rich in protein.

Chapter 10

Solution Concentrations

*D*oug Carson, a 23-year-old college student, was brought to the emergency room of the Good Samaritan Hospital by his friends. Doug was in an extremely agitated state: he was unable to sit still, was talking in a rapid-fire, nonstop fashion—insisting to everyone that he had just been appointed secretary to the governor—and seemed confused by even simple questions. His friends told the doctor that Doug had been staying up with little or no sleep for the last three weeks, had been eating little, and had been abnormally cheerful and carefree. That evening, he had shocked them all by throwing a deskful of books out of his dorm window "to get rid of alien influences."

A physical examination and several blood tests ruled out the possibility that Doug was suffering from a physical disease or from hallucinations that might be caused by drugs. The emergency-room doctor then called in a psychiatrist, who diagnosed mania (that is, Doug was suffering from the agitated phase of manic–depressive illness). The psychiatrist recommended that Doug be admitted to the hospital and treated with lithium carbonate (600 mg three times a day). Lithium has been used for many years for its calming effects; it is thought to help regulate the balance between ions inside and outside cells in the body.

Two days after beginning lithium therapy, Doug became confused and uncoordinated and began vomiting. A blood serum test for lithium was quickly performed. Lithium is effective therapeutically at blood serum levels of 0.6 to 1.5 milliequivalents/liter (mEq/L). Doug's blood, however, had a Li^+ concentration of 1.8 mEq/L. High levels of lithium can cause toxic effects as severe as kidney damage, coma, and death.

Doug was immediately given extra fluids and sodium to reduce the serum lithium level in his blood, and his new symptoms soon disappeared. When Doug's original agitated symptoms again began, lithium therapy was resumed at 300 mg four times daily. Four days later, a blood test revealed Doug's serum lithium level to be at 0.75 mEq/L, and Doug was noticeably calmer and more rational. After nine days of treatment in the hospital, Doug was discharged and returned to his college studies. Each Tuesday he met with the psychiatrist, and his friends reported that he was "back to normal." Doug continued to take lithium carbonate for six months. During this time, his serum lithium concentration was carefully monitored, and the drug dosage adjusted to maintain the lithium level within the therapeutic range.

As you can see, knowing the exact concentration of a component of the blood, or a drug in a prescription, can be critical to the diagnosis and treatment of disease. In this chapter we discuss some commonly used units for measuring solution concentration.

10.1 Saturated and Unsaturated Solutions

Define saturated and unsaturated solutions.

We may say that a solution is **dilute** if only a few solute particles are dissolved in it, or **concentrated** if many solute particles are dissolved. The terms "dilute" and "concentrated," however, are not particularly precise, so scientists have developed methods to describe more accurately the concentration of a solute in solution. A **saturated** solution is one that contains all the solute particles the solvent can normally hold at that temperature (if it contained fewer particles the solution would be **unsaturated**). A state of equilibrium exists, in a saturated solution, between the dissolved solute and the undissolved solute remaining in the container.

$$\text{Undissolved solute} \underset{\text{rate}_r}{\overset{\text{rate}_f}{\rightleftharpoons}} \text{dissolved solute}$$

Explain what makes a solution supersaturated, and give an example.

When a saturated solution is cooled, the solubility of the solute decreases. The rate of the crystallizing process (rate_r) is greater than the rate of the dissolving process (rate_f). By Le Chatelier's principle, the solute continues to form crystals until a new equilibrium is established with fewer solute particles in solution. Occasionally, however, the solution initially contains no crystals on which the dissolved particles can deposit. In such a case, no crystals form when the temperature of the solution is lowered, and the solution is then said to be **supersaturated.** A supersaturated solution, however, is very unstable. If a small crystal is added, onto which the solute can deposit, a very rapid formation of crystals often occurs (Fig. 10.1). Honey and jellies are often supersaturated sugar solutions that, after long storage, may be found to contain sugar crystals.

Figure 10.1 In the left photo, a seed crystal has been added to the supersaturated solution. The solute rapidly crystallizes. The solution surrounding the crystal in the right photo is now a saturated solution.

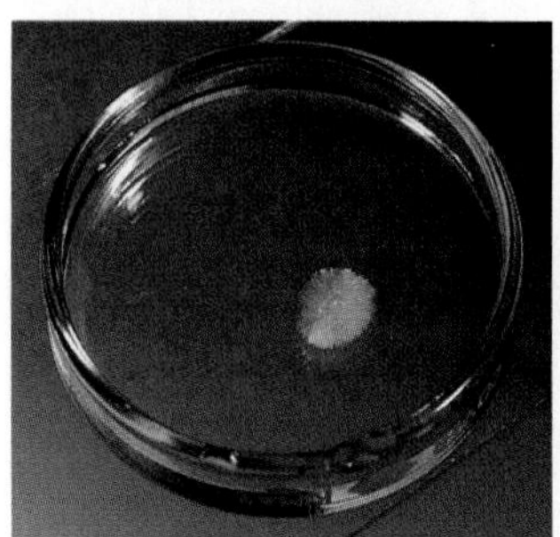

(1)

(2)

(3)

(4)

(5)

Figure 10.2 The steps in preparing a solution of known concentration. (1) Add a known weight of solute to a volumetric flask. (2) Add some of the solvent. (3) Swirl to dissolve the solute. (4) Add solvent carefully, until the solution level reaches the mark on the volumetric flask. (5) Shake the stoppered flask to mix well.

10.2 Molar Concentration

When working with solutions, we often need to know exactly how much solute is present. The **concentration** of a solution is a numerical measure of the relative amount of solute in the solution. This measure is always expressed as a ratio.

One of the most useful units of concentration is molarity. The **molarity (*M*)** of a solution is defined as the number of moles of solute in a liter of solution.

$$\text{Molarity } (M) = \frac{\text{mol of solute}}{\text{L of solution}}$$

◆ **Explain how to prepare a solution of a given molarity.**

A one molar, or 1 *M*, solution contains 1 mole of solute in enough water to make 1 liter of solution. Note that it is the total volume of the solution that is measured, not the volume of the water that is added. A bottle labeled 6.0 *M* HCl, therefore, contains 6.0 mol of HCl in each liter of solution. Figure 10.2 shows the procedure for making a solution of known concentration.

Molarity indicates the number of moles of a substance in a liter. Concentrations of blood electrolytes such as potassium ions are very small, however, so blood analyses from medical laboratories report the concentrations of electrolytes in millimoles per liter (mmol/L). If the concentration of potassium ions is reported as 4.3 mmol/L, then 1 L of blood serum contains 4.3 mmol (or 0.0043 mol) of potassium ions.

Example 10-1

1. How would you make 1.00 L of 0.100 *M* NaCl solution?

See the Question

This problem asks how many grams of NaCl to weigh out and dissolve in enough water to make a 0.100 *M* solution with a final volume of 1.00 L.

Think It Through

To determine the weight of 0.100 mol of NaCl, calculate the weight of 1 mol (the formula weight) of NaCl and multiply it by 0.100.

Execute the Math

$$\text{Formula weight of NaCl} = 23.0 + 35.5 = 58.5$$

$$1 \text{ mol of NaCl} = 58.5 \text{ g}$$

$$0.100\ \cancel{\text{mol NaCl}} \times \frac{58.5\ \text{g}}{1\ \cancel{\text{mol NaCl}}} = 5.85\ \text{g}$$

Prepare the Answer Because the initial concentration has three significant figures, our answer must also have three significant figures. To make a 0.100 *M* NaCl solution you therefore first dissolve 5.85 g of NaCl in a small amount of water. Then add enough water to make the final volume of the solution equal to 1.00 L and mix well.

2. To carry out a particular reaction in the laboratory, you need 1 mol of hydrochloric acid (HCl) dissolved in water. The only solution available to you is a 2 *M* HCl solution. How many milliliters of this stock solution should you use?

S The problem asks: 1 mol = (?) mL of 2 *M* HCl

T From the molarity of the stock solution and the fact that 1 L = 1000 mL, two conversion factors can be written:

$$\frac{2\ \text{mol HCl}}{1000\ \text{mL}} \quad \text{and} \quad \frac{1000\ \text{mL}}{2\ \text{mol HCl}}$$

E Using the second conversion factor, we can calculate the number of milliliters that contain 1 mol.

$$1\ \cancel{\text{mol HCl}} \times \frac{1000\ \text{mL}}{2\ \cancel{\text{mol HCl}}} = 500\ \text{mL}$$

P Our answer should have one significant figure. So, 500 mL of 2 *M* HCl stock solution contains 1 mol of HCl.

Check Your Understanding

1. Describe how to prepare each of the following solutions:

(a) 250 mL of 0.200 *M* Na_2CO_3

(b) 1.5 L of 0.75 *M* H_3PO_4

(c) 150 mL of 0.600 *M* $KMnO_4$

2. A student requires 0.250 mol of KOH for a reaction, but the only reagent available is a stock solution marked 0.400 *M* KOH. How many milliliters of this stock solution should she use?

3. The normal concentration range for sodium ions in blood serum is 138 to 148 mmol/L. If 1 dL of serum contains 320 mg of sodium, does the concentration of sodium in this sample fall within the normal range?

10.3 Percentage Concentration

Another way of describing the concentration of a solution—a way that does not take into account the formula weight of the solute—is by percentage concentration. We shall discuss two different types of percentage concentration that are used extensively in clinical reports and the biological sciences.

Weight/Volume Percent

Weight/volume percent is the concentration often found on clinical reports. **Weight/volume (w/v) percent** gives the number of grams of solute per 100 mL of solution.

$$\text{Weight/volume (w/v) percent} = \frac{\text{g of solute}}{\text{100 mL of solution}} \times 100\%$$

A 10% (w/v) NaOH solution, therefore, contains 10 g of NaOH in 100 mL of solution. For example, to make a 20% (w/v) NaOH solution, weigh out 20 g of NaOH. Dissolve this in a small amount of water in a 100-mL volumetric flask, and then add enough water to make 100 mL of solution (see Fig. 10.2).

Explain how to prepare a solution of a given percentage of solute.

Milligram Percent

Milligram percent is a unit of concentration often used in clinical reports to describe extremely low solute concentrations. **Milligram percent (mg%)** gives the number of milligrams of solute per 100 mL, or 1 dL, of solution.

$$\text{Milligram percent (mg\%)} = \frac{\text{mg of solute}}{\text{100 mL of solution}} \times 100\%$$

For example, blood urea nitrogen (BUN) is measured in milligram percent. An infant suffering from dehydration might have a blood urea nitrogen level of 32 mg%, which means that 32 mg of urea is present in 1 dL of blood. Blood chemistry laboratory reports also list the serum levels of calcium and phosphate ions in milligrams per deciliter (mg/dL), or milligram percent (see Fig. 10.3 on page 235).

Example 10-2

1. How is 150 mL of 0.4% (w/v) $NaHCO_3$ prepared?

See the Question

The problem is asking how many grams of $NaHCO_3$ need to be added to water to make 150 mL of a 0.4% (w/v) solution.

Think It Through

To solve problems involving percentage solutions, start by writing conversion factors given by the definition of that unit of concentration. In this case, 0.4% (w/v) $NaHCO_3$ gives us the conversion factors

$$\frac{0.4\text{ g } NaHCO_3}{100\text{ mL solution}} \quad \text{and} \quad \frac{100\text{ mL solution}}{0.4\text{ g } NaHCO_3}$$

Execute the Math

Use the conversion factor on the left to calculate the number of grams of $NaHCO_3$ that are needed.

$$150\text{ }\cancel{\text{mL solution}} \times \frac{0.4\text{ g } NaHCO_3}{100\text{ }\cancel{\text{mL solution}}} = 0.6\text{ g } NaHCO_3$$

Prepare the Answer

To prepare 150 mL of 0.4% (w/v) $NaHCO_3$ you dissolve 0.6 g $NaHCO_3$ in a small amount of water and add enough water to make 150 mL of solution.

2. How is 20 mL of 9.0 mg% sodium phosphate made?

S

The problem is asking you to determine how many milligrams of sodium phosphate must be added to 20 mL of solution to make a 9.0 mg% solution.

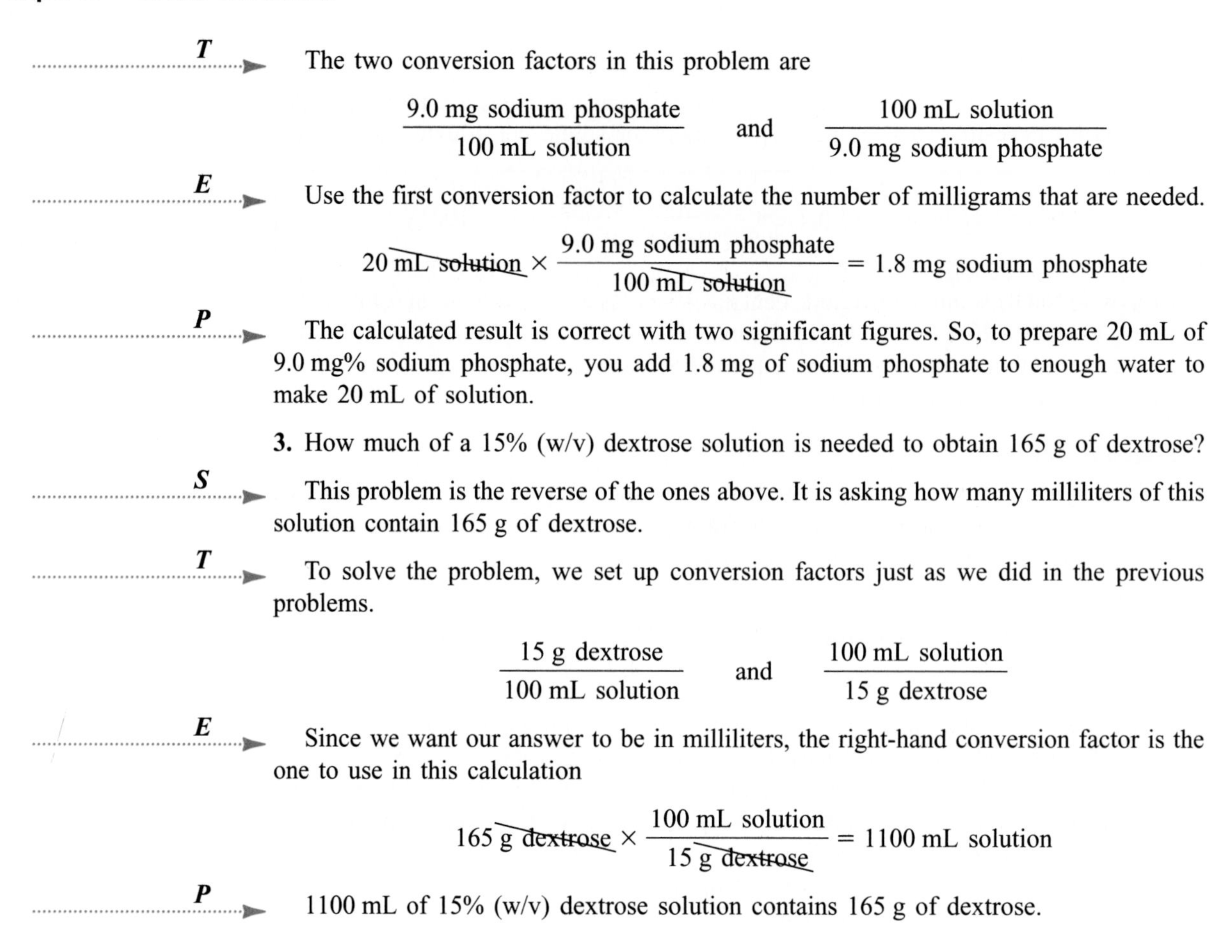

T The two conversion factors in this problem are

$$\frac{9.0 \text{ mg sodium phosphate}}{100 \text{ mL solution}} \quad \text{and} \quad \frac{100 \text{ mL solution}}{9.0 \text{ mg sodium phosphate}}$$

E Use the first conversion factor to calculate the number of milligrams that are needed.

$$20 \text{ mL solution} \times \frac{9.0 \text{ mg sodium phosphate}}{100 \text{ mL solution}} = 1.8 \text{ mg sodium phosphate}$$

P The calculated result is correct with two significant figures. So, to prepare 20 mL of 9.0 mg% sodium phosphate, you add 1.8 mg of sodium phosphate to enough water to make 20 mL of solution.

3. How much of a 15% (w/v) dextrose solution is needed to obtain 165 g of dextrose?

S This problem is the reverse of the ones above. It is asking how many milliliters of this solution contain 165 g of dextrose.

T To solve the problem, we set up conversion factors just as we did in the previous problems.

$$\frac{15 \text{ g dextrose}}{100 \text{ mL solution}} \quad \text{and} \quad \frac{100 \text{ mL solution}}{15 \text{ g dextrose}}$$

E Since we want our answer to be in milliliters, the right-hand conversion factor is the one to use in this calculation

$$165 \text{ g dextrose} \times \frac{100 \text{ mL solution}}{15 \text{ g dextrose}} = 1100 \text{ mL solution}$$

P 1100 mL of 15% (w/v) dextrose solution contains 165 g of dextrose.

Check Your Understanding

1. Describe how to prepare each of the following:

(a) 350 mL of 4.60% (w/v) NaCl

(b) 55.0 mL of 0.220% (w/v) K_2CO_3

(c) 25 mL of 80 mg% glucose

2. How many milliliters of a blood sample contain 5.6 mg of uric acid if the concentration of uric acid in the blood is 7.0 mg%?

3. How many grams of glucose are in 275 mL of a 12.0% (w/v) glucose solution?

10.4 Parts per Million and Parts per Billion

When a solution is extremely dilute, the concentration units of **parts per million (ppm)** or **parts per billion (ppb)** may be used. One part per million is equivalent to 1 mg of solute per liter of solution. One part per billion is 1 μg of solute per liter of solution.

$$1 \text{ ppm} = \frac{1 \text{ mg of solute}}{1 \text{ L of solution}} \qquad 1 \text{ ppb} = \frac{1\ \mu\text{g of solute}}{1 \text{ L of solution}}$$

It is hard to imagine concentrations this small, but the following comparison might help: If you were to earn a yearly salary of $100,000, then 1 ppm of your salary would be 10 cents, and 1 ppb would be only one one-hundredth of a penny! But it would be wrong to think that such concentrations are too small to even care about. Some pollutants found in industrial waste water can be extremely harmful in concentrations as small as 1 ppm.

◆ **Determine the concentration of a solution in parts per million and parts per billion.**

Example 10-3

1. Infant formula is often prepared from evaporated milk. That's why it was alarming in 1972 to learn that canned evaporated milk contained up to 3.2 ppm of lead. The toxic lead had dissolved out of the lead solder used in manufacturing the cans. At this concentration, how many grams of lead are present in 8 oz (470 mL) of evaporated milk?

See the Question

This question asks: 470 mL milk = (?) grams of lead

Think It Through

To solve this problem, we need to write the two conversion factors given by the unit of concentration.

$$\frac{3.2 \text{ mg Pb}}{1000 \text{ mL milk}} \quad \text{and} \quad \frac{1000 \text{ mL milk}}{3.2 \text{ mg Pb}}$$

Execute the Math

Using the conversion factor on the left, 470 mL of lead contains

$$470 \text{ mL milk} \times \frac{3.2 \text{ mg Pb}}{1000 \text{ mL milk}} = 1.504 \text{ mg Pb}$$

The question asks "how many grams," so we must convert our answer to grams.

$$1.504 \text{ mg Pb} \times \frac{1 \text{ g}}{1000 \text{ mg}} = 0.001504 \text{ g Pb}$$

Prepare the Answer

The value given for parts per million of lead in the problem has two significant figures, therefore the answer is 0.0015 g Pb.

2. After four months of drinking evaporated milk containing 1 ppm lead, an infant was found to have a lead concentration in her blood of 55 μg Pb/100 mL of blood. What was the lead concentration in the infant's blood in parts per billion and parts per million?

S

This question asks us to determine (a) the number of micrograms of Pb per liter (ppb) and (b) the number of milligrams of Pb per liter (ppm) in the infant's blood.

T

To solve this problem, we use the following equalities to write conversion factors: 1 L = 1000 mL and 1 mg = 1000 μg.

E

(a) $$\frac{55\ \mu\text{g Pb}}{100 \text{ mL}} \times \frac{1000 \text{ mL}}{1 \text{ L}} = \frac{550\ \mu\text{g Pb}}{1 \text{ L}}$$

(b) $$55\ \mu\text{g Pb} \times \frac{1 \text{ mg}}{1000\ \mu\text{g}} = 0.055 \text{ mg Pb}$$

Then, $\dfrac{0.055 \text{ mg Pb}}{100 \text{ mL}} \times \dfrac{1000 \text{ mL}}{1 \text{ L}} = \dfrac{0.55 \text{ mg Pb}}{1 \text{ L}}$

P

The initial concentration contains two significant figures, so our answers must have two significant figures. Thus, the concentration of lead in the infant's blood was (a) 550 ppb and (b) 0.55 ppm.

Check Your Understanding

The purity of drinking water is of increasing concern around the world as underground water supplies become contaminated with environmental pollutants. A 250-mL water sample from a private well was tested for the four contaminants listed in the table. Were any of the contaminants in the well water above the safe limits set by the U.S. Environmental Protection Agency (EPA) in 1986?

Contaminant	*Amount in a 250-mL Sample (μg)*	*EPA Safe Water Limits*
Carbon tetrachloride (CCl_4)	0.40	0.4 ppb
Selenium	1.3	0.010 ppm
Lead	11	0.050 ppm
Mercury	0.21	2.0 ppb

10.5 Equivalents

Doctors often refer to the important ionic components of the blood in terms of their ionic charge. The unit used to describe the concentration of ionic components is the **equivalent (Eq),** which corresponds to Avogadro's number (one mole) of electric charges.

1 Eq of an ion = 1 mol of charge (either + or −)

For example, 1 mol of the sodium ion Na^+ contains 1 mol of positive charge and is equal to 1 Eq of sodium ions. But 1 mol of Mg^{2+} ions contains 2 mol of positive charge; therefore, 1 Eq of magnesium ions equals $\frac{1}{2}$ mol of Mg^{2+}. For the HCO_3^- ion, 1 mol contains 1 mol of negative charge and equals 1 Eq of HCO_3^- ions.

To determine the number of equivalents of an ion in a solution, we need to know the weight of one equivalent of that ion. The **equivalent weight** of an ion is calculated by dividing the formula weight of the ion by the charge on the ion.

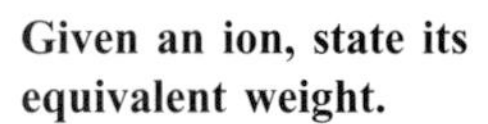

Given an ion, state its equivalent weight.

$$\text{Equivalent weight of an ion} = \frac{\text{formula weight}}{\text{charge of the ion}}$$

For the ions discussed here, we calculate the equivalent weight as follows:

$$Na^+: \frac{23 \text{ g}}{1} = 23 \text{ g} \qquad Mg^{2+}: \frac{24 \text{ g}}{2} = 12 \text{ g} \qquad HCO_3^-: \frac{61 \text{ g}}{1} = 61 \text{ g}$$

Because the concentration of ions in the blood is very dilute, medical reports often state ion concentrations in **milliequivalents,** or **mEq** (1000 mEq = 1 Eq). For example, it is

Trig	8		mg/dL		
Na^+	9	135—148	mEq/L	122	L
K^+	10	3.5—5.3	mEq/L	3.2	L
Cl^-	11	96—109	mEq/L	83	L
Bicarb	12		mEq/L		
Ca^{2+}	13	8.1—10.7	mg/dL	9.1	
Phos	14	2.6—4.8	mg/dL	2.9	

Figure 10.3 Portion of a blood chemistry laboratory report, with normal serum levels at the left and values for the patient at the right. (Courtesy Good Samaritan Hospital, Corvallis, Oregon)

important to check regularly the concentration of sodium ions (Na^+) and potassium ions (K^+) in the blood serum of a person with congestive heart failure who is also taking diuretics (Fig. 10.3). A diuretic is a substance that increases the amount of water released into the urine by the kidneys. Because sodium and potassium also pass into the urine along with the water, it is critical to keep track of these ions. Patients need to add sodium or potassium ions to their diets if their concentrations in the blood fall below the normal range of 135 to 148 mEq/L for Na^+ and 3.5 to 5.3 mEq/L for K^+.

Example 10-4

Calculate the number of milliequivalents of an ion in a given weight per volume percentage solution.

The calcium ion (Ca^{2+}) serves many important functions in the body. How many milliequivalents of calcium ions are present in 100 mL of a 0.1% (w/v) Ca^{2+} solution?

See the Question

The problem asks: 100 mL of a 0.1% (w/v) Ca^{2+} solution = (?) mEq

Think It Through

To solve this problem we must determine the equivalent weight of the calcium ion. Using that value we can determine the number of equivalents of calcium ion in 100 mL of our solution. Then, we can calculate the number of milliequivalents.

Execute the Math

(a) Equivalent weight of Ca^{2+}:

$$\text{Formula weight of Ca} = 40 \text{ g}$$

$$\text{Equivalent weight of } Ca^{2+} = \frac{40 \text{ g}}{2} = 20 \text{ g}$$

(b) Number of equivalents of Ca^{2+} in 100 mL of a 0.1% (w/v) Ca^{2+} solution:

From Section 10.3 we know that

$$0.1\% \text{ (w/v) } Ca^{2+} = \frac{0.1 \text{ g } Ca^{2+}}{100 \text{ mL}}$$

Therefore, 100 mL of 0.1% (w/v) Ca^{2+} contains 0.1 g Ca^{2+}.

If we use the result of (a), 1 Eq of Ca^{2+} equals 20 g, then

$$0.1 \cancel{\text{g}}\ Ca^{2+} \times \frac{1 \text{ Eq of } Ca^{2+}}{20 \cancel{\text{g}}} = 0.005 \text{ Eq of } Ca^{2+}$$

(c) Number of milliequivalents:

We know that 1000 mEq equals 1 Eq, so

$$0.005\ \text{Eq Ca}^{2+} \times \frac{1000\ \text{mEq}}{1\ \text{Eq}} = 5\ \text{mEq Ca}^{2+}$$

Prepare the Answer

100 mL of 0.1% (w/v) Ca^{2+} solution contains 5 mEq of Ca^{2+}.

Check Your Understanding

1. What is the equivalent weight of the following ions:
   (a) K^+ **(c)** HPO_4^{2-}
   (b) Cl^- **(d)** SO_4^{2-}
2. A normal value for the concentration of HPO_4^{2-} in a liter of body fluid is 140 mEq. How many grams of HPO_4^{2-} are in 1 L of body fluid?
3. A patient's chart records his serum chloride ion concentration as 94 mEq/L. What is his chloride ion concentration in millimoles per liter?

10.6 Dilutions

A very common task in a chemical or medical laboratory is preparing a needed solution using a concentrated stock solution (see Fig. 10.4). You can use the following relationship to find the amount of stock solution required for a desired solution. (*Note:* Be sure to use the same units of concentration and units of volume on both sides of the equation.)

$$\underset{\text{final concentration}}{C_2} \times \underset{\text{final volume}}{V_2} = \underset{\text{initial concentration}}{C_1} \times \underset{\text{initial volume}}{V_1} \quad (1)$$

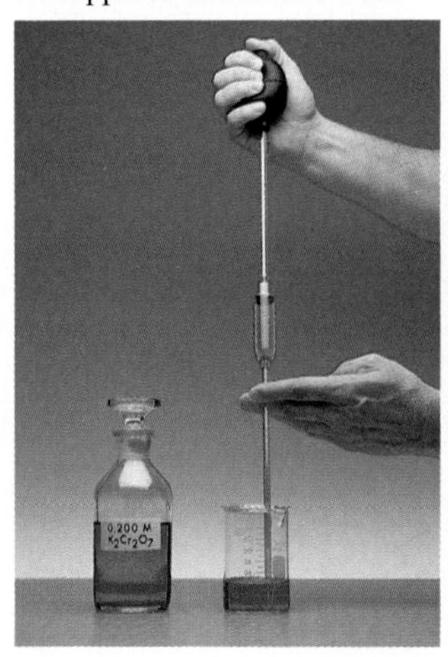

(a)

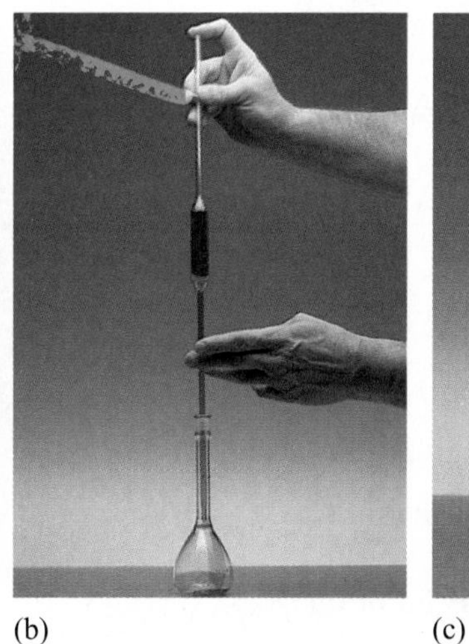

(b)

(c)

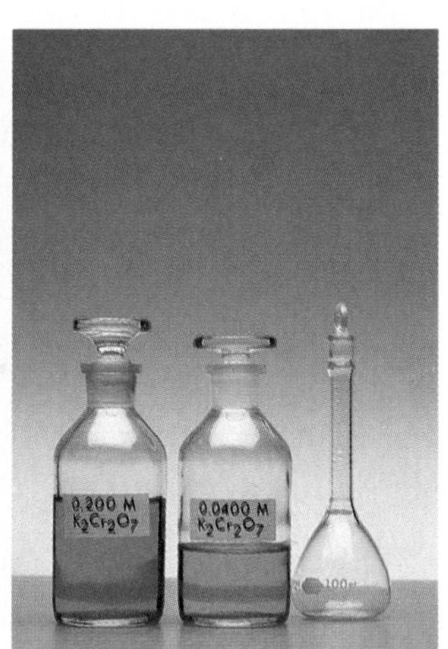

(d)

Figure 10.4 Preparing a solution by the dilution of a stock solution. (a) The calculated amount of the concentrated stock solution is drawn out using a pipet. (b) The solution is transferred to the volumetric flask. (c) Water is carefully added to the flask up to the mark on the volumetric flask. (d) The flask is stoppered and mixed well.

You can also use this equation to determine the final concentration of a solution made by adding solvent to increase the volume of the solution. If the volume of the solution is doubled, you have a 1:2 dilution; if the volume is increased by five times, you have a 1:5 dilution. For example, instructions on frozen lemonade concentrate tell you to add four cans of water to one can of frozen lemonade concentrate. This increases the total volume from one can to five cans, making a 1:5 dilution of the lemonade concentrate. Looking again at Equation (1), we can write

$$C_a \times V_a = C_b \times V_b \qquad (2)$$

where the subscripts a and b stand for "after dilution" and "before dilution," respectively.

Example 10-5

1. How do you prepare 100 mL of 1.5 *M* HCl from a 6.0 *M* HCl stock solution?

The question is asking how many milliliters of the 6.0 *M* stock solution should be added to water to make 100 mL of 1.5 *M* HCl. — **See the Question**

To determine the number of milliliters of the stock solution to use, we can use Equation (1) and substitute the values stated in the problem. — **Think It Through**

Execute the Math

$$(1.5\ M\ \text{HCl}) \times (100\ \text{mL}) = (6.0\ M\ \text{HCl}) \times V_1$$

$$\frac{1.5\ \cancel{M\ \text{HCl}} \times 100\ \text{mL}}{6.0\ \cancel{M\ \text{HCl}}} = V_1$$

$$25\ \text{mL} = V_1$$

To make 100 mL of 1.5 *M* HCl, you take 25 mL of the 6.0 *M* HCl stock solution and add enough water to make 100 mL of solution. — **Prepare the Answer**

2. What is the final concentration of a 1:5 dilution of 100 mL of 15% NaCl?

The question asks us to determine the percentage concentration after the dilution. — **S**

A 1:5 dilution means that the volume has been increased five times by adding more solvent. — **T**

$$\text{Volume after dilution} = 100\ \text{mL (volume before dilution)} \times 5 = 500\ \text{mL}$$

We use Equation (2) to find the final concentration.

Substituting the known values in Equation (2) we have — **E**

$$C_a \times 500\ \text{mL} = 15\%\ \text{NaCl} \times 100\ \text{mL}$$

$$C_a = \frac{15\%\ \text{NaCl} \times 100\ \cancel{\text{mL}}}{500\ \cancel{\text{mL}}}$$

$$= 3\%\ \text{NaCl}$$

The value for the concentration given in the problem has two significant figures, so our answer should also have two significant figures: 3.0% NaCl. — **P**

....

Check Your Understanding

1. How do you prepare 1.65 L of 2.75 *M* NaCl
   (a) Using solid NaCl
   (b) Using a 4.12 *M* NaCl stock solution
2. Describe how to make a 1 : 6 dilution of the solution prepared in Question 1. What is the final concentration of the solution?

10.7 Colligative Properties

Describe three colligative properties of solutions.

Solutions have certain properties, called **colligative properties,** that depend only on the number of solute particles that are dissolved, regardless of the chemical nature of the solute. Such properties include lowering of the freezing point and raising of the boiling point of water by solute particles. The more solute particles in the solution, the lower the freezing point and the higher the boiling point. For example, automotive antifreeze is added to the water in a car's radiator to lower the freezing point of the water, preventing it from freezing (and cracking the radiator) in the winter. The antifreeze also raises the boiling point of the water, preventing the radiator from boiling over in the summer. Nature has given some insects their own antifreeze. They have a high concentration of the compound glycerol in their body fluids, which helps them survive the harsh winter temperatures by lowering the freezing point of the water in their bodies.

10.8 Osmosis

Define osmosis.

Another colligative property is osmotic pressure. **Osmosis** is the flow of water molecules through a differentially permeable membrane from a region of lower solute concentration to a region of higher solute concentration. (A differentially permeable membrane is a barrier that allows the passage of water but not solute particles.) For example, if pure water and a sugar solution are separated by a differentially permeable membrane, water molecules flow through the membrane into the sugar solution at a higher rate than they return. The volume of the sugar solution therefore increases (Fig. 10.5).

Osmotic pressure is defined as the amount of pressure that must be applied to prevent the flow of water, or osmosis, through the membrane from a solution of lower solute concentration to a solution of higher solute concentration. Osmotic pressure depends only on the number of solute particles in the solution: the greater the difference in solute concentration between the solutions, the greater the pressure that must be applied to prevent osmosis (Fig. 10.5).

A good example of the way in which nature uses osmotic pressure is the method by which trees get water to their leaves. A tree loses water (through transpiration) from its leaves to the atmosphere; as a result, the solute concentration within the leaves increases. This generates osmotic pressure within the tree that forces water from the soil into the roots, up the trunk, and through the branches to the leaves. A coast redwood tree can grow to be 364 ft tall. To force water up a pipe as tall as a coast redwood, you would need a pump capable of generating 12 atm of pressure!

Chemical Perspective / Peter Piper's Pickles

If Peter Piper picked a peck of pickled peppers, how many pickled peppers did Peter Piper pick? Actually, it is hard to believe that Peter Piper was able to pick *any* pickled peppers. Usually we have to pick our peppers unpickled and then process them ourselves. In fact, however, it's more likely that we would be pickling cucumbers, which are what we call pickles.

Making pickles is just one of many interesting applications of osmosis. Salt is the major ingredient in the fermentation process used to make pickles. In this process, fresh cucumbers are soaked for 2 to 6 weeks in a 10% salt solution kept at 86°F. The salt solution draws both moisture and certain natural sugars from the cucumbers, and these combine to form an acid bath that "cures" the pickles. The cucumbers shrivel while sitting in this brine as water flows out of the cucumbers into the more concentrated brine solution.

Before they are packaged or eaten, pickles are washed several times with water to remove most of the salt. Dill pickles are fermented with dill herb and are packaged in vinegar after the washing process. Sweet pickles are packaged in a sweet, spiced vinegar.

Your Perspective: "Peter Piper" is a tongue-twister, a sentence that is difficult to say quickly. (Another well-known tongue-twister is "She sells sea shells by the seashore.") Write your own tongue-twister based on chemical principles from this textbook.

Given the molar concentration of two solutions, give the osmolarity of each solution and tell which has the higher osmotic pressure.

The total concentration of particles in the blood or urine is often recorded in terms of osmolarity. Because every particle in solution (whether an ion or molecule) contributes to the osmotic pressure of that solution, **osmolarity** describes the total number of moles of all particles in a liter of solution. One **osmole (osm)** is defined to be 1 mol of any combination of particles. For example, 1 mol of KCl dissolves in a liter of solution and produces 1 mol of K^+ and 1 mol of Cl^-, or 2 mol of particles—that is, 2 osm. A liter of 1 M KCl therefore has an osmolarity of 2 osm per liter. Solutions containing different solute par-

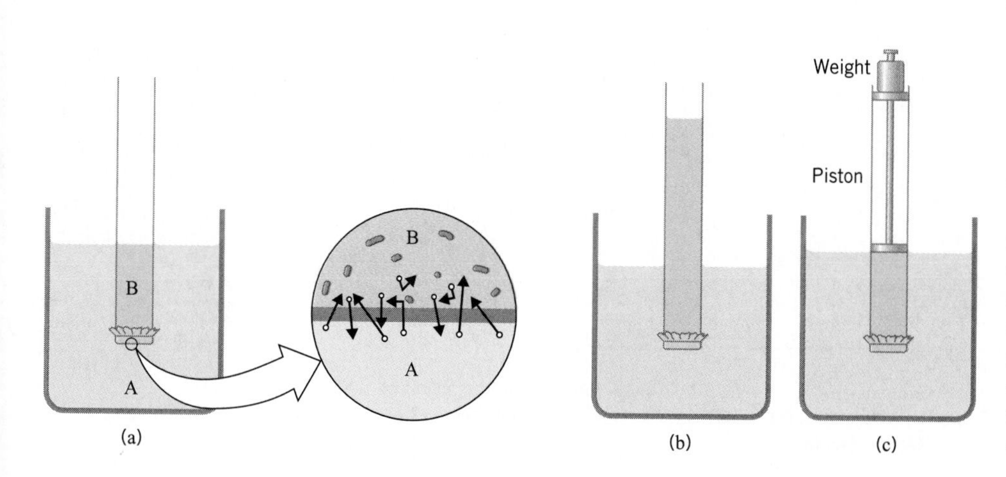

Figure 10.5 Osmosis. (a) A tube containing a sugar solution B is placed in a beaker of water (A). The differentially permeable membrane on the end of the tube permits water molecules to pass through, but not sugar molecules. Water moves from the region of lower solute concentration to the region of higher concentration. (b) When equilibrium is established, the side with the sugar solution has more material. (c) Osmotic pressure equals the pressure applied to the sugar solution to prevent osmosis.

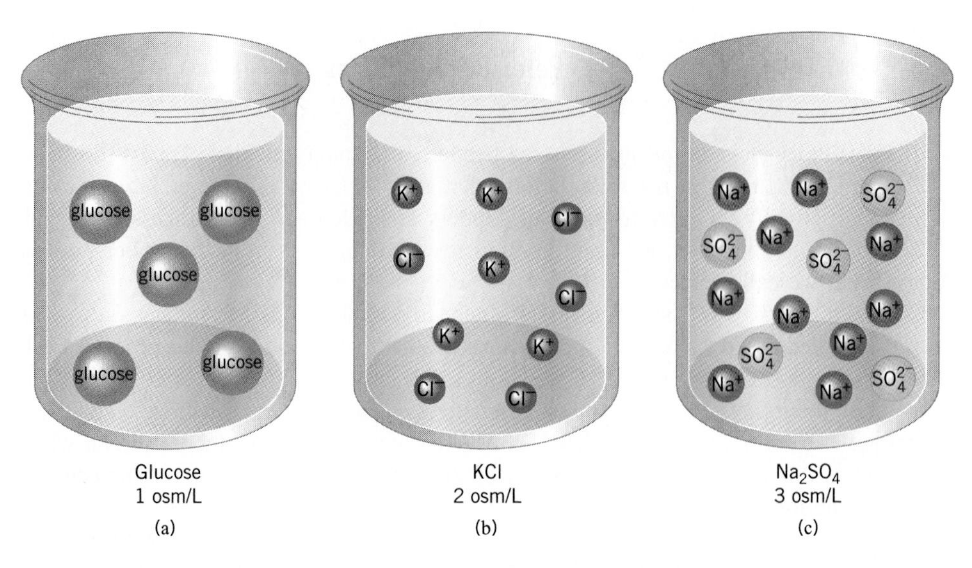

Figure 10.6 Each of these beakers contains a 1 molar solution. (a) This 1 *M* sugar solution has an osmolarity of 1 osm/L. (b) This 1 *M* KCl solution has an osmolarity of 2 osm/L. (c) This 1 *M* Na_2SO_4 solution has an osmolarity of 3 osm/L.

ticles but having the same osmolarity have the same osmotic pressure (Fig. 10.6). A 1 *M* KCl solution has the same osmolarity and, therefore, the same osmotic pressure as a 1 *M* NaBr solution. A comparison of the osmolarity of blood and urine is often a good indication of kidney function. Normal serum osmolarity is 275 to 295 milliosmols per liter (mOsm/liter), and normal urine osmolarity is 300 to 1000 mOsm/liter.

10.9 Isotonic Solutions

Define isotonic, hypertonic, and hypotonic.

The passage of water in and out of living cells is an important biological process. If the concentration of solute is equal on both sides of the cell membrane, the osmotic pressures are equal, and the solutions are then called **isotonic.** The usual concentration of the salts in blood plasma is isotonic to a 0.9% (w/v) NaCl solution. This concentration of NaCl is called **normal saline** or **physiological saline.**

A normal saline solution is isotonic to red blood cells. When red blood cells are placed in the saline, no change is observed. If red blood cells are placed in a solution of lower solute concentration, called a **hypotonic solution** (for example, distilled water), water

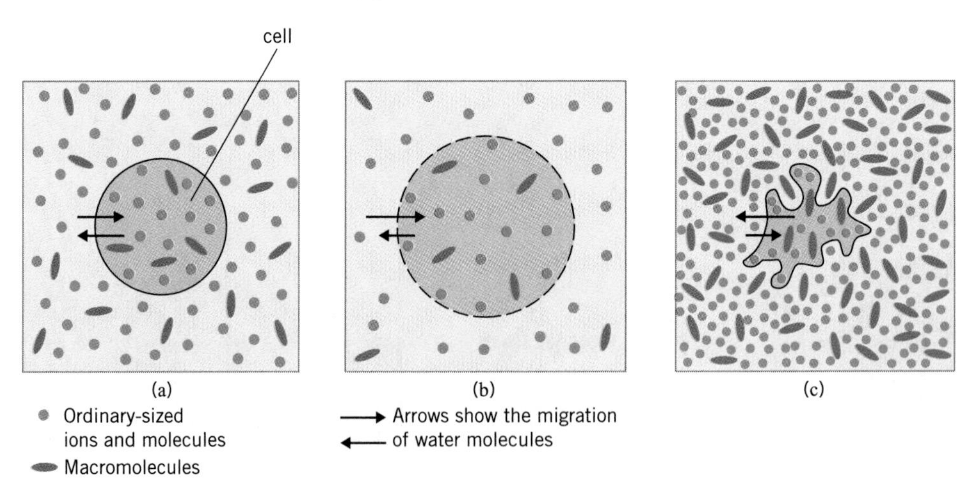

Figure 10.7 (a) A red blood cell in an isotonic solution. (b) This red blood cell is about to undergo hemolysis because the cell is swollen by water entering from the surrounding hypotonic solution. (c) A red blood cell undergoes crenation when it is placed in a hypertonic solution.

Health Perspective / Intravenous Solutions and Electrolyte Balance

Although there is always some danger in injecting fluids and electrolytes directly into the bloodstream, there are many medical situations for which the benefits of ''an IV'' (intravenous solution) outweigh the risk. These situations involve people who

- can't swallow safely, such as patients in a coma or those under anesthesia, or patients suffering from seizures, paralysis, or severe vomiting
- can't drink enough to keep up with their loss of fluids or electrolytes, such as patients suffering from major burns, hemorrhage, diabetic ketoacidosis, or severe diarrhea
- require medications that are destroyed by gastric juices or are poorly absorbed from the gastrointestinal tract
- must rapidly increase the concentration of a medication or electrolyte in their blood

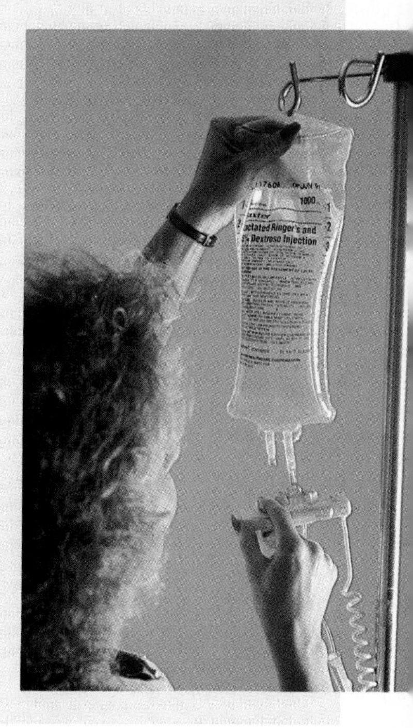

The type of IV solution that is used may be hypertonic, hypotonic, or isotonic, depending on the patient's particular fluid or electrolyte imbalance. A patient who has lost a great deal of water compared with the loss of salts requires hypotonic IV fluids. In some cases, the opposite situation might occur. In general, however, most IV solutions are isotonic—having essentially the same osmolarity as body fluids. Examples of isotonic solutions include 0.9% NaCl (normal saline) and Ringer's lactate.

Complications can occur when IV solutions are not isotonic. Consider the case of a person who is badly dehydrated and needs increased water in the bloodstream. If this person were given a rapid IV infusion of pure water, the red blood cells would swell and rupture (hemolysis). Instead, a 5% dextrose solution (D5W) is administered, which is nearly isotonic and easily tolerated by the patient. As the dextrose is metabolized, it leaves behind in the blood the ''free'' water that the patient needed. Another example might be a person who is suffering from a low level of Na^+ in the blood. Although the patient needs a hypertonic infusion to correct the low electrolyte balance, such a solution could cause inflammation of the vein around the IV site. Instead, an isotonic ''normal saline'' solution is administered, which still results in raising the patient's very low electrolyte level.

moves into the cells. The cells then swell and rupture in a process called **hemolysis.** If red blood cells are placed in a solution of higher solute concentration, called a **hypertonic solution** (for example, blood plus 3% NaCl solution), the water moves from the cells to the solution. The cells then shrink in a process called **crenation** (Fig. 10.7).

Give examples of solutions that are isotonic, hypertonic, and hypotonic to blood plasma.

Some solutes in the bloodstream cannot pass through the walls of the capillaries, which gives the blood a higher solute concentration than the surrounding extracellular fluid. As a result, water moves from the extracellular fluid into the capillaries, keeping the veins full and preventing the collapse of the blood vessels. In some diseases, such as malnutrition or kidney failure, the concentration of particles in the blood goes down, decreasing the osmotic pressure and decreasing the flow of water from the tissues into the veins. This results in a condition called **edema,** which is a swelling of tissues such as in the hands or legs.

Health Perspective / Laxative Effect of Magnesium Ions

Magnesium compounds like $MgCO_3$ and $Mg(OH)_2$ act as antacids when taken in small doses, but in larger doses they act as laxatives. A person suffering from a severe case of acid indigestion might be unpleasantly surprised if she consumes large amounts of one of these compounds to alleviate the problem! The apparent change in medicinal properties is caused by the magnesium ion. When these compounds are used as antacids, magnesium is a ''spectator'' ion; it is only along for the ride. The actual neutralization of acid is carried out by the anion portion of the salt. The laxative effect comes about because magnesium ions are poorly absorbed in the intestine. An increase in the concentration of magnesium ions in the large intestine creates a hypertonic condition. The resulting osmotic pressure causes water to flow from the neighboring tissues into the large intestine, diluting the stool and causing diarrhea. For this reason, over-the-counter drugs like milk of magnesia (a suspension of magnesium hydroxide and water) have printed on their labels both the antacid dosage and the laxative dosage.

Your Perspective: You want to request that the U.S. Food and Drug Administration include mandatory labels on magnesium-containing antacids to warn of their possible laxative effects. Write a letter to the FDA, explaining the importance of this request.

KEY CONCEPTS

Several general terms can be used to describe concentration: dilute, concentrated, saturated, unsaturated, and supersaturated.

Units of concentration that determine precisely the number of solute particles in a specific amount of solvent are molarity, weight/volume percent, milligram percent, parts per million, and parts per billion. Equivalents and milliequivalents describe the amount of ionic charge in a solution.

Colligative properties are those properties of a solution that depend only on the number of particles in the solution, not on the nature of the particle. These properties include raising the boiling point and lowering the freezing point of the solvent, and osmotic pressure.

Osmosis is the flow of water through a membrane from a region of lower solute concentration to a region of higher solute concentration. Osmotic pressure is the pressure that must be applied to prevent such a flow. Osmolarity describes the total concentration of all particles in a solution. Isotonic solutions are ones with the same osmolarity. A hypotonic solution has fewer solute particles and a hypertonic solution more solute particles than a given solution.

Important Equations

$$\text{Molarity } (M) = \frac{\text{mol of solute}}{\text{L of solution}}$$

$$\text{Weight/volume \%} = \frac{\text{g of solute}}{\text{100 mL of solution}} \times 100\%$$

$$\text{Milligram \%} = \frac{\text{mg of solute}}{\text{100 mL of solution}} \times 100\%$$

$$\text{1 ppm} = \frac{\text{1 mg of solute}}{\text{1 L of solution}}$$

$$\text{1 ppb} = \frac{1\ \mu\text{g of solute}}{\text{1 L of solution}}$$

$$\text{1 Equivalent} = \text{1 mol of charge}$$

$$\text{Equivalent weight of an ion} = \frac{\text{formula weight}}{\text{charge of the ion}}$$

$$\text{Dilutions:} \quad C_2 \times V_2 = C_1 \times V_1$$

$$\text{Osmolarity} = \frac{\text{osm}}{\text{1 L}}$$

Review Problems

Section 10.1

1. State the difference between a dilute and a concentrated solution.

2. State the difference between an unsaturated, a saturated, and a supersaturated solution.

3. What is the result of cooling a saturated solution?

4. Sugar is added to a beaker containing a saturated solution of sugar in water. Will any of this sugar go into solution? Explain your answer.

Section 10.2

5. Describe how to prepare each of the following aqueous solutions:

(a) 50 mL of 0.20 *M* NaOH **(c)** 0.10 L of 0.50 *M* NH_4Cl

(b) 250 mL of 0.11 *M* Na_2SO_4

6. Describe how to prepare each of the following aqueous solutions:

(a) 55 mL of 1.0 *M* KCl **(c)** 0.25 L of 0.010 *M* $(NH_4)_2SO_4$

(b) 10 mL of 0.20 *M* K_2HPO_4

7. If you dissolve 4.9 g of H_2SO_4 in 250 mL of solution, what is the molarity of the solution?

8. What is the concentration of each of the following in molarity?

(a) 11 μg $HgCl_2$ in 10 mL of solution **(c)** 0.039 g KCN in 100 mL of solution

(b) 5.0 mg NaF in 250 mL of solution **(d)** 0.99 mg SeO_2 in 750 mL of solution

Section 10.3

9. pHisoHex soap contains 3.00% (w/v) hexachlorophene. How many grams of hexachlorophene is in a 148-mL bottle of pHisoHex soap?

10. How many milliliters of 0.10% (w/v) sodium carbonate solution contain 2.5 g of Na_2CO_3?

11. How do you prepare 25 mL of a 2.5 mg% fructose solution?

12. Describe how to prepare each of the following aqueous solutions:

    (a) 0.50 L of 5.0% (w/v) KOH **(b)** 125 mL of 8.25% (w/v) glucose

13. Describe how to prepare each of the following aqueous solutions:

    (a) 225 mL of 22.0 mg% lactic acid **(b)** 35 mL of 0.832 mg% KCl

14. What is the concentration of each of the following in milligram percent:

    (a) 11 μg $HgCl_2$ in 10 mL of solution
    (b) 5.0 mg NaF in 250 mL of solution
    (c) 0.039 g KCN in 100 mL of solution
    (d) 0.99 mg SeO_2 in 750 mL of solution

15. What is the molarity of NaCl in normal saline [0.900% (w/v) NaCl]?

16. You have available three stock solutions to use in performing an experiment: 6.00 *M* HCl, 0.100 *M* NaOH, and 5.00% (w/v) Na_2CO_3. State the amount of each stock solution required to supply the following amounts of reactants needed for the experiment:

    (a) 0.300 mol HCl
    (b) 0.150 mol NaOH
    (c) 2.50 g Na_2CO_3
    (d) 40.0 mg NaOH
    (e) 21.9 g HCl
    (f) 0.250 mol Na_2CO_3

Section 10.4

17. A 150-mL water sample contains 0.26 μg of cadmium. What is this concentration in parts per billion?

18. What is the concentration of each of the following in parts per million?

    (a) 11 μg $HgCl_2$ in 10 mL of solution
    (b) 5.0 mg NaF in 250 mL of solution
    (c) 0.039 g KCN in 100 mL of solution
    (d) 0.99 mg SeO_2 in 750 mL of solution

Section 10.5

19. Give the equivalent weight for the following ions.

    (a) PO_4^{3-} **(b)** Br^- **(c)** Ba^{2+} **(d)** CO_3^{2-}

20. If the concentration of Mg^{2+} in extracellular fluid (the fluid found between body cells) is 2.0 mEq/L, how many milligrams of Mg^{2+} are present in 10 mL of extracellular fluid?

21. Determine the number of milliequivalents of the listed ion in each of the following solutions:

    (a) 100 mL of 0.1% (w/v) PO_4^{3-}
    (b) 250 mL of 0.2% (w/v) CO_3^{2-}
    (c) 125 mL of 0.35% (w/v) SO_4^{2-}

22. The normal range for the concentration of potassium ion in the blood of an adult is 3.5 to 5.3 mEq/L. Does a person taking diuretics need a potassium supplement if a 5.0-mL blood sample contains 0.39 mg K^+?

Section 10.6

23. What is the final concentration of a 1:5 dilution of

    (a) 50 mL of 6.0 *M* H_2SO_4 **(b)** 150 mL of 2 ppm Cd^{2+}

24. What is the final concentration of a 1:5 dilution of
    (a) 30 mL of 3% (w/v) KCl
    (b) 10 mL of 32 mg% urea
25. How do you prepare each of the following aqueous solutions from the given stock solution?
    (a) 250 mL of 8.00% (w/v) NaCl from 10.0% (w/v) NaCl
    (b) 150 mL of 0.030 *M* NaOH from 0.10 *M* NaOH
    (c) 1:10 dilution of 75 mL of 2.5 *M* H_2SO_4
26. How do you prepare each of the following aqueous solutions from the given stock solution?
    (a) 55 mL of 6.4 mg% uric acid from 7.8 mg% uric acid
    (b) 650 mL of 4.0 ppm Pb^{2+} from 10 ppm Pb^{2+}

Section 10.7

27. Why is covering a sidewalk with salt (NaCl) an effective way to prevent ice from forming on the sidewalk in the winter?
28. Compare the freezing point and boiling point of the ocean with that of fresh water.

Section 10.8

29. Referring to Figure 10.5a, state whether the level in A or B rises for each of the following:
    (a) 0.1 *M* HCl in A and 0.01 *M* HCl in B
    (b) 1% (w/v) dextrose in A and 5% (w/v) dextrose in B
    (c) 1 *M* NaCl in A and 1% (w/v) NaCl in B
30. What is the osmolarity of an isotonic saline solution [0.9% (w/v) NaCl]?
31. Celery stored for long periods in a refrigerator often goes limp. It can be made crisp again by putting it in a container of water. Explain the principle behind the celery's return to its original crispness.
32. A dried prune is placed in a glass of water. After a period of time, the prune is observed to have increased in size considerably, and the water in the glass tastes sweet. Explain these changes.
33. Explain why the osmotic pressure of a solution depends only on the concentration of its solute particles and is not affected by their chemical properties.
34. The salinity of the ocean is equivalent to a 3.5% (w/v) aqueous NaCl solution. What is the osmolarity of the ocean?
35. Potassium sulfate (K_2SO_4) is a strong electrolyte, and alcohol is a nonelectrolyte. Explain why the osmotic pressure of a 1 *M* solution of K_2SO_4 is three times that of a 1 *M* alcohol solution.
36. Which of the following solutions has the same osmotic pressure (osmolarity) as a 0.1 *M* NaCl solution?
    (a) 0.1 *M* $CaCl_2$ solution
    (b) 0.05 *M* Na_2CO_3 solution
    (c) 0.1 *M* NaOH solution

Section 10.9

37. What happens to red blood cells if they are placed in the following solutions?

(a) 1.5% (w/v) NaCl solution

(b) 0.154 *M* NaCl solution

(c) 0.15% (w/v) NaCl solution

38. Identify each of the solutions in Problem 37 as isotonic, hypertonic, or hypotonic to red blood cells.

APPLIED PROBLEMS

39. Novocain is sometimes administered as an aqueous solution of its hydrochloride salt, procaine hydrochloride ($C_{13}H_{20}N_2O_2HCl$). How would you prepare 250 mL of a 5.0% (w/v) solution of procaine hydrochloride? What is the molarity of this solution?

40. Nitrosamines are potent carcinogens (chemicals that cause cancer). In 1979, the average level of dimethylnitrosamine in beer was found to be 5.90 ppb.

(a) How many micrograms of dimethylnitrosamine were in a 12-oz can of beer in 1979?

(b) By 1981, the average concentration of dimethylnitrosamine in beer had been reduced to 0.200 ppb. By how many micrograms was the nitrosamine content of a 12-oz can of beer reduced?

41. The blood of a patient being treated with lithium carbonate was found to contain 1.4 mg of Li^+ in a 100-mL sample. The blood concentration of Li^+ should not exceed 1.5 mEq/L. Should the patient stop taking lithium carbonate?

42. A blood chemistry laboratory report shows a patient's serum sodium at 142 mEq/L. What is the total mass of Na^+ in the patient's blood (5.5 L)?

43. Some hospital laboratories are reporting blood serum levels of Na^+, K^+, and Cl^- in millimoles per liter (mmol/L). What is the blood serum level of Na^+, K^+, and Cl^- in millimoles per liter for the patient whose laboratory report is shown in Figure 10.3?

44. In cases of severe burns, protein molecules often enter the tissues. What effect does this increase in protein have on the fluid levels of the tissues?

45. Most freshwater fish do not survive when placed in salt water. Why is this so?

46. The packet of oral rehydration salts (ORS) given to the Ethiopian refugees discussed at the beginning of Chapter 9 contains the following:

Glucose	20.0 g	Sodium bicarbonate	2.5 g
Sodium chloride	3.5 g	Potassium chloride	1.5 g

The instructions on the packet say to dissolve the contents in enough water to make 1 L of solution.

(a) What is the concentration in weight/volume percent of each of the components of ORS?

(b) What is the concentration in molarity of each of the components of ORS?

(c) What is the osmolarity of the ORS solution?

47. When a houseplant is grown in a pot without holes, salt dissolved in the tap water builds up in the soil and appears as white crystals on the soil surface.

(a) Explain why, as time goes on, the plant's growth slows and the plant eventually dies.

(b) Why is it that some plants are able to live very successfully in saltwater marshes?

Chapter 11

Acids and Bases

*B*ob Knight was a middle-aged insurance agent with a very sensitive stomach. He kept a roll of antacids handy in his pocket to control the acid indigestion he constantly suffered, and drank large amounts of milk to keep an ulcer from forming. One afternoon, Bob began suffering some very unusual symptoms. He became extremely dizzy, had high and irregular blood pressure, felt nauseated, and had an acid stomach that was worse than usual. His doctor sent Bob to the hospital for a series of tests but was unable to determine what was wrong. The symptoms soon disappeared, and Bob went home. But not long after he returned to work, the symptoms reappeared. Again he was hospitalized, and again the doctors could find nothing wrong. On Bob's third visit to the hospital, his doctor even called in a psychiatrist for consultation. The breakthrough came when a very determined intern, after long discussions with Bob about his daily habits, found a reference to a condition called the ''milk-alkali'' syndrome in a medical textbook. This syndrome, which included the symptoms suffered by Bob, is caused by a high intake of calcium-containing antacids and large quantities of milk.

To understand what was happening to Bob, we must take a closer look at the work done by the stomach. Cells in our stomach lining secrete hydrochloric acid (HCl), which is very corrosive. Normally the stomach lining secretes about 2 liters of HCl solution over a 24-hour period, most coming after meals. The proteins in our food, and drugs like alcohol and caffeine, stimulate the production of HCl. Our emotions also affect the secretion of stomach acid. In the stomach this acid works to kill bacteria in foods, to soften foods, and to convert the inactive enzyme pepsinogen into its active form, pepsin, to begin the digestion of protein.

Hydrochloric acid is secreted into the stomach fluid in the form of hydronium ions (H_3O^+, or hydrated protons) and chloride ions. It is the presence of the hydronium ion that makes the stomach contents acidic. A jelly-like layer of mucous lines the stomach to protect it from the

acid. The esophagus, however, does not have such a mucous membrane. If the stomach's contents back up into the esophagus, the result is "heartburn"—a stinging sensation behind the breastbone. Swallowing one of the various products sold as antacids can neutralize the stomach's contents and reduce the occurrence of heartburn.

The "active ingredient" in the antacid tablets Bob was taking is calcium carbonate ($CaCO_3$). Dissolved in the water mixture in the stomach, calcium carbonate dissociates to form the calcium ion and the carbonate ion.

$$CaCO_3(s) \underset{}{\overset{H_2O}{\rightleftharpoons}} Ca^{2+}(aq) + CO_3^{2-}(aq)$$

The carbonate ions react with and neutralize hydronium ions in the stomach fluid, thus relieving the indigestion and heartburn.

$$CO_3^{2-}(aq) + 2H_3O^+ \longrightarrow 3H_2O + CO_2(g)$$

Calcium carbonate is an excellent antacid for many reasons: It acts rapidly, can neutralize a large amount of acid, and has a long-lasting effect at low cost. As Bob discovered, however, it can also have serious side effects when used in large amounts over long periods. Ingesting too much calcium can lead to constipation and may cause "acid rebound," a condition in which the stomach produces even more acid—and more heartburn.

An even greater danger, however, is caused by the high level of calcium ions that can occur in the blood when a person repeatedly takes calcium-based antacids. The possibility of a high blood calcium concentration is significantly increased in persons who also drink large amounts of calcium-rich milk or who have kidney problems. Although calcium ions are essential for normal body functioning, excessively high blood calcium levels disrupt reactions in the body and cause the unusual symptoms suffered by Bob. Such elevated calcium levels can also impair kidney function and lead to the formation of kidney stones. Bob's symptoms cleared up about a week after he reduced his intake of milk and switched to an aluminum–magnesium antacid to relieve his indigestion.

A critical balance between acids and bases must be maintained in all parts of the living organism to permit normal physiological reactions to occur. In this chapter we discuss the nature of acids and bases and the ways in which their concentrations are regulated in the human body.

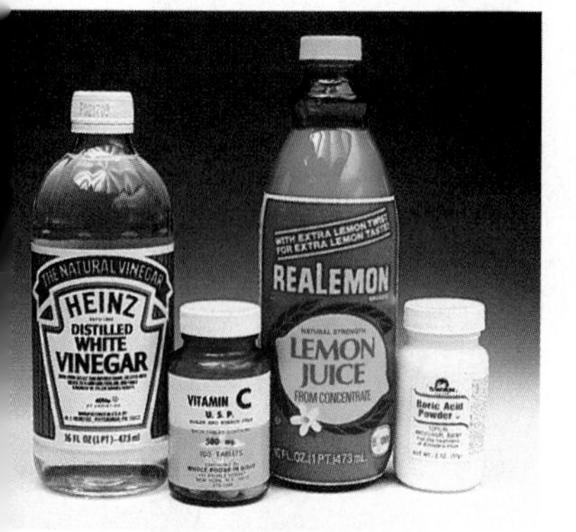

Common household acids

Common household bases

11.1 Acids and Bases: The Brønsted–Lowry Definition

Acids and bases play important roles in living organisms. Strong acids and bases can be quite harmful; they damage tissue by destroying protein material and drawing out water. For example, concentrated sulfuric acid is a strong dehydrating (water-removing) agent that rapidly injures tissues on contact. Concentrated bases react with the fats that make up the protective membranes of cells, destroying such membranes and causing even more widespread destruction to tissues than acids do. Strong laundry soaps and detergents contain bases. Clothes containing wool and silk (which are animal proteins) cannot be washed in such soaps because the base in the soap causes the fibers of these materials to shrink and partially dissolve.

Acids are compounds that, when dissolved in water, produce solutions that taste sour, turn litmus paper from blue to red, and react with metals such as zinc or magnesium to produce hydrogen gas. **Bases** form solutions that taste bitter, feel slippery to the touch, and

Table 11.1 **Some Common Acids**

Name	*Formula*	*Name*	*Formula*	*Name*	*Formula*
Hydrofluoric acid	HF	Nitric acid	HNO_3	Acetic acid	$HC_2H_3O_2$
Hydrochloric acid	HCl	Nitrous acid	HNO_2	Carbonic acid	H_2CO_3
Hydrobromic acid	HBr	Sulfuric acid	H_2SO_4	Oxalic acid	$H_2C_2O_4$
Hydroiodic acid	HI	Sulfurous acid	H_2SO_3	Phosphoric acid	H_3PO_4

Litmus paper in acid (left) and base (right)

turn litmus paper from red to blue. Bases react with acids to neutralize each others' properties. For example, in the chapter-opening story we saw that one popular antacid contains the base carbonate (CO_3^-), which neutralizes the hydrochloric acid secreted into the stomach.

Acids composed of hydrogen and a nonmetal are called binary acids. They are named by using the prefix *hydro-,* adding an *-ic* to the name of the nonmetal, then adding the word *acid.* For example, HBr is hydrobromic acid. Not all binary hydrogen compounds are acids, however. When the binary compound has the hydrogen listed first (HCl, H_2S), it is an acid; but compounds such as CH_4 and NH_3 are not acidic. Table 11.1 lists the names and formulas of some common acids.

Define a Brønsted–Lowry acid and a Brønsted–Lowry base.

What makes a compound an acid or a base? Although there are several definitions for these terms, we'll find it convenient for our study of biochemistry to use the definitions proposed independently in 1923 by the Danish chemist Johannes N. Brønsted and the English chemist Thomas M. Lowry.

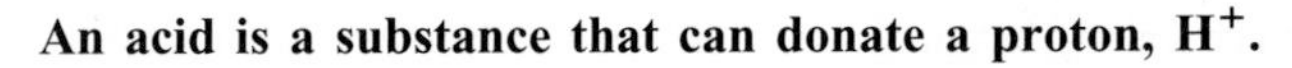

An acid is a substance that can donate a proton, H^+.

A base is a substance that can accept a proton, H^+.

Note that a hydrogen atom has just one proton in its nucleus, and one electron. The hydrogen ion (H^+, a hydrogen atom that has lost one electron) is therefore simply a proton.

Let's look at some examples. Consider the following reaction:

$$\underset{\text{nitrous acid}}{HNO_2} + \underset{\text{water}}{H_2O} \rightleftharpoons \underset{\text{nitrite ion}}{NO_2^-} + \underset{\text{hydronium ion}}{H_3O^+} \qquad (1)$$

In this reaction, nitrous acid is donating a proton to water. Nitrous acid is therefore the acid (the proton donor), and water acts as the base (the proton acceptor).

$$\underset{\text{ammonia}}{NH_3} + \underset{\text{water}}{H_2O} \rightleftharpoons \underset{\text{ammonium ion}}{NH_4^+} + \underset{\text{hydroxide ion}}{OH^-} \qquad (2)$$

Chemical Perspective / Acid–Base Components of Drain Cleaners

Clogged drains are an unpleasant part of home living. Two types of substances usually cause such a clog. In the kitchen the culprit is most likely fat; in the bathroom it is probably hair.

As you will learn in Chapter 16, strong bases like sodium hydroxide react with fat to produce soap. This is why solid sodium hydroxide or concentrated solutions of sodium hydroxide are used in drain cleaners. When sodium hydroxide is added to the drain, some of the fat reacts with the NaOH to form soap. The newly formed soap is soluble in water and can easily be washed away. The soap can also act as an emulsifier, breaking down the remaining fat and permitting it to be washed away. When solid sodium hydroxide is used as a drain cleaner, it is often mixed with powdered aluminum. The solid sodium hydroxide liberates a considerable quantity of heat when it contacts the water in the drain. This heat alone can often be enough to melt the fat and break up the clot. The aluminum metal reacts with water in the presence of the hydroxide ion (OH^-) to produce hydrogen gas. The gas creates pressure in the drain that helps break up the clog. The combination of the two ingredients delivers a one–two punch that is very effective.

Sodium hydroxide also works on the hair that blocks bathroom drains. It dissolves the hair and clears the drain. Sulfuric acid is another powerful hair-dissolving substance that is used in some drain cleaners.

Your Perspective: Using language that young viewers would understand, write the script for a short segment of ''Sesame Street'' explaining how a sodium hydroxide drain cleaner works.

In this reaction, ammonia (NH_3) accepts a proton from a water molecule. Ammonia is therefore the base, and water (as the proton donor) is the acid. You can see that water may function as either an acid or a base depending on the substance with which it is reacting. A substance that can act as either an acid or a base is called **amphoteric.**

Consider for a moment the reverse reaction in Equation (1). Here, the hydronium ion (H_3O^+) is the acid, and the nitrite ion (NO_2^-) is the base. Similarly, in the reverse of Equation (2), the ammonium ion (NH_4^+) is the acid, and the hydroxide ion (OH^-) is the base.

$$\underset{\text{acid}_1}{HNO_2} + \underset{\text{base}_2}{H_2O} \rightleftharpoons \underset{\text{base}_1}{NO_2^-} + \underset{\text{acid}_2}{H_3O^+}$$

$$\underset{\text{base}_1}{NH_3} + \underset{\text{acid}_2}{H_2O} \rightleftharpoons \underset{\text{acid}_1}{NH_4^+} + \underset{\text{base}_2}{OH^-}$$

Identify the conjugate acid–base pairs in a chemical reaction.

You might notice that HNO_2 and NO_2^- differ only by a proton. The same is true of H_2O and H_3O^+. They are known as **conjugate acid–base pairs.** The nitrite ion is the conjugate base of nitrous acid, and the hydronium ion is the conjugate acid of water.

Some acids, called **polyprotic acids,** can donate more than one proton in a reaction with a base. For example, sulfuric acid (H_2SO_4) and carbonic acid (H_2CO_3) each have two ionizable hydrogens that can be donated in an acid–base reaction. Phosphoric acid (H_3PO_4) has three ionizable hydrogens.

$$H_3PO_4 + H_2O \rightleftharpoons H_3O^+ + H_2PO_4^-$$

$$H_2PO_4^- + H_2O \rightleftharpoons H_3O^+ + HPO_4^{2-}$$

$$HPO_4^{2-} + H_2O \rightleftharpoons H_3O^+ + PO_4^{3-}$$

11.2 Strength of Acids and Bases

We define strong and weak acids in much the same way that we defined strong and weak electrolytes. A strong acid is one that completely, or almost completely, ionizes to donate all its protons. The result of adding a strong acid to water is a large increase in the concentration of hydronium ions. For example, nitric acid is a strong acid. In 0.1 M HNO_3, 100% of the nitric acid molecules are ionized to hydronium ions and nitrate ions.

State in your own words the difference between a strong and weak acid, and a strong and weak base.

$$HNO_3 + H_2O \longrightarrow H_3O^+ + NO_3^-$$

Other strong acids are hydrochloric acid (HCl), hydrobromic acid (HBr), hydroiodic acid (HI), and sulfuric acid (H_2SO_4). Most acids are weak acids, however. A weak acid only partially dissociates in water to donate protons. The addition of a weak acid to water therefore results in only a small increase in the concentration of hydronium ions. Acetic acid is an example of a weak acid. In a 0.1 M CH_3COOH solution, only 1.3% of the molecules are ionized.

$$CH_3COOH + H_2O \rightleftharpoons H_3O^+ + CH_3COO^-$$

Other weak acids are nitrous acid (HNO_2), carbonic acid (H_2CO_3), and boric acid (H_3BO_3).

Similarly, a strong base has a very large attraction for protons. A weak base has a weak attraction, and only a few of its molecules accept protons. The hydroxide ion is a strong base, whereas ammonia is a weak base.

If an acid is strong and, therefore, has a strong tendency to donate protons, its conjugate base is weak and has a low attraction for protons. The reverse is true for a weak acid; a weak acid has a conjugate base with a strong attraction for protons. Table 11.2 lists the relative strengths of some conjugate acid–base pairs. (Note that the number of hydrogens in the formula of an acid is not an indication of its strength as an acid.)

From looking at the polyprotic acids in Table 11.2, we see that the neutral acid donating the first proton is a stronger acid than the negative ion donating the second proton. For example, H_2CO_3 is a stronger acid than HCO_3^- because it is easier to remove a positive ion (H^+) from a neutral molecule than from a negatively charged ion.

11.3 Neutralization Reactions

Acids react with bases to form water and the dissolved ions of an ionic compound, commonly called a **salt.** For example,

$$\underset{\text{acid}}{HCl(aq)} + \underset{\text{base}}{NaOH(aq)} \longrightarrow H_2O(\ell) + \underset{\text{salt}}{NaCl(aq)}$$

Write the balanced equation for a neutralization reaction between an acid and a base.

If equal amounts of hydronium ions and hydroxide ions react, the resulting solution has neither acidic nor basic properties and is said to be **neutral.** A **neutralization reaction,**

For generations, families had looked forward to spending their summer vacations fishing on Darts Lake in the Adirondack Mountains. But now the fish have disappeared, leaving biologists and resort owners wondering why. A clue to this mystery lies in data showing that the lake has slowly become more acidic over the last 20 years. Scientists know that most living plants and animals are able to tolerate only very small changes in acidity. So the increasing acidification of lakes in the northeastern United States and around the world may be producing long-term harmful effects on fish and other aquatic life, such as clams, crayfish, and insects.

One possible cause of this acidification may be industrial air pollution. Coal-burning electric power companies and exhaust-spewing automobiles release many gases into the atmosphere. Two of these polluting gases are nitrogen monoxide (NO) and nitrogen dioxide (NO_2)—together represented as NO_X. Another polluting gas is sulfur dioxide (SO_2), which is formed when sulfur-containing coal or oil is burned. Once they enter the atmosphere, both NO_X and SO_2 can be transformed into acids. When they are later washed out of the atmosphere by rain, these compounds acidify the rainwater. Acidic rainwater falling on streams and lakes has the ability to make these bodies of water more acidic.

But this "acid-rain" theory doesn't explain all of the evidence that has been collected. Data show that acidification of lakes and streams began in the early 1900s, long before air pollution from automobiles, power plants, and other industries became a problem in the eastern United States. In fact, scientists have found that lakes in the western United States—where the rain is not particularly acidic—are also showing a slow acidification. What, then, could be other sources of acids affecting the lakes? It turns out that many natural biological processes, such as plant decay and the natural oxidation of organic nitrogen in the soil, produce acids. Such acids are leached out of the ground as rainwater passes through the rocks and soil on its way to streams and lakes. These naturally produced acids might also be playing a role in the acidification of lakes and streams.

As usual in the modern world, things may be more complicated than they first appear. Although acid rain does contribute to the acidification of lakes and streams, the soil on which the rain falls can change the chemistry of the acidic rainwater. At the same time, natural processes in the soil may be producing other acids that contribute to the acidification of streams and lakes. The acidification of streams and lakes is just one example of the need to understand how environmental processes can be affected by both human actions and natural causes.

Your Perspective: Suppose that the owner of Dart's Lake Resort wrote an angry letter to the editor demanding that the state impose restrictions that would result in lowering the pH of rainwater all the way to zero. Using language that a nonscientist would understand, write a letter to the editor responding to the resort owner's letter.

Table 11.2 **Relative Strengths of Conjugate Acid–Base Pairs**

Acid			*Base*
Sulfuric acid	H_2SO_4	HSO_4^-	Hydrogen sulfate ion
Hydrochloric acid	HCl	Cl^-	Chloride ion
Nitric acid	HNO_3	NO_3^-	Nitrate ion
Hydronium ion	H_3O^+	H_2O	Water
Sulfurous acid	H_2SO_3	HSO_3^-	Hydrogen sulfite ion
Hydrogen sulfate ion	HSO_4^-	SO_4^{2-}	Sulfate ion
Phosphoric acid	H_3PO_4	$H_2PO_4^-$	Dihydrogen phosphate ion
Nitrous acid	HNO_2	NO_2^-	Nitrite ion
Acetic acid	$HC_2H_3O_2$	$C_2H_3O_2^-$	Acetate ion
Carbonic acid	H_2CO_3	HCO_3^-	Hydrogen carbonate ion
Hydrogen sulfite ion	HSO_3^-	SO_3^-	Sulfite ion
Dihydrogen phosphate ion	$H_2PO_4^-$	HPO_4^{2-}	Hydrogen phosphate ion
Ammonium ion	NH_4^+	NH_3	Ammonia
Hydrogen carbonate ion	HCO_3^-	CO_3^{2-}	Carbonate ion
Hydrogen phosphate ion	HPO_4^{2-}	PO_4^{3-}	Phosphate ion
Water	H_2O	OH^-	Hydroxide ion
Ammonia	NH_3	NH_2^-	Amide ion

Increasing Acid Strength (↑) — Increasing Base Strength (↓)

therefore, is one in which either an acidic or basic solution is converted to a neutral solution.

The reaction between, for example, hydrochloric acid and sodium hydroxide occurs between aqueous ions. The complete ionic equation for the reaction is

$$H^+(aq) + Cl^-(aq) + Na^+(aq) + OH^-(aq) \longrightarrow H_2O(\ell) + Na^+(aq) + Cl^-(aq)$$

The net ionic equation for this reaction is

$$H^+(aq) + OH^-(aq) \longrightarrow H_2O(\ell)$$

Example 11-1

1. Write the balanced equation for the neutralization of sulfuric acid by potassium hydroxide.

See the Question

The problem is asking us to write the correct formulas for the reactants of the neutralization reaction, then to predict the products of that reaction, and finally to balance the equation for the reaction between sulfuric acid and potassium hydroxide.

Think It Through

Write the correct formulas for the reactants.

Sulfuric acid: H_2SO_4 Potassium hydroxide: KOH

The products of this neutralization reaction are water and the salt potassium sulfate.

Water: H_2O Potassium sulfate: K_2SO_4

Knowing the reactants and the products, we can write the unbalanced equation.

$$H_2SO_4 + KOH \longrightarrow H_2O + K_2SO_4$$

Execute the Math

To balance the equation, we begin by balancing the potassium atoms so that two are on each side of the equation.

$$H_2SO_4 + 2KOH \longrightarrow H_2O + K_2SO_4$$

The sulfate ions are already balanced (one on each side of the arrow), so next we balance the hydrogens. Four hydrogen atoms are on the left of the arrow, so we need two waters (four hydrogens) on the right side.

$$H_2SO_4 + 2KOH \longrightarrow 2H_2O + K_2SO_4$$

Now we check the total number of oxygen atoms on both sides: 6

Prepare the Answer

The number of atoms of each element is the same on both sides of the equation, so the equation is balanced.

2. Write the net ionic equation for the preceding reaction.

S

We must take the balanced equation from Problem 1 and write the net ionic equation for the reaction.

T

Ionic compounds dissociate in aqueous solution, and a neutralization reaction is a reaction between aqueous ions. So, take the balanced equation and rewrite it showing these ions.

$$H_2SO_4 + 2KOH \longrightarrow 2H_2O + K_2SO_4$$

becomes

$$2H^+(aq) + SO_4^{2-}(aq) + 2K^+(aq) + 2OH^-(aq) \longrightarrow 2H_2O(\ell) + 2K^+(aq) + SO_4^{2-}(aq)$$

E

Crossing out the "spectator ions" that appear on both sides of the equation, we have:

$$2H^+(aq) + \cancel{SO_4^{2-}}(aq) + \cancel{2K^+}(aq) + 2OH^-(aq) \longrightarrow 2H_2O(\ell) + \cancel{2K^+}(aq) + \cancel{SO_4^{2-}}(aq)$$

P

This gives us the net ionic equation for the reaction.

$$2H^+(aq) + 2OH^-(aq) \longrightarrow 2H_2O(\ell)$$

Check Your Understanding

Write the balanced equation and net ionic equation for the neutralization of sodium hydroxide by nitric acid.

11.4 Ionization of Water

In Chapter 9 we discussed water's ability to dissolve ionic compounds and to ionize some highly polar molecular compounds. To a very small extent, water molecules also can ionize other water molecules. At 25°C, one water molecule out of every 550 million is

Chemical Perspective / Chemical Leavening

We are so accustomed to light breads and cakes that we seldom give thought to the role leavenings play in their formation. If the dough didn't expand, most baked goods would be heavy, dense, unappetizing pastries.

What causes dough to rise? Part of the expansion (anywhere from 30%–80%) results from steam that is formed from the moisture in the dough. But when we think of leavens, what usually comes to mind are baking powders and yeast. Yeasts are living organisms that feed on sugars and produce alcohol and carbon dioxide gas. As the carbon dioxide expands, it causes the bread to rise.

Because it takes so long for yeast to form sufficient quantities of carbon dioxide, we use chemical leavenings for making many baked goods. Cakes, cookies, doughnuts, pancakes, and even some pizza are made with chemical leavening agents called baking powders. These powders are mixtures of several components, with sodium bicarbonate ($NaHCO_3$) as the active ingredient. In fact, sodium bicarbonate is also called baking soda because of its use in baking powders. Sodium bicarbonate reacts with acid to form carbon dioxide according to the following equations:

$$NaHCO_3(aq) + H^+(aq) \longrightarrow Na^+(aq) + H_2CO_3(aq)$$

$$H_2CO_3(aq) \text{ (unstable)} \longrightarrow CO_2(g) + H_2O$$

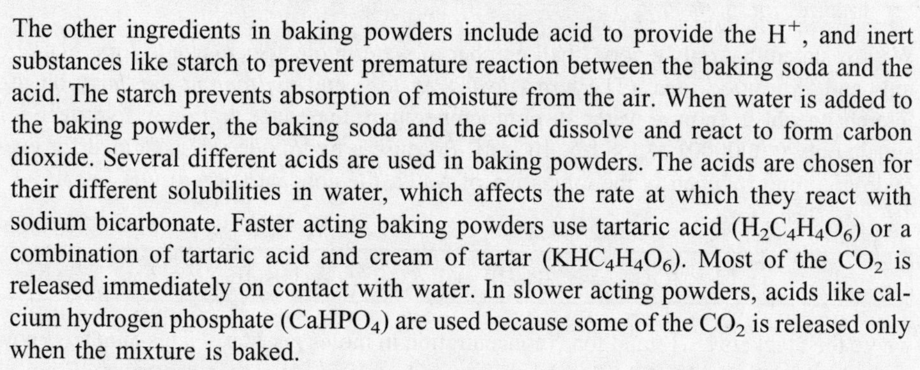

The other ingredients in baking powders include acid to provide the H^+, and inert substances like starch to prevent premature reaction between the baking soda and the acid. The starch prevents absorption of moisture from the air. When water is added to the baking powder, the baking soda and the acid dissolve and react to form carbon dioxide. Several different acids are used in baking powders. The acids are chosen for their different solubilities in water, which affects the rate at which they react with sodium bicarbonate. Faster acting baking powders use tartaric acid ($H_2C_4H_4O_6$) or a combination of tartaric acid and cream of tartar ($KHC_4H_4O_6$). Most of the CO_2 is released immediately on contact with water. In slower acting powders, acids like calcium hydrogen phosphate ($CaHPO_4$) are used because some of the CO_2 is released only when the mixture is baked.

Your Perspective: A televised science show for children is planning a segment using an animated yeast cell who explains how yeast and chemical leavening agents work. Write the script for the yeast cell's explanation.

dissociated to form a hydronium ion, H_3O^+, and a hydroxide ion, OH^-. In 1 L of water, the concentration of H_3O^+ and OH^- are both $1 \times 10^{-7}\ M$.

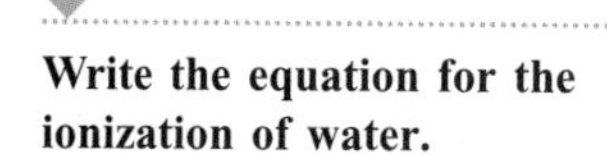

Write the equation for the ionization of water.

$$H_2O + H_2O \rightleftharpoons OH^- + H_3O^+$$

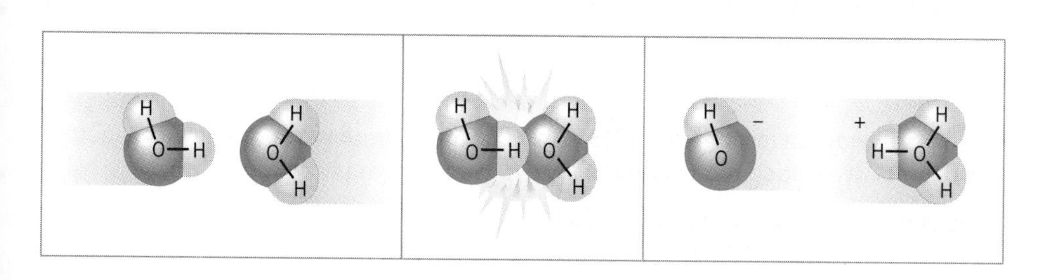

For convenience, we often write the equation for the ionization of water as follows:

$$\underset{}{H_2O} \rightleftharpoons \underset{\substack{\text{hydrogen ion} \\ \text{(proton)}}}{H^+} + \underset{\text{hydroxide ion}}{OH^-}$$

It is important to understand, however, that although we often use the terms *protons* or *hydrogen ions* when talking about aqueous solutions, single hydrogen ions never exist in water. They are always associated with water molecules in the form of the hydronium ion (hydrated proton).

$$H^+ + :\overset{\displaystyle H}{\underset{\cdot\cdot}{\overset{\cdot\cdot}{O}}}:H \longrightarrow \left[H:\overset{\displaystyle H}{\underset{\cdot\cdot}{\overset{\cdot\cdot}{O}}}:H\right]^+$$

In the covalent bond formed between the oxygen and the hydrogen ion, the oxygen molecule donates both of the electrons. This type of bond, in which one atom donates both of the electrons, is known as a **coordinate covalent bond.**

11.5 Ion Product of Water, K_w

We just stated that only a very small number of water molecules ionize to form hydrogen ions and hydroxide ions. (The terms *hydrogen ions* and *hydronium ions* may be used interchangeably.) In pure water at room temperature, the concentration of hydrogen ions equals only 0.0000001 *M*, or 1×10^{-7} *M*. Because a hydroxide ion is formed for every hydrogen ion produced in the ionization of water, the concentration of OH^- is also 1×10^{-7} *M*. Therefore,

$$[H^+] \times [OH^-] = (1 \times 10^{-7})(1 \times 10^{-7}) = 1 \times 10^{-14} = K_w$$

where the brackets [ ] stand for, ''concentration in moles per liter.'' This number, known as the **ion product of water, K_w,** always equals 1×10^{-14} at room temperature. From this equation we can calculate the concentration of either H^+ or OH^- if we know the concentration of the other ion.

$$[H^+] = \frac{1 \times 10^{-14}}{[OH^-]} \qquad [OH^-] = \frac{1 \times 10^{-14}}{[H^+]}$$

Example 11-2

The concentration of hydrogen ions in a slightly acidic solution is 1×10^{-5} *M*. What is the hydroxide ion concentration of this solution?

See the Question The question asks: $[OH^-] = (?)$ mol/L

Think It Through The problem statement gives us $[H^+]$, which we can use in the preceding equation to calculate the hydroxide ion concentration.

Execute the Math Substituting the known value, we have

$$[OH^-] = \frac{1 \times 10^{-14}}{[H^+]} = \frac{1 \times 10^{-14}}{1 \times 10^{-5}} = 1 \times 10^{-9}$$

(If you are unsure about how to divide numbers in exponential form, see Appendix 1.)

Prepare the Answer

The answer should have one significant figure.

$$[OH^-] = 1 \times 10^{-9} \text{ mol/L}$$

Check Your Understanding

Calculate the concentration of hydrogen ions in moles per liter for a solution whose hydroxide ion concentration is

(a) 1×10^{-3} *M*

(b) 5×10^{-6} *M*

(c) 0.0004 mol in 200 mL

(d) 0.00040 *M*

(e) 0.01 *M*

(f) 8.00×10^{-8} mol in 500 mL

11.6 The pH Scale

Small changes in hydrogen ion concentration can be of great importance to living cells, and are critical in many fields of scientific investigation. As a result, scientists are constantly measuring hydrogen ion concentrations. In 1909, a Danish biochemist named Sören Sörenson developed the pH scale to provide a way of showing the hydrogen ion concentration more conveniently than by using a negative exponent (10^{-7}) or a decimal fraction (0.0000001). pH is the negative power to which the number 10 must be raised to express the concentration of hydrogen ions in moles per liter. Mathematically,

$$[H^+] = 1 \times 10^{-pH} \quad \text{or} \quad pH = -\log [H^+]$$

At room temperature, the $[H^+]$ in pure water is 1×10^{-7}. The pH of pure water is therefore 7. Because in pure water the $[H^+]$ equals $[OH^-]$, pure water is considered to be neutral. This means that the pH of a neutral solution is 7. Acidic solutions are ones in which the hydrogen ion concentration is greater than the hydroxide ion concentration. From Table 11.3 we see that for an acidic solution the exponent of 10 is less negative than -7 (-2 for example). Thus, the pH of an acidic solution is less than 7. A basic solution is one in which the hydrogen ion concentration is less than the hydroxide ion concentration, and the pH is greater than 7 (Fig. 11.1).

The measurement of pH is an important laboratory procedure because the pH of a solution affects the activity of biological molecules. This means that pH can influence the behavior of cells and even entire organisms. For example, bacteria grow best in a very small range of pH. The pH of culture media must therefore be carefully controlled. Enzymes, the biological catalysts, work best in narrow limits of pH that can vary from an optimum pH range of 1 to 4 for pepsin (an enzyme in the stomach) to an optimum pH range of 8 to 9 for trypsin (an enzyme in the small intestine). Most body fluids are maintained in a very narrow range of pH that, if changed, can be toxic to the organism (Table 11.4).

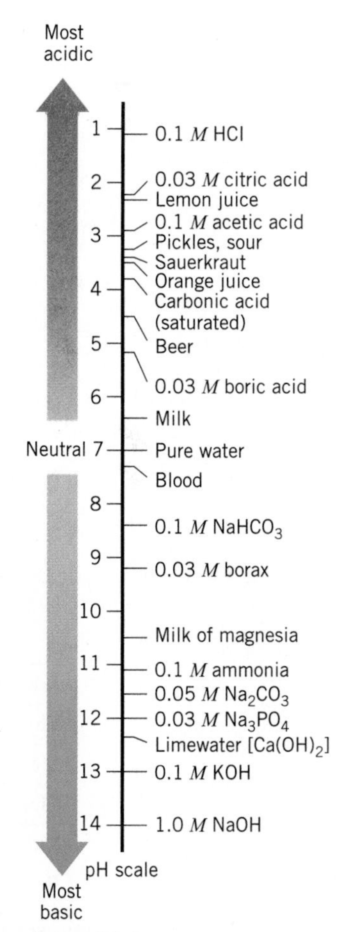

Figure 11.1 A pH scale showing the pH values of some common substances.

Table 11.3 The pH Scale and Corresponding Concentrations of Hydrogen and Hydroxide Ions

$[H^+]$	pH	$[OH^-] = \dfrac{1 \times 10^{-14}}{[H^+]}$
$10^0 = 1$	0	10^{-14}
$10^{-1} = 0.1$	1	10^{-13}
$10^{-2} = 0.01$	2	10^{-12}
$10^{-3} = 0.001$	3	10^{-11}
$10^{-4} = 0.0001$	4	10^{-10}
$10^{-5} = 0.00001$	5	10^{-9}
$10^{-6} = 0.000001$	6	10^{-8}
$10^{-7} = 0.0000001$	7	10^{-7}
$10^{-8} = 0.00000001$	8	10^{-6}
$10^{-9} = 0.000000001$	9	10^{-5}
$10^{-10} = 0.0000000001$	10	10^{-4}
$10^{-11} = 0.00000000001$	11	10^{-3}
$10^{-12} = 0.000000000001$	12	10^{-2}
$10^{-13} = 0.0000000000001$	13	10^{-1}
$10^{-14} = 0.00000000000001$	14	10^0

Table 11.4 Normal pH Range of Body Fluids

Fluid	*pH*	*Fluid*	*pH*
Gastric juice	1.0–3.0	Blood	7.35–7.45
Vaginal secretion	3.8	Intestinal secretions	7.7
Urine	5.5–7.0	Bile	7.8–8.8
Saliva	6.5–7.5	Pancreatic juice	8

Table 11.5 Colors of Acid–Base Indicators at Various pH Levels

Indicator	*0*	*1*	*2*	*3*	*4*	*5*	*6*	*7*	*8*	*9*	*10*	*11*	*12*	*13*	*14*
Thymol blue[a]		Red	Transition	Yellow →											
Methyl orange		←		Red	Transition	Yellow →									
Methyl red		←			Red	Transition		Yellow →							
Litmus		←				Red	Transition	Blue →							
Bromothymol blue		←				Yellow	Transition		Blue →						
Metacresol purple		←						Yellow	Transition	Purple →					
Thymol blue[a]		←							Yellow	Transition	Blue →				
Phenolphthalein		←							Colorless	Transition	Pink →				

[a]Thymol blue indicator undergoes two color changes, one in the acid range and one in the base range.

The pH of a solution is best measured using a pH meter. These instruments make use of the fact that the voltage of an electric current passing through a solution changes according to the pH of the solution. A second, but less accurate method of measuring pH is by a colorimetric indicator. Such methods use chemical dyes, called **acid–base indicators,** which change color at certain hydrogen ion concentrations (Table 11.5). For example, paper containing the dye nitrazine is yellow at a pH of 4.5 and blue at a pH of 7.5. Such paper is used to test the pH of urine. Acidic urine—urine with a pH of 4.5 or lower—turns the paper yellow and often indicates a serious disorder.

A **titration** uses a neutralization reaction to determine the amount of acid or base in a solution of unknown concentration. In a titration (Fig. 11.2), you use a buret to add a solution with a known concentration of base to a solution of unknown concentration of acid (or a known acid can be added to an unknown base). The neutralization reaction is monitored by a pH meter or an acid–base indicator. When the number of hydrogen ions donated by the acid equals the number of hydrogen ions that can be accepted by the base, you have reached the **equivalence point** of the titration. The acid–base indicator used in a titration must be carefully chosen so that the color change, or **endpoint,** of the titration occurs at the pH of the equivalence point. Some important uses of titrations are to determine the alkali (or basic) constituents of blood, the acidity of the stomach, or the acidity of urine.

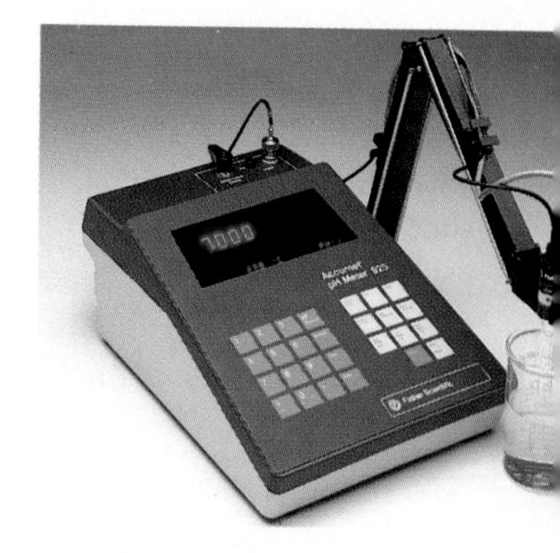

pH meter

◆

Describe how a titration can be used to determine an unknown concentration of an acid solution.

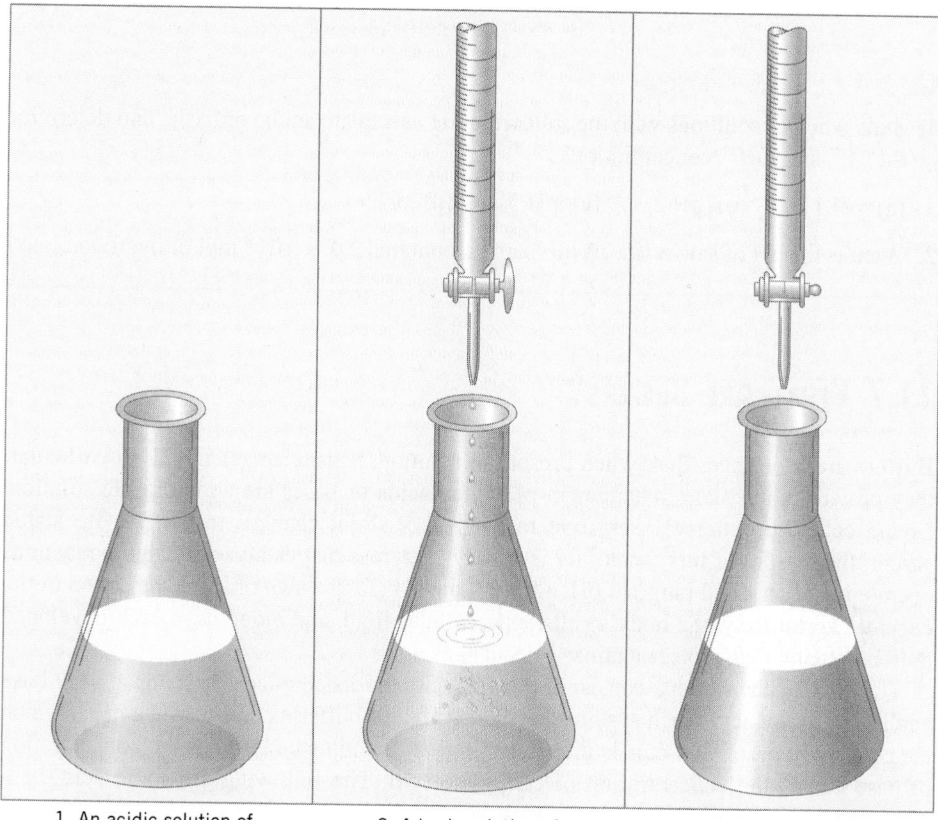

1. An acidic solution of unknown concentration is placed in a flask with an indicator.

2. A basic solution of known concentration is added slowly from the buret.

3. The flow of base from the buret is stopped when the indicator changes color, showing that just enough base has been added to react with all the acid in the flask.

Figure 11.2 A titration.

Example 11-3

Given the pH of a solution, state whether the solution is acidic, basic, or neutral.

As we saw in an earlier chapter, the air pollutants sulfur dioxide and nitrogen dioxide react with rainwater to form acids. A sample of rainwater from Sweden was found to have a pH of 4.

(a) Is the sample acidic or basic?

The pH is less than 7, so the sample is acidic.

(b) Determine the concentrations of H^+ and OH^- in the sample.

See the Question → The problem asks: $[H^+] = (?)$ mol/L and $[OH^-] = (?)$ mol/L

Think It Through → The pH tells us the concentration of H^+. Once we have that value, we can use the expression for the ion product of water to calculate the concentration of OH^-.

Execute the Math →

$$[H^+] = 1 \times 10^{-pH} = 1 \times 10^{-4}$$

$$[OH^-] = \frac{1 \times 10^{-14}}{[H^+]} = \frac{1 \times 10^{-14}}{1 \times 10^{-4}} = 1 \times 10^{-10}$$

Prepare the Answer → Both answers should have one significant figure.

$$[H^+] = 1 \times 10^{-4}\ \text{mol/L} \qquad [OH^-] = 1 \times 10^{-10}\ \text{mol/L}$$

Check Your Understanding

1. State whether solutions with the following pH values are acidic or basic, and determine the H^+ and OH^- concentrations.

   (a) pH 11 **(b)** pH 2 **(c)** pH 5 **(d)** pH 9

2. What is the pH of saliva if a 20-mL sample contains 2.0×10^{-9} mol of hydrogen ions?

11.7 What Are Buffers?

Define buffer.

Buffers are substances that, when present in solution, resist changes in pH. In particular, they protect against large changes in pH when acids or bases are added to the solution. Living cells are extremely sensitive to even very slight changes in pH. As we stated earlier, the reason for this sensitivity is that the enzymes that catalyze metabolic reactions operate in only a small range of pH. Altering the pH slows down or stops the action of the enzyme. Fortunately, the body's cells, extracellular fluid, and blood have each developed buffer systems that protect against pH changes.

The best buffer systems consist of a weak acid and its conjugate base, or a weak base and its conjugate acid. Such systems have their highest buffering capacity at a pH at which the concentration of acid equals the concentration of conjugate base, or the concentration of base equals the concentration of conjugate acid. The following are some acid–base pairs that can be used in buffer systems.

$$\underset{\text{carbonic acid}}{H_2CO_3} + H_2O \rightleftharpoons \underset{\text{bicarbonate ion}}{HCO_3^-} + H_3O^+$$

$$CH_3COOH + H_2O \rightleftharpoons CH_3COO^- + H_3O^+$$
acetic acid — acetate ion

$$H_2PO_4^- + H_2O \rightleftharpoons HPO_4^{2-} + H_3O^+$$
dihydrogen phosphate ion — monohydrogen phosphate ion

$$NH_3 + H_2O \rightleftharpoons NH_4^+ + OH^-$$
ammonia — ammonium ion

How do buffer systems protect body fluids? The major buffer system in the blood is the carbonic acid–bicarbonate system. Consider the following equilibrium equation:

$$H_2CO_3 \rightleftharpoons HCO_3^- + H^+$$

Give an example of the way in which a buffer protects a solution against large changes in pH.

Adding a strong acid to the system increases the concentration of H^+, driving the reaction to the left and forming more carbonic acid.

$$H_2CO_3 \leftharpoondown\!\!\rightharpoonup HCO_3^- + H^+$$

Carbonic acid is unstable, however, and decomposes to form carbon dioxide and water.

$$H_2CO_3 \longrightarrow CO_2(g) + H_2O$$

The carbon dioxide so formed can be removed from the blood and exhaled by the lungs. This buffer system continues to protect against the pH change until all the bicarbonate has reacted.

Various factors, such as heart failure or diabetes, can cause an abnormal increase in acid levels in the blood (lowering the blood pH), resulting in a condition known as acidosis. Other factors, such as vomiting or hyperventilation, can raise the blood pH, causing a condition called alkalosis. The mechanisms by which the body corrects these changes in blood pH, as well as the treatment for acidosis and alkalosis, are discussed in Section 21.6.

KEY CONCEPTS

The Brønsted–Lowry definition of acids and bases states that an acid is a substance that can donate protons (H^+), and a base is a substance that can accept protons. When dissolved in water, acids produce solutions that react with metals to produce hydrogen, taste sour, and turn blue litmus paper red. Bases form solutions that taste bitter, feel slippery, and turn red litmus paper blue.

A weak acid added to water causes a much smaller increase in the hydrogen ion concentration than the same amount of strong acid. A strong acid is one that has a large tendency to donate protons and, when dissolved in water, is almost completely ionized, or dissociated. Only a few molecules of a weak acid donate their protons. A strong base has a very large attraction for protons, whereas a weak base has a weak attraction.

A neutralization reaction occurs when equal amounts of acid and base react. A titration uses a neutralization reaction to determine an unknown concentration of acid or base.

In water, the hydrogen ion concentration $[H^+]$ times the hydroxide ion concentration $[OH^-]$ always equals 1×10^{-14}, or K_w, the ion product of water. The pH scale is a convenient way to show the hydrogen ion concentration of a solution. Pure water, which is neutral, has a pH of 7. Solutions with a pH less than 7 are acidic, and those with a pH greater than 7 are basic.

Buffers are systems that protect solutions against changes in pH caused by the addition of acid or base. Buffer systems consist of either a weak acid and its conjugate base, or a weak base and its conjugate acid. Because living organisms are sensitive to changes in pH, they include buffer systems within the cells, the extracellular fluid, and the blood to protect against such changes.

Important Equations

$$\text{Acid} + \text{base} \longrightarrow \text{salt} + \text{water}$$

$$[H^+] \times [OH^-] = 1 \times 10^{-14} = K_w$$

$$[H^+] = 1 \times 10^{-pH} \quad \text{or} \quad pH = -\log [H^+]$$

Review Problems

Section 11.1

1. Identify the acids and bases in the following reactions:
   (a) $H_2O + HCl \rightleftharpoons H_3O^+ + Cl^-$
   (b) $H_2O + CH_3NH_2 \rightleftharpoons CH_3NH_3^+ + OH^-$
2. Identify the acids and bases in the following reactions:
   (a) $HSO_4^- + H_2O \rightleftharpoons H_3O^+ + SO_4^{2-}$
   (b) $NH_2^- + H_2O \rightleftharpoons NH_3 + OH^-$
3. Describe three properties of solutions formed by dissolving an acid in water.
4. Describe three properties of solutions formed by dissolving a base in water.
5. Write the formula for the conjugate acid of each of the following:
   (a) CO_3^{2-} (c) HSO_3^- (e) NH_2^-
   (b) NH_3 (d) HPO_4^{2-} (f) OH^-
6. Write the formula for the conjugate base of each of the following:
   (a) $H_2PO_4^-$ (d) NH_4^+
   (b) H_2SO_3 (e) H_2O
   (c) $HClO$ (f) NH_3
7. Write equations showing sulfuric acid donating each of its two ionizable protons.
8. Write equations showing carbonic acid donating each of its two ionizable protons.

Section 11.2

9. Using Table 11.2, state which is the stronger base in each of the following pairs:
   (a) Cl^- or OH^- (c) NO_3^- or NH_3
   (b) PO_4^{3-} or HPO_4^{2-} (d) HCO_3^- or CO_3^{2-}

10. Using Table 11.2, state which is the stronger acid in each of the following pairs:

(a) H_3O^+ or HSO_4^- **(c)** H_3PO_4 or $H_2PO_4^-$

(b) H_2O or NH_4^+ **(d)** HNO_2 or HSO_3^-

Section 11.3

11. For the neutralization reaction between ammonium chloride and sodium hydroxide,

(a) Write a balanced equation for the reaction.

(b) Write the net ionic equation for the reaction.

(c) Is the odor of ammonia greater before or after the sodium hydroxide is added to a solution of ammonium chloride? Give the reason for your answer.

12. Write the balanced equation and the net ionic equation for the complete neutralization of arsenic acid (H_3AsO_4) by sodium hydroxide.

Section 11.4–11.5

13. Calculate the concentration of hydrogen ions in moles per liter for a solution whose hydroxide ion concentration is:

(a) $1 \times 10^{-8}\ M$ **(c)** $0.0030\ M$

(b) 0.0030 mol in 450 mL **(d)** $2.6 \times 10^{-3}\ M$

14. Calculate the concentration of hydroxide ions in moles per liter for a solution whose hydrogen ion concentration is:

(a) $1 \times 10^{-6}\ M$ **(c)** $0.0025\ M$

(b) 0.0030 mol in 550 mL **(d)** 3.6×10^{-5}

Section 11.6

15. At room temperature, what is the concentration of H^+ and OH^- in a solution of

(a) pH 1, **(b)** pH 6, **(c)** pH 12?

16. Arrange the following pH values in order of increasing acid strength (from the least acidic to the most), and indicate which values represent basic, neutral, and acidic solutions.

4, 6.3, 9.5, 1.4, 7, 5.5, 8.4, 12, 7.4

17. At room temperature, what is the pH of a solution for which

(a) $[H^+] = 0.01$ **(b)** $[H^+] = 1 \times 10^{-8}$ **(c)** $[OH^-] = 1 \times 10^{-4}$

18. Indicate whether each of the solutions in Problem 17 is acidic or basic.

19. Two 5-mL samples of fog water collected in Southern California had pH values of 3.0 and 4.0. How many moles of hydrogen ions and hydroxide ions are in each sample?

20. Solution A has a pH of 3 and solution B a pH of 5.

(a) Which solution is more acidic, A or B?

(b) What is the hydrogen ion concentration in each solution?

(c) By what factor do the two hydrogen ion concentrations differ?

Section 11.7

21. Use words and equations to explain how an acetic acid–acetate ion buffer system can protect against pH changes **(a)** when acid is added to the system, **(b)** when base is added.

22. What is the buffer system that controls the acidity in the blood plasma and red blood cells?
23. Use equations and words to show how the carbonate buffer system responds to the addition of acid.
24. Use equations and words to show how the carbonate buffer system responds to the addition of base.

Applied Problems

25. Before the development of modern detergents, washing soda (Na_2CO_3) was often added to water to reduce its acidity and allow soap to stay in solution. The carbonate in washing soda is the conjugate base of carbonic acid (H_2CO_3). Write the net ionic equation showing how washing soda neutralizes acid.
26. One of the uses of baking soda ($NaHCO_3$) described on the box is to relieve acid indigestion. Explain how baking soda acts in the relief of indigestion (use words or an equation).
27. Baking powder is made up of baking soda ($NaHCO_3$) and some source of H^+ to react with it. One example is cream of tartar, $KHC_4H_4O_6$, the potassium acid salt of tartaric acid. When water is added to this combination and heat is applied, carbon dioxide gas is produced—which makes cakes rise. Write the net ionic equation for the reaction between baking soda and cream of tartar to produce CO_2 gas.
28. In some areas, the increasing acidity of lakes has led conservationists to drop lime (calcium hydroxide) into the lakes. Explain how this helps to lower the acidity of a lake.
29. You may have noticed that when you add lemon juice to your tea, the tea changes color. Suggest a reason for the color change.
30. Rainwater has a large amount of CO_2 dissolved in it. Explain why the pH of normal rain is 5.6.
31. A large amount of lactic acid is produced in the muscles during hard exercise. The lactic acid must be transported by the blood to the liver, where it is broken down. Why doesn't the pH of the blood change drastically after hard exercise?

Integrated Problems

1. Years ago, travelers often kept a canvas bag full of water attached to their car's front fender. Even when traveling in extremely hot weather, the water in the canvas bag was always cool. Why was this so?
2. In hot, humid weather, there is nothing more satisfying than an ice-cold beverage. You may have noticed that if you are drinking from an aluminum cup, condensation forms almost immediately on the outside of the cup. When using a cup made from glass or ceramic, however, this doesn't happen. What accounts for this difference?

3. **(a)** Aspartame (the sweetening ingredient Nutrasweet found in Equal sweetener) is an artificial sweetener used in soft drinks and fruit yogurts. Because it decomposes at high temperatures, aspartame cannot be used in cooked foods. If a soft-drink manufacturer puts 1.00 g of aspartame in each kilogram of soft drink, what is the concentration of aspartame (in weight/volume percent) in a 350-mL bottle with a density of 0.985 g/mL?

   (b) If the formula for aspartame is $C_{14}H_{18}N_2O_5$, what is the molarity of the aspartame solution?

4. Dextran can be administered in normal saline solution to help maintain the blood pressure of persons in shock. The recommended dosage for the first 24 h is 2.00 g of dextran per kilogram of body weight. How many milliliters of 10.0% (w/v) dextran in normal saline should be administered to a 165-lb man during the first 24 h?

5. Bile is produced in the liver and stored in the gallbladder, from which it is released after each meal to aid in digestion of fats. The capacity of the gallbladder is only 40 mL. To store sufficient fluid between each meal, the bile produced by the liver is concentrated to 10% of its original volume. The cells of the gallbladder actively transport sodium and chloride ions out of the bile and into the intercellular fluid. Explain how this action works to concentrate the bile.

6. The following equilibrium is formed between ammonia and the ammonium ion:

$$NH_3 + H^+ \rightleftharpoons NH_4^+$$

   Use the following statements and your knowledge of Le Chatelier's principle to explain how ammonia is removed from the blood into the kidney tubules that carry urine to the bladder.
   (a) NH_3 can pass through cell membranes, but NH_4^+ cannot.
   (b) The normal pH of the urine in the kidney tubules is between 5.5 and 6.5.

7. Barium sulfate is used to coat the gastrointestinal tract of patients having CT scans.
   (a) Does a precipitate form if a solution of barium chloride is added to a solution of sodium sulfate? Write a balanced equation and a net ionic equation for the reaction that occurs.
   (b) How many moles of barium sulfate are produced when 50 mL of 0.1 *M* barium chloride is added to 100 mL of 0.05 *M* sodium sulfate?

8. A 100-mL sample of well water was found to contain the following levels of contaminants: arsenic, 40 ppb; cadmium, 5 ppb; and mercury, 5 ppb. Do any of these contaminants exceed the following maximum contaminant levels for drinking water as set by the EPA: arsenic, 0.05 mg/L; cadmium, 0.01 mg/L; and mercury, 0.002 mg/L?

9. Acetic acid is the active ingredient in vinegar. A solution of acetic acid forms the following equilibrium:

$$HC_2H_3O_2 + H_2O \rightleftharpoons H_3O^+(aq) + C_2H_3O_2^-(aq)$$

   (a) What happens to the equilibrium if sodium acetate is added to the solution?
   (b) What effect does adding NaOH to the solution have on the equilibrium?

10. The trace element fluorine can be absorbed into the body both as the fluoride ion and as hydrofluoric acid. Which form is mainly absorbed in the stomach and which in the small intestine? Explain your answer.

11. Although a sodium intake of less than 2400 mg/day is recommended, the average American consumes 6 to 8 g of sodium daily. Sodium is a factor in causing high blood

pressure in many people. Explain why high daily dietary concentrations of sodium might cause high blood pressure.

12. An intravenous solution contains 5.0% (w/v) dextrose ($C_6H_{12}O_6$) in 0.45% (w/v) NaCl.

   (a) What is the osmolarity of the solution? Is the solution isotonic, hypotonic, or hypertonic to blood serum?

   (b) Should this IV solution be given to a postoperative patient that is suffering from edema and low blood pressure?

13. People who are frequent users of hot tubs run the risk of getting an itchy rash caused by the bacterium *Pseudomonas aeruginosa.* These bacteria are brought into the hot tub on people's skin and are kept in check by using halogen disinfectants (chlorine or bromine). A halogen disinfectant loses its effectiveness, however, if its concentration falls too low or if the water becomes too alkaline (basic). The Centers for Disease Control recommend that the free chlorine level in a heated tub be kept above 2 ppm and the pH between 7.2 and 7.8. What measures, if any, should you take if you test your hot tub and find the free Cl_2 concentration to be 0.1 mg% and the metacresol purple indicator you use turns purple?

Section 2

The Elements Necessary for Life

When something is alive, the atoms and molecules from which it is formed are organized into elaborate structures and interact in precise ways. It is sometimes useful to think of these structures as having several levels of organization. The most basic level consists of the atoms common to all living things. The atoms are organized into relatively simple molecules, and these simple molecules are arranged in repeating units that form much more complex molecules.

These complex molecules join together in living organisms to form organelles that make up cells, which are the next level in structural organization. Cells and groups of cells, in turn, form tissues. Different tissues make up organs, which act together to form the organ systems of living organisms. Finally, the organisms themselves come together to form populations and biological communities.

Let's look at an example of this organizational structure. Atoms of carbon, hydrogen, oxygen, nitrogen, and magnesium can be combined to form the chemical structure of the chlorophyll molecule. Chlorophyll, a vital component of plant cells, absorbs energy from sunlight. Later we will see how this energy, through the process of photosynthesis, is used to produce glucose in the leaves of plants.

Many other different molecules are required for photosynthesis to occur. These molecules, along with the chlorophyll, are located in organelles called chloroplasts. In turn, other types of organelles work together with the chloroplasts to perform the functions of a plant cell. Plant cells are arranged to form tissues. The cells containing the chloroplasts help make up the tissues that form a leaf, which is one of the organs of a plant. Such organs form complex organ systems that together can make up a tree. A tree is also a member of a biological community that contains many other types of organisms.

In Section 2, we study how the atoms that compose the majority of living matter (carbon, hydrogen, oxygen, and nitrogen) join together to form the simple molecules that are the basic building blocks of life. We see how specific arrangements of these atoms called functional groups give these molecules characteristic physical and chemical properties. Understanding the chemistry of these organic compounds is the foundation for understanding the chemistry of living organisms.

Chapter 12

Carbon and Hydrogen: The Hydrocarbons

*H*ow did life on earth begin? There are many theories about the beginning of life, but little hard evidence exists to support the theories. Because the oldest rocks on earth are not much older than the earliest organisms, there is no geologic record of the conditions on early earth. In one theory proposed by the Russian chemist A. I. Oparin, the planet earth was born 4.5 billion years ago, having formed from a ball of hot gases in space. Over endless centuries the surface of the earth slowly cooled and became solid, forming the rocks known as the earth's mantle. As the cooling continued, steam condensed to form water that filled the low places in the mantle, creating earth's oceans. The infant earth was a lifeless landscape of rock and water, constantly pounded by storms and heavy rains. Oparin proposed that the first atmosphere, enriched by gases pouring from the interior of the earth in volcanic eruptions, consisted mostly of methane (CH_4) and nitrogen, with hydrogen, ammonia, carbon monoxide, and water vapor in smaller quantities. There was no free oxygen.

In this very earliest (or primordial) atmosphere, chemical reactions were started by ultraviolet radiation, ionizing radiation, and lightning. Many products of the reactions were simple carbon compounds and water. The carbon compounds formed in the atmosphere were brought to earth with the rains. They settled in the pools and oceans, forming a warm, dilute aqueous solution of carbon compounds—a ''primordial soup'' rich in the building blocks for the earliest life.

Another theory suggests that, in addition to basic carbon compounds formed in the earth's atmosphere, many of the key molecules were formed in outer space, in the vast clouds of gas and dust that collapsed to form our solar system. These molecules were brought to earth by the meteorites and comets that constantly bombarded the planet until about 4 billion years ago. Evidence for this theory has been mounting since the late 1960s when Nobel laureate Charles Townes and his

research group identified ammonia and water molecules in interstellar space. (Until then, the gaseous clouds of interstellar space were thought to contain only hydrogen and helium atoms.) Since that first discovery, over 75 different molecules have been identified in interstellar space. A 1986 study of Halley's Comet showed that its icy dust contained organic carbon compounds, including the first polymer (a large molecule of repeating units) ever identified in space. In 1987, Samuel Epstein of the California Institute of Technology and John Cronin of Arizona State University identified amino acids (the building blocks of proteins) in pieces of the Murchison Meteorite.

Over 100,000 comets may have collided with the earth during its first billion years, bringing many tons of organic matter and water to the earth's surface (most of a comet's mass is water in the form of ice). Comets not exploding in the atmosphere would have hit the earth's surface, forming a warm pool with a concentration of the molecules needed to form the building blocks of life.

Oparin suggested that as the centuries passed the carbon compounds in the pools began to clump together to form droplets (or coacervates). The coacervates developed a surface layer, or boundary, with properties different from the droplet itself. The coacervates would absorb through this boundary only selected materials from the surrounding environment. This allowed various molecules to be held very close to one another within the coacervate, causing their concentration to become much higher than in the surrounding environment. Oparin considered this the first major step toward the beginning of life.

Oparin believed that, as time passed, increasingly complex coacervate systems developed that were capable of growth as well as simple maintenance. Such growth required the development of complicated chemical pathways to trap chemical energy obtained from other molecules in the primordial soup. Oparin gave the name protobionts to these coacervate droplets with improved internal organization. He suggested that life truly began about 3.5 billion years ago as the protobionts developed methods for controlling their chemical reactions and reproduction. The protobionts, therefore, represent the link between the original droplets and the most primitive of living organisms.

12.1 Organic Chemistry: An Overview

Define organic chemistry.

Organic chemistry is the study of the compounds of carbon. The unique properties of these compounds, as we will see, make them the building blocks of life. We include the study of organic chemistry in this text because of the importance of many organic compounds to living organisms. For example, vitamins are organic compounds that must be included in our diet to maintain health. Other organic compounds can be dangerous. For example, the organic polychlorinated biphenyls (PCBs), which are used in insulating materials, are toxic to living organisms.

Our discussion of organic chemistry can cover only a very small part of this field. We will emphasize, therefore, those compounds and reactions that prepare you for the study of carbohydrates, lipids, proteins, and nucleic acids—substances critical to life.

Describe four properties of carbon that make it unique among the elements.

What is it about carbon that makes it unique? Looking at the periodic table, you will see that carbon is the first member of Group IVA. As a member of this group, carbon has four valence electrons. The octet rule (see Section 5.1) states that an atom achieves stability by having eight electrons in its outermost energy level. It is unlikely that carbon would gain

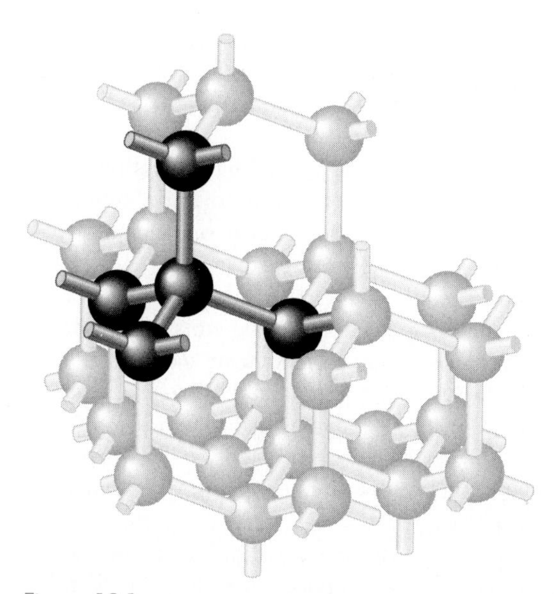

Figure 12.1 Bonding in diamond. Each carbon atom sits at the center of a tetrahedron and is bonded to four other carbon atoms.

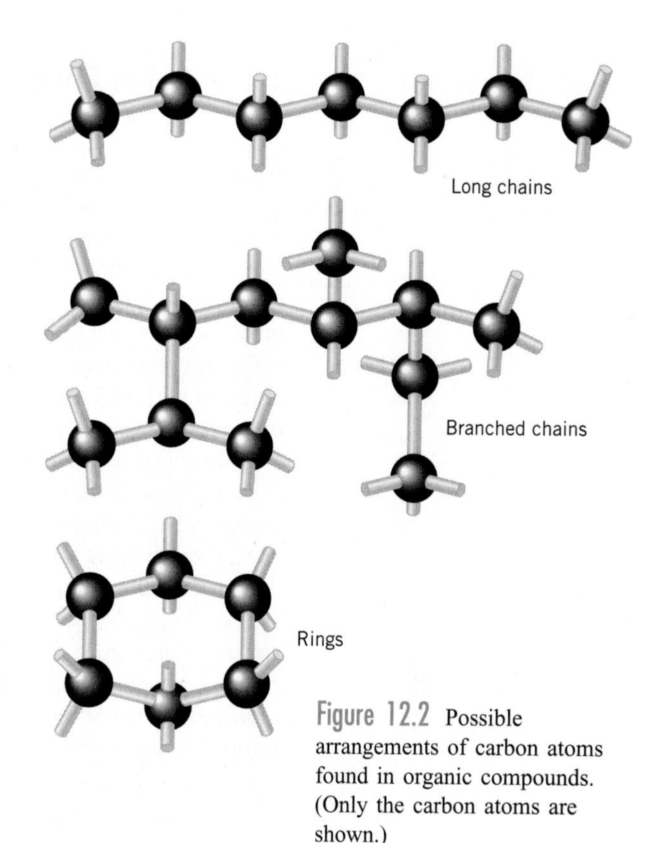

Figure 12.2 Possible arrangements of carbon atoms found in organic compounds. (Only the carbon atoms are shown.)

or lose four electrons to become a 4− or a 4+ ion. By sharing four more electrons, however, it can achieve an octet in its outermost energy level. Thus, carbon achieves stability by forming four single, covalent bonds. If these four bonds are to other carbons and if these other carbons in turn each bond to three more carbons, and each of these to three more carbons, and so on, the molecule that results is the very stable diamond (Fig. 12.1).

Carbon can bond to other carbons to form chains. In some compounds these chains are branched, whereas in others they loop into rings (Fig. 12.2). Carbon can also bond to many other elements. We will see compounds in which carbon bonds to hydrogen, oxygen, nitrogen, sulfur, phosphorous, and the halogens. Because of carbon's unique bonding ability, there are more than 3 million known natural and synthetic organic compounds. Compare this with the 500,000 known inorganic compounds. Inorganic compounds with more than 12 atoms are rare, but complex organic compounds, such as proteins, may contain more than a million atoms.

Diamonds are a crystalline form of pure carbon.

You might find the thought of learning about millions of compounds rather intimidating. Fortunately, the chemical properties of many organic compounds are very similar, allowing them to be grouped into several classes of compounds. As you examine the structures of compounds within a particular class, you will find that they each have a group of atoms arranged in a particular way. These arrangements of atoms, which give rise to certain chemical properties, are called **functional groups.** For example, the three compounds in Figure 12.3 have similar chemical properties because they contain the same functional group, the alcohol group, $-\overset{|}{\underset{|}{C}}-O-H$.

Environmental Perspective / Carbon Dioxide and the Greenhouse Effect

We are used to hearing about the dangers of air pollution caused by the burning of fossil fuels. From the smoke released by burning wood to the sulfur dioxide, sulfuric acid, and other harmful compounds released in the combustion of coal, oil, and gas, we have been filling the atmosphere with toxic substances. Global concerns over this problem are having some effect. In recent years, air quality has improved greatly in many parts of the world. We have done much to reduce particulate emissions, especially in urban areas. Levels of sulfur oxides and carbon monoxide are down in most cities. In all this, however, we have overlooked two very important pollutants: carbon dioxide and water. No matter how clean we make our engines and factories, carbon dioxide and water will always be produced when the source of energy is fossil fuels. Because 90% of our energy comes from fossil fuels, we continue to spew a tremendous amount of carbon dioxide and water into the air.

Carbon dioxide and water have generally not been considered pollutants because they are natural components of our environment. But scientists throughout the world are becoming alarmed over the possible long-term effects of their buildup in the atmosphere. Both carbon dioxide and water vapor absorb infrared radiation (heat) given off by the earth. Somewhat like the glass in a greenhouse, they act as a one-way filter—allowing the sun's rays to enter the atmosphere but preventing the longer wavelength infrared radiation from leaving. This is called the "greenhouse effect." Because water vapor quickly falls back to the earth as rain, it plays less of a role in the greenhouse effect. Carbon dioxide, on the other hand, stays around much longer. Part of it is consumed by plants in photosynthesis or absorbed by the oceans, but at least half remains in the atmosphere to affect the earth's climate. Every gram of fossil fuel burned releases nearly 3 g of carbon dioxide. Some scientists predict that a doubling of carbon dioxide in the atmosphere will raise the average global temperature 3° to 8°F by the middle of the next century. This could result in widespread climatic change, increasing the frequency and severity of both droughts and destructive storms, and causing sea levels to rise.

Your Perspective: Some scientists estimate that the greenhouse effect could cause sea level to rise by as much as several feet. Explain why this might occur, and describe the possible results of such a rise in sea level.

Combustion of fossil fuels releases air pollutants.

Our study of organic chemistry is structured around functional groups. As we discuss each functional group, you will learn to recognize the particular arrangement of atoms that makes up that functional group. We then discuss several important members of the class of compounds containing that functional group and study some of the reactions they undergo.

As is true with any field of study, you need to know something about the language of organic chemistry in order to study organic compounds. We already saw how to name some inorganic compounds when we learned that NaCl is called sodium chloride. The nomenclature of organic chemistry is more complex because so many organic compounds exist. The system of nomenclature we use was developed by the International Union of Pure and Applied Chemistry and is referred to as the IUPAC system. In this system, a set of rules was developed to ensure that each compound has its own distinct name. It is confusing enough if two people have the same name. It would be even more confusing if two organic compounds shared the same name. For example, one of the compounds might be beneficial to a living organism, whereas the other is toxic. On the other hand, just as we often have nicknames, some organic compounds have more than one name. In addition to their IUPAC name, they may have a trivial (or common) name. Often, these other names are based on where the compound is found in nature or on some property of the compound. It is not hard to guess where to look for cinnamaldehyde or cadaverine. Later in this chapter, we learn a few of the IUPAC naming rules so that we can name and identify some of the simpler organic compounds.

```
    H
    |
H—C—O—H
    |
    H
```
methanol

```
    H  H
    |  |
H—C—C—O—H
    |  |
    H  H
```
ethanol

```
       H
       |
    H  O  H
    |  |  |
H—C—C—C—H
    |     |
    H     H
```
2-propanol

Figure 12.3 These three compounds all contain the alcohol functional group and have similar properties.

12.2 Bonding with Hybrid Orbitals

In Chapter 5 we learned that two or more atoms can share some of their valence electrons, thereby forming covalent bonds. Earlier, in Chapter 4, we discussed how these valence electrons are located in atomic orbitals (or probability regions) of various shapes, surrounding the nucleus. A covalent bond is formed when two atoms are close enough to each other that an atomic orbital of one atom overlaps with an atomic orbital of the second atom. The strength of the covalent bond is determined by the amount of overlap of the two orbitals (differently shaped orbitals allow for different amounts of overlap).

This "orbital overlap" explanation of covalent bonding helps us understand the shapes taken on by the resulting molecules. In particular, it is used to explain the angles formed between different bonds in a molecule. Experimental observation of the bond angles formed in some molecules, however, cannot be explained by the simple overlap of *s*, *p*, or *d* atomic orbitals. It seems that in forming some bonds, the atomic orbitals "mix together" to form **hybrid orbitals.** The shapes of these hybrid orbitals allow for greater overlap, which results in the formation of stronger bonds and thus more stable molecules.

How does the theory of hybridization relate to our study of organic chemistry? The theory can be used to explain the shapes of some organic compounds. For example, it has been shown experimentally that carbon can form four equal single covalent bonds that are directed to the corners of a tetrahedron (Fig. 12.4). Carbon has four valence electrons and a valence electron configuration of $2s^2 2p^2$. Given these two types of orbitals, there ought to be two different types of bonds. How can there be four equal covalent bonds?

To form four equal covalent bonds, one of the 2*s* electrons moves to the empty 2*p* orbital. These four orbitals (one *s* and three *p*) then "mix," or hybridize, to form four equal sp^3 hybrid orbitals. Each hybrid orbital contains one electron, and each is directed toward a corner of the tetrahedron. In methane (CH_4), for example, carbon uses four sp^3

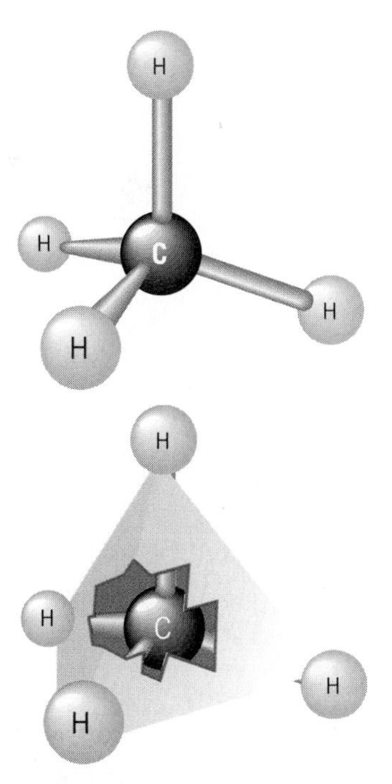

Figure 12.4 The carbon in methane sits at the center of a tetrahedron.

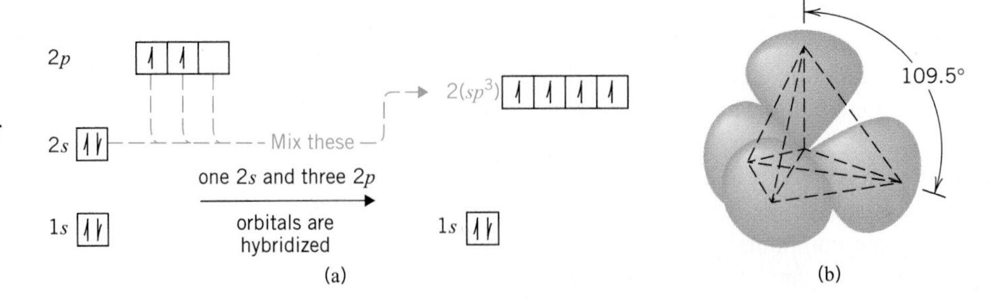

Figure 12.5 The sp^3 hybrid orbitals. (a) In carbon, one 2*s* and three 2*p* orbitals mix to form four sp^3 hybrid orbitals. (b) Tetrahedral arrangement of the four sp^3 orbitals.

orbitals to form four single covalent bonds with hydrogen. Methane has the expected tetrahedral shape, with the four bond angles all being the same 109.5° (Fig. 12.5). Carbon can also form other types of hybrid orbitals (Fig. 12.6).

12.3 Structural Formulas and Isomers

Describe the difference between the molecular and structural formula of a compound.

The molecular formula of a compound indicates the type and number of elements present in the compound. Several examples of molecular formulas are C_2H_6, $C_5H_{10}O$, and C_5H_{12}. Because it is common in organic chemistry to find more than one compound with the same molecular formula, we need to use **structural formulas** to differentiate among compounds of similar formula. Structural formulas are no more than chemical diagrams that

Figure 12.6 Hybrid orbitals.

Hybrid	Orbitals mixed	Orbital shape
sp	$s + p$	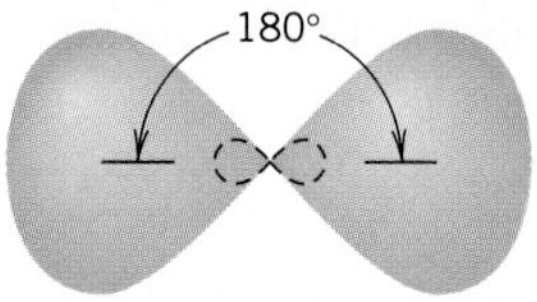 180°; 2 orbitals, 180° apart, linear shape
sp^2	$s + p + p$	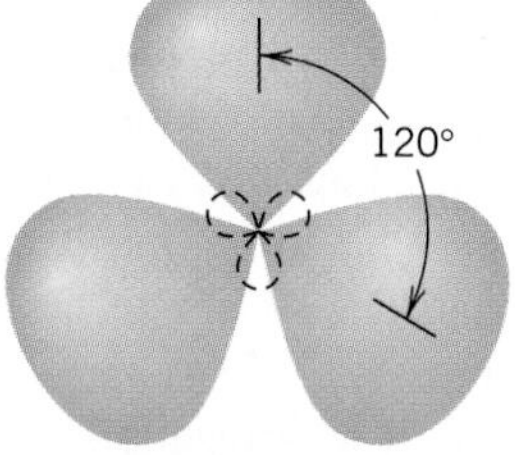 120°; 3 orbitals, 120° apart, planar–triangular shape
sp^3	$s + p + p + p$	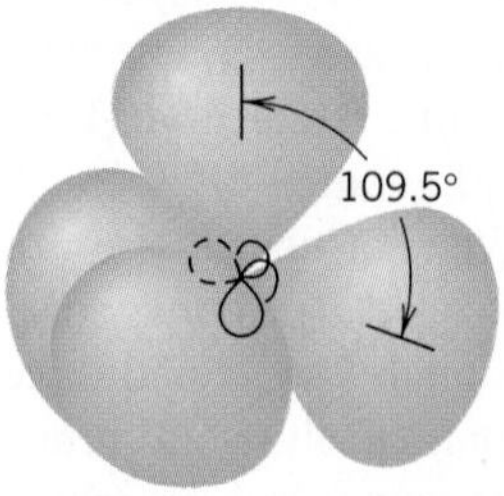 109.5°; 4 orbitals, 109.5° apart, tetrahedral shape

make organic chemistry easier to follow in the same way that diagrams help in the assembly of a bicycle or in the sewing of a dress. Although chemical compounds have three-dimensional structures, it is often very difficult to draw them that way. Instead, we use two-dimensional diagrams to represent the three-dimensional structures. For example,

```
                            H  H  H
                            |  |  |
instead of  [model]  we write H—C—C—C—H
                            |  |  |
                            H  H  H
```

In the structural formula of a molecule, the shared pair of electrons in a covalent bond is indicated by a line drawn between the symbols of the two atoms connected by that bond.

Consider the molecular formula C_5H_{12}. We can draw three different molecules with that same molecular formula (Fig. 12.7). The two-dimensional structural formulas for these three compounds are written as follows:

```
                                                        H
                                                        |
                                                      H—C—H
   H  H  H  H  H            H    H    H  H          H   |   H
   |  |  |  |  |            |    |    |  |          |   |   |
 H—C—C—C—C—C—H            H—C————C————C—C—H       H—C———C———C—H
   |  |  |  |  |            |    |    |  |          |   |   |
   H  H  H  H  H            H    |    H  H          H   |   H
                               H—C—H                  H—C—H
                                 |                      |
                                 H                      H
       (a)                       (b)                    (c)
```

We know that molecules (a), (b), and (c) are different compounds because they have different structural formulas, yet they all have the same molecular formula. Molecules that have the same molecular formula but different structural formulas are called **isomers.**

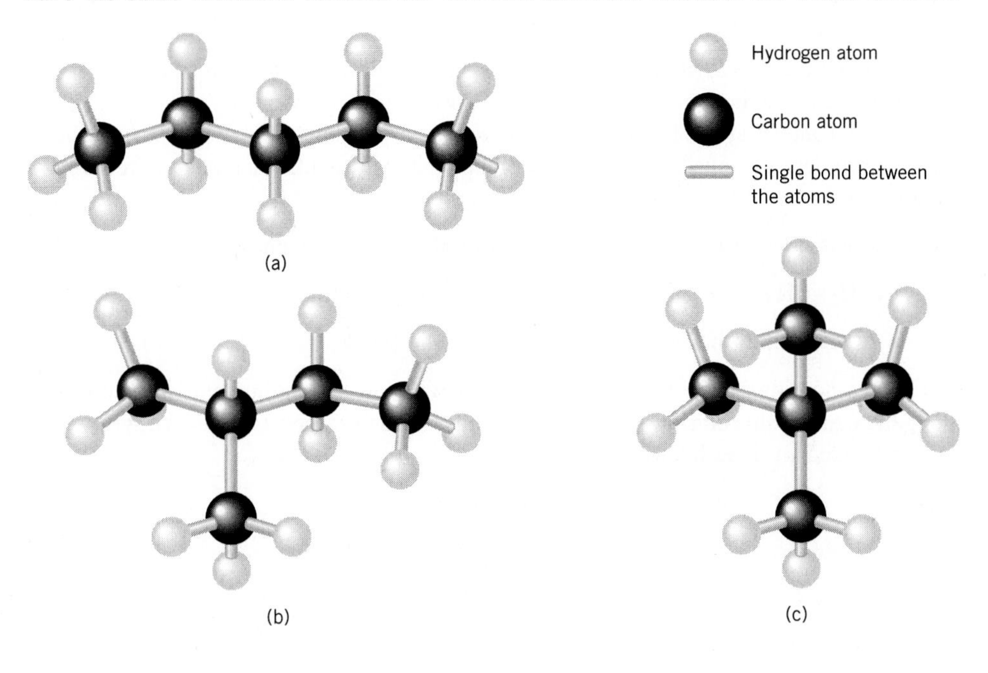

Figure 12.7 Molecules (a), (b), and (c) all have the same molecular formula, C_5H_{12}. They are constitutional isomers because they have different structural formulas.

Because drawing the structures of large molecules can be rather time-consuming, organic chemists have developed several ways of shortening the procedure. Here are some examples of condensed structural formulas.

```
    H   H   H   H
    |   |   |   |
H—C—C—C—C—H
    |   |   |   |
    H   H   H   H
```

becomes $CH_3—CH_2—CH_2—CH_3$

or $CH_3(CH_2)_2CH_3$

```
              H
              |
          H—C—H
H   H   H     |     H
|   |   |     |     |
H—C—C—C—C—C—H
|   |   |     |     |
H   H   H     H     H
```

becomes

```
                               CH3
                               |
CH3—CH2—CH2—CH—CH3
```

or $CH_3(CH_2)_2CH(CH_3)_2$

Quite often, we are only interested in a small part of a larger molecule. In this case we draw the extended structural formula for that part of the molecule and represent the rest of the molecule either as a condensed formula or as the **letter R.**

```
    H   H   H   H
    |   |   |   |
H—C—C—C—C—O—H
    |   |   |   |
    H   H   H   H
```

becomes

```
                         H
                         |
CH3CH2CH2—C—O—H
                         |
                         H
```

or

```
     H
     |
R—C—OH
     |
     H
```

*E*xample 12-1

In this example the *STEP* strategy is used, but the ***Execute the Math*** step is not necessary. This is the case in a number of problems that do not require mathematical calculations. For clarity, the ***T*** and ***E*** symbols are combined for these problems. As you ***Think It Through*** for each problem you must decide whether mathematical calculations are required for the solution or not.

Write the extended structural formula for the following compounds:

1. $CH_3(CH_2)_5CH_3$

See the Question ▶ The structure is to be written showing all of the bonds.

Think It Through (Execute the Math) ▶ Each carbon must have four bonds. We start with the first carbon and work our way along the main chain connecting four lines (bonds) to each carbon symbol. We start with a carbon bonded to three hydrogens

```
    H
    |
H—C—
    |
    H
```

The fourth bond is then connected to the next carbon. This second carbon is shown as $(CH_2)_5$, which means we have five CH_2 groups in a row. Our structure now looks like this:

```
     H   H   H   H   H   H
     |   |   |   |   |   |
  H—C—C—C—C—C—C—
     |   |   |   |   |   |
     H   H   H   H   H   H
```

Finally, we attach the last carbon and its three hydrogens:

Prepare the Answer

```
     H   H   H   H   H   H   H
     |   |   |   |   |   |   |
  H—C—C—C—C—C—C—C—H
     |   |   |   |   |   |   |
     H   H   H   H   H   H   H
```

2. Write the extended structural formula for the following molecule:

$$CH_3CH_2CH(CH_3)CH_2CH(CH_3)_2$$

The structure is to be written showing all bonds.

S

This structure is somewhat more complex than the previous one. The groups in parentheses are bonded to the carbon that precedes them in the condensed formula. The first carbon in the chain is attached to three hydrogens and to the second carbon.

T • E

```
     H
     |
  H—C—C
     |
     H
```

That carbon, in turn, must be attached to three more atoms: two H's and the next carbon on the chain.

```
     H   H
     |   |
  H—C—C—C
     |   |
     H   H
```

This carbon must now be bonded to three other atoms: the next carbon on the chain, a hydrogen, and a methyl group (shown in parentheses in the condensed formula).

```
             H
             |
          H—C—H
     H   H   |
     |   |   |
  H—C—C—C—C
     |   |   |
     H   H   H
```

The last carbon we added is now bonded to two H's and the next carbon on the chain.

```
             H
             |
          H—C—H
     H   H   |   H
     |   |   |   |
  H—C—C—C—C—C
     |   |   |   |
     H   H   H   H
```

P To complete the structural formula, a hydrogen and the two methyl groups shown in parentheses at the end of the condensed formula are bonded to the carbon that we added in the previous step. This results in the complete extended structural formula:

```
            H
            |
          H—C—H
     H  H   |   H  H  H
     |  |   |   |  |  |
  H—C—C—C—C—C—C—H
     |  |   |   |  |  |
     H  H   H   H  |  H
                 H—C—H
                   |
                   H
```

Check Your Understanding

1. Write the extended structural formula for each of the following compounds:

(a) $CH_3CH_2CH(CH_3)CH_2C(CH_3)_2CH_3$

(b) $CH_3C(CH_3)_3$

Now see if you can reverse this process by writing condensed structural formulas from extended structural formulas.

2. Write the condensed structural formula for each of the following compounds:

(a)

```
     H  H  H  H  H  H
     |  |  |  |  |  |
  H—C—C—C—C—C—C—H
     |  |  |  |  |  |
     H  H  H  H  H  H
```

(b)

```
                  H
                  |
                H—C—H
     H  H  H      |   H
     |  |  |      |   |
  H—C—C—C—C—C—H
     |  |  |   |  |
     H  H  |   H  H
         H—C—H
           |
           H
```

12.4 Alkanes: The Saturated Hydrocarbons

Define alkane and saturated hydrocarbon.

The simplest organic compounds are the **hydrocarbons,** which contain only carbon and hydrogen. Our study of organic chemistry begins with the **alkanes**—a group of hydrocarbons whose molecular formulas all fit the general pattern C_nH_{2n+2} (where n is any whole number). We recognize the alkanes by the fact that in each molecule the carbons are singly bonded to four other atoms. Because each carbon forms four single bonds, no additional atoms can be added to the molecule. For this reason, the alkanes are referred to as being **saturated.** (*Saturated* in organic chemistry means something entirely different from its use in the phrase "saturated solution" in Chapter 9.) Because of the strength and stability of the single bonds, the alkanes are the least reactive class of hydrocarbons.

The simplest hydrocarbon is the gas methane, CH_4. This compound is the major component of natural gas, which is used to heat our homes and cook our food. Methane

Health Perspective / Physiological Properties of Alkanes

Although the alkanes are generally quite unreactive, they have a variety of effects on the human body. Gaseous alkanes, as well as vapors from the liquid alkanes, act as anesthetics. They affect the protective coatings of nerve cells and disrupt the transmission of nerve impulses to and from the brain.

Liquid alkanes can do extensive harm if they get into the lungs. They dissolve lipid molecules in the cell membranes of the alveoli, which decreases the ability of the alveoli to expel fluids and leads to pneumonia-like symptoms. People accidentally swallowing a liquid alkane should *never* be induced to vomit because this might result in some alkane entering their lungs. Liquid alkanes have a similar effect when they contact the skin. Naturally occurring body oils are dissolved by the alkanes, causing the skin to dry out. For this reason, gloves should be worn whenever handling liquid alkanes.

The heavier liquid alkanes have properties more like the naturally occurring body oils. When applied to the skin, mixtures such as mineral oil act as softeners by replacing the natural skin oils lost in repeated contact with water or other solvents. Petroleum jelly, a mixture of high-molecular-weight liquid and solid alkanes, is used as both a skin softener and protective film. Because water and water solutions like urine cannot dissolve alkanes, petroleum jelly protects the skin from excessive contact with water. A good example is the use of petroleum jelly to protect a baby's skin from diaper rash caused by prolonged contact of the baby's skin with urine.

is a symmetrical nonpolar molecule. It is insoluble in water, which is a very polar solvent. (Recall that "like dissolves like.") Methane has a very low boiling point: −161.5°C (Fig. 12.8).

The next member of the alkane family is ethane, CH_3CH_3. Ethane is followed by propane ($CH_3CH_2CH_3$) and butane ($CH_3CH_2CH_2CH_3$, also written as $CH_3(CH_2)_2CH_3$). Propane and butane can be liquefied in tanks under pressure, allowing them to be easily stored and transported. This gives them wide use as fuels for lighters, torches, and furnaces in rural homes. The first 10 compounds in the unbranched alkane series are listed in Table 12.1. Note that all of these compounds fit the general formula of C_nH_{2n+2}.

Table 12.1 **Alkanes**

Number of Carbons	*Molecular Formula*	*IUPAC Prefix*	*Name*	*Structural Formula*	*Boiling Point in °C*
1	CH_4	meth-	Methane	CH_4	−162
2	C_2H_6	eth-	Ethane	CH_3CH_3	−89
3	C_3H_8	prop-	Propane	$CH_3CH_2CH_3$	−42
4	C_4H_{10}	but-	Butane	$CH_3CH_2CH_2CH_3$	0
5	C_5H_{12}	pent-	Pentane	$CH_3CH_2CH_2CH_2CH_3$	36
6	C_6H_{14}	hex-	Hexane	$CH_3CH_2CH_2CH_2CH_2CH_3$	69
7	C_7H_{16}	hept-	Heptane	$CH_3CH_2CH_2CH_2CH_2CH_2CH_3$	98
8	C_8H_{18}	oct-	Octane	$CH_3CH_2CH_2CH_2CH_2CH_2CH_2CH_3$	126
9	C_9H_{20}	non-	Nonane	$CH_3CH_2CH_2CH_2CH_2CH_2CH_2CH_2CH_3$	151
10	$C_{10}H_{22}$	dec-	Decane	$CH_3CH_2CH_2CH_2CH_2CH_2CH_2CH_2CH_2CH_3$	174

Figure 12.8 Methane.

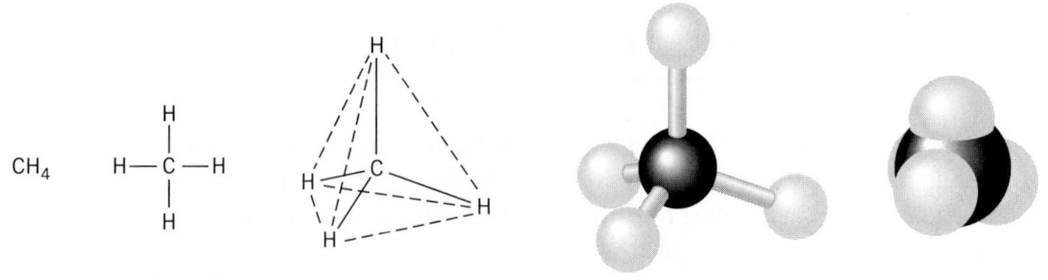

CH_4

H—C—H (with H above and below C)

12.5 Constitutional Isomers

The fourth member of the alkanes has a molecular formula of C_4H_{10}. Let's take a closer look. We can write structural formulas for two different compounds having this molecular formula:

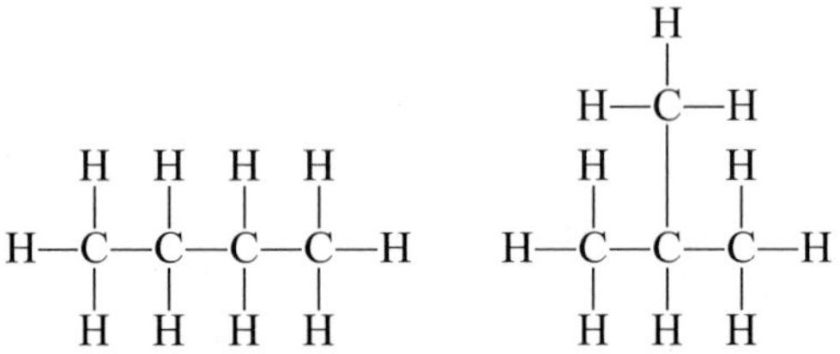

butane: H—C—C—C—C—H (each C bearing H above and below)

2-methylpropane: H—C—C—C—H (central C bearing an H—C—H methyl group above and H below; end carbons bearing H above and below)

These two compounds have their carbons arranged in a different order in the three-dimensional structure of the molecule. In one, the carbon atoms lie in a straight (unbranched) chain. In the other, the carbon atoms are arranged in a branched chain. These compounds are called **constitutional isomers.** Because the two compounds have their atoms connected in different ways, they have different structural formulas. They are, in fact, different compounds with different physical and chemical properties. If you make models of these two compounds, no amount of twisting and turning of the bonds will convert one of these isomers into the other. The only way to convert one constitutional isomer into another is to break bonds and rearrange the order in which the atoms are bonded to each other.

Define constitutional isomer, and, given the molecular formula of a compound, draw the structural formulas of its isomers.

Although there are only two constitutional isomers with the molecular formula C_4H_{10}, the number of possible isomers increases as the number of carbon atoms in a molecule increases. There are 18 different constitutional isomers with the formula C_8H_{18}. Each of these isomers has slightly different properties. One of them, ''isooctane,'' burns very well in car engines and is used as a standard in determining the octane rating of gasolines. A gasoline with an octane rating of 100 burns as smoothly in a car engine as ''isooctane.''

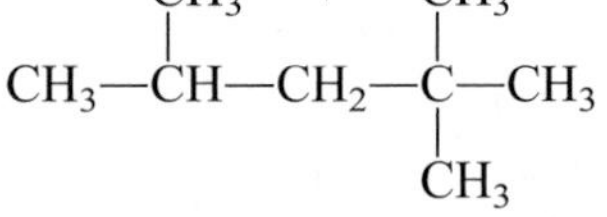

$$CH_3-CH(CH_3)-CH_2-C(CH_3)_2-CH_3$$

''isooctane'' (2, 2, 4-trimethylpentane)

12.6 Nomenclature

Because two different compounds have the formula C_4H_{10}, they must have different names. For this reason, IUPAC has established rules for naming such compounds. The nomenclature rules themselves must be quite complex to provide a different name for

```
    H              H              H  H              H  H
    |              |              |  |              |  |
 H—C—H          H—C—Br        H—C—C—H          H—C—C—Cl
    |              |              |  |              |  |
    H              H              H  H              H  H
 methane       bromomethane       ethane         chloroethane
```

The names of several important substituents are shown in the table.

Group	*Name*	*Group*	*Name*
—F	fluoro-	—I	iodo-
—Cl	chloro-	$—NH_2$	amino-
—Br	bromo-	$—NO_2$	nitro-

Rule 5. If the organic compound contains more than one of the same type of substituent, the number is indicated by a prefix: di- indicates two, tri- indicates three, and tetra- indicates four. For example, dibromomethane is a methane molecule containing two atoms of bromine, and tetrachloromethane is a methane molecule containing four atoms of chlorine.

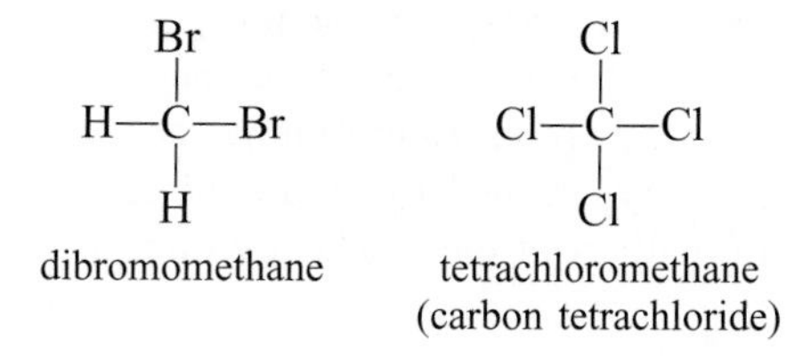

dibromomethane tetrachloromethane (carbon tetrachloride)

Rule 6. If two or more different substituents are attached to the carbon chain, they are written in alphabetical order (for example, fluoroiodomethane).

Rule 7. To indicate the carbon atoms to which substituents are attached, numbers corresponding to the carbon atoms precede the name of the substituent. The carbon atoms are numbered from the end of the chain nearest the first substituent. For example, 1,2-dichloroethane is an ethane molecule containing two atoms of chlorine—one attached to carbon number 1 and one attached to carbon number 2.

```
     Cl Cl                 Br Cl
     |  |                  |  |
  H—C—C—H              Br—C—C—H
    1| 2|                 1| 2|
     H  H                  H  H
1,2-dichloroethane   1,1-dibromo-2-chloroethane
```

This might seem very complicated to you right now, but after a little practice these rules will fall into place.

Example 12-2

Let's see how these rules work by naming the following compounds:

1.

```
              CH3 CH3
               |   |
CH3—CH2—CH—CH—CH3
```

See the Question — The IUPAC rules for naming compounds must be applied to this compound.

Think It Through (Execute the Math) —

(a) Compounds are actually named by working backwards, starting with the last part of the name. This compound is a hydrocarbon in which all of the bonds are single bonds. It is therefore an alkane and, based on Rule 1, its name must end in -ane.

(b) Next, draw a circle around the longest continuous carbon chain. This is considered to be the main chain. In our compound there are five carbons in the longest chain and so, from Rule 2, we know that the correct prefix is pent-. This compound is a pentane.

(c) There are two —CH_3 groups attached to the main chain. Table 12.2 tells us that the —CH_3 group is the methyl group, and Rule 5 tells us that the correct prefix is di-. At this point the name is dimethylpentane.

(d) We must now indicate the positions of the two methyl groups on the main chain. We number the main chain from the end closer to the first attached group. The methyls are therefore attached to carbon 2 and 3.

Prepare the Answer — The correct name for this compound is 2,3-dimethylpentane.

We should notice several things about the naming process. First, ''left'' or ''right'' have no significance in the name. We numbered this compound from the right only because the first attached group was closer to the right. If it had been closer to the left, we would have numbered from that end. The attached groups are indicated in two ways. Both the quantity (prefix) and position (number) of the attached groups are found in the name.

2. Let's try another one.

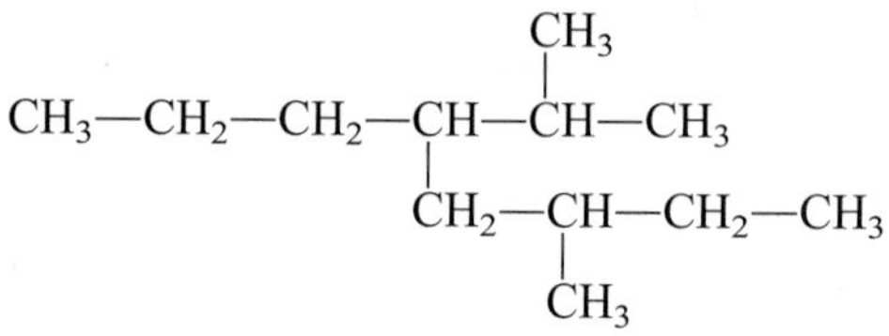

$$
\begin{array}{llllll}
 & & & & CH_3 & \\
 & & & & | & \\
CH_3— & CH_2— & CH_2— & CH— & CH— & CH_3 \\
 & & & | & & \\
 & & & CH_2— & CH— & CH_2—CH_3 \\
 & & & & | & \\
 & & & & CH_3 &
\end{array}
$$

S — Once again, the IUPAC nomenclature rules must be used to name this compound.

T • E —

(a) We have a hydrocarbon that has all single bonds. Therefore, it is an alkane and its name ends in -ane.

(b) Checking carefully, we find that the longest continuous chain we can circle is eight carbons long. Note that the longest chain has not been drawn in a straight line in this compound. The prefix for eight is oct-, so this compound is an octane. Here is the same compound redrawn to more clearly show the eight-carbon longest chain. Although it may look different, it is exactly the same compound.

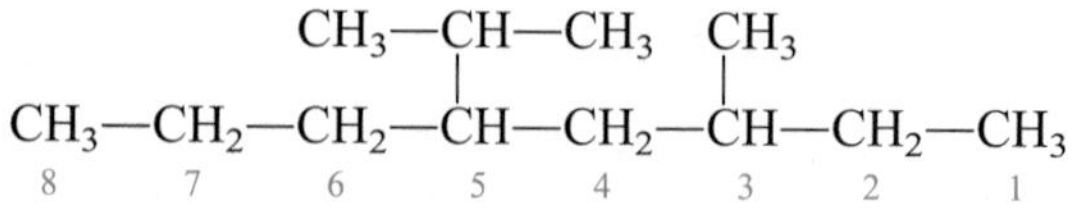

$$
\begin{array}{llllllll}
 & & CH_3— & CH—CH_3 & & CH_3 & & \\
 & & & | & & | & & \\
CH_3— & CH_2— & CH_2— & CH— & CH_2— & CH— & CH_2— & CH_3 \\
8 & 7 & 6 & 5 & 4 & 3 & 2 & 1
\end{array}
$$

(c) There are two groups attached to the eight-carbon main chain: one is a methyl group and the other is an isopropyl group (see Table 12.2).

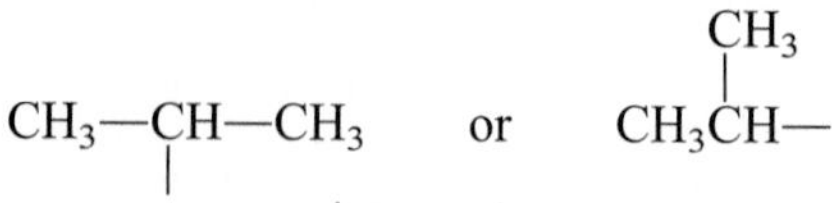

$$
\begin{array}{c}
CH_3—CH—CH_3 \\
|
\end{array}
\quad \text{or} \quad
\begin{array}{c}
CH_3 \\
| \\
CH_3CH—
\end{array}
$$

isopropyl

(d) The methyl group is attached to carbon 3 and the isopropyl group is attached to carbon 5. If we had numbered from the other end, the isopropyl group would be at carbon 4 and the methyl at carbon 6, which would be incorrect (remember Rule 7).

(e) We name the groups attached to the main chain in alphabetical order.

The name of this compound is 5-isopropyl-3-methyloctane.

P

Check Your Understanding

Try to name the following compounds yourself:

1.
$$\begin{array}{l} \quad\quad\;\; CH_3 \quad\quad\quad\;\; CH_3 \\ \quad\quad\quad | \quad\quad\quad\quad\quad | \\ CH_3-CH-CH_2-CH-CH_3 \end{array}$$

2.
$$\begin{array}{l} \quad\quad\quad\quad\quad\quad\;\; CH_3 \\ \quad\quad\quad\quad\quad\quad\quad | \\ \quad\quad\quad\quad\quad\quad\;\; CH_2 \;\; CH_3 \\ \quad\quad\quad\quad\quad\quad\quad | \quad\quad\;\; | \\ CH_3-(CH_2)_2-CH-CH-CH_3 \end{array}$$

3.
$$\begin{array}{l} \quad\quad\quad\quad\quad\quad\;\; CH_3 \\ \quad\quad\quad\quad\quad\quad\quad | \\ \quad\quad\quad\quad\quad\quad\;\; CH_2 \\ \quad\quad\quad\quad\quad\quad\quad | \\ CH_3-(CH_2)_4-CH-CH_2-CH_3 \end{array}$$

12.7 Reactions of the Alkanes

Oxidation

You are probably most familiar with alkanes as the fuels that heat your home, power your car, and light your camping lantern. Because they burn so easily, you might be surprised to learn that alkanes are actually quite unreactive compounds. They are not affected by strong acids or strong bases, nor by most strong oxidizing or reducing agents. But alkanes do react well with oxygen, and when this reaction takes place rapidly, we say that the compound is burning.

Combustion is another word for burning. As alkanes undergo combustion, the carbon and hydrogen atoms from the hydrocarbon each bond with oxygen atoms from the air. If this highly exothermic (heat-producing) reaction is allowed to go to completion in the presence of sufficient oxygen, the only products of the combustion are carbon dioxide (CO_2) and water (H_2O). For example, the equation for the combustion of methane is

Write the chemical equations for the combustion and substitution reactions of a given alkane.

$$CH_4 + 2O_2 \longrightarrow CO_2 + 2H_2O + \text{energy}$$

If not enough oxygen is present when alkanes burn, incomplete combustion occurs. In such a case, the end products are carbon monoxide (CO) and water. For example, if methane burns in an oxygen-poor environment, the following reaction occurs:

$$2CH_4 + 3O_2 \longrightarrow 2CO + 4H_2O + \text{energy}$$

Carbon monoxide produced by incomplete combustion in factories and cars is the largest source of carbon monoxide pollution in our society. Thankfully, the use of catalytic converters on automobiles since the late 1970s has significantly decreased atmospheric carbon monoxide levels.

Figure 12.9 Both photographs illustrate oxidation reactions, but the reactions are occurring at vastly different rates.

Combustion is an example of a rapidly occurring oxidation reaction. The process of oxidation in, say, a forest fire is easily seen and dramatic in its force (Fig. 12.9). But even the results of very slow oxidation can be observed by noticing the yellowing of old newspapers or the rusting of farm equipment left out in the rain. Oxidation reactions also occur continuously in our bodies and are the source of energy for our cells. If this oxidation occurred in the same way as methane burns, our cells would be destroyed by heat. Oxidation in the body, however, occurs at a very controlled rate, allowing cells to trap and use a portion of the energy that is released. Our bodies do not use alkanes as fuel, but rather use derivatives of alkanes: carbohydrates, fats, and proteins. Red blood cells supply the tissues with the oxygen needed for this reaction. The oxygen is picked up in the lungs and distributed throughout the body. The red blood cells then return the waste product, carbon dioxide, to the lungs to be exhaled. The water produced in the oxidation reaction either remains in the cells and tissues or is excreted through sweat and urine. Just as a fire can be smothered by a blanket that cuts off the oxygen supply, we can be smothered by anything that cuts off our supply of oxygen.

Halogenation

The other type of reaction that alkanes commonly undergo is **halogenation,** in which the alkane reacts with fluorine, chlorine, or bromine. For example, if a mixture of methane and chlorine is exposed to light or heat, the reaction begins immediately. The products of this reaction are chloromethane and hydrogen chloride.

$$CH_4 + Cl_2 \xrightarrow{\text{light}} CH_3Cl + HCl$$

Notice that in this reaction, one of the chlorine atoms *replaces* one of the hydrogens that had been bonded to the carbon. This is an example of a **substitution reaction**—a reaction in which one atom (or group of atoms) replaces another. We will be seeing substitution reactions throughout our study of organic chemistry.

Example 12-3

We just saw that methane reacts with chlorine to produce chloromethane. If the chloromethane from this reaction is allowed to react with additional chlorine, the substitution of chlorine for hydrogen proceeds until all of the hydrogens have been replaced by chlorine atoms. Write the equation for each of these reactions, and name the products of each.

See the Question

The problem asks for an equation for each reaction as chlorine replaces, one by one, the three hydrogens on chloromethane. It also asks for the name of each product using IUPAC nomenclature rules.

Think It Through (Execute the Math)

This reaction is an example of a halogenation reaction. The equation for the reaction of chlorine with methane when the reactants are exposed to heat or light is

$$CH_4 + Cl_2 \longrightarrow CH_3Cl + HCl$$

Prepare the Answer

If we substitute one more chlorine for a hydrogen, the reaction is

$$CH_3Cl + Cl_2 \longrightarrow CH_2Cl_2 + HCl$$

If we continue the process,

$$CH_2Cl_2 + Cl_2 \longrightarrow CHCl_3 + HCl$$

$$CHCl_3 + Cl_2 \longrightarrow CCl_4 + HCl$$

With the help of Rules 4 and 5 in Section 12.6, we can determine that

CH_3Cl is chloromethane (common name: methyl chloride)

CH_2Cl_2 is dichloromethane (also known as methylene chloride)

CCl_4 is tetrachloromethane (also known as carbon tetrachloride)

Each of these compounds is widely used as a solvent although their use is decreasing because they can cause liver damage.

Check Your Understanding

Write equations for the following reactions:

1. Write a balanced equation for the complete combustion of hexane.
2. Chloroethane, a fast-acting anesthetic, can be formed from ethane. Write the equation for this reaction.

12.8 Unsaturated Hydrocarbons

In addition to its ability to form chains, branched chains, and even rings, carbon can also share more than two electrons with another carbon. If two carbons share four electrons, they form a double bond. The sharing of six electrons results in the formation of a triple bond.

·C· ·C·   :C: :C: ·C: :C·

C—C   C=C   —C≡C—

The class of hydrocarbons containing double bonds between carbons is known as the **alkenes;** those compounds containing a triple bond are known as the **alkynes.**

Carbon–carbon double and triple bonds are more reactive than carbon–carbon single bonds. Reactions in the alkenes and alkynes therefore usually occur at the double or triple bond. We used the term *saturated* to describe the alkanes because that class of compounds has four single bonds to each carbon, so additional atoms cannot be added. The alkenes and alkynes are said to be **unsaturated** because additional atoms *can* be added to these molecules. If a compound contains more than one double or triple bond between carbon atoms, it is said to be **polyunsaturated.**

◆ **Indicate whether a hydrocarbon is saturated or unsaturated, given its structural formula.**

$$H-\underset{H}{\overset{H}{C}}-\underset{H}{\overset{H}{C}}-\underset{H}{\overset{H}{C}}-H \qquad CH_3-\overset{CH_3}{CH}-CH_3$$

saturated hydrocarbons

$$H-\underset{H}{\overset{H}{C}}-\overset{H}{C}=\overset{H}{C}-H \qquad CH_3-\overset{CH_3}{C}=CH_2 \qquad H-\underset{H}{\overset{H}{C}}-\underset{H}{\overset{H}{C}}-C\equiv C-H$$

unsaturated hydrocarbons

$$H-\overset{H}{C}=\overset{H}{C}-\overset{H}{C}=\overset{H}{C}-H \qquad CH_3-CH=CH-\overset{CH_3}{C}=CH_2$$

polyunsaturated hydrocarbons

12.9 The Structure and Nomenclature of the Alkenes

◆ **Describe the differences between alkanes and alkenes and give examples of compounds found in each class.**

The simplest alkene is ethene, which is also commonly called ethylene (Fig. 12.10). In ethene, all the atoms are located in the same plane. Notice also that the bond angles are 120°.

The rules for naming the alkenes are much like those for naming the alkanes, with a few modifications. IUPAC nomenclature is based on a set of priorities. Because the double

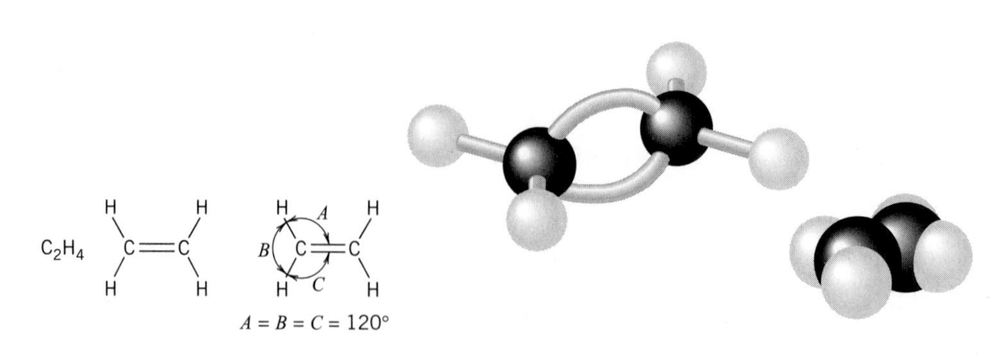

Figure 12.10 Ethene (ethylene).

Table 12.3 Alkenes

Number of Carbons	*Number of Double Bonds*	*Molecular Formula*	*Name*	*Structural Formula*
3	1	C_3H_6	Propene	$CH_2{=}CHCH_3$
4	1	C_4H_8	1-Butene	$CH_2{=}CHCH_2CH_3$
4	1	C_4H_8	2-Butene	$CH_3CH{=}CHCH_3$
4	1	C_4H_8	2-Methylpropene	CH_3 | $CH_2{=}C{-}CH_3$
5	2	C_5H_8	1,3-Pentadiene	$CH_2{=}CHCH{=}CHCH_3$
5	2	C_5H_8	2-Methyl-1,3-butadiene	CH_3 | $CH_2{=}C{-}CH{=}CH_2$

bond takes precedence over the single bond, we must base the names for alkenes on the position of the double bond. The rules for naming alkenes are

Rule 1. The names of alkenes end in -ene.

Rule 2. The longest carbon chain **containing the double bond** is selected.

Rule 3. The carbons in the longest chain are numbered, starting from the end closest to the double bond. The position of the double bond is indicated by a prefix giving the number of the double-bonded carbon nearest the beginning of the chain. For example, $CH_3CH_2CH{=}CHCH_3$ is 2-pentene, not 3-pentene or 2,3-pentene.

Rule 4. If the molecule has two or more double bonds, the prefixes di-, tri-, and so on are used to indicate the number of double bonds.

Carefully study the names and structures in Table 12.3 and in the following examples to see how these rules are actually applied. Notice especially the placement of commas and hyphens. Numbers are separated from each other by commas, whereas numbers and words are separated from one another by hyphens.

Example 12-4

1. Name the following compound:

$$\begin{array}{ccccccccc} & & CH_3 & & & & CH_3 & & \\ & & | & & & & | & & \\ CH_3 & - & CH & - & CH & = & C & - & CH_3 \end{array}$$

Apply the IUPAC rules to the naming of this alkene.

See the Question

(a) The compound contains a double bond. It is an alkene and, therefore, its name ends in -ene.

Think It Through (Execute the Math)

(b) Find the longest chain that contains the double bond, and number it from the end closer to the double bond. The name is now 2-pentene.

$$\begin{array}{ccccccccc} & & CH_3 & & & & CH_3 & & \\ & & | & & & & | & & \\ CH_3 & - & CH & - & CH & = & C & - & CH_3 \\ 5 & & 4 & & 3 & & 2 & & 1 \end{array}$$

(c) Now, name and number the substituents. Two methyl groups are bonded to carbons 2 and 4.

Prepare the Answer The name of the compound is 2,4-dimethyl-2-pentene.

2. Convert the name of the following compound into a structural formula: 4-methyl-2-hexene.

S In this problem the IUPAC name must be converted into the structural formula.

T • E **(a)** The last part of the name tells us we have a six-carbon chain with a double bond starting at the second carbon.

$$-\overset{|}{\underset{|}{C}}-\overset{|}{\underset{|}{C}}-\overset{|}{\underset{|}{C}}-\underset{|}{C}=\underset{|}{C}-\overset{|}{\underset{|}{C}}-$$

(b) The "4-methyl" tells us that there is a methyl on the fourth carbon.

$$\begin{array}{c} H \\ | \\ H-C-H \\ | \\ -\overset{|}{\underset{6|}{C}}-\overset{|}{\underset{5|}{C}}-\underset{4|}{C}-\underset{3|}{C}=\underset{2|}{C}-\overset{|}{\underset{1|}{C}}- \end{array}$$

P The structure is completed by putting in the correct number of hydrogens so that each carbon has exactly four bonds.

$$\begin{array}{c} H \\ | \\ H-C-H \\ \quad\; H\;\; H \quad | \quad H\;\; H\;\; H \\ H-\overset{|}{\underset{\underset{H}{|}}{C}}-\overset{|}{\underset{\underset{H}{|}}{C}}-\underset{\underset{H}{|}}{C}-\overset{|}{C}=\overset{|}{C}-\overset{|}{\underset{\underset{H}{|}}{C}}-H \end{array}$$

4-methyl-2-hexene

3. Let's try a compound that has more than one double bond. Draw the structure of 2-methyl-1,3-hexadiene.

S This is another conversion from the IUPAC name to the structural formula.

T • E **(a)** The -diene ending tells us that there are two double bonds.

(b) The "1,3-hexadiene" tells us that the main chain is six carbons long and that the double bonds start at carbons 1 and 3.

(c) "2-methyl" tells us to attach a methyl group at the second carbon.

(d) Attach the correct number of hydrogens to each carbon.

P The structure is

$$CH_3-CH_2-CH=CH-\overset{\overset{CH_3}{|}}{C}=CH_2$$

2-methyl-1,3-hexadiene

Chemical Perspective / Ethylene: The Workhorse of Organic Chemicals

In addition to being a trace component of crude oil and natural gas, ethylene is produced by plants to control the ripening of fruit, the germination of seeds, and the maturing of flowers. Fruits and vegetables naturally give off ethylene when they start to ripen. You can speed up the ripening process by placing fruit in a paper bag, thereby surrounding the fruit with the ethylene it gives off. Food processing companies now add ethylene to unripe fruit to ready it for market. Two pounds of green tomatoes can be ripened in 24 hours by exposure to only 0.08 mL of ethylene. Ethylene is also a flammable, anesthetic gas that is nontoxic to tissues even in high concentrations. Its anesthetic effects are rapid; a patient is ready for surgery 2 to 4 minutes after administration.

Ethylene is the primary starting material for the manufacture of such important industrial chemicals as ethyl alcohol; ethylene glycol (an antifreeze); and many plastics, such as polyethylene, polyvinyl chloride (PVC), and polystyrene. No other organic chemical is produced in larger quantities than ethylene. In the United States, 15 million tons of ethylene is produced each year. In the past, supplies of ethylene have come from the distillation of petroleum. But in 1989, scientists cloned the gene for the enzyme that regulates the production of ethylene in plants. Using this gene, researchers are working on ways to control fruit ripening and to produce large quantities of ethylene for industrial purposes using bacteria or yeast.

Your Perspective: The owner of a health food store claims in a newspaper article that tomatoes ripened through exposure to ethylene are dangerous to eat. Write a letter to the editor of your local paper either agreeing or disagreeing with this opinion, clearly stating your reasons.

Check Your Understanding

1. Name the following compounds.

(a) $CH_3CH{=}CHCHCH_3$ with $CH_2CH_2CH_3$ on the fourth carbon

$$\begin{array}{l} CH_3CH{=}CH\underset{\displaystyle |}{C}HCH_3 \\ \qquad\qquad\quad CH_2CH_2CH_3 \end{array}$$

(b) $Cl_2C{=}CHCH_3$

2. Draw the structural formula for the following compounds.

(a) 3-isopropyl-4-nonene

(b) 2,4,6-trimethyl-1-octene

12.10 Bonding in the Alkenes

A single bond is formed by the overlap of two atomic orbitals (*s*, *p*, or hybrid) directly along the bond axis—the imaginary line that connects the two nuclei (Fig. 12.11). This is sometimes also called frontwise overlap of atomic orbitals. This type of bond is called a **sigma bond,** or **σ bond.**

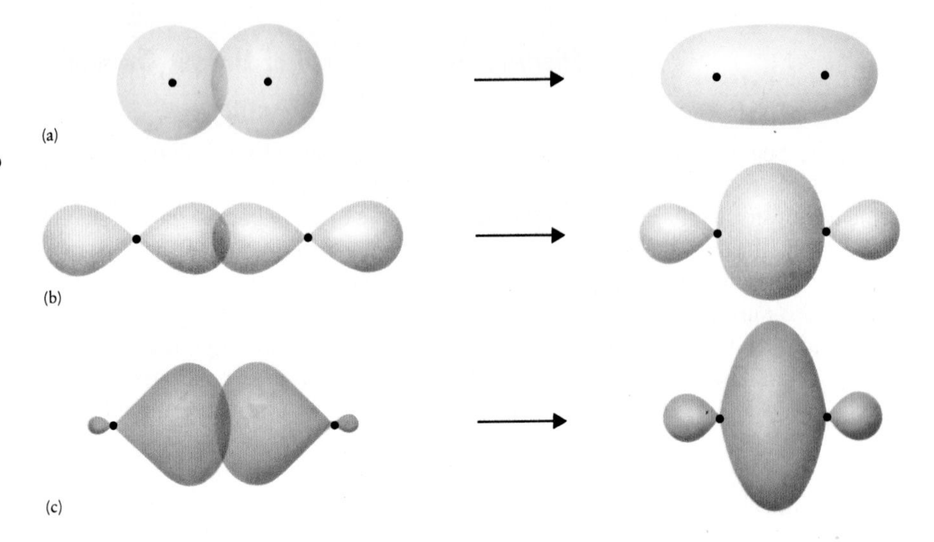

Figure 12.11 A sigma (σ) bond is formed by the overlap of atomic orbitals along the bond axis. (a) Overlap of *s* orbitals. (b) End-to-end overlap of *p* orbitals. (c) Overlap of hybrid orbitals.

A double bond consists of two kinds of bonds. One is a σ bond; the other is a **pi bond,** or **π bond.** The π bond is formed by the overlap of a *p* orbital from each carbon. Notice in Figure 12.12 that the overlap in the π bond is sideways rather than frontwise as in the σ bond. Also note that the π bond has two parts: two regions that lie on opposite sides of the bond axis like two hamburger buns covering the bond.

The formation of π bonds allows two atoms to share more than one pair of electrons. Let's look at ethene.

```
H       H
 \     /
  C=C
 /     \
H       H
```

ethene

Each carbon forms three σ bonds using sp^2 hybrid orbitals.

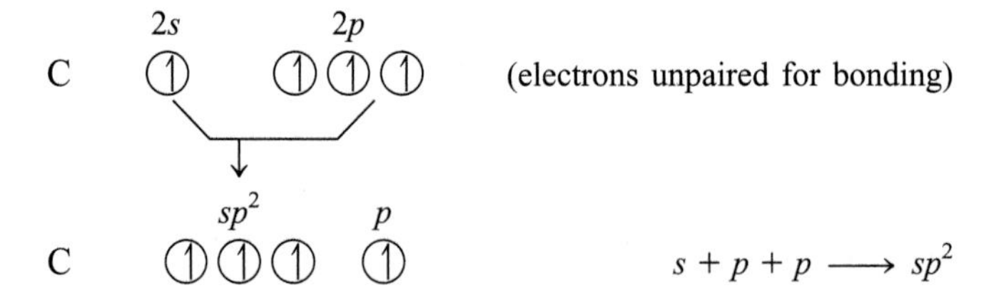

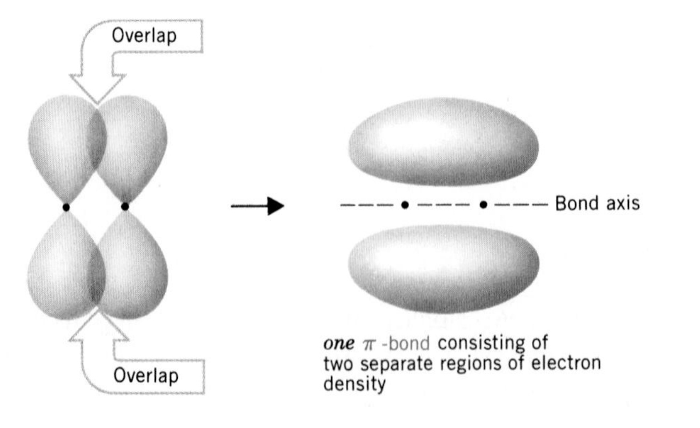

Figure 12.12 A pi (π) bond is formed when *p* orbitals overlap on opposite sides of the bond axis.

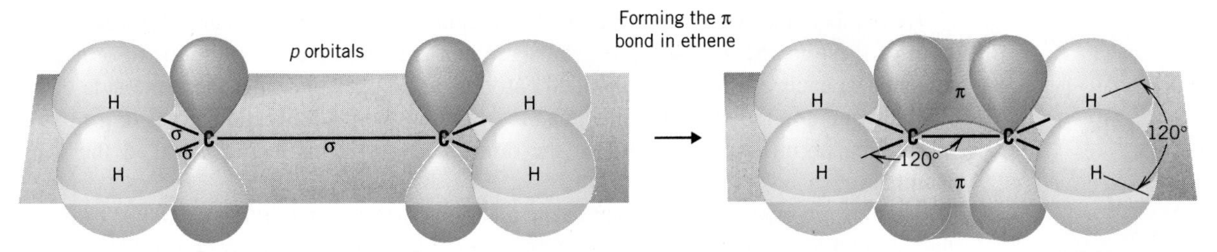

Figure 12.13 Ethene contains one π bond formed by the overlap of one p orbital from each carbon atom.

The formation of three sp^2 hybrid orbitals leaves one free p orbital. This orbital is used to form the π bond. Because the sp^2 orbitals form a triangular-planar shape (Fig. 12.13), all of the atoms in ethene lie in the same plane, with the π bond forming above and below the plane. The triple bond of ethyne consists of one σ and two π bonds. Because carbon-to-carbon triple bonds do not naturally occur in living organisms, we omit the study of their orbitals.

12.11 *Cis-Trans* Isomerism

The carbon-to-carbon single bond found in the alkanes does not restrict the rotation of atoms around the bond. The carbons can be viewed as being similar to two balls connected by a string. You can freely twist the two balls. In the same way, the three structures shown in Figure 12.14 for 1,2-dibromoethane result from rotation about the carbon-to-carbon single bond. They are, in fact, the same compound.

When two carbons are connected by a double bond, the situation is very different. The double bond is relatively rigid and it prevents free rotation. In this case, the carbons can be viewed as similar to two balls glued to the ends of a rigid rod. Because free rotation does not exist around a double bond, the two structures shown for 1,2-dibromoethene are different compounds.

```
Br       H          Br       Br
  \     /             \     /
   C=C                 C=C
  /     \             /     \
H        Br          H       H
```

trans-1,2-dibromoethene   *cis*-1,2-dibromoethene

◆ **Define geometric isomerism, and draw the structures of *cis* and *trans* isomers of an alkene and a cyclic hydrocarbon.**

Because these two different compounds share the same molecular formula ($C_2H_2Br_2$), they are isomers. But the atoms in the two molecules are connected in the same order, so they are not constitutional isomers. They differ only in the arrangement in space, or geometry, of the atoms. Compounds that differ only in the way their atoms are arranged in space are called **geometric isomers.** The particular kind of geometric isomerism we see here is called ***cis-trans* isomerism.**

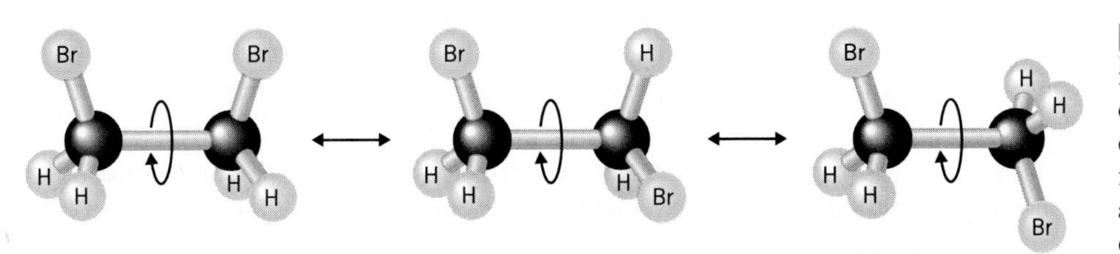

Figure 12.14 *Cis-trans* isomerism does not exist in 1,2-dibromoethane because the carbon-to-carbon single bond is free to rotate. As a result, the atoms bonded to the carbons can be in any position.

Remember that two different compounds cannot have the same name. We cannot call both of these isomers 1,2-dibromoethene. When specified atoms or groups of atoms are attached to each of the double-bonded carbons, and both appear on the same side of the double bond, the molecule is called the ***cis* isomer.** When the specified atoms or groups of atoms appear on opposite sides of the double bond, the molecule is called the ***trans* isomer.**

Isomers are critically important in the reactions that occur in living cells. For example, maleic acid and fumaric acid are *cis* and *trans* isomers that look very similar.

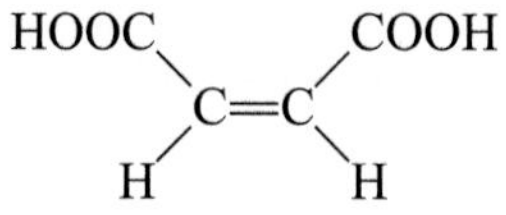

$$\begin{array}{c}\text{HOOC}\quad\quad\text{COOH}\\ \text{C=C}\\ \text{H}\quad\quad\quad\text{H}\end{array}\qquad\qquad\begin{array}{c}\text{HOOC}\quad\quad\text{H}\\ \text{C=C}\\ \text{H}\quad\quad\quad\text{COOH}\end{array}$$

maleic acid 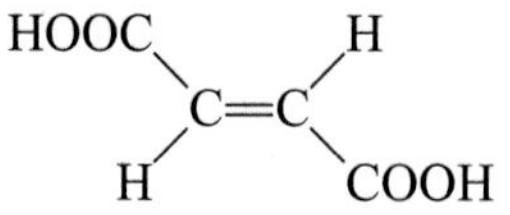 — fumaric acid

Maleic acid, however, is poisonous to living cells, whereas fumaric acid is produced in the cell as part of the citric acid cycle (the series of reactions that produces energy in the cell).

Because *cis* and *trans* isomers can have such different effects on the body, it is important that we be able to tell them apart. How can we tell if a molecule exhibits *cis-trans* isomerism? If a compound contains a double bond and also contains two or more of the same specified atoms or groups of atoms (which we will label G), then we determine the existence of *cis-trans* isomerism using the following two criteria:

1. *Cis* isomers do not exist when the compound fits the following general formulas:

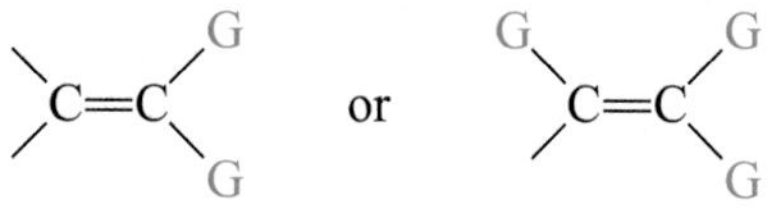

$$\begin{array}{c}\quad\quad\quad\text{G}\\ \text{C=C}\\ \quad\quad\quad\text{G}\end{array}\quad\text{or}\quad\begin{array}{c}\text{G}\quad\quad\text{G}\\ \text{C=C}\\ \quad\quad\quad\text{G}\end{array}$$

For example:

$$CH_3\text{—}CH\text{=}\overset{\overset{\displaystyle CH_3}{|}}{C}\text{—}CH_3 \qquad CH_3\text{—}CH_2\text{—}CH\text{=}\overset{\overset{\displaystyle Cl}{|}}{C}\text{—}Cl \qquad Br\text{—}CH\text{=}\overset{\overset{\displaystyle Br}{|}}{C}\text{—}Br$$

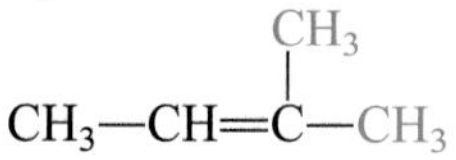

2-methyl-2-butene 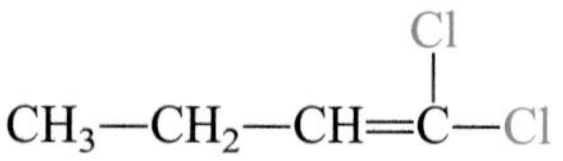 — 1,1-dichloro-1-butene 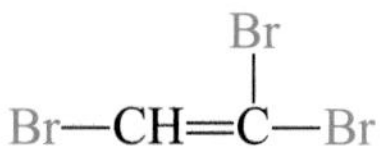 — 1,1,2-tribromoethene

2. *Cis-trans* isomers do exist when the compound fits these general formulas:

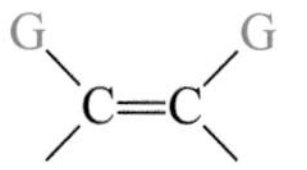

$$\begin{array}{c}\text{G}\quad\quad\text{G}\\ \text{C=C}\\ \end{array}\qquad\begin{array}{c}\text{G}\quad\quad\quad\\ \text{C=C}\\ \quad\quad\quad\text{G}\end{array}$$

cis isomer 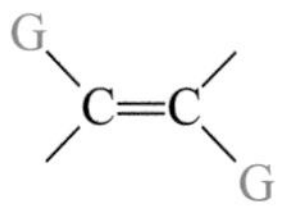 — *trans* isomer

For example:

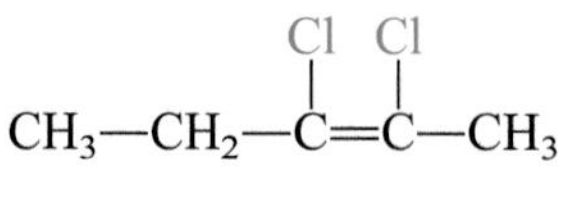

$$CH_3\text{—}CH_2\text{—}\overset{\overset{\displaystyle Cl}{|}}{C}\text{=}\overset{\overset{\displaystyle Cl}{|}}{C}\text{—}CH_3 \qquad CH_3\text{—}\overset{\overset{\displaystyle Cl}{|}}{C}\text{=}\underset{\underset{\displaystyle Br}{|}}{C}\text{—}CH_3$$

cis-2,3-dichloro-2-pentene 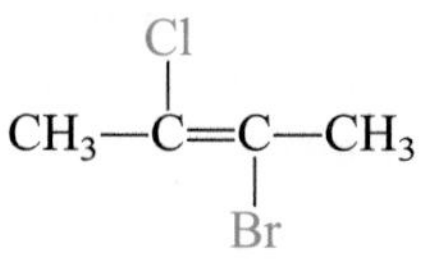 — *trans*-2-bromo-3-chloro-2-butene

Example 12-5

Let's see if we can identify compounds that exhibit *cis-trans* isomerism. Which of the following compounds can exist in *cis* and *trans* forms? If *cis* and *trans* isomers do exist, draw the structural formulas of each.

(a) 2-methyl-1-butene

See the Question

We are asked to determine whether or not this compound fits the general formula for a molecule that exhibits *cis-trans* isomerism. To do this, we must first draw the extended structural formula for the compound.

Think It Through (Execute the Math)

Drawing the extended structural formula for this compound, we get

$$\begin{array}{ccccccc} & & & \text{H} & & & \\ & & & | & & & \\ & & \text{H}- & \text{C} & -\text{H} & & \\ & & & | & & & \\ & \text{H} & \text{H} & | & & \text{H} & \\ & | & | & | & & | & \\ \text{H}- & \text{C}- & \text{C}- & \text{C} & = & \text{C} & \\ & | & | & & & | & \\ & \text{H} & \text{H} & & & \text{H} & \end{array} \qquad \begin{array}{c} \diagdown \quad\quad \diagup \text{G} \\ \text{C}=\text{C} \\ \diagup \quad\quad \diagdown \text{G} \end{array}$$

2-methyl-1-butene

Because there are two hydrogens on one of the carbons of the double bond, this compound fits the general pattern in which two of the groups (G) are on the same carbon.

Prepare the Answer

The compound does not exhibit *cis-trans* isomerism.

(b) 2-methyl-2-butene

S

This problem asks for the same information as the previous example.

T • E

The structural formula for this compound is

$$\begin{array}{c} \text{CH}_3 \\ | \\ \text{CH}_3-\text{CH}=\text{C}-\text{CH}_3 \end{array} \qquad \begin{array}{c} \diagdown \quad\quad \diagup \text{G} \\ \text{C}=\text{C} \\ \text{G} \diagup \quad\quad \diagdown \text{G} \end{array}$$

2-methyl-2-butene

Again in this compound we see that one of the double-bonded carbons has two identical groups attached to it.

P

This compound does not exhibit *cis-trans* isomerism.

(c) 3-methyl-2-hexene

S

This example also asks us to compare a structural formula with the general rules for *cis-trans* isomerism.

T • E

The structural formula is

$$\begin{array}{c} \text{CH}_3 \\ | \\ \text{CH}_3-(\text{CH}_2)_2-\text{C}=\text{CH}-\text{CH}_3 \end{array}$$

3-methyl-2-hexene

Each of the carbons connected by the double bond has two different groups attached to it: carbon 2 has a hydrogen and a methyl group, and carbon 3 has a methyl and a propyl group. The fact that carbons 2 and 3 each have a methyl is not important, what is essential is that neither double-bonded carbon has two groups on it that are both the same.

$$\begin{array}{c} \text{G} \diagdown \quad\quad \diagup \text{G} \\ \text{C}=\text{C} \\ \diagup \quad\quad \diagdown \end{array} \qquad \begin{array}{c} \text{CH}_3 \diagdown \quad\quad \diagup \text{CH}_3 \\ \text{C}=\text{C} \\ \text{CH}_3(\text{CH}_2)_2 \diagup \quad\quad \diagdown \text{H} \end{array}$$

P

The molecule can have *cis* and *trans* isomers. That is, *cis*-3-methyl-2-hexene and *trans*-3-methyl-2-hexene both exist.

$$\begin{array}{ccc} CH_3(CH_2)_2 & & CH_3 \\ & C{=}C & \\ CH_3 & & H \end{array} \qquad \begin{array}{ccc} CH_3 & & CH_3 \\ & C{=}C & \\ CH_3(CH_2)_2 & & H \end{array}$$

trans-3-methyl-2-hexene　　*cis*-3-methyl-2-hexene

Check Your Understanding

Draw the structural formulas of the *cis* and *trans* isomers of 2-pentene.

12.12 Reactions of the Alkenes

Compare the chemical reactivity of alkanes and alkenes.

Because a carbon-to-carbon double bond is more reactive than a single bond, alkenes have many commercial uses. They are especially important in the chemical industry, where they are used in the production of many other compounds. We cover only two of the many types of reactions of the alkenes: oxidation and addition reactions.

Oxidation

Write the equations for the combustion and addition reactions of a given alkene.

Early in this chapter, we discussed the oxidation of the alkanes by combustion. Burning, however, is not the only kind of oxidation reaction. A wide range of reactions are oxidation reactions, or more precisely, **oxidation–reduction (redox)** reactions. Because oxidation involves the loss of electrons by a reactant, and reduction involves the gain of electrons by another reactant, oxidation and reduction must always occur together. In many of the oxidation reactions of organic chemistry, a reactant molecule gains oxygen atoms or loses hydrogen atoms. Similarly, the reduction reaction is the one in which a reactant loses oxygen or gains hydrogen.

As with the alkanes, alkenes can be completely oxidized to carbon dioxide and water.

$$2CH_3{-}CH{=}CH_2 + 9O_2 \longrightarrow 6CO_2 + 6H_2O$$

In this reaction the propene loses hydrogen and gains oxygen, so it undergoes oxidation. Oxygen gains hydrogen, so it is reduced. Under certain circumstances the oxidation of an alkene may be incomplete, resulting in the formation of an alcohol rather than carbon dioxide.

$$\begin{array}{c} \text{H H H} \\ \text{H—C—C=C—H} \\ \text{H} \end{array} \xrightarrow{KMnO_4} \begin{array}{c} \text{H} \\ \text{H O H} \\ \text{H—C—C—C—O—H} \\ \text{H H H} \end{array}$$

propene　　1,2-propanediol

Addition Reactions

In an **addition reaction,** atoms react with the double bond and convert it to a single bond. The addition of bromine to an alkene is called bromination. The bromination reaction can be used to test compounds to see if they have unsaturated bonds (that is, double or triple

Environmental Perspective / Pheromones

Honeybees use pheromones to mark trails to food and to warn of danger.

One of the most interesting approaches to insect control involves chemicals called *pheromones.* These compounds are excreted by insects to mark a trail, send an alarm, or attract a mate. Pheromones are specific for each species; they affect receptors on insects only of the same species as the sender. Pheromones that are used to mark a trail are large molecules that remain in the environment for some time. Social insects like bees, wasps, ants, and termites use trail pheromones to locate food. Alarm pheromones are smaller, much more volatile compounds released as a warning of danger. Because of its high volatility, the alarm pheromone isoamyl acetate produced by honeybees evaporates rapidly after the alert is over. Female insects excrete sex attractant pheromones to attract males. Males of the species can detect extremely low concentrations of the compounds. For example, the sex attractant of the gypsy moth has been effective in insect traps at concentrations of 0.0000000000001 g (also written as 1×10^{-13} g). The use of sex attractants to control insect pests in the mobile stage of their life cycles has become increasingly popular. It is often a safe alternative to the use of environmentally harmful insecticides.

A pheromone can be a simple compound, but its three-dimensional structure is very important. The molecules must be constructed so that only the appropriate species is attracted. The sex attractant for the common housefly is made up of two compounds, one of which, 9-tricosene, is excreted as the *cis* isomer only.

$$\begin{array}{ccc} CH_3(CH_2)_6CH_2 & & CH_2(CH_2)_{11}CH_3 \\ & C{=}C & \\ H & & H \end{array}$$

The *trans* isomer has no effect on the male housefly. Such examples of geometry-specific structure are seen throughout nature. In almost all situations in which several isomers are possible, only one is produced naturally.

Your Perspective: On behalf of a client, write a letter to a Congressman urging that the federal government spend more money on pheremone research to reduce the need for chemical pesticides. Clearly explain your reasoning for this request.

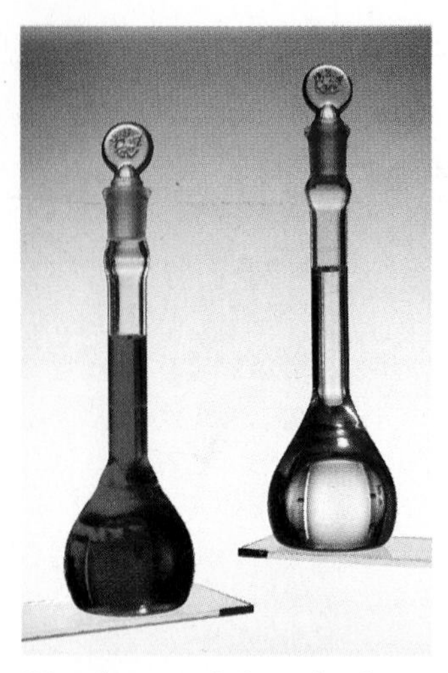

The characteristic red color of bromine disappears when an alkene is added.

bonds). Bromine in CCl_4 forms a reddish brown solution, but the bromides formed from the bromination reaction are colorless. If the bromine solution reacts with a hydrocarbon and the result is a colorless solution, the hydrocarbon may therefore have contained unsaturated bonds.

$$\underset{\text{propene}}{\begin{array}{ccccccccc} & & H & & H & & H & & \\ & & | & & | & & | & & \\ H & - & C & - & C & = & C & - & H \\ & & | & & & & & & \\ & & H & & & & & & \end{array}} + Br_2 \longrightarrow \underset{\text{1,2-dibromopropane}}{\begin{array}{ccccccccc} & & H & & H & & H & & \\ & & | & & | & & | & & \\ H & - & C & - & C & - & C & - & H \\ & & | & & | & & | & & \\ & & H & & Br & & Br & & \end{array}}$$

The following are two other addition reactions of the alkenes:

1. **Hydrogenation** is the addition of hydrogen to a carbon–carbon double bond to form an alkane.

$$\underset{\text{propene}}{\begin{array}{ccccccccc} & & H & & H & & H & & \\ & & | & & | & & | & & \\ H & - & C & - & C & = & C & - & H \\ & & | & & & & & & \\ & & H & & & & & & \end{array}} + H_2 \longrightarrow \underset{\text{propane}}{\begin{array}{ccccccccc} & & H & & H & & H & & \\ & & | & & | & & | & & \\ H & - & C & - & C & - & C & - & H \\ & & | & & | & & | & & \\ & & H & & H & & H & & \end{array}}$$

2. **Hydration** is the addition of water to a carbon–carbon double bond to form an alcohol.

$$\underset{\text{ethene}}{\begin{array}{ccccccc} & & H & & H & & \\ & & | & & | & & \\ H & - & C & = & C & - & H \end{array}} + HOH \longrightarrow \underset{\text{ethanol}}{\begin{array}{ccccccccc} & & H & & H & & & & \\ & & | & & | & & & & \\ H & - & C & - & C & - & O & - & H \\ & & | & & | & & & & \\ & & H & & H & & & & \end{array}}$$

Liquid vegetable oils generally contain numerous carbon-to-carbon double bonds. Hydrogenation of some of these double bonds converts these oils to solid fats, such as margarine and vegetable shortening.

Polymerization

You are probably familiar with the word ''ethylene,'' having seen it in the name of a type of plastic, polyethylene. Polyethylene is formed by an addition reaction involving thousands of ethylene units. This is how three ethylene units bond together:

$$\cdots \begin{array}{ccc} H & & H \\ & C{=}C & \\ H & & H \end{array} + \begin{array}{ccc} H & & H \\ & C{=}C & \\ H & & H \end{array} + \begin{array}{ccc} H & & H \\ & C{=}C & \\ H & & H \end{array} \cdots$$

yields

$$\cdots \begin{array}{ccccccccccccc} & H & & H & & H & & H & & H & & H & \\ & | & & | & & | & & | & & | & & | & \\ - & C & - & C & - & C & - & C & - & C & - & C & - \\ & | & & | & & | & & | & & | & & | & \\ & H & & H & & H & & H & & H & & H & \end{array} \cdots$$

polyethylene

Bottles made from polyethylene.

The process of joining many simple units, called **monomers,** together to form very large molecules, called **polymers,** is **polymerization.** Thus, polyethylene is a polymer of the monomer ethylene. Synthetic (artificial) polymers have wide use in our daily lives, ranging from the plastic containers, bags, and wrapping we use, to the synthetic fibers such as

Orlon (polyacrylonitrile), rayon, and acrylics that we wear, to the synthetic rubber that we use for tires and for parts in appliances.

Check Your Understanding

Write the equations for the following reactions:

1. The addition of bromine to 1-pentene
2. The addition of hydrogen to *cis*-2-butene
3. The addition of water to *cis*-2-butene

12.13 Alkynes

The alkynes are a class of compounds containing the carbon-to-carbon triple bond. Like the double bond, the triple bond provides a reactive area on the molecule. Alkynes are named following the same rules as alkenes, except that the names of alkynes end in -yne.

The simplest alkyne is ethyne, more commonly known as acetylene. This compound is a flammable gas that, when burned with oxygen in an oxyacetylene torch, provides enough energy to cut and weld metals. Acetylene is a convenient starting material for many industrial products. But, as we mentioned earlier, the human body doesn't naturally contain any compounds with triple bonds in their structures.

12.14 Cycloalkanes and Cycloalkenes

In addition to forming chains and branched chains, carbon can also form rings. Such **cyclic hydrocarbons** are found in a variety of sizes and may contain only single bonds or both double and single bonds. Although the most common cyclic hydrocarbons have rings containing five or six carbon atoms, both larger and smaller rings can be found (Table 12.4).

The smallest number of carbons possible in a cyclic compound is three. This compound, cyclopropane, is a highly potent anesthetic, which must be handled carefully because it is both flammable and explosive. Many hydrocarbons besides cyclopropane act as anesthetics when inhaled. These include methane, acetylene, ethylene, and cyclobutane. An anesthetic is a compound that decreases a person's sensitivity to pain and, in most cases, also causes the person to lose consciousness. The anesthetic property of the hydrocarbons is due to the nonpolar nature of these molecules and the absence of a polar hydrogen atom for hydrogen bonding.

Anesthetics act on the nerves, preventing nerve impulses from moving along the nerve fibers. This happens, in part, because each nerve is surrounded by a protective coating composed of molecules that are nonpolar. When an anesthetic is inhaled, it enters the bloodstream from the lungs and travels to the nerve tissues. It then dissolves in the nonpolar coating around the nerves. As the anesthetic accumulates in this coating, it disrupts the transmission of nerve impulses. Because the nerve impulses no longer reach the brain, there is no sensation of pain. You may have heard about deaths from inhaling methane or sniffing the solvents in glue. These deaths occurred because nonpolar hydrocarbons dissolved in the nonpolar protective coatings of the nerves and disrupted the critical nerve impulses traveling from the brain to the lungs and heart.

Table 12.4 Cyclic Hydrocarbons

Molecular Formula	*Name*	*Structural Formula*
C_3H_6	Cyclopropane	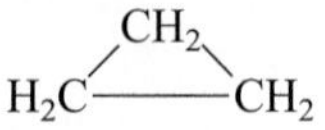 CH_2 / H_2C—CH_2
C_4H_8	Cyclobutane	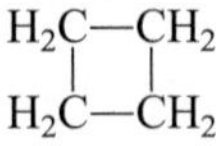 H_2C—CH_2 / H_2C—CH_2
C_5H_{10}	Cyclopentane	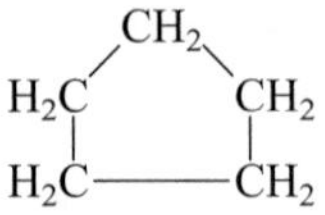 CH_2 / H_2C CH_2 / H_2C—CH_2
C_5H_6	1,3-Cyclopentadiene	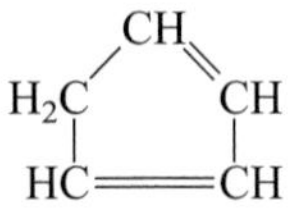 CH / H_2C CH / HC═CH
C_6H_{12}	Cyclohexane	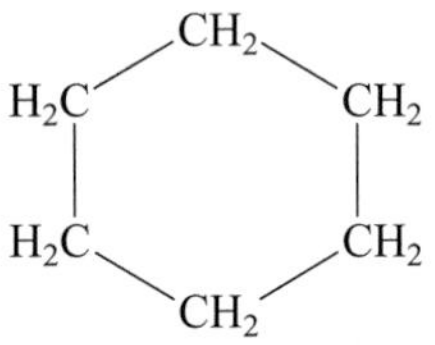 CH_2 / H_2C CH_2 / H_2C CH_2 / CH_2

The cyclic hydrocarbons are named following the same rules given for the straight-chain hydrocarbons, except that the unsubstituted name begins with cyclo-. If a ring has only one substituent, it is not necessary to give it a number.

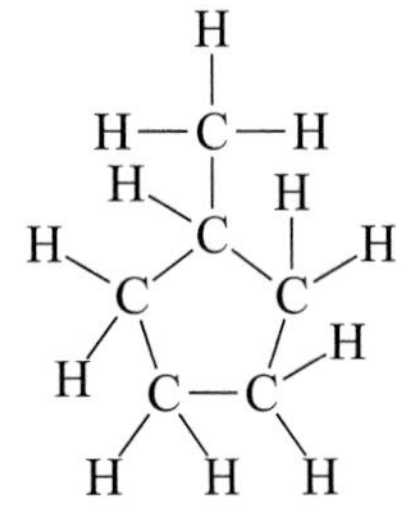

methylcyclopentane

can be abbreviated as shown. Each corner of the ring is a carbon and the hydrogens on the ring are not shown.

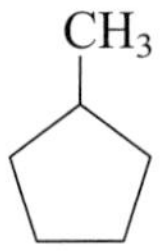

CH_3

If the ring has more than one substituent, the ring is numbered so that one of the substituents is at carbon number 1 and the other substituent is given the lowest possible number. In the following example, the correct name is 1,3-dichlorocyclohexane, not 1,5-dichlorocyclohexane. You can number the ring going either clockwise or counterclockwise, whichever gives the lower number for the substituent.

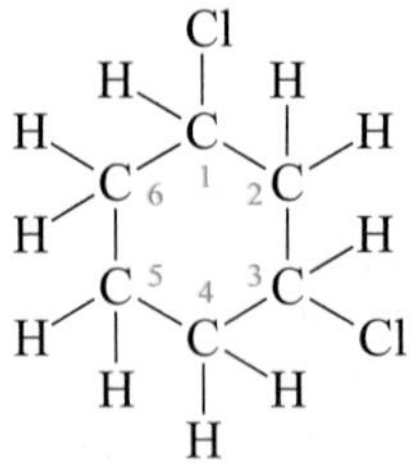

or

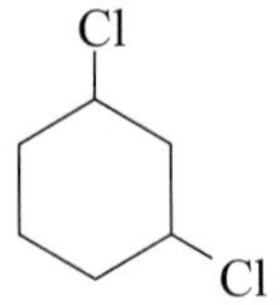

Cl Cl

1,3-dichlorocyclohexane

When carbons are arranged in rings, free rotation is restricted even around single bonds. This situation is very similar to what we observed in the alkenes, and it has the same effect on molecular structure. *Cis* and *trans* isomerism occurs in cyclic compounds. In a *cis* isomer, the specified atoms or groups of atoms lie on the same side of the ring.

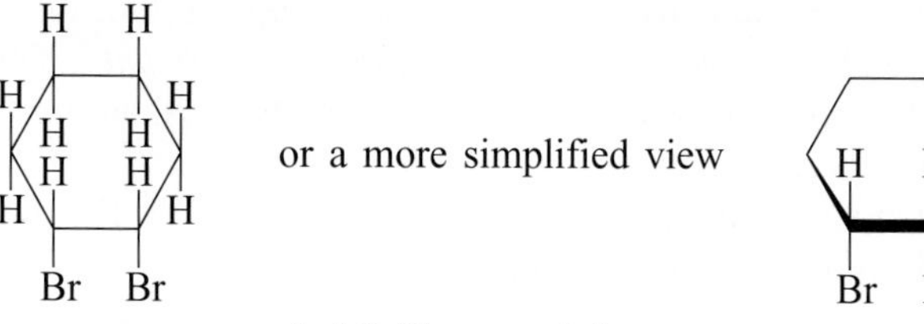

cis-1,2-dibromocyclohexane

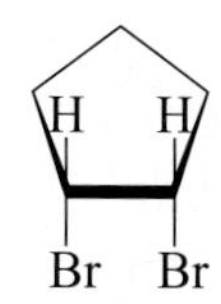

cis-1,2-dibromocyclopentane

In *trans* isomers, the specified atoms or groups of atoms lie on opposite sides of the ring.

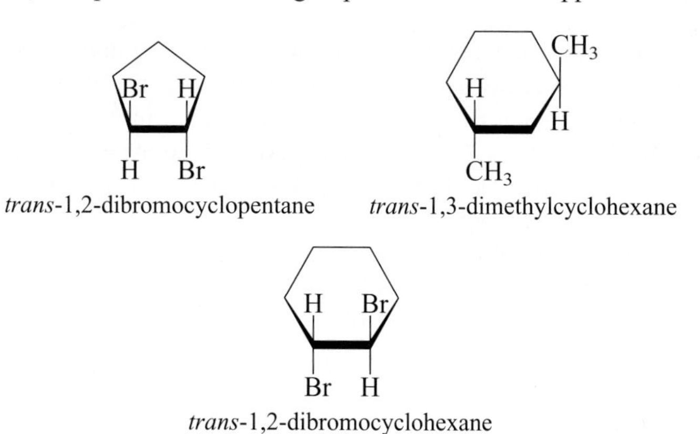

trans-1,2-dibromocyclopentane *trans*-1,3-dimethylcyclohexane

trans-1,2-dibromocyclohexane

Check Your Understanding

1. Name the following compounds:

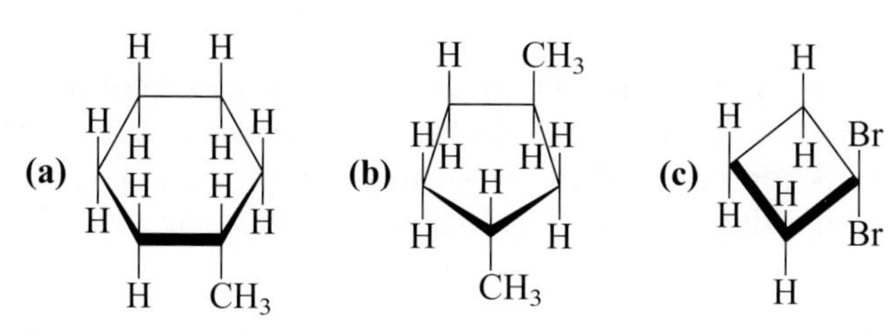

2. Draw structures for the following compounds:

(a) 1,2-dichlorocyclohexane

(b) *cis*-1,3-dibromocyclopentane

(c) 1,1-dimethylcyclohexane

12.15 Aromatic Hydrocarbons

The alkanes, alkenes, alkynes, cycloalkanes, and cycloalkenes belong to a larger class of compounds called the **aliphatic hydrocarbons.** A second large class of hydrocarbons is called the **aromatic hydrocarbons.** The first natural compounds identified as part of this class had a definite and fairly pleasant odor, hence the name *aromatic.* This historical fact notwithstanding, the aromatic hydrocarbons are generally speaking no ''smellier'' than the aliphatic hydrocarbons.

The aromatic compounds include benzene and its derivatives. The molecular formula for benzene is C_6H_6, the structural formula of which is

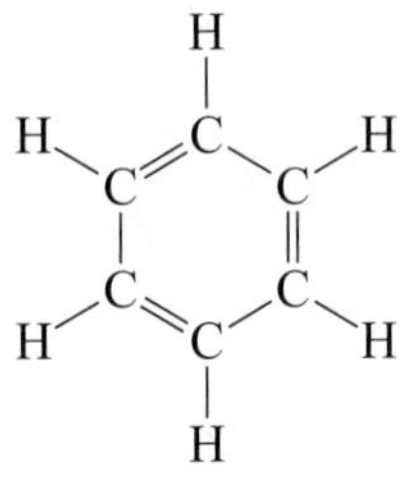

benzene

Describe the structure of the benzene molecule, and explain why the molecule is extremely stable.

Benzene comprises six carbon atoms in a ring, with alternating double and single bonds. Each carbon is also bonded to a hydrogen. The problem is that this structure does not adequately represent benzene. Benzene is more stable than might be expected based on this structure. Moreover, the carbon-to-carbon bonds in benzene all have the same length and the same strength. This wouldn't be the case if some of the carbon-to-carbon bonds were double and some were single. Finally, as we'll soon see, benzene does not undergo some of the reactions that are typical of alkenes.

If the preceding simple structural formula does not adequately describe benzene, then what is its structure? Benzene is

neither 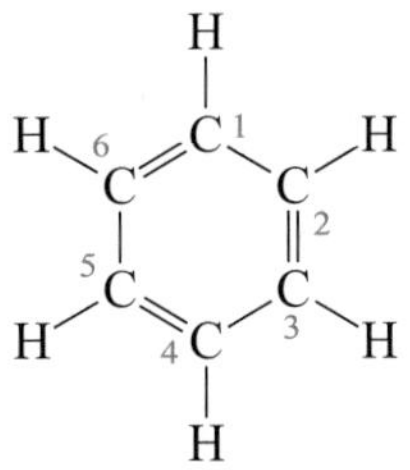 nor

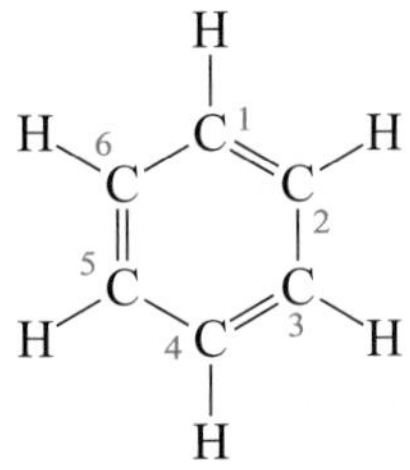

but rather a structure in which the electrons from the three double bonds are spread out, or delocalized, over all six carbon atoms. Such delocalization of electrons is called **resonance,** and the true structure for benzene is called a **resonance hybrid.** We can represent this structure for benzene by drawing a circle in the center of the benzene ring.

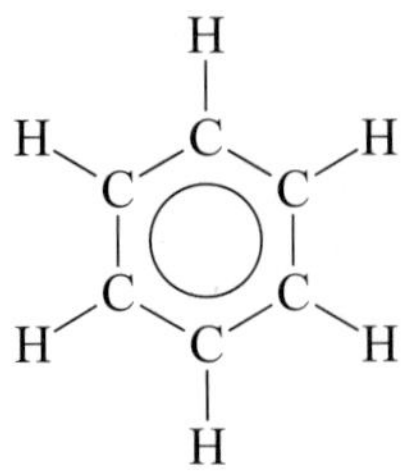

or more typically

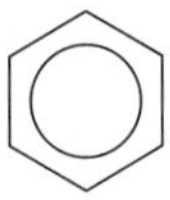

Benzene rings can fuse together to form such compounds as naphthalene (which gives moth balls their characteristic odor) and anthracene (which is an important starting material in the production of dyes.)

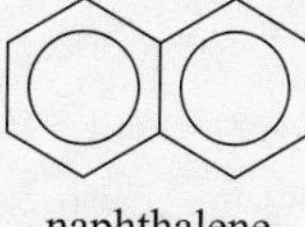

naphthalene

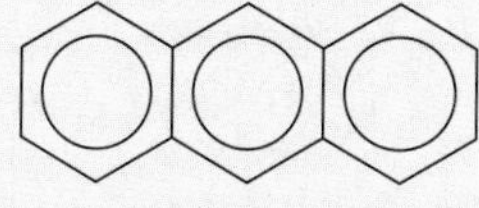

anthracene

Additional benzene rings can fuse to these molecules to form structures that look like honeycombs. Such complicated compounds are created in the production of coal tar, the harmful effects of which became evident when workers in European coal tar factories developed skin cancer. It has since been shown that several fused-ring aromatic compounds can cause cancer in mice. Other compounds with fused-ring structures similar to those found in coal tar are also formed in the partial combustion of large organic molecules. Only a few milligrams of one such compound, 3,4-benzpyrene, is enough to cause cancer in laboratory animals. Compounds that cause cancer in animals are known as **carcinogens.** The compound 3,4-benzpyrene is one of the major carcinogens found in cigarette smoke.

Although scientists still don't know the exact way in which these rather inert compounds produce cancer, recent studies have shed some light on the subject. Certain enzymes located mostly in the liver and kidneys, but also found in the lungs and other tissues, work to detoxify foreign chemicals that enter the body. One of the ways in which they do this is to make nonpolar compounds (such as 3,4-benzpyrene) polar. This increased polarity makes these compounds more water-soluble and, therefore, more easily excreted by the kidneys. In most cases, the products of these enzyme reactions are less harmful to the cells than the original reactants (they have been detoxified). In some cases, however, the products of these reactions turn out to be more toxic—and sometimes very carcinogenic. So in their attempt to detoxify foreign chemicals, our body's enzymes may be contributing to the synthesis of carcinogens from cigarette smoke. The manner in which such carcinogens turn a normal cell into a cancer cell is still being studied.

Your Perspective: Imagine that the mayor of your town has developed lung cancer from years of smoking and has filed a lawsuit against a cigarette company. Write a letter to the editor of your local newspaper either defending or disagreeing with the mayor's lawsuit, clearly explaining your reasons for doing so.

All the atoms in benzene are located in one plane, with bond angles of 120°. As a result, *cis* and *trans* isomers do not exist. Resonance therefore makes benzene very stable. (Note that benzene and cyclohexane are very different compounds. When you draw the structure of benzene, be sure to include the circle to avoid confusion.)

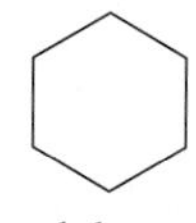

cyclohexane (C_6H_{12})

is not the same as

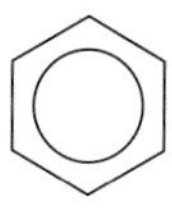

benzene (C_6H_6)

Table 12.5 **Aromatic Compounds**

Name	*Structural Formula*	*Name*	*Structural Formula*
Benzene	(benzene ring)	Toluene	CH_3 on benzene ring
Phenol	OH on benzene ring	Nitrobenzene	NO_2 on benzene ring
Benzoic acid	C(=O)—OH on benzene ring	Aniline	NH_2 on benzene ring

Benzene and other aromatics can be obtained from coal tar, which is produced by heating soft coal in the absence of air. Benzene is widely used in the chemical industry as a solvent and as the starting material in the production of many other compounds. Benzene, however, is toxic. Its fumes can cause nausea or death from respiratory and heart failure. Prolonged exposure to benzene can interfere with red blood cell production in a condition known as aplastic anemia.

The benzene ring is found in many compounds that are important to the life processes of living organisms. Plants synthesize benzene from carbon dioxide, water, and inorganic materials. Animals cannot synthesize aromatic rings. This means that they must obtain the essential aromatic compounds from their diets.

Some important aromatic compounds are listed in Table 12.5. The names of several important aromatic compounds appear in our discussion of their reactions.

12.16 Reactions of the Aromatic Compounds

We saw earlier that the alkenes undergo addition reactions. For example, the reaction of cyclohexene with bromine involves the addition of bromine atoms across the double bond of cyclohexene.

$$\text{cyclohexene} + Br_2 \longrightarrow \text{1,2-dibromocyclohexane}$$

Write the equation for the substitution reaction of benzene with a halogen.

The situation is very different for benzene and other aromatic hydrocarbons. Very special conditions are required for benzene to react with bromine. When this does occur, the benzene remains intact and one bromine atom substitutes for a hydrogen atom.

$$C_6H_6 + \underset{\substack{\text{bromine} \\ (Br_2)}}{Br{-}Br} \xrightarrow{\text{catalyst}} C_6H_5Br + HBr$$

Health Perspective / Cancer

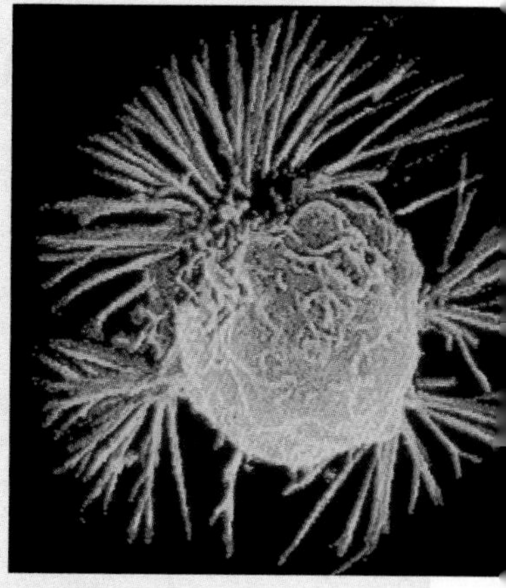

Breast cancer cell

Although it ranks behind heart disease as the most common cause of death in the United States, cancer is perhaps the most feared human disease. For many people, the word *cancer* brings a vision of death, pain, disfigurement, and dependency. Cancer has been known since ancient times—Hippocrates introduced the term *carcinoma* to describe a tumor that spread and destroyed the host—but cancer appears to be increasing in modern societies. Cancer is a disease that becomes more common with increased age, so part of the increase in cancer cases can be explained by the doubling of life expectancy in the last century.

Cancer is not just one disease. It is a group of more than 100 diseases, all of which result from unregulated cell growth. Once an organism has reached a certain age, its cells normally maintain a state of dynamic equilibrium; cells divide at about the same rate as other cells die. Healthy cells divide only 20 to 60 times before they die. Also, they do not invade the space occupied by other cells touching them. Nearly all tissues of the human body can develop disorders of cell growth, however. When this results in unrestricted cell growth, tumors are formed. Tumors are abnormal cellular masses that perform no useful function in the body and grow at the expense of healthy neighboring tissue. They are classified as either benign (nonmalignant) tumors, such as warts and moles, or malignant tumors. Malignant tumors are commonly called cancerous. Unlike cells in benign tumors, cells in malignant tumors can divide an unlimited number of times and have no regard for cellular boundaries. These cells grow on top of one another, as well as on top of and between normal cells. They rapidly invade nearby tissues and organs, disrupting them and causing blood vessels to bleed. In addition, cancer cells are less cohesive than normal cells. They have a tendency to break off from the tumor and move (or metastasize) to other parts of the body, where they form secondary tumors (or metastases).

Many different conditions can cause the uncontrolled cell growth of cancer. Ionizing radiation, viruses, and a number of synthetic or natural chemicals called carcinogens can all damage the DNA of cells or interfere with the mechanisms responsible for controlling cell division. The development of cancer is often a two-step process. Exposure of tissues to a carcinogen can prime their cells for transformation, but the malignant tumor may not form until many years later when one of the cells suffers a final transforming event and begins to grow uncontrollably.

Your Perspective: Imagine you are a pediatric nurse at a hospital where a six-year-old boy has just been found to have leukemia (a cancer affecting white blood cells). Suppose that his 10-year-old sister asks you about cancer. In words she would understand, explain what cancer is.

KEY CONCEPTS

Living organisms are made up of compounds containing only 25 of the 90 naturally occurring elements. Of these elements, the four most abundant are carbon, hydrogen, oxygen, and nitrogen.

Carbon-containing compounds are the building blocks of life. The carbon atom is especially suited for this purpose because it forms strong stable bonds with up to four other carbon atoms. Molecules can contain carbon atoms in long chains, branched chains, or rings—resulting in a great variety of molecules.

The field of chemistry that studies the compounds of carbon is called organic chemistry.

Hydrocarbons are a large group of organic compounds composed only of carbon and hydrogen. Compounds that have the same molecular formula but a different ordering of the carbon and hydrogen atoms within the molecule are called constitutional isomers. Hydrocarbons are nonpolar and dissolve only in other nonpolar substances.

Hydrocarbons are either saturated or unsaturated. They are saturated when the molecules contain only carbon-to-carbon single bonds. The alkanes are saturated hydrocarbons with the general formula C_nH_{2n+2}. Unsaturated hydrocarbons contain one or more double or triple bonds. The alkenes are the class of unsaturated hydrocarbons containing carbon-to-carbon double bonds. The alkynes are unsaturated hydrocarbons containing carbon-to-carbon triple bonds. Cyclic hydrocarbons are compounds that contain carbon atoms bonded together to form rings. They may be either saturated or unsaturated.

There are reactivity differences among the hydrocarbon families. The alkanes are relatively unreactive, but they do undergo oxidation and substitution reactions. The alkenes and alkynes are more reactive than the alkanes. The alkenes undergo oxidation and addition reactions. This text does not study reactions of the alkynes.

Because of restricted rotation about carbon-to-carbon double bonds, alkenes may exhibit *cis-trans* isomerism (geometric isomerism). *Cis-trans* isomers are molecules having the same order of carbon atoms but a different arrangement in space. Because a ring structure also restricts rotation about carbon-to-carbon bonds, *cis* and *trans* isomers are also found in cyclic hydrocarbons.

Benzene and its derivatives make up a large class of compounds known as the aromatic hydrocarbons. The ring structure of benzene is unusually stable and remains intact through most chemical reactions. Although the aromatic hydrocarbons are unsaturated, they do not generally undergo addition reactions. They undergo substitution reactions in which other atoms or groups of atoms substitute for a hydrogen on the benzene ring.

Important Equations

Oxidation of alkanes and alkenes

Complete oxidation

$$CH_4 + 2O_2 \longrightarrow CO_2 + 2H_2O$$

$$CH_2{=}CH_2 + 3O_2 \longrightarrow 2CO_2 + 2H_2O$$

Partial oxidation of alkanes

$$CH_2{=}CH_2 \xrightarrow{KMnO_4} H{-}O{-}\underset{\displaystyle H}{\overset{\displaystyle H}{\underset{|}{\overset{|}{C}}}}{-}\underset{\displaystyle H}{\overset{\displaystyle H}{\underset{|}{\overset{|}{C}}}}{-}O{-}H$$

Addition reaction of alkenes

```
                              H  H
                              |  |
CH2=CH2 + Br2  ---->   Br—C—C—Br
                              |  |
                              H  H
```

Substitution reaction of benzene

```
       H                             Br
  (C6H5)—   + Br2  --catalyst-->  (C6H5)—    + HBr
```

Review Problems

Section 12.1

1. Define *organic chemistry* in your own words.
2. Carbon is often said to form the ''backbone'' of life molecules. What does this mean, and what special properties of carbon make this possible?
3. Why are there so many more organic compounds than inorganic compounds?
4. Define *functional group* in your own words.

Section 12.2

5. What are hybrid orbitals, and why do they form?
6. List three types of hybrid orbitals. What type of hybrid orbital does carbon have in a molecule of methane?

Section 12.3

7. What is the difference between the molecular formula and the structural formula of a compound?
8. Write the structural formulas for the five constitutional isomers of C_6H_{14}.
9. Write extended structural formulas for the following:
   (a) $(CH_3)_3CCH_2CH(CH_3)CH_2CH_3$ (c) $(CH_3)_2CHCH(CH_3)CH(CH_3)_2$
   (b) $(CH_3)_2CHC(CH_3)_2CH(CH_3)_2$
10. Write condensed structural formulas for the following:

```
        H
        |
      H—C—H
    H   |   H  H
    |   |   |  |
(a) H—C—C—C—C—H
    |   |   |  |
    H   H   H  H
```

```
               H
               |
             H—C—H
    H  H       |       H  H  H
    |  |       |       |  |  |
(b) H—C—C——————C——————C—C—C—H
    |  |       |       |  |  |
    H  |       H       |  H  H
     H—C—H         H—C—H
       |               |
       H               H
```

```
          H
          |
        H—C—H
      H   |   H
      |   |   |
(c) H—C—C—C—H
      |   |   |
      H   |   H
        H—C—H
          |
          H
```

Section 12.4

11. How do we recognize that a compound is an alkane?

12. How does the meaning of the term *saturated,* as used in organic chemistry, differ from its meaning when we refer to solutions?

Section 12.5

13. State whether the following pairs or structures are (1) the same compound, (2) constitutional isomers, or (3) unrelated compounds. To be the same compound, the structural formulas must have the identical sequence (or order) of atoms, no matter how they are arranged on the paper. If the molecular formulas are the same, but the sequence of atoms is different, then the compounds are constitutional isomers.

(a) $CH_3—CH_2—CH_3$

```
          H
          |
    H  H  C  H  H
     \ | / | \ | /
      C    H    C
      |         |
      H         H
```

(b) $CH_3—CH_2—CH_2—CH_3$ $\qquad$ $CH_3—CH(CH_3)—CH_3$

```
        CH3
        |
  CH3—CH—CH3
```

(c) $CH_3—CH(CH_3)—CH_2—CH_3$ $\qquad$ $CH_3—CH(CH_3)—CH_3$

```
        CH3                      CH3
        |                        |
  CH3—CH—CH2—CH3           CH3—CH—CH3
```

(d)

```
        CH3             CH3
        |               |
  CH3—CH—CH3            CH—CH3
                        |
                        CH3
```

(e)

```
        OH                   OH
        |                    |
  CH3CHCH2CH3          CH3CH2CHCH3
```

(f) $CH_3CH(OH)CH_3$ $\qquad$ $CH_3CH_2CH_2OH$

```
       OH
       |
  CH3CHCH3
```

(g) $CH_3—O—CH_3$ $\qquad$ $CH_3CH_2—OH$

(h) $CH_3CH_2—\overset{\overset{O}{\|}}{C}—CH_3$ $\qquad$ $CH_3CH_2CH_2—\overset{\overset{O}{\|}}{C}—H$

(i) $HO—\overset{\overset{O}{\|}}{C}—CH_2CH_2CH_3$ $\qquad$ $CH_3CH_2CH_2—\overset{\overset{O}{\|}}{C}—OH$

(j) $CH_3CH_2—\overset{\overset{O}{\|}}{C}—O—CH_3$ $\qquad$ $CH_3CH_2—\overset{\overset{O}{\|}}{C}—CH_3$

(k) $HO\overset{\overset{O}{\|}}{C}CH_2CH_2CH_3$ $\qquad$ $CH_3CH_2CH_2\overset{\overset{O}{\|}}{C}OCH_3$

(l) $CH_3CH_2NH_2$ $\qquad$ CH_3NHCH_3

(m)
$$\begin{array}{ccccccccc} & H & & OH & & H & & H & O \\ & | & & | & & | & & | & \| \\ H- & C & - & C & - & C & - & C & -C-H \\ & | & & | & & | & & | & \\ & H & & H & & CH_3 & & H & \end{array} \qquad \begin{array}{ccccccccccc} & O & & H & & CH_3 & & H & & H & H \\ & \| & & | & & | & & | & & | & | \\ H- & C & - & C & - & C & - & C & - & C & -C-H \\ & & & | & & | & & | & & | & | \\ & & & H & & H & & OH & & OH & H \end{array}$$

(n)
$$\begin{array}{c} CH_3 \qquad\qquad O \\ |\qquad\qquad\quad \| \\ CH_3-CH-CH_2-C-CH_3 \end{array} \qquad \begin{array}{c} CH_3 \qquad\qquad OH \\ |\qquad\qquad\quad | \\ CH_3-CH-CH_2-CH-CH_3 \end{array}$$

(o)
$$\begin{array}{c} O \\ \| \\ CH_3-CH-C-NH_2 \\ | \\ CH_3 \end{array} \qquad \begin{array}{c} O \\ \| \\ CH_3-NH-C-CH-CH_3 \\ | \\ CH_3 \end{array}$$

(p)
$$\begin{array}{c} CH_3 \qquad\qquad\qquad CH_3 \\ CH_3-CH-CH_2-CH-CH-CH_3 \\ CH_2CH_3 \end{array} \qquad \begin{array}{c} CH_2CH_3 \qquad CH_3 \\ CH_3-CH-CH-CH_2-CH-CH_3 \\ CH_3 \end{array}$$

(In the first structure of (p), the CH_3 groups are on the 2nd and 5th carbons and CH_2CH_3 is on the 4th carbon; in the second structure, CH_3 is below the 2nd carbon, CH_2CH_3 above the 3rd carbon and CH_3 above the 5th carbon.)

14. Do the following pairs of structural formulas represent the same compound or different compounds?

(a)
$$\begin{array}{l} CH_3CH_2CH_2 \\ \qquad\quad | \\ \qquad\quad CH_2CH_2CH_3 \end{array} \qquad \begin{array}{l} CH_3CH_2 \\ \qquad | \\ \qquad CH_2CH_2 \\ \qquad\qquad | \\ \qquad\qquad CH_2 \\ \qquad\qquad | \\ \qquad\qquad CH_3 \end{array}$$

(b) $CH_3CH_2CH_2CH(CH_3)CH_2CH_3$ $\qquad$ $CH_3(CH_2)_2CH(CH_3)CH_2CH_3$

(c) $(CH_3)_2CHCH_2CH(CH_3)CH_2CH_3$ $\qquad$ $CH_3CH(CH_3)CH_2CH_2CH(CH_3)_2$

Section 12.6

15. Give the IUPAC name for the following alkanes:

(a) $CH_3CH_2CH_3$

(b) $CH_3(CH_2)_7CH_3$

(c)
$$\begin{array}{l} CH_3CHCH_2CH_3 \\ \qquad | \\ \qquad CH_3 \end{array}$$

(d)
$$\begin{array}{c} CH_3-CH-CH_2-CH-CH_3 \\ |\qquad\qquad\quad | \\ CH_3 \qquad\quad CH_3 \end{array}$$

(e) $(CH_3)_3CH$

(f)
$$\begin{array}{l} \qquad\qquad CH_2CH_3 \\ \qquad\qquad | \\ CH_3CH_2CHCHCH_2CH_2CH_3 \\ \qquad\quad | \\ \qquad\quad CH_3 \end{array}$$

(g)
$$\begin{array}{l} CH_3CH_2-CH-CH_2CH_2CH_3 \\ \qquad\qquad | \\ \qquad CH_3CHCH_3 \end{array}$$

(h)
$$\begin{array}{l} CH_3CH_2CH_2CHCH_3 \\ \qquad\qquad\quad | \\ \qquad\qquad\quad CH_2CH_2CH_3 \end{array}$$

(i)
$$\begin{array}{c} CH_2CH_2CH_3 \\ | \\ CH_3CH_2CH_2-C-CH_2CH_2CH_2CH_3 \\ | \\ CH_3CHCH_3 \end{array}$$

16. Write the structural formula for the following compounds:

(a) 5-ethyldecane

(b) 2,2-dimethylpropane

(c) 4-isopropyloctane

(d) 1,1,2-trichlorobutane

(e) 2-iodo-2,4,6-trimethylheptane

(f) 5-butyl-2,4-dimethylnonane

17. Give the IUPAC name for each of the following compounds:

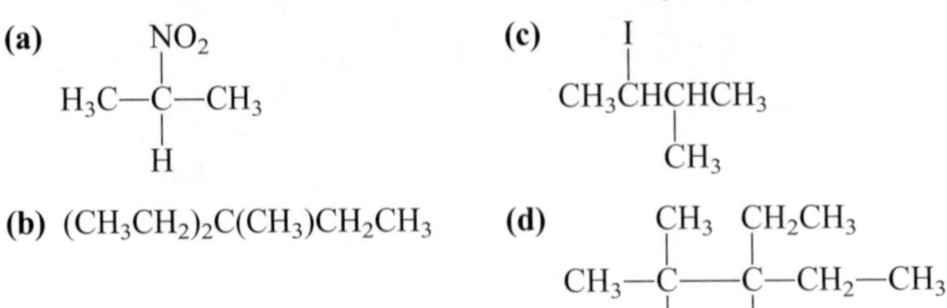

18. Write the structural formula for the following compounds:

(a) 1,1,2,2-tetrabromoethane
(b) 2-aminoheptane
(c) 3-chloro-2,3-dimethylhexane
(d) 3,3-diethylpentane
(e) 1,4-diaminobutane
(f) 1-fluoro-4-isobutyloctane

19. Write the structural formulas for the nine constitutional isomers of C_7H_{16}. Name each according to the IUPAC system.

20. Write the structural formula for the following compounds or alkyl groups:

(a) *sec*-butyl group
(b) isopropyl group
(c) isobutyl group
(d) *tert*-butyl group
(e) *n*-propyl group
(f) *n*-pentane
(g) isopentyl group
(h) isopentane

21. Each of the following are incorrect IUPAC names. Explain why they are incorrect, and write the correct IUPAC name for each compound.

(a) 4-methylpentane
(b) 2-ethyl-3-methylhexane
(c) 4,4-dimethylhexane
(d) 2,2-diethylpentane
(e) 2-*sec*-butylheptane

22. The following structure was incorrectly named as 1-bromo-2-ethyl-2-methylpropane:

$$Br{-}CH_2{-}\overset{\displaystyle CH_3}{\underset{\displaystyle CH_2CH_3}{C}}{-}CH_3$$

Explain why this name is incorrect, and give the correct name for this compound.

Section 12.7

23. Write the balanced equation for the complete combustion of

(a) propane **(b)** octane **(c)** 3-methylpentane

24. Write the equation for the monosubstitution reaction (the reaction in which only one hydrogen on the alkane is replaced) of

(a) methane and bromine
(b) propane with bromine (*Note:* Two different products are possible; write the structural formulas for both.)

Section 12.8

25. How many electrons are shared by carbon atoms joined by **(a)** a single bond, **(b)** a double bond, and **(c)** a triple bond?

26. Why are the alkenes and alkynes described as being unsaturated?

Section 12.9

27. Write the structural formula for the following compounds:

(a) 2-methyl-3-heptene

(b) 2,3-dimethyl-2-butene

(c) 1,3,5-octatriene

(d) 3,3,4,4-tetrabromo-1-hexene

(e) 5,5-dimethyl-2-hexene

(f) 2,3,5-trimethyl-4-propyl-1-heptene

28. Determine the IUPAC name for each of the following compounds:

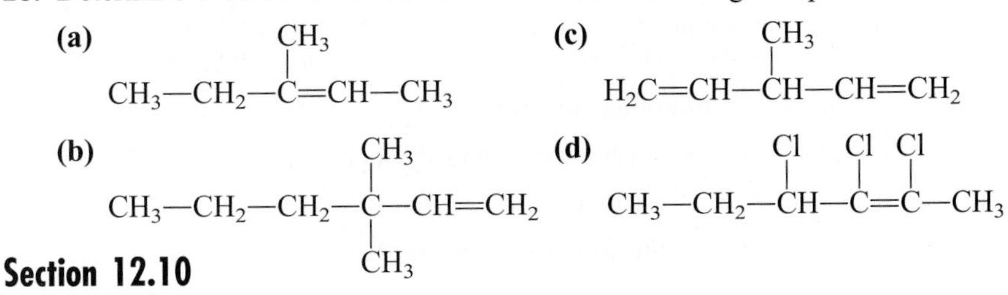

Section 12.10

29. What two types of bonds form **(a)** the double bond in an ethene molecule, **(b)** the triple bond in the ethyne molecule?

30. Indicate the type of hybrid orbitals found in each carbon atom in the following compounds:

(a) $CH_2{=}CHCH{=}C{=}CHCH_3$

(b) $CH_2{=}\overset{\overset{\large Cl}{|}}{C}CH{=}CHCH_3$

Section 12.11

31. Draw the *cis* and *trans* isomers for the following compounds:

(a) 2-pentene

(b) 1,2-dichloroethene

(c) 1,2-dichlorocyclopentane

32. Write the structural formula for the following compounds:

(a) *cis*-1,2-dichloropropene

(b) *trans*-4-phenyl-2-pentene

(c) *cis*-3-hexene

(d) *trans*-4,5-diethyl-6-butyl-4-decene

Section 12.12

33. Write the equations for the following reactions:

(a) the complete oxidation of butene

(b) the addition reaction between 2-pentene and hydrogen

34. Show the product for each of the following reactions:

(a)

$$CH_3-CH=\overset{\overset{\displaystyle CH_3}{|}}{C}-CH_3 + Br_2 \longrightarrow$$

(b) $CH_3-CH_2-CH=CH-CH_2-CH_3 + H_2O \longrightarrow$

(c)

$$CH_3-CH=\overset{\overset{\displaystyle CH_3}{|}}{C}-CH_3 + KMnO_4 \longrightarrow$$

Section 12.14

35. Write the structural formula for the following compounds:

(a) 1,5-dimethylcyclopentene (*Hint:* In a cyclic compound with a double or triple bond, always number the carbons beginning with the multiple bond.)

(b) 1,2-diethylcyclohexene

(c) 3-isopropylcyclopentene

36. Draw all the structural formulas, including the *cis-trans* isomers, for the cycloalkanes with the molecular formula C_5H_{10}. Name each according to IUPAC rules.

37. Write the equation for the following reactions:

(a) the addition reaction between hydrogen and cyclopentene

(b) the complete oxidation of cyclohexane

38. Show the product for each of the following reactions:

(a) 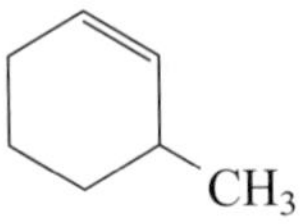 [cyclohexene ring with CH_3] $+ H_2O \longrightarrow$

(b) 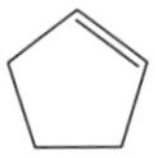 [cyclopentene ring] $+ Br_2 \longrightarrow$

(c) 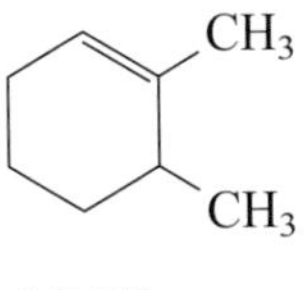 [cyclohexene ring with two CH_3 groups] $+ H_2 \longrightarrow$

Section 12.15

39. Which of the following are aromatic compounds?

(a) 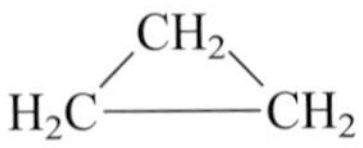 [ring: CH_2, H_2C, CH_2]

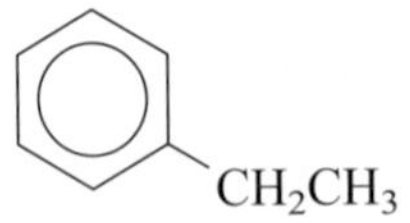

(b) 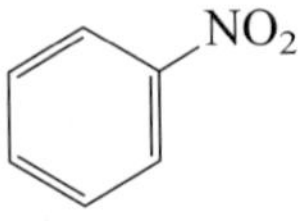 [ring with NO_2]

(c) [benzene ring with CH_2CH_3]

(d) 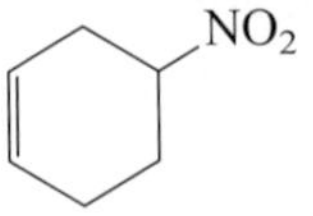 [ring with NO_2]

40. Aromatic compounds are classified as unsaturated compounds, just like the alkenes and alkynes. Why then do aromatic compounds undergo substitution reactions rather than addition reactions like the alkenes and alkynes?

41. Identify each of the following (1) as an alkane, alkene, alkyne, or aromatic hydrocarbon, and (2) as saturated or unsaturated.

(a) $CH_3CH_2CH_2CH_3$

(b) 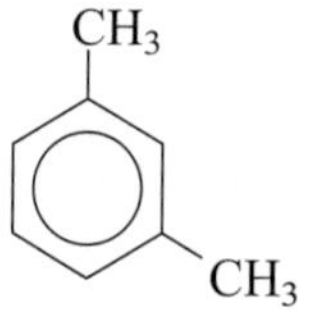 (benzene ring with two CH_3 substituents)
CH_3
CH_3

(c) 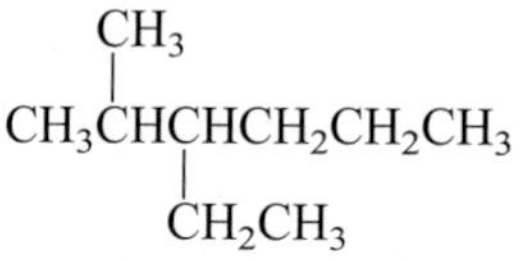
$$\begin{array}{l} \quad\;\; CH_3 \\ \quad\;\; | \\ CH_3CHCHCH_2CH_2CH_3 \\ \qquad\;\; | \\ \qquad\;\; CH_2CH_3 \end{array}$$

(d) $CH_3C{\equiv}CH$

(e) 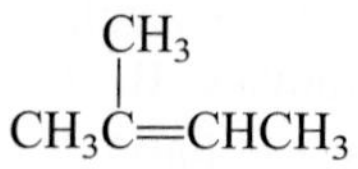
$$\begin{array}{l} \quad\;\; CH_3 \\ \quad\;\; | \\ CH_3C{=}CHCH_3 \end{array}$$

(f) $CH_3(CH_2)_6CH_3$

(g) 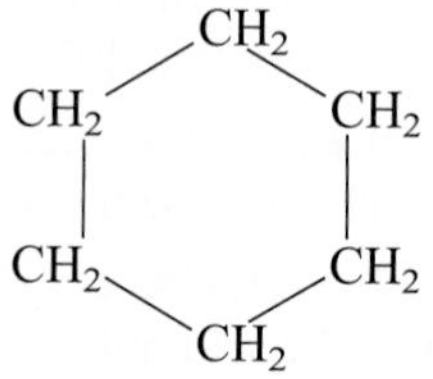 (ring of six CH_2 groups)
CH_2
CH_2 CH_2
CH_2 CH_2
CH_2

(h) 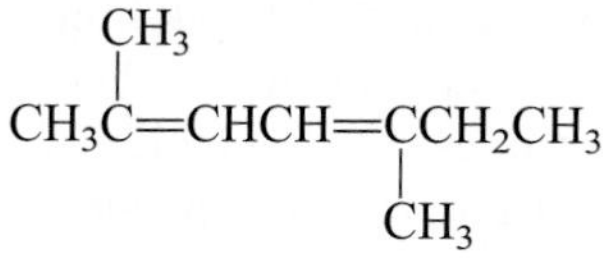
$$\begin{array}{l} \quad\;\; CH_3 \\ \quad\;\; | \\ CH_3C{=}CHCH{=}CCH_2CH_3 \\ \qquad\qquad\qquad\; | \\ \qquad\qquad\qquad\; CH_3 \end{array}$$

42. Identify each of the following (1) as an alkane, alkene, alkyne, or aromatic hydrocarbon, and (2) as saturated or unsaturated.

(a) 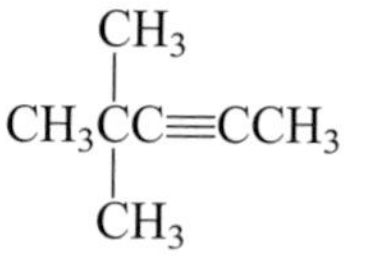
$$\begin{array}{l} \quad\;\; CH_3 \\ \quad\;\; | \\ CH_3CC{\equiv}CCH_3 \\ \quad\;\; | \\ \quad\;\; CH_3 \end{array}$$

(b) 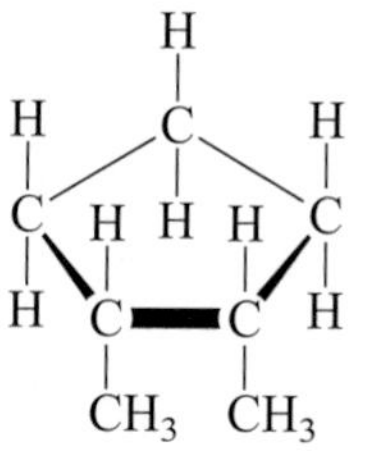 (five-membered ring)
H
H C H
C H H H C
H C=C H
CH_3 CH_3

(c)
$$\begin{array}{l} \quad\;\; CH_3 \\ \quad\;\; | \\ CH_3CCH_2CH_3 \\ \quad\;\; | \\ \quad\;\; CH_2CH_2CH_2CH_3 \end{array}$$

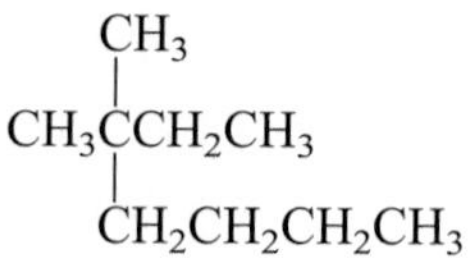

(d) 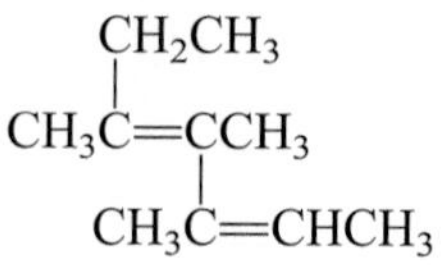
$$\begin{array}{l} \quad\;\; CH_2CH_3 \\ \quad\;\; | \\ CH_3C{=}CCH_3 \\ \qquad\;\; | \\ \quad CH_3C{=}CHCH_3 \end{array}$$

(e) 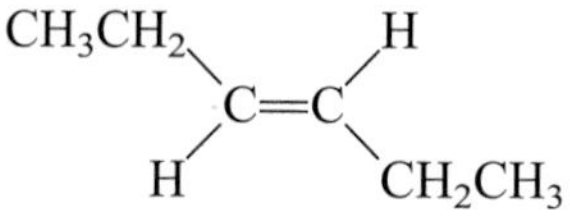
$$\begin{array}{ccc} CH_3CH_2 & & H \\ & C{=}C & \\ H & & CH_2CH_3 \end{array}$$

Applied Problems

43. Many reactions produce more than one organic product. If these products are difficult to separate, the reaction is not very useful for synthesis purposes. Draw the structural formulas for all the monochlorination products formed when propane reacts with chlorine in the presence of heat or light. Name each product.

$$CH_3CH_2CH_3 + Cl_2 \longrightarrow \text{monochloropropanes} + HCl$$

44. When propane reacts with chlorine as shown in Problem 43, some dichloropropanes also form along with the monochloropropanes. Draw the structural formulas for and name all of the possible dichloropropanes formed.

45. Methylene chloride, CH_2Cl_2, is a major component of liquid paint strippers and is a common propellant in spray paints, pesticides, and lubricants. What is the IUPAC name for methylene chloride? What is the chemical formula of methylene bromide?

46. Double bonds that alternate with single bonds, such as in 1,3-butadiene, are said to be conjugated. Conjugated systems play an important role in living organisms. The color of many compounds, for example, is due to conjugated systems. Six possible diene isomers have the formula C_5H_8. Draw the structural formula for each of the dienes, and draw a circle around those that have conjugated double bonds.

47. *Cis* alkenes are less stable than their *trans* isomers. For example *cis*-2-methyl-3-hexene is much less stable than its isomer *trans*-2-methyl-3-hexene. Use their structural formulas to explain this difference in stability.

48. Electric switches in a hospital operating room are not allowed to be turned on or off while gaseous anesthetic is being administered. Suggest a reason for this regulation.

49. The labels from two bottles have been removed. One bottle contained pentane and the other cyclohexene. What simple test can you use to correctly relabel the bottles?

50. Benzenes that have alkyl substituents are called arenes. Many of these compounds, like benzene, are obtained by heating coal in the absence of air. Eight possible benzene derivatives have the molecular formula C_9H_{12}. Draw structures for these derivatives, and name each compound.

Chapter 13

Organic Compounds Containing Oxygen

There were no classes that Friday, and 18-year-old Jane was bored. She felt a bit guilty about wasting the day at her friend's apartment, drinking beer and watching MTV, but after the beer ran out and they switched to vodka and Seven-Up, the time seemed to pass more quickly. As 4:30 PM came around, Jane knew she had to leave. She realized she was a little drunk, but decided it would be OK as long as she was a bit more cautious than usual when driving home.

The young boy on the skateboard never knew what hit him when Jane's car rounded the corner and swerved right into his path. Incredibly, although the boy was thrown onto the hood and bounced off the car's windshield, Jane never realized that she had hit anything. Her only memory was a fuzzy impression of being pulled from the car by a police officer after hitting a tree a bit farther down the road. Jane's blood alcohol level was measured at 0.22%, way above the legal limit for driving (0.10% in most states). She was arrested for both drunk driving and hit-and-run.

The specific substance that the police measured in Jane's blood—the chemical responsible for Jane's lack of judgment and her poor muscle coordination—was ethanol. Ethanol, or ethyl alcohol, is just one of many alcohols. These are all substances whose molecules contain a special group of atoms called the hydroxyl group, which consists of an oxygen atom and a hydrogen atom connected by a single bond: —O—H.

What happened as Jane drank her beer and vodka? Even as ethanol is swallowed, it is being absorbed by the mucous membranes in the nose and throat. Further absorption of ethanol into the body begins immediately after it is swallowed, with about 20% of the alcohol being absorbed through the stomach and the remainder through the small intestines. It takes only an hour for the alcohol in a drink to be absorbed when the stomach is empty, but it can take up to 6 hours when the stomach is full. Also, research has shown that women do not metabolize alcohol in the stomach as well as men. This means that relatively more alcohol reaches a woman's bloodstream than a man's when both have consumed the same amount of alcohol.

Ethanol is completely soluble in water. Once in the bloodstream it moves rapidly into the tissues, especially into organs with large blood supplies such as the brain. The movement into the tissues continues until the concentration of ethanol in the tissues equals the concentration of ethanol in the blood. Because the ethanol becomes equally distributed throughout the tissues of the body, the concentration of ethanol in a person's breath or urine is a fairly accurate indicator of the level of ethanol in the blood. That is why a breath test can be used to determine the blood alcohol level of a driver.

A small amount of ethanol in the blood acts as a stimulant to most organs and body systems, but as the level of ethanol increases, it acts as a depressant. This is especially true in the brain, where ethanol has a disruptive effect on the nerve cell membranes, making the brain less responsive to stimuli. As the blood alcohol level rises toward 0.10%, a person is generally in a mood of pleasant relaxation; tension and anxieties are eased. At this level, inhibitions become decreased as the control center in the brain becomes less active. At higher concentrations, the increased disruption of the nervous system results in a lack of muscular coordination, slurred speech, and difficulty in understanding what is seen and heard. A concentration of alcohol in the blood above 0.36% can result in delirium, anesthesia, coma, and even death.

Table 13.1 Alcohol's Effect on the Body and Behavior

Blood Alcohol Content (%)	*Effect*
0.05	Relaxed state; judgment not as sharp
0.08	Everyday stress lessened
0.10[a]	Movements and speech become clumsy
0.20	Very drunk; loud and difficult to understand; emotions unstable
0.40	Difficult to wake up; incapable of voluntary action
0.50	Coma and/or death

[a]Most states use 0.10 as the lowest indicator of driving while intoxicated. A few states use 0.08, and some go as high as 0.12.

In addition to being found in alcohols, atoms of oxygen appear in many other organic molecules. In this chapter we study the various classes of oxygen-containing compounds.

13.1 Functional Groups Containing Oxygen

You already know that our lives depend critically on a supply of oxygen. Without oxygen (O_2) in the air to breathe and without water (H_2O) to drink, we would die. Oxygen is also essential, however, as an element in many of the compounds that are vital to other life processes. One such group of compounds is the alcohols. In the previous chapter we saw that alcohols all contain the —OH group bonded to a carbon. The specific arrangement of atoms that gives certain characteristic properties to a molecule is called a **functional group.** The study of organic chemistry is greatly simplified by grouping compounds according to their functional group. Look at the structure of the alcohol ethanol:

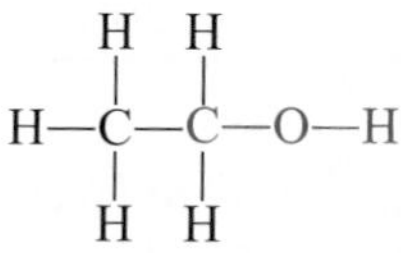

```
    H  H
    |  |
H—C—C—O—H
    |  |
    H  H
```

ethanol

The alcohol functional group is indicated in color. Functional groups are reactive areas within a molecule that give the molecule certain specific chemical properties. In this chapter we study the major oxygen-containing functional groups. When studying complex molecules containing oxygen we often look on the hydrocarbon (alkane) portion of the molecule as a "skeleton" to which the oxygen-containing functional group is attached. Consider, for example, the following complicated molecule:

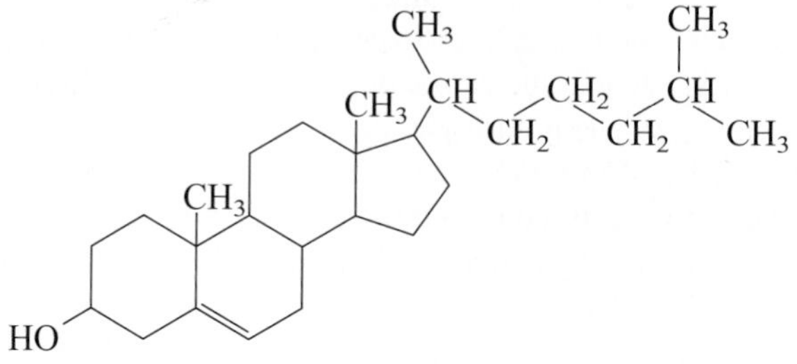

CH_3 CH_3 CH_3 CH CH_2 CH CH_2 CH_2 CH_3 CH_3 HO

cholesterol

Although this molecule, with its four rings, is quite complex, it is easily recognizable as an alcohol.

When discussing a general reaction for a class of compounds, we will often use the letter "R" as we did in the previous chapter to refer to an alkyl group, such as a methyl or ethyl group, for example. We use R′ and R″ to represent other alkyl groups that are different from R. This should help you understand Table 13.2, which lists the oxygen-containing functional groups that we study in this chapter.

◆ **Classify the organic compounds having oxygen-containing functional groups.**

The functional groups in Table 13.2 play important roles in the chemistry of life. For example, the alcohol group includes many compounds that participate in the chemical reactions of living organisms. Such compounds are said to be **physiologically active.** The

Table 13.2 **Functional Groups Containing Oxygen**

Functional Group	*Class of Compound*	*Typical Compound*	
R—OH	Alcohol	CH_3CH_2—OH	Ethanol
R—O—R′	Ether	CH_3—O—CH_2CH_3	Ethyl methyl ether
R—C(=O)—H	Aldehyde	CH_3CH_2—C(=O)—H	Propanal
R—C(=O)—R′	Ketone	CH_3—C(=O)—CH_3	Propanone (acetone)
R—C(=O)—OH	Carboxylic acid	CH_3CH_2—C(=O)—OH	Propanoic acid
R—C(=O)—O—R′	Ester	CH_3CH_2—C(=O)—O—CH_3	Methyl propanoate

glycerol molecule, which contains three alcohol functional groups, is one such physiologically active compound. It is a component of fats and oils that the cell can use to produce energy.

H—O—CH₂—CH(OH)—CH₂—O—H

$$\text{H—O—}\underset{\text{H}}{\overset{\text{H}}{\text{C}}}\text{—}\underset{\text{H}}{\overset{\text{O—H}}{\text{C}}}\text{—}\underset{\text{H}}{\overset{\text{H}}{\text{C}}}\text{—O—H}$$

glycerol

The functional groups that we study in this chapter will show up again in the chapters on biochemistry. We will see that the carbohydrates contain the alcohol functional group as well as an aldehyde or ketone group. The fats and oils are examples of esters, and the amino acids are derived from carboxylic acids. As we study each of the oxygen-containing functional groups we learn the structure of the group, see how to name simple members of the group, and discuss important reactions of that class of compounds. We then focus on a few members of each class that are particularly important to living organisms.

13.2 Ethers

Ethers are a class of compounds having an oxygen atom bonded singly to two carbon atoms, —C—O—C— (general formulas ROR or ROR′). The ethers are only slightly more soluble in water than are the alkanes, but they are generally soluble in nonpolar solvents. They vaporize at relatively low temperatures. The simplest ether, dimethyl ether (molecular weight 46), has a boiling point of −23°C. We will see that the boiling points of the ethers are more similar to the alkanes than to other oxygen-containing compounds. Low-molecular-weight ethers are extremely flammable and must be used with great care.

◆ **Given the structural formula of an ether, state its IUPAC name.**

Simple ethers are frequently given common names that list the two alkyl groups attached to the oxygen. Because you learned the names of the simple alkyl groups in the previous chapter, you should be able to name these ethers. For example:

CH_3OCH_3 is dimethyl ether

$CH_3OCH_2CH_3$ is ethyl methyl ether

$CH_3CH_2CH_2CH_2OCH_2CH_3$ is butyl ethyl ether

Notice that the alkyl groups are each written as a separate word and are listed alphabetically.

Most ethers have some anesthetic properties. Diethyl ether is the anesthetic we commonly call ether or ethyl ether (now rarely used because of its flammability).

$$CH_3{-}CH_2{-}O{-}CH_2{-}CH_3$$

diethyl ether

$$H_2C{=}CH{-}O{-}CH{=}CH_2$$

divinyl ether

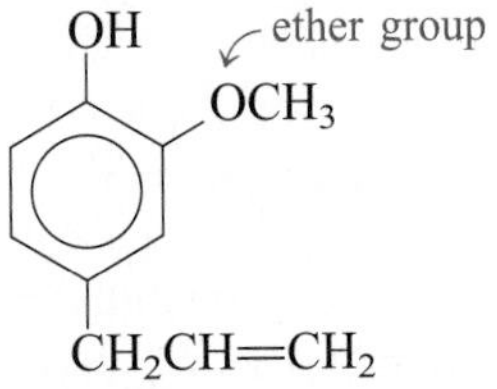

Eugenol: benzene ring with OH, OCH_3 (ether group), and $CH_2CH{=}CH_2$ substituents

eugenol

Divinyl ether is a fast-acting anesthetic that is also little used because of its flammability. Notice that divinyl ether is both an ether and a diene. Because of the double bonds in its structure, it is highly reactive and must be handled with care to prevent decomposition. Eugenol, which is obtained from cloves, is a mild anesthetic used by dentists to lessen the pain of the injection when filling cavities. A small amount of eugenol is applied to the gum before the injection.

Check Your Understanding

1. Name the following ethers:

(a) $CH_3CH_2OCH_2CH_2CH_3$

(b)

$$CH_3{-}CH_2{-}O{-}\underset{\substack{|\\CH_3}}{\overset{\substack{CH_3\\|}}{CH}}$$

2. Draw the structures of the following ethers:

(a) cyclohexyl methyl ether

(b) butyl methyl ether

13.3 Alcohols

Alcohols are organic compounds that contain the hydroxyl group attached to a carbon atom

$$-\overset{|}{\underset{|}{C}}-OH$$

The general formula for the alcohols is ROH. The —OH group in the alcohols should not be confused with the basic hydroxide ion (OH^-) discussed in Chapter 11. The hydroxide

Health Perspective / Ether: An Explosive Anesthetic

In 1846, Oliver Wendell Holmes (physician and father of the Supreme Court justice of the same name) introduced the word *anesthetic* to describe any drug that brought on the condition of anesthesia (an insensibility to pain and sensations). Before that time, there were no effective anesthetics for surgical or dental procedures. Such procedures could be extremely painful. Old movie scenes of a patient consuming huge amounts of whiskey before having a tooth extracted or a bullet removed were fairly accurate. In the 1840s, three substances—ether, nitrous oxide, and chloroform—were introduced as general anesthetics. When inhaled, these compounds depress the activity of the brain and cause the total loss of pain sensations.

Ether (correctly called diethyl ether) was the first of these inhalation anesthetics to be used successfully for surgery. Although it takes a long time for diethyl ether to take full effect on a patient (10–15 min), it is useful in long operations because there is a large safety margin in its use. That is, there is a large difference between the concentration that causes anesthesia and the concentration that can kill the patient. Often a chemical having a more rapid anesthetic effect was used first, and then the patient was switched to diethyl ether.

Ether is used less often today as an anesthetic because of the great danger of explosion when ether mixes with oxygen in the air. Ether vapors are denser than air. If allowed to escape from an unsealed container, ether vapors can flow across the room along the floor. If they encounter a flame or spark, they explode. When working with ether in the laboratory, be sure there are no open flames. If an ether fire should occur, never use water to extinguish it. Ether is less dense than water, and it will simply float on top of the water while continuing to burn.

Even the long-term storage of ether can be dangerous. Ether reacts with oxygen from the air to form peroxides. Such peroxides are extremely explosive. Simply jarring the container can result in an explosion. For this reason, old, half-empty containers of ether can become time bombs. When stored, ether containers should be tightly closed to minimize exposure to air and should be kept away from strong light (which increases the rate of peroxide formation).

Your Perspective: You are inspecting the storeroom in a small rural clinic and notice some bottles of ether that appear to be fairly old. In a formal memorandum, urge the head of the clinic to have the ether removed. Give reasons for your request, and offer suggestions for how this might be done.

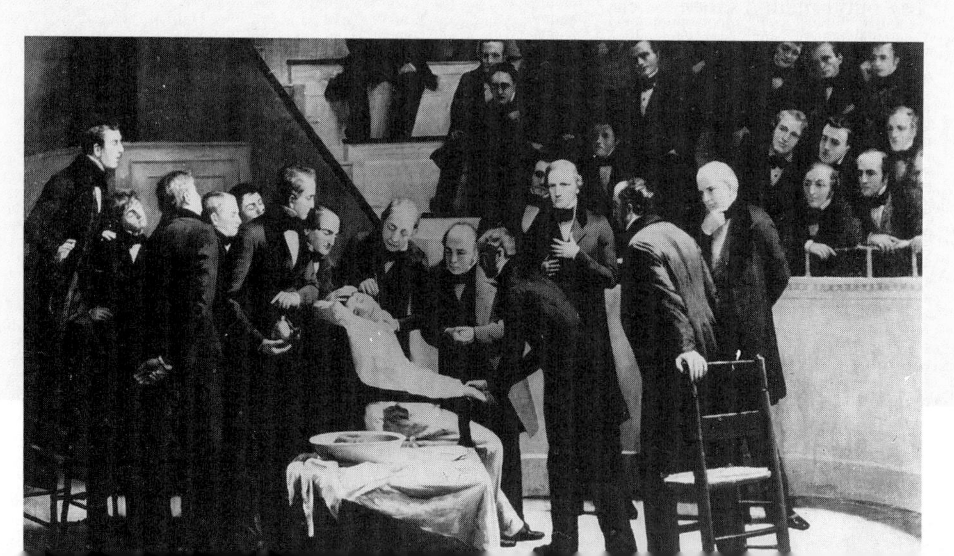

The first public demonstration of the use of ether as a surgical anesthetic took place at Massachusetts General Hospital in Boston on October 16, 1846.

ion bonds through an ionic bond to a metal ion to form such compounds as sodium hydroxide (NaOH), whereas the organic hydroxyl group forms a covalent bond with an adjacent carbon. Unlike the hydroxide ion, the hydroxyl group is not basic.

The hydroxyl group is quite reactive and is easily formed, which makes the alcohols good starting materials for the synthesis of many other compounds. This is just as true in living organisms as it is in the laboratory. In later chapters we will see the importance of the alcohol functional group in our study of carbohydrates, fats, and oils.

Given the structural formula of an alcohol, state its IUPAC name.

Alcohols are named by changing the final -e in the name of the parent alkane to -ol. To identify the parent alkane, find the longest carbon chain to which the hydroxyl group is attached. Number the carbons on that chain starting at the end nearer to the hydroxyl group. Indicate the position of the hydroxyl group itself by adding a number in front of the name. If there is more than one hydroxyl group in the molecule, indicate how many by a prefix before the -ol ending. A compound with two hydroxyl groups is a diol, so 1, 2-ethanediol is a two-carbon alcohol with a hydroxyl group on carbon 1 and on carbon 2.

State the difference between a primary, secondary, and tertiary alcohol.

Alcohols are also classified according to the placement of the hydroxyl group within the molecule. This becomes important when we discuss the reactions of alcohols. **Primary alcohols** have the hydroxyl group attached to a carbon atom that is bonded to no more than one other carbon.

```
   H  H  H                    CH3
   |  |  |                    |
H—C—C—C—OH          CH3—CH—CH2—OH
   |  |  |
   H  H  H
  1-propanol            2-methyl-1-propanol
```

Secondary alcohols have the hydroxyl group attached to a carbon atom that is bonded to two other carbon atoms.

```
   H  OH H                        OH
   |  |  |                        |
H—C—C—C—H          CH3—CH2—CH—CH3
   |  |  |
   H  H  H
  2-propanol                2-butanol
```

Tertiary alcohols have the hydroxyl group attached to a carbon atom that is bonded to three other carbon atoms.

```
      H
      |
   H  O  H                          OH
   |  |  |                          |
H—C—C—C—H          CH3—CH2—C—CH2—CH3
   |  |  |                          |
   H  |  H                          CH3
   H—C—H
      |
      H
2-methyl-2-propanol          3-methyl-3-pentanol
```

The hydroxyl group in an alcohol forms a polar area on the otherwise nonpolar hydrocarbon chain. Oxygen is more electronegative than hydrogen, so it has a stronger attraction for their shared electrons than does the hydrogen. That makes the bond between oxygen and hydrogen a polar covalent bond. For this reason, the hydrogen in the hydroxyl group can form hydrogen bonds with other alcohol molecules or with water (Fig. 13.1).

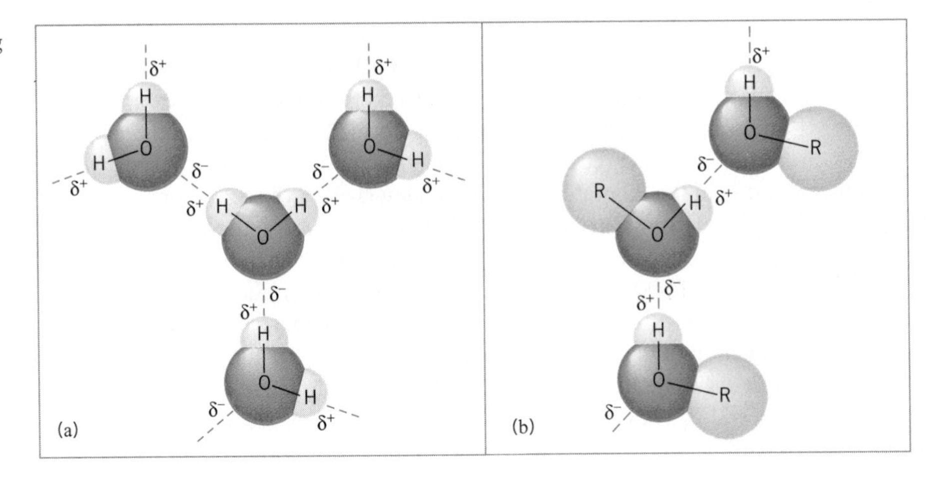

Figure 13.1 Hydrogen bonding in (a) water and (b) alcohols. The polar hydroxyl group on alcohols can enter into hydrogen bonding in much the same way as the —OH group in water. Hydrogen bonds are indicated by dashed lines.

You might want to review the concept of hydrogen bonds (see Section 5.8), because this important feature of many organic functional groups is critical to the shape and properties of many biological molecules.

Because of hydrogen bonding, alcohols with short carbon chains are soluble in water. As the length of the carbon chain increases, however, the nonpolar nature of the hydrocarbon chain becomes more important than the attraction of the hydroxyl group for water. This means that larger alcohol molecules are less and less soluble in water and more and more soluble in nonpolar solvents, like fats, benzene, and carbon tetrachloride. If the number of hydroxyl groups on the molecule is increased, however, the number of sites where hydrogen bonding can occur increases and the solubility of the molecule in water increases (Table 13.3). We will see this same pattern repeat as we continue our study of organic compounds. In general, the solubility of compounds in water *decreases* as the number of carbons in the molecule increases, but also, the solubility of compounds in water *increases* as the number of polar functional groups in the molecule increases.

◆ **Compare the polarity and water solubility of alcohols, ethers, aldehydes, ketones, and carboxylic acids.**

The polar hydroxyl group also affects the melting point and boiling point of the alcohol molecule. Hydrogen bonding increases the amount of energy needed to pull the molecules apart, which increases both the melting and boiling points. Alcohols have much higher boiling points than alkanes or ethers with comparable molecular weights. For example, ethanol has a molecular weight of 46 and a boiling point of 78°C. But propane (with a molecular weight of 44) has a boiling point of −45°C, and dimethyl ether (with a molecular weight of 46) has a boiling point of −23°C.

Table 13.3 Solubility of Alcohols (in Water)

Compound	*Formula*	*Solubility*
Ethanol	CH_3CH_2OH	Completely soluble
1-Pentanol	$CH_3CH_2CH_2CH_2CH_2OH$	Slightly soluble
1,2-Pentanediol	$CH_3CH_2CH_2CHOHCH_2OH$	Soluble
2-Hexanol	$CH_3(CH_2)_3CHOHCH_3$	Very slightly soluble
2,3-Hexanediol	$CH_3(CH_2)_2CHOHCHOHCH_3$	Soluble
1-Decanol	$CH_3(CH_2)_8CH_2OH$	Insoluble

Example 13-1

1. Name the following compound:

$$\begin{array}{c} \qquad\qquad\qquad\qquad\qquad\quad OH \\ \qquad\qquad\qquad\qquad\qquad\quad | \\ CH_3—CH_2—CH_2—CH—CH_3 \end{array}$$

See the Question

The IUPAC nomenclature rules must be applied, in the correct order, to the naming of this compound.

Think It Through (Execute the Math)

(a) Since this is an alcohol, the name ends in -ol.

(b) The only chain in the molecule is five carbons long, so this is a pentanol.

(c) We number the chain from the end closer to the hydroxyl group, so the hydroxyl group is on carbon 2.

Prepare the Answer

The name of the compound is 2-pentanol.

2. Now try to name the following compound:

$$\begin{array}{c} \qquad\quad CH_3\;\; OH \\ \qquad\quad |\qquad\; | \\ CH_3—CH—CH—CH_3 \end{array}$$

S

Again, you must apply the rules of nomenclature to this compound.

T • E

(a) The compound has a four-carbon chain to which the hydroxyl group is attached, so it is a butanol.

(b) There is a methyl group on the chain, so it is a methylbutanol.

(c) We must number the chain from the end closer to the hydroxyl group.

P

The name of the compound is 3-methyl-2-butanol.

3. Write the structural formula for 2-methyl-2-hexanol.

S

The name of the compound must be used to determine its structure.

T • E

(a) We must break the name down into its parts to determine the structure of the compound.

(b) The compound has a six-carbon chain. The hydroxyl group is attached at the second carbon.

(c) There is also a methyl attached to the second carbon.

P

The structure of the compound is

$$\begin{array}{c} \qquad\qquad\qquad\qquad\qquad\quad OH \\ \qquad\qquad\qquad\qquad\qquad\quad | \\ CH_3—CH_2—CH_2—CH_2—C—CH_3 \\ \qquad\qquad\qquad\qquad\qquad\quad | \\ \qquad\qquad\qquad\qquad\qquad\quad CH_3 \end{array}$$

2-methyl-2-hexanol

4. State whether the alcohols in questions 1 through 3 are primary, secondary, or tertiary.

S

This question asks us to determine how many carbons are bonded to the carbon to which the hydroxyl group is attached.

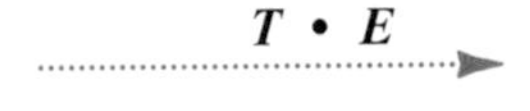

(a) In 2-pentanol, the carbon bonded to the hydroxyl group has two carbons attached to it.

(b) The carbon bonded to the hydroxyl group in 3-methyl-2-butanol is also bonded to two other carbons.

(c) Carbon 2 in 2-methyl-2-hexanol is bonded to the hydroxyl group and to three other carbons.

(a) 2-Pentanol is a secondary alcohol.

(b) 3-Methyl-2-butanol is also a secondary alcohol.

(c) 2-Methyl-2-hexanol is a tertiary alcohol.

Check Your Understanding

1. Name the following alcohols, and identify them as primary, secondary, or tertiary:

(a) $CH_3—CH_2—CH_2—CH_2—C(OH)(CH_3)—CH_3$

(c) (cyclohexane ring)—OH

(b) $CH_3—CH_2—C(CH_3)_2—CH_2—OH$

2. Draw the structural formula for each of the following:

(a) 3, 3-dimethyl-2-butanol

(b) 1, 2-propanediol

(c) 5-methyl-3-isopropyl-1-hexanol

13.4 Some Important Alcohols

Methanol

Methanol (CH_3OH) is sometimes called wood alcohol because it was first obtained by heating wood in the absence of air. Perhaps you did this experiment in a science lab. Methanol is very poisonous. Ingestion of small quantities can cause blindness, and consumption of larger quantities can be fatal. Industrially, methanol has many important uses as a solvent and as a starting material for manufacturing other products.

Ethanol

Ethanol, often referred to simply as "alcohol," can be obtained from the fermentation of grains, such as wheat or barley (thus the term *grain alcohol*). Its common name is ethyl alcohol. In fermentation, yeast cells in the grain or fruit use the nutrients found there to supply themselves with energy. The waste products of this process are ethanol and carbon dioxide.

$$\underset{\text{glucose (a sugar)}}{C_6H_{12}O_6} \xrightarrow{\text{yeast}} 2CH_3CH_2OH + 2CO_2 + \text{energy}$$

Ethanol has many uses. It is an excellent solvent, so it has widespread use in such products as medicines and perfumes. Ethanol in a 70% aqueous solution is useful as an antiseptic because it destroys bacteria by precipitating their protein. When used for industrial purposes, ethanol is often mixed with foul-smelling (and often poisonous) chemicals that make the alcohol undrinkable. Ethanol that is treated in this manner is said to be *denatured.* Ethanol can also be mixed with gasoline to form a fuel known as gasohol. In fact, a large proportion of the automobiles in Brazil are fueled by pure ethanol or a gasohol mixture.

2-Propanol

2-Propanol is often referred to as isopropyl alcohol. Because of its rapid evaporation, it is used in astringents to cool the skin and constrict blood vessels.

1, 2-Ethanediol

Dihydric alcohols are compounds that contain two hydroxyl groups. 1,2-Ethanediol, often referred to as ethylene glycol, is such a compound. This thick, colorless, and sweet-tasting liquid is about as toxic as methanol. Its combination of sweetness and toxicity has led to tragic results when small children or animals have consumed this compound. Ethylene glycol is the major ingredient in automobile antifreeze.

```
                                  H
                                  |
       H  H                    H  O  H
       |  |                    |  |  |
  H—O—C—C—O—H            H—O—C—C—C—O—H
       |  |                    |  |  |
       H  H                    H  H  H
   1, 2-ethanediol        1, 2, 3-propanetriol
  (ethylene glycol)            (glycerol)
```

1, 2, 3-Propanetriol

Although glycerol (1, 2, 3-propanetriol) looks very similar to ethylene glycol, some major differences exist. Glycerol, also called glycerin, is not toxic. It is a component of fats and oils and is used commercially as a humectant (to prevent water loss) in cream-filled candies, tobacco, inks, and plastic clays. It is used as a sweetening agent, as a solvent for medicines, and as an ingredient in some liqueurs. Because it protects the skin from drying out, glycerol is used in many hand lotions and cosmetics.

Menthol

Menthol is an important cyclic alcohol. It has been used in folk medicine for thousands of years. It causes an unusual cooling and refreshing sensation when rubbed on the skin, leading to its use in after-shave lotions and cosmetics. It is also used in nose and throat sprays, in cough drops, and in cigarettes because of its ability to sooth inflamed mucous tissues, such as those in the nose and throat.

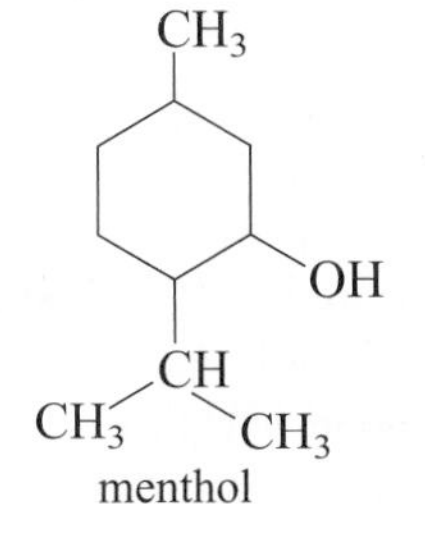

Menthol is used in cough drops to soothe mucous membranes.

Historical Perspective / Alcohol: The Oldest Synthetic Chemical

It is likely that alcohol was the first substance synthesized by humans. Alcohol was originally produced by mixing honey, fruits, berries, cereals, or other plant materials with water and leaving them in the sun. This created a liquid prized as food, as a ceremonial or religious potion, and as medicine. The first known brewery dates from about 3700 BC in Egypt.

Grapes had considerable significance for the ancient peoples of the Mediterranean and Middle East. Fresh grapes are a high-energy food, containing about 20% to 25% sugar. When grapes are dried properly, they can contain up to 75% sugar. In those days, therefore, grapes constituted one of the few sources of sweet food that could be easily stored and transported. Freshly harvested grapes were placed in large vessels and stored in cellars to be eaten at a later time. Entering the cellars to retrieve the grapes could sometimes be very hazardous. When yeast on the skins of the grapes contacted juice that had escaped from the fruit, fermentation began. Carbon dioxide produced from the fermentation could build up in the closed cellars to deadly levels.

There is an interesting story from ancient Mesopotamia about the discovery of wine. The story goes that a young concubine had fallen out of favor with the king. Deciding to take her life rather than face his rejection, she ran off to the cellar where the grapes were stored. While sitting there waiting to die from carbon dioxide asphyxiation, she decided to hasten the process by drinking the juice that had formed among the grapes. After drinking a considerable amount of the strange-tasting juice, she became so drunk that she forgot why she had gone to the cellar. She skipped happily up the stairs and back to the company of the king. Her dramatic change in attitude fascinated the king. When the young lady explained what she had done, he hurriedly led the way to the cellar to try the drink himself.

Early wines must have been of poor quality. They were probably drunk during or soon after the primary fementation, before they turned to vinegar. Although wine was probably the first alcoholic beverage discovered, beer was the beverage that was first produced on a large scale. When barley grain was wetted, allowed to germinate, and then dried, it had a much sweeter taste and was less perishable. This is what we know today as *malt.* When flour made from the ''malted'' barley was mixed with water and allowed to sit for a while, yeast in the air fermented the sugars and starches present in the mixture. Eventually this fermentation produced an alcoholic soup that had much the same intoxicating effect as early wines. Although the scientific processes responsible for creating beer and wine were not understood, drinking these beverages did make people feel good. It is not surprising that, during a period when life was so difficult for the average person, alcoholic beverages became an important part of their lives. In fact, beer and wine, with their mild alcoholic content, were often the only liquids available that were safe to drink.

Phenols

When one or more hydroxyl groups are bonded to benzene, a new class of compounds is produced. The simplest of these is phenol, which has led to calling the entire class ''the phenols.'' In 1865, the English surgeon Joseph Lister was the first to apply chemicals to a wound to prevent infection. For this purpose he used phenol, also known as carbolic

Example 13-2

Determine the product of the oxidation of cyclohexanol.

See the Question

We need to write the reaction for the oxidation of this alcohol.

Think It Through (Execute the Math)

(a) The structure of cyclohexanol is

OH

cyclohexanol

(b) The carbon to which the hydroxyl group is attached is bonded to two other carbons, so this is a secondary alcohol. Oxidation of a secondary alcohol involves the loss of the hydrogen on the oxygen and on the carbon attached to the hydroxyl group.

Prepare the Answer

This oxidation yields

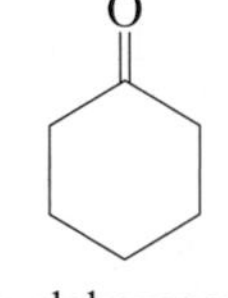

cyclohexanone

Because cyclohexanol is a secondary alcohol, the product is a ketone.

Check Your Understanding

Now see if you can determine the product of each of the following reactions:

1. dehydration of 2-propanol
2. hydration of 2-butene
3. oxidation of methanol
4. oxidation of 3-methyl-2-hexanol

13.7 Aldehydes and Ketones

The aldehydes and ketones are two functional groups that contain the **carbonyl group:** a carbon double bonded to an oxygen.

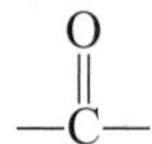

The carbonyl group appears in the remaining functional groups we cover in this chapter. It is also a part of lipids, carbohydrates, proteins, and nucleic acids—compounds that we cover in the next section of the text. The chemistry of the compounds containing the carbonyl group is very much at the center of our life processes.

State the difference in structure between an aldehyde and a ketone.

In aldehydes, the carbon of the carbonyl group always has at least one hydrogen attached to it. Stated another way, the carbonyl group in an aldehyde can have no more than one carbon bonded to it. In the ketones, the carbonyl group has two carbons bonded to it.

$$\begin{array}{c} H \\ | \\ -C{=}O \end{array} \qquad \begin{array}{c} \quad O \quad \\ |\quad \| \quad | \\ -C-C-C- \\ | \qquad\quad | \end{array}$$

aldehyde — ketone

$$\begin{array}{c} H \\ | \\ R-C{=}O \end{array} \qquad \begin{array}{c} O \\ \| \\ R'-C-R \end{array}$$

aldehyde — ketone

Given the structural formula of an aldehyde or ketone, state its IUPAC name.

To name an aldehyde, select the longest carbon chain that contains the carbonyl group. Replace the -e ending of the parent alkane with -al. Because the carbonyl group in an aldehyde is always at the end of the chain, you don't need to indicate its position (it is always at carbon 1). If the carbonyl group is attached to a benzene ring, the name of the compound is benzaldehyde. Many aldehydes have common names in addition to their IUPAC names. The name formaldehyde for methanal and acetaldehyde for ethanal are more typically used in the health sciences.

To name a ketone, select the longest chain that contains the carbonyl group and replace the -e ending with -one. Number the carbon chain from the end closer to the carbonyl group, and indicate the position of the carbonyl group by a number placed before the name.

Here are some examples of aldehydes and ketones:

$$\begin{array}{c} H \\ | \\ H-C{=}O \end{array} \qquad \begin{array}{c} H \quad H \\ | \quad\; | \\ H-C-C{=}O \\ | \\ H \end{array} \qquad C_6H_5-\overset{O}{\overset{\|}{C}}-H$$

methanal (formaldehyde) — ethanal (acetaldehyde) — benzaldehyde

$$CH_3-\overset{O}{\overset{\|}{C}}-CH_3 \qquad CH_3-CH_2-\overset{O}{\overset{\|}{C}}-CH_3 \qquad O{=}C_6H_{10}$$

2-propanone (acetone) — 2-butanone — cyclohexanone

Compare the polarity and water solubility of alcohols, ethers, aldehydes, ketones, and carboxylic acids.

Although the aldehydes and ketones are both highly reactive groups, the aldehydes are the more reactive. The presence of the carbonyl group creates a polar region in the aldehydes and ketones. Because of hydrogen bonding between the carbonyl group and water, low-molecular-weight aldehydes and ketones are water-soluble.

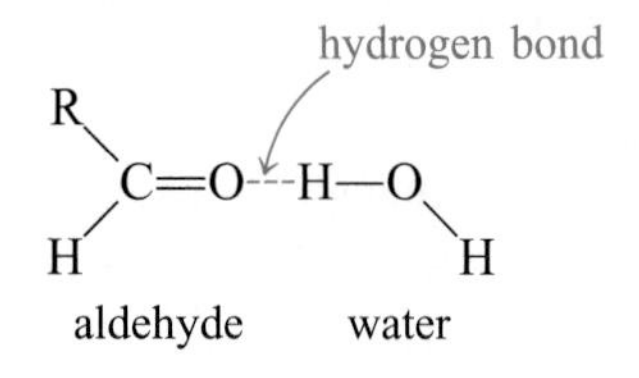

Example 13-3

1. Name this compound:

$$CH_3—CH_2—\overset{\overset{\displaystyle O}{\|}}{C}H$$

Apply the IUPAC nomenclature rules. ◄ *See the Question*

(a) This compound has a three-carbon chain, so its name is based on propane. ◄ *Think It Through (Execute the Math)*

(b) Because it has a carbonyl group bonded to one hydrogen and only one carbon, it is an aldehyde. The final -e is replaced with -al.

The name of the compound is propanal. ◄ *Prepare the Answer*

2. Name this compound:

$$CH_3—CH_2—CH_2—\overset{\overset{\displaystyle O}{\|}}{C}—CH_2—CH_3$$

Apply the IUPAC nomenclature rules. ◄ *S*

(a) This compound has a six-carbon chain, so its name is based on hexane. ◄ *T • E*

(b) Because the carbonyl group is bonded to two other carbons, it is a ketone. The final -e is dropped, and -one is added.

(c) So far, that makes the name hexanone.

(d) The chain is numbered from the end closer to the carbonyl group, which makes the carbonyl carbon number 3.

The name of the compound is 3-hexanone. ◄ *P*

3. Name this compound:

$$CH_3—CH_2—\overset{\overset{\displaystyle CH_3}{|}}{C}H—\overset{\overset{\displaystyle O}{\|}}{C}—CH_3$$

Apply the IUPAC nomenclature rules. ◄ *S*

(a) Because this is a ketone whose longest carbon chain contains five carbons, it is a pentanone. ◄ *T • E*

(b) Numbering from the end closer to the carbonyl, we find it is located on the second carbon. That makes the compound 2-pentanone.

(c) There is a methyl group at carbon 3.

The name is 3-methyl-2-pentanone. ◄ *P*

4. Name this compound:

$$CH_3—\overset{\overset{\displaystyle CH_3}{|}}{C}H—CH_2—\overset{\overset{\displaystyle O}{\|}}{C}—CH_2—CH_3$$

Apply the IUPAC nomenclature rules. ◄ *S*

(a) The longest carbon chain has six carbons. ◄ *T • E*

(b) The carbonyl group of this ketone is on the third carbon from the end closer to the carbonyl.

(c) There is a methyl group at carbon 5.

P The name is 5-methyl-3-hexanone.

Check Your Understanding

1. Name the following compounds:

(a)

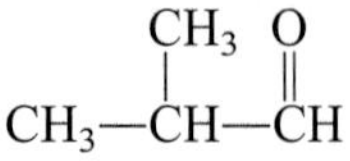

(c)

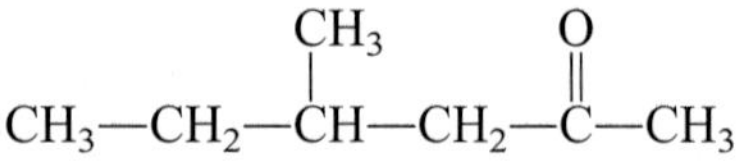

(b)

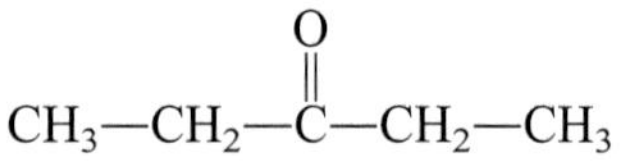

2. Write the structural formula for each of the following compounds:

(a) 4-ethylheptanal

(b) 4-ethyl-2-methyl-3-octanone

(c) 2-methylcyclohexanone

Nail polish removers contain the solvent acetone.

13.8 Some Important Aldehydes and Ketones

Many aldehydes and ketones play important roles in our lives. Formaldehyde is the simplest aldehyde. At room temperature it is a gas with an irritating odor. It is commonly prepared as a 37% aqueous solution known as formalin. You may have used formalin as a disinfectant or as a preservative for biological specimens. If you have ever come in contact with formaldehyde, you know that it can irritate the membranes in your eyes, nose, and throat. Concerns over health hazards associated with formaldehyde have led to a decline in its use.

Acetone is the simplest ketone and is used mainly as a solvent because of its solubility in water and its ability to dissolve a wide range of organic compounds. For example, it is the main ingredient in some nail polish removers. Acetone can be produced in the body through a reaction that occurs only when there is some metabolic disorder. An example is the disease diabetes mellitus, in which the absence of certain metabolic reactions causes acetone to be produced in large quantities. This compound builds up in the tissues and appears in the urine as well as in the breath of untreated diabetics.

Some aldehydes and ketones have pleasant odors or tastes, which makes them useful in perfumes or as flavoring agents.

These candies contain the flavoring agent, cinnamaldehyde.

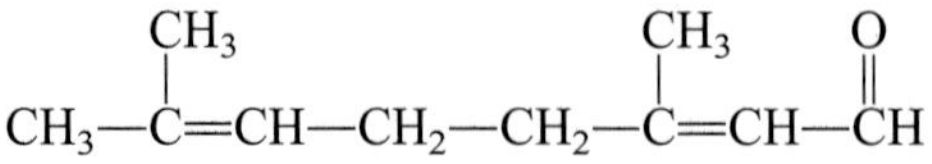

citral
(a fragrance found in lemon grass oil)

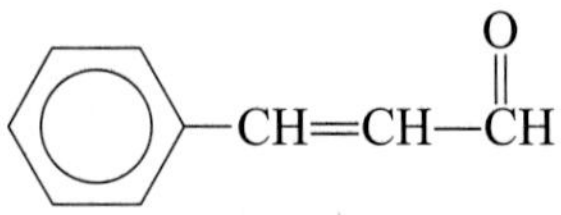

cinnamaldehyde

Aldehydes and ketones are also important in the metabolic reactions of living organisms. For example, the pyruvate ion plays a role in the oxidation of carbohydrates to produce energy in the cell. Progesterone is a complex diketone that is the female sex hormone produced during pregnancy.

$$CH_3\overset{O}{\overset{\|}{C}}CO_2^-$$

pyruvate ion

CH_3 H_3C $C{=}O$ CH_3 O

progesterone

13.9 Synthesis of Aldehydes and Ketones

Aldehydes and ketones can be produced by the oxidation of primary and secondary alcohols in the presence of oxidizing agents such as potassium permanganate (see Section 13.6). As we will see when we study the chemical reactions in living organisms, these same types of reactions occur in the body under much milder conditions, made possible by the action of enzymes.

Describe one method of preparing aldehydes and ketones.

$$H-\overset{H}{\underset{H}{C}}-\overset{H}{\underset{H}{C}}-\overset{H}{\underset{H}{C}}-\overset{H}{\underset{H}{C}}-O-H \xrightarrow{-2H} H-\overset{H}{\underset{H}{C}}-\overset{H}{\underset{H}{C}}-\overset{H}{\underset{H}{C}}-\overset{H}{C}{=}O$$

1-butanol → butanal

$$H-\overset{H}{\underset{H}{C}}-\overset{H}{\underset{H}{C}}-\overset{\overset{H}{|}}{\overset{O}{\underset{H}{C}}}-\overset{H}{\underset{H}{C}}-H \xrightarrow{-2H} H-\overset{H}{\underset{H}{C}}-\overset{H}{\underset{H}{C}}-\overset{O}{\overset{\|}{C}}-\overset{H}{\underset{H}{C}}-H$$

2-butanol → 2-butanone

Under stronger oxidizing conditions, the oxidation of a primary alcohol yields a carboxylic acid rather than an aldehyde.

$$H-\overset{H}{\underset{H}{C}}-\overset{H}{\underset{H}{C}}-O-H \longrightarrow H-\overset{H}{\underset{H}{C}}-\overset{O}{\overset{\|}{C}}-O-H$$

ethanol → acetic acid

13.10 Reactions of Aldehydes and Ketones

Reduction

Describe a method for preparing an alcohol in the laboratory.

The primary alcohols can be synthesized by the reduction of aldehydes; the secondary alcohols, on the other hand, are formed through the reduction of ketones. If it sounds like we are going in circles, it's true. Aldehydes and ketones are one step up in oxidation state from the alcohols. To go up in oxidation state to the aldehydes and ketones, we oxidize an alcohol. To go down one step from an aldehyde or a ketone, we form alcohols. Reduction of aldehydes and ketones can be accomplished in the laboratory by allowing the compound to react with hydrogen in the presence of a catalyst such as platinum (Pt).

$$\underset{\text{acetaldehyde}}{H{-}CH_2{-}CH{=}O} + H{-}H \xrightarrow{Pt} \underset{\text{ethanol}}{H{-}CH_2{-}CH_2{-}O{-}H}$$

$$\underset{\text{2-butanone}}{H{-}CH_2{-}CH_2{-}\overset{O}{\overset{\|}{C}}{-}CH_2{-}H} + H{-}H \xrightarrow{Pt} \underset{\text{2-butanol}}{H{-}CH_2{-}CH_2{-}\overset{\overset{H}{|}}{\underset{H}{C}}(O){-}CH_2{-}H}$$

Oxidation

Compare the ease with which aldehydes and ketones can be oxidized.

Aldehydes are very easily oxidized to produce carboxylic acids. Ketones, on the other hand, can be oxidized only under very strong conditions. This difference in ease of oxidation is one of the ways we can distinguish in the laboratory between these two classes of compounds. As we can see in the following reaction, oxidation of an aldehyde involves the gain of an oxygen atom.

$$\underset{\text{propanal}}{H{-}CH_2{-}CH_2{-}CH{=}O} \xrightarrow{Cu^{2+}} \underset{\text{propanoic acid}}{H{-}CH_2{-}CH_2{-}\overset{O}{\overset{\|}{C}}{-}O{-}H}$$

The blue Benedict's solution changes to brick red in a positive test.

In addition to being more easily oxidized than ketones, aldehydes are more easily oxidized than the alcohols from which they can be formed. For example, propanal can be oxidized under milder conditions than 1-propanol. The ease of oxidation of the aldehydes is the basis of **Benedict's test,** which is used to determine their presence. The reagent for this test is an aqueous solution of copper(II) sulfate, $CuSO_4$. Aldehydes are oxidized by this solution when heated, whereas ketones and alcohols are not. A positive test is indicated by the formation of a brick-red precipitate of copper(I) oxide, Cu_2O.

The Formation of Hemiacetals and Acetals, Hemiketals and Ketals

Aldehydes react with alcohols under certain conditions to produce a hemiacetal.

$$\underset{\text{ethanal}}{H{-}CH_2{-}CH{=}O} + \underset{\text{methanol}}{H{-}O{-}CH_2{-}H} \rightleftharpoons \underset{\text{a hemiacetal}}{H{-}CH_2{-}\overset{OH}{\overset{|}{\underset{H}{C}}}{-}O{-}CH_2{-}H}$$

This reaction can be seen as the addition of the alcohol across the carbonyl bond.

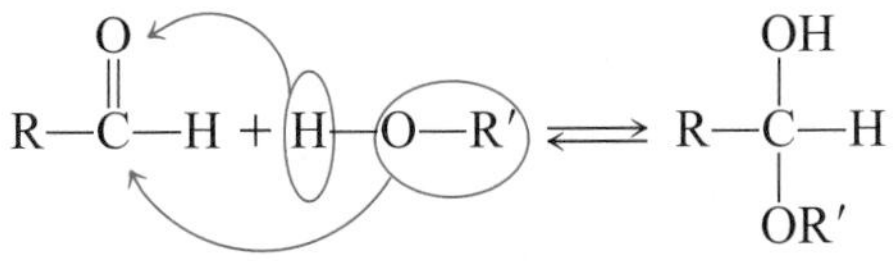

$$\mathrm{R{-}\overset{\overset{\displaystyle O}{\|}}{C}{-}H + H{-}O{-}R' \rightleftharpoons R{-}\underset{\underset{\displaystyle OR'}{|}}{\overset{\overset{\displaystyle OH}{|}}{C}}{-}H}$$

Write the equation for the preparation of a hemiacetal and an acetal.

The prefix *hemi-* means half. A hemiacetal is ''halfway'' to being an acetal. If the hemiacetal reacts with another molecule of alcohol, an acetal is formed.

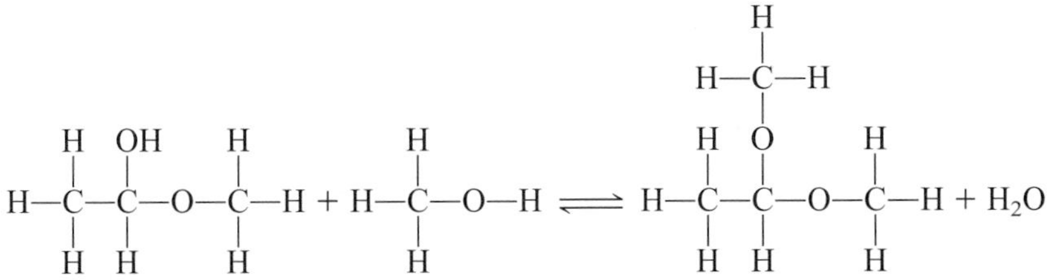

$$\underset{\text{a hemiacetal}}{\mathrm{CH_3{-}CH(OH){-}O{-}CH_3}} + \underset{\text{methanol}}{\mathrm{CH_3{-}O{-}H}} \rightleftharpoons \underset{\text{an acetal}}{\mathrm{CH_3{-}CH(OCH_3){-}O{-}CH_3}} + H_2O$$

The generalized equation for this reaction is

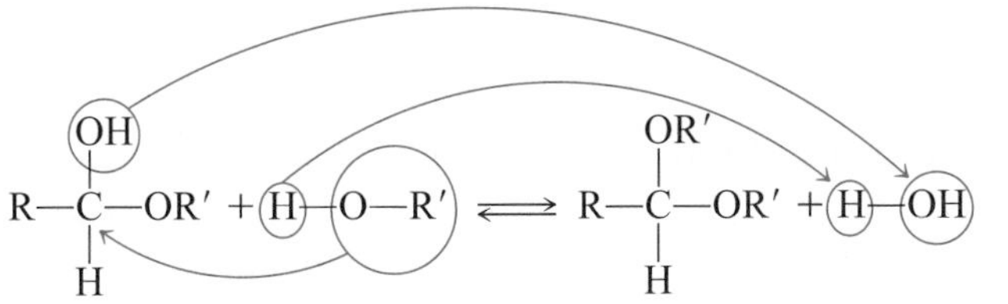

$$\mathrm{R{-}\underset{\underset{\displaystyle H}{|}}{\overset{\overset{\displaystyle OH}{|}}{C}}{-}OR' + H{-}O{-}R' \rightleftharpoons R{-}\underset{\underset{\displaystyle H}{|}}{\overset{\overset{\displaystyle OR'}{|}}{C}}{-}OR' + H{-}OH}$$

When an aldehyde and an alcohol group on the same molecule react with each other, a cyclic hemiacetal is formed. Much of the chemistry of glucose and other sugars is the chemistry of cyclic hemiacetals and acetals. If you look carefully at the cyclic form of glucose, you will see that it is a hemiacetal.

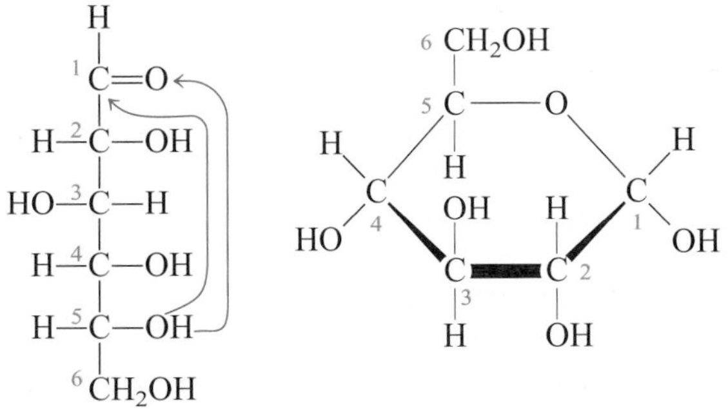

Open-chain form: H–C¹=O; H–C²–OH; HO–C³–H; H–C⁴–OH; H–C⁵–OH; C⁶H₂OH. Cyclic form: ring of C1–C5 and O, with $^6CH_2OH$ on C5; OH on C1, C2 (below), C3 (above), C4 HO.

glucose

Ketones can form hemiketals and ketals in reactions very similar to those of the aldehydes.

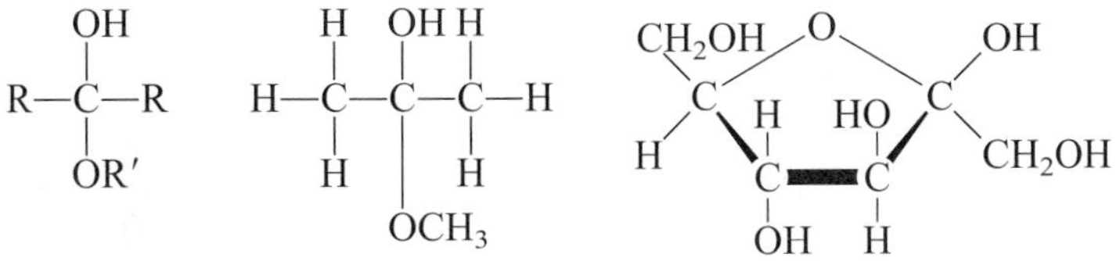

$$\underset{\text{hemiketal}}{\mathrm{R{-}\underset{\underset{\displaystyle OR'}{|}}{\overset{\overset{\displaystyle OH}{|}}{C}}{-}R}} \qquad \underset{\text{a hemiketal}}{\mathrm{CH_3{-}C(OH)(OCH_3){-}CH_3}}$$

Fructose ring: CH_2OH, O, OH, H, HO, H, CH_2OH, OH, H

fructose, a hemiketal

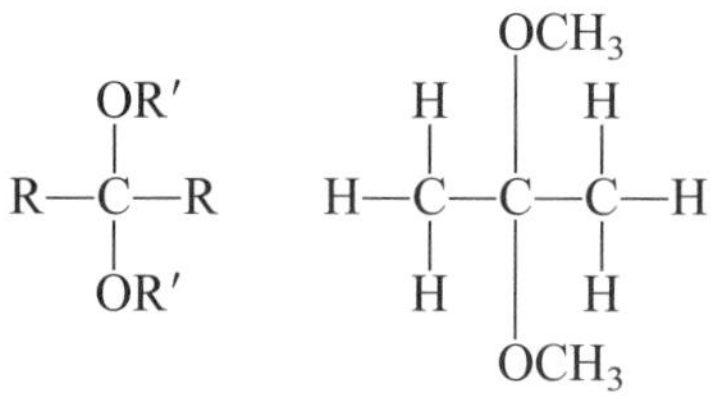

$$\underset{\text{ketal}}{R-\overset{\overset{OR'}{|}}{\underset{\underset{OR'}{|}}{C}}-R} \qquad \underset{\text{a ketal}}{H-\overset{\overset{H}{|}}{\underset{\underset{H}{|}}{C}}-\overset{\overset{OCH_3}{|}}{\underset{\underset{OCH_3}{|}}{C}}-\overset{\overset{H}{|}}{\underset{\underset{H}{|}}{C}}-H}$$

Check Your Understanding

Write the formula for the product of each of the following reactions:

1. reduction of pentanal
2. reduction of cyclohexanone
3. oxidation of 1-pentanol
4. oxidation of 2-methyl-3-hexanol
5. formation of a hemiacetal between butanal and ethanol
6. formation of an acetal between propanal and two molecules of methanol
7. formation of a hemiketal between acetone and ethanol

13.11 Carboxylic Acids

One step up in oxidation from the aldehydes is the functional group known as the carboxyl group

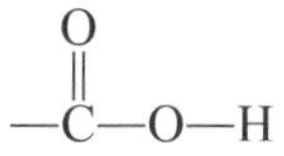

$$-\overset{\overset{O}{\|}}{C}-O-H$$

This group contains a hydroxyl group attached to the carbon of the carbonyl group. The class of compounds containing the carboxyl functional group is called the **carboxylic acids,** the general formula for which is RCOOH, or RCO_2H.

Compare the polarity and water solubility of alcohols, ethers, aldehydes, ketones, and carboxylic acids.

Carboxylic acids form hydrogen bonds. They have higher boiling points than alcohols with comparable molecular weights because they have a higher degree of intermolecular hydrogen bonding. Hydrogen bonding also explains their solubility in water. The lower molecular-weight carboxylic acids are completely soluble in water. Those of higher molecular weight are less soluble but are still more soluble than alcohols, ketones, or aldehydes with comparable molecular weights.

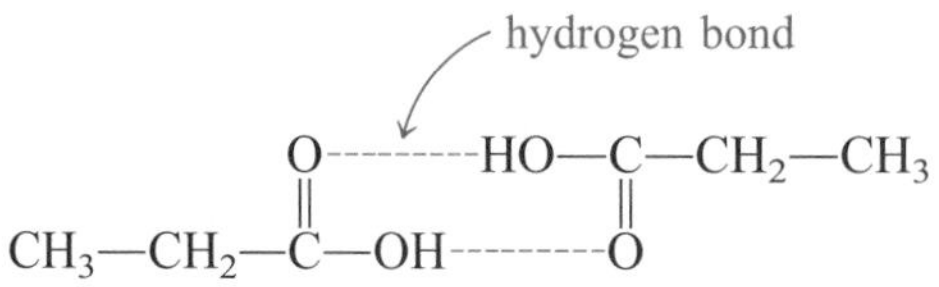

hydrogen bond

$$CH_3-CH_2-\overset{\overset{O}{\|}}{C}-OH \cdots\cdots O=\overset{}{C}(-OH\cdots)-CH_2-CH_3$$

propanoic acid

hydrogen bond

$CH_3-CH_2-C(=O\cdots H-O-H)-O-H\cdots OH_2$, with $H-O-H\cdots O$ (water molecules hydrogen bonded)

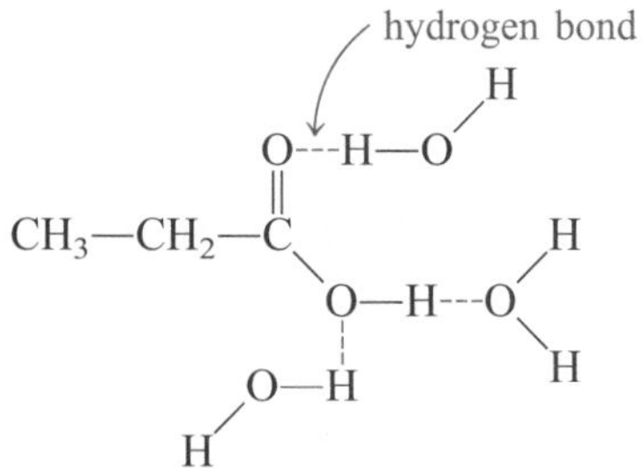

propanoic acid and water

Table 13.4 Common Carboxylic Acids

Common Name	*IUPAC Name*	*Formula*
Formic acid	Methanoic acid	$H{-}\overset{\overset{O}{\parallel}}{C}{-}OH$
Acetic acid	Ethanoic acid	$CH_3{-}\overset{\overset{O}{\parallel}}{C}{-}OH$
Lactic acid	2-Hydroxypropanoic acid	$CH_3\overset{\overset{OH}{\mid}}{C}H{-}\overset{\overset{O}{\parallel}}{C}{-}OH$
Butyric acid	Butanoic acid	$CH_3CH_2CH_2{-}\overset{\overset{O}{\parallel}}{C}{-}OH$
Capric acid	Decanoic acid	$CH_3(CH_2)_8{-}\overset{\overset{O}{\parallel}}{C}{-}OH$
Tartaric acid	2,3-Dihydroxybutanedioic acid	$HO{-}\overset{\overset{O}{\parallel}}{C}{-}\overset{\overset{OH}{\mid}}{C}H{-}\overset{\overset{OH}{\mid}}{C}H{-}\overset{\overset{O}{\parallel}}{C}{-}OH$
Benzoic acid	Benzoic acid	$C_6H_5{-}\overset{\overset{O}{\parallel}}{C}{-}OH$
Salicylic acid	2-Hydroxybenzoic acid	$C_6H_4(OH){-}\overset{\overset{O}{\parallel}}{C}{-}OH$ (OH ortho)

Carboxylic acids are named by dropping the -e ending of the parent alkane and replacing it with -oic acid. The carboxylic acid group is always located at carbon 1. Compounds having two carboxylic acid groups in the molecule also exist. Their names end in -dioic acid. Many carboxylic acids occur naturally and have common names in addition to their IUPAC names. The name of the simplest carboxylic acid (HCOOH) is methanoic acid, but it is usually called formic acid. CH_3COOH is ethanoic acid in IUPAC nomenclature, but it is typically referred to as acetic acid. The simplest dicarboxylic acid (HOOCCOOH) is ethanedioic acid or, more commonly, oxalic acid. Table 13.4 lists the formulas and names of many common acids.

Given the structural formula of a carboxylic acid, state its IUPAC name.

Example 13-4

1. Name the following compound:

$$CH_3{-}CH_2{-}\overset{\overset{CH_3}{\mid}}{C}H{-}CH_2{-}\overset{\overset{CH_3}{\mid}}{C}H{-}\overset{\overset{O}{\parallel}}{C}{-}OH$$

Apply the IUPAC nomenclature rules. ← *See the Question*

(a) The compound has a carboxyl group, so it is a carboxylic acid. Thus its name ends in -oic acid. ← *Think It Through (Execute the Math)*

(b) It has six carbons in the chain containing the carboxyl group, so it is a hexanoic acid.

(c) It has two methyl groups on the main chain, and they are at carbons 2 and 4. (Remember that the carboxyl group carbon is always carbon number 1.)

Prepare the Answer ➤ The name is 2,4-dimethylhexanoic acid.

2. Write the structural formula of 2-phenylbutanoic acid.

S ➤ Convert the IUPAC name into the structural formula.

T • E ➤ (a) Remember that the phenyl group is a benzene ring that is a substituent on another chain. We must attach the benzene ring to carbon 2 of our main chain.

(b) Butanoic acid has a four-carbon chain.

P ➤ The structural formula is therefore

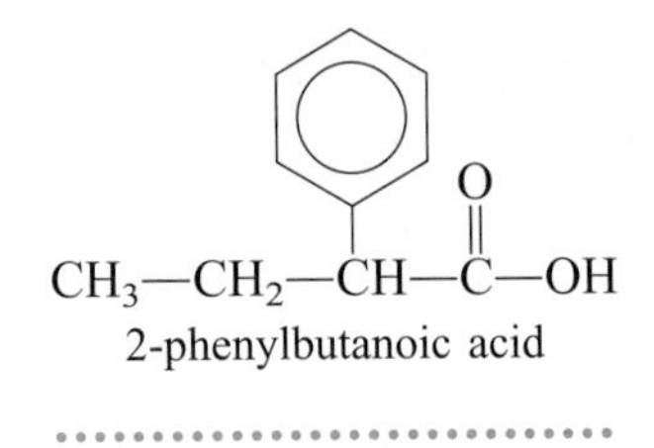

2-phenylbutanoic acid

Check Your Understanding

1. Name the following compounds:

(a)

$$C_6H_5-CH_2-CH_2-\overset{\overset{\displaystyle O}{\|}}{C}-OH$$

(b)

$$CH_3-\underset{}{\overset{\overset{\displaystyle CH_3}{|}}{C}H}-CH_2-CH_2-\overset{\overset{\displaystyle CH_3}{|}}{C}H-CH_2-CH_2-\overset{\overset{\displaystyle O}{\|}}{C}-OH$$

2. Draw the structural formula for each of the following:

(a) hexanedioic acid

(b) 4-ethyl-2-methylheptanoic acid

13.12 Some Important Carboxylic Acids

The name formic acid is derived from the Latin word for ant (*formica*) because ants are a source of this compound. Formic acid (methanoic acid) is very irritating to tissues. In humans, it can be formed in the liver by the oxidation of methanol. When this formic acid enters the bloodstream it can cause severe acidosis, a condition that disrupts the blood's ability to transport oxygen.

Vinegar is a dilute solution of acetic acid (ethanoic acid). You may have heard that spoiled wine can taste like vinegar. This happens because vinegar is formed by the oxidation of the alcohol ethanol. In later chapters we will see that acetic acid is used by living cells in making fatty acids. In fact, a salt of acetic acid (acetyl CoA) plays a key role in the metabolism of carbohydrates, proteins, and lipids.

Health Perspective / Body Odor

Have you ever wished that your blind date had put on a little more Arrid Extra Strength antiperspirant before he came to pick you up? Scientists are just beginning to understand what causes the unpleasant smell we call "body odor." The production of body odor begins with secretions from glands called apocrine glands, located most commonly in the armpit but also in the groin and in the breast. Two types of skin bacteria that live in the armpit use these secretions to produce energy and give off several dozen compounds as waste products. Scientists have isolated one of these compounds, 3-methyl-2-hexenoic acid, as being largely responsible for body odor.

$$CH_3{-}CH_2{-}CH_2{-}\overset{\overset{\displaystyle CH_3}{|}}{C}{=}CH{-}\overset{\overset{\displaystyle O}{\|}}{C}{-}OH$$

3-methyl-2-hexenoic acid

The bacteria (lipophilic diphtheroid) that produce the most 3-methyl-2-hexenoic acid are found in about 90% of male armpits but in only about 60% of females'. This fact has led to the development of different deodorant products for men and women.

Oxalic acid (ethanedioic acid), which is found in many plants, is toxic. In fact, even a small amount of rhubarb leaves contains enough oxalic acid to be poisonous if consumed raw. The cooked stalks of rhubarb are safe to eat, however, because they contain only a very small amount of oxalic acid. Despite the toxicity of oxalic acid, one of its salts is a normal product of the body's metabolism.

Benzoic acid and its sodium salt are used to prevent spoilage from the growth of mold or bacteria in cheese and bread. Salicylic acid, although very similar in structure to benzoic acid, cannot be used in foods because it is too irritating to tissues. Its irritating properties, however, can destroy horny growths like corns and warts and it is used in products to treat these conditions.

Numerous carboxylic acids and their derivatives are involved in the metabolic processes of living organisms. For example, when muscle tissues metabolize glucose, under certain conditions they synthesize lactic acid. This acid is also found in sour milk. Proteins are formed from certain carboxylic acids that also have a nitrogen group attached. The digestion of fats and oils produces fatty acids, which are then oxidized to carbon dioxide and water.

Sodium benzoate is added to these products to retard spoilage.

13.13 Synthesis of Carboxylic Acids

Carboxylic acids can be prepared by the oxidation of primary alcohols or aldehydes in the presence of an oxidizing agent.

Write the equation for the preparation of a carboxylic acid.

$$\underset{\text{1-butanol}}{CH_3{-}(CH_2)_2{-}CH_2OH} \xrightarrow{KMnO_4} \underset{\text{butanoic acid}}{CH_3{-}(CH_2)_2{-}\overset{\overset{\displaystyle O}{\|}}{C}{-}OH}$$

$$\underset{\text{propanal}}{CH_3{-}CH_2{-}\overset{\overset{\displaystyle O}{\|}}{C}H} \xrightarrow{KMnO_4} \underset{\text{propanoic acid}}{CH_3{-}CH_2{-}\overset{\overset{\displaystyle O}{\|}}{C}{-}OH}$$

The oxidation of a primary alcohol to form a carboxylic acid is a two-step process. The first step is the oxidation of the alcohol to form an aldehyde (see Section 13.9). Once the aldehyde is formed, it rapidly oxidizes to form the carboxylic acid. (If, for some reason, the aldehyde is your desired end product, the reaction must be carefully controlled to prevent the rapid oxidation of that aldehyde).

$$\underset{\text{ethanol}}{CH_3{-}CH_2{-}O{-}H} \xrightarrow{-2H} \underset{\text{ethanal}}{CH_3{-}CH{=}O}$$

$$\underset{\text{ethanal}}{CH_3{-}CH{=}O} \longrightarrow \underset{\text{acetic acid}}{CH_3{-}C({=}O){-}O{-}H}$$

13.14 Reactions of Carboxylic Acids

The carboxylic acids are called acids because they have acidic properties. The aqueous solutions of those carboxylic acids that are soluble in water taste sour (for example, vinegar), they turn litmus red, neutralize bases, and have other characteristics we associate with acids. In the case of acetic acid, the reaction in water is

Write the equation for the dissociation of a carboxylic acid.

$$\underset{\text{acetic acid}}{CH_3{-}C({=}O){-}OH} + H_2O \rightleftharpoons \underset{\text{acetate ion}}{CH_3{-}C({=}O){-}O^-} + \underset{\text{hydronium ion}}{H_3O^+}$$

Carboxylic acids are considered to be weak acids. They do not donate a proton as readily as such strong acids as hydrochloric acid and sulfuric acid. In a 0.1 M aqueous solution of acetic acid, only a little more than 1% of the acetic acid molecules donate a proton and exist as the acetate ion.

Write the equation for the preparation of an ester.

Carboxylic acids react with alcohols in the presence of an acid catalyst to yield esters. This reaction, called **esterification,** involves the removal of a molecule of water from a molecule of carboxylic acid and a molecule of alcohol. In general, a reaction in which a molecule of water is removed from two molecules that combine is called a **condensation** reaction. Many of the reactions of biological systems which we will study are condensation reactions.

$$\underset{\text{propanoic acid}}{CH_3{-}CH_2{-}C({=}O){-}OH} + \underset{\text{ethanol}}{HO{-}CH_2{-}CH_3} \overset{H^+}{\rightleftharpoons} \underset{\text{ethyl propanoate}}{CH_3{-}CH_2{-}C({=}O){-}O{-}CH_2{-}CH_3} + H_2O$$

The general esterification reaction is

$$\underset{\text{carboxylic acid}}{R{-}C({=}O){-}OH} + \underset{\text{alcohol}}{HO{-}R'} \underset{\text{acid catalyst}}{\overset{H^+}{\rightleftharpoons}} \underset{\text{ester}}{R{-}C({=}O){-}O{-}R'} + \underset{\text{water}}{H_2O}$$

13.15 Esters

The ester functional group is

$$\overset{\displaystyle O}{\overset{\|}{-C}}-O-C- \quad \text{(ester linkage)}$$

Esters are named by listing the alcohol portion first as one word, with the -anol ending replaced by -yl. For example, if the alcohol portion is ethanol we use ethyl. (Another way to state this is that we use the alkyl group name to indicate the alcohol: methyl, ethyl, isopropyl, and so on.) The acid portion is listed by dropping the -ic acid from its name and replacing it with -ate. For example, acetic acid gives us acetate, propanoic acid gives us propanoate, and so on.

Given the structural formula of an ester, state its IUPAC name.

$$[CH_3-CH_2-CH_2-\overset{\displaystyle O}{\overset{\|}{C}}]-[O-CH_3]$$

carboxylic acid portion — alcohol portion

methyl butanoate

$$\left[H_3C-\underset{}{\overset{\displaystyle CH_3}{\overset{|}{CH}}}-CH_2-O\right]\left[\overset{\displaystyle O}{\overset{\|}{C}}-C_6H_5\right]$$

alcohol portion — carboxylic acid portion

isobutyl benzoate

Example 13-5

Write the structural formula of the ester formed by the reaction of octanoic acid and 2-propanol.

See the Question

We must use nomenclature rules to draw the structures of the carboxylic acid and the alcohol and then apply the general reaction for the formation of esters to this particular case.

Think It Through (Execute the Math)

1. First we write out the structures of the two reactants.
2. We then remove water from the two molecules and bond the oxygen on the alcohol to the carbon of the carbonyl group
3. After drawing the structure we name the alcohol portion first. The alkyl group in this case is the isopropyl group.
4. Then we name the acid portion by dropping -ic acid and replacing it with -ate.
5. The name of the ester is isopropyl octanoate.

Prepare the Answer

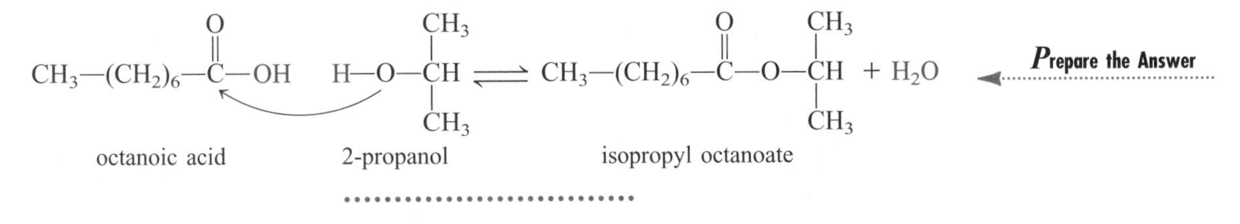

Check Your Understanding

1. Write the structural formula of the carboxylic acid formed by the oxidation of 2-methylbutanal with potassium permanganate.
2. Write the structural formula of the ester formed by the reaction of 3-methylpentanoic acid and 1-propanol.

13.16 Some Important Esters

Many esters have pleasant odors that we associate with flowers and fruits (Table 13.5). Nitroglycerin, the active substance in dynamite, is also an ester. It is formed by the condensation reaction between an inorganic acid (nitric acid) and an alcohol (glycerol). Nitroglycerin is also used to relieve pain from the heart disorder angina pectoris. It acts to enlarge (dilate) smaller blood vessels and to relax smooth muscles in the arteries, which lowers blood pressure and alleviates the pain.

Salicylic acid, found in the bark of the willow tree, is a very irritating substance with a disagreeable taste. An interesting feature of the salicylic acid molecule is that it contains both an acid and an alcohol group. The acid group can form esters with alcohols, and the alcohol group can form esters with other acids. Two of these esters are acetylsalicylic acid and methyl salicylate.

salicylic acid   methyl salicylate (wintergreen)   acetylsalicylic acid (aspirin)

Aspirin, the common name for acetylsalicylic acid, is an analgesic (pain reliever), antipyretic (fever reducer), and an antiinflammatory agent. Although various brands of

Table 13.5 Esters Used as Flavorings

IUPAC Name (Common Name)	*Formula*	*Flavor*
Ethyl methanoate (ethyl formate)	$CH_3CH_2{-}O{-}\overset{O}{\overset{\Vert}{C}}{-}H$	Rum
Pentyl ethanoate (amyl acetate)	$CH_3(CH_2)_4{-}O{-}\overset{O}{\overset{\Vert}{C}}{-}CH_3$	Banana
Octyl ethanoate (octyl acetate)	$CH_3(CH_2)_7{-}\overset{O}{\overset{\Vert}{C}}{-}CH_3$	Orange
Pentyl butanoate (amyl butyrate)	$CH_3(CH_2)_4{-}O{-}\overset{O}{\overset{\Vert}{C}}{-}CH_2CH_2CH_3$	Apricot

aspirin are heavily advertised, repeated testing has shown that there is no difference among competing brands except for their price. Regular aspirin tablets each contain only 5 grains (0.33 g) of acetylsalicylic acid. The rest of the tablet is binder used to put the aspirin in tablet form.

Aspirin is by far the most widely used medicine in the United States. The amount produced each year is equivalent to 100 tablets for every man, woman, and child in the country. Although aspirin has been used since 1899, only recently have scientists begun to understand how it works. Its major effects come from blocking the production of a family of chemicals called prostaglandins (which we discuss in a later chapter). Aspirin is relatively safe, but its continuous use in high doses can cause stomach ulcers and gastrointestinal bleeding. The use of aspirin by children suffering from the viral infections influenza or chicken pox has been connected with the development of a rare, but often fatal disorder known as Reye's syndrome. For this reason, children suffering from these diseases should be given one of the nonaspirin pain relievers.

Methyl salicylate, which is formed by a condensation reaction between salicylic acid and methanol, is also known as oil of wintergreen. It is used in perfumes and candies. It is also used in ointments that cause a mild burning sensation on your skin in an attempt to take your mind off your sore muscles.

13.17 Reactions of Esters

Hydrolysis

The hydrolysis of esters is the reverse of the esterification reaction. So, having learned to make esters, we need only turn that reaction around and we have hydrolysis. Water is a product in esterification, so it is a reactant in hydrolysis. But if the two reactions are the reverse of each other, how do we control which reaction takes place? Notice that the esterification reaction is always written as an equilibrium; therefore, we can use Le Chatelier's principle. If we want to make an ester (esterification), we use a small amount of concentrated acid as a catalyst. If we want to hydrolyze an ester, we use lots of water—that is, a dilute acid. Fats and oils are esters of glycerol and fatty acids. The first step in digestion of the fats and oils we eat is their hydrolysis in the stomach. We study this important process in Chapter 19.

◆ **Write the equation for the hydrolysis of an ester.**

$$\underset{\text{ethyl acetate}}{CH_3-\overset{\overset{O}{\|}}{C}-O-CH_2-CH_3} + HOH \overset{H^+}{\rightleftharpoons} \underset{\text{acetic acid}}{CH_3-\overset{\overset{O}{\|}}{C}-OH} + \underset{\text{ethanol}}{CH_3-CH_2-OH}$$

$$\underset{\text{methyl benzoate}}{C_6H_5-\underset{\underset{O}{\|}}{C}-O-CH_3} + HOH \overset{H^+}{\rightleftharpoons} \underset{\text{benzoic acid}}{C_6H_5-\underset{\underset{O}{\|}}{C}-OH} + \underset{\text{methanol}}{CH_3-OH}$$

Saponification

Esters can also be hydrolyzed when heated in an aqueous solution of strong base such as NaOH or KOH. This reaction, known as saponification, has been used since ancient times. It produces an alcohol and the sodium or potassium salt of the carboxylic acid.

$$CH_3-\overset{\overset{\displaystyle O}{\|}}{C}-O-CH_2-CH_3 + NaOH(aq) \longrightarrow CH_3-\overset{\overset{\displaystyle O}{\|}}{C}-O^- \; Na^+ + CH_3-CH_2-OH$$

ethyl acetate — sodium acetate — ethanol

$$C_6H_5-\underset{\underset{\displaystyle O}{\|}}{C}-O-CH_3 + KOH(aq) \longrightarrow C_6H_5-\underset{\underset{\displaystyle O}{\|}}{C}-O^- \; K^+ + CH_3-OH$$

methyl benzoate — potassium benzoate — methanol

The saponification of fats and oils produces soap. If you spill a basic solution on your skin, you will notice that it has a soapy, slippery feeling. In that case, you are actually saponifying the fats and oils (esters) on your skin. Many centuries ago, people discovered that a solution obtained by letting wood ashes settle in water had this same slippery feeling. Later it was discovered that the combination of animal fats with this solution produced what we now know as soap. In Chapter 16 we learn more about the structure of soap and how it works.

KEY CONCEPTS

Oxygen-containing functional groups create a reactive area on the molecule. Each oxygen-containing functional group adds its own special properties to the molecule.

The hydroxyl group, carbonyl group, and carboxylic acid group are all polar. Because of hydrogen bonding, low-molecular-weight compounds containing these functional groups are soluble in water. The following are the general formulas for the chemical families with these oxygen-containing functional groups:

Family Name	*General Formula*	*Specific Example*
Ether	ROR or ROR′	$CH_3-CH_2-O-CH_2-CH_3$ diethyl ether
Alcohol	ROH	CH_3-CH_2-OH ethanol
Aldehyde	$R\overset{\overset{\displaystyle O}{\|}}{C}H$ or RCHO	$CH_3-\overset{\overset{\displaystyle O}{\|}}{C}H$ ethanal
Ketone	$R\overset{\overset{\displaystyle O}{\|}}{C}R'$ or RCOR′	$CH_3-CH_2-\overset{\overset{\displaystyle O}{\|}}{C}-CH_3$ 2-butanone
Carboxylic acid	$R\overset{\overset{\displaystyle O}{\|}}{C}OH$ or RCOOH	$CH_3-CH_2-\overset{\overset{\displaystyle O}{\|}}{C}-OH$ propanoic acid
Ester	$R\overset{\overset{\displaystyle O}{\|}}{C}OR'$ or RCOOR′ or RCO_2R'	$CH_3-CH_2-\overset{\overset{\displaystyle O}{\|}}{C}-O-CH_3$ methyl propanoate

The relationships between the oxygen-containing functional groups are very important to the study of the carbohydrates, lipids, and proteins. We will see these reactions over and over again as we study biochemistry and the metabolic pathways used by living organisms. The chart summarizes the relationships between compounds containing these functional groups.

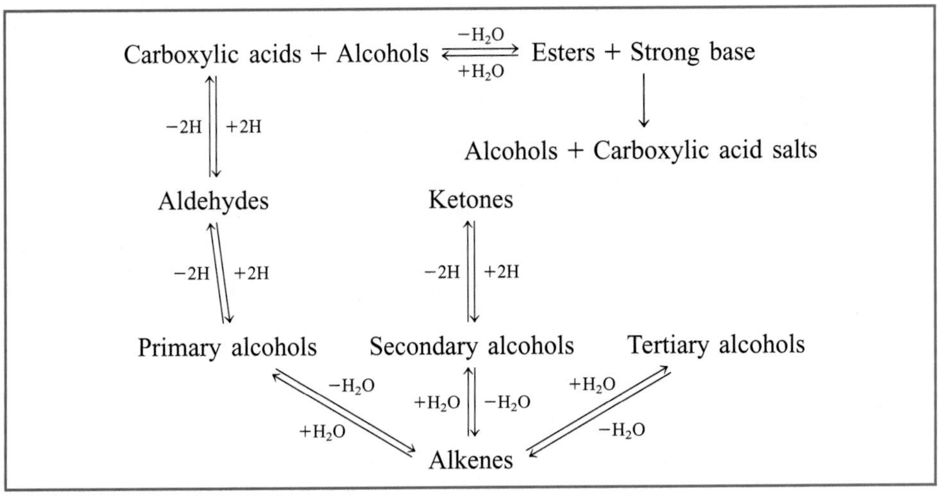

$$\text{Carboxylic acids} + \text{Alcohols} \underset{+H_2O}{\overset{-H_2O}{\rightleftharpoons}} \text{Esters} + \text{Strong base} \longrightarrow \text{Alcohols} + \text{Carboxylic acid salts}$$

$$\text{Carboxylic acids} \underset{+2H}{\overset{-2H}{\rightleftharpoons}} \text{Aldehydes} \underset{+2H}{\overset{-2H}{\rightleftharpoons}} \text{Primary alcohols}$$

$$\text{Ketones} \underset{+2H}{\overset{-2H}{\rightleftharpoons}} \text{Secondary alcohols}$$

$$\text{Primary alcohols} \underset{+H_2O}{\overset{-H_2O}{\rightleftharpoons}} \text{Alkenes}$$

$$\text{Secondary alcohols} \underset{+H_2O}{\overset{-H_2O}{\rightleftharpoons}} \text{Alkenes}$$

$$\text{Tertiary alcohols} \underset{+H_2O}{\overset{-H_2O}{\rightleftharpoons}} \text{Alkenes}$$

Important Equations

The chemical reactions that we discussed in this chapter can be summarized as follows:

Oxidation

Alcohols

primary alcohol ⟶ aldehyde

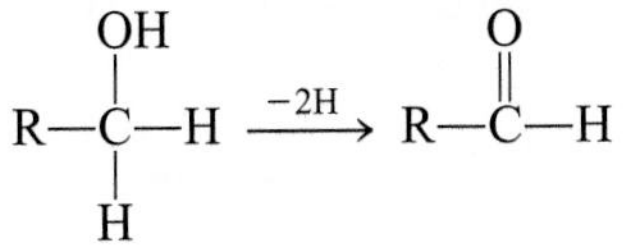

$$R{-}\underset{\displaystyle H}{\overset{\displaystyle OH}{\underset{|}{\overset{|}{C}}}}{-}H \xrightarrow{-2H} R{-}\overset{\displaystyle O}{\overset{\|}{C}}{-}H$$

secondary alcohol ⟶ ketone

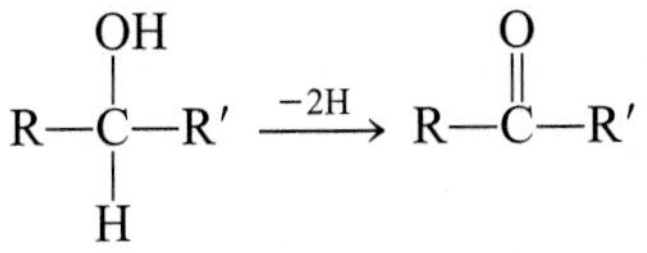

$$R{-}\underset{\displaystyle H}{\overset{\displaystyle OH}{\underset{|}{\overset{|}{C}}}}{-}R' \xrightarrow{-2H} R{-}\overset{\displaystyle O}{\overset{\|}{C}}{-}R'$$

primary alcohol ⟶ carboxylic acid

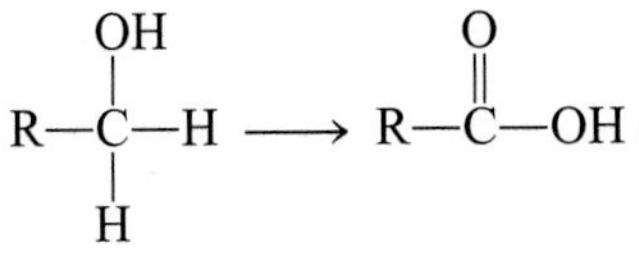

$$R{-}\underset{\displaystyle H}{\overset{\displaystyle OH}{\underset{|}{\overset{|}{C}}}}{-}H \longrightarrow R{-}\overset{\displaystyle O}{\overset{\|}{C}}{-}OH$$

Aldehydes

$$\text{aldehyde} \longrightarrow \text{carboxylic acid}$$

$$R{-}\overset{\overset{O}{\|}}{C}{-}H \longrightarrow R{-}\overset{\overset{O}{\|}}{C}{-}OH$$

Reduction

Aldehydes

$$\text{aldehyde} \longrightarrow \text{primary alcohol}$$

$$R{-}\overset{\overset{O}{\|}}{C}{-}H \xrightarrow{+2H} R{-}\underset{\underset{H}{|}}{\overset{\overset{OH}{|}}{C}}{-}H$$

Ketones

$$\text{ketone} \longrightarrow \text{secondary alcohol}$$

$$R{-}\overset{\overset{O}{\|}}{C}{-}R' \xrightarrow{+2H} R{-}\underset{\underset{H}{|}}{\overset{\overset{OH}{|}}{C}}{-}R'$$

Dehydration

$$\text{alcohol} \longrightarrow \text{alkene} + \text{water}$$

$$RCH_2CH_2OH \longrightarrow RCH{=}CH_2 + H_2O$$

Condensation (Esterification)

$$\text{alcohol} + \text{carboxylic acid} \longrightarrow \text{ester} + \text{water}$$

$$ROH + R'\overset{\overset{O}{\|}}{C}OH \longrightarrow RO\overset{\overset{O}{\|}}{C}R' + H_2O$$

Hydrolysis

$$\text{ester} + \text{water} \longrightarrow \text{alcohol} + \text{carboxylic acid}$$

$$RO\overset{\overset{O}{\|}}{C}R' + H_2O \longrightarrow ROH + HO\overset{\overset{O}{\|}}{C}R'$$

Saponification

$$\text{ester} + \text{strong base} \longrightarrow \text{alcohol} + \text{carboxylic acid salt}$$

$$RO\overset{\overset{O}{\|}}{C}R' + NaOH \longrightarrow ROH + R'\overset{\overset{O}{\|}}{C}O^-\ Na^+$$

REVIEW PROBLEMS

Section 13.1

1. Alcohols contain the —OH group attached to a carbon. What do we mean by the term *functional group?*
2. What does the term *physiologically active* mean when applied to an organic compound?

Section 13.2

3. Write the structural formulas for the following ethers:
   (a) dipropyl ether
   (b) isobutyl methyl ether
   (c) cyclohexyl ethyl ether
4. Name the following ethers:
   (a) $CH_3—CH_2—O—CH_3$
   (b) 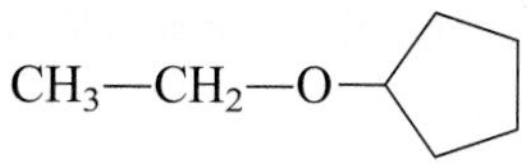 $CH_3—CH_2—O—$(cyclopentyl)
   (c) $CH_3—CH_2—CH_2—CH_2—CH_2—O—CH_3$
   (d) $CH_3—CH_2—O—CH_2—CH_3$

Section 13.3

5. State the IUPAC name for each of the following compounds, and identify each as either a primary, secondary, or tertiary alcohol:

(a) 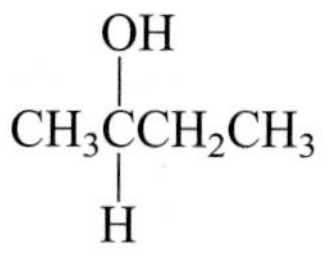 $$CH_3\overset{\overset{\displaystyle OH}{|}}{\underset{\underset{\displaystyle H}{|}}{C}}CH_2CH_3$$

(b) (benzene ring)$—CH_2OH$

(c) $$CH_3CH_2—\overset{\overset{\displaystyle CH_3}{|}}{\underset{\underset{\displaystyle CH_3}{|}}{C}}—OH$$

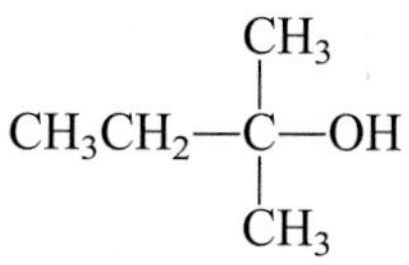

(d) 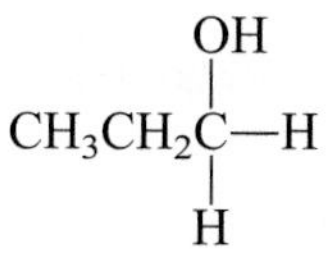 $$CH_3CH_2\overset{\overset{\displaystyle OH}{|}}{\underset{\underset{\displaystyle H}{|}}{C}}—H$$

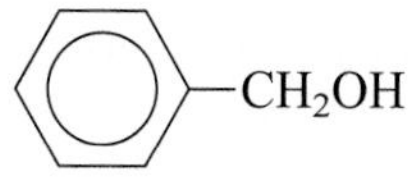

(e) 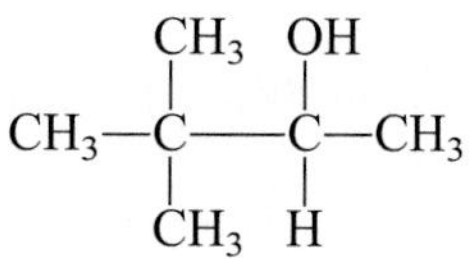 $$CH_3—\overset{\overset{\displaystyle CH_3}{|}}{\underset{\underset{\displaystyle CH_3}{|}}{C}}—\overset{\overset{\displaystyle OH}{|}}{\underset{\underset{\displaystyle H}{|}}{C}}—CH_3$$

(f) 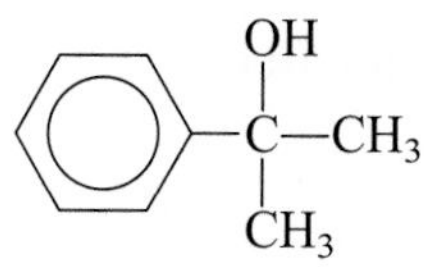 $$(\text{benzene ring})—\overset{\overset{\displaystyle OH}{|}}{\underset{\underset{\displaystyle CH_3}{|}}{C}}—CH_3$$

6. Draw the structures for each of the following alcohols:
   (a) a tertiary alcohol of 5 carbons
   (b) a secondary alcohol of formula C_3H_8O
   (c) a primary alcohol of four carbons
7. Write the structural formulas for the following alcohols:
   (a) 3-chloro-1-propanol
   (b) 4,4-dimethyl-2-pentanol
   (c) 1,3-pentanediol
   (d) 2-methylcyclohexanol

8. Name the following alcohols:

(a) 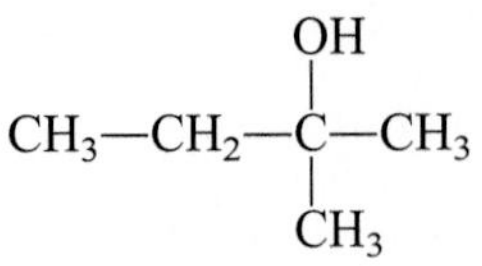 $$\begin{array}{ccccccc} & & & & OH & & \\ & & & & | & & \\ CH_3 & - & CH_2 & - & C & - & CH_3 \\ & & & & | & & \\ & & & & CH_3 & & \end{array}$$

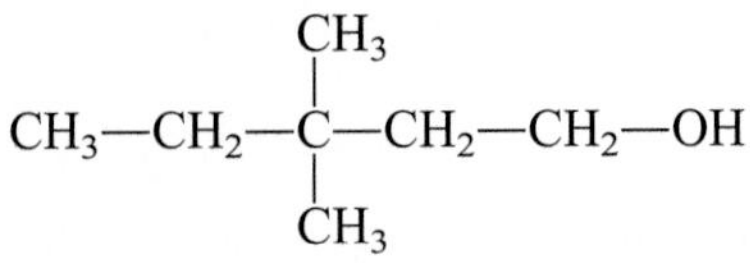

(b) 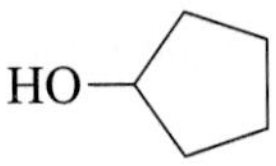 HO—(cyclopentane ring)

(c) $$\begin{array}{ccccccccccc} & & & & CH_3 & & & & & & \\ & & & & | & & & & & & \\ CH_3 & - & CH_2 & - & C & - & CH_2 & - & CH_2 & - & OH \\ & & & & | & & & & & & \\ & & & & CH_3 & & & & & & \end{array}$$

(d) 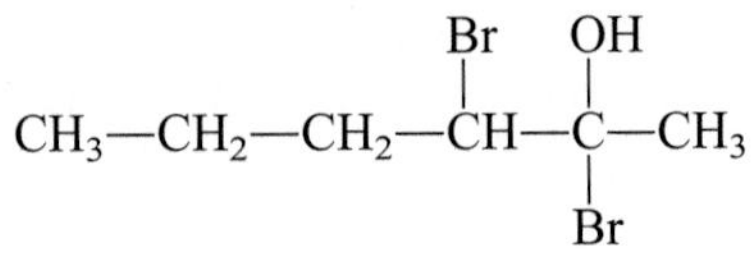 $$\begin{array}{ccccccccccc} & & & & & & Br & & OH & & \\ & & & & & & | & & | & & \\ CH_3 & - & CH_2 & - & CH_2 & - & CH & - & C & - & CH_3 \\ & & & & & & & & | & & \\ & & & & & & & & Br & & \end{array}$$

9. Explain why 2-hexanol is only slightly soluble in water, whereas 2,3-hexanediol is water-soluble.

10. Why is it that propane boils at −45°C and methyl ether at −23°C, but ethanol boils at 78°C?

11. Eight isomeric alcohols have the molecular formula $C_5H_{12}O$.

(a) Draw the structural formulas for the eight alcohols, and name each compound.

(b) Classify each of the alcohols as primary, secondary, or tertiary.

12. The molecular formula $C_4H_{10}O$ has seven possible structural formulas, some of which are alcohols and some ethers. Draw the seven structural formulas, and give their names.

Section 13.4

13. Why is methanol used to denature ethanol?

Section 13.5

14. Draw the structural formula of the alcohol that is the product of the hydration of each of the following alkenes:

(a) 2-butene **(c)** cyclohexene

(b) 3-hexene

15. Draw the structural formula of an alkene that can be converted by hydration to each of the following alcohols:

(a) 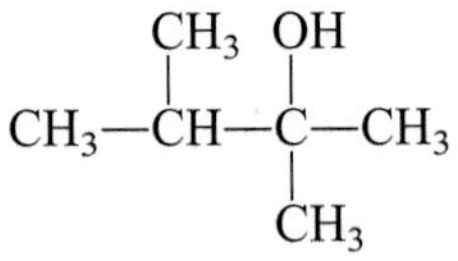 $$\begin{array}{ccccccc} & & CH_3 & & OH & & \\ & & | & & | & & \\ CH_3 & - & CH & - & C & - & CH_3 \\ & & & & | & & \\ & & & & CH_3 & & \end{array}$$

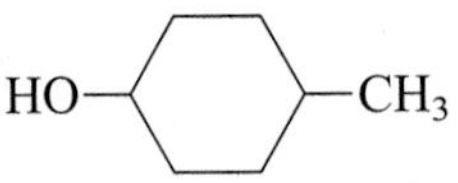

(b) 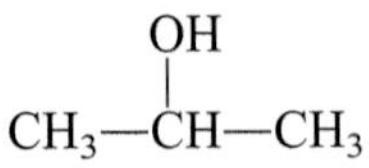 $$\begin{array}{ccccc} & & OH & & \\ & & | & & \\ CH_3 & - & CH & - & CH_3 \end{array}$$

(c) HO—(cyclohexane ring)—CH_3

Section 13.6

16. Write the structure of the product of the dehydration of **(a)** 1-butanol, **(b)** 3-hexanol.

17. Write the structure of the aldehydes or ketones that form from the oxidation of the following alcohols:

(a) 1-propanol **(e)** cyclopentanol

(b) 2-propanol
(c) 3-methyl-1-butanol
(d) 3-methyl-2-butanol
(f) ethanol
(g) 1-pentanol
(h) 2-hexanol

18. Draw the structural formula of an alcohol that, when dehydrated, forms each of the following alkenes:

(a) 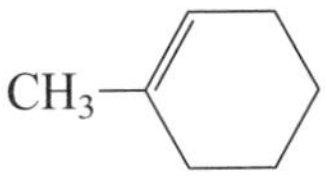 CH_3—(cyclohexene ring)

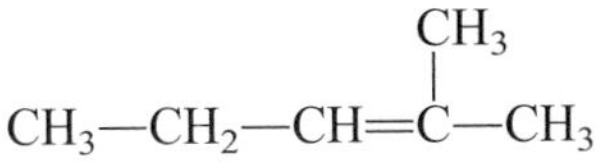

(b) 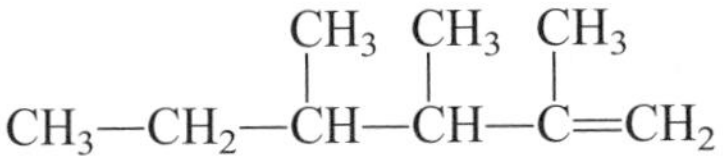 $CH_3—CH_2—\underset{|}{\overset{CH_3}{CH}}—\underset{|}{\overset{CH_3}{CH}}—\overset{CH_3}{C}=CH_2$

(c) $CH_3—CH_2—CH=\overset{CH_3}{C}—CH_3$

19. Draw the structural formula of the alcohol that, when oxidized, forms each of the following aldehydes or ketones:

(a) 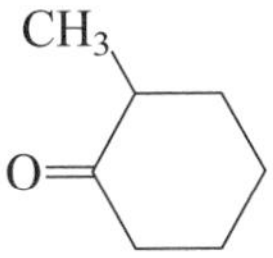 O=(cyclohexane ring) with CH_3

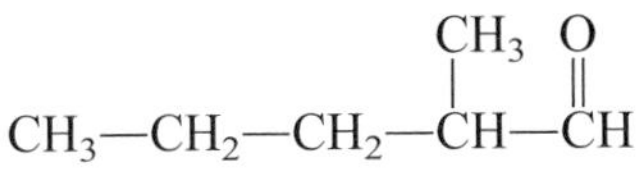

(b) 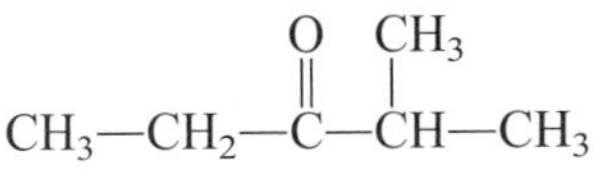 $CH_3—CH_2—\overset{O}{\overset{\|}{C}}—\overset{CH_3}{CH}—CH_3$

(c) $CH_3—CH_2—CH_2—\overset{CH_3}{CH}—\overset{O}{\overset{\|}{CH}}$

(d) 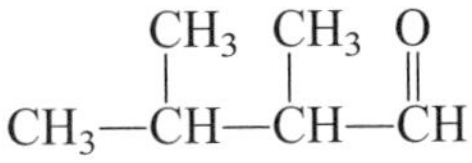 $CH_3—\overset{CH_3}{CH}—\overset{CH_3}{CH}—\overset{O}{\overset{\|}{CH}}$

Section 13.7

20. Seven aldehydes and ketones have the molecular formula $C_5H_{10}O$. Draw the formulas for the seven, and name each structure.

21. Name the compounds in problem 19.

Section 13.8

22. The presence of acetone in the urine may be an indication of what disease?

23. What weight of formaldehyde is present in 450 mL of formalin?

Section 13.10

24. Write the structure of the carboxylic acids that form from the oxidation of the following aldehydes:

(a) acetaldehyde
(b) butanal
(c) hexanal
(d) benzaldehyde

25. Draw the structural formula of the aldehyde that, when oxidized, forms each of the following carboxylic acids:

(a) 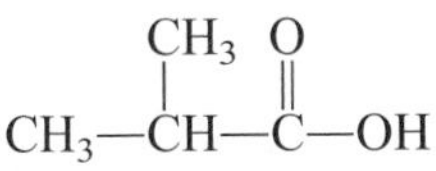 $CH_3—\overset{CH_3}{CH}—\overset{O}{\overset{\|}{C}}—OH$

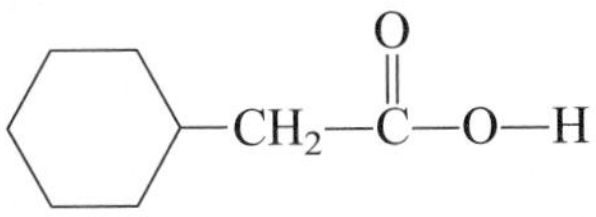

(b) 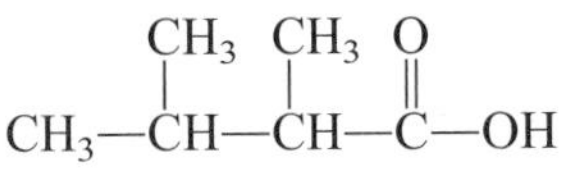 $CH_3—\overset{CH_3}{CH}—\overset{CH_3}{CH}—\overset{O}{\overset{\|}{C}}—OH$

(c) (cyclohexane ring)—$CH_2—\overset{O}{\overset{\|}{C}}—O—H$

26. Write the structure of the hemiacetals or hemiketals that form between the following pairs of compounds:
    - (a) acetaldehyde and methanol
    - (b) acetone and ethanol
    - (c) butanone and methanol
    - (d) formaldehyde and ethanol
    - (e) propanal and methanol
    - (f) acetone and 1-propanol
27. For each of the following hemiacetals and hemiketals, draw the structures of the alcohol and the aldehyde or ketone from which it can be formed:

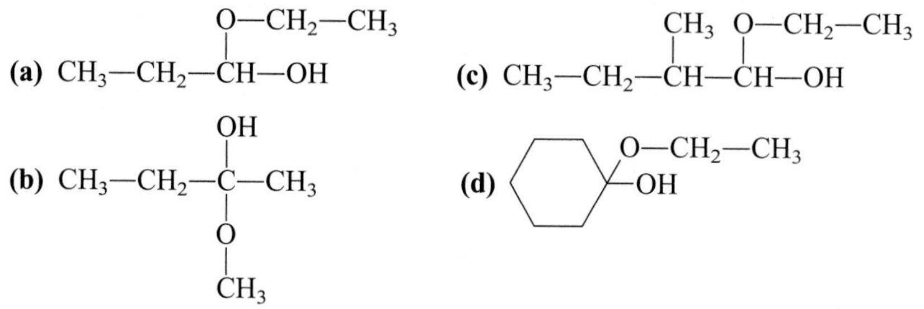

(a)
$$\begin{array}{ccccc} & & O{-}CH_2{-}CH_3 & & \\ & & | & & \\ CH_3{-} & CH_2{-} & CH & {-}OH & \end{array}$$

(b)
$$\begin{array}{cccc} & & OH & \\ & & | & \\ CH_3{-} & CH_2{-} & C & {-}CH_3 \\ & & | & \\ & & O & \\ & & | & \\ & & CH_3 & \end{array}$$

(c)
$$\begin{array}{ccccc} & & CH_3 & O{-}CH_2{-}CH_3 & \\ & & | & | & \\ CH_3{-} & CH_2{-} & CH{-} & CH & {-}OH \end{array}$$

(d) [cyclohexane ring] with $O{-}CH_2{-}CH_3$ and $-OH$ on the same ring carbon

Section 13.11

28. Draw the structural formula for each of the following carboxylic acids:
    - (a) octanoic acid (caprylic acid)
    - (b) 3-chloro-2,3-dimethylbutanoic acid
    - (c) 3-phenylpentanoic acid
    - (d) butanedioic acid
29. Name the following carboxylic acids:

(a)
$$\begin{array}{ccc} & O & \\ & \| & \\ CH_3{-}(CH_2)_5{-} & C & {-}OH \end{array}$$

(b)
$$\begin{array}{cccccc} & & CH_3{-}CH{-}CH_3 & & O & \\ & & | & & \| & \\ CH_3{-}CH_2{-} & CH_2{-} & CH{-} & CH_2{-} & C & {-}OH \end{array}$$

(c)
$$\begin{array}{cccccc} & & CH_3 & & & \\ & & | & & & \\ & & CH_2 & & O & \\ & & | & & \| & \\ CH_3{-} & CH_2{-} & CH{-} & CH_2{-} & C & {-}OH \end{array}$$

(d)
$$\begin{array}{cccccc} & & Cl & Cl & O & \\ & & | & | & \| & \\ CH_3{-} & CH_2{-} & CH{-} & CH{-} & C & {-}OH \end{array}$$

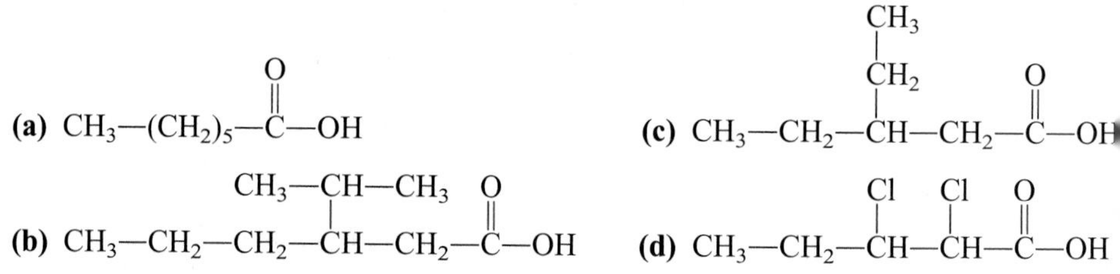

Section 13.12

30. Name four carboxylic acids that can be found in the home.
31. What carboxylic acid is produced in the muscles by the metabolism of glucose?

Section 13.13

32. Draw the structural formula for the carboxylic acid that is produced when each of the following compounds is oxidized:
    - (a) butanal
    - (b) 1-pentanol
    - (c) 2,3-dimethylhexanal
    - (d) 3,4-dimethyl-1-hexanol
33. Draw the structural formulas for both the alcohol and the aldehyde that can be oxidized to form each of the following carboxylic acids:
    - (a) pentanoic acid
    - (b) benzoic acid
    - (c) 2,3-dichloropropanoic acid
    - (d) 2,2-dimethylbutanoic acid

Section 13.14

34. Write the equation for the reaction of each of the following acids with water:

(a) benzoic acid **(b)** hydrochloric acid **(c)** 2-methylpropanoic acid

35. Compare the strength of benzoic acid with that of hydrochloric acid.

Section 13.15

36. Draw the structural formula for each of the following esters, and identify the acid and the alcohol portions of the molecule:

(a) benzyl acetate **(c)** phenyl 2-methylpentanoate **(e)** acetylsalicylic acid

(b) methyl formate **(d)** methyl salicylate

37. Write the equation for the esterification reaction between each of the following, and name the products that are formed:

(a) butanoic acid and methanol **(d)** acetic acid and 2-butanol

(b) benzoic acid and ethanol **(e)** salicylic acid and ethanol

(c) formic acid and 1-propanol **(f)** salicylic acid and propanoic acid

38. Identify each of the following as either an alcohol, ether, carboxylic acid, ester, aldehyde, or ketone:

(a) $CH_3\overset{\overset{CH_3}{|}}{C}H-\overset{\overset{O}{\|}}{C}-OH$

(h) $CH_3CH_2\underset{\underset{CH_2CH_3}{|}}{\overset{\overset{OH}{|}}{C}}CH_2CH_2CH_3$

(b) $CH_3CH_2CH_2OCH_2CH_3$

(c) $CH_3CH_2O\overset{\overset{O}{\|}}{C}CH_2CH_2CH_3$

(i) $CH_3CH_2CH_2CH_2\overset{\overset{O}{\|}}{C}H$

(d) $CH_3\overset{\overset{CH_3}{|}}{C}H-\overset{\overset{O}{\|}}{C}H$

(j) $CH_3\underset{\underset{CH_3}{|}}{C}H-\overset{\overset{O}{\|}}{C}-\underset{\underset{CH_3}{|}}{C}HCH_3$

(e) $CH_3CH_2CH_2\overset{\overset{O}{\|}}{C}CH_2CH_3$

(k) $C_6H_5-O\overset{\overset{O}{\|}}{C}CH_2CH_2CH_2CH_3$ (benzene ring drawn)

(f) $CH_3\overset{\overset{OH}{|}}{C}HCH_2\overset{\overset{O}{\|}}{C}OH$

(g) $C_6H_5-O-C_6H_5$ (benzene rings drawn)

(l) $CH_3CH_2\underset{\underset{\underset{\underset{CH_3}{|}}{CHCH_3}}{|}}{\overset{\overset{CH_3}{|}}{C}H}\overset{\overset{OH}{|}}{C}HCH_2CH_2CH_2CH_3$

39. Name each of the compounds shown in problem 38.

Section 13.16

40. Name three esters that have medicinal uses, and list what they are used for.

41. How many moles of acetylsalicylic acid are in each regular (5-grain) aspirin tablet?

Section 13.17

42. For each of the following esters write (1) the structural formula and (2) the equation for the hydrolysis of the ester, naming each of the products formed.

(a) ethyl benzoate
(b) isopropyl butanoate
(c) octyl acetate
(d) *tert*-butyl propanoate

43. Write the equations for the saponification of the following esters by sodium hydroxide, and write the formulas (condensed or structural) for each of the products formed.

(a) methyl acetate
(b) ethyl propanoate
(c) *sec*-butyl benzoate
(d) diethyl oxalate

Applied Problems

44. The formation of a hemiacetal appears to come about when the oxygen on the alcohol attaches to the carbon on the carbonyl group of the aldehyde. Why do you think the alcohol is attracted to this group?

45. When two functional groups that can react with each other are on the same compound, they may undergo an intramolecular reaction, resulting in a cyclic compound. 5-Hydroxyhexanal readily forms a stable six-member cyclic hemiacetal. Draw the structural formula for this cyclic hemiacetal.

46. A student is handed a vial containing a colorless liquid and is told that the liquid is either 2-pentanol or pentanal. Using chemical reactions, how can the student determine which of these two compounds is in the vial?

47. Many flavoring agents are esters of carboxylic acids. How do you prepare the flavoring agent for wintergreen candies in a laboratory?

48. Hydrocortisone is a steroid used to reduce swelling due to such causes as exposure to poison ivy. Identify the functional groups present in the hydrocortisone molecule.

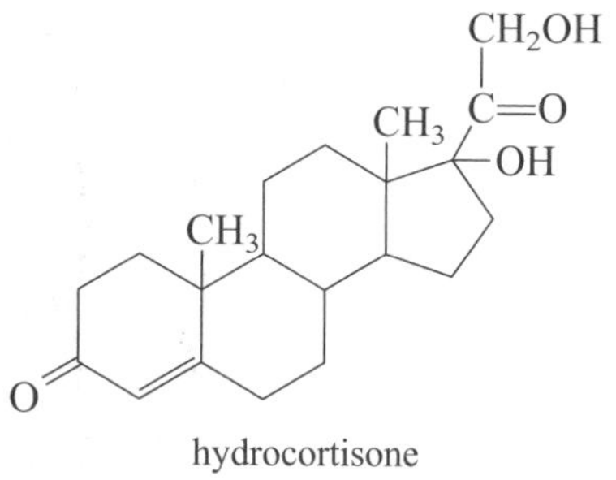

hydrocortisone

49. Estrone and estradiol are two important female sex hormones. How is estradiol converted into estrone?

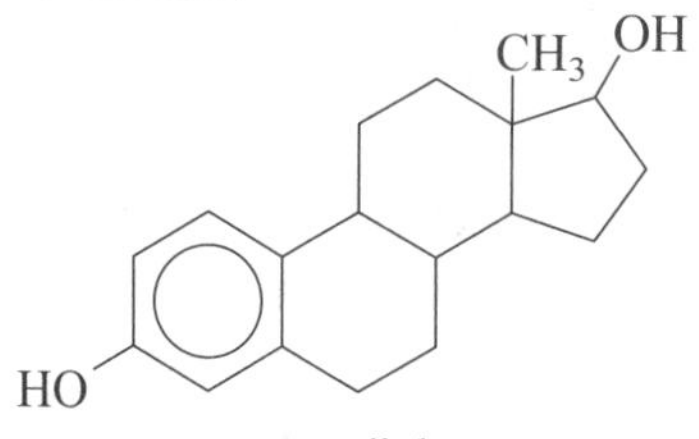

estradiol

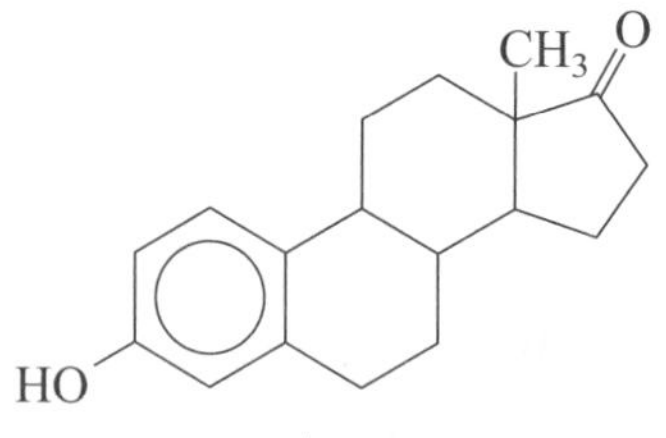

estrone

Chapter 14

Organic Compounds Containing Nitrogen

*K*athy had been staring at the patient's chart for minutes without really registering what it said. The young man had been brought up to the intensive care unit from the emergency room suffering from a brain hemorrhage. The staff in the emergency room had worked for 2 hours to stabilize the patient. Kathy knew that her careful monitoring of the dopamine infusion rate and the cardiac monitor was crucial to his survival. So why was she having so much trouble concentrating?

She had come to work 3 hours early because the day shift was a nurse short, and with two heart-attack patients and the accident victim they were really stretched to the limit. Kathy hadn't had time for her dinner break. She had simply grabbed a cheese pastry and some coffee between taking patient blood pressures and replacing several IVs. Now she felt as if she were working in a fog. She ate a candy bar to see if it would help.

At the nurses' station, Kathy told her friend Ann about the trouble she was having concentrating. To Kathy's surprise, Ann suggested that her problem might be due to what she had been eating. Ann told Kathy about the research being conducted by Dr. Judith Wurtman of the Massachusetts Institute of Technology (MIT) on the effect of foods on brain neurotransmitters. Dr. Wurtman's research showed that you could influence your mind and mood with the foods that you eat. You could use foods to help you stay awake and alert even during hours when you normally would be sleeping.

The brain is made up of millions of nerve cells. For them to work together, bits of information (or messages) must pass from one cell to another. The messages are moved through the cell by electric impulses and are passed between cells by chemicals called neurotransmitters. Three of the chemical neurotransmitters are synthesized by nerve cells in the brain from the foods we eat: dopamine, norepinephrine, and serotonin. Research by Dr. Wurtman and others has shown that dopamine and norepinephrine are alertness chemicals, and serotonin is a calming chemical. Increased production of dopamine and norepinephrine by the brain helps you think quickly and react more rapidly, but increased production of serotonin makes you more relaxed and sluggish and helps you to fall asleep.

These neurotransmitters are synthesized from nitrogen-containing compounds called amino acids. Amino acids are the building blocks of the proteins in our food. Dopamine and norepinephrine are synthesized from the amino acid tyrosine, and serotonin from the amino acid tryptophan.

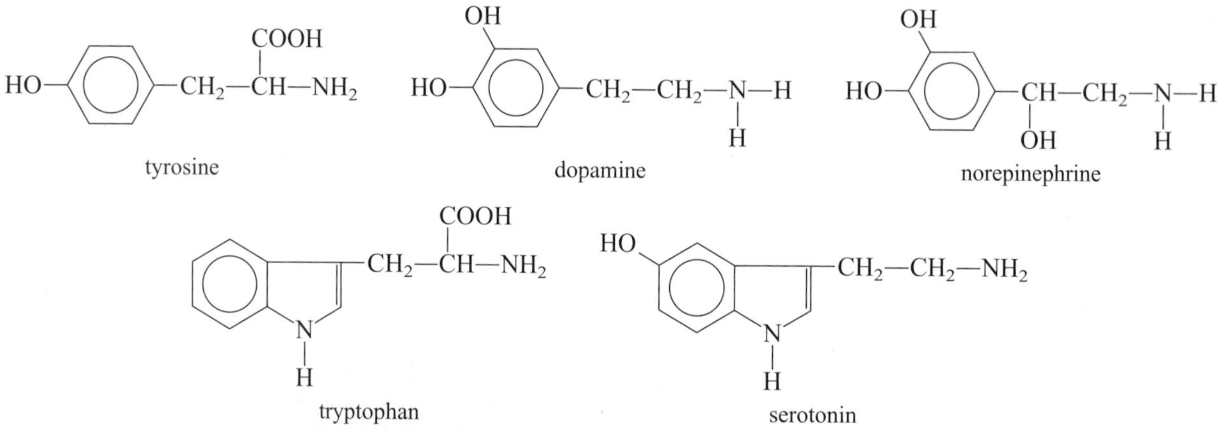

tyrosine dopamine norepinephrine

tryptophan serotonin

When you eat a piece of fish, for example, the protein in the fish is broken down by the digestive process into amino acids that are absorbed into the blood. Because of the way in which amino acids are absorbed by the brain, more tyrosine in comparison to tryptophan enters the brain after eating protein and, therefore, the production of norepinephrine and dopamine increase. A person feels more motivated and mentally alert after a meal of protein, thus, in this instance, endorsing the old saw that "fish is brain food."

But what if you are too stimulated and your mind is racing? What can you do to slow your brain down or get to sleep after a mentally demanding day? You need to increase the amount of tryptophan getting into the brain and, therefore, the synthesis of serotonin. You can do this by eating carbohydrates, foods that are sweet or starchy like candy or pasta. A meal of carbohydrate causes the concentration of tryptophan in the blood to increase relative to the concentration of tyrosine. This causes the amount of tryptophan entering the brain to increase, resulting in a rise in the production of serotonin. By eating a cheese pastry and a candy bar, both high in carbohydrates, Kathy was increasing the production of serotonin in her brain, producing a sluggish, sleepy feeling.

Dr. Wurtman suggests that individuals who must remain alert in the late evening hours use food to counteract the body's natural tendency to wind down and become sluggish and sleepy. By eating a small snack high in protein in the late afternoon, and having a late dinner, one can remain mentally alert until very late in the evening. The evening meal should be low in calories; it should contain protein and carbohydrate

but very little fat. The key to increasing the brain's production of dopamine and norepinephrine, the alertness neurotransmitters, is to start by eating the protein part of the meal before the carbohydrates. On the other hand, when you want to wind down and go to sleep, Dr. Wurtman suggests a meal high in carbohydrate (such as pasta or a muffin with jelly) or a snack of fig newtons, popcorn, or gumdrops.

The next evening, Kathy tried Dr. Wurtman's advice. She had a light meal of broiled chicken breast and a salad, and had a carton of low-fat yogurt at break. She was pleasantly surprised to find herself more alert and much more able to cope with the demands of the intensive care unit.

Tyrosine, tryptophan, dopamine, norepinephrine, and serotonin are all very important compounds whose structures include functional groups that contain nitrogen. In the previous few chapters we discussed important organic compounds composed of carbon, hydrogen, and oxygen. In this chapter, we examine the classes of compounds that have nitrogen-containing functional groups.

14.1 Functional Groups Containing Nitrogen

Nitrogen is the fourth most abundant element in the human body, making up 1.4% of the total number of atoms. As a member of Group VA on the periodic table, it has five valence electrons. To achieve an octet in its outermost energy level, nitrogen forms three covalent bonds, consisting of either three single bonds, a double and a single bond, or one triple bond (Fig. 14.1). Because nitrogen can form strong bonds to carbon, oxygen, and hydrogen, it is found in many different arrangements in organic molecules. We limit our study to only two of the nitrogen-containing functional groups: the amines and the amides (Table 14.1).

Classify organic compounds having nitrogen-containing functional groups.

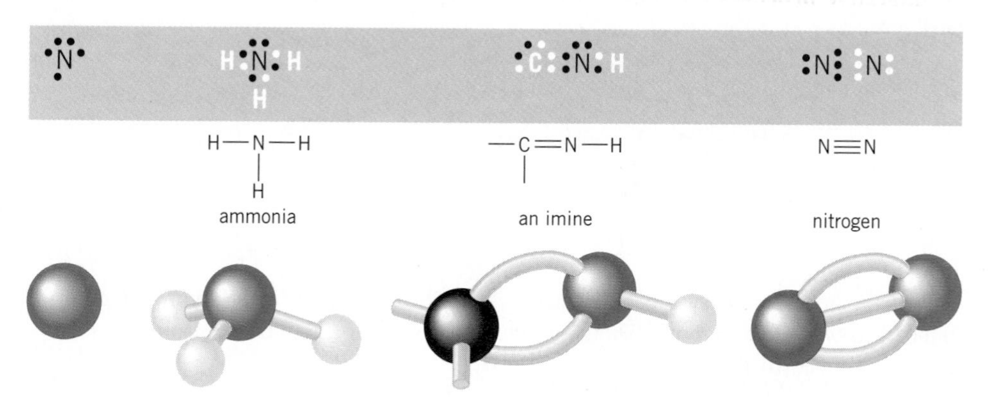

Figure 14.1 Nitrogen can form single, double, and triple covalent bonds.

Table 14.1 **Nitrogen-Containing Functional Groups**

Functional Group	*Class of Compound*	*Typical Compound*
$R—NH_2$	Amine	Methylamine $H_3C—NH_2$
$R—C(=O)—NH_2$	Amide	Ethanamide $CH_3—C(=O)—NH_2$

14.2 Amines

Amines are organic compounds that contain the **amino** functional group: $—NH_2$. The amines are known for their strong, pungent, and often fishy odors. The amine methylamine (CH_3NH_2) smells like spoiled fish. The natural decay of dead organisms produces amines called ptomaines.

$$H_2N—CH_2—CH_2—CH_2—CH_2—NH_2$$
1,4-diaminobutane
(putrescine)

$$H_2N—CH_2—CH_2—CH_2—CH_2—CH_2—NH_2$$
1,5-diaminopentane
(cadaverine)

The common names of these compounds should give you a clue about their smells. It was once thought that such amines were responsible for the vomiting and diarrhea that results from eating spoiled food, so these symptoms were referred to as ptomaine poisoning. It is now known that our reactions to spoiled food result from more complicated causes.

Given the structural formula of an amine, identify it as a primary, secondary, or tertiary amine.

Amines are classified as primary, secondary, or tertiary—depending on the number of carbon atoms bonded directly to the nitrogen atom. In a primary amine, one carbon atom and two hydrogen atoms are directly bonded to the nitrogen. In a secondary amine, two carbons and one hydrogen bond directly to the nitrogen. In a tertiary amine, all three atoms bonded to the nitrogen are carbons.

$$H—\overset{\overset{\displaystyle H}{|}}{N}—CH_3$$
methylamine
(a primary amine)

$$CH_3—\overset{\overset{\displaystyle H}{|}}{N}—CH_3$$
dimethylamine
(a secondary amine)

$$CH_3—\overset{\overset{\displaystyle CH_3}{|}}{N}—CH_3$$
trimethylamine
(a tertiary amine)

Although we classify both alcohols and amines as primary, secondary, and tertiary, notice that the rules for classifying compounds containing these two functional groups are quite different.

$$CH_3—CH_2—\underset{\underset{\displaystyle H}{|}}{\overset{\overset{\displaystyle OH}{|}}{C}}—CH_3$$
2-butanol

$$CH_3—CH_2—\overset{\overset{\displaystyle H}{|}}{N}—CH_2—CH_3$$
diethylamine

2-Butanol is a secondary alcohol because the carbon bonded to the hydroxyl group is bonded to two other carbons. Diethylamine is a secondary amine because there are two carbons bonded to the nitrogen.

$$CH_3—\overset{\overset{\displaystyle OH}{|}}{CH}—CH_3$$
isopropyl alcohol
(2-propanol)

$$CH_3—\overset{\overset{\displaystyle NH_2}{|}}{CH}—CH_3$$
isopropylamine

Even though these two compounds look very similar, isopropyl alcohol is a secondary alcohol, whereas isopropyl amine is a primary amine. The carbon to which the hydroxyl group in isopropyl alcohol is bonded is itself bonded to two carbons, so this is a secondary alcohol. Only one carbon is bonded to the nitrogen in isopropylamine, so isopropylamine is a primary amine.

Primary amines are named by first identifying the group attached to the nitrogen and then adding the suffix -amine. For examine, $CH_3CH_2NH_2$ is ethylamine, and $CH_3CH_2CH_2CH_2NH_2$ is butylamine. Secondary and tertiary amines are named in the same way. Each group on the nitrogen is identified and then listed alphabetically, using the prefixes di- and tri- where appropriate. For example:

Write the structural formula of an amine when given its IUPAC name, and state the name when given its structural formula.

$$CH_3{-}CH_2{-}NH{-}CH_3$$
ethylmethylamine

$$CH_3CH_2{-}NH{-}CH_2CH_3$$
diethylamine

$$CH_3CH_2{-}\underset{}{\overset{\overset{\displaystyle CH_2CH_3}{|}}{N}}{-}CH_2CH_3$$
triethylamine

Amines containing other functional groups are often named by using the prefix amino- to indicate the $-NH_2$ group. For example:

$$H_2N{-}CH_2{-}CH_2{-}\overset{\overset{\displaystyle O}{\|}}{C}{-}OH$$
3-aminopropanoic acid

$$CH_3{-}CH_2{-}CH_2{-}\overset{\overset{\displaystyle NH_2}{|}}{CH}{-}\overset{\overset{\displaystyle O}{\|}}{CH}$$
2-aminopentanal

$$H_2N{-}CH_2{-}(CH_2)_4{-}CH_2{-}NH_2$$
1,6-diaminohexane

An *N* appearing before the name of a substituent group indicates that the group named after the *N* is attached to the nitrogen of the amino group. In Table 14.2, the name *N*-methylaniline indicates that the methyl is attached to the nitrogen rather than to the benzene ring.

Table 14.2 Amines

Name	*Formula*	*Type*
Propylamine	$CH_3CH_2CH_2NH_2$	Primary
Isopropylamine	$CH_3{-}\overset{\overset{\displaystyle CH_3}{|}}{CH}{-}NH_2$	Primary
Ethylmethylamine	$CH_3CH_2{-}\underset{\underset{\displaystyle CH_3}{|}}{N}{-}H$	Secondary
Dimethylethylamine	$CH_3CH_2{-}\underset{\underset{\displaystyle CH_3}{|}}{N}{-}CH_3$	Tertiary
Aniline	$C_6H_5{-}NH_2$	Primary
N-Methylaniline	$C_6H_5{-}\underset{\underset{\displaystyle H}{|}}{N}{-}CH_3$	Secondary
N,N-Dimethylaniline	$C_6H_5{-}\underset{\underset{\displaystyle CH_3}{|}}{N}{-}CH_3$	Tertiary

Health Perspective / Amphetamines

The structure of the secondary amine epinephrine (adrenalin) was discovered in the early 20th century (see Section 14.9). Using this structure, researchers were able to prepare a series of chemically related compounds with very similar drug effects. This group of compounds became known as *amphetamines.* Their powerful action in stimulating the central nervous system has made them the standard by which all other stimulants are measured.

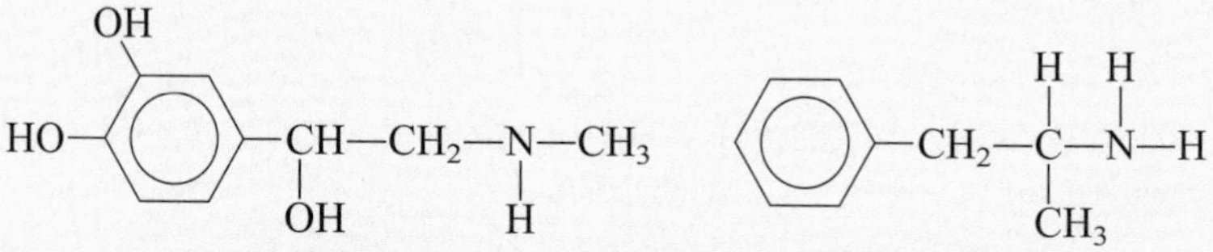

$$\text{HO}-C_6H_3(\text{OH})-\underset{|}{\text{CH}}(\text{OH})-CH_2-\text{N}(\text{H})-CH_3 \qquad C_6H_5-CH_2-\text{C}(\text{H})(CH_3)-\text{N}(\text{H})-\text{H}$$

epinephrine amphetamine

The amphetamine Benzedrine was first used in 1932 in inhalers. These inhalers cleared blocked nasal passages by shrinking mucous membranes. Unfortunately, as the effect of the inhalers wore off, some users experienced even more severe blockage. Perhaps of even greater concern was the discovery that Benzedrine was a euphoriant, a substance that creates a feeling of well-being and buoyancy. Drug users quickly found a new way to get high by extracting the amphetamine from the inhaler paper. This was done by using carbonated beverages as a source of carbonic acid to remove the basic amphetamine.

Amphetamines next found use in the treatment of two disorders that were complete opposites. The first was an epileptic seizure disorder called narcolepsy in which a person falls asleep for no apparent reason. Patients were able to stay awake and function while taking amphetamines because of their stimulant effect. The second application came in the treatment of hyperkinetic children. For some reason, amphetamines acted to calm down the hyperactive child.

A side effect of amphetamines reported by narcolepsy patients, the loss of appetite, led to the use of amphetamines in diet pills. As sales of amphetamines increased, more of their side effects became recognized. These included heart palpitations, headaches, increased blood pressure, and wakefulness. This last side effect gave rise to the extensive use of amphetamines by soldiers in World War II to stay awake and reduce fatigue. After the war, amphetamines were sold by the millions to everyone from students to truck drivers.

On continued use, tolerance to amphetamines develops. The user eventually needs up to 300 mg daily, a huge increase over the initial 10 to 20 mg/day. Many people who regularly use amphetamines discover that they can't sleep at night and have to take ''downer'' drugs like barbiturates. A vicious cycle often develops in which the person needs to take amphetamines in the morning to come up from the downers, and barbiturates at night to come down from the stimulant.

Because of widespread abuse of these drugs, the U.S. Food and Drug Administration in 1965 began to regulate the manufacture and distribution of amphetamines. Prescriptions for these drugs must be written on special forms and are not refillable. Illegal laboratories have continued to produce amphetamines in large quantities, however, especially for those who use them for euphoric effects. The preferred illegal drug is methamphetamine, often called ''crystal,'' which is injected (''mainlined'') directly into the bloodstream. These illegal laboratories are called ''meth labs'' because of the methamphetamine and methylmethamphetamine they produce.

Example 14-1

1. Name and classify the following compound:

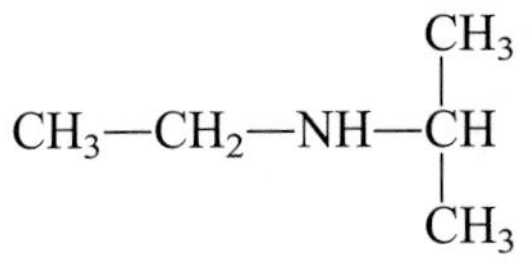

$$CH_3-CH_2-NH-\underset{\underset{\displaystyle CH_3}{|}}{\overset{\overset{\displaystyle CH_3}{|}}{CH}}$$

We are asked to apply the IUPAC nomenclature rules by determining how many groups are attached to the nitrogen. ◄ *See the Question*

(a) This compound contains nitrogen in an amine functional group, so it is an amine. ◄ *Think It Through (Execute the Math)*

(b) The alkyl groups bonded to the nitrogen of the amine functional group are the ethyl group and the isopropyl group.

(c) The groups are named alphabetically: ethyl before isopropyl.

The name is ethylisopropylamine. Because two carbons are bonded to the nitrogen, this is a secondary amine. ◄ *Prepare the Answer*

2. Name and classify the following compound:

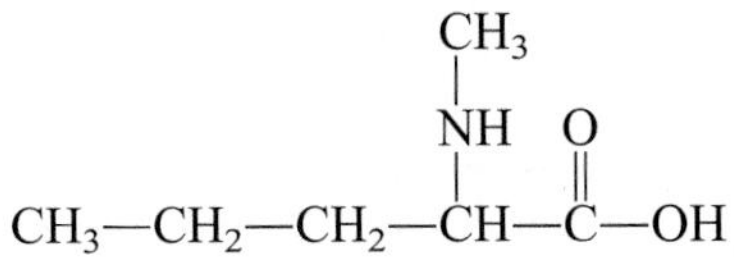

$$CH_3-CH_2-CH_2-\underset{}{\overset{\overset{\displaystyle CH_3}{|}}{\overset{\displaystyle NH}{\overset{|}{CH}}}}-\overset{\overset{\displaystyle O}{\|}}{C}-OH$$

This problem asks for the same information as the previous example. ◄ *S*

(a) This compound contains two functional groups: an amino group and a carboxylic acid group. ◄ *T • E*

(b) The carbon chain containing the carboxylic acid group is five carbons long, so this is a pentanoic acid.

(c) The alkyl group attached to the amine nitrogen is a methyl group, so this substituent is named *N*-methylamino-.

(d) Because the carboxylic acid carbon is always carbon 1, the *N*-methylamino group is on carbon 2.

The name of this compound is 2-(*N*-methylamino)pentanoic acid. Because two carbons are bonded to the nitrogen, this is a secondary amine. ◄ *P*

3. Draw the structure of ethylmethylisopropylamine.

We must use the nomenclature rules to develop and draw the structure of this compound. ◄ *S*

(a) The name indicates that three alkyl groups are attached to the nitrogen of the amino group. So this is a tertiary amine. ◄ *T • E*

(b) The three alkyl groups are: ethyl ($—CH_2CH_3$), methyl ($—CH_3$), and isopropyl

$$\begin{array}{c} CH_3 \\ | \\ CH_3—CH— \end{array}$$

P The structure of this compound is

$$\begin{array}{l} \quad\;\; CH_3\;\; CH_3 \\ \quad\;\;\; | \qquad\; | \\ CH_3—CH—N—CH_2CH_3 \end{array}$$

Check Your Understanding

1. Name and classify the following amines:

(a) $H_2N—CH_2—(CH_2)_6—CH_3$

(b) $CH_3—CH_2—N(—CH_2CH_3)—CH_2—CH_3$

$$\begin{array}{c} CH_3 \\ | \\ CH_2 \\ | \\ CH_3—CH_2—N—CH_2—CH_3 \end{array}$$

(c) $H_2N—CH_2—CH_2—CH_2—OH$

(d)

$$\begin{array}{l} \qquad\quad NH_2\;\; CH_3\;\; O \\ \qquad\quad\;\; | \qquad\; | \qquad \| \\ CH_3—CH—CH—CH \end{array}$$

(e)

$$\begin{array}{r} O \\ \| \\ CH_3—NH—CH_2—CH_2—C—OH \end{array}$$

2. Write the structural formula for each of the following amines:

(a) diethylisobutylamine

(b) cyclohexylmethylamine

(c) 1,4-diaminohexane

(d) *N*-ethyl-*N*-methylaniline

14.3 The Amines as Bases

When placed in water, ammonia reacts as follows:

$$NH_3 + H_2O \rightleftharpoons NH_4^+ + OH^-$$

The ammonia acts as a base and accepts a proton, whereas the water acts as an acid and donates a proton. Because ammonia is a weak base, only a small percentage of the ammonia molecules actually accept a proton. The amines can be viewed as derivatives of ammonia in which one or more of the protons on the ammonia are replaced by alkyl groups. It should not be surprising to learn that amines are also weak bases. In water,

Health Perspective / Drugs in the Form of Their Salts

Many drugs used today are amines, which are bases. Most of these drugs are large molecules with high molecular weights, and they are insoluble in aqueous body fluids. For these drugs to be absorbed into the body and carried to the target organs or cells, they must be given in a water-soluble form. As you know, when a base combines with an acid, they react to form a salt. The salt formed is usually water-soluble. For this reason, many of the amine drugs are converted to their salts by the addition of an acid. As an example,

$$\underset{\text{(a base)}}{\text{Amphetamine}} + \text{sulfuric acid} \longrightarrow \underset{\text{(a salt)}}{\text{amphetamine sulfate}}$$

In addition to making the drug water-soluble, the salt form is usually more stable. Many different acids are used for salt formation, including sulfuric, hydrochloric, nitric, tartaric, benzoic, and salicylic acids. When hydrochloric acid is used, the resulting salt is called a hydrochloride, such as procaine hydrochloride (Novocain).

The term *free* base refers to the base after it is converted back to basic form from the salt. It is now "free" of its salt form. The term is most often heard in reference to the "freebasing" of cocaine. The volatile base cocaine is brought into this country as the more stable hydrochloride salt and is sold on the streets in that form. Users often convert it back to the more volatile free base form in order to inhale the fumes by smoking it.

methylamine accepts a proton and becomes the methylammonium ion. The aqueous solutions of amines are basic and have a pH greater than 7.

Write equations for the reactions of amines with water and with a carboxylic acid.

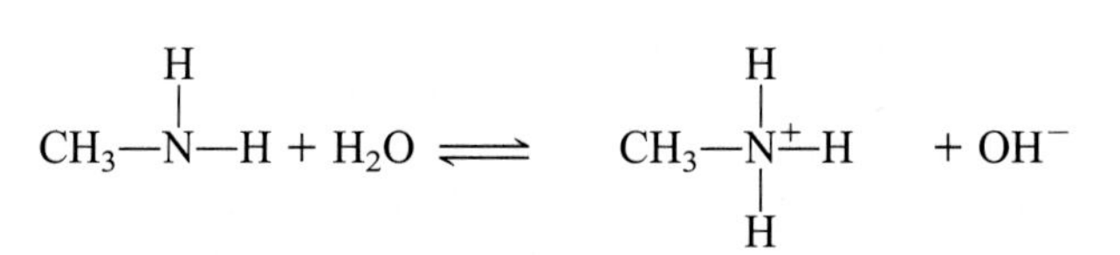

methylamine methylammonium ion

In Chapter 13, we saw that carboxylic acids can act as acids in water. Carboxylic acids and amines can react in a typical acid–base reaction. For example, methylamine and acetic acid can react to give a salt, methylammonium acetate.

$$\underset{\text{acetic acid}}{CH_3COOH} + \underset{\text{methylamine}}{CH_3NH_2} \rightleftharpoons \underset{\text{methylammonium acetate}}{CH_3COO^- \; H_3N^+CH_3}$$

14.4 Important Amine Derivatives

Write equations for the reactions of amines with inorganic acids and with alkyl halides.

Amines can also react with inorganic acids to produce salts.

$$\underset{\text{methylamine}}{H_2N{-}CH_3} + HCl \longrightarrow \underset{\text{methylammonium chloride}}{CH_3NH_3^+\ \ Cl^-}$$

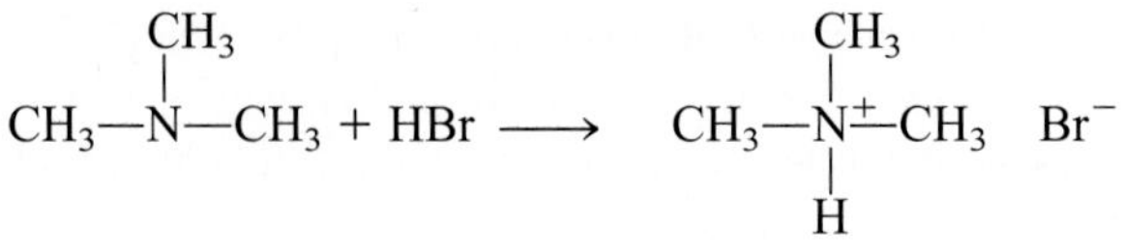

$$\underset{\text{trimethylamine}}{CH_3{-}\overset{\displaystyle CH_3}{\overset{|}{N}}{-}CH_3} + HBr \longrightarrow \underset{\text{trimethylammonium bromide}}{CH_3{-}\underset{H}{\underset{|}{\overset{\displaystyle CH_3}{\overset{|}{N^+}}}}{-}CH_3\ \ Br^-}$$

Organic ammonium ions, known as quaternary ammonium salts, are formed when tertiary amines react with alkyl halides such as methyl iodide.

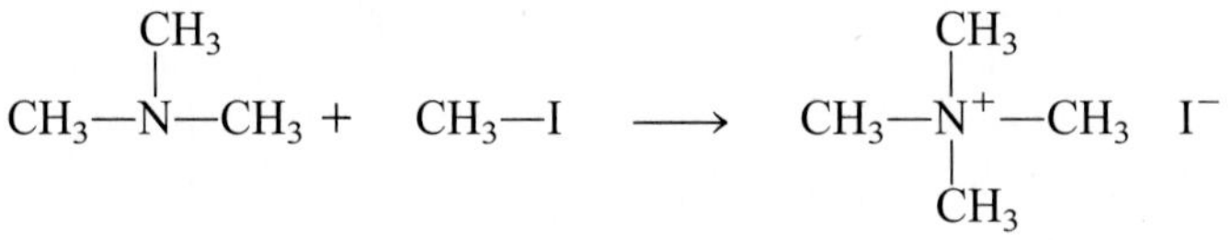

$$\underset{\text{trimethylamine}}{CH_3{-}\overset{\displaystyle CH_3}{\overset{|}{N}}{-}CH_3} + \underset{\text{methyl iodide}}{CH_3{-}I} \longrightarrow \underset{\text{tetramethylammonium iodide}}{CH_3{-}\underset{CH_3}{\underset{|}{\overset{\displaystyle CH_3}{\overset{|}{N^+}}}}{-}CH_3\ \ I^-}$$

Choline is an important compound that contains an organic ammonium ion. Choline and its derivatives, such as acetylcholine, are critical to the structure of the cell and the transmission of nerve impulses. Release of acetylcholine by a nerve triggers muscle contraction. For muscle contraction to occur again, the acetylcholine must be removed. This is done by the hydrolysis of acetylcholine to choline and acetic acid. The enzyme acetylcholinesterase is the catalyst for this reaction.

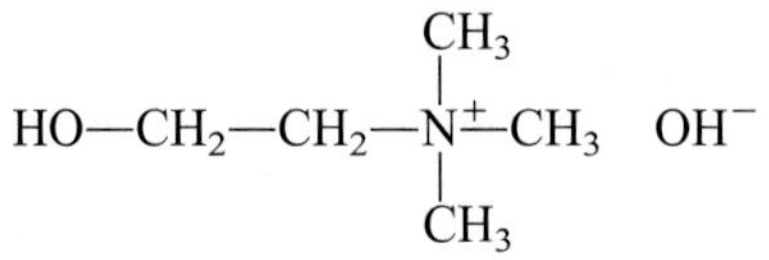

$$\underset{\text{choline}}{HO{-}CH_2{-}CH_2{-}\underset{CH_3}{\underset{|}{\overset{\displaystyle CH_3}{\overset{|}{N^+}}}}{-}CH_3\ \ OH^-} \qquad \underset{\text{acetylcholine}}{CH_3{-}\overset{\displaystyle O}{\overset{\|}{C}}{-}O{-}CH_2{-}CH_2{-}\underset{CH_3}{\underset{|}{\overset{\displaystyle CH_3}{\overset{|}{N^+}}}}{-}CH_3\ \ OH^-}$$

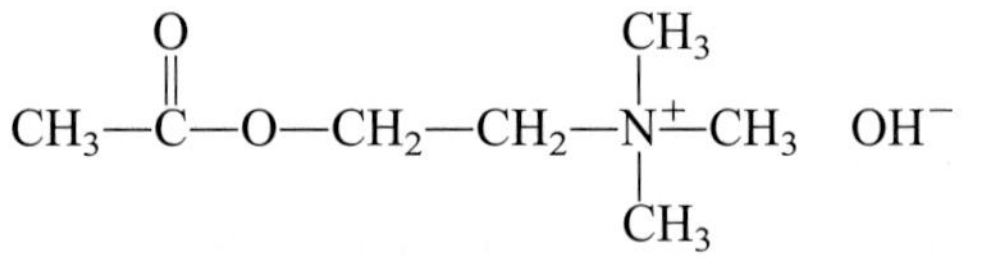

Nitrates, which are naturally present in high levels in such foods as celery and beets, can be converted to nitrites by the bacteria in saliva. Nitrites are also commonly added to food commercially to enhance color and to preserve poultry, other meats, and fish. No matter what their source, nitrites may react in the stomach with amines naturally present in foods, drugs, and cigarette smoke to produce compounds called nitrosamines. Nitrosamines have been shown to cause cancer in all types of laboratory animals and are suspected of causing human cancers. The general formula for this class of highly potent carcinogens is

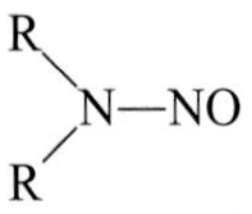

$$\underset{\text{nitrosamine}}{R_2N{-}NO} \qquad \underset{N\text{-nitrosopyrrolidine}}{(CH_2)_4N{-}NO}$$

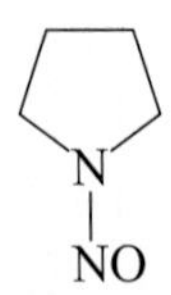

N-Nitrosopyrrolidine has been found in cooked bacon. Cigarette smoke has been shown to contain at least one *N*-nitrosamine.

14.5 Amides

Amides are derivatives of carboxylic acids and amines, in which the amino group ($—NH_2$) replaces the hydroxyl (—OH) group. One or both of the hydrogens on the amino group can be replaced by alkyl groups (Table 14.3). The general formula for the amides can be written as:

$$\overset{O}{\overset{\|}{RC}}NH_2 \quad \text{or} \quad \overset{O}{\overset{\|}{RC}}NHR' \quad \text{or} \quad \overset{O}{\overset{\|}{RC}}NR'R''$$

In Section 13.14, we saw that an ester is formed by the condensation reaction in which a molecule of water is removed as an alcohol combines with a carboxylic acid. An amide is formed by a similar reaction between an amine (or ammonia) and a carboxylic acid. The single covalent bond between the nitrogen and the carbon of the carbonyl group is called the **amide linkage.** This very stable bond is also found in proteins, where it is called the peptide bond.

$$CH_3—\overset{O}{\overset{\|}{C}}—OH \qquad CH_3—\overset{O}{\overset{\|}{C}}—NH_2$$

acetic acid — acetamide (arrow to C—N bond: amide linkage)

The solubility of the amides follows the same pattern we saw with the other carbonyl groups. The polar amide group causes the lower molecular-weight amides to be soluble in water. Amides of increasing molecular weight are less and less water-soluble. The melting and boiling points of the amides are higher than that of alkanes with comparable molecular weights because of hydrogen bonding between the polar amide groups on adjacent molecules. Remember that a hydrogen bonded to a nitrogen, oxygen, or fluorine can form a hydrogen bond to another nitrogen, oxygen, or fluorine.

Table 14.3 Amides

Name	*Structural Formula*	*Name*	*Structural Formula*
Formamide	$H—\overset{O}{\overset{\|}{C}}—NH_2$	Acetanilide	$CH_3\overset{O}{\overset{\|}{C}}—\overset{H}{\overset{\|}{N}}—C_6H_5$
Acetamide	$CH_3\overset{O}{\overset{\|}{C}}—NH_2$	Benzamide	$C_6H_5—\overset{O}{\overset{\|}{C}}—NH_2$
Propanamide	$CH_3CH_2\overset{O}{\overset{\|}{C}}—NH_2$		
Butanamide	$CH_3CH_2CH_2\overset{O}{\overset{\|}{C}}—NH_2$	Urea	$H_2N—\overset{O}{\overset{\|}{C}}—NH_2$

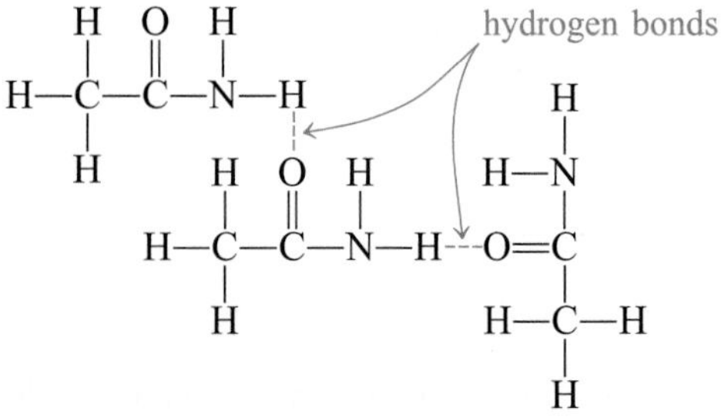

$$\begin{array}{c} \text{H-C(H)(H)-C(=O)-N(H)-H} \cdots \text{O=C(CH}_3\text{)-N(H)-H} \cdots \text{O=C(CH}_3\text{)-NH}_2 \end{array}$$

hydrogen bonds

hydrogen bonding in acetamide

Write the structural formula of an amide when given its IUPAC name, and state the name when given its structural formula.

Amides that have two hydrogens attached to the nitrogen are named by dropping the -ic or -oic ending from the name of the acid and adding -amide. For example:

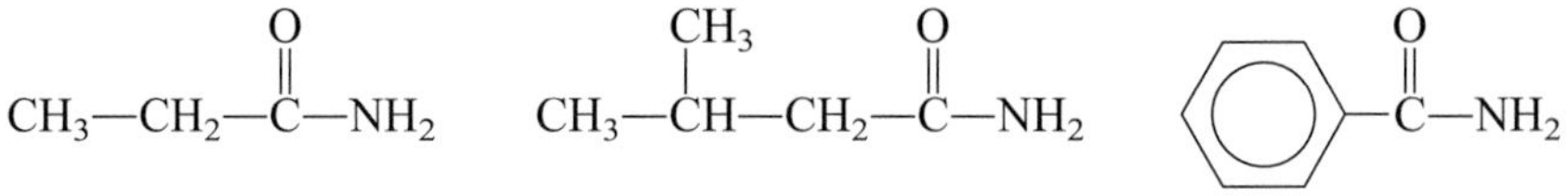

$$CH_3-CH_2-\overset{\overset{\displaystyle O}{\|}}{C}-NH_2 \qquad CH_3-\overset{\overset{\displaystyle CH_3}{|}}{C}H-CH_2-\overset{\overset{\displaystyle O}{\|}}{C}-NH_2 \qquad C_6H_5-\overset{\overset{\displaystyle O}{\|}}{C}-NH_2$$

propanamide — 3-methylbutanamide — benzamide

Naming amides in which one or both hydrogens in the amino group are replaced by an alkyl group is only slightly more complicated. Remember that a capital *N* means that a group is bonded to the nitrogen rather than somewhere else in the molecule. To name these *N*-substituted amides we follow the same procedure as previously, but we indicate the group or groups on the nitrogen by using the *N*. Notice the following example of *N*-methylbutanamide and compare it with 3-methylbutanamide.

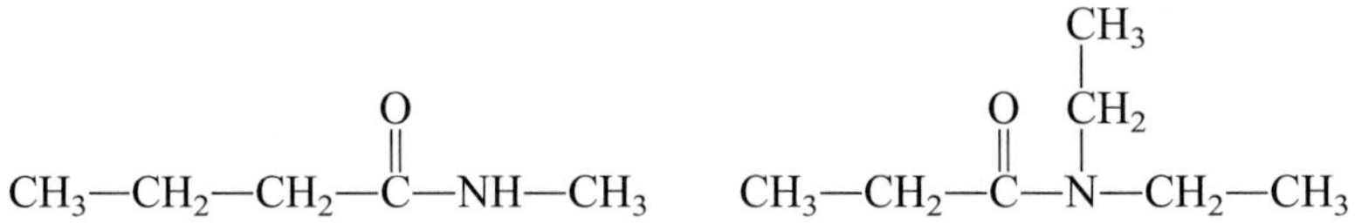

$$CH_3-CH_2-CH_2-\overset{\overset{\displaystyle O}{\|}}{C}-NH-CH_3 \qquad CH_3-CH_2-\overset{\overset{\displaystyle O}{\|}}{C}-\overset{\overset{\displaystyle CH_2-CH_3}{|}}{N}-CH_2-CH_3$$

N-methylbutanamide — *N*,*N*-diethylpropanamide

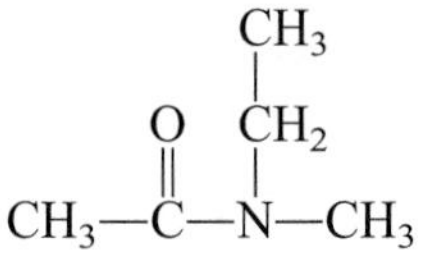

$$CH_3-\overset{\overset{\displaystyle O}{\|}}{C}-\overset{\overset{\displaystyle CH_2-CH_3}{|}}{N}-CH_3$$

N-ethyl-*N*-methylacetamide
(*N*-ethyl-*N*-methylethanamide)

Example 14-2

1. Name the following compound:

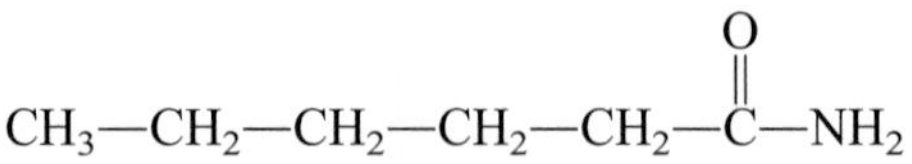

$$CH_3-CH_2-CH_2-CH_2-CH_2-\overset{\overset{\displaystyle O}{\|}}{C}-NH_2$$

See the Question We must apply IUPAC nomenclature rules to name this compound.

Think It Through (Execute the Math)

(a) The acid from which this amide is derived has six carbons, so it is hexanoic acid.

(b) The amide is unsubstituted (there are only hydrogens on the nitrogen), so we drop -oic acid and add -amide.

Prepare the Answer The name is hexanamide.

2. Now let's try an *N*-substituted amide. Name this compound:

$$CH_3{-}CH_2{-}CH_2{-}\overset{\overset{\large O}{\|}}{C}{-}NH{-}CH_2{-}CH_3$$

Apply IUPAC nomenclature rules as in the previous example. ◄ ***S***

(a) The acid from which this amide is derived has four carbons, so it is butanoic acid. ◄ ***T • E***

(b) The amide must be a butanamide.

(c) The alkyl group attached to the nitrogen is an ethyl group ($—CH_2CH_3$).

(d) The alkyl group is indicated by writing *N*-ethyl before the name of the amide.

The name is *N*-ethylbutanamide. ◄ ***P***

3. Now let's convert a name into a structural formula. Write the structural formula for *N,N*-dimethylpropanamide.

We are being asked to use the nomenclature rules to convert a name into an extended structural formula. ◄ ***S***

(a) The acid from which propanamide is derived is propanoic acid, the three-carbon acid. ◄ ***T • E***

$$CH_3{-}CH_2{-}\overset{\overset{\large O}{\|}}{C}{-}OH$$

(b) Because this is an amide, we replace the hydroxyl group with an amino group.

$$CH_3{-}CH_2{-}\overset{\overset{\large O}{\|}}{C}{-}NH_2$$

(c) The *N,N*-dimethyl indicates that we must replace the two hydrogens on the nitrogen with methyl groups.

The structure of *N,N*-dimethylpropanamide is ◄ ***P***

$$CH_3{-}CH_2{-}\overset{\overset{\large O}{\|}}{C}{-}\overset{\overset{\large CH_3}{|}}{N}{-}CH_3$$

Check Your Understanding

1. Name the following compounds:

(a)

$$CH_3{-}\overset{\overset{\large O}{\|}}{C}{-}NH_2$$

(b)

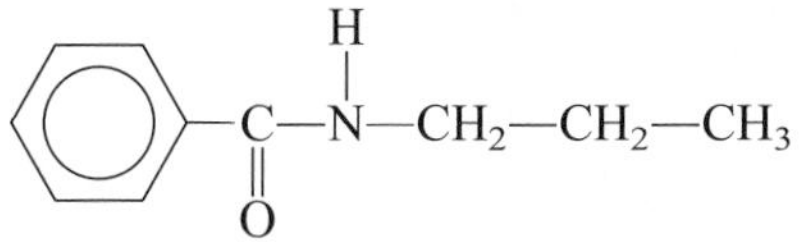

$$C_6H_5{-}\underset{\underset{\large O}{\|}}{C}{-}\overset{\overset{\large H}{|}}{N}{-}CH_2{-}CH_2{-}CH_3$$

(c)

$$CH_3{-}CH_2{-}CH_2{-}CH_2{-}\overset{\overset{\large O}{\|}}{C}{-}NH{-}CH_2{-}CH_2{-}CH_3$$

(d)
$$\begin{array}{c} CH_3 \\ | \\ O\ \ CH_2 \\ \|\ \ \ | \\ CH_3-C-N-CH_2-CH_3 \end{array}$$

2. Write structural formulas for the following compounds:
   (a) pentanamide
   (b) *N*-isopropylbutanamide
   (c) *N,N*-diethylpentanamide
   (d) *N*-butyl-*N*-ethylbenzamide

14.6 Some Important Amides

Many amides are important to living organisms. Urea is a waste product of the body's metabolism (see Table 14.3). When the body breaks down compounds containing carbon, hydrogen, and oxygen, it produces carbon dioxide and water as waste products. The water is eliminated through the breath, sweat, and urine, and the carbon dioxide is eliminated through the lungs. When the body breaks down compounds containing nitrogen, the additional waste product ammonia (NH_3) is also produced. Ammonia, however, is toxic to living tissues. Fish and other aquatic animals can eliminate the ammonia directly into the water, but land animals must convert the ammonia into less toxic substances that can be transported through the body and eliminated. Our bodies convert ammonia into urea, which can accumulate in the blood without harm before being eliminated in the urine by the kidneys.

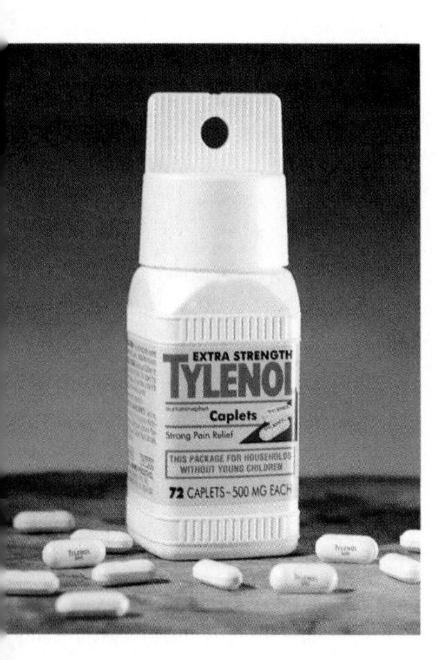

We saw in Chapter 13 that continuous use of aspirin in high doses may cause stomach upset, gastrointestinal bleeding, and stomach ulcers. This has led to the search for other pain relievers. One of these, acetaminophen, contains an amide group rather than the ester group found in aspirin. Acetaminophen (best known under the brand name Tylenol) is as effective as aspirin in relieving pain and reducing fever, but it does not reduce inflammation.

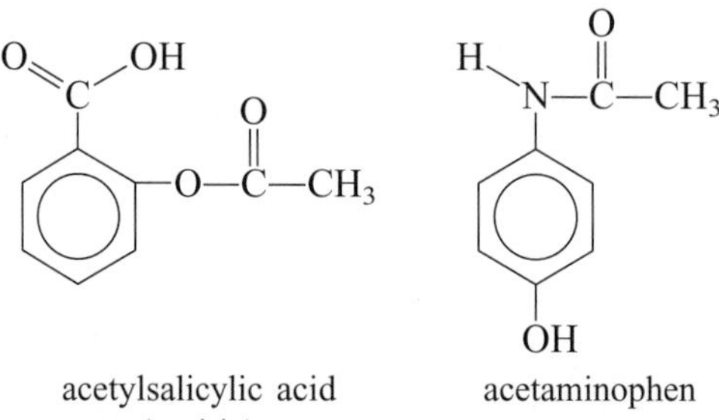

acetylsalicylic acid (aspirin) — acetaminophen

A number of anesthetics contain the amide group. Compare the structures of lidocaine and procaine. Lidocaine is made up of an amide and an amine; procaine, on the other hand, is made up of an ester and an amine. Procaine is better known as Novocain and is used extensively in dentistry. Of the two, procaine is the stronger anesthetic.

$$C_6H_5-\overset{O}{\overset{\|}{C}}-O-CH_2CH_2N\begin{matrix} CH_2CH_3 \\ CH_2CH_3 \end{matrix}$$

procaine

$$(CH_3)_2C_6H_3-NH-\underset{O}{\underset{\|}{C}}-CH_2N\begin{matrix} CH_2CH_3 \\ CH_2CH_3 \end{matrix}$$

lidocaine

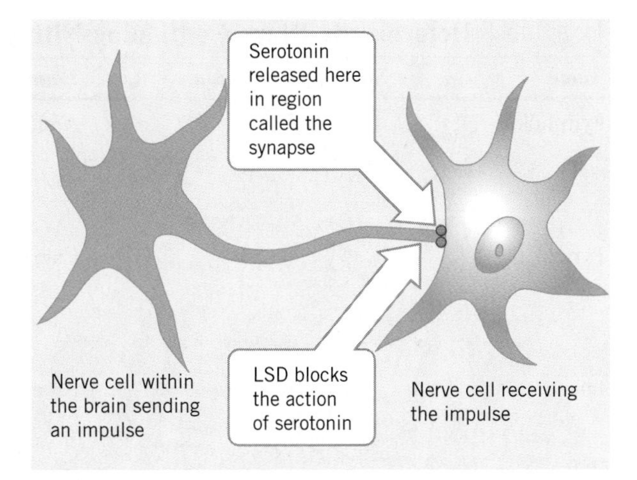

Figure 14.2 The hallucinogen LSD probably affects brain cells by blocking the action of serotonin.

The hallucinogenic drug LSD (lysergic acid diethylamide) is also an amide. LSD seems to disrupt the transmission of nerve impulses in the brain. The structure of LSD has certain features that resemble serotonin, one of the neurotransmitters. Nerve cells in the brain probably confuse LSD with serotonin, which prevents serotonin from carrying out its control function (Fig. 14.2). This results in uncontrolled nerve impulses in the brain, causing hallucinations and other behavioral abnormalities.

LSD

serotonin

14.7 Reactions of Amides

Proteins are complex compounds containing many amide linkages. The stability of the amide bond partially explains the stability of the proteins in our bodies. Amides ordinarily hydrolyze (are split apart by water) very slowly. But under certain conditions and in the presence of certain enzymes, they hydrolyze rather rapidly to form a carboxylic acid and an amine. These conditions are found in the digestive tract.

Write equations for the hydrolysis of amides.

$$\underset{\text{propanamide}}{CH_3{-}CH_2{-}\overset{\overset{\large O}{\|}}{C}{-}NH_2} + H{-}OH \xrightarrow{\text{catalyst}} \underset{\text{propanoic acid}}{CH_3{-}CH_2{-}\overset{\overset{\large O}{\|}}{C}{-}OH} + \underset{\text{ammonia}}{NH_3}$$

$$\underset{N\text{-methylacetamide}}{CH_3{-}\overset{\overset{\large O}{\|}}{C}{-}NH{-}CH_3} + H_2O \longrightarrow \underset{\text{acetic acid}}{CH_3{-}\overset{\overset{\large O}{\|}}{C}{-}OH} + \underset{\text{methylamine}}{H_2N{-}CH_3}$$

Table 14.4 **Heterocyclic Rings Containing Nitrogen**

Name	*Structural Formula*	*Found in the Structure of*
Pyrrolidine		Amino acids
Pyrrole		Chlorophyll, hemoglobin, and vitamin B_{12}
Indole		An amino acid
Pyrimidine		Nucleic acids
Purine		Nucleic acids

14.8 Heterocyclic Compounds

Define heterocyclic compound, and give several examples.

So far, the cyclic compounds we have studied have had rings made up only of carbon atoms. But some cyclic compounds have rings that also contain nitrogen, oxygen, or sulfur. Cyclic compounds containing an atom other than carbon as part of the ring are known as **heterocyclic compounds,** or **heterocycles.** An enormous number of naturally occurring compounds are heterocyclic. Table 14.4 shows some heterocyclic compounds that contain nitrogen as part of the ring.

14.9 Alkaloids

The **alkaloids** are a large class of nitrogen-containing organic compounds that include nitrogen as part of a heterocyclic ring. Many complex alkaloids come from plants, in which they serve to defend the plant against predators. Because they can produce strong physiological effects when administered to animals, some alkaloids have been used as drugs for centuries. Many have become part of what we now call ''folk medicine.''

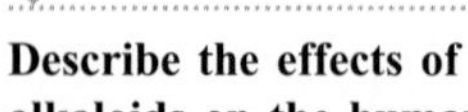

Describe the effects of alkaloids on the human body.

Some alkaloids are strongly addictive. Cocaine is an alkaloid that is extracted from the coca plant of South America; morphine and codeine are alkaloids that can be extracted from opium poppies. Cocaine and morphine have somewhat similar medicinal effects. Both are narcotics (drugs that induce a profound sleep), analgesics (pain-relievers), and antitussives (cough suppressants). Both are also highly addictive. Codeine, although less effective than morphine, is also less likely to cause addiction. Another natural alkaloid, atropine, is chemically similar to cocaine. In dilute solution, atropine can be used for

Health Perspective / Cocaine and Crack

"Crack" vials

If our most serious drug problem is with alcohol—affecting the lives of more people than all other drugs combined—then cocaine is our most insidious. Cocaine is a complex alkaloid compound that is extracted from coca leaves (see Table 14.5). Once a person has used cocaine, the urge to use it again is almost irresistible. It is a seductive drug that brings on pleasurable, overwhelmingly ecstatic experiences. Despite engendering feelings of overwhelming ecstasy, however, cocaine never produces satisfaction. When we eat a meal, our hunger disappears and our appetite is satisfied. With cocaine, however, the feeling of satisfaction never comes—only the desire for more. The user is left with an overpowering craving to experience the joy of cocaine over and over again, which leads to tolerance and an ever-increasing dose. Laboratory research has shown that animals prefer cocaine to food, water, and sex. If given free access to it, they continually take the drug until they overdose and die. Ken Liska, in his book *Drugs and the Human Body,* describes a study in which monkeys pressed a bar 12,800 times to get a dose of cocaine.

Until recently, the form of cocaine most readily available on the streets was the water-soluble hydrochloride salt. Users either sniffed this into their nose or injected it intravenously. Over the last few years, however, the free base form of cocaine (called *crack* because it makes a crackling sound when heated) has appeared with ever-increasing frequency. Crack looks like small pieces of white rock and can be placed in a pipe, heated, and inhaled. The amount necessary to make a dose is small and sells for less than $15. Because of the low cost of crack, it has become available to almost everyone, hugely increasing the drug problems in this country.

The initial effects of cocaine are similar to any central nervous system stimulant: constriction of the blood vessels, dilation of the pupils, and increase in heart rate, breathing rate, blood pressure, and body temperature. Users normally lose their appetite and have tremendous difficulty sleeping. When the drug is sniffed, the severe vasoconstriction that occurs can damage the nasal septum and mucosa. Those who snort cocaine often have a runny nose. Frequent users suffer a large number of other symptoms, including irritability, weight loss, poor nutrition, depression, and paranoia. The mental depression that follows the euphoric high is so intolerable that the user helplessly returns to the drug. In addition to all this, cocaine forces the heart to beat rapidly while slowing the flow of blood to it. This can result in a lack of synchronism between heartbeat and pulse, called fibrillation, in which the heart literally quivers just like Jell-O. The resulting muscle tissue damage can be severe, leading to heart attack or sudden cardiac death.

Your Perspective: Some reputable scientists have proposed legalizing drugs as a means of reducing the criminal behavior associated with drug dealing and drug acquisition. Write a brief essay commenting on such proposals from a biochemical (as opposed to a societal) viewpoint.

dilating the pupils of the eyes, or for keeping the mouth dry during dental work. This extract from the deadly nightshade plant (*Atropa belladonna*) also serves as an antidote for nerve gas poisoning (Table 14.5).

Nicotine is an alkaloid found in the leaves of the tobacco plant. In pure form, it is a rapid-acting, extremely toxic drug. Nicotine stimulates the central nervous system (causing irregular heartbeat and increased blood pressure), induces vomiting and diarrhea, and first stimulates and then inhibits glandular secretions. Absorption of only about 4 mg by a

Table 14.5 Alkaloids with Nitrogen-Containing Rings

CH_3 N O C—OCH_3 O O—C

Cocaine

HO N—CH_3 O HO

Morphine

CH_3 N CH—C—O H H_2COH O

Atropine

HO N—CH_3 O CH_3O

Codeine

nonsmoker can lead to nausea, vomiting, diarrhea, and weakness. Smokers build up a tolerance to nicotine and may be able to absorb twice as much as a nonsmoker before suffering any noticeable effects. Although the smoke from one cigarette may contain up to 6 mg of nicotine, only about 0.2 mg is absorbed into the body. A solution of nicotine sulfate is also used as an insecticide. There have been reports of gardeners who have died from respiratory failure only a few minutes after handling a nicotine-based insecticide.

N CH_3 N

nicotine

Some naturally occurring alkaloids have been synthesized in the laboratory. By studying and replacing certain functional groups on these alkaloids, chemists have been able to produce compounds with desirable properties (such as relieving pain) without such accompanying side effects of the natural alkaloid as dependence (addiction) and tolerance (the need to keep increasing the dose to produce the same effect). Let's take a look at one group of relatively simple alkaloids to see how changing the structure of an alkaloid can change its physiological effects.

Epinephrine (also called adrenalin) is produced in a pair of hat-shaped organs, known as the adrenals, which lie on top of the kidneys. Epinephrine is a hormone, one of a number of chemical "messengers" produced by special glands in the body.

HO— —CH—CH_2—N—CH_3 OH H HO

epinephrine

Health Perspective / Designer Drugs

One way that unscrupulous chemists have tried to get around the Controlled Substance Act is by designing drugs with structures similar to known narcotics. Often these "designer drugs" turn out to be more powerful than the narcotics from which they are designed. One example is the narcotic analgesic fentanyl (marketed under several brand names, including Sublimaze), which is used in a majority of the surgeries performed in the United States. It is 100 times more powerful than morphine and is a controlled substance. Fentanyl has become the chemical model for the creation of designer drugs. The first analog of fentanyl, α-methylfentanyl, hit the streets in 1979 and proved to be 200 times more powerful than morphine. Since then, a series of analogs have followed, each more powerful than the one before. Among these is 3-methylfentanyl, called "China White," which is 3000 times more powerful a narcotic than morphine. Because of the extraordinary potency of these designer drugs, and because they are often sold on the streets as synthetic heroin, the potential for accidental overdose is tremendous.

One danger of designer drugs stems from the lack of chemical knowledge among the underground chemists who make them. Many of these illegitimate chemists have no experience other than introductory high school or college chemistry courses. The synthesis of illegal drugs by these poorly trained opportunists has enormous potential for disaster. And a disaster is just what happened when one of them attempted to synthesize MPPP (methylphenylpropionoxypiperidine), a designer drug modeled after the controlled opiate meperidine (Demerol). Through an error in the synthesis, the unexpected compound MPTP (methylphenyltetrahydropyridine) was produced, which causes a crippling condition closely resembling Parkinson's disease. Hundreds of heroin users unknowingly used MPTP thinking it was MPPP, and the results were tragic. They became violent and hallucinatory as the MPTP caused permanent brain damage and paralysis.

Notice that epinephrine is a secondary amine. Under normal conditions, small amounts of this compound are released into the blood to help control blood pressure and to maintain the level of sugar in the blood. Epinephrine is also one of the hormones used by the body to meet emergencies and deal with such stress as emotional excitement, strenuous exercise, or extreme temperature changes. The epinephrine produced by the body results in increased blood sugar, faster heart rate, accelerated blood clotting, and increased muscular strength. These changes enable the body to meet the initial challenge of the stress.

A compound similar to epinephrine, but lacking the methyl group on the nitrogen, is called norepinephrine. The prefix "nor" means something has been removed from the parent compound—in this case, the methyl group.

$$\text{HO}-\text{C}_6\text{H}_3(\text{OH})-\underset{\text{OH}}{\underset{|}{\text{CH}}}-\text{CH}_2-\underset{\text{H}}{\underset{|}{\text{N}}}-\text{H} \qquad \text{HO}-\text{C}_6\text{H}_3(\text{OH})-\text{CH}_2-\text{CH}_2-\underset{\text{H}}{\underset{|}{\text{N}}}-\text{H}$$

norepinephrine — dopamine

Norepinephrine acts as a chemical messenger between nerve cells. Its main function in the body is to maintain muscle tone in the blood vessels, which helps to control blood pressure. The drug reserpine (which is given to help reduce blood pressure in persons suffering from hypertension) works by greatly reducing the amount of norepinephrine in the nerve endings.

The body synthesizes norepinephrine from dopamine, which is also a neurotransmitter. One of the drugs used to treat schizophrenia hinders the ability of dopamine to participate in the transmission of nerve impulses. Synthetically produced dopamine is used to treat congestive heart failure. As you can see, neurotransmitters, such as norepinephrine and dopamine, and hormones such as epinephrine play important roles in regulating various chemical and electric processes in the body. Yet it is only the smallest of differences in the structures of these three compounds that gives them such greatly different functions in the body.

KEY CONCEPTS

Nitrogen is the fourth most common element in the human body. It is found in many types of compounds.

Amines are weak organic bases that can be classified as primary, secondary, or tertiary according to the groups attached to the nitrogen. This classification is different from that used for the alcohols.

Amines form salts in neutralization reactions with acids and in reactions with alkyl halides.

Amides are derivatives of carboxylic acids, in which a $—NH_2$, —NHR, or —NRR′ replaces the —OH on the carboxyl group. The covalent bond between the carbonyl carbon and the nitrogen is very stable and is known as an amide linkage. This same bond in proteins is called a peptide bond. Amides ordinarily undergo hydrolysis very slowly.

Nitrogen atoms can form rings with carbon atoms. Compounds containing rings that include atoms other than carbon are known as heterocyclic compounds.

Alkaloids are a large class of nitrogen-containing compounds in which the nitrogen is often part of a ring. Most complicated alkaloids are produced in plants as part of their defense system. Many have been used for centuries as drugs because of their physiological effects.

Important Equations

Salt Formation

Amines with acids

$$\text{Amine} + \text{organic or inorganic acid} \longrightarrow \text{alkylammonium salt}$$

$$RNH_2 + HCl \longrightarrow RNH_3^+ \ Cl^-$$

Amines with alkyl halides

$$\text{Tertiary amine} + \text{alkyl halide} \longrightarrow \text{quaternary ammonium halide}$$

$$(CH_3)_3N + CH_3I \longrightarrow (CH_3)_4N^+ \ I^-$$

Hydrolysis

$$\text{Amide} + \text{water} \longrightarrow \text{carboxylic acid} + \text{amine}$$

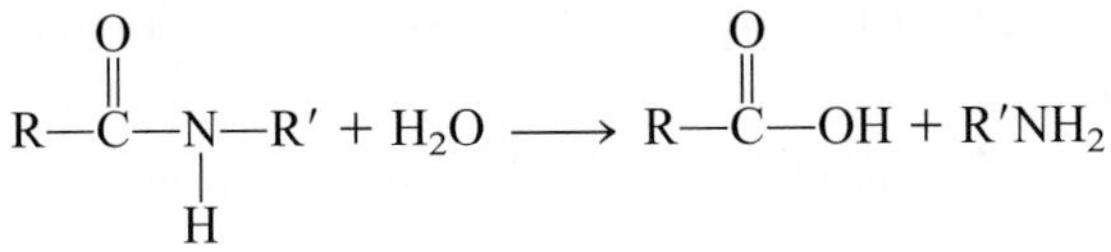

$$R-\overset{\overset{\displaystyle O}{\|}}{C}-\underset{\underset{\displaystyle H}{|}}{N}-R' + H_2O \longrightarrow R-\overset{\overset{\displaystyle O}{\|}}{C}-OH + R'NH_2$$

Review Problems

Section 14.1

1. How many covalent bonds must nitrogen form to achieve an octet in its outermost energy level?
2. Draw the structure of the amine and the amide groups.

Section 14.2

3. Identify the following compounds as primary, secondary, or tertiary amines:

(a) 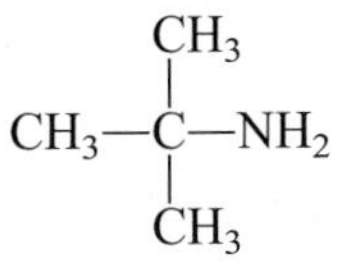 $CH_3-\overset{\overset{\displaystyle CH_3}{|}}{\underset{\underset{\displaystyle CH_3}{|}}{C}}-NH_2$

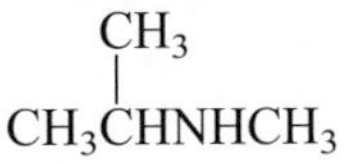

(b) 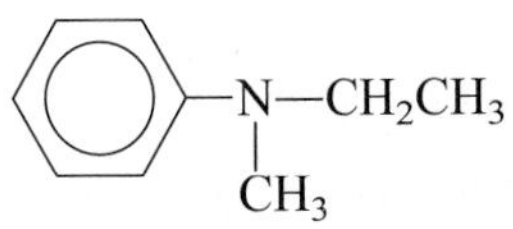 $C_6H_5-\underset{\underset{\displaystyle CH_3}{|}}{N}-CH_2CH_3$ (benzene ring)

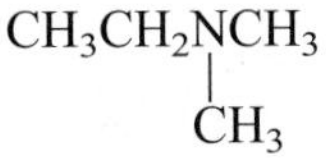

(c) $CH_3CH_2CH_2NH_2$

(d) $CH_3\overset{\overset{\displaystyle CH_3}{|}}{C}HNHCH_3$

(e) $CH_3CH_2\underset{\underset{\displaystyle CH_3}{|}}{N}CH_3$

(f) 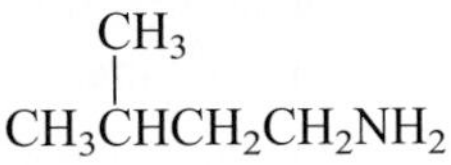 $CH_3\overset{\overset{\displaystyle CH_3}{|}}{C}HCH_2CH_2NH_2$

4. Name each of the amines in Problem 3.

Section 14.3

5. Draw the structures of the products of the following reactions:

(a) $CH_3-CH_2-NH-CH_2-CH_3 + H_2O \rightleftharpoons$

(b) 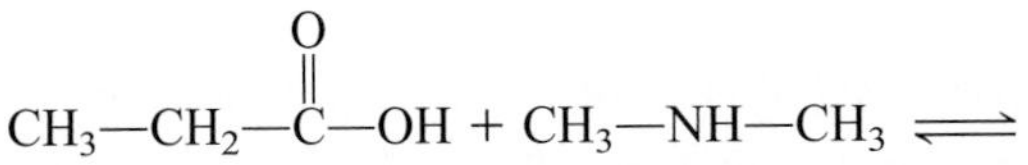 $CH_3-CH_2-\overset{\overset{\displaystyle O}{\|}}{C}-OH + CH_3-NH-CH_3 \rightleftharpoons$

(c) $CH_3-CH_2-CH_2-NH_2 + CH_3-\overset{\overset{\displaystyle O}{\|}}{C}-OH \rightleftharpoons$

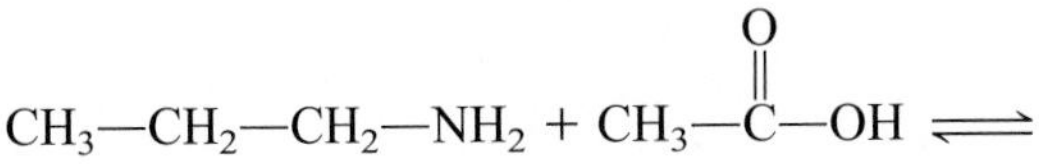

6. Draw the structures of the amine and the acid that form each of the following salts:

(a) 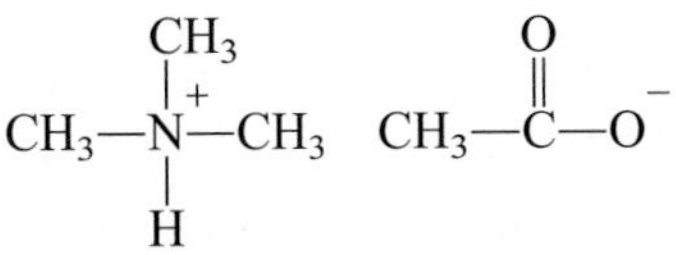 $CH_3-\overset{\overset{\displaystyle CH_3}{|}}{\underset{\underset{\displaystyle H}{|}}{N^+}}-CH_3 \quad CH_3-\overset{\overset{\displaystyle O}{\|}}{C}-O^-$

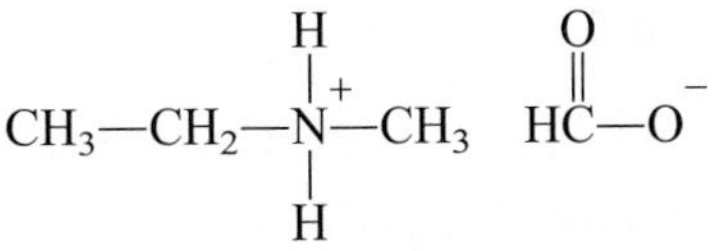

(b) 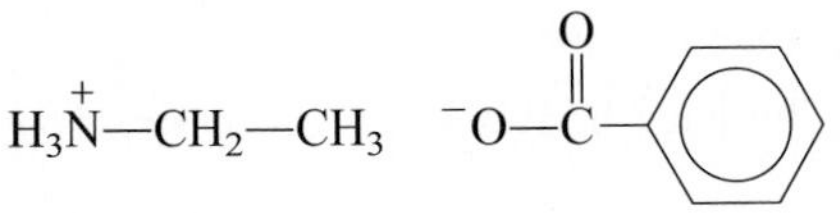 $H_3\overset{+}{N}-CH_2-CH_3 \quad {}^-O-\overset{\overset{\displaystyle O}{\|}}{C}-C_6H_5$ (benzene ring)

(c) $CH_3-CH_2-\overset{\overset{\displaystyle H}{|}}{\underset{\underset{\displaystyle H}{|}}{N^+}}-CH_3 \quad H\overset{\overset{\displaystyle O}{\|}}{C}-O^-$

Section 14.4

7. Write the equations for the following reactions, and name the products:

(a) isobutylamine with HCl

(b) triethylamine with HCl

(c) triethylamine with ethylbromide

8. Draw the structure of the product from each of the following reactions:

(a) $CH_3—CH_2—NH—CH_2—CH_3 + HBr \longrightarrow$

(b) $H_2N—C_6H_{11}$ (cyclohexyl) $+ HCl \longrightarrow$

(c) $CH_3—CH_2—NH—CH_3 + CH_3—CH_2—Br \longrightarrow$

Section 14.5

9. Name the following amides:

(a) $CH_3CH_2CH_2CH_2\overset{O}{\overset{\|}{C}}NH_2$

(b) $CH_3\underset{CH_3}{\underset{|}{N}}\overset{O}{\overset{\|}{C}}CH_3$

(c) $C_6H_5—\underset{H}{\underset{|}{N}}—\overset{O}{\overset{\|}{C}}CH_2CH_3$

(d) $CH_3CH_2\underset{CH_3}{\underset{|}{N}}\overset{O}{\overset{\|}{C}}CH_2CH_2CH_2CH_2CH_3$

10. Write the structural formula for each of the following compounds:

(a) methylbenzylamine

(b) pentanamide

(c) triisopropylamine

(d) *N,N*-diethylhexanamide

(e) 2-(*N*-methylamino)-1-propanol

(f) *N*-methyl-*N*-phenylpropanamide

Section 14.6

11. What functional groups are present in LSD?

12. What functional groups are present in procaine and in lidocaine?

Section 14.7

13. Write the equation for the hydrolysis of each of the amides in problem 9.

14. Name all of the hydrolysis products in problem 13.

Section 14.8

15. Draw the structure of two different heterocyclic rings that contain nitrogen.

Section 14.9

16. Of what benefit to plants are the alkaloids they synthesize?

17. The following descriptions refer to specific alkaloids. Identify each one.

(a) controls blood pressure

(b) used to relieve coughing

(c) found in tobacco

(d) initiates body's reactions to stress

Applied Problems

18. People suffering from Parkinson's disease lack dopamine-producing cells in an area of the brain called the substantia nigra. Is dopamine a primary, secondary, or tertiary amine?

19. In our study of the hydrocarbons, we saw that increasing molecular weight is accompanied by an increase in boiling point. Even though trimethylamine has a higher molecular weight than dimethylamine, it boils at a lower temperature (3° versus 7°C). Why do you think this occurs?

20. A common household remedy for a bee sting is the application of aqueous ammonia. The chemical in a bee sting causing the pain is formic acid (HCOOH). Ammonia neutralizes this acid. Write the equation for the neutralization reaction.

21. Morphine is a very effective pain reliever. Why must its continued use in a patient be avoided?

22. Nylon 66 is an important fiber that was first developed at the E.I. duPont de Nemours Company by Wallace Carothers. It can be made into fibers that are stronger than natural fibers and more inert. It is a polymer prepared from adipic acid (hexanedioic acid) and 1,6-diaminohexane. Draw the structure of nylon.

23. Concerns about smoking usually relate to the carcinogens present in cigarette smoke, but nicotine also has profound effects on the body.
(a) Describe three specific effects of nicotine on the human body.
(b) Suggest a reason why smokers can tolerate higher concentrations of nicotine than can nonsmokers before feeling ill effects.

24. As we saw in the Health Perspective in Section 14.2, the amphetamines are powerful central nervous system stimulants. Amphetamine is never dispensed medically as the pure amine. Explain why it is usually dispensed in the form of its sulfate salt (Benzedrine).

25. Morphine and codeine (see Table 14.5) have similar medicinal effects. Although codeine is less effective than morphine, it is also less addictive. What change must be made in the morphine molecule to convert it to codeine?

Integrated Problems

1. In laboratory experiments, compounds in several commercially available products have been found to be carcinogenic in mice.
(a) What does it mean for a compound to be carcinogenic?
(b) Should humans continue to use such products? Why or why not?

2. A solution containing equal amounts of acetic acid (CH_3COOH) and acetate ion (CH_3COO^-) has a pH that remains quite constant when small amounts of strong acid or base are added. Explain why this happens.

3. **(a)** Write the balanced equation for the complete combustion of isooctane (2,2,4-trimethylpentane).
(b) If 1000 g of isooctane is burned as a fuel, how many liters of carbon dioxide gas form at STP?

4. A chemist uses bromine to test an unknown organic compound with the molecular formula C_6H_{10} for unsaturation. A 2.05-g sample of the compound requires the addition of 41.7 mL of a 1.2 *M* Br_2/CCl_4 solution before the reddish brown color of bromine remains. What is a reasonable structure for the original hydrocarbon?

5. **(a)** Write the structural formula for 1,2-ethanediol.
   (b) In 1985 this compound was added to some Austrian wines to increase their sweetness. Why should this have resulted in the banning of all sales and export of this wine?

6. In 1986, all wine exports from Italy were suspended when it was discovered that methanol had been added to some of the wine to increase its alcohol content to 12%. Twenty Italians died after drinking this wine, and other consumers were blinded after drinking it. How does methanol act on the body to cause blindness and death?

7. Arrange the following compounds in order of increasing boiling points:
   (a) $CH_3CH_2{-}O{-}CH_2CH_3$ **(c)** CH_3CH_2OH **(e)** $CH_3CH_2CH_2NH_2$
   (b) $CH_3{-}O{-}CH_3$ **(d)** $CH_3CH_2CH_2OH$

8. **(a)** Explain why an amine can act as a base in a chemical reaction.
   (b) Write the equation describing what occurs when propylamine is added to water.
   (c) Which of the following is the strongest base: aniline, dipropylamine, or propylamine? Why?

9. 1,2-Cyclopentanediol exists as *cis* and *trans* isomers. The boiling point for the *trans* isomer is slightly higher than that of the *cis* isomer.
   (a) Draw structures for the two isomers and show why intramolecular (within the same molecule) hydrogen bonding is greater in the *cis* isomer.
   (b) Why does this result in a lower boiling point for the *cis* isomer?

10. When an aldehyde or ketone is dissolved in water, a water molecule can add to the $-\overset{O}{\overset{\|}{C}}-$ much as an alcohol does in the formation of hemiacetals and hemiketals. These compounds are often referred to as *gem*-diols. Normally these products are very unstable and cannot be isolated. One exception is chloral hydrate, which is formed when water is added to chloral (trichloroacetaldehyde). Chloral hydrate is better known as "knockout drops." Chloral hydrate, when added to alcohol, forms the infamous "Mickey Finn." Write the equation showing the formation of chloral hydrate from water and chloral.

11. **(a)** What is the alcohol level of a suspect if 1 mL of her blood contains the same amount of alcohol as 7.5 mL of a 0.020% standard alcohol solution.
    (b) If the legal limit for drunk driving in the state is 0.08% blood alcohol, should the district attorney prosecute this suspect?

12. Ants living in the Amazon jungle produce formic acid both for defense and for communication. About 10% to 15% of the ant's body weight is stored formic acid. It is estimated that the ants release 2×10^{13} g of formic acid per year. The formic acid provides most of the acidity for the acid rain that falls in regions of the Amazon.
    (a) Write the structural formula for formic acid.
    (b) What is the IUPAC name for formic acid?
    (c) How many moles of formic acid are released each year by the ants? How many tons?

Section 3

The Compounds of Life

We saw at the beginning of the previous section that living organisms have a complex structure with several levels of organization. The cell is the simplest structure in an organism that can perform all the activities of life. All living organisms are made up of cells. Some organisms consist of a single cell; others are multicellular organisms with tissues and organs made up of many different types of cells.

The existence of cells was first discovered in 1665. Robert Hooke, an English scientist, was looking at slices of cork under a microscope that magnified the sample 30 times. He thought that the tiny boxes, or "cells," that he saw were something that existed only in cork. At about the same time, the Dutch scientist Antonie van Leeuwenhoek was looking through a microscope at pond water. He was astonished to observe the great variety of single-celled organisms in the water sample. But it was not until 160 years later, with the development of high-quality microscope lenses, that the importance of the cell became understood. In the late 1830s, the German scientists Matthias Schleiden (working with plants) and Theodor Schwann (working with animals) separately concluded that all plants and animals were made up of cells. This idea, called the cell theory, was later expanded to include the recognition that cells cannot arise from nonliving matter, but rather that all cells come from the division of other cells.

Two kinds of cells make up living organisms. One kind, called a prokaryotic cell, does not contain a nucleus or any organelles (other than ribosomes). For example, the cells of bacteria are prokaryotic. The other kind contains a nucleus and organelles and is called a eukaryotic cell. All organisms other than bacteria and viruses are made up of eukaryotic cells.

The nucleus of eukaryotic cells contains DNA and various proteins—all of which play a role in regulating the growth and reproduction of the cell. The nucleus is surrounded by the cytoplasm, a thick colloidal liquid in which are suspended the organelles that perform most of the cell's functions. Depending on the type of cell, these organelles might include mitochondria (which produce energy), ribosomes (which make proteins), lysosomes (which break down large molecules), vacuoles (which store various substances), and chloroplasts (in which photosynthesis takes place in plants). All cells are surrounded by a plasma membrane that controls the passage of materials in and out of the cell. In addition to the membrane, plant cells have a rigid cell wall surrounding the cell. Animal cells have a plasma membrane but no cell wall.

All of these cell parts are constructed from four types of large molecules: carbohydrates, lipids, proteins, and nucleic acids. In Section 3, we study these four classes of molecules and the chemical reactions essential to the living cell.

Chapter 15

Carbohydrates

Although Ed had been fighting a cold, he couldn't take time to rest—his report for the manager's meeting had to be ready by Monday! He knew he'd been pushing himself too hard, and now he was paying for it with a fever and a terrible cough. On the way home Ed thought he'd better stop at the immediate care center.

As part of Ed's examination, the doctor ordered a chest X ray, sputum culture, blood count, and a routine urine analysis. The chest X ray showed no signs of pneumonia, so Ed was sent home with a supply of antibiotics, cough medicine, and some strong advice to take it easy for a few days. The next afternoon, however, Ed received a call from the doctor saying that the urine analysis had found sugar in the urine. Ed made an appointment to return for a fasting chemical profile of his blood.

By the time of his appointment, Ed had recovered completely from his bronchitis and was feeling fine. He couldn't imagine that anything could be wrong with him.

Figure 15.1 This giant sequoia tree is the largest living object in the world. It is as tall as a six-story building, wider at the base than an average city street, and is estimated to be between 2500 and 3000 years old. The support structure of trees is composed of the carbohydrate cellulose.

The doctor did a complete physical and then asked Ed to join him in his office. Ed's chemical profile showed elevated levels of glucose (325 mg%), serum cholesterol (350 mg%), triglycerides (475 mg%), and uric acid (8.2 mg%). When the doctor examined the retina of Ed's eyes, he saw thickening and irregularities in the arteries. When he had felt the arteries in Ed's feet, the pulses were not as strong as they should be. Both of these symptoms indicated the beginning of atherosclerotic changes unusual in a person as young as Ed. These findings, and the fact that Ed's grandmother and uncle suffered from diabetes, suggested a diagnosis of noninsulin-dependent diabetes with early complications of atherosclerosis. Ed was distressed by the diagnosis (since he couldn't even pronounce it!) and wanted more information.

The doctor explained that there are two forms of diabetes mellitus: insulin-dependent and noninsulin-dependent. Insulin-dependent diabetes often strikes children, adolescents, or young adults; it has a rapid onset that can result in ketosis and diabetic coma. It is caused by a low level or total lack of the hormone insulin, which is critical in controlling the level of the sugar glucose in the blood. The initial treatment requires daily doses of rapid-acting insulin and a strictly controlled diet. Careful regulation of the blood glucose level in these patients reduces the occurrence of such complications as atherosclerosis, visual problems, coronary heart disease, strokes, and circulatory problems in the legs and feet.

The noninsulin-dependent diabetes found in Ed results from the body's inability to use the insulin that it produces. This form of diabetes usually begins later in life, commonly in people who are overweight. It can usually be controlled by a high-fiber, low-cholesterol diet with 50% to 60% carbohydrates and less than 30% total fat. The purpose of the diet is to reduce the patient's weight and to lower the blood sugar to the normal range of 80 mg% to 120 mg%. If Ed's blood sugar didn't return to normal after six to eight weeks of the diet, the doctor would prescribe an oral antidiabetic drug such as glyburide, which stimulates the beta cells of the pancreas to produce more insulin. In addition to the diet, the doctor recommended that Ed attend the hospital's diabetic training course. The course would help him with a weight reduction and exercise program and inform him about oral antidiabetic drugs, insulin therapy, and good foot care. Luckily, Ed was able to lose weight and lower his blood glucose to normal by diet alone.

Diabetes is the most common disease associated with the class of compounds called carbohydrates. In this chapter we study the structure, properties, and functions of carbohydrates.

15.1 Carbohydrates

◆

Define carbohydrate, and write the general formula for a carbohydrate.

Carbohydrates, which are essential components of all living organisms, are the most abundant class of biological molecules. For example, sugars, glycogen, starches, cellulose, dextrins, and gums are all carbohydrates. The name carbohydrate literally means "carbon hydrate," which reflects the general formula of these compounds: $C_x(H_2O)_y$. Carbohydrates are found mainly in plants, where they make up about 75% of the solid plant material. They function both as part of the structure that supports the plant and as storehouses for the plant's energy supply. For example, the carbohydrate cellulose is the most important component of the supporting tissue of plants—such as the wood in trees (Fig.

15.1)—and the carbohydrate starch is the energy storage molecule in plants. Carbohydrates also perform a variety of functions in animals, such as forming the chitin that makes up lobster shells and the fluid that lubricates the joints of our bodies.

Carbohydrates are classified according to the size of the molecule. **Monosaccharides** are carbohydrates that cannot be broken into smaller units on hydrolysis. **Disaccharides** produce two monosaccharides on hydrolysis, and **polysaccharides** produce three or more monosaccharides on hydrolysis. In fact, polysaccharides can contain as many as 3000 monosaccharide units. An -ose ending on the name of a compound indicates that the compound is a carbohydrate.

Define monosaccharide, disaccharide, and polysaccharide.

Monosaccharides, also called simple sugars, can be further classified by the number of carbons in the molecule.

three carbons—triose	five carbons—pentose
four carbons—tetrose	six carbons—hexose

Another way of classifying monosaccharides is by the carbonyl functional group found in the molecule.

Given the structure of a monosaccharide, identify the compound as an aldose or a ketose.

aldose—aldehyde functional group

ketose—ketone functional group

For example, ribose is an aldopentose (a five-carbon sugar molecule containing an aldehyde group), and fructose is a ketohexose (a six-carbon sugar molecule containing a ketone group).

Monosaccharides are white crystalline solids that are highly soluble in water because of their polar hydroxyl groups. This also means that monosaccharides are insoluble in nonpolar solvents. For reasons not totally understood, most monosaccharides have a sweet taste. The most common monosaccharides are the hexoses. Table 15.1 shows the structures of some important monosaccharides.

15.2 Monosaccharides: Glucose

Glucose (also known as blood sugar, grape sugar, and dextrose) is the most common of the hexoses. It is the only aldose that is commonly found in nature as a monosaccharide. It occurs in the juices of fruits (especially grape juice), in the saps of plants, and in the blood and tissues of animals. It is the immediate source of energy for many energy-requiring cellular reactions, such as tissue repair and synthesis, muscle contraction, and nerve transmission. The average adult has 5 to 6 g of glucose in the blood (about 1 tsp). This amount of glucose supplies the body's energy needs for only about 15 minutes, so the body must continuously replenish the glucose in the blood from compounds stored in the liver. The level of glucose in the blood of a normal adult is fairly constant, although it rises after each meal and falls during periods of fasting.

Glucose is a part of many polysaccharides and can be produced by the hydrolysis of these polysaccharides (it is produced commercially by the hydrolysis of cornstarch). Because glucose is found in most living cells, its chemistry is an important part of the carbohydrate chemistry of the body.

Table 15.1 Important Monosaccharides

Pentoses			
Aldopentoses			*Ketopentose*
H	H	H	
C=O	C=O	C=O	CH_2OH
H—C—OH	HO—C—H	H—C—OH	C=O
H—C—OH	H—C—OH	HO—C—H	H—C—OH
H—C—OH	H—C—OH	H—C—OH	H—C—OH
CH_2OH	CH_2OH	CH_2OH	CH_2OH
D-ribose	D-arabinose	D-xylose	D-ribulose

Hexoses			
Aldohexoses			*Ketohexose*
H	H	H	
C=O	C=O	C=O	CH_2OH
H—C—OH	HO—C—H	H—C—OH	C=O
HO—C—H	HO—C—H	HO—C—H	HO—C—H
H—C—OH	H—C—OH	HO—C—H	H—C—OH
H—C—OH	H—C—OH	H—C—OH	H—C—OH
CH_2OH	CH_2OH	CH_2OH	CH_2OH
D-glucose	D-mannose	D-galactose	D-fructose

Structure of Glucose

The structure of glucose can be drawn as a straight chain (see Table 15.1). The straight chain structure, however, does not explain some of the properties of glucose. In fact, glucose is found in three forms in water solution. The three forms exist in equilibrium and are easily converted one into another. The straight-chain form of glucose makes up only 0.02% of the molecules. The two other forms are ring compounds that result when an internal hemiacetal forms on the molecule (review Section 13.10). In glucose, the hemiacetal forms between the aldehyde group on carbon 1 and the alcohol group on carbon 5.

Draw the linear and ring structure of glucose.

H		OH	
1 C=O		H—1 C	(ring bond to O)
H—2 C—OH		H—2 C—OH	
HO—3 C—H	→	HO—3 C—H	O
H—4 C—OH		H—4 C—OH	
H—5 C—OH		H—5 C	(ring bond to O)
6 CH_2OH		6 CH_2OH	

Other structural formulas are often used to simplify the drawing of sugar ring structures. In the Haworth structural formula, the ring formed by the sugar molecule is shown as though we were looking at it from the side (rather than looking down from above the ring). The thickened side of the ring is the one closest to us, and the groups attached to the carbon atoms are then shown either above or below the ring.

In going from the straight-chain structure of glucose to the Haworth structure, any atom or group of atoms to the right of carbons 2, 3, and 4 in the straight-chain structure are placed below the ring, and any atom or group of atoms to the left are placed above the ring. Notice that in the shorthand form on the far right, a straight line indicates a bond to an OH, and a right-angled line a bond to a CH_2OH group.

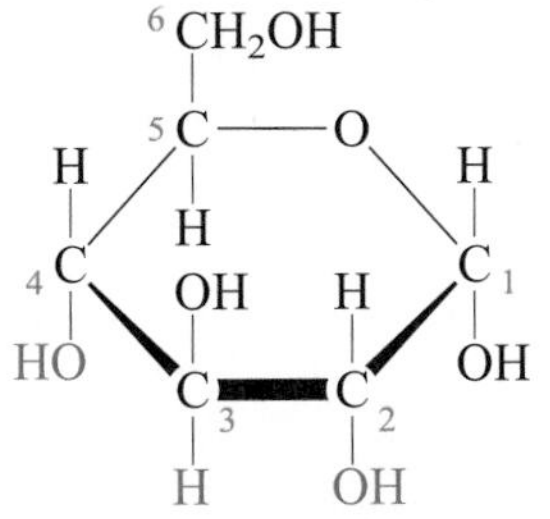

Haworth formula

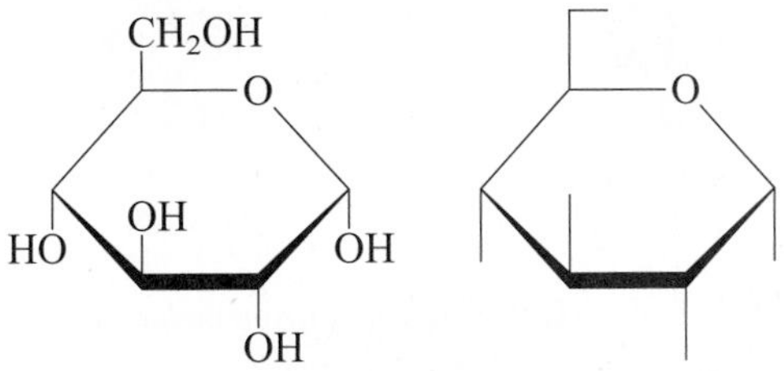

shorthand forms

The two ring forms of glucose depend on the placement of the hydrogen and hydroxyl groups on carbon 1. A hydroxyl group on carbon 1 below the plane of the ring is the alpha (α) form; one that appears above the ring is the beta (β) form.

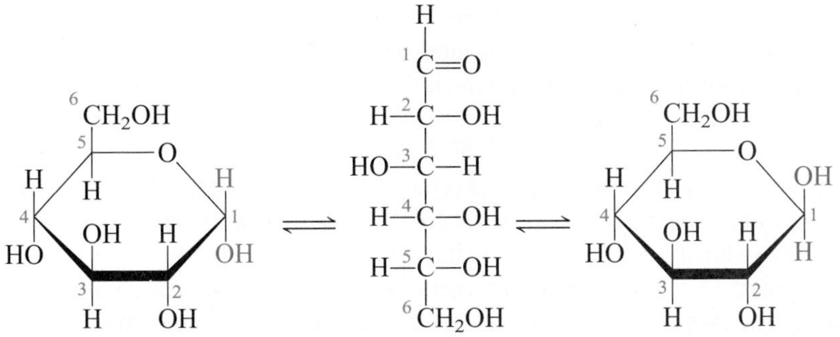

α form

β form

You may wonder what difference the position of that one hydroxyl group on the ring could possibly make. As we study the metabolism of living organisms, we shall see that such small differences can determine whether or not a cell can use a molecule. The difference between starch (a digestible glucose polymer) and cellulose (an indigestible glucose polymer) is the position of the hydroxyl group on carbon 1 of the glucose molecule.

Although Haworth formulas are easy to draw (and we use them throughout the rest of this book), they do not accurately represent the true shape of the ring. In other readings, you might instead see the ring structures drawn in a "puckered" form, as shown here for glucose.

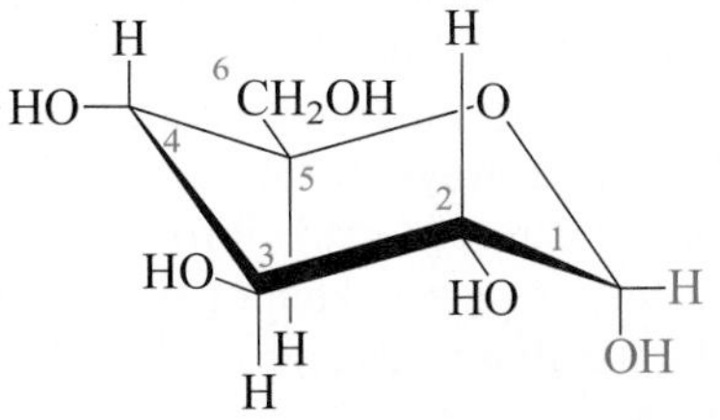

α-glucose

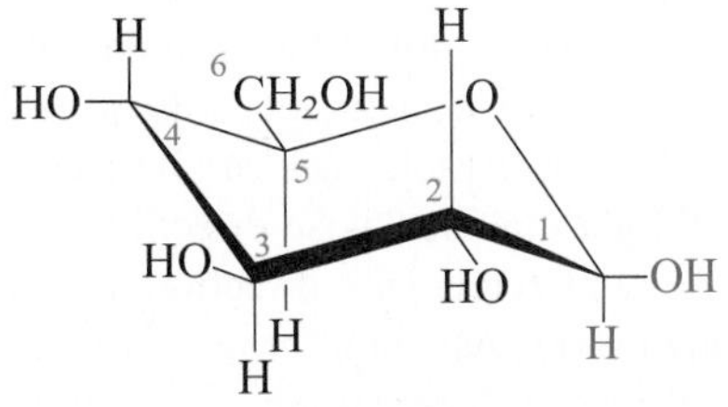

β-glucose

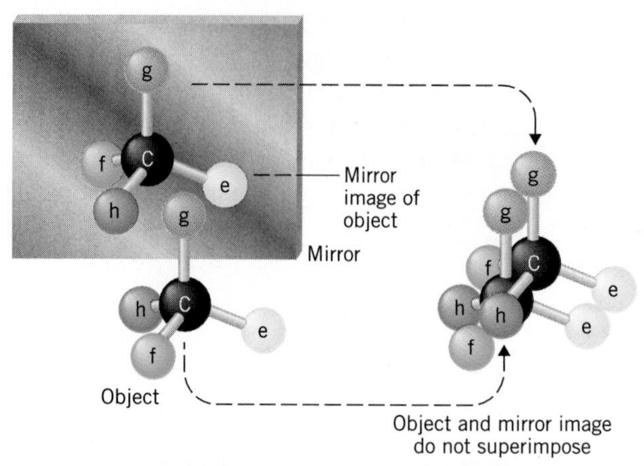

Figure 15.2 This molecule and its mirror image cannot be superimposed. They are optical isomers.

15.3 Optical Isomerism

Figure 15.3 Your hands are familiar examples of asymmetrical, or chiral, objects. There is no way you can put a left-hand glove on your right hand.

In Chapter 12, we discussed various forms of isomerism in compounds having the same molecular formulas. Another type of isomerism, resulting only from the arrangement of atoms in space (not from the order in which they are arranged), is called **optical isomerism.** We have seen that the carbon atom can have four groups attached to it, each directed toward one corner of a tetrahedron. If the four groups are all different, the carbon atom possesses a property called **chirality** and is then called a **chiral center.** A **chiral molecule** is a molecule that cannot be superimposed on its mirror image. In other words, if you were to take a three-dimensional model of a chiral molecule and a model of its mirror image, you could not exactly superimpose the two models so that all of the groups matched up (Fig. 15.2). The word *chiral* comes from the Greek *cheir* meaning "hand," and chiral objects are said to possess "handedness." Your hands are familiar examples of chiral, or asymmetrical, objects. Your right and left hands cannot be superimposed on each other; a left-hand glove cannot possibly fit correctly on your right hand—except by turning the glove inside out (Fig. 15.3). A chiral molecule and its mirror image are called **enantiomers.**

What difference does a chiral center make in a molecule? A chiral molecule and its mirror image have the same melting and boiling points, density, vapor pressure, and solubility in common solvents. They differ, however, in the way they interact with polarized light—light in which the waves vibrate in only a single plane. When placed in solution, chiral molecules interact with polarized light by rotating the plane of vibration of such light in either a clockwise or counterclockwise direction. Molecules having this property are said to be optically active. **Optical isomers** are optically active compounds that share the same molecular formula but rotate the plane of polarized light in opposite directions. Optical isomers are critical in the chemistry of the living organism.

◆

Define optical isomers, and describe why they are important in the chemistry of living organisms.

Optically active molecules can be classified in several ways. A substance that rotates polarized light in a clockwise direction is said to be dextrorotatory. If it rotates polarized light in a counterclockwise direction, it is levorotatory. A second kind of classification is based on the arrangement of groups around the chiral carbon. This type of classification is illustrated by the three-carbon sugar glyceraldehyde.

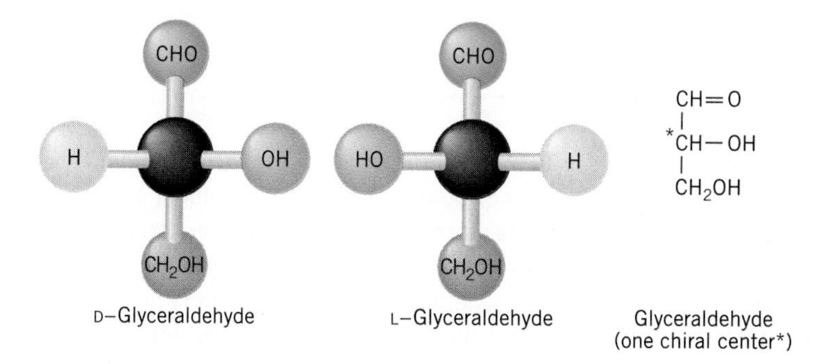

D-Glyceraldehyde has the hydroxyl group to the right of the chiral carbon, whereas L-glyceraldehyde has the hydroxyl group to the left. The D-family of monosaccharides contains compounds whose highest numbered chiral center has the same configuration as D-glyceraldehyde (Fig. 15.4). The L-family has the same configuration as L-glyceralde-

Figure 15.4 The D-family of aldoses. The configuration of atoms around the carbon in red shows that the compound belongs to the D-family. The members of each pair differ from each other in the arrangement of the H and OH on carbon 2. (blue shaded)

```
        H
      1 |
        C=O
      2 |
    H—C—OH
      3 |
        CH2OH
  D-glyceraldehyde
```

```
        H                        H
      1 |                        |
        C=O                      C=O
      2 |                        |
    H—C—OH                  HO—C—H
      3 |                        |
    H—C—OH                   H—C—OH
      4 |                        |
        CH2OH                    CH2OH
     D-erythrose               D-threose
```

```
      H              H              H              H
    1 |              |              |              |
      C=O            C=O            C=O            C=O
    2 |              |              |              |
  H—C—OH        HO—C—H         H—C—OH        HO—C—H
    3 |              |              |              |
  H—C—OH         H—C—OH        HO—C—H        HO—C—H
    4 |              |              |              |
  H—C—OH         H—C—OH         H—C—OH         H—C—OH
    5 |              |              |              |
      CH2OH          CH2OH          CH2OH          CH2OH
   D-ribose      D-arabinose      D-xylose       D-lyxose
```

```
      H           H           H           H           H           H           H           H
    1 |           |           |           |           |           |           |           |
      C=O         C=O         C=O         C=O         C=O         C=O         C=O         C=O
    2 |           |           |           |           |           |           |           |
  H—C—OH     HO—C—H      H—C—OH     HO—C—H      H—C—OH     HO—C—H      H—C—OH     HO—C—H
    3 |           |           |           |           |           |           |           |
  H—C—OH      H—C—OH    HO—C—H     HO—C—H       H—C—OH      H—C—OH    HO—C—H     HO—C—H
    4 |           |           |           |           |           |           |           |
  H—C—OH      H—C—OH     H—C—OH      H—C—OH    HO—C—H     HO—C—H     HO—C—H     HO—C—H
    5 |           |           |           |           |           |           |           |
  H—C—OH      H—C—OH     H—C—OH      H—C—OH     H—C—OH      H—C—OH     H—C—OH      H—C—OH
    6 |           |           |           |           |           |           |           |
      CH2OH       CH2OH       CH2OH       CH2OH       CH2OH       CH2OH       CH2OH       CH2OH
  D-allose    D-altrose   D-glucose   D-mannose   D-gulose    D-idose   D-galactose  D-talose
```

Chemical Perspective / Birth of Stereochemistry

On a fine spring day in 1848, the French chemist Louis Pasteur was trying to precipitate crystals of tartaric acid salt. When he looked at the crystals under a microscope, he discovered that there were two crystal forms: one that he saw as ''right-handed'' and the other as ''left-handed.'' Using tweezers, he sorted the two types of crystals into two piles. When he dissolved the piles in separate flasks and shined rays of plane-polarized light through each solution, he found that one solution rotated the light plane toward the right and the other rotated it an equal angle to the left. This discovery led to the birth of stereochemistry, an area of organic chemistry that studies the three-dimensional configuration of molecules and its effects on the properties of those molecules.

Stereochemistry is a very active area of chemical research because drug and pesticide manufacturers are finding that mirror-image twins, called enantiomers, can cause vastly different responses in organisms. In nature, usually only one of these twins is found. That is because the enzymes produced in living organisms can recognize, bind to, and produce only one enantiomer. So when drugs are extracted from plants, they contain only one of the twins and are pure. This is not true when chemists synthesize a drug. In this case the synthesis produces a racemate—a 50:50 mixture of enantiomers that are often extremely difficult to separate. In fact, as much as 75% of the drugs sold on the market are racemates. This means that each dose contains an equal weight of isomers, half of which may have little therapeutic value or may even have harmful side effects.

The importance of being able to produce pure enantiomers from a racemate mixture is illustrated by the drug thalidomide. In the early 1960s, this drug was prescribed to thousands of patients as a sleeping pill. Thalidomide was considered safe until it was discovered that women taking it in the first trimester of pregnancy gave birth to babies with malformed or shortened limbs. Research now suggests that if pregnant women could have taken only the drug's ''right-handed'' enantiomer, thalidomide might have been safe. The ''left-handed'' enantiomer appears to have been responsible for the birth defects. There are other well-known examples of racemate problems. One enantiomer of ethambutol fights tuberculosis, whereas the other causes blindness. One enantiomer of naproxen reduces arthritic inflammation, whereas the other poisons the liver.

Until recently, chemists were not able to produce many drugs in forms other than as racemates. Now, however, they have learned to use microorganisms to produce large amounts of one enantiomer of vitamin C, as well as the desired enantiomer of aspartame (an artificial sweetener). Chemists are also producing synthetic enzymes that allow the production of almost pure enantiomers of such complex molecules as beetle pheromone, a sex attractant used in insect traps. As these techniques become perfected, drug manufacturers will increasingly be able to produce pure enantiomers that avoid the possible problems of racemate contamination.

Your Perspective: For some important drugs synthesized in the laboratory, scientists do not yet know how to separate the enantiomers. Thus, the only way to make the drug available to large numbers of people is in the form of a racemic mixture. If the undesirable enantiomer has potentially dangerous side effects, how should the drug company and the government go about deciding whether to distribute the drug? Write a brief essay proposing some guidelines for making such a decision.

hyde around its highest numbered chiral center. Most naturally occurring carbohydrates belong to the D-family. Of the 16 possible optical isomers of the aldohexoses (8 in the D-family and 8 in the L-family), the three most abundant in nature are D-glucose, D-mannose, and D-galactose.

Again, the small difference in the arrangement of atoms between the D- and L-isomers may seem unimportant to you, but to the cells of a living organism the difference is critical. Cells can recognize the difference and often can use only one of the isomers. For example, yeast can ferment D-glucose to produce alcohol but not L-glucose. As we'll see in Chapter 17, our cells can use only L-amino acids to build proteins. This is because the enzymes that catalyze reactions in cells are asymmetrical compounds themselves, just as your shoes are asymmetrical. To catalyze a reaction, the enzyme must fit the reactant—just as your right shoe can fit only your right foot.

15.4 Fructose

Fructose (also called levulose, or fruit sugar) is a ketohexose found in many fruit juices and in honey (see Table 15.1). It is the sweetest sugar known, much sweeter than table sugar. Fructose is a part of the disaccharide sucrose; it is produced by hydrolysis of the polysaccharide inulin. Fructose molecules form internal hemiketals and exist in five-member ring structures such as the following:

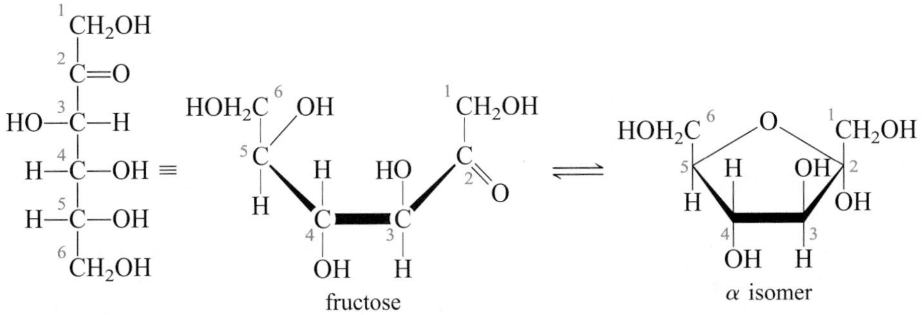

fructose

α isomer

15.5 Galactose

Galactose is not found naturally as a free monosaccharide, but it can be formed from the hydrolysis of larger carbohydrates. It is a part of lactose, the sugar found in milk; thus, it is a part of our diets from birth.

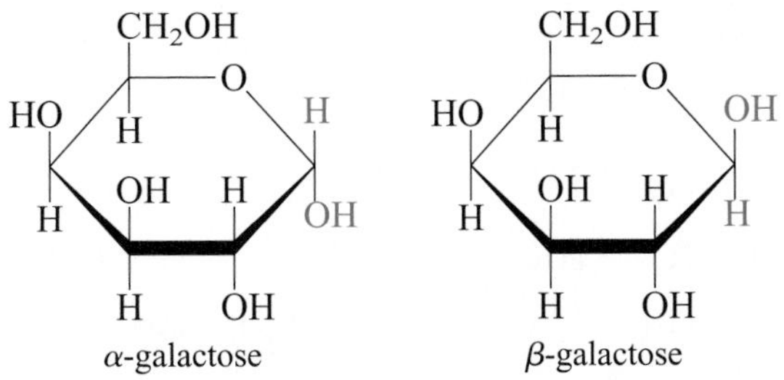

α-galactose

β-galactose

Galactose is also found in glycolipids, fat-like substances that are components of the brain and nervous system. Agar, a polysaccharide extracted from certain types of seaweed, is a polymer of galactose that humans cannot digest. Nevertheless, it is used as a thickener in sauces and ice creams, as well as in nutrient broths and microorganism culture media plates used in microbiology.

15.6 Pentoses

Some important five-carbon sugars are arabinose, which is formed by the hydrolysis of gum arabic and the gum of a cherry tree; and xylose, which is a component of wood, straw, corncobs, and bran (see Fig. 15.4). Pentoses that play a role in human chemistry are ribose and deoxyribose, which are the sugars found in the nucleic acids (the subject of Chapter 20). Both the alpha and beta forms of ribose exist in solution, but only the beta form is found in nucleic acids and other metabolically active compounds.

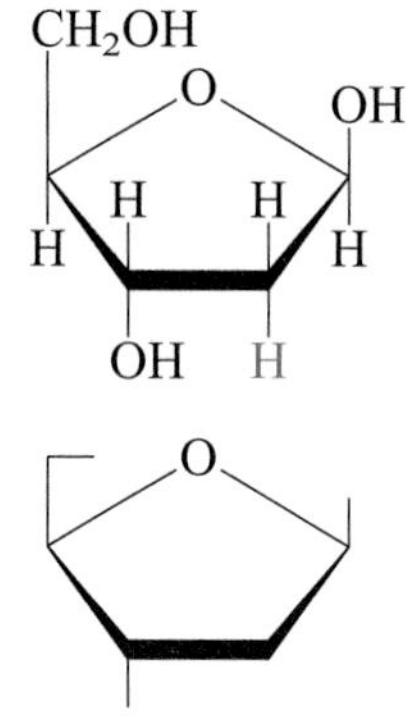

β-deoxyribose

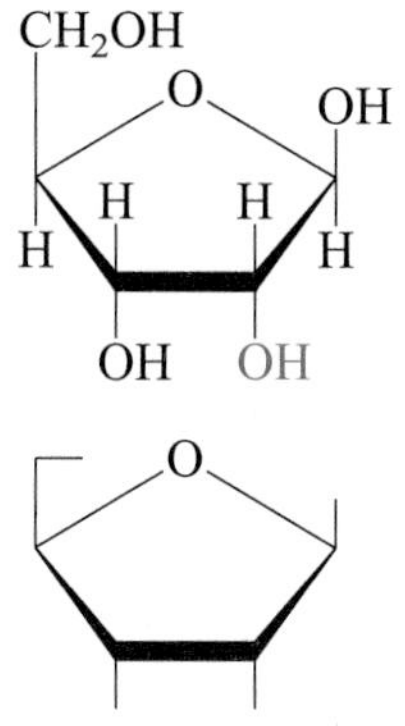

β-ribose

15.7 Reducing Sugars

State what makes a substance a reducing sugar, and describe the test used to identify reducing sugars.

Carbohydrates that contain a free, or potentially free, aldehyde or ketone group reduce alkaline (basic) solutions of mild oxidizing agents, such as Cu^{2+} (Benedict's solution, Fehling's solution) or Ag^{+} (Tollens's solution). Benedict's solution is widely used for the detection of reducing sugars. The reagent for this test contains an alkaline solution of copper(II) sulfate, $CuSO_4$. This blue reagent solution is mixed with the unknown solution and heated. If a reducing sugar is present, an oxidation–reduction reaction takes place. The aldehyde group on the sugar is oxidized, and the Cu^{2+} ions are reduced to Cu^{+} ions, forming a brick-red precipitate of copper(I) oxide (Cu_2O). The amount of precipitate formed indicates the amount of reducing sugar present. Glucose is a reducing sugar, and if a solution of glucose is mixed with Benedict's solution, the following reaction occurs:

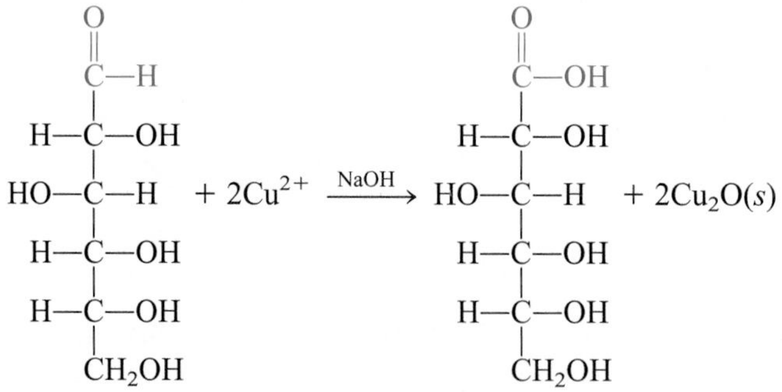

$$\underset{\text{D-glucose}}{\text{CHO–(CHOH)}_4\text{–CH}_2\text{OH}} + 2Cu^{2+} \xrightarrow{\text{NaOH}} \underset{\text{D-gluconic acid}}{\text{COOH–(CHOH)}_4\text{–CH}_2\text{OH}} + 2Cu_2O(s)$$

All monosaccharides yield a positive Benedict's test. Clinitest, which is a widely used brand of tablets for testing for sugar in the urine, is based on the same principle as Benedict's test. In this case, a green color indicates very little sugar in the urine, whereas a brick-red color indicates more than 2 g of reducing sugar per 100 mL of urine.

Note, however, that a false-positive Benedict's test for glucose (and an erroneous diagnosis of diabetes mellitus) can occur with patients having a rare condition called pentosuria. These persons have the ketopentose xylulose, a reducing sugar, in their urine but suffer no ill effects from it. To guard against such an erroneous diagnosis, Tollens's pentose test, which identifies pentose sugars, can be used to determine if the sugar found in the urine is glucose or xylulose.

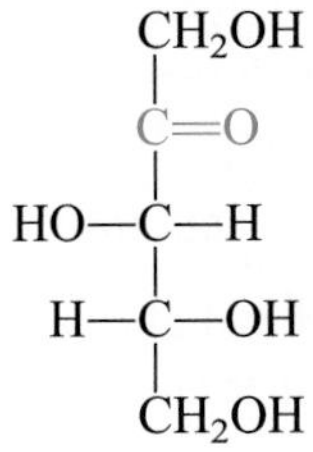

$$
\begin{array}{c}
CH_2OH \\
| \\
C{=}O \\
| \\
HO{-}C{-}H \\
| \\
H{-}C{-}OH \\
| \\
CH_2OH
\end{array}
$$

D-xylulose

Check Your Understanding

Classify each of the following by its carbonyl functional group and by the number of carbons in the molecule. For example, glucose is an aldohexose.

1. ribose

2. fructose

3. threose

4. galactose

5. mannose

6. ribulose

15.8 Disaccharides: Maltose

Maltose (malt sugar) is a disaccharide made up of two glucose units. It is produced by the incomplete hydrolysis of starch, glycogen, or dextrins. Maltose that is produced from grains germinated under controlled conditions is called malt and is used in the manufacture of beer.

List three disaccharides, and explain the difference in their structures.

Disaccharides are formed by a condensation reaction between two monosaccharides. The reaction involves the formation of an acetal (called a glycoside) from a hemiacetal and an alcohol (see Section 13.10).

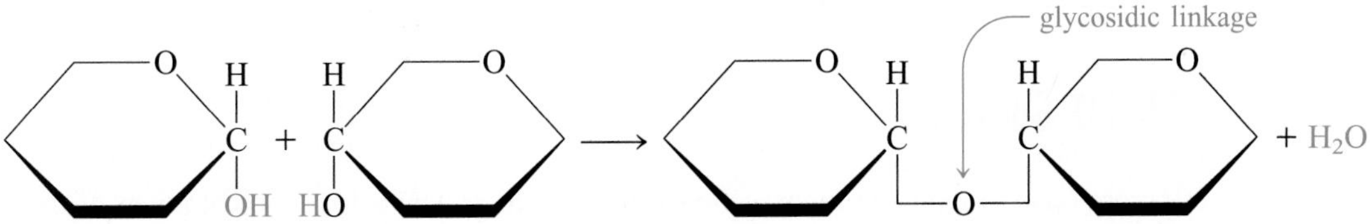

glycosidic linkage

H—C—OH + HO—C—H → H—C—O—C—H + H_2O

In the condensation reaction, one monosaccharide unit acts as the hemiacetal and the other as the alcohol. The linkage that is formed is called a glycosidic linkage (or acetal linkage) and is more stable than the hemiacetal. This linkage does not react with bases; only acids or specific enzymes can break the bond. The glycosidic linkage in maltose occurs between carbon 1 of a glucose molecule in the alpha form and carbon 4 on the other glucose. Such a bond is called an α,1:4 linkage. The bond is also sometimes called an $\alpha(1 \rightarrow 4)$ linkage or an α-1,4 linkage.

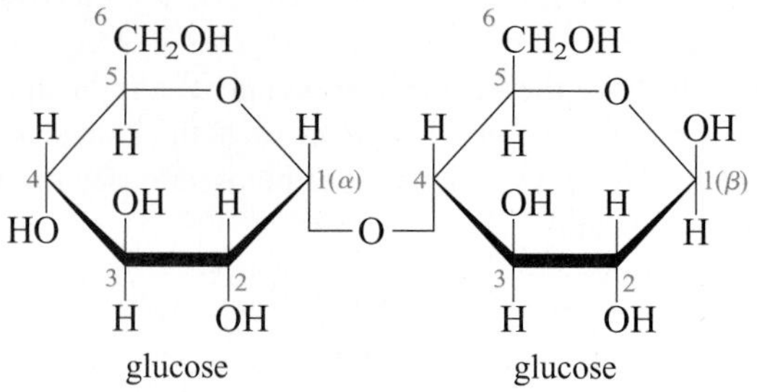

glucose glucose

maltose

Because the aldehyde group of the second glucose molecule is not involved in the glycosidic linkage, maltose can exist in either an alpha or beta form and is a reducing sugar.

15.9 Lactose

Lactose (milk sugar) is found only in the milk of mammals. Its synthesis within the mammary glands from glucose and galactose is regulated by hormones produced after giving birth. Four percent to five percent of cow's milk is lactose, whereas human milk contains 6% to 8%. Lactose itself is a colorless powder that is nearly tasteless. It can, therefore, be used in large amounts in special high-calorie diets.

Lactose is formed by a condensation reaction between glucose and galactose. The bond is formed between carbon 1 of galactose in the beta form and carbon 4 of glucose, resulting in a β,1:4 linkage.

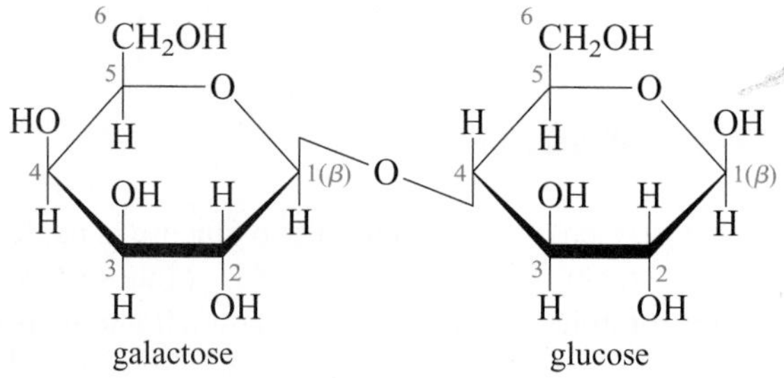

galactose glucose

lactose

As in the case with maltose, lactose has a potentially free aldehyde group in the glucose unit and is a reducing sugar.

15.10 Sucrose

Sucrose (also called table sugar, cane sugar, and beet sugar) is found in the juices of fruits and vegetables. It is produced commercially from sugar cane or sugar beets and is the sugar that we use in cooking. It is estimated that we each consume an average of 100 lb of sucrose a year.

Health Perspective / Lactose Intolerance and Galactosemia

The disaccharide lactose is an important dietary carbohydrate that is found in the milk of all mammals. Lactose digestion occurs in the small intestine, where the enzyme lactase (β-galactosidase) hydrolyzes the molecule to form galactose and glucose. Many adults, including most blacks and almost all Asians, have a low level of this enzyme. The lactose in the milk they drink travels through the small intestine to the colon mostly undigested. In the colon, bacterial fermentation of the lactose produces large quantities of CO_2, H_2, and irritating organic acids. This results in bloating, flatulence, abdominal pain, and diarrhea from drinking milk. The diarrhea results from fluids moving into the small intestine in response to the osmotic pressure from the undigested lactose. Such an inability to metabolize lactose is called lactose intolerance, or milk intolerance. People who suffer from lactose intolerance are able to eat cheese, yogurt, and other cooked foods that contain milk because the lactose is hydrolyzed when milk is fermented or cooked. Food scientists have now produced milk in which the lactose is enzymatically hydrolyzed to glucose and galactose. Such milk can be consumed without a problem by individuals with lactose intolerance.

A more serious disease involving lactose is galactosemia. More than one out of every 20,000 children is born with a deficiency in certain enzymes necessary for the metabolism of galactose. Their bodies can hydrolyze lactose to form glucose and galactose, but they are unable to metabolize the galactose. As the galactose accumulates in the blood and urine, the body converts it to other compounds that can lead to cataract formation and mental disorders. In more severe cases, some children have died from liver damage resulting from the buildup of galactose.

Your Perspective: You work for a public health clinic that commonly publishes a number of brochures explaining general health issues to city residents. In language that can be understood by a local population having only basic skills in reading English, write a short description of lactose intolerance, describing its symptoms and causes.

A sucrose molecule contains one unit of glucose and one unit of fructose. The linkage occurs between the hemiacetal group of glucose and the hemiketal group of fructose and is called an α,1:2 linkage.

glucose — fructose

sucrose

Because both the aldehyde group of glucose and the ketone group of fructose are involved in the linkage, sucrose does not have a potentially free aldehyde or ketone group and is not a reducing sugar. Sucrose can be hydrolyzed by acids or enzymes found in the intestines and in yeast. Hydrolysis of sucrose produces a mixture of fructose and glucose called invert sugar.

Check Your Understanding

Complete the chart.

Disaccharide	*Source*	*Monosaccharide Subunits*	*Reducing Sugar*	*Type of Linkage*
Sucrose				
Lactose				
Maltose				

15.11 Polysaccharides: Starch

Describe the metabolic function of starch.

Describe how the structure of amylose differs from amylopectin.

Polysaccharides are complex carbohydrate polymers containing three or more monosaccharides. They are used both for storing energy and as part of the structural fibers of the organism. Starch is the storage form of glucose used by plants. Granules of starch are found in organelles in their leaves, roots, and seeds. The granules are insoluble in water; their coating must be broken open for the starch to mix with water. Heat breaks open the granules, producing a colloidal suspension whose thickness increases with heating. For this reason, cornstarch is widely used as a thickening agent in cooking.

Natural starches are a mixture of two types of polysaccharides: amylose and amylopectin. Amylose is a large linear polysaccharide (molecular weight 150,000–600,000) whose glucose units are connected by α,1:4 linkages.

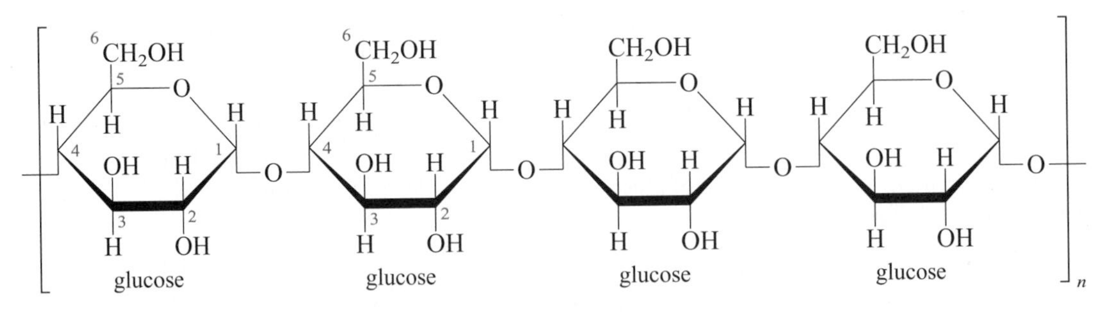

amylose

Amylopectin (molecular weight of 1–6 million) is a highly branched glucose polymer. The nonbranching portion of the molecule has glucose units connected by α,1:4 linkages. The branching occurs every 20 to 24 glucose units and is a result of α,1:6 linkages between carbon 1 on one glucose in the alpha form and carbon 6 on another glucose (Figs. 15.5 and 15.6).

Dextrins are polysaccharides formed by the partial hydrolysis of starch by acids, enzymes, or dry heat. The golden color of bread crust results from the formation of dextrins. Dextrins get sticky when wet and are therefore used as adhesives on stamps and envelopes and in wallpaper paste.

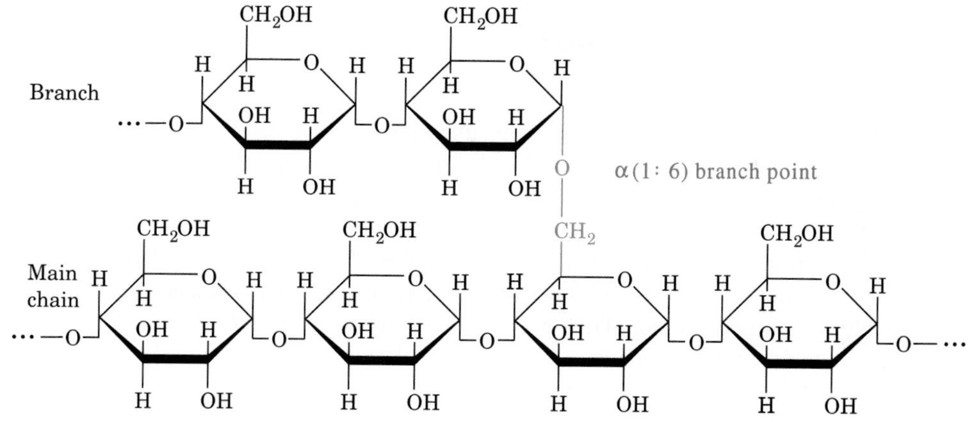

Figure 15.5 Structure of amylopectin and glycogen. The glucose units are connected by α,1:4 and α,1:6 linkages.

15.12 Glycogen

Glycogen is a heavily branched molecule that is the storage form of glucose in animals. It accounts for about 5% of the weight of the liver and 0.5% of the weight of the muscles in the body. Enough glucose is stored as glycogen in a well-nourished body to supply it with energy for about 18 hours.

The structure of a glycogen molecule is similar to that of amylopectin (see Figs. 15.5 and 15.6). It consists of straight chains of glucose units connected by α,1:4 linkages. The branching that results from α,1:6 linkages between glucose units in a glycogen molecule is more frequent than in amylopectin, occurring every 8 to 12 glucose units.

State the metabolic function of glycogen, and describe how its structure compares with that of starch.

The ability of the body to form glycogen from glucose (and glucose from glycogen) is extremely important because glucose is the main source of energy for all cells. When we eat a meal, glucose resulting from the breakdown of carbohydrates enters the bloodstream. If this large amount of glucose were to remain in the blood, the osmotic balance between the blood and the extracellular and intracellular fluids would be completely disrupted. The excess glucose, however, does not circulate in the blood but is instead converted to

Figure 15.6 The highly branched structure of the amylopectin and glycogen molecules is a result of the α,1:6 linkages. The red glucose units indicate branch points.

Chemical Perspective / Starches: How Do They Differ?

Plant starches are a mixture of two molecules: the straight-chain amylose and the branched-chain amylopectin. Both molecules are insoluble in cold water, but form colloidal dispersions in hot water. The branched chains in the amylopectin keep the molecules from packing together too tightly and allow for easier penetration by water molecules. Once the branched chains are dispersed in the hot water, they have a tendency to remain that way. On the other hand, the straight-chain molecules in amylose easily align with one another. They pack much closer together and are not so easily penetrated by water. When amylose molecules are dispersed in hot water, they eventually join back together and precipitate.

This explains how bread becomes stale. Bread is made from flour, which contains a mixture of both kinds of starch. Over time the straight-chain amylose molecules find their way back together and form the firmer, rigid texture of stale bread. You may have observed that if you microwave a stale piece of bread or a roll, it regains the soft texture it had when it was fresh. When microwaved, the water molecules in the bread are excited to high energies and can again disperse the starch molecules. This softness doesn't last long, however, since much of the water is lost from the bread when it is microwaved, and the starches quickly come back together.

By separating the two types of starch molecules, food chemists can take advantage of their differing properties. When amylose is heated and then cooled, it forms an opaque, rigid gel that is useful for making chewy candies, puddings, and digestible films. When amylopectin is heated and later cooled, it forms a clear, pastelike substance that does not gel. This makes it particularly useful for making gravies and sauces, and in the preparation of adhesives. The major source of starch in the United States is cornstarch, which contains 70% to 80% amylopectin and 20% to 30% amylose. This is a much higher percentage of branched-chain amylopectin than in ordinary flour, which is why many cooks prefer cornstarch for thickening gravies.

Anyone who has ever made real pudding knows how long you have to stand at the stove, constantly stirring until the pudding reaches the boiling point and starts to thicken. By inserting phosphate ester groups into straight-chain amylose molecules, food chemists can disrupt the tight molecular alignment and obtain quicker dispersion in water. Starches with these phosphate ester groups are used in ''instant'' puddings that do not need to be heated to disperse the starch and thicken the pudding.

glycogen in the liver. The large, branched glycogen molecule is ideally suited for storage because it cannot pass through cell membranes. Later, as glucose is used by the cells, the blood glucose is maintained at its normal level by the hydrolysis of glycogen in the liver and the resulting release of glucose into the blood. In this way, the blood glucose level remains relatively constant even though we eat at widely spaced intervals during the day.

15.13 Cellulose

Cellulose is a glucose polymer (molecular weight 150,000–1,000,000) produced by plants. It makes up the main structural support for plants, whose cells produce this compound to form the exterior cell wall. The glucose units in cellulose are held together by a glycosidic linkage between carbon 1 in the beta position on the first glucose and carbon 4 on the second glucose.

glucose glucose glucose glucose

cellulose

Cellulose fibers are composed of many long cellulose chains tightly held together by many hydrogen bonds. This hydrogen-bonded structure gives the cellulose fibers exceptional strength and rigidity and makes them insoluble in water (Fig. 15.7).

Humans do not possess enzymes that can break the glycosidic (β,1:4) linkage that occurs between glucose units in cellulose. Therefore, any cellulose we eat passes through the digestive tract undigested, supplying the roughage we need for proper elimination. Some microorganisms, however, can digest cellulose. Grass-feeding animals such as cows have extra stomachs to hold the grass for long periods while these microorganisms hydrolyze the cellulose into glucose.

Explain why we can digest starch but not cellulose.

Over 50% of the total organic matter in the living world is cellulose. For example, wood is about 50% cellulose, and cotton is almost pure cellulose. When treated with a wide variety of chemicals, cellulose forms many useful products: celluloid; rayon; guncotton (an explosive); cellulose acetate (used in plastics, food wrapping films, and fingernail polish); methyl cellulose (used in fabric sizing, pastes, and cosmetics); and ethyl cellulose (used in plastic coatings and films).

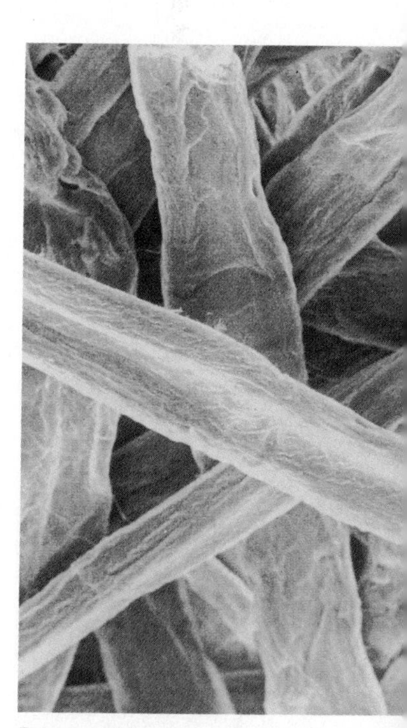

Figure 15.7 Electron micrograph of a cell wall showing the cellulose fibers.

15.14 Iodine Test

The iodine test is used to detect small amounts of starch in a solution. Starch produces a dark blue-black color when mixed with the iodine test reagent, a solution of potassium iodide containing iodine. The test can be used to monitor the hydrolysis of starch—the color slowly changes as the starch continues to be hydrolyzed into shorter carbon chain products.

Hydrolysis:	starch $\xrightarrow{\text{hydrolysis}}$	dextrins $\xrightarrow{\text{hydrolysis}}$	glucose
Iodine Test:	*blue-black* $\longrightarrow$	*reddish* $\longrightarrow$	*colorless*

15.15 Photosynthesis

We have said that polysaccharides are energy storage molecules. But where does the energy come from? It comes originally from the sun. Nuclear reactions occurring on the sun produce energy that radiates out into space. Plants growing on the earth trap this radiant energy and use it to produce carbohydrates and certain amino acids. **Photosynthesis,** the process by which plants capture and use this energy, is quite complex and not

Describe photosynthesis.

Chemical Perspective / Hydrolysis of Starch

Because of the huge size of starch molecules (molecular weights ranging from 150,000 to 6,000,000), an enormous number of glycosidic bonds must be broken to obtain the individual glucose units. This can often be a problem in digesting the starches in our foods. By eating slowly, and carefully chewing our foods, we can aid digestion by providing a larger surface area for the hydrolysis of starches. Cooked foods are much easier to digest because many of the starches and celluloses are hydrolyzed by the heat to maltose. This also accounts for the much sweeter taste of cooked vegetables such as carrots compared with their taste when raw. Even the partial hydrolysis of starches to dextrins makes them easier to digest. This is one reason for the age-old treatment of stomach upset with tea and toast. Toast (and even bread crusts) is easier to digest because of the large amounts of dextrin. Infant formulas often contain mixtures of dextrins and maltose rather than starch for easier digestion.

Corn syrup is another example of the hydrolysis of starch. It is made by the hydrolysis of cornstarch, producing a mixture of dextrins, maltose, and glucose. This mixture is then treated with the enzyme glucose isomerase, which catalyzes the conversion of some of the glucose to fructose. Since fructose is about 1.5 times as sweet as sucrose, the resulting syrup is often used as a substitute for table sugar. Corn syrup is similar to honey, which is mostly a mixture of glucose and fructose (invert sugar) that bees make from flower nectar. Both products find wide use as sweeteners in products like pancake syrups, candies, and cake frostings.

Write the overall equation for the reactions of photosynthesis.

totally understood. We do know that the reactions of photosynthesis require the presence of light and molecules of chlorophyll. The reactions of photosynthesis take place in regions of the plant cell called chloroplasts. These reactions can be divided into two categories: the light reactions and the dark reactions. The light reactions require the presence of chlorophyll to absorb radiant light energy and to use this energy in the production of oxygen and energy-rich molecules. The dark reactions then use the energy-rich molecules to reduce carbon dioxide to glucose and other organic products. The overall equation, summarizing the many reactions in this process, shows the formation of glucose from carbon dioxide and water.

$$\text{Energy} + 6CO_2 + 6H_2O \longrightarrow \underset{\text{glucose}}{C_6H_{12}O_6} + 6O_2$$

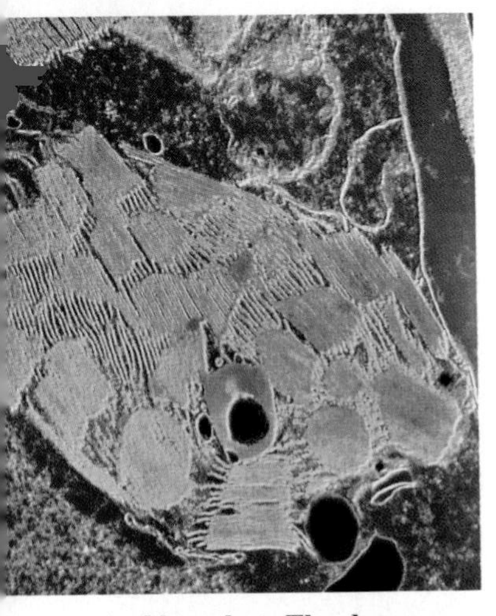

A chloroplast. The dense green stacks are the grana, the main location of the chlorophyll molecule

The reverse of this equation, in which cells convert glucose to carbon dioxide and water to produce the energy necessary for life (a process called cellular respiration) is the subject of Chapter 19. As the critical component of organic molecules, carbon is continuously exchanged among the earth, its atmosphere, and living organisms in a cyclic process called the **carbon cycle** (Fig. 15.8). Green plants absorb carbon (in the form of carbon dioxide) from the atmosphere and, through photosynthesis, use it to form glucose and other organic molecules. Carbon dioxide is released back into the atmosphere through cellular respiration, decomposition of dead organisms, and the burning of fossil fuels, formed over time by deposits of nondecayed organic materials.

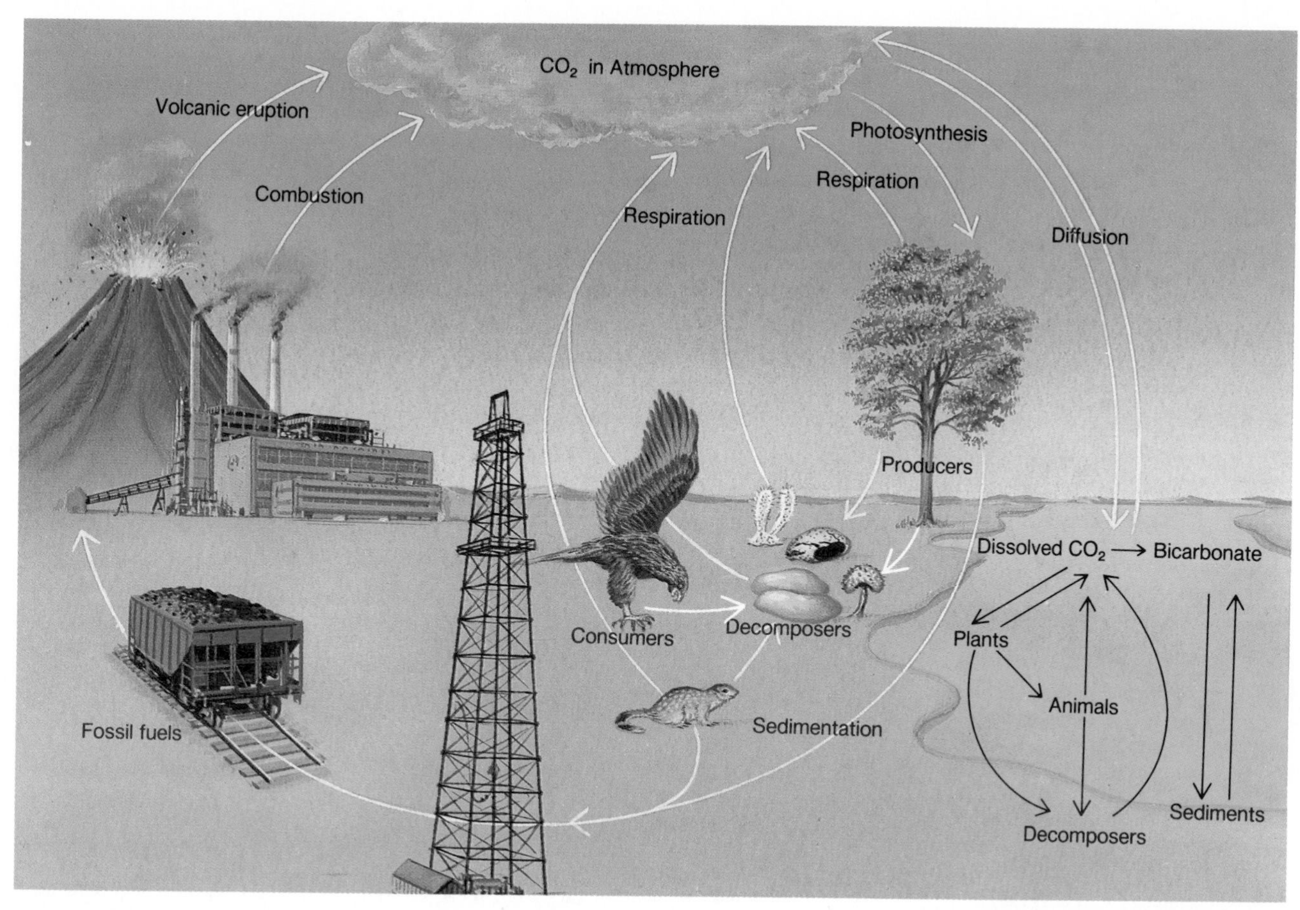

Figure 15.8 The carbon cycle involves the cycling of carbon in different chemical forms from the atmosphere to living organisms and back into the atmosphere.

Check Your Understanding

Identify the following carbohydrates:

1. glucose storage molecule in animals
2. product of the partial hydrolysis of starch
3. glucose storage molecule in plants
4. main structural molecule in plants
5. end product of photosynthesis

KEY CONCEPTS

Carbohydrates form a large class of compounds that includes monosaccharides, disaccharides, and polysaccharides.

Monosaccharides are carbohydrates that cannot be converted into smaller units on hydrolysis. Important examples are glucose, fructose, galactose, and ribose. Monosaccharides exist mainly as five- or six-sided ring structures that result from the formation of an internal hemiacetal or hemiketal. These rings have alpha and beta forms that make a great difference in how the molecule can be used by an organism. All monosaccharides are reducing sugars and yield a positive Benedict's test.

Disaccharides produce two monosaccharides on hydrolysis. The linkage formed between the monosaccharide units is a glycosidic linkage and is quite stable. Important examples of disaccharides are sucrose (fructose + glucose), maltose (glucose + glucose), and lactose (glucose + galactose).

Polysaccharides produce more than two monosaccharides on hydrolysis. Polysaccharides can be linear polymers of glucose, such as amylose (glucose units connected by α,1:4 linkages) or cellulose (glucose units connected by β,1:4 linkages), or can be branched polymers of glucose, such as glycogen and amylopectin (glucose units connected by α,1:4 linkages, with branches formed by α,1:6 linkages).

Important Equations

Formation of Cyclic Hemiacetals

CHO
H—C—OH
HO—C—H
H—C—OH
H—C—OH
CH_2OH

⟶

CH_2OH, O, H, OH, HO, OH, OH + CH_2OH, O, OH, OH, HO, H, OH

Photosynthesis

$$\text{Energy} + 6CO_2 + 6H_2O \longrightarrow C_6H_{12}O_6 + 6O_2$$

Review Problems

Section 15.1

1. Describe the differences among monosaccharides, disaccharides, and polysaccharides.
2. Identify each of the following as (a) an aldose or ketose, (b) a triose, tetrose, pentose, hexose, or heptose.

(a)
H
C=O
H—C—OH
H—C—OH
H—C—OH
CH_2OH

(b)
CH_2OH
C=O
HO—C—H
H—C—OH
H—C—OH
H—C—OH
CH_2OH

(c)
CH_2OH
C=O
HO—C—H
H—C—OH
H—C—OH
CH_2OH

(d)
H
C=O
H—C—OH
CH_2OH

Section 15.2

3. Draw the following:
   (a) the linear structure of glucose
   (b) the ring structure of α-glucose
   (c) the ring structure of β-glucose
4. What is one way in which the position of the hydroxyl group on carbon 1 in glucose is important in living organisms?

Section 15.3

5. Draw the open-chain configuration for the following L-sugars. (An L-sugar is the mirror image of the corresponding D-sugar.)
   (a) L-erythrose (b) L-xylose (c) L-mannose
6. How many different aldoheptoses are theoretically possible? Of these sugars, how many belong to the D-family and how many to the L-family?

Section 15.4

7. Draw the hemiketal formed by a molecule of fructose.
8. Because fructose forms a hemiketal, it can exist in an alpha and a beta form. The structure shown in Section 15.4 is the alpha form. Draw the beta form.

Section 15.5

9. Draw the equilibrium that shows the interconversion of galactose between the open-chain form, the alpha form, and the beta form.
10. Name two foods you ate recently that contain lactose.

Section 15.6

11. Examine the following Haworth structure for a simple sugar:

CH_2OH
O H
C C
H H
H C C OH
OH OH

   (a) Is this a hexose or a pentose?
   (b) Is it an alpha or beta form?
   (c) Draw the open-chain form for this molecule.
   (d) What is the name for the open-chain molecule?
12. What structural difference do you observe between β-ribose and β-deoxyribose?

Section 15.7

13. Which of the following sugars yields a positive Benedict's test? Give the reason for your answer.

(a) fructose (d) maltose
(b) ribose (e) sucrose
(c) lactose

14. Write the equation for the reaction that occurs when a solution of galactose is mixed with Benedict's solution.

Section 15.8

15. Write the equation for the hydrolysis of maltose, and name the product(s).
16. Draw the beta form of maltose. Circle the acetal and hemiacetal groups.

Section 15.9

17. Write the equation for the hydrolysis of lactose, and name the resulting product(s).
18. Draw the alpha form of lactose. Circle the acetal and the hemiacetal groups.

Section 15.10

19. Write the equation for the hydrolysis of sucrose, and name the resulting product(s).
20. Draw sucrose, and circle the acetal and ketal groups.
21. Draw D-xylulose (see Section 15.7) in its five-member cyclic β-hemiketal form.
22. Draw D-mannose (see Fig. 15.1) in its six-member cyclic α-hemiacetal form.
23. Ribulose, a ketopentose, has the open-chain form shown here. Draw ribulose in its five-member cyclic alpha and beta forms.

$$
\begin{array}{c}
CH_2OH \\
| \\
C{=}O \\
| \\
H{-}C{-}OH \\
| \\
H{-}C{-}OH \\
| \\
CH_2OH
\end{array}
$$

24. The prefix deoxy- means "minus an oxygen." Draw the structure of both the ring form and the open-chain form of 2-deoxyglucose.

Section 15.11

25. (a) Name the two polysaccharides making up starch.
    (b) Describe the difference in their structures.
26. What is the role of starch in plants?

Section 15.12

27. What is the storage form of glucose in animals? How does its structure compare with that of starch?
28. Compare the structure of glycogen with that of (a) amylose, and that of (b) amylopectin. How do the structures differ? How are they similar?

Section 15.13

29. Compare the structure of cellulose with that of amylose. How do the structures differ? How are they similar?

30. Why are humans unable to digest cellulose?

Section 15.14

31. Design an experiment for the hydrolysis of starch. How would you check the progress of the hydrolysis?

32. What monosaccharide is obtained when starch is completely hydrolyzed?

Section 15.15

33. **(a)** Name the reactants that must be present for photosynthesis to occur.

(b) Where do the reactions of photosynthesis take place?

34. **(a)** Write the overall equation for the reactions of photosynthesis.

(b) What is the energy source for these reactions?

Applied Problems

35. Suppose that a sample of urine from an infant gives a positive test with a Clinitest tablet. Is it correct for the analyst to report glucose in the urine? Why or why not?

36. Although table sugar is sucrose, this carbohydrate is never administered intravenously, whereas glucose is. Why is this so?

37. Both celery and potato chips are composed of molecules that are polymers of glucose. Explain why celery is a good snack for people on a diet, whereas potato chips are not.

38. The disaccharide called melibiose is found in some plant juices.

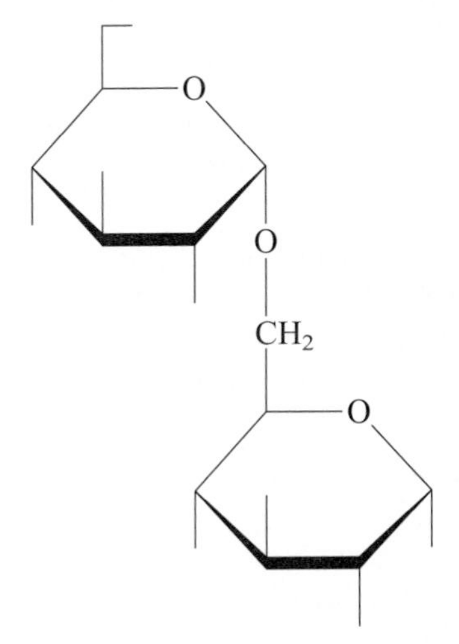

(a) What are the two monosaccharides that make up this disaccharide?

(b) Is this disaccharide a reducing sugar? Why or why not?

(c) What type of linkage connects the two monosaccharides?
(d) What polysaccharide also contains this linkage?

39. Raffinose is a trisaccharide that is widely distributed in nature.

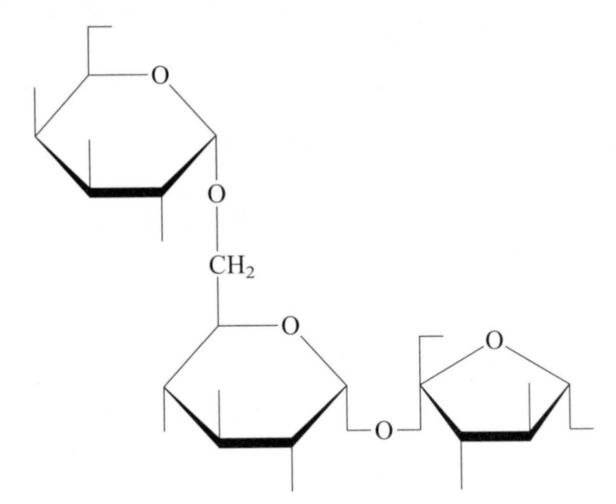

(a) Is raffinose a reducing sugar? Explain your answer.
(b) Identify the types of glycosidic linkages that exist in this trisaccharide.
(c) Identify the monosaccharides produced by the hydrolysis of raffinose, and draw the structure of each.

40. We saw in Section 15.8 that disaccharides form by a condensation reaction in which one monosaccharide acts as the hemiacetal and the other acts as the alcohol to form an acetal. The linkage between the two monosaccharides is called the glycosidic linkage. Monosaccharides can also react with other alcohols that are not monosaccharides. When a monosaccharide is treated with an alcohol in a strong acidic solution, a glycoside forms. Write the equation for the reaction of α-glucose with ethanol to form ethyl-α-glucoside, showing the product in its Haworth structural formula.

41. Maltose is a reducing sugar formed by the condensation of two glucose units. Explain how to join two molecules of α-glucose so that the resulting disaccharide is not a reducing sugar.

42. Because of the way glucose and fructose link together to form the disaccharide sucrose, it is not a reducing sugar. Show how these two monosaccharides can be joined together to form a reducing sugar.

43. Trehalose is a carbohydrate found in mushrooms and yeast as well as in some insects. It is a disaccharide formed when two glucose units in the alpha-ring form are joined together by an α,1 : 1 linkage.
(a) Draw the structure for this disaccharide.
(b) Is trehalose a reducing sugar?

44. The disaccharide gentiobiose is a rare sugar found in saffron. It can also be called β-glucose–β,1 : 6-glucose.
(a) Draw the structure for gentiobiose.
(b) Is gentiobiose a reducing sugar?

Chapter 16

Lipids

Suzy was found lying in bed. She was snuggled up with her teddy bear, and the alarm on her clock radio was ringing. This would have been her sixth birthday, but the morning never came for her. An autopsy the next day revealed that Suzy's arteries were clogged like those of a 60-year-old man. She had suffered a massive heart attack.

What would make a young child's arteries look like those of a much older person? This was the question that sent Michael Brown and Joseph Goldstein of the University of Texas Health Science Center on a 13-year investigation—a scientific journey that ended in Stockholm, Sweden, where they received the 1985 Nobel Prize in Physiology or Medicine. Along the way they made new biochemical discoveries that not only brought hope to people like Suzy but completely revolutionized the thinking about many chemical processes in the human body.

Suzy suffered from a rare and severe form of familial hypercholesterolemia (FH), in which the level of cholesterol in the bloodstream can be as much as 10 times higher than that in healthy individuals. Cholesterol is a fascinating molecule. It is essential for the formation of cell membranes and several body hormones, yet too much cholesterol in the blood can cause atherosclerosis. In this disease, cholesterol accumulates in the walls of arteries and forms bulges, or plaques. Over time such plaques can become so thick

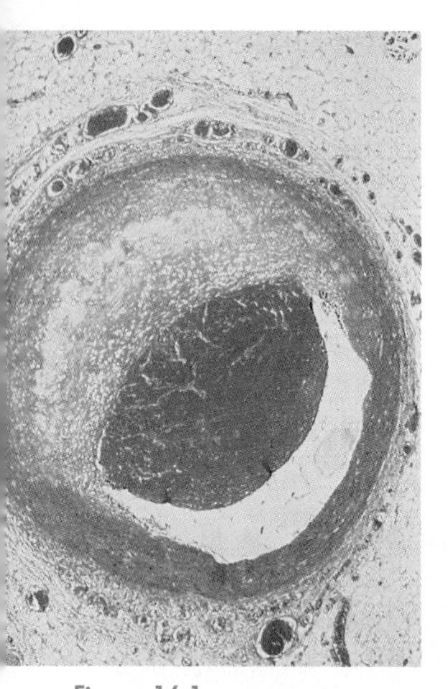

Figure 16.1 An atherosclerotic plaque has thickened the wall of this artery, severely restricting the blood flow to the point that a blood clot has formed.

that they severely restrict the flow of blood, to the point that a blood clot can eventually block the artery completely (Fig. 16.1). A heart attack occurs when such obstruction takes place in an artery feeding the heart muscle itself.

Because cholesterol is completely insoluble in water, it is carried in the bloodstream by special particles called low-density lipoproteins, or LDLs. The more LDL there is in the bloodstream, the greater the chance of suffering atherosclerosis. Sadly, more than half the people in the industrialized countries of the West have blood LDL levels putting them in the high-risk category. Brown and Goldstein, however, wanted to know why some people have much higher levels of LDL in their bloodstream than do others.

By comparing cells of patients suffering from familial hypercholesterolemia with those of normal people, the two scientists discovered that most normal body cells contain special molecules called LDL receptors. The receptors stick out from the plasma membrane surface and "capture" LDL particles circulating in the blood. These LDL receptors tend to cluster on the plasma membrane in indented or dimpled areas called coated pits. Every few minutes the dimpled areas pouch inward and pinch off from the surface to form a sac inside the cell. The LDL particles bound to the LDL receptors in the coated pits are then released and broken down inside the cell to provide the needed cholesterol. The receptors, meanwhile, return to the cell surface to bind to more LDL particles. Brown and Goldstein showed that the cells of people suffering from FH either have no functional LDL receptors at all or have only very few, and therefore do poorly at removing the cholesterol-carrying LDL from the bloodstream. Individuals having the most severe form of FH, like Suzy, have no working LDL receptors at all.

But the work of these two scientists went far beyond explaining the cause and developing a treatment for familial hypercholesterolemia. Brown and Goldstein went on to show that cells have a complex feedback mechanism that regulates the amount of cholesterol they contain. As the cholesterol level in the cell rises, the cell stops synthesizing cholesterol and produces fewer LDL receptors, thus decreasing the cholesterol taken in from the blood. Based on this finding, other investigators showed that the "receptor" method of incorporating LDL into the body's cells is also used for a variety of other large molecules (such as insulin) that are vital to body function.

The investigative work of Brown and Goldstein has been remarkable in its use of methods from many areas of science: genetics, medicine, cell biology, molecular biology, biochemistry, pharmacology, nuclear medicine, and immunology. This type of interdisciplinary biochemical study is typical of the investigations of a large number of scientists who are using their knowledge of chemistry to discover new facts about living organisms.

16.1 Saponifiable Lipids: Simple Lipids

◆ **Describe the function of lipids in living organisms.**

The cholesterol and fats that make up the plaques so important in atherosclerosis belong to a class of compounds called lipids. **Lipids** include all biological compounds that are not soluble in water but are soluble in organic solvents, such as chloroform, methanol, ether, or benzene. This class of compounds includes fats, oils, some vitamins and hormones, and most nonprotein components of cell membranes. In living cells, lipids function as part of the structure of membranes, as energy storage molecules for the cell, and as starting materials for the synthesis of prostaglandins, vitamins, and hormones.

◆ **State the difference between a saponifiable and a nonsaponifiable lipid, and a simple and a compound lipid.**

Lipids are a varied group of compounds that can be categorized in several ways. We can divide the lipids into two major classes: those that can be **saponified** (hydrolyzed by

Table 16.1 **Classification of Lipids**

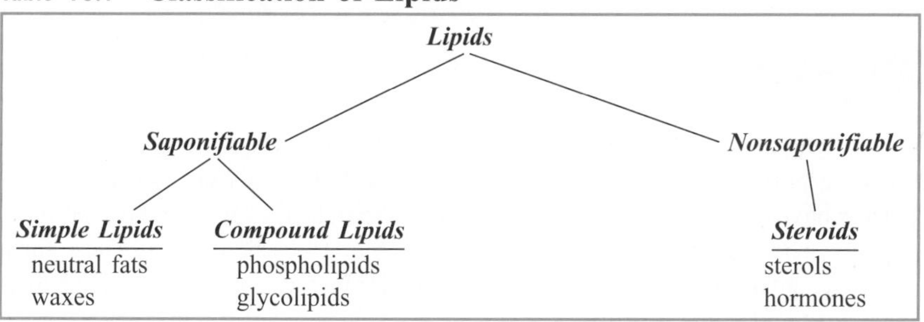

Fat cells

a base) and those that are **nonsaponifiable.** The saponifiable lipids can be further subdivided into **simple lipids,** which yield fatty acids and an alcohol on hydrolysis, and **compound lipids,** which yield fatty acids, alcohol, and other compounds on hydrolysis (Table 16.1).

16.2 Fats and Oils

The simplest and most abundant lipids are the neutral fats, which are also called **triacylglycerols,** or **triglycerides.*** These compounds are esters of glycerol and three fatty acids. They are the main form of fat storage in plants and in the adipose cells (or fat cells) of animals. An average man's body is 21% fat (26% for women), enough fat to supply his body's energy needs for 2 to 3 months.

Draw the general structure of a triacylglycerol.

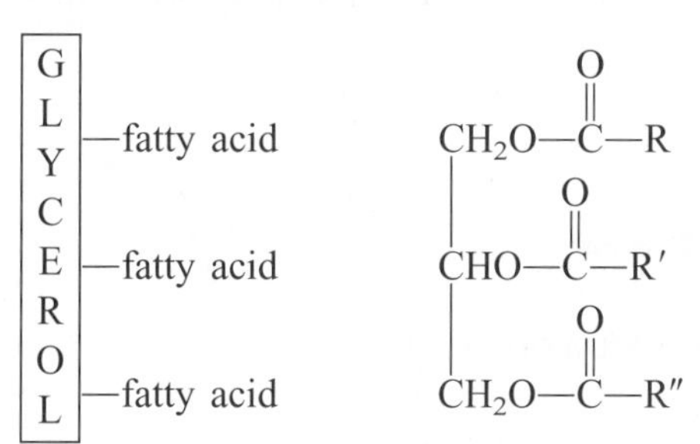

fat, or triacylglycerol

Simple triacylglycerols contain the same fatty acid in all three positions on the glycerol molecule, whereas **mixed triacylglycerols** contain two or more different fatty acids. Natural fats are a mixture of simple and mixed triacylglycerols. Triacylglycerols and their derivatives have very wide commercial use in the production of soaps, varnishes, printing inks, ointments, and creams.

State the difference between a simple and a mixed triacylglycerol.

$$\begin{array}{l} CH_2O\overset{O}{\overset{\|}{C}}(CH_2)_7CH{=}CH(CH_2)_7CH_3 \\ |\\ CHO\overset{O}{\overset{\|}{C}}(CH_2)_7CH{=}CH(CH_2)_7CH_3 \\ |\\ CH_2O\overset{O}{\overset{\|}{C}}(CH_2)_7CH{=}CH(CH_2)_7CH_3 \end{array} \qquad \begin{array}{l} CH_2O\overset{O}{\overset{\|}{C}}(CH_2)_{14}CH_3 \\ |\\ CHO\overset{O}{\overset{\|}{C}}(CH_2)_{16}CH_3 \\ |\\ CH_2O\overset{O}{\overset{\|}{C}}(CH_2)_7CH{=}CH(CH_2)_7CH_3 \end{array}$$

simple triacylglycerol (triolein) — mixed triacylglycerol

*The IUPAC nomenclature committee has recommended that the commonly used (but chemically inaccurate) term *triglyceride* be replaced by the term *triacylglycerol.*

16.3 Fatty Acids

Fatty acids are long-chain carboxylic acids that are formed from the hydrolysis of triacylglycerols. The carbon chains of the naturally occurring fatty acids are generally nonbranching. They have an even number of carbon atoms because they are synthesized from two-carbon units. The most common fatty acids have 16 or 18 carbon atoms. These fatty acids are palmitic, stearic, linoleic, and oleic acids, with oleic acid making up more than half the total fatty acid content of many fats.

◆ **State the difference between a saturated and an unsaturated fatty acid.**

Saturated fatty acids have only single bonds between carbon atoms, are unreactive, and are waxy solids at room temperature. Unsaturated fatty acids have one or more carbon-to-carbon double bonds and are liquids at room temperature (Table 16.2). Stearic and oleic acids are both fatty acids with 18 carbons. Oleic acid is unsaturated—it contains one double bond in the *cis* configuration, which gives the molecule a rigid bend.

$$CH_3CH_2CH_2CH_2CH_2CH_2CH_2CH_2CH_2CH_2CH_2CH_2CH_2CH_2CH_2CH_2CH_2COOH$$

stearic acid

$$\begin{array}{ccccc} & H & & H & \\ & \diagdown & & \diagup & \\ & & C{=}C & & \\ & \diagup & & \diagdown & \\ CH_3CH_2CH_2CH_2CH_2CH_2CH_2CH_2 & & & & CH_2CH_2CH_2CH_2CH_2CH_2CH_2COOH \end{array}$$

oleic acid

We can group unsaturated fatty acids by the number and placement of the double bonds. The *n*-3 family of fatty acids has the first double bond on the third carbon from the

Table 16.2 Common Fatty Acids

Common Name (IUPAC Name)	*Symbol*[a]	*Structure*	*Melting point in °C*
Saturated			
Butyric acid (butanoic acid)	4:0	$CH_3(CH_2)_2COOH$	−4.2
Myristic acid (tetradecanoic acid)	14:0	$CH_3(CH_2)_{12}COOH$	52
Palmitic acid (hexadecanoic acid)	16:0	$CH_3(CH_2)_{14}COOH$	63.1
Stearic acid (octadecanoic acid)	18:0	$CH_3(CH_2)_{16}COOH$	69.6
Arachidic acid (eicosanoic acid)	20:0	$CH_3(CH_2)_{18}COOH$	75.4
Unsaturated			
Oleic acid (9-octadecanoic acid)	18:1*n*-9	$CH_3(CH_2)_7CH{=}CH(CH_2)_7COOH$	13.4
Linoleic acid (9,12-octadecadienoic acid)	18:2*n*-6	$CH_3(CH_2)_4CH{=}CHCH_2CH{=}CH(CH_2)_7COOH$	−9
Linolenic acid (9,12,15-octadecatrienoic acid)	18:3*n*-3	$CH_3CH_2CH{=}CHCH_2CH{=}CHCH_2CH{=}CH(CH_2)_7COOH$	−17
Arachidonic acid (5,8,11,14-eicosatetraenoic acid)	20:4*n*-6	$CH_3(CH_2)_4(CH{=}CHCH_2)_4(CH_2)_2COOH$	−49.5
EPA (5,8,11,14,17-eicosapentaenoic acid)	20:5*n*-3	$CH_3CH_2(CH{=}CHCH_2)_5(CH_2)_2COOH$	−54

[a]Number of carbon atoms: number of double bonds and position of the first double bond from the methyl end of the molecule.

methyl end of the molecule, and the *n*-6 family of fatty acids has the first double bond on the sixth carbon from the methyl end. To indicate the length of the fatty acid chain, as well as the number and position of double bonds, we use the following shorthand code:

$$(A\!:\!Bn\text{-}C)$$

where A = number of carbon atoms; B = number of double bonds; and C = position of the first double bond from the methyl end of the molecule. You can think of "*n*" as standing for the words "double bonds," the first of which will be C carbons from the methyl end of the molecule. For example, DHA is docosahexaenoic acid (22:6*n*-3). It has 22 carbons and 6 double bonds, the first of which is on the third carbon from the methyl end of the molecule.

$$CH_3CH_2CH{=}CHCH_2CH{=}CHCH_2CH{=}CHCH_2CH{=}CHCH_2CH{=}CHCH_2CH{=}CH(CH_2)_2COOH$$

The difference between fats and oils comes from the number of unsaturated fatty acids present. Animal fats, such as lard, tallow, and butter, are mixed fats containing more saturated fatty acids than unsaturated fatty acids. They are waxy, white solids at room temperature. Vegetable oils, such as olive oil, corn oil, and cottonseed oil, contain a higher concentration of unsaturated fatty acids and are liquids at room temperature (Table 16.3).

16.4 Essential Fatty Acids

◆ **Define essential fatty acid.**

Linoleic acid (18:2*n*-6) is called an **essential fatty acid** because it is not synthesized by the body and is essential for the health and growth of tissues, especially in infants. Linoleic acid is found in large amounts in vegetable oils. Dietary linoleic acid lowers the level of LDL (low-density lipoprotein or "bad cholesterol") in the blood. Arachidonic acid is

Table 16.3 Common Fats and Oils and Their Fatty Acid Composition[a]

		Percentage Composition of the Most Abundant Fatty Acids								
		Saturated				*Unsaturated*				
	Melting Point in °C	Myristic	Palmitic	Stearic	Arachidic	Palmitoleic	Oleic	Linoleic	Linolenic	Iodine[b] Number
Animal fats										
Butter	32	11	29	9	2	5	27	4	—	36
Lard	30	1	28	12	—	3	48	6	—	59
Tallow	N/A	6	27	14	—	—	50	3	—	50
Human fat	15	3	24	8	—	5	47	10	—	68
Plant oils										
Corn	−20	1	10	3	—	2	50	34	—	123
Cottonseed	−1	1	23	1	1	2	23	48	—	106
Linseed	−24	—	6	2	1	—	19	24	47	179
Olive	−6	—	7	2	—	—	84	5	—	81
Peanut	3	—	8	3	2	—	56	26	—	93
Safflower	N/A	←	7		→	—	19	70	3	145
Soybean	−16	—	10	2	—	—	29	51	6	130

[a]Values are averages. Extreme variation may occur in the values depending on the source, treatment, and age of the fat or oil.
[b]The higher the iodine number, the more unsaturated the fat.

Health Perspective / Fish Oils and Heart Disease

It has long been known that a diet high in fat and cholesterol can lead to atherosclerosis and heart disease. Over recent decades, millions of Americans have changed their eating habits. They are eating less saturated fat and cholesterol, and instead are substituting the polyunsaturated fats in vegetables. Meanwhile, scientists are continuing their research to see if other dietary changes can further reduce the risk of heart disease.

One such finding comes from studies of the Alaskan Inuit. These people eat a diet rich in fats and as a result, have high average serum cholesterol levels. Yet they experience very low rates of atherosclerosis and heart disease. Although their diet is high in fat, this fat comes from fish and sea mammals. Scientists have found that the fat from fish has a different composition from that of land animals or even that of vegetables. Fish oil contains a high concentration of polyunsaturated fatty acids (PUFA), but these fatty acids belong to the omega-3 (or n-3) family rather than the n-6 family of PUFA found in vegetable oils.

To understand how these fish oils might affect a person's health, researchers have studied how atherosclerosis and heart attacks occur. Atherosclerosis results from the formation of bulges, or plaques, on the walls of blood vessels. As a plaque grows, it can cause the formation of blood clots that block the artery. Blood platelets play an important role in the formation of both plaques and blood clots. Using the fatty acid arachidonic acid ($20{:}4n$-6), platelets produce chemicals called eicosanoids. These particular eicosanoids cause blood pressure to increase and stimulate blood platelets to stick together to form blood clots. High blood pressure and blood clot formation can lead to heart disease and death from heart attacks. Scientists tested the blood of Inuits to see if it differed from the blood of people who eat other diets. Inuit blood had much lower platelet function—that is, a lower tendency for platelets to stick together and form blood clots.

This study was confirmed in examinations of other populations. For example, Japanese fishermen who ate mainly fish showed lower platelet function and lower incidence of atherosclerosis and heart disease than Japanese farmers who ate little fish. Greenland Eskimos had reduced platelet activity and fewer deaths from heart attack than their relatives who had moved from Greenland to Denmark and no longer ate the traditional diet of fish.

Fish oils contain linolenic acid ($18{:}3n$-3), EPA ($20{:}5n$-3), and DHA ($22{:}6n$-3), all members of the n-3 family of polyunsaturated fatty acids. These PUFAs are synthesized by plankton and seaweed, which are eaten by fish and sea mammals and incorporated into their fatty tissue. Research has shown that EPA inhibits the formation of the particular eicosanoids that cause the blood platelets to stick together. In this way, high amounts of EPA in the blood inhibit the formation of blood clots and reduce the chance of a heart attack. Thus, it appears that eating fish regularly, especially fatty fish such as cod and salmon (which have greater amounts of EPA and DHA than other fish), can help lower your risk of developing atherosclerosis or heart disease.

Your Perspective: Suppose that the owner of a chain of supermarkets has read about the benefits of fish oils, and thinks this might help her sell more salmon from the fish department. She has asked you to write a 30-second radio advertisement that will convince listeners to buy salmon at her stores to obtain better health. Write the script for such a radio ad, making sure that the ad will be easily understood by listeners of all educational levels.

also often referred to as an essential fatty acid, but it can be synthesized in the liver from linoleic acid. Adequate amounts of arachidonic acid are crucial to good health because it is used by most mammalian cells to synthesize the large class of compounds called eicosanoids. Eicosanoids, which include prostaglandins and leukotrienes, have a wide range of physiological effects (see Section 18.19). For example, prostaglandins are powerful inducers of fever and inflammation. Aspirin functions as an antipyretic (fever reducer) by controlling the production of prostaglandins in the temperature-regulating cells of the brain.

16.5 Waxes

Waxes are esters of long-chain fatty acids and long-chain monohydric alcohols (alcohols with one hydroxyl group). For example, beeswax is largely an ester of myricyl alcohol and palmitic acid.

Describe the general structure of a wax.

$$CH_3(CH_2)_{28}CH_2O{-}\overset{\overset{\displaystyle O}{\|}}{C}CH_2CH_2CH_2CH_2CH_2CH_2CH_2CH_2CH_2CH_2CH_2CH_2CH_2CH_2CH_3$$

beeswax

Because of their two long hydrocarbon chains, waxes are completely insoluble in water. Waxes are also flexible and nonreactive, which makes them excellent protective and water-repellant coatings on skin, fur, feathers, leaves, and fruits. Waxes have commercial use in cosmetics, floor waxes, furniture and car polishes, ointments, and creams (Table 16.4).

Check Your Understanding

1. Use your own words to define each of the following terms:
   - **(a)** simple and compound lipid
   - **(b)** essential fatty acid
   - **(c)** triacylglycerol
   - **(d)** wax
   - **(e)** saturated and unsaturated fatty acid
2. Write the structural formula for a triacylglycerol formed between glycerol, palmitic acid, stearic acid, and oleic acid. Is this a simple or mixed triacylglycerol?
3. Write the formula of the wax that is formed between a straight-chain carboxylic acid with 26 carbons and a straight-chain primary alcohol with 26 carbons.

Table 16.4 Common Waxes

Name	*Melting Point in °C*	*Source*	*Uses*
Beeswax	61–69	Honeycomb	Candles, polishes
Carnauba	83–86	Carnauba palm	Floor waxes, polishes
Lanolin	36–43	Wool	Cosmetics, skin ointments
Spermaceti	42–50	Sperm whale	Cosmetics, candles

16.6 Iodine Number

Given the iodine number of a lipid, describe the physical properties and most likely source of the lipid.

The unsaturated bonds in a fatty acid react to add iodine, giving us a useful tool for determining unsaturation (see Fig. 16.2 and Section 12.12). Chemists have defined the iodine number of a simple lipid to be the number of grams of iodine that react with 100 g of fat or oil. The higher the iodine number, the more unsaturated the fat. Fats generally have an iodine number below 70, and oils have an iodine number above 70 (see Table 16.3).

Figure 16.2 Iodine test. A few drops of iodine are added to peanut oil (left) and sunflower oil (right), producing a red color. The color disappears in the sunflower oil because sunflower oil is more polyunsaturated than peanut oil.

16.7 Hydrogenation

Oils can be converted to solid fats by hydrogenation, the addition of hydrogen to the double bonds of the molecule in the presence of a catalyst.

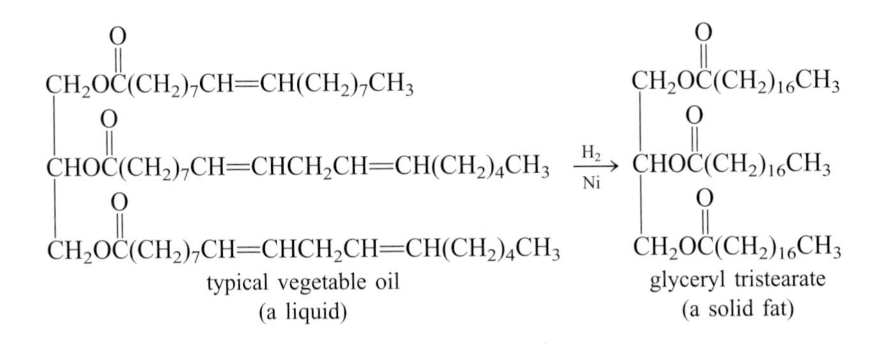

For example, vegetable shortenings and margarines are commercially produced by the partial hydrogenation of soybean, corn, or cottonseed oil. (The complete hydrogenation of these oils would produce a hard, brittle product.)

16.8 Rancidity

When fats and oils start to develop a disagreeable odor and taste, they are then said to be rancid. There are two causes of rancidity: hydrolysis and oxidation. For example, butter becomes rancid when left too long at room temperature. This occurs because some of the fat in the butter undergoes hydrolysis, which is accelerated by enzymes produced by microorganisms in the air. This hydrolysis produces the fatty acid butyric acid, which causes the odor of rancid butter.

Describe the process by which butter becomes rancid.

Oxygen in the air reacts with the unsaturated chains of triglycerides, breaking them apart and converting them into a variety of smaller foul-smelling and foul-tasting acids and aldehydes. To slow down such oxidation in commercial cooking oils and manufactured products, such as bread or crackers, manufacturers add chemicals that react preferentially with the oxygen. Examples of these antioxidant additives are the synthetic compounds BHA and BHT and the natural antioxidant vitamin E. Use of such chemical additives greatly extends the shelf life of these products.

Health Perspective / *Trans* Fatty Acid

For years we have been warned about the need to lower the amount of saturated fat in our diets. Even some of the largest fast-food chains have switched to vegetable oils to fry their foods. Saturated fats—fats such as butter, which remain solid at room temperature—increase blood levels of cholesterol and low-density lipoproteins (LDL). It is the LDL that deposits cholesterol in the arteries. Saturated fats are found in meat, dairy products, and tropical oils such as palm oil. Many people have switched from butter to margarine, and now cook with vegetable oil rather than animal fat to lower their intake of saturated fats. Manufacturers have recently substituted partially hydrogenated vegetable oils for tropical oils in baked goods, cereals, and other products.

It might seem that we are well on our way to winning the war against cholesterol. Recent studies in the United States and the Netherlands, however, have shown that the partially hydrogenated oils used in margarine and solid vegetable shortenings may actually be worse than saturated fat in raising cholesterol levels! Hydrogenation of liquid vegetable oils is necessary to produce margarine or solid shortening with the firmness and plasticity that both food manufacturers and consumers desire. Natural vegetable oils contain only the *cis* isomer of fatty acids, but in the hydrogenation process a mixture of *cis* and *trans* isomers is produced. Partially hydrogenated margarines and shortenings may contain up to 40% of their fats in the *trans* configuration.

These latest studies show that *trans* fatty acids produced during hydrogenation not only raise the level of serum cholesterol, they also increase the levels of LDL and decrease HDL levels. (High levels of LDL are associated with atherosclerosis, whereas high levels of HDL protect against heart disease.) This apparent double-barreled effect has researchers concerned about the consumption of *trans* fatty acids, and additional studies are being conducted. In the meantime, persons worried about their blood cholesterol levels should avoid eating *trans* fatty acids by using liquid vegetable oil for cooking (rather than hard margarine or shortening) and soft margarine on their bread (instead of stick margarine).

16.9 Hydrolysis

Given the formula of a triacylglycerol, write the equation for its hydrolysis.

The hydrolysis of fats can occur in the presence of superheated steam, hot mineral acids, or specific enzymes. Hydrolysis under such conditions produces glycerol and three fatty acids. In general,

$$\underset{\text{fat}}{\begin{matrix} CH_2OCOR \\ | \\ CHOCOR' \\ | \\ CH_2OCOR'' \end{matrix}} + 3H_2O \longrightarrow \underset{\text{glycerol}}{\begin{matrix} CH_2OH \\ | \\ CHOH \\ | \\ CH_2OH \end{matrix}} + \underset{\text{fatty acids}}{\begin{matrix} RCOOH \\ R'COOH \\ R''COOH \end{matrix}}$$

For example,

$$
\begin{array}{l}
CH_2O\overset{\overset{\displaystyle O}{\|}}{C}(CH_2)_{16}CH_3 \\
| \\
CHO\overset{\overset{\displaystyle O}{\|}}{C}(CH_2)_7CH{=}CH(CH_2)_7CH_3 + 3H_2O \\
| \\
CH_2O\overset{\overset{\displaystyle O}{\|}}{C}(CH_2)_{14}CH_3
\end{array}
\longrightarrow
\left\{
\begin{array}{l}
\begin{array}{l} CH_2OH \\ | \\ CHOH \; + \\ | \\ CH_2OH \end{array} \\
\text{glycerol} \\
CH_3(CH_2)_{16}\overset{\overset{\displaystyle O}{\|}}{C}OH + \\
\text{stearic acid} \\
CH_3(CH_2)_7CH{=}CH(CH_2)_7\overset{\overset{\displaystyle O}{\|}}{C}OH + \\
\text{oleic acid} \\
CH_3(CH_2)_{14}\overset{\overset{\displaystyle O}{\|}}{C}OH \\
\text{palmitic acid}
\end{array}
\right.
$$

16.10 Saponification

Write the equation for the saponification of a specific triacylglycerol.

When hydrolysis is carried out in the presence of a strong base such as sodium hydroxide, glycerol and the sodium salts of the fatty acids are produced. In general,

$$
\begin{array}{l}
CH_2O\overset{\overset{\displaystyle O}{\|}}{C}R \\
| \\
CHO\overset{\overset{\displaystyle O}{\|}}{C}R' \\
| \\
CH_2O\overset{\overset{\displaystyle O}{\|}}{C}R'' \\
\quad\text{fat}
\end{array}
+ 3NaOH \longrightarrow
\begin{array}{l}
CH_2OH \\
| \\
CHOH \\
| \\
CH_2OH \\
\text{glycerol}
\end{array}
+
\begin{array}{l}
R\overset{\overset{\displaystyle O}{\|}}{C}O^-Na^+ \\
R'\overset{\overset{\displaystyle O}{\|}}{C}O^-Na^+ \\
R''\overset{\overset{\displaystyle O}{\|}}{C}O^-Na^+ \\
\text{sodium salts} \\
\text{(soap)}
\end{array}
$$

For example,

$$
\begin{array}{l}
CH_2O\overset{\overset{\displaystyle O}{\|}}{C}(CH_2)_{16}CH_3 \\
| \\
CHO\overset{\overset{\displaystyle O}{\|}}{C}(CH_2)_{16}CH_3 \\
| \\
CH_2O\overset{\overset{\displaystyle O}{\|}}{C}(CH_2)_{16}CH_3 \\
\quad\text{tristearin}
\end{array}
+ 3NaOH \longrightarrow
\begin{array}{l}
CH_2OH \\
| \\
CHOH \\
| \\
CH_2OH \\
\text{glycerol}
\end{array}
+ 3CH_3(CH_2)_{16}\overset{\overset{\displaystyle O}{\|}}{C}O^-Na^+
\quad \text{sodium stearate}
$$

Explain how soap functions to remove grease from your hands.

Such salts are called **soaps.** Sodium salts of fatty acids are used in bar soaps, and potassium salts are used in liquid soaps. A hard soap contains a larger number of saturated fatty acids than a soft soap. (Remember that saturated fatty acids have higher melting points than unsaturated fatty acids, and are solids at room temperature.) Commercial soaps con-

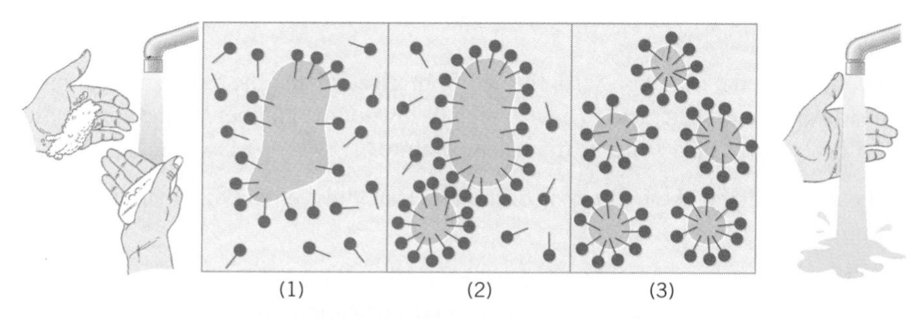

Figure 16.3 Cleansing action of soap. (1) The nonpolar tails of the soap dissolve in the nonpolar grease. (2) Small colloidal grease particles break off and are surrounded by the negatively charged polar head of the soap molecules. This prevents the grease from reforming into large droplets. (3) In this manner, the grease can be completely broken up, and the colloidal droplets washed away with water.

tain various additives to give them their colors and scents. In addition, floating soaps contain air bubbles, and scouring soaps contain abrasives.

Cleansing Action of Soap

Water does a poor job of removing grease and oil because water molecules tend to stick together rather than penetrate the nonpolar grease. Soap greatly improves the cleansing power of water. A soap molecule has two portions: a nonpolar hydrophobic (water-repelling) ''tail'' formed by the hydrocarbon chain, and a polar hydrophilic (water-attracting) ''head'' formed by the carboxyl group.

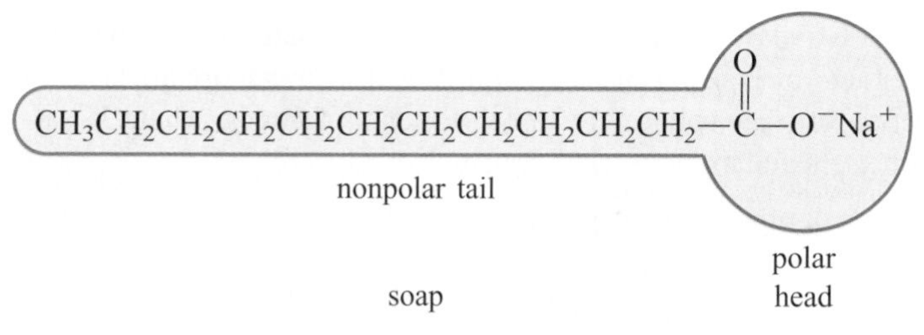

Water's surface tension causes it to bead on the surface of fabrics.

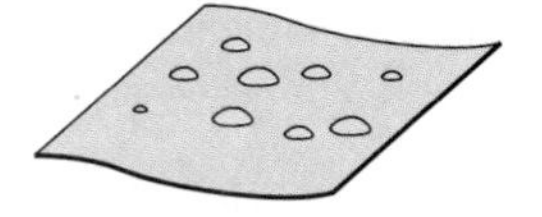

Adding detergent lowers water's surface tension and allows it to . . .

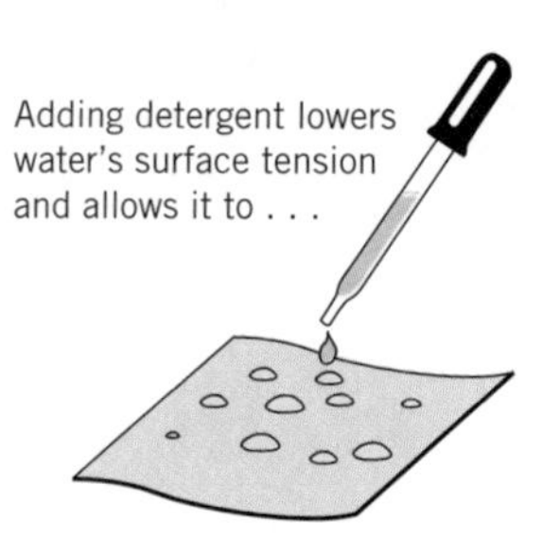

. . . penetrate into the fabric.

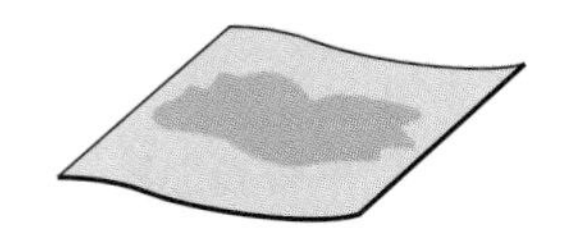

The nonpolar tail readily dissolves in the nonpolar grease, whereas the polar head remains dissolved in the water. This allows the soap to break up the grease into small colloidal droplets (that is, it emulsifies the grease), which can then be washed away (Fig. 16.3).

Several factors affect the cleansing power of soap. For example, ''hard'' water contains one or more of the metallic ions Ca^{2+}, Mg^{2+}, Fe^{2+}, or Fe^{3+}, which form insoluble salts with soap. These salts precipitate out (forming soap scum and bathtub ring), leaving less soap in the water to do the cleaning. Water softeners work by replacing such metallic ions with other ions, such as sodium, which do not interfere with the action of soap. Lowering the pH of the water also decreases the cleansing action of soap by neutralizing the charge on the fatty acid ion. Detergents, which are mixtures of sodium salts of sulfuric acid esters, have cleansing properties similar to soaps but also have important advantages over them. Their calcium and magnesium salts are water-soluble, and they are not affected by pH.

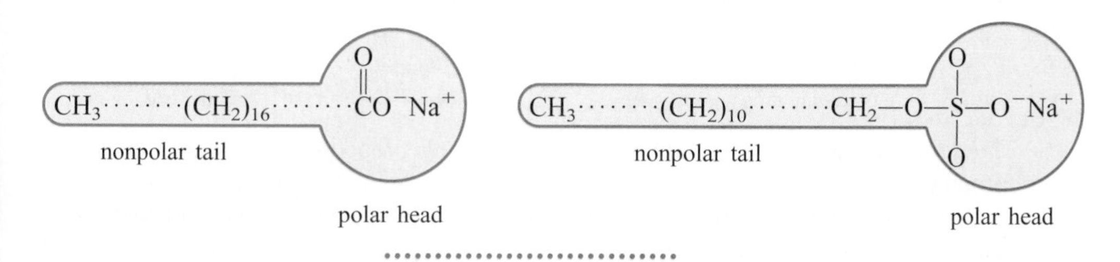

Check Your Understanding

1. Would you expect the triacylglycerol formed from glycerol and the fatty acids palmitic acid, stearic acid, and oleic acid to have a high or low iodine number? Is the compound likely to be a liquid or a solid at room temperature?
2. Write the equations for the hydrolysis of the two triacylglycerols whose structures are shown at the end of Section 16.2.
3. Write the equations for the saponification by sodium hydroxide of the two triacylglycerols whose structures are shown at the end of Section 16.2.

16.11 Compound Lipids

Draw the general structure, and describe the function of two types of compound lipids.

Compound lipids are lipids that, upon hydrolysis, yield fatty acids, an alcohol (either glycerol or sphingosine), and other compounds (such as the phosphate group or a sugar).

Phospholipids

Phospholipids are a class of lipids that contain a phosphate group. They are waxy solids that are used to form cell plasma membranes and to transport lipids in the body. The general structure of a phospholipid can be written as follows, where X can be the compounds choline, ethanolamine, serine, inositol, or glycerol.

GLYCEROL —fatty acid (saturated)
—fatty acid (unsaturated)
—phosphate—X

phosphoglyceride
(a phospholipid)

The phosphate group forms a polar hydrophilic (water-attracting) head on the molecule, and the two fatty acids form two nonpolar hydrophobic (water-repelling) tails on the molecule.

nonpolar tails — polar head

This structure gives phospholipids good emulsifying properties and good membrane-forming properties.

One important phospholipid is phosphatidylcholine (abbreviated as PC and commonly called lecithin).

Health Perspective / Multiple Sclerosis and the Myelin Sheath

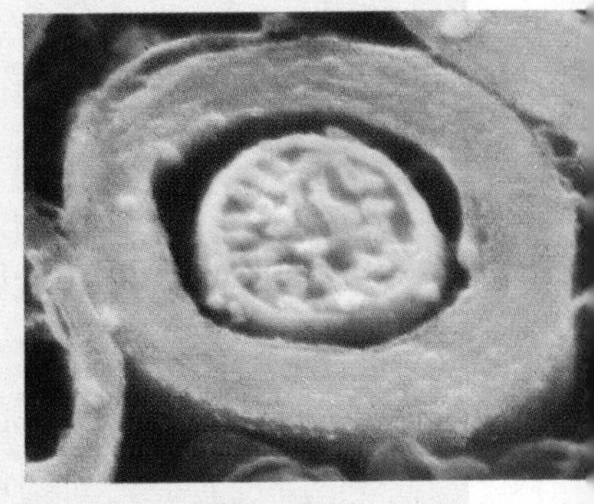

The gray myelin sheath surrounds a nerve fiber.

For many of us, the study of chemistry would be a breeze if we could just store all these facts in our gray matter. The term *gray matter* refers to the brain, which really *is* composed of gray matter. This gray matter contains the nerve cell bodies. A smaller percentage of the brain is composed of *white matter,* which contains the nerve fibers called axons. The white color of the axons comes from a white lipid substance called myelin, which acts as an insulator for the conduction of nerve impulses. The axons are coated, or sheathed, in this myelin. This coating, therefore, is called the *myelin sheath.*

Multiple sclerosis (MS) is a widespread neurological disease in which the nerve fibers slowly lose their insulating myelin sheaths. This disease affects approximately 250,000 people in the United States. It is about twice as common in women as in men. Although the exact cause of multiple sclerosis is not known, it is believed to result from a childhood infection that somehow triggers an autoimmune response later in life. The body produces antibodies that attack and destroy the myelin sheaths of nerve fibers in the central nervous system. As the myelin is destroyed, it is replaced by hard scar tissue called sclerotic plaques. The name ''multiple sclerosis'' comes from these multiple patches of hardened tissue. Without the myelin coating, nerve impulses slow down and in some instances are totally blocked. Depending on the part of the brain or spinal cord that is affected, MS produces a wide variety of symptoms, including numbness, visual problems, fatigue, and loss of coordination and balance. In a small percentage of MS cases, the destruction of the myelin sheath is progressive, eventually leading to paralysis and death.

Your Perspective: You are a volunteer at a local hospital where a 20-year-old woman has just been diagnosed with multiple sclerosis. The woman is afraid of the doctor and asks you to explain this disease to her. You want to explain the disease in terms she would understand, without unnecessarily alarming her. Write down the words that you would say, just as if you were talking directly to her.

GLYCEROL —fatty acid
—fatty acid
—phosphate—choline

$$
\begin{array}{l}
CH_2\text{—}O\overset{\overset{\displaystyle O}{\|}}{C}(CH_2)_{16}CH_3 \\
| \\
CH\text{—}O\overset{\overset{\displaystyle O}{\|}}{C}(CH_2)_7CH\text{=}CH(CH_2)_7CH_3 \\
| \\
CH_2\text{—}O\text{—}\overset{\overset{\displaystyle O}{\|}}{\underset{\underset{\displaystyle O^-}{|}}{P}}\text{—}O\text{—}CH_2\text{—}CH_2\text{—}\overset{\overset{\displaystyle CH_3}{|}}{\underset{\underset{\displaystyle CH_3}{|}}{N^+}}\text{—}CH_3
\end{array}
$$

phosphatidylcholine, lecithin
(PC)

Phosphatidylcholine plays an important role in the metabolism of fats in the liver and in the transport of fats from one part of the body to another. It serves as a source of inorganic phosphate for tissue formation and is an excellent emulsifying agent. PC finds commercial use as an emulsifying agent in such products as chocolate candies, margarine, and medicines. Egg yolks contain a large amount of PC and are used to emulsify salad oil and vinegar to make mayonnaise. The removal of one fatty acid from PC forms lysolecithin, a compound that causes the destruction of red blood cells and spasmodic muscle contractions. The venom of poisonous snakes contains enzymes to catalyze the formation of lysolecithin from PC.

Another important class of phospholipids are the sphingolipids, which contain the alcohol sphingosine.

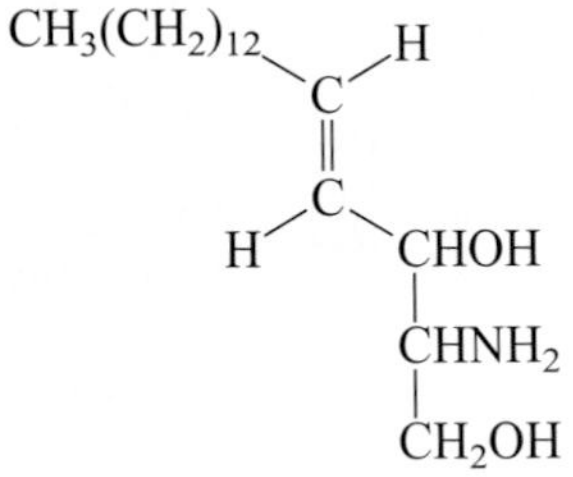

$CH_3(CH_2)_{12}CH{=}CHCH(OH)CH(NH_2)CH_2OH$

sphingosine

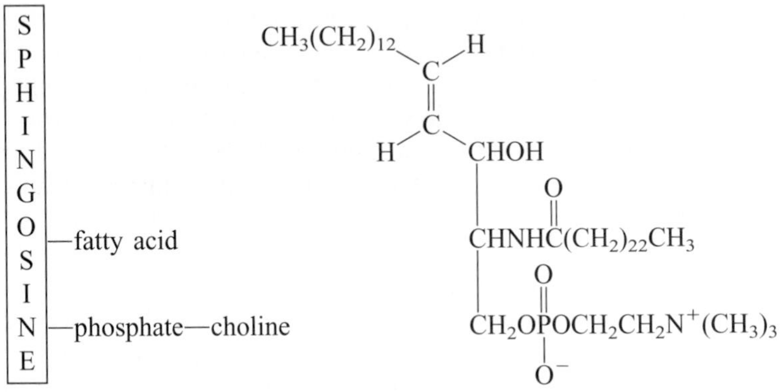

SPHINGOSINE —fatty acid —phosphate—choline

$CH_3(CH_2)_{12}CH{=}CHCH(OH)CHNHC(O)(CH_2)_{22}CH_3$ with $CH_2OP(O)(O^-)OCH_2CH_2N^+(CH_3)_3$

sphingomyelins

The most common sphingolipid is sphingomyelin. Large amounts of sphingomyelins are found in brain and nervous tissue. They form part of the myelin sheath, the protective coating of nerves.

Glycolipids

The main difference between glycolipids and phospholipids is that a glycolipid contains a sugar group rather than a phosphate group. The sugar group is usually galactose but may also be glucose. The alcohol is either glycerol or sphingosine.

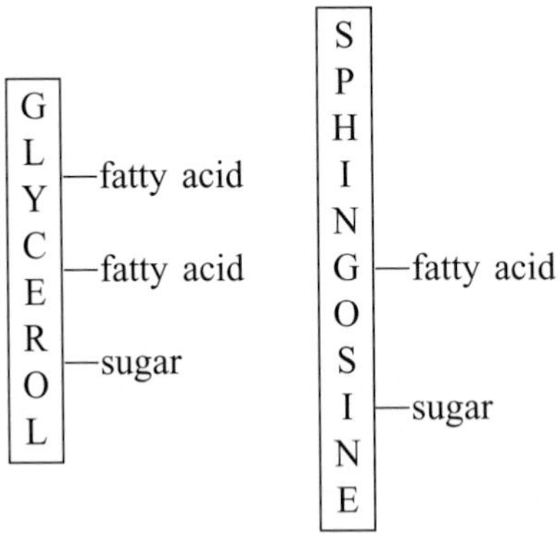

GLYCEROL —fatty acid —fatty acid —sugar

SPHINGOSINE —fatty acid —sugar

glycolipids

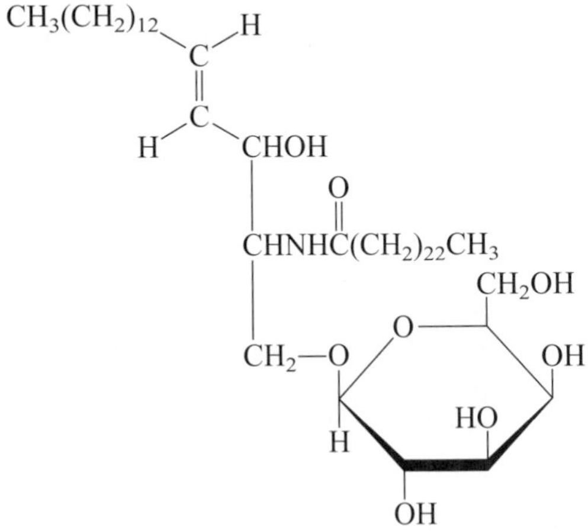

$CH_3(CH_2)_{12}CH{=}CHCH(OH)CHNHC(O)(CH_2)_{22}CH_3$ with $CH_2{-}O$–galactose (ring with CH_2OH, OH, HO, OH, H)

cerebroside

Cerebrosides are glycolipids that contain sphingosine. They are formed in high concentrations in the brain and nerve cells, especially in the myelin sheath.

Several hereditary fat metabolism diseases have been linked to errors in the enzymes that hydrolyze glycolipids. In Gaucher's disease, glucocerebroside, which contains glucose instead of galactose, is not hydrolyzed and collects in the spleen and kidneys. An infant born with Tay–Sachs disease lacks one of the first enzymes required in the hydrolysis of glycolipids. In such infants the glycolipid GM_2 collects in the tissues of the brain and eyes, causing muscular weakness, mental retardation, seizures, blindness, and in severe cases, death by the age of 3 or 4.

16.12 Nonsaponifiable Lipids: Steroids

Nonsaponifiable lipids are those that are not broken apart by alkaline hydrolysis. **Steroids** are nonsaponifiable lipids whose structure contains a complicated four-ring framework consisting of three cyclohexane rings and one cyclopentane ring.

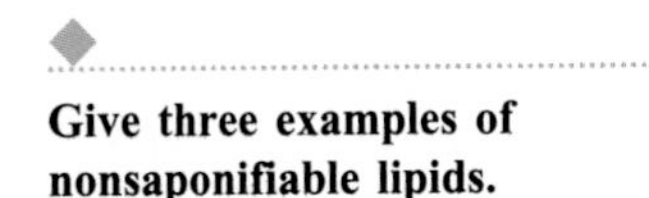

Give three examples of nonsaponifiable lipids.

steroid nucleus

The steroid nucleus is found in the structure of several vitamins, hormones, drugs, poisons, bile acids, and sterols. The structure and function of some familiar steroids are shown in Table 16.5.

Cholesterol

Sterols are steroid alcohols, the most common of which is cholesterol.

cholesterol

Describe how cholesterol is both essential and harmful to living organisms.

Cholesterol is synthesized by animal cells from a compound called acetyl coenzyme A (acetyl-CoA) and forms part of all cell membranes. Specialized cells in the liver, adrenal glands, and ovaries also use cholesterol to synthesize bile acids, steroid hormones, and vitamin D. The liver plays a key role in the body's cholesterol balance because it synthesizes 90% of the 3 to 5 g of cholesterol the body makes each day.

The specific amount of cholesterol produced by the liver at any time, however, is controlled by the amount of cholesterol already circulating in the blood. Cholesterol circulates in the blood in several forms. These forms all have a nonpolar core that contains triacylglycerols and cholesterol esters surrounded by a coating of protein, phospholipid, and cholesterol. One of these forms consists of chylomicrons, which are particles formed in the small intestine. Via the lymph and blood, chylomicrons carry dietary triacylglycerols to the tissues and dietary cholesterol to the liver. In the chapter opening story we mentioned low-density lipoproteins (LDL), which are the major particles that transport

Table 16.5 **Structure and Function of Steroids**

Steroid	*Structure*	*Function*
Cortisone	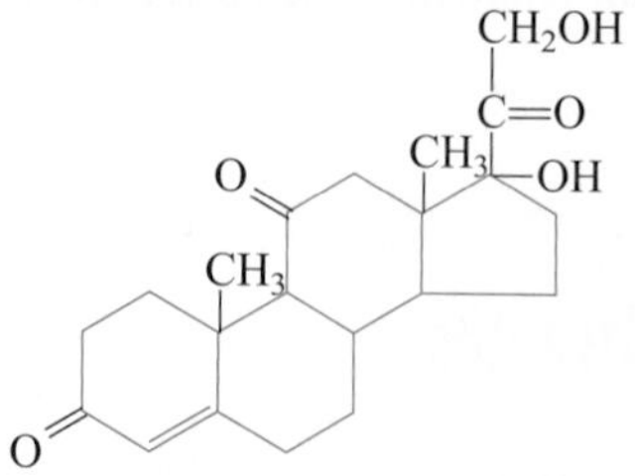	One of many hormones produced in the adrenal glands, cortisone is important in controlling carbohydrate metabolism and is used therapeutically to relieve symptoms of inflammation, especially in rheumatoid arthritis.
Vitamin D_2	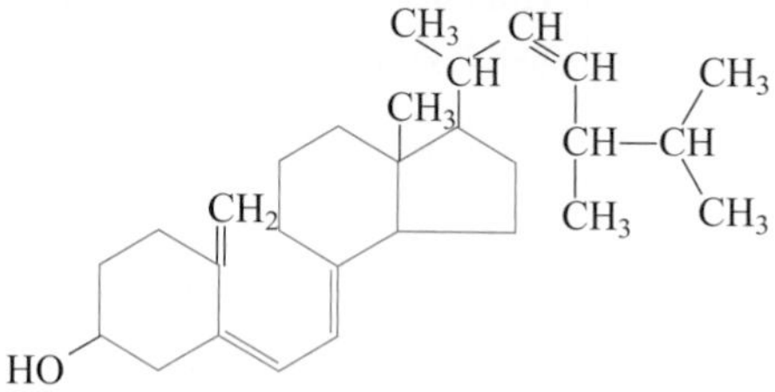	Irradiation of the steroid hormone ergosterol with ultraviolet light breaks open one of the rings in the steroid nucleus, producing vitamin D_2. This vitamin is essential in preventing rickets, a disease of calcium metabolism.
Digitoxigenin	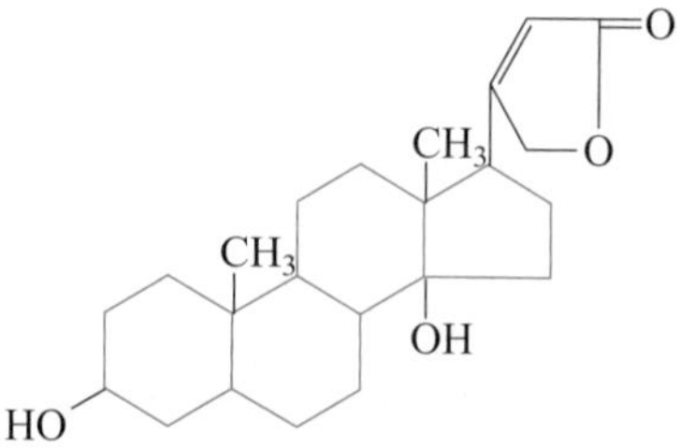	Extracted from the digitalis plant, digitoxigenin is used in small doses to regulate a diseased heart. In large doses it causes death.
Testosterone	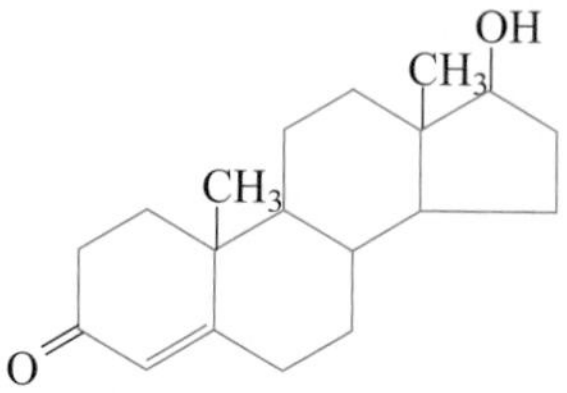	The male sex hormone testosterone regulates the development of the male reproductive organs.
Progesterone	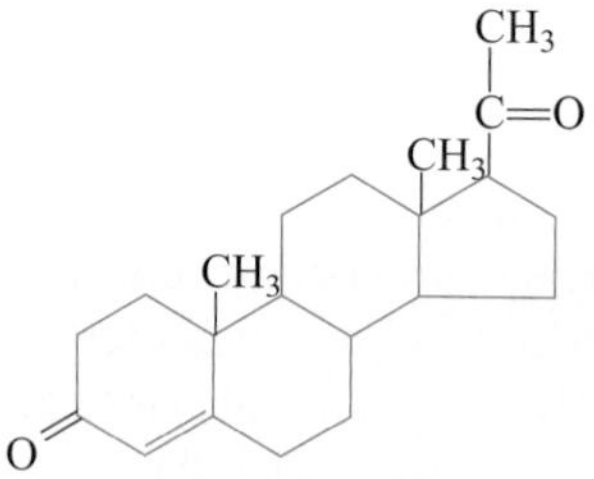	Progesterone is the female sex hormone that is produced in pregnancy and acts on the uterine lining, preparing it to receive the embryo.

cholesterol and triacylglycerols produced by the liver to the tissues (Fig. 16.4). Of equal importance are high-density lipoproteins (HDL), which act as cholesterol scavengers. They remove cholesterol from tissue cell surfaces and carry it back to the liver, where it is removed from circulation in the blood. Heart disease has been correlated with high LDL and low HDL levels in the blood.

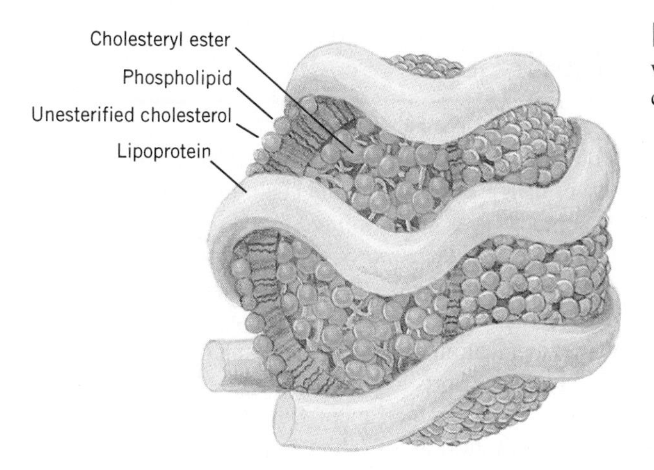

Figure 16.4 An LDL particle, which is the major cholesterol carrier in the bloodstream.

16.13 Cellular Membranes

Membranes perform many specific functions in living organisms. All cells have a plasma membrane, and organelles in eukaryotic cells are surrounded by membranes. Membranes control the chemical environment of the space they enclose by keeping out certain compounds and selectively transporting others through the membrane. Plasma membranes maintain the shape of the cell and control cellular movement. The chemical composition of the membrane allows cell-to-cell recognition, and the membrane contains receptors for specific substances, such as hormones or nutrients. Many biochemical processes occur on the surface of membranes; examples include electron transport on the mitochondrial membrane and photosynthesis on the inner membrane of chloroplasts.

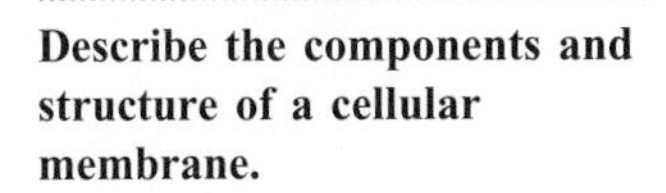

Describe the components and structure of a cellular membrane.

The two major components of membranes are lipids and proteins, with the proportion of the two types of compounds varying among different kinds of membranes. For example, the myelin sheath is about 70% lipid, whereas the nuclear membrane is only about 40% lipid. Membranes also contain small amounts of carbohydrate found in glycolipids and glycoproteins. These carbohydrates serve as the biological markers for cellular recognition. Membrane glycoproteins also form the structure for receptor sites, which are areas on the membrane where specific molecules such as hormones attach to the cell.

When viewed in cross section through an electron microscope, biological membranes reveal two dark bands on either side of a light band. These bands show us that the molecular structure of a membrane is a lipid bilayer—two rows of phospholipids, each with their polar hydrophilic heads toward the outside of the membrane and their nonpolar hydrophobic tails toward the water-free interior (Fig. 16.5). The protein components of the

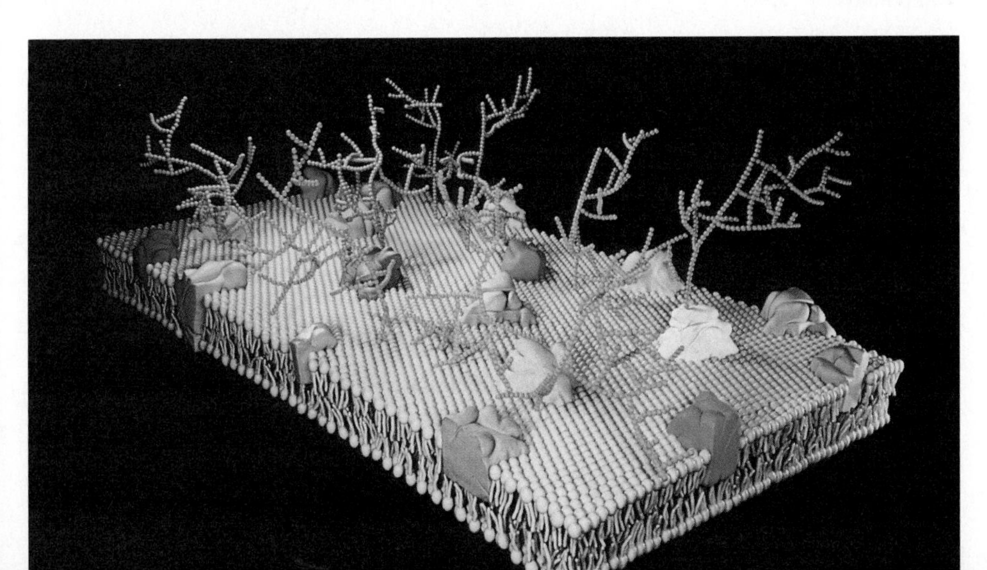

Figure 16.5 Model of a membrane. The membrane is composed of a phospholipid bilayer (yellow base). Colored proteins are imbedded in the bilayer. Carbohydrates joined to the membrane proteins appear as tree-like structures on the external surface of the membrane.

Sprinter Ben Johnson

In today's win-at-all-costs sports worlds, some players and coaches have turned to the use of dangerous drugs in the hope of improving performance. Some athletes continue to compete even when injured by taking analgesics to kill the pain and antiinflammatory drugs to reduce the swelling in damaged joints and tissues. To mask symptoms of fatigue and to create a feeling of increased stamina, they may take stimulants like amphetamines and cocaine. News stories have recorded the sudden deaths from cocaine intoxication of both amateur and professional athletes. The sports community has increased its efforts to eliminate the use of cocaine by suspending athletes from competition or even banning them from the sport if caught using the drug.

Anabolic steroids are illegal drugs popular among athletes in sports requiring muscle building and strength. Like cortisone, which has many important medical uses, anabolic steroids are hormones that are naturally produced by the body. They differ from other steroids in promoting the buildup of muscle tissue. Men normally have a greater muscle mass than women because the male hormone testosterone is an anabolic steroid. Many athletes, both male and female, have taken testosterone to increase their muscular development. Because testosterone is not very effective when taken orally, it is injected for best results. This has led scientists to develop synthetic anabolic steroids similar to testosterone that can be taken orally. Examples of these steroids are clostebol, kabolin, methandienone, methenolone, nandrolone, decanoate, oxymesterone, and stanozolol. It was stanozolol that was found in the urine of Canadian sprinter Ben Johnson shortly after he won the 100-meter dash in the 1988 Olympics. Because of this drug use, he lost the gold medal and was banned for several years from competition.

All anabolic steroids have the potential for serious side effects. In men these side effects include atrophy of the testicles, difficulty in urination, decreased sperm production, growth of breasts, impotence, increased cholesterol levels, edema, high blood pressure, heart attack, kidney disease, and liver cancer. Women who use these steroids have increased growth of facial hair, loss of body curves, irreversible baldness, deepening of the voice, and menstrual problems. Many users also suffer from ''steroid psychosis,'' which causes violent mood swings, hallucinations, and delusions of grandeur. It may even be the delusions of grandeur that help convince athlete users that anabolic steroids work in the first place.

Your Perspective: You are told by a friend that her son, a high-school wrestler, is using anabolic steroids. Using language appropriate for a high-school student, write a letter to your friend's son, telling him what you think he should know about the use of these compounds.

membrane may be on the surface of, embedded in, or even extending completely through the lipid bilayer. The lipids found most abundantly in membranes are phosphoglycerides, sphingolipids, and cholesterol.

Check Your Understanding

1. List four steroids that are important to living organisms.
2. What lipids are the major components of cellular membranes? What other classes of compounds are found in the membrane structure?

3. Draw the structure for the following lipids:

(a) cephalin (phosphatidylethanolamine): a phosphoglyceride in which the fatty acids bonded to the glycerol are stearic acid and linolenic acid, and the phosphate group is bonded to ethanolamine ($—CH_2CH_2N^+H_3$)

(b) a cerebroside in which the fatty acid unit is linoleic acid

KEY CONCEPTS

Lipids are waxy or oily substances that are not soluble in water. Their main functions are to form part of the structure of biological membranes and to store energy for the cell. They are also the starting material for the formation of prostaglandins, hormones, vitamins, and bile acids. Lipids can be classified as saponifiable and nonsaponifiable. The saponifiable lipids are either simple lipids (which are esters of glycerol and fatty acids) or compound lipids (which yield fatty acids, glycerol, and other compounds upon hydrolysis).

Fatty acids are long-chain carboxylic acids. They are formed from the hydrolysis of fats and oils. Two polyunsaturated fatty acids—linoleic and arachidonic acids—are essential fatty acids. They are used in the synthesis of prostaglandins.

Triacylglycerols are simple lipids that may be either solid (fats) or liquid (oils) at room temperature. Oils contain more unsaturated fatty acids than do fats. Waxes are esters of monohydric alcohols and long-chain fatty acids.

Triacylglycerols can undergo hydrogenation, hydrolysis, and saponification. Unsaturated fatty acids, or unsaturated fatty acid side chains of triacylglycerols, can be hydrogenated to produce saturated fatty acids or saturated fats. Fats and oils can become rancid by hydrolysis of the triacylglycerols or by oxidation of their unsaturated fatty acid side chains. Triacylglycerols can be hydrolyzed, producing three fatty acids and glycerol. If the hydrolysis occurs in the presence of a strong base, glycerol and the salts of fatty acids (or soaps) are formed. Soaps are good emulsifying agents because their molecules contain a nonpolar region that dissolves in the fat or grease and a polar region that dissolves in water.

Phospholipids are compound lipids that form the structure of cell membranes and that help to transport other lipids in the body. All phospholipids contain fatty acids, an alcohol (glycerol or sphingosine), and a phosphate group bound to a polar compound such as choline.

Glycolipids are compound lipids that contain a sugar group rather than a phosphate group. They are found in high concentrations in brain and nervous tissue.

Steroids are nonsaponifiable lipids with a complicated four-ring structure. Vitamin D, sterols, bile acids, and many hormones, drugs, and poisons contain this structure. Cholesterol is the most common sterol. It is a component of all cells and is used by the cell as the starting material for the synthesis of many other compounds.

Important Equations

Hydrolysis

$$\begin{array}{l} \overset{\text{O}}{\overset{\|}{\text{CH}_2\text{OCR}}} \\ | \\ \overset{\text{O}}{\overset{\|}{\text{CHOCR}'}} \\ | \\ \overset{\text{O}}{\overset{\|}{\text{CH}_2\text{OCR}''}} \end{array} + 3\text{H}_2\text{O} \longrightarrow \begin{array}{l} \text{CH}_2\text{OH} \\ | \\ \text{CHOH} \\ | \\ \text{CH}_2\text{OH} \end{array} + \begin{array}{l} \overset{\text{O}}{\overset{\|}{\text{RCOH}}} \\ \overset{\text{O}}{\overset{\|}{\text{R}'\text{COH}}} \\ \overset{\text{O}}{\overset{\|}{\text{R}''\text{COH}}} \end{array}$$

Saponification

$$\underset{\text{fat}}{\begin{array}{l} \overset{\text{O}}{\overset{\|}{\text{CH}_2\text{OCR}}} \\ | \\ \overset{\text{O}}{\overset{\|}{\text{CHOCR}'}} \\ | \\ \overset{\text{O}}{\overset{\|}{\text{CH}_2\text{OCR}''}} \end{array}} + 3\text{NaOH} \longrightarrow \underset{\text{glycerol}}{\begin{array}{l} \text{CH}_2\text{OH} \\ | \\ \text{CHOH} \\ | \\ \text{CH}_2\text{OH} \end{array}} + \underset{\substack{\text{sodium salts} \\ \text{(soap)}}}{\begin{array}{l} \overset{\text{O}}{\overset{\|}{\text{RCO}^-\text{Na}^+}} \\ \overset{\text{O}}{\overset{\|}{\text{R}'\text{CO}^-\text{Na}^+}} \\ \overset{\text{O}}{\overset{\|}{\text{R}''\text{CO}^-\text{Na}^+}} \end{array}}$$

Review Problems

Section 16.1

1. State the difference between a saponifiable lipid and a nonsaponifiable lipid.
2. State the difference between a simple lipid and a compound lipid.

Section 16.2

3. State the difference between a simple triacylglycerol and a mixed triacylglycerol.
4. Write the formulas for the following triacylglycerols:
   (a) tripalmitin
   (b) the mixed triacylglycerol containing glycerol, arachidic, stearic, and oleic acids
   (c) triolein

Section 16.3

5. State the differences among a saturated fatty acid, an unsaturated fatty acid, and a polyunsaturated fatty acid.
6. What is the structural difference between the triacylglycerols in animal fats and vegetable oils?

7. **(a)** Identify the following as saturated fatty acids or unsaturated fatty acids:
   (1) myristic acid **(3)** linolenic acid
   (2) oleic acid **(4)** lauric acid

   (b) Which of the fatty acids in part (a) would you most likely find in olive oil?
8. Write the structural formulas for the following:
   (a) *cis, cis, cis*-linolenic acid
   (b) *cic, cis, cis, cis*-arachidonic acid
9. Draw structures of the fatty acids that have the following shorthand codes:
   (a) 18:2*n*-3
   (b) 20:3*n*-6
10. What are the shorthand codes for the following fatty acids:
    (a) $CH_3(CH_2)_8COOH$
    (b) $CH_3(CH_2)_5CH{=}CH(CH_2)_6COOH$

Section 16.4

11. State the difference between an essential fatty acid and a nonessential fatty acid.
12. Why is a diet lacking in linoleic acid bad for a person's health?

Section 16.5

13. Write the structural formula for the wax that could be formed between myricyl alcohol and arachidic acid.
14. Why are waxes insoluble in water?

Section 16.6

15. What information about a lipid do you obtain from its iodine number?
16. Which triacylglycerol in problem 4
    (a) has the highest melting point
    (b) is most likely to be found in animal fat
    (c) has the highest iodine number

Section 16.7

17. Describe how margarine is produced from corn oil.
18. Write the equations for the complete hydrogenation of the triacylglycerols in parts (b) and (c) of problem 4.

Section 16.8

19. **(a)** Why does butter turn rancid?
    (b) How does refrigeration slow this process?
    (c) What causes the odor of rancid butter?
20. What classes of compounds are produced by the oxidation of the double bonds in the fatty acids of unsaturated triacylglycerols as they become rancid?

Section 16.9

21. Write the equations for the hydrolysis of the three triacylglycerols in problem 4.

22. State the difference between hydrolysis and hydrogenation.

Section 16.10

23. **(a)** Describe what is happening, on the molecular level, when you use soap to wash salad oil from your hands.

(b) Suggest a possible reason for the antibacterial action of soap, using the fact that bacterial cell membranes are made up of lipids.

24. In what ways are detergents superior to soaps?

Section 16.11

25. State the difference between each of the following pairs:

(a) a phospholipid and a triacylglycerol

(b) phosphatidylcholine and sphingomyelin

(c) a phospholipid and a glycolipid

26. What physical property of a phospholipid clearly distinguishes it from a fat?

27. Write the general formula for the following compounds, and describe the function of each in the human body:

(a) phospholipid

(b) phosphatidylcholine

(c) sphingolipid

(d) sphingomyelin

(e) glycolipid

28. To what class of lipids does the following compound belong?

$$
\begin{array}{l}
\mathrm{CH(OH)CH{=}CH(CH_2)_{12}CH_3} \\
\;| \qquad \mathrm{O} \\
\;| \qquad \| \\
\mathrm{CHNHC(CH_2)_{22}CH_3} \\
\;| \qquad \mathrm{O} \\
\;| \qquad \| \\
\mathrm{CH_2OPOCH_2CH_2N^+(CH_3)_3} \\
\qquad\;\; | \\
\qquad\;\; \mathrm{O^-}
\end{array}
$$

Section 16.12

29. **(a)** The hormones testosterone and progesterone belong to what class of lipids?

(b) From what compound are they synthesized?

(c) What is the difference between the structures of the two compounds?

30. What steroid is a component of cell membranes?

31. What functional groups are present in vitamin D_2 (see Table 16.5)?

32. What functional groups are present in digitoxigenin (see Table 16.5)?

Section 16.13

33. **(a)** Describe the structure of cellular membranes.

(b) Explain how the structure of phospholipids allows them to form a bilayer.

34. **(a)** What is the function of the lipid bilayer in cell membranes?
    (b) What type of compound most easily diffuses into the cell through this membrane?

APPLIED PROBLEMS

35. Elaidic acid is a *trans* isomer of oleic acid. It is produced in the manufacture of margarine by the partial hydrogenation of corn oil.
    (a) Draw the structural formulas of oleic and elaidic acids.
    (b) Why should there be increasing concern about the production and consumption of *trans* fatty acids in shortenings and margarines?
36. Glycerol molecules that are esterified by one or two fatty acid units are called, respectively, monoglycerides and diglycerides. They are used along with lecithin as the main emulsifying agents in margarine (a water in oil dispersion). They are also used to form water–oil emulsions in cake batters. This allows more water to be added to the batter and increases the amount of sugar that can be incorporated. Draw the structural formulas for all the diglycerides that can be formed from glycerol, oleic acid, and stearic acid.
37. In Section 16.3 you learned that at room temperature saturated fatty acids are solid and unsaturated fatty acids are liquids. Why does the addition of double bonds into a fatty acid lower its melting point?
38. Most margarines made from hydrogenated soybean oil have an iodine number between 78 and 90. The iodine number for peanut oil is very close to this, with an average value of 93. Why are margarines solid at room temperature whereas peanut oil is a liquid? (*Hint:* The hydrogenation process produces a mixture of *cis* and *trans* isomers.)
39. Why do you think olive oil (iodine number 81) is better suited for use as a cooking oil than safflower oil (iodine number 145)?
40. Why can companies that produce vegetable oils advertise their products as cholesterol-free?
41. The following triacylglycerol is found in lard:

$$\begin{array}{l} CH_2-O-\overset{\overset{\large O}{\|}}{C}(CH_2)_{14}CH_3 \\ | \\ CH-O-\overset{\overset{\large O}{\|}}{C}(CH_2)_{16}CH_3 \\ | \\ CH_2-O-\overset{\overset{\large O}{\|}}{C}(CH_2)_7CH{=}CH(CH_2)_7CH_3 \end{array}$$

Before commercially produced soap was available, people made soap from lard and lye that was extracted from wood ashes with a small amount of water. This lye solution contained basic substances such as KOH, Na_2CO_3, and K_2CO_3. Write the equation for the formation of soap from this triacylglycerol and KOH.

42. The following triacylglycerol is typical of those found in animal fats. Write the chemical equations for the following reactions using as a reactant the triacylglycerol shown here.

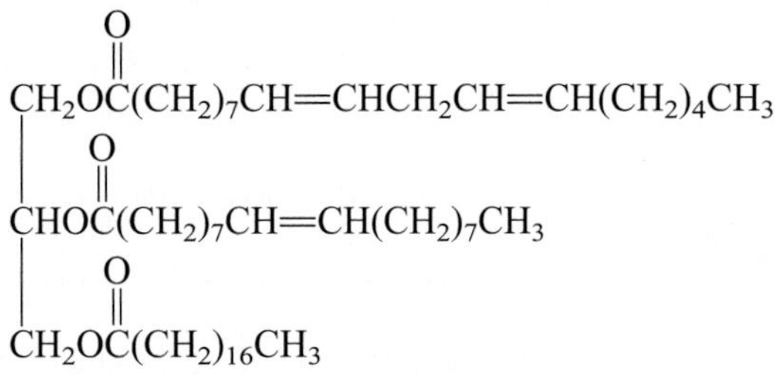

$$\begin{array}{l} CH_2OC(=O)(CH_2)_7CH{=}CHCH_2CH{=}CH(CH_2)_4CH_3 \\ | \\ CHOC(=O)(CH_2)_7CH{=}CH(CH_2)_7CH_3 \\ | \\ CH_2OC(=O)(CH_2)_{16}CH_3 \end{array}$$

(a) hydrogenation
(b) hydrolysis
(c) saponification
(d) the reaction with Br_2 in CCl_4

Chapter 17

Proteins

*I*t had not been a good year for George and Fran Bobo. They had spent a very frustrating winter working with the contractor who was renovating their recently purchased turn-of-the-century victorian house. And now their six-year-old son seemed to be suffering from some mysterious ailment. It had started when Carson came home from school with a bad cold, complaining that he felt tired and had pains in his stomach. Later that evening Carson had begun vomiting uncontrollably and acting very lethargic. When George took his son to the town's small clinic, the doctors found nothing requiring immediate attention (although blood tests did suggest a mild case of anemia).

At the local doctor's suggestion, however, George took Carson to the city hospital a few hours away. In the hospital's pediatric clinic, the doctor reviewed the results of the laboratory tests and noticed the anemia that had been found the previous week. She also observed that certain blood cells appeared unusual. This signaled to the doctor the possibility of lead poisoning, and a test for lead in the blood confirmed her suspicion. Carson's blood contained a high concentration of lead: 70 μg/dL, compared with a normal level of less than 10 μg/dL.

But how could a child living in a small town be exposed to dangerously high concentrations of lead? The culprit turned out to be the renovation of the house. Carson had been playing in the house as the

remodelers were sanding the walls. As a result, the dust that he had been breathing contained high concentrations of lead from the old paint. To remedy the situation, the doctor simply ordered Carson to stay with friends or relatives until all the house remodeling was finished, and to have his blood lead level checked weekly until it returned to the normal range. The blood test would then be repeated at regular intervals for a few years to make certain lead levels didn't increase again.

Lead is just one of many heavy metals that are toxic to the body. Children are much more sensitive to lead exposure than are adults. The body absorbs lead in the form of inorganic lead salts (such as those found in paints and pottery glazes) or as tetraethyl lead (found in leaded gasoline). Most absorbed lead is stored in the bones, but it is the lead in the blood that causes the effects of lead poisoning. Any amount of lead in the blood can be harmful. At low blood concentrations, the Pb^{2+} ion inhibits several enzymes containing sulfhydryl groups (—SH). These enzymes affect the synthesis of the heme molecule that is necessary for the production of hemoglobin. The resulting decrease in the concentration of hemoglobin in the blood causes anemia. At high blood concentrations, lead disrupts kidney function and affects the central nervous system, causing permanent brain damage. At blood concentrations of 100 μg/dL, lead can cause convulsions, coma, and death.

Until recently, high blood lead levels were always treated with chelating agents such as ethylenediaminetetraacetic acid (EDTA), which bind to the lead in the blood and allow it to be excreted in the urine. This treatment has become controversial, however, and may prove to be harmful. Rapid removal of the blood lead disrupts the equilibrium between the lead in the bones and that in the blood. The sudden lowering of the concentration of Pb^{2+} in the blood causes large amounts of lead to be released from the bones, momentarily raising the concentration of lead in the blood and central nervous system, where even a temporary rise can result in immediate and permanent brain damage. The best way to lower blood lead concentrations, therefore, is to remove the patient immediately from any further exposure to lead so that the body can slowly rid itself of the accumulated lead.

In this chapter, we discuss proteins and enzymes. We also see why substances such as the lead ion, which can interfere with the function of proteins and enzymes, can have profound effects on living organisms.

17.1 Amino Acids: Building Blocks of Proteins

Proteins are the most complex and varied class of molecules found in living organisms. They are found in all cells, and their biological importance cannot be overemphasized. The Dutch chemist G. J. Mulder recognized this fact in 1838 when he gave this class of compounds the name **protein,** which means "of prime importance."

All proteins are composed of the elements carbon, nitrogen, oxygen, and hydrogen. Most proteins also contain sulfur, and some have phosphorus and other elements, such as iron, zinc, or copper. Proteins are large polymers that, on hydrolysis, produce monomer units called **amino acids.** The molecular weight of most proteins ranges from 12,000 to 1 million or more (as compared with other groups of compounds whose molecular weights are generally under 1000). This large size gives protein molecules colloidal properties. For example, they cannot pass through differentially permeable membranes such as the membranes of a cell. This is why the presence of proteins in the urine warns doctors of the possibility of damage to the membranes of the kidneys.

The sequence, or order, of amino acids in the protein molecule determines the function of that protein. To study proteins, therefore, we must first discuss amino acids. **Amino**

acids are carboxylic acids that have an amino group on the alpha carbon, the carbon next to the carboxyl group. The general structure of an amino acid is

◆ **Write the general formula for an amino acid.**

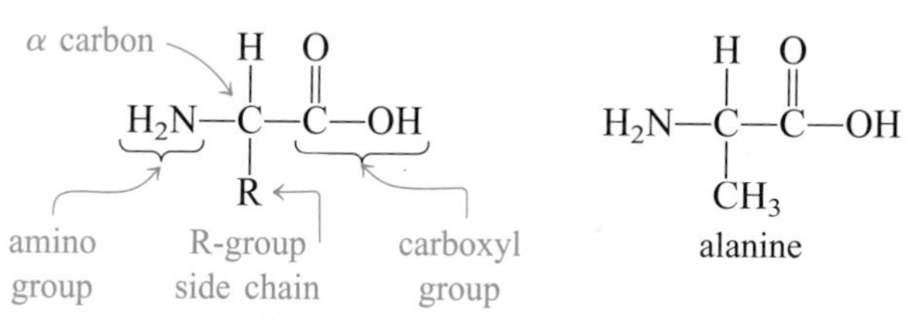

It is the different R-group side chains on the amino acids that make one amino acid different from another. Most naturally occurring proteins are composed of the 20 amino acids shown in Table 17.1, but a few specialized types of proteins contain other rarer amino acids. Amino acids that are not found in proteins can exist in a free or combined form; these are, in general, derivatives of the 20 amino acids found in proteins.

17.2 The L-Family of Amino Acids

All amino acids found in proteins, except for glycine, are optically active and belong to the L-family. D-Family isomers of amino acids can be found in nature but never occur in proteins.

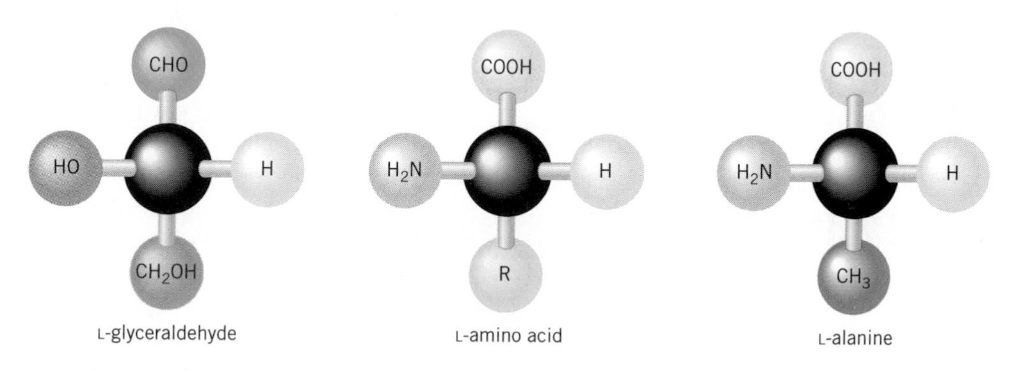

17.3 Essential Amino Acids

The body can synthesize 10 of the 20 amino acids found in proteins. The 10 amino acids not adequately synthesized by the body are called the **essential amino acids.** Nine of the essential amino acids must be supplied in the diet throughout life. The tenth, arginine, is synthesized by the body but not in sufficient quantities to meet the needs of children during periods of rapid growth (Table 17.2).

Proteins that contain all 10 of the essential amino acids are known as **adequate proteins.** Animal protein and milk are adequate proteins, but many vegetable proteins are missing one or more of the essential amino acids and, therefore, are said to be inadequate proteins. The protein in soybeans is adequate, but the protein in corn is too low in lysine and tryptophan to support growth in young children. Rice is low in lysine and threonine, and wheat is low in lysine. For this reason, people on vegetarian diets must eat a combination of vegetables to obtain all the essential amino acids. This is why protein deficiency diseases, which cause both slowed growth and reduced resistance to disease in children, are commonly found in parts of the world where a single plant (such as corn) is the major source of dietary protein.

◆ **Describe what makes a protein an adequate protein.**

Table 17.1 Structure of the R-Group Side Chains of the 20 Most Common Amino Acids.

	Structure of R in $H_3N^+-\overset{\overset{H}{|}}{\underset{\underset{R}{|}}{C}}-\overset{\overset{O}{\|}}{C}-O^-$ [a]	*Name of Amino Acid*	*Abbreviation*
Nonpolar R Group	$-H$	Glycine	Gly
	$-CH_3$	Alanine	Ala
	$-\underset{\underset{CH_3}{|}}{CH}-CH_3$	Valine	Val
	$-CH_2-\underset{\underset{CH_3}{|}}{CH}-CH_3$	Leucine	Leu
	$-\underset{\underset{CH_3}{|}}{CH}-CH_2-CH_3$	Isoleucine	Ile
	$-CH_2-CH_2-S-CH_3$	Methionine	Met
	$-CH_2-C_6H_5$	Phenylalanine	Phe
	$-CH_2-C_6H_4-OH$	Tyrosine	Tyr
	$-CH_2-$(indole ring, $\underset{\underset{H}{|}}{N}$)	Tryptophan	Trp
	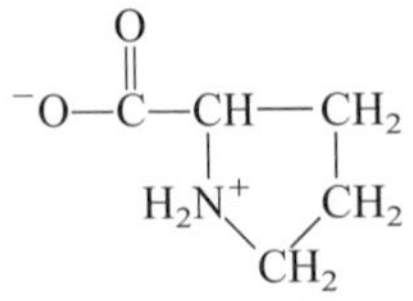 $^{-}O-\overset{\overset{O}{\|}}{C}-CH-CH_2$ / H_2N^+, CH_2, CH_2 (ring) (complete structure)	Proline	Pro

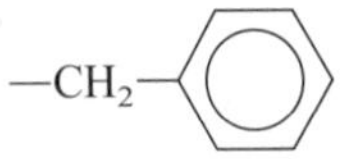

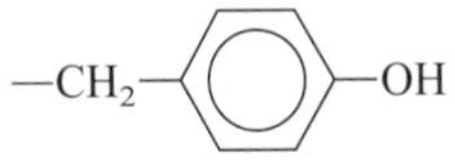

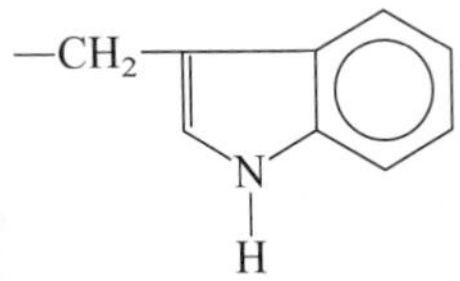

17.4 Acid–Base Properties

Define zwitterion, and give an example.

Amino acids are soluble in water and have very high melting points. This suggests that they do not exist as uncharged molecules but are found as a highly polar **zwitterion,** or dipolar ion.

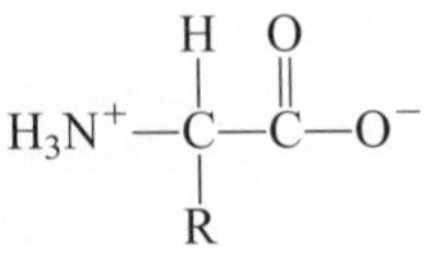

$$H_3N^+-\overset{\overset{H}{|}}{\underset{\underset{R}{|}}{C}}-\overset{\overset{O}{\|}}{C}-O^- \qquad H_3N^+-\overset{\overset{H}{|}}{\underset{\underset{\underset{\underset{CH_3}{|}}{CH-CH_3}}{|}}{C}}-\overset{\overset{O}{\|}}{C}-O^-$$

zwitterion, or dipolar ion 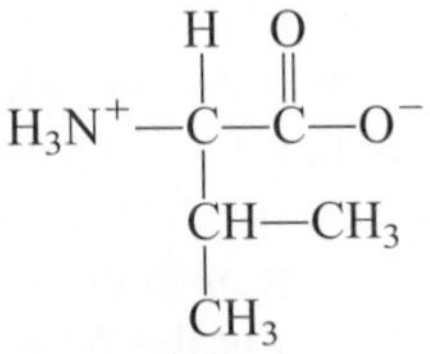 — valine

Although we may write the structure of the amino acid in the uncharged form, keep in mind that in the pH range of most human tissues, amino acids always exist as the dipolar ion.

Table 17.1 **continued**

	Structure of R in $H_3N^+—C(H)(R)—C(=O)—O^-$	Name of Amino Acid	Abbreviation
Polar R Group	$—CH_2—OH$	Serine	Ser
	$—CH(CH_3)—OH$	Threonine	Thr
	$—CH_2—SH$	Cysteine	Cys
	$—CH_2—C(=O)—NH_2$	Asparagine	Asn
	$—CH_2—CH_2—C(=O)—NH_2$	Glutamine	Gln
Acidic R group	$—CH_2—C(=O)—OH$	Aspartic acid	Asp
	$—CH_2—CH_2—C(=O)—OH$	Glutamic acid	Glu
Basic R group	$—CH_2—CH_2—CH_2—CH_2—NH_2$	Lysine	Lys
	$—CH_2—CH_2—CH_2—NH—C(=NH)—NH_2$	Arginine	Arg
	$—CH_2—C{=}CH$ (imidazole ring: HN—CH=N)	Histidine	His

[a]The complete structure of the amino acid can be written by substituting the formula of the R group in the general formula. For example, the structure of serine can be written:

$$H_3N^+—\underset{R}{\overset{H}{C}}—\overset{O}{\overset{\|}{C}}—O^- \qquad H_3N^+—\underset{CH_2—OH}{\overset{H}{C}}—\overset{O}{\overset{\|}{C}}—O^-$$

Because proteins can often be viewed as very large amino acids, a knowledge of the acid–base properties of amino acids can help us understand some of the properties of proteins. Amino acids contain both an acidic group (the carboxyl group) and a basic group (the amino group). In water, amino acids can act as either acids or bases. Molecules

Table 17.2 **Essential and Nonessential Amino Acids in Humans**

Essential		*Nonessential*	
Arginine[a]	Methionine	Alanine	Glutamine
Histidine	Phenylalanine	Asparagine	Glycine
Isoleucine	Threonine	Aspartic acid	Proline
Leucine	Tryptophan	Cysteine	Serine
Lysine	Valine	Glutamic acid	Tyrosine

[a]Essential for children but synthesized by the body in sufficient quantities for adults.

having this property are called **amphoteric.** Because amino acids (and proteins) can act as either acids or bases, they are effective buffers in an aqueous solution.

Explain how a protein can act as a buffer.

$$H_2O + H_2N-\underset{R}{\overset{H}{C}}-\overset{O}{\overset{\|}{C}}-O^- \xleftarrow{+OH^-} H_3N^+-\underset{R}{\overset{H}{C}}-\overset{O}{\overset{\|}{C}}-O^- \xrightarrow{+H^+} H_3N^+-\underset{R}{\overset{H}{C}}-\overset{O}{\overset{\|}{C}}-OH$$

One important function of proteins in the blood is to act as buffers, helping to keep the blood pH within a very narrow normal range (pH 7.35–7.45).

17.5 Isoelectric Point

Explain how proteins can be separated on the basis of their isoelectric points.

The specific pH at which each amino acid and protein is electrically neutral and does not move in an electric field is called the **isoelectric point** for that molecule. The isoelectric point is indicated by the symbol p*I*. At a pH more basic than the isoelectric point, the amino acid has a net negative charge and moves toward the positive electrode. At a pH more acidic than the isoelectric point, the amino acid carries a net positive charge and moves toward the negative electrode. Because some amino acids have an ionizable R group, each amino acid and protein has a specific isoelectric point (Table 17.3). Proteins can be separated from one another based on their charge at different pH levels through a process called electrophoresis. In this technique, each protein migrates toward the positive or negative electrode at a different speed depending on the pH and voltage applied. Paper electrophoresis is a useful tool in analyzing the proteins in human blood serum (Fig. 17.1). At the isoelectric point the protein also has minimum solubility. At this pH, the electrically neutral protein molecules cluster together and can be easily removed from solution.

Table 17.3 **Isoelectric Points of Amino Acids and Proteins**

Compound	*Isoelectric Point (pI)*	*Compound*	*Isoelectric Point (pI)*
Amino acid		Protein	
Glutamic acid	3.2	Egg albumin	4.6
Phenylalanine	5.5	Urease	5.0
Alanine	6.0	Hemoglobin	6.8
Leucine	6.0	Myoglobin	7.0
Lysine	9.7	Chymotrypsin	9.5
Arginine	10.8	Lysozyme	11.0

Figure 17.1 Paper electrophoresis allows proteins and amino acids to be separated by their rate of migration in an electric field. (a) A schematic diagram of the electrophoresis apparatus. (b) The paper strip after electrophoresis is completed. The positive ions have migrated toward the negative pole and the negative ions toward the positive pole. Neutral amino acids and proteins remain at the point of sample application.

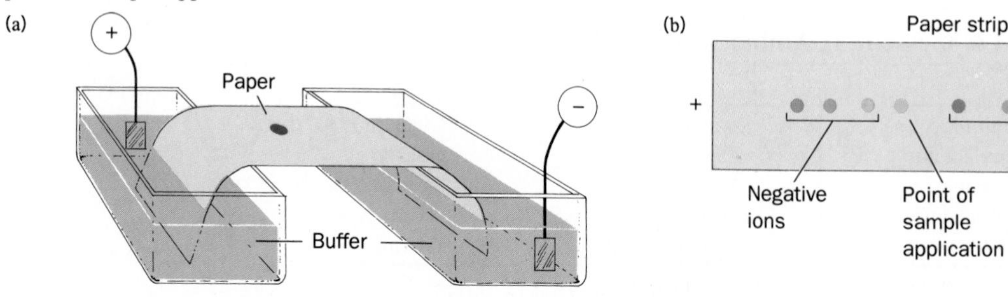

Example 17-1

1. Write the equation for the reaction of valine with KOH.

See the Question

We need to write a balanced equation for this acid–base reaction.

Think It Through (Execute the Math)

First we must write the correct formula for the amino acid valine. In Table 17.1 we find the R-group for valine and substitute it in the general formula for an amino acid.

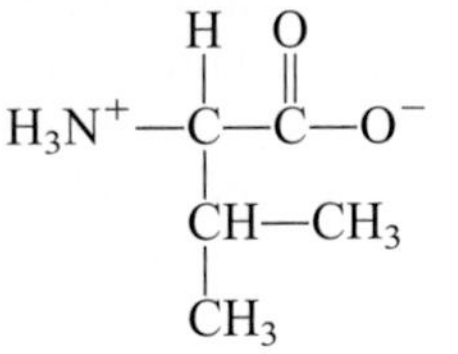

$$H_3N^+-\underset{\underset{\underset{CH_3}{|}}{CH-CH_3}}{\overset{\overset{H}{|}}{\underset{|}{C}}}-\overset{\overset{O}{\|}}{C}-O^-$$

Since KOH is a base, the amino acid acts as an acid and donates a proton to the hydroxide ion to form water.

Prepare the Answer

Writing the equation, we have

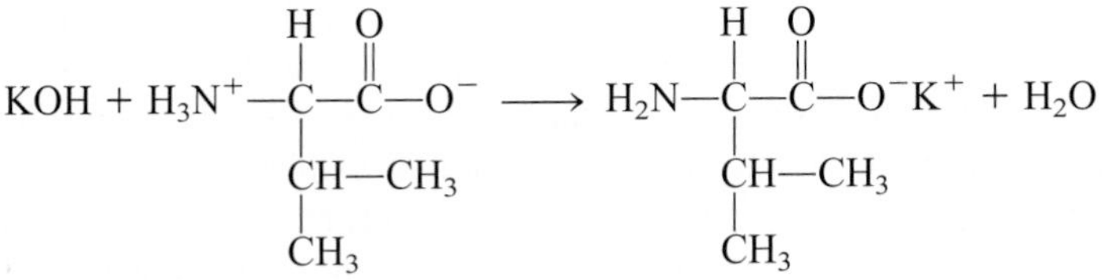

$$KOH + H_3N^+-\underset{\underset{\underset{CH_3}{|}}{CH-CH_3}}{\overset{\overset{H}{|}}{\underset{|}{C}}}-\overset{\overset{O}{\|}}{C}-O^- \longrightarrow H_2N-\underset{\underset{\underset{CH_3}{|}}{CH-CH_3}}{\overset{\overset{H}{|}}{\underset{|}{C}}}-\overset{\overset{O}{\|}}{C}-O^-K^+ + H_2O$$

2. Suppose that myoglobin is added to a solution having a pH of 6.0. This solution is placed into a paper electrophoresis apparatus. Toward what pole does the myoglobin migrate?

S

We must determine toward which electric pole a molecule of myoglobin migrates.

T • E

To answer the question, we must know the isoelectric point of myoglobin (that is, the pH at which the molecule is electrically neutral). From Table 17.3, we see that the p*I* for myoglobin is 7.0. The pH of the solution is 6.0, which is more acidic than myoglobin's isoelectric point. The protein therefore carries a net positive charge.

P

Since myoglobin carries a net positive charge at pH 6.0, it migrates toward the negative pole.

Check Your Understanding

1. **(a)** How can an amino acid act as a buffer?

   (b) Write the equation for the reaction of leucine with HCl.

   (c) Write the equation for the reaction of phenylalanine with NaOH.

2. **(a)** What is the p*I* of phenylalanine?

   (b) If a solution containing phenylalanine at a pH of 7 is placed in an electric field, toward what pole does the phenylalanine migrate?

17.6 Proteins

◆ **Describe the difference between a simple and a conjugated protein.**

Because of their varied natures, proteins can be classified in several different ways. They may be divided into two major classes: **simple proteins,** which produce only amino acids on hydrolysis, and **conjugated proteins,** which produce amino acids and other organic or

Chemical Perspective / Cheese Making

"Little Miss Muffet sat on her tuffet, eating her curds and whey. Along came a spider and sat down beside her, and frightened Miss Muffet away." It is interesting that many people who recite this nursery rhyme don't really know what Miss Muffet was eating.

If you have ever seen badly spoiled milk or cream, you are familiar with curdled milk. The lumpy material floating in the liquid is the *curd.* Curd is nothing more than proteins in milk, called casein, that have precipitated out of solution. Casein has an isoelectric point of pH 4.7. The normal pH of milk is 6.7. When milk begins to sour, bacteria in the milk produce lactic acid that lowers the pH of the milk. This lowers the solubility of the casein, causing the milk to curdle. (The same principle is used in the production of soft cheese such as cottage cheese.) As the curd precipitates, it traps most of the milk fat and some of the lactose in the milk. The liquid that remains is called *whey.*

A different process for the precipitation of casein is used to produce hard cheeses, such as cheddar or swiss. Casein is actually a mixture of protein molecules (called micelles), the most important of which are alpha-, beta-, and kappa-casein. The alpha and beta forms of casein are insoluble in the presence of the calcium ions found in milk. The kappa-casein, however, positions itself around the alpha- and beta-casein, with the hydrophilic (water-attracting) ends of the micelle on the outside of the colloidal particle that is formed. In this way the kappa-casein stabilizes the milk proteins. Cheesemakers add a mixture of enzymes called rennet to the milk. The rennet breaks off the hydrophilic portions of the kappa-casein micelle, causing immediate precipitation of the curd. When first formed, this curd does not have the sour taste associated with lactic acid. By adding lactic acid bacteria to the cheese milk, however, the lactic acid that is produced helps give the cheese a desirable flavor. Following cooking, the cheesemakers separate the curd from the whey and treat the curd in different ways to produce different cheese types. Because the cheese milk is inoculated with bacteria or mold cultures and the curd is allowed to age (or ripen), some of the milk fat is broken down into fatty acids during the aging process. The fatty acids give various cheeses their distinctive flavors and aromas.

Your Perspective: Imagine that you work at the Tillamook Cheese Factory and have been invited to speak to a fifth-grade class at the local elementary school to explain how cheese is made. Develop a two-minute presentation on cheese making, using language that a fifth grader would understand.

inorganic substances on hydrolysis. These other substances are called **prosthetic groups** (Table 17.4).

Describe the difference in structure and function of a globular and fibrous protein.

A second classification is based on the physical characteristics of the protein molecule. **Globular proteins** are generally soluble in water, are quite fragile, and have an active function, such as catalyzing reactions (in the case of enzymes) or transporting other substances (as in the case of hemoglobin). **Fibrous proteins** are insoluble in water, are physically tough, and have a structural or protective function. The keratins of hair, skin, and fingernails; the collagen in tendons and hides; and the silks are examples of fibrous proteins.

The biological importance of proteins results from their wide variety of functions (Table 17.5). Proteins are the body's main dietary source of nitrogen and sulfur. Besides

Table 17.4 **Conjugated Proteins**

Class	*Prosthetic Group*	*Examples*
Glycoproteins	Carbohydrate	Connective tissue, mucins, heparin, immunoglobulins
Lipoproteins	Lipid	High- and low-density lipoproteins in the blood, membrane-bound lipoproteins
Nucleoproteins	Nucleic acid	Viruses, chromosomes
Metalloproteins	Metal ions	Ferritin (Fe), alcohol dehydrogenase (Zn)
Chromoproteins	Colored groups: riboflavin, heme, etc.	Hemoglobin, chlorophyll, luciferase, cytochromes

Table 17.5 **Classification of Proteins Based on Their Functions in Living Organisms**

Class	*Function*	*Example*
Enzymes	Catalyze biological reactions	Pepsin catalyzes the hydrolysis of proteins in the stomach
Structural proteins	Provide structural support	Collagen is the major extracellular support in tendons and bones
Storage proteins	Store nutrients	Ferritin stores Fe in the spleen
Transport proteins	Bind and transport specific molecules in the blood	Hemoglobin transports oxygen
Membrane proteins	Carry out dynamic membrane function and catalyze chains of reactions such as the citric acid cycle	Membrane glycoproteins facilitate the transport of glucose into the cell
Hormones	Regulate body metabolism	Insulin regulates glucose metabolism
Contractile proteins	Perform contraction and movement	Actin and myosin form the contraction system in muscles
Protective proteins	Protect against foreign substances	Antibodies inactivate foreign proteins in the blood
Toxins	Defend organisms	Botulinus toxin is poisonous to organisms other than *Clostridium botulinum*

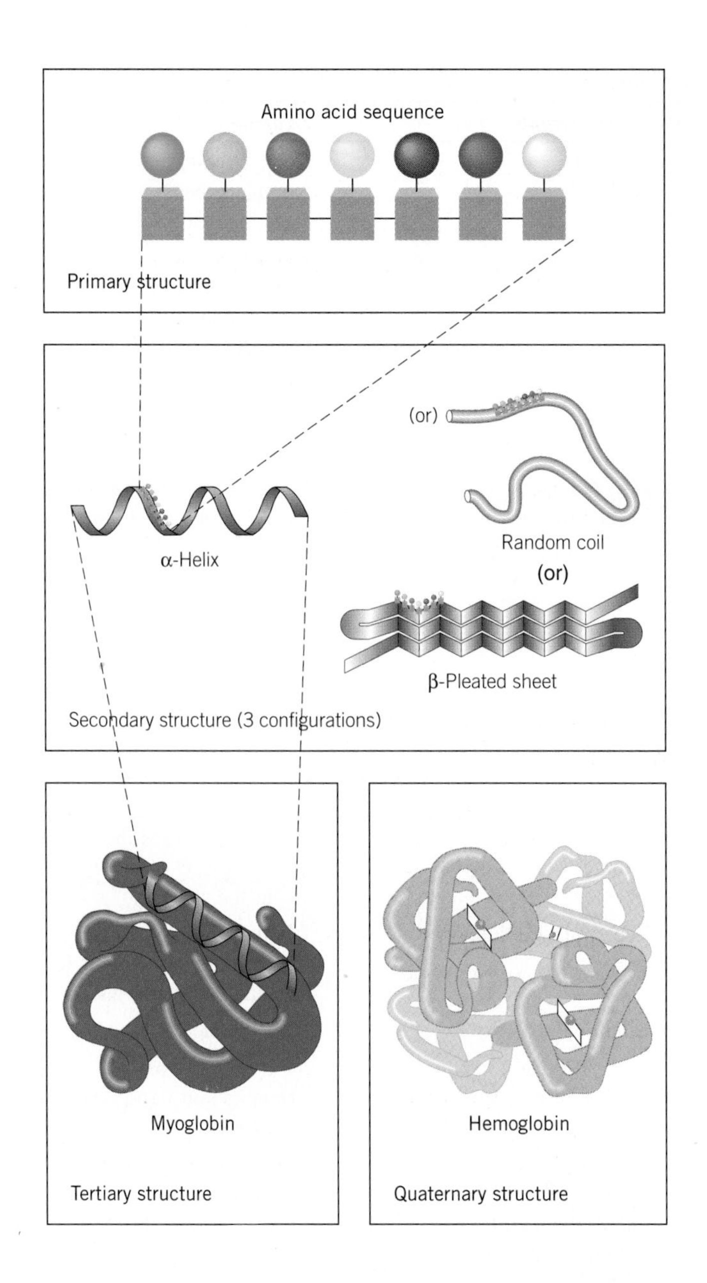

Figure 17.2 Four levels of protein structure.

their catalytic and structural functions, they make up the contractile system of muscles. As antibodies they are the defense system of the body, and as hormones they regulate the body's glandular activity. In the blood, proteins maintain fluid balance, are part of the clotting process, and transport oxygen and lipids. As a key part of cellular membranes, proteins are responsible for such processes as the transport of molecules and ions across the membrane and communication between cells. They can act as poisons, such as the venoms in animal bites and stings, or toxins, such as the toxin that produces botulism. Some antibiotics that are secretions of bacteria and fungi also are protein in nature.

The way a protein functions in a living organism is determined by its structure. The structure of a large complex protein can be described in terms of four levels of organization (Fig. 17.2).

1. *Primary structure* describes the sequence of amino acids in a chain.
2. *Secondary structure* describes the spatial arrangement of atoms in various sections of the amino acid chain.
3. *Tertiary structure* describes the three-dimensional structure of the entire amino acid chain.
4. *Quaternary structure* describes the spatial arrangement of all the amino acid chain subunits that make up a large protein molecule.

17.7 Primary Structure

The **primary structure** of a protein is given by the sequence of amino acids in the protein molecule. The amino acids are joined by amide linkages called **peptide bonds** (see Section 14.5). The peptide bond is formed by joining the carboxyl group of one amino acid to the amino group of a second amino acid through the elimination of water (a condensation reaction).

Define the primary structure of a protein, and describe the type of bonding in the primary structure.

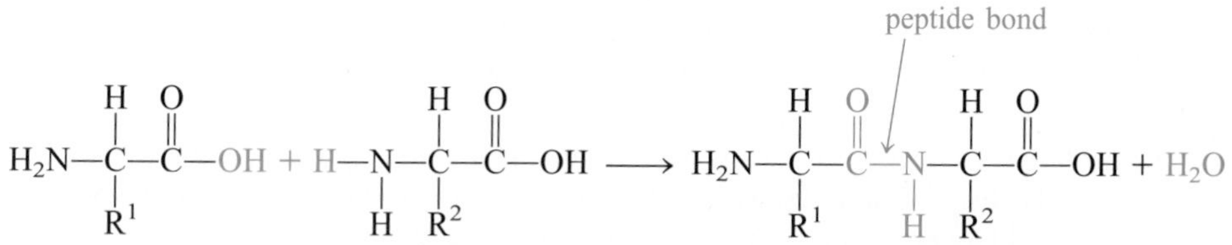

$$\underset{\text{amino acid}_1}{H_2N-\underset{R^1}{\overset{H}{C}}-\overset{O}{\overset{\|}{C}}-OH} + \underset{\text{amino acid}_2}{H-\underset{H}{N}-\underset{R^2}{\overset{H}{C}}-\overset{O}{\overset{\|}{C}}-OH} \longrightarrow \underset{\text{dipeptide}}{H_2N-\underset{R^1}{\overset{H}{C}}-\overset{O}{\overset{\|}{C}}-\underset{H}{N}-\underset{R^2}{\overset{H}{C}}-\overset{O}{\overset{\|}{C}}-OH} + H_2O$$

peptide bond

The peptide bond is unaffected by changes in pH, in solvents, or in salt concentrations. It can be broken only by acid or base hydrolysis, or by specific enzymes.

Two amino acids held together by a peptide bond are called a **dipeptide;** three amino acids form a **tripeptide;** 4 to 10 an **oligopeptide;** and more than 10 a **polypeptide.** There is no precise dividing line between polypeptides and proteins. For example, insulin, with 51 amino acids in its primary structure, is a very small protein. Glucagon, with 21 amino acids, is considered a large polypeptide. Many small polypeptides have important functions in biological systems. Vasopressin, a diuretic hormone produced by the pituitary, is an oligopeptide containing nine amino acids. The antibiotic bacitracin is a polypeptide with 11 amino acids in its sequence. Leucine-enkephalin, a hormone that plays a role in controlling pain and emotional states, is a pentapeptide.

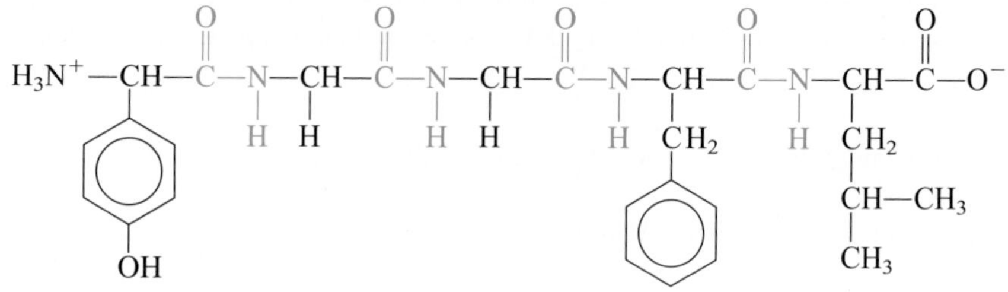

$$H_3N^+-\underset{CH_2C_6H_4OH}{CH}-\overset{O}{\overset{\|}{C}}-\underset{H}{N}-\underset{H}{CH}-\overset{O}{\overset{\|}{C}}-\underset{H}{N}-\underset{H}{CH}-\overset{O}{\overset{\|}{C}}-\underset{H}{N}-\underset{CH_2C_6H_5}{CH}-\overset{O}{\overset{\|}{C}}-\underset{H}{N}-\underset{CH_2-CH(CH_3)-CH_3}{CH}-\overset{O}{\overset{\|}{C}}-O^-$$

N-terminal end — leucine-enkephalin — C-terminal end

The leucine-enkephalin molecule contains four peptide bonds. The end of a peptide molecule with the free amino group is called the N-terminal end and is written first. The end with the free carboxyl group is called the C-terminal end.

Two different dipeptides can be formed between two amino acids. For example, for the amino acids glycine and alanine we have

$$\underset{\substack{\text{glycine}\\(\text{Gly})}}{H_2N-\overset{H}{\underset{H}{C}}-\overset{O}{\overset{\|}{C}}-OH} + \underset{\substack{\text{alanine}\\(\text{Ala})}}{H-\underset{H}{N}-\overset{H}{\underset{CH_3}{C}}-\overset{O}{\overset{\|}{C}}-OH} \longrightarrow \underset{\substack{\text{glycylalanine}\\(\text{Gly-Ala})}}{H_2N-\overset{H}{\underset{H}{C}}-\overset{O}{\overset{\|}{C}}-\underset{H}{N}-\overset{H}{\underset{CH_3}{C}}-\overset{O}{\overset{\|}{C}}-OH} + H_2O$$

$$\underset{\substack{\text{alanine}\\(\text{Ala})}}{H_2N-\overset{H}{\underset{CH_3}{C}}-\overset{O}{\overset{\|}{C}}-OH} + \underset{\substack{\text{glycine}\\(\text{Gly})}}{H-\underset{H}{N}-\overset{H}{\underset{H}{C}}-\overset{O}{\overset{\|}{C}}-OH} \longrightarrow \underset{\substack{\text{alanylglycine}\\(\text{Ala-Gly})}}{H_2N-\overset{H}{\underset{CH_3}{C}}-\overset{O}{\overset{\|}{C}}-\underset{H}{N}-\overset{H}{\underset{H}{C}}-\overset{O}{\overset{\|}{C}}-OH} + H_2O$$

Considering that 20 different amino acids make up proteins, the number of different possible proteins is enormous. This gives us a good reason for discussing the standard rules for naming peptides and proteins. Because proteins contain very large numbers of amino acids, we use three-letter abbreviations of the amino acid names when writing the amino acid sequence (see Table 17.1 for the abbreviations). A dash or dot between the amino acid names represents the peptide bond. One end of the protein consists of an amino acid having a free amino group. This is the N-terminal end of the protein, and the N-terminal amino acid is listed first in the sequence of amino acids. At the other end of the protein is an amino acid having a free carboxyl group. This is the C-terminal amino acid, and it is the last amino acid listed in the sequence. The following are two examples of these naming rules:

lysine-aspartic acid-serine-asparagine-glutamic acid (1)
Lys-Asp-Ser-Asn-Glu

valine · phenylalanine · alanine · tryptophan · leucine (2)
Val · Phe · Ala · Trp · Leu
(N-terminal end) (C-terminal end)

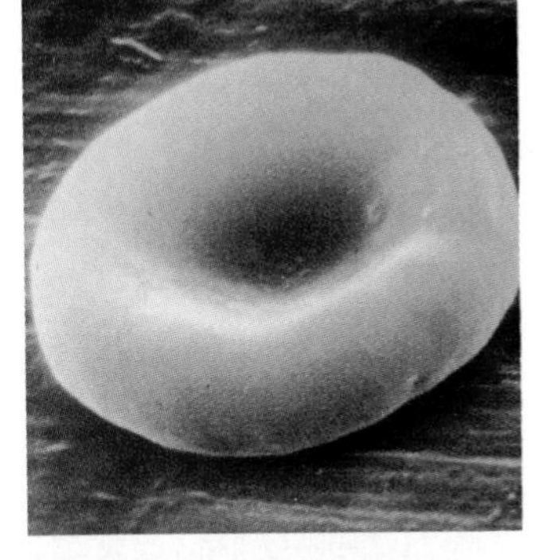

Normal red blood cell

The order of amino acids in a protein determines its structure and function, which are critical to its biological activity. A change of just one amino acid in the sequence can disrupt the entire protein molecule. For example, hemoglobin, the molecule in the blood that carries oxygen, has four polypeptide chains (two alpha chains and two beta chains) with a total of 574 amino acid units. Changing just one specific amino acid in one of the beta chains results in the defective hemoglobin molecule found in patients with sickle cell anemia.

Adult hemoglobin (Hb-A)	Val-His-Leu-Thr-Pro-Glu-Glu-Lys- . . .
Sickle cell hemoglobin (Hb-S)	Val-His-Leu-Thr-Pro-Val-Glu-Lys- . . .

Sickled red blood cell

Similarly, in their study of familial hypercholesterolemia (FH), Brown and Goldstein showed that in the chain of approximately 822 amino acids making up the LDL receptor molecule, a change of just one amino acid (a cysteine instead of a tyrosine) causes a defect in the receptor. Such defective LDL receptors bind to LDL but fail to cluster in the coated pits; therefore, they cannot carry the cholesterol-containing LDL particles into the cell.

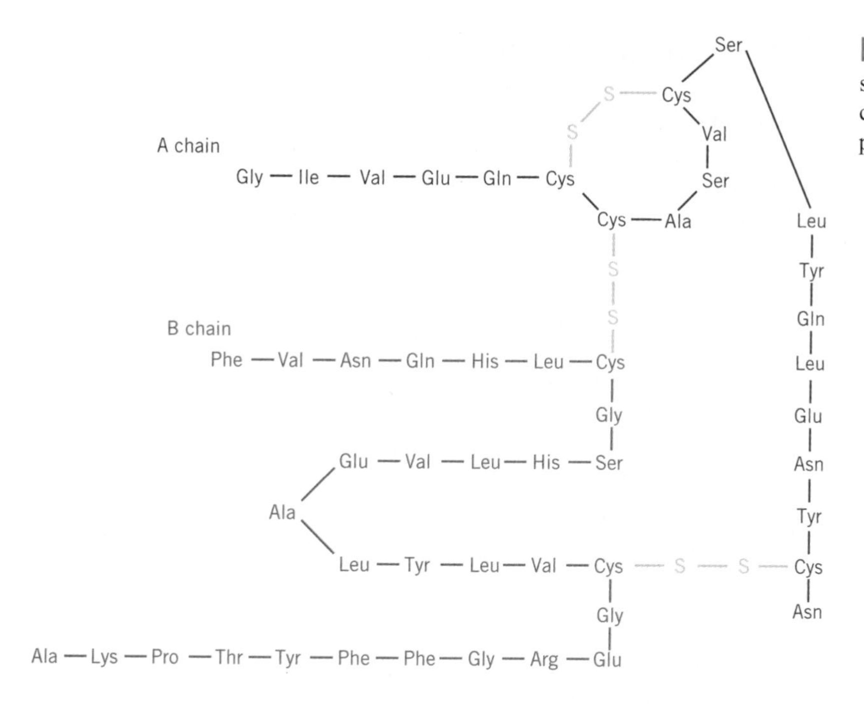

Figure 17.3 The amino acid sequence of bovine insulin contains 51 amino acids in two polypeptide chains.

Determining the amino acid sequence in a protein is a complex procedure. The technique was first developed by Frederick Sanger in 1953 when he determined the amino acid sequence of the protein insulin obtained from beef pancreas (Fig. 17.3). For many years, bovine (beef) insulin—along with insulin from pigs and sheep—was used to treat diabetes mellitus in humans. But there are small differences between animal insulin and human insulin in their sequence of amino acids, and these small differences have caused serious side effects for some patients. In 1978, however, human insulin was successfully synthesized in the laboratory. In 1982, human insulin became the first genetically engineered product to be approved for human use.

The insulin proteins from closely related species differ very little in their primary structures. Small variations in the sequence of amino acids may occur in regions of the protein that are not essential to its function. Consider the protein cytochrome *c*, which is essential for the production of energy in all cells. Researchers have studied the amino acid sequence of cytochrome *c* in many different species of plants and animals. They have determined that, of the 100 different amino acids in its structure, 38 positions contain the same amino acids in all species. These 38 amino acids, therefore, are thought to be critical to the structure and function of cytochrome *c*.

Check Your Understanding

1. Give the amino acid sequence and then draw the structure for the tripeptide containing the amino acids glutamine, arginine (C-terminal end), and serine (N-terminal end).

2. Enkephalins, produced in the brain, are neurotransmitters that act like opiates. Methionine enkephalin, which acts to inhibit the sense of pain, is a pentapeptide with the following amino acid sequence:

   Tyr-Gly-Gly-Phe-Met

   Draw the structural formula for methionine enkephalin, boxing in the peptide bonds.

17.8 Secondary Structure

◆ **Describe the secondary structure of a protein.**

The **secondary structure** of a protein is the shape of specific regions of the polypeptide backbone. Parts of the polypeptide backbone may take on regular folding patterns, such as helices, pleated sheets, loops, and turns. These structures were first explained by Linus Pauling on the basis of X-ray diffraction data. The folding patterns are held in place by hydrogen bonding between the hydrogen on the nitrogen of the peptide bond and the oxygen on the carbon of another peptide bond.

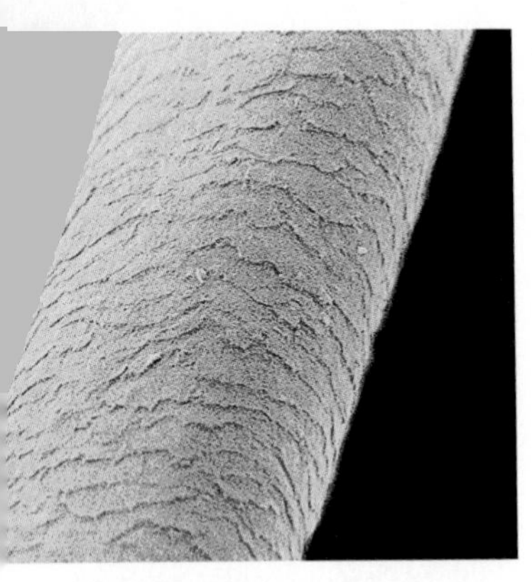

Human hair

(a) (b)

These hydrogen bonds can form (a) between amino acids on the same polypeptide chain, forming a loop in the molecule, or (b) between amino acids on different polypeptide chains.

Fibrous Proteins

Secondary structure is very important to fibrous proteins. **Keratins** are the fibrous proteins that make up hair, fur, wool, claws, hooves, and feathers. Pauling determined that the polypeptide chains in the keratin protein were curled in an arrangement called an **alpha helix (α helix).** In this arrangement, the amino acids form loops in which the hydrogen on the nitrogen atom in the peptide bond is hydrogen-bonded to the oxygen attached to the carbon atom of a peptide bond farther down the chain (Fig. 17.4a). There are 3.6 amino acids in each turn of the α helix, with the R groups on these amino acids extending outward from the helix.

Collagen fibrils from skin

In hair, pairs of polypeptides arranged in alpha helices coil around one another. Then these coils themselves twist around one another—much like the fibers in a rope—to form a protofibril that is held together by disulfide bridges (see Section 17.9). A single hair consists of many of these protofibrils arranged parallel to one another and embedded in an insoluble protein framework.

Collagen is the most abundant protein in the body; it is found in skin, bones, teeth, tendons, cartilage, blood vessels, and connective tissue. It is a protein formed of long insoluble fibers of great strength. Collagen has an unusual amino acid composition, containing about one-third glycine and one-third proline and 4-hydroxyproline. The arrangement of the amino acids in the polypeptide chain give collagen its strong secondary structure: three left-handed polypeptide helices twisted together to form a triple-helix structure called tropocollagen (Fig. 17.4b).

Domestic silk worm

Silks contain the fibrous protein fibroin held together by an amorphous protein called sericin. In the secondary structure of silks, hydrogen bonding occurs between amino acids on adjacent polypeptide chains. In fibroin, several polypeptide chains running in opposite

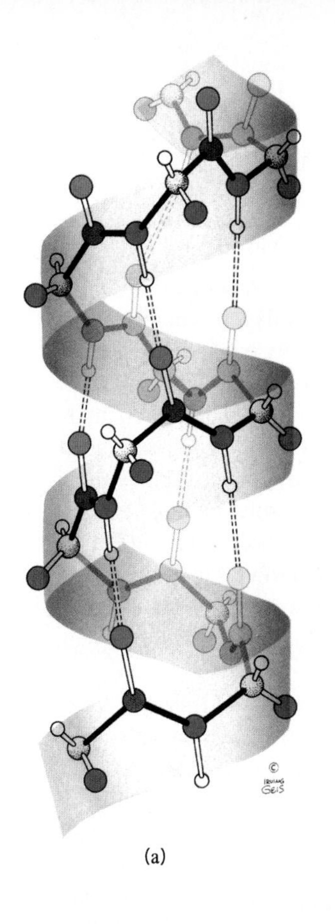

(a)

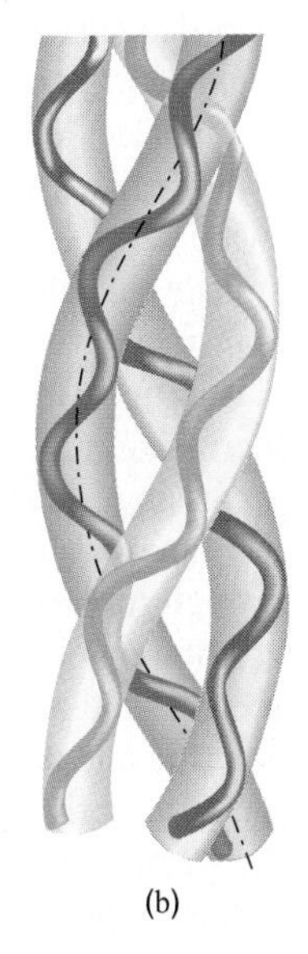

(b)

Figure 17.4 (a) Polypeptide chain in an α helix configuration. The amino acids are coiled in a manner resembling a circular staircase, with loops held together by hydrogen bonds. (b) The triple-stranded helix of collagen.

directions are located next to each other in an arrangement called a **beta configuration (β configuration),** or **beta pleated sheet.** This gives the protein a zigzag appearance, from which we get the name *pleated sheet.* The polypeptide chains in silks are held together only by extensive hydrogen bonding. Note that the R groups extend above and below the sheet (Fig. 17.5).

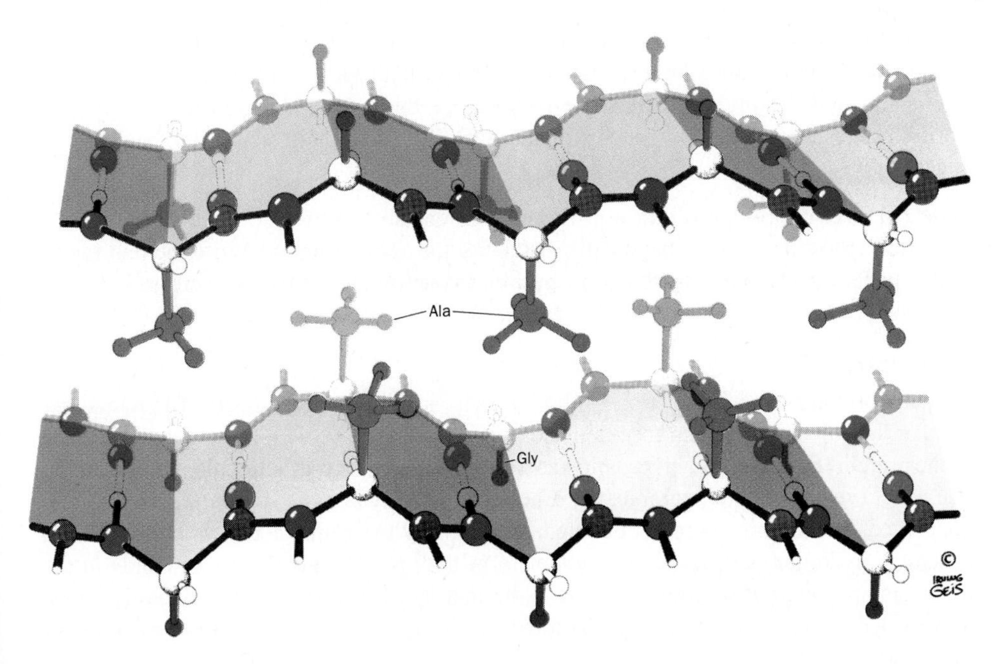

Figure 17.5 Two polypeptide chains in the β pleated-sheet configuration. Dashed lines represent hydrogen bonds.

Chemical Perspective / Collagen

Because collagen is the main protein making up connective tissue, the condition of our skin depends a great deal on the condition of our collagen. Collagen contains three separate protein chains or strands, each of which winds around its own axis to form a left-handed spiral (or helix). The intertwining of these three twisted strands results in triple-helix structures called tropocollagen. The tropocollagen fibrils themselves intertwine to make up the collagen that is the structural support for the skin. The collagen strands are held together, or cross-linked, by hydrogen bonds and some covalent bonds. As we grow older, the amount of cross linking increases. This results in a decrease in elasticity, and drying and wrinkling of the skin. You can also observe this effect of increased cross linking in the meat you buy. As an animal grows older, the meat increasingly becomes tougher. Older chickens are sold as stewing chickens because their meat requires the longer cooking process of stewing for it to become tender enough to eat.

Although increased cross linking between collagen fibrils may be undesirable when it comes to aging, it can be of considerable importance in other ways. Collagen can be attacked and broken down by bacteria, resulting in the destruction of the collagen and the skin. Early humans found this out when they used animal skins for protection from the weather. In a matter of days the skins began to decompose and had to be discarded. It was in the Mediterranean region about 3500 years ago that a process called tanning was developed to preserve hides, skins, and furs. The term "tanning" comes from one of the first materials used to convert skin to leather: the crushed bark of an oak tree called "tan." A number of chemicals found in the bark, roots, wood, leaves, and fruit of many trees and plants can cause the tanning process. These chemicals, called tannins, are mixtures of large polyphenols with molecular weights ranging from 500 to 3000.

In the tanning process, a skin or hide is placed in a dilute acid solution. The acid quickly penetrates the open regions of the collagen, breaking the hydrogen bonds. The collagen then swells, permitting the tannins to penetrate its structure. Polyphenols in the tannins form hydrogen bonds. The new, stronger hydrogen bonds formed between the tannins and the protein chains greatly increase the cross linking in the collagen fibrils. The resulting leather is much stronger, water-repellant, and resistant to bacterial attack.

Your Perspective: Suppose you are trying to invent an "antiwrinkle" cream. What does your cream have to do chemically to reverse the aging process? Write the text for a newspaper ad that you might use to promote your revolutionary new cream.

Globular Proteins

Globular proteins can have sections of both α helices and β pleated sheets. The remaining parts of the polypeptide chain form loops or turns that are most often located at the surface of the protein (Fig. 17.6). Each protein has a specific secondary structure depending on the amino acid sequence. For example, the α helix is formed when the R groups are small and uncharged. This arrangement is disrupted, however, by the amino acid proline, which has no amide hydrogen and, therefore, cannot form hydrogen bonds. Kinks or bends in the molecule are found at proline positions. Note that hydrogen bonding is relatively weak, noncovalent bonding and is easily disrupted by changes in pH, temperature, solvents, or salt concentrations.

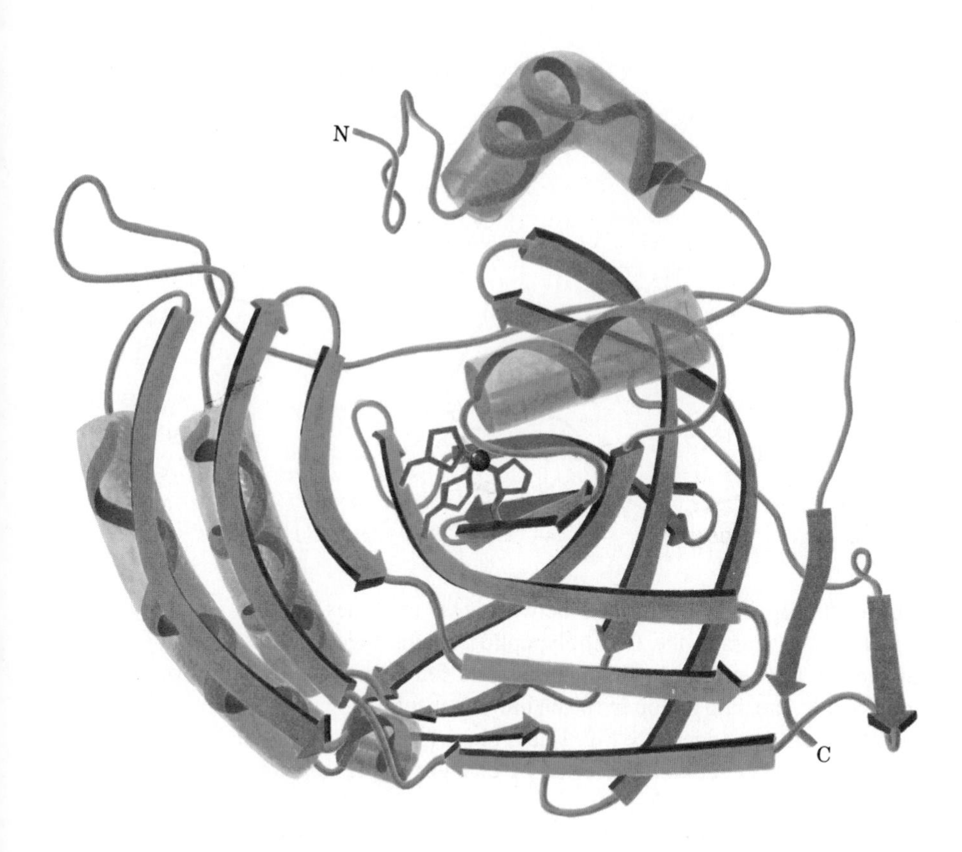

Figure 17.6 Model of the enzyme carbonic anhydrase. The α helix portions of the molecule are represented by cylinders and the β pleated sheet by arrows. The ball in the middle is a Zn^{2+} ion.

17.9 Tertiary Structure

◆ **Describe the tertiary structure of a protein.**

Tertiary structure is the three-dimensional structure of globular proteins. At normal pH and temperature, each protein takes on a shape that is energetically the most stable given the specific sequence of amino acids and the various types of interactions that may be involved. This shape is called the **native state,** or **native configuration,** of the protein. In general, globular proteins are very tightly folded into a compact, spherical (globular) form. This folding results from interactions between the R-group side chains of amino acids and may involve some combination of hydrogen bonding, disulfide bridges, hydrophobic forces, or salt bridges (Fig. 17.7). Hydrogen bonding is a major factor in the tertiary structure of a protein. It is internal hydrogen bonds that help fold the protein into its native state.

◆ **Give four examples of interactions that hold a protein in its native configuration.**

Disulfide bridges occur between molecules of the amino acid cysteine. Sulfhydryl groups (—SH) on the side chains of cysteine molecules are easily oxidized to form a disulfide (—S—S—) linkage or bond.

$$\underset{\text{cysteine}}{HO-\overset{O}{\overset{\|}{C}}-\underset{NH_2}{\underset{|}{\overset{H}{\overset{|}{C}}}}-CH_2-SH} + \underset{\text{cysteine}}{HS-CH_2-\underset{NH_2}{\underset{|}{\overset{H}{\overset{|}{C}}}}-\overset{O}{\overset{\|}{C}}-OH} \xrightarrow{-2H} HO-\overset{O}{\overset{\|}{C}}-\underset{NH_2}{\underset{|}{\overset{H}{\overset{|}{C}}}}-CH_2-\underbrace{S-S}_{\text{disulfide linkage}}-CH_2-\underset{NH_2}{\underset{|}{\overset{H}{\overset{|}{C}}}}-\overset{O}{\overset{\|}{C}}-OH$$

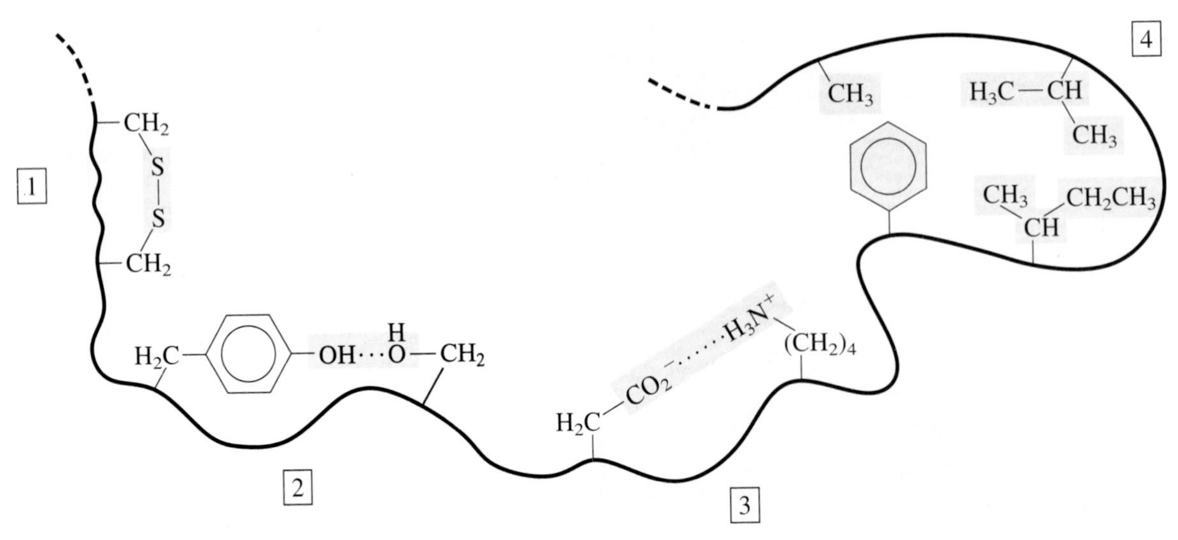

Figure 17.7 Interactions between R groups in proteins. (1) disulfide bridge (2) hydrogen bonding (3) salt bridge (4) hydrophobic interaction.

The disulfide linkage can be formed between two cysteines on different polypeptide chains, in this way linking the two polypeptide chains together to form tough, strong material such as keratin. The linkage can also be found between cysteines on the same polypeptide chain, creating a loop in the chain. The insulin molecule contains examples of both types of disulfide bridges (see Fig. 17.3). These bridges are covalent linkages that can be broken by reduction but are unaffected by changes in pH, solvents, or salt concentrations.

Hydrophobic interactions occur between the nonpolar side chains of the amino acids in the protein molecule, and they are perhaps the most important interactions in determining protein shape. Because they are repelled by water, the nonpolar R groups tend to be found on the inside of the molecule with other hydrophobic groups.

Salt bridges result from ionic interactions between charged carboxyl or amino side chains found on amino acids such as aspartic acid, glutamic acid, lysine, and arginine (see Table 17.1). These linkages are easily broken by changes in pH.

17.10 Quaternary Structure

◆

Describe the quaternary structure of a protein.

The **quaternary structure** of a protein is the manner in which separate polypeptide chains and prosthetic groups fit together in proteins containing more than one polypeptide chain. Hydrogen bonding, hydrophobic interactions, and salt bridges may be involved in holding the chains in position. Hemoglobin is just one example of a protein that contains more than one polypeptide chain. It has four polypeptide chains—two alpha chains and two beta chains—each arranged around an iron-containing heme group (Fig. 17.8). All four chains must be present to have a biologically active molecule.

17.11 Denaturation

Various changes in the surroundings of a protein can disrupt the complex secondary, tertiary, or quaternary structure of the molecule. Disruption of the native state of the protein is called **denaturation.** This process involves the uncoiling of the protein molecule into a random state and causes the protein to lose its biological activity.

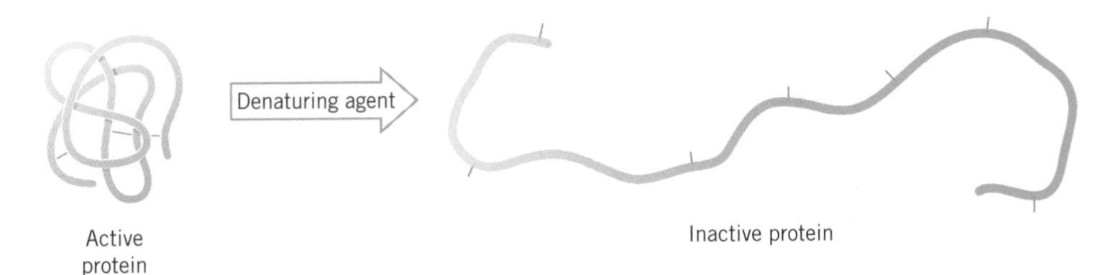

Denaturation may or may not be permanent; sometimes the protein returns to its native state when the denaturing agent is removed. Denaturation also may result in coagulation, with the protein precipitating out of the solution. Denaturing agents come in many forms, a few of which are discussed here.

◆

List four changes in the environment of a protein that can result in its denaturation.

pH

Changes in pH have their greatest disruptive effect on hydrogen bonding and salt bridges. For example, the polypeptide polylysine is composed entirely of the amino acid lysine, which has an amino group on its side chain. In acidic pH, all the side chains are positively charged and repel each other, causing the molecule to uncoil. In basic pH, however, the side chains are neutral. They do not repel, and the molecule coils into an α helix. Exposing proteins to strong acids or bases for long periods leads to the hydrolysis of the peptide chain.

Heat and Radiation

Heat causes an increase in the thermal vibration of the molecules, disrupting hydrogen bonding and salt bridges. After gentle heating, the protein can usually regain its native state. Violent heating, however, results in irreversible denaturation and coagulation of the protein (as, for example, in cooking an egg). Similarly, heat used to sterilize equipment coagulates the protein of microorganisms, thereby destroying the bacteria. Heat can also

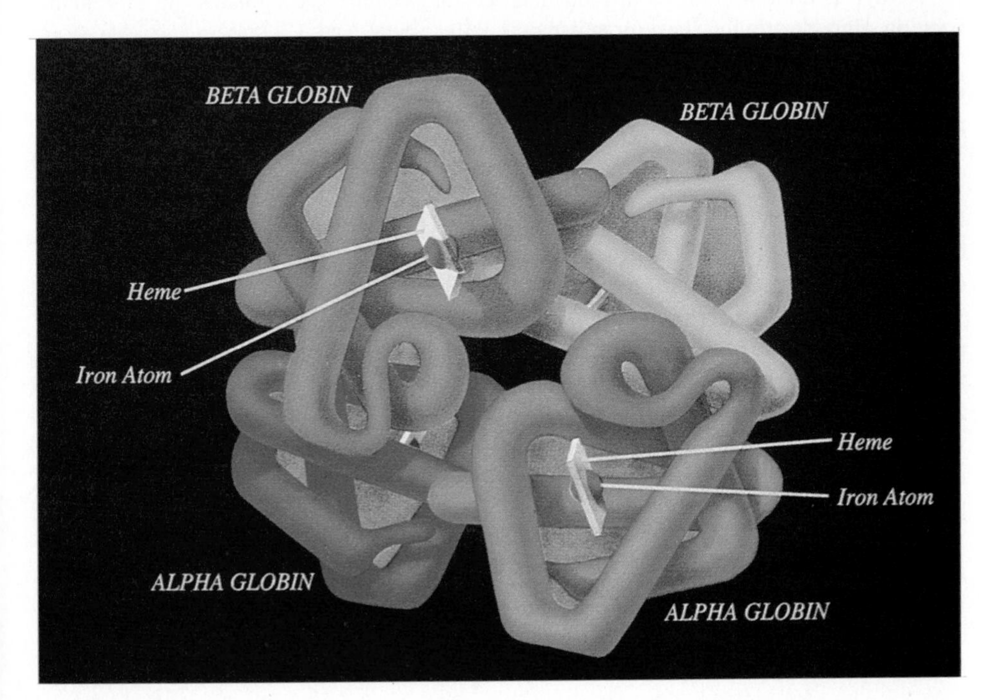

Figure 17.8 Quaternary structure of hemoglobin.

Health Perspective / Treating Burns

Over 2 million burn injuries are reported annually in the United States, with most occurring in the home from carelessness or ignorance. One-third of the victims are children. To minimize the effects of any burn, it is important to act quickly. The injured area should be cooled with cold water or ice. You can minimize the depth of the burn injury in the skin if it is treated within one minute.

The greatest initial threat to the victim of a major burn is the loss of body fluids. The barrier against fluid loss is the outermost layer of skin. When a burn is severe enough to open large portions of the skin, severe reactions from fluid loss can occur. For this reason, it is crucial to treat the burn with something to prevent the loss of body fluids. Alkaloid reagents, like tannic or picric acid, are an especially effective treatment. When applied to the burn they precipitate surface proteins, creating a protective barrier that stops the fluid loss. A household source of tannic acid is tea. Applying dampened tea bags to a burn precipitates protein and forms a protective crust over the wound, thus slowing the loss of body fluids and promoting healing.

be used to detect the presence of protein in urine; urine that turns cloudy when heated indicates the presence of protein. Nonionizing radiation is similar to heat in its effect on proteins. For example, ultraviolet light denatures proteins in the skin, causing sunburn.

Organic Reagents

Organic solvents, such as alcohol or acetone, interfere with the hydrophobic interactions of the nonpolar side chains of the amino acids in the protein molecule. This change in hydrophobic interactions with solvent causes the protein to denature and coagulate. A 70% ethyl alcohol solution is a good disinfectant because it coagulates the protein in bacteria, thus destroying these organisms.

Reagents such as tannic and picric acid affect salt bridges and hydrogen bonding, causing proteins to precipitate. Tannic acid is sometimes applied to severe burns. The precipitated protein forms a crust over the wound and slows the loss of body fluids.

Heavy-Metal Ions

Heavy-metal ions, such as Pb^{2+}, Hg^{2+}, and Ag^{+}, may disrupt the natural salt bridges of a protein by forming salt bridges of their own with the protein. This usually causes coagulation of the protein. Organomercury ions, such as methyl mercury (CH_3Hg^+), which are present in mercury pollution, coagulate proteins in this way. Heavy-metal ions also bond with sulfhydryl groups, disrupting the disulfide linkages in the protein molecule, thereby denaturing the protein. Such heavy metals are therefore toxic to living organisms. As an antidote to heavy-metal poisoning, patients are given substances high in protein, such as milk or egg whites. These substances bond with the metals while they are still in the stomach. The victim must then be made to vomit before the metals are again released through the processes of digestion.

Reducing Agents

Reducing agents disrupt disulfide bridges formed between cysteine molecules. Thioglycolic acid, a component of permanent-wave hair preparations, breaks the disulfide bridges in hair and reduces them to —SH groups.

$$HS—CH_2—\overset{\overset{\large O}{\|}}{C}—OH$$

thioglycolic acid

The hair is then curled and an oxidizing agent such as hydrogen peroxide (H_2O_2) is applied. This causes the formation of new disulfide bridges.

Check Your Understanding

1. Describe the types of bonding or attractive interactions that occur in the following protein structures:

   (a) the primary structure of a protein

   (b) the secondary structure

   (c) the tertiary structure

   (d) the quaternary structure

2. Define native state and denaturation.

3. Which structure or structures of a protein are disrupted by

   (a) a strong acid **(b)** sunlight **(c)** lead (Pb^{2+}) ions

KEY CONCEPTS

Proteins are large polymers of amino acids and have many functions in the body. They may be simple proteins that contain only amino acids, or complex proteins that contain other nonprotein or inorganic components in addition to amino acids. Enzymes, hormones, antibodies, and transport proteins are all globular proteins. Keratin, muscle fibers, and collagen are examples of fibrous proteins.

Proteins are constructed out of 20 amino acids. The amino acids each contain an amino group and a carboxyl group. As a result, they are amphoteric and can act as buffers. Each amino acid and protein has a characteristic isoelectric point. The isoelectric point is also the pH at which the protein is the least soluble in water.

Adequate proteins contain all 10 essential amino acids. Essential amino acids are those amino acids not synthesized adequately by the body.

Proteins have complex structures that can be analyzed at four levels: primary, secondary, tertiary, and quaternary. The primary structure is the sequence of amino acids held together by peptide bonds in the polypeptide chain. The secondary structure is the shape of specific regions of the polypeptide backbone, which result from hydrogen bonding between the amide groups. Alpha helix, beta configuration, and triple helix are all examples of different types of secondary structures. Tertiary structure refers to the three-dimensional structure of globular proteins, including the folding of the secondary structure and the arrangement of the amino acid side groups. The interactions between side groups include hydrogen bonding, disulfide bridges, salt bridges, and hydrophobic interactions. Most globular proteins contain more than one polypeptide chain. The way in which these multiple chains fit together (often with prosthetic groups) is called the quaternary structure of the protein.

The native state of a protein is the most stable structural arrangement for that protein. Denaturation is the disruption, whether permanent or temporary, of the native state of a protein. Denaturation can be caused by heat, radiation, changes in pH, organic solvents, heavy-metal ions, alkaloid reagents, or reducing agents.

Review Problems

Section 17.1

1. What is the difference between an amino acid and a protein?
2. Give the common name for the following amino acids:
   (a) aminoacetic acid
   (b) 2-aminopropanoic acid
   (c) 2-amino-4-methylpentanoic acid
   (d) 2-amino-3-methylbutanoic acid

Section 17.2

3. Write the structural formulas for the following amino acids. Mark with an asterisk each carbon that is a chiral center.
   (a) L-valine **(b)** D-glutamic acid
4. Write the structural formulas for the following amino acids. Mark with an asterisk each carbon that is a chiral center.
   (a) L-threonine **(b)** L-isoleucine

Section 17.3

5. **(a)** What is meant by an adequate protein?
   (b) What are the nutritional problems of eating a strictly vegetarian diet?
6. Is a diet consisting mainly of rice an adequate diet? Why or why not?

Section 17.4

7. What fact about the structure of amino acids explains their high melting points?
8. Draw the structure for the zwitterion form of the following amino acids:
   (a) valine
   (b) threonine
   (c) phenylalanine
   (d) tyrosine
9. Write the reactions that show how alanine is able to act as an acid and as a base.
10. Describe how blood proteins help protect against pH changes.

Section 17.5

11. If placed in an electric field, how does each of the following proteins migrate when the pH of the solution is 7?
    (a) urease **(b)** myoglobin **(c)** chymotrypsin

12. Why is a protein the least soluble at its isoelectric point?

13. Draw the structural formula for each of the following amino acids in the form in which it exists at the given pH.

(a) phenylalanine at pH 3.0 **(c)** alanine at pH 8.0

(b) phenylalanine at pH 6.0 **(d)** alanine at pH 5.0

14. **(a)** At what pH is the amino acid leucine electrically neutral? Draw the structural formula of leucine at this pH.

(b) Why does the solubility of leucine increase as the pH is increased? Write the equation for the addition of base to the solution in part (a).

(c) Why does the solubility of leucine increase as the pH is decreased? Write the equation for the addition of acid to the solution in part (a).

Section 17.6

15. Describe the difference between each of the following:

(a) a simple protein and a conjugated protein

(b) a globular protein and a fibrous protein

(c) a glycoprotein and a lipoprotein

16. List four functions of proteins in living organisms.

Section 17.7

17. Write structural formulas for the two possible dipeptides that can be formed from tyrosine and aspartic acid.

18. Using three-letter abbreviations for the amino acids, show the six tripeptides that contain phenylalanine, serine, and isoleucine.

19. Explain what is meant by the N-terminal end and the C-terminal end of a protein or polypeptide molecule.

20. A pentapeptide is made up of amino acids in the following sequence:

Ser-Gly-Tyr-Ile-Cys

Draw the structural formula for this pentapeptide, boxing in the peptide bonds. Identify the N-terminal and C-terminal amino acids.

21. Write the structures and identify each of the products of the acid hydrolysis of the following polypeptide:

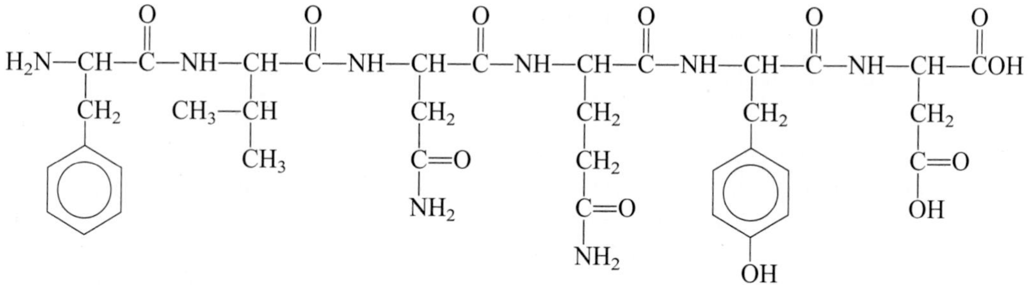

```
           O              O              O              O              O              O
           ||             ||             ||             ||             ||             ||
H2N—CH—C—NH—CH—C—NH—CH—C—NH—CH—C—NH—CH—C—NH—CH—COH
     |           |              |              |              |              |
    CH2    CH3—CH            CH2            CH2            CH2            CH2
     |           |              |              |              |              |
  (C6H5)        CH3            C=O            CH2          (C6H4)          C=O
                                |              |              |              |
                               NH2            C=O             OH             OH
                                               |
                                              NH2
```

22. Using the three-letter abbreviations for the amino acids, write the amino acid sequence for the polypeptide in problem 21.

Section 17.8

23. Describe the difference between each of the following:

(a) the primary structure and the secondary structure of a protein

(b) a dipeptide and a polypeptide

(c) the alpha helix configuration and the beta configuration

24. What type of bonding occurs in each of the following?

(a) α-helix configuration

(b) globular proteins

(c) β configuration

(d) triple-helix configuration

25. What structural difference makes the keratins of wool flexible and those of hooves rigid?

26. Which is more completely disrupted by heat: the structure of keratin or that of silk? Give the reason for your answer.

Section 17.9

27. (a) What is meant by the native state of a protein?

(b) Name four types of interactions between amino acids in a protein that are important in determining the native state.

28. One of the amino acids that is most important in determining the three-dimensional structure of globular proteins is cysteine. Explain.

Section 17.10

29. Describe the difference between the following: the tertiary structure and the quaternary structure of a protein

30. What types of interactions hold the protein in the shape of its quaternary structure?

Section 17.11

31. (a) What happens to a protein when it is denatured?

(b) Is it possible to reverse this process?

32. By means of a chemical reaction, show the effect of a change in pH on the side chain amino group of lysine.

APPLIED PROBLEMS

33. Why does protein in the urine indicate a possible kidney disorder?

34. Suggest several ways in which a diet consisting mainly of corn can be supplemented to provide enough adequate protein.

35. The following is the amino acid sequence of the hormone vasopressin:

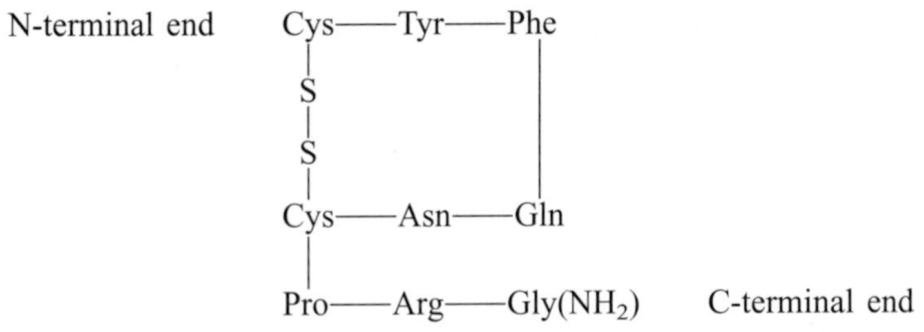

```
N-terminal end    Cys——Tyr——Phe
                   |         |
                   S         |
                   |         |
                   S         |
                   |         |
                  Cys——Asn——Gln
                   |
                  Pro——Arg——Gly(NH2)    C-terminal end
```

(a) Draw the structural formula of vasopressin. The notation Gly(NH_2) means that the carboxyl group of glycine has reacted to form an amide, $CONH_2$.

(b) What kind of linkage other than a peptide bond is present in the molecule?

36. Albumin, the most abundant protein in the blood plasma, transports slightly soluble molecules, such as aspirin, bilirubin, and barbiturates, through the plasma and extracellular fluid. Albumin has an isoelectric point of 4.8. What charge does this molecule have at physiological pH (pH 7.4)?

37. Explain, on the molecular level, the process of hair straightening.

38. One effective home remedy for the treatment of severe burns is the application of a moist tea bag. (Tea contains tannic acid.) Can you explain how this affects the burn?

39. A 1% solution of silver nitrate is placed in the eyes of newborn infants as a disinfectant against the bacteria causing gonorrhea. How is this effective in killing the bacteria?

40. A 70% alcohol solution is an effective disinfectant. Interestingly, the more commonly available 95% alcohol solution is not nearly as effective at killing the bacteria. Suggest a reason why this might be the case.

41. If a child consumes a poison containing heavy metals, eggs can be administered as an antidote.

(a) Why are eggs a good antidote for lead poisoning?

(b) What additional treatment is required when an egg is used as an antidote?

Chapter 18

Enzymes, Vitamins, and Hormones

*K*athy Phipps felt she hadn't a care in the world. She loved her job teaching high school mathematics, her two children were both doing well in school, and her husband had just been promoted to manager of his department.

Well, she actually did have one little worry. When she had examined her breasts the previous month, she thought she had detected a small lump. She hadn't been sure, and couldn't always find it again, but lately there was no question that the lump was there. This is what worried Kathy. Her mother had developed breast cancer and

had undergone a radical mastectomy about 15 years earlier. Kathy cringed at the thought of losing a breast and looking the way her mother did. She promised that if the lump was still there in a month, she would go see her doctor.

Dr. Huntington felt the lump in Kathy's breast and was dismayed to learn that Kathy had first discovered it more than three months ago. He sent Kathy for a mammogram and telephoned the next day to tell her the lump appeared to be cancerous. He had already scheduled her for a biopsy of the lump and the lymph nodes under her arm. If the pathology reports from the biopsy were positive, he said they would discuss the various options for her treatment.

Three days later Kathy learned that the biopsy had found malignant (cancerous) cells both in the lump and in the lymph nodes under her arm. This meant that the malignant cells had begun to spread, or metastasize, into her body. Dr. Huntington explained that she could choose between two courses of treatment: she could undergo surgical removal of the breast (a modified radical mastectomy) followed by chemotherapy to destroy the metastatic cells, or she could be treated with chemotherapy followed by irradiation of her breast to destroy any cancerous cells remaining in the area of the biopsy incision. The choice was hers—both treatments had shown similar survival rates in clinical trials. Because of her strong negative feelings about mastectomy, Kathy chose to be treated with chemotherapy and radiation.

The following week Kathy began a six-month program of chemotherapy treatment. During this period she was given three drugs in combination: methotrexate, 5-fluorouracil, and cyclophosphamide. Used together, these drugs work to destroy cells that are rapidly dividing. Unfortunately, cancer cells are not the only cells in a person's body that divide rapidly. Kathy's treatment disrupted and damaged some normal tissues as well as the cancer cells.

Dr. Huntington had warned Kathy about the well-known side effects of the chemotherapy. She had been told, for example, that because bone marrow cells reproduce very rapidly to continually replace the cells in the blood, the chemotherapy would disrupt the function of her bone marrow. That is why Kathy had her blood tested weekly to monitor the decrease in her white blood cell and platelet count. Too severe a decrease in these blood cells would require that the chemotherapy be stopped. Similarly, the body is constantly sloughing off and replacing cells from the mucous linings of the mouth and intestines (the mucous lining of the small intestine replaces itself every 48 hours!). The cells producing this lining are among those disrupted by the chemotherapy, commonly producing sores in the mouth, nausea, and vomiting. Another side effect that is much more visible is due to the use of cyclophosphamide, which causes the hair to fall out. Although she often felt tired and listless, Kathy bore up well under the side effects of the chemotherapy and was able to return to her classroom and finish the term.

In this chapter we learn about enzymes, which are critical to the functioning of living cells. In particular, we discuss how the inhibition of enzyme function can lead to disruption and death of the cell. 5-Fluorouracil and methotrexate are antimetabolites, chemicals that obstruct enzyme function. In this case, both compounds inhibit the function of enzymes that are needed for the synthesis of DNA in the cell. But the synthesis of DNA is a necessary step in the division of a cell. By acting to prevent cell reproduction, these two chemicals have especially damaging effects on rapidly dividing cells such as cancer cells. This gives physicians a way to destroy cancer cells that have broken off from a tumor and are circulating in the blood.

Extensive research into the structure and function of enzymes has led not only to the development of chemotherapy drugs that inhibit enzyme function but also to the

identification of specific enzymes released into the blood by damaged cells. The level of these enzymes in the blood can reveal how much damage has been done to the heart muscle after a heart attack, or whether a patient is suffering from infectious hepatitis.

18.1 What Is Metabolism?

Metabolism can be defined as all the enzyme-catalyzed reactions in the body. These reactions are of two types: **catabolic reactions,** in which molecules are broken down to produce smaller molecules and energy, and **anabolic reactions,** in which the cell uses energy to produce the molecules it needs for growth and repair (these are also called biosynthetic reactions). Anabolic and catabolic reactions occur continuously in the cell, and each involves a complicated series of carefully controlled enzyme-catalyzed reactions.

◆ **Define metabolism.**

◆ **Explain how the anabolic and catabolic reactions of metabolism differ.**

18.2 Enzymes

Enzymes are the largest and most highly specialized class of proteins. They function as biological catalysts. Everything that goes on within a cell is either catalyzed by or directed by enzymes. Earlier we described catalysts as substances that increase the rate of a chemical reaction by lowering the activation energy of that reaction. The reactions of metabolism would ordinarily occur at extremely slow rates at normal body temperature and pH. Without the enzymes in our digestive tract, for example, it would take us about 50 years to digest a single meal! Enzymes greatly increase this reaction rate, allowing cells to function under normal body conditions.

Because a catalyst is not consumed in the reaction, it can be used over and over again. Such compounds, therefore, need to be present in only very small amounts. Enzymes are water-soluble globular proteins that carry out their functions in body fluids or bound to the membranes of the cell. Enzymes can vary in molecular weight from 12,000 to over 1 million. The enzyme molecule may be a simple single polypeptide chain or a more complex molecule composed of several polypeptide chains and other nonprotein parts.

For example, the enzyme ribonuclease consists of a single polypeptide chain of 124 amino acids and has a molecular mass of about 13,700 daltons (remember, 1 dalton = 1 amu). The enzyme pyruvate dehydrogenase, on the other hand, is a macromolecular complex containing 42 individual molecules and having a molecular mass of about 10 million daltons.

Before continuing with our discussion of enzymes, some special terms used in the study of these complex molecules must be defined (Fig. 18.1).

◆ **Identify the parts of an enzyme molecule.**

Apoenzyme The apoenzyme is the protein part of the enzyme molecule.

Cofactor Cofactors are additional chemical groups that are required for enzyme activity. Cofactors may consist of metal ions or complex organic molecules. Some enzymes require both types of cofactors.

Coenzyme When the cofactor is a complex organic molecule other than a protein, the cofactor is called a coenzyme.

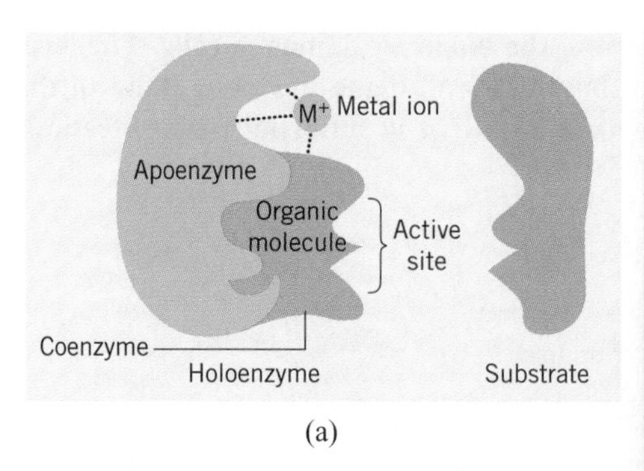

(a)

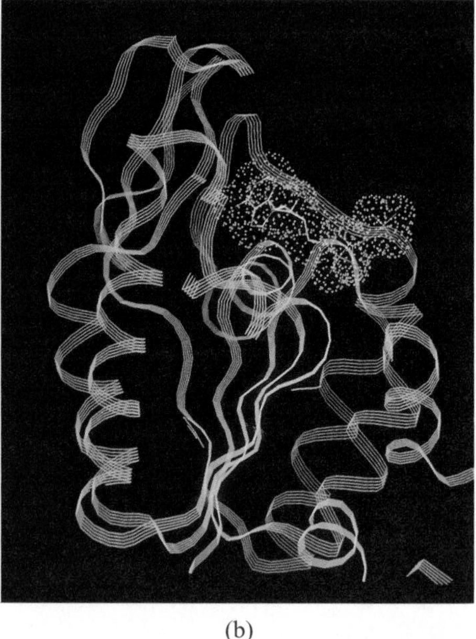

(b)

Figure 18.1 (a) Important terms used in the study of enzymes. (b) A computer model of the substrate ATP binding to the enzyme phosphoglycerate kinase.

Prosthetic group If the cofactor is a permanent part of the enzyme, bound to the protein portion with covalent bonds, it is called a prosthetic group.

Holoenzyme A holoenzyme is an entire active enzyme, which consists of an apoenzyme and one or more cofactors.

Proenzyme, or zymogen A proenzyme (or zymogen) is an enzyme in its inactive form. Enzymes (especially digestive enzymes) are often synthesized in an inactive form, transported to the place where activity is desired, and then converted to their active forms.

Substrate The substrate is the chemical substance or substances on which the enzyme acts.

◆ **Describe the function of the active site.**

Active site The active site is the specific area of the enzyme to which the substrate attaches during the reaction. An enzyme molecule can have several active sites.

18.3 Enzyme Nomenclature and Classification

Because enzymes are the largest class of proteins, scientists have been able to isolate and describe a great number of individual enzymes. Originally, enzymes were named simply by ending the enzyme name with the suffix -in to indicate a protein. Names such as trypsin, renin, and pepsin are examples of this nomenclature and are the common names still used today. However, such names give no indication of the reaction being catalyzed by the enzyme or the substrate involved.

In 1961, the Commission on Enzymes of the International Union of Biochemistry proposed a standard classification of enzymes. They recommended that enzymes be divided into six major classes (each with several subclasses) based on the reactions that they catalyze (Table 18.1). Enzymes are now named by adding an -ase ending on the name of the substrate (or on a phrase describing the enzyme's activity). For example, ribonuclease

Table 18.1 Classes of Enzymes

Example	*Reaction Catalyzed*
Hydrolases: Enzymes that catalyze hydrolysis reactions	
Carbohydrases	Polysaccharides $\xrightarrow{+H_2O}$ Monosaccharides and dissaccharides
Esterases	Ester $\xrightarrow{+H_2O}$ Acid + alcohol
Proteases	Protein $\xrightarrow{+H_2O}$ Peptides and amino acids
Nucleases	Nucleic acids $\xrightarrow{+H_2O}$ Pyrimidines + purines + sugars + phosphoric acid
Oxidoreductases: Enzymes that catalyze oxidation–reduction reactions	
Oxidases	Addition of oxygen to a substrate
Dehydrogenases	Removal of hydrogen from a substrate
Transferases: Enzymes that catalyze reactions involved in the transfer of functional groups	
Transaminases	Transfer of $—NH_2$
Transmethylases	Transfer of $—CH_3$
Transacylases	Transfer of $—\overset{\overset{\large O}{\|}}{C}—R$
Transphosphatases (Kinases)	Transfer of $—O—\overset{\overset{\large O}{\|}}{\underset{\underset{\large OH}{\vert}}{P}}—OH$
Lyases: Enzymes that catalyze the elimination of groups to form double bonds	
Isomerases: Enzymes that catalyze the interconversion of isomers	
Ligases: Enzymes that, in conjunction with ATP, catalyze the formation of new bonds	

hydrolyzes ribonucleic acids, and pyruvate dehydrogenase catalyzes the removal of hydrogen from pyruvate.

18.4 Method of Enzyme Action

We stated that enzymes catalyze reactions in cells by lowering the activation energy. They do this by forming a complex with the substrate (that is, by binding to the substrate), thereby increasing the probability that the reaction will occur. We can outline the steps of an enzyme-catalyzed reaction as follows:

Describe the method by which enzymes catalyze reactions.

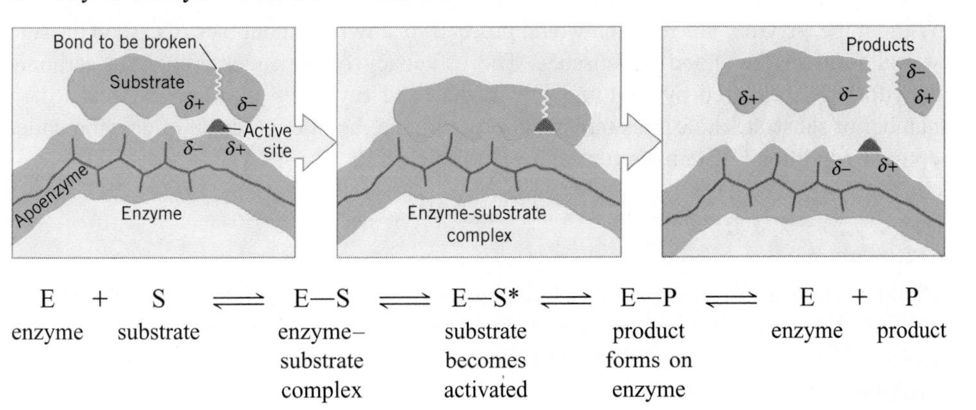

$$\underset{\text{enzyme}}{E} + \underset{\text{substrate}}{S} \rightleftharpoons \underset{\text{enzyme–substrate complex}}{E—S} \rightleftharpoons \underset{\text{substrate becomes activated}}{E—S^*} \rightleftharpoons \underset{\text{product forms on enzyme}}{E—P} \rightleftharpoons \underset{\text{enzyme}}{E} + \underset{\text{product}}{P}$$

Health Perspective / The Value of a Balanced Diet

Everybody remembers a parent, at one time or another, telling them to eat their vegetables. For centuries people have known that a balanced diet containing vegetables and fruits promotes health. But scientists are just now beginning to discover the wealth of chemicals in fruits and vegetables, called phytochemicals, that protect against cancer. These chemicals evolved in plants as a means of protecting them from sunlight. There is growing evidence, however, that these phytochemicals also block many of the processes that lead to cancer in humans.

Here are some examples. Broccoli and other members of the cruciferous family of vegetables (such as brussel sprouts, cauliflower, and kale) contain sulforaphane. This compound stimulates animal and human cells to produce enzymes that help whisk carcinogens out of the cell before they can do any damage. Citrus fruits and berries contain flavenoids, which keep cancer-causing hormones from latching onto a cell. Tomatoes, strawberries, pineapples, and green peppers are rich in *p*-coumaric acid and chlorogenic acid, which help prevent nitric oxide and protein amines from forming the carcinogenic nitrosamines. Soybeans contain genistein, a phytochemical that keeps tiny tumors from connecting to capillaries. Without a source of oxygen and nutrition, the tumor never grows, spreads, or kills. Garlic and onions contain phytochemicals called allylic sulfides that protect against stomach cancer by stimulating enzymes that detoxify cancer-causing chemicals in the stomach. Chili peppers contain capsaicin, which keeps carcinogens from cigarette smoke from binding with DNA and triggering the changes that lead to lung cancer and other cancers.

Your Perspective: Imagine that you are a nutritional consultant for a large school district and the district has asked you to teach elementary school children the health benefits of eating fruits and vegetables. Compose a song or a poem that the children can use to help them understand why eating fruits and vegetables is so beneficial to their health.

The rate at which an enzyme catalyzes a reaction varies with different cellular conditions, and from enzyme to enzyme. Some enzymes, such as carbonic anhydrase (carbonate dehydratase), are extremely efficient in catalyzing a reaction. Carbonic anhydrase catalyzes the formation of bicarbonate from carbon dioxide in red blood cells.

$$CO_2 + H_2O \xrightleftharpoons{\text{carbonic anhydrase}} H^+ + HCO_3^-$$

Without the enzyme, this reaction would proceed so slowly that bubbles of carbon dioxide would form in the blood and tissues. The efficiency of an enzyme such as carbonic anhydrase is measured by its turnover number. The **turnover number** is equal to the number of substrate molecules transformed per minute by one molecule of enzyme under optimal conditions of temperature and pH (Table 18.2).

Example 18-1

Cholinesterase is the enzyme that works at nerve endings to catalyze the hydrolysis of acetylcholine to acetate and choline. This enzyme has a turnover number of 1.5×10^6 molecules/min. How many molecules of acetylcholine can one molecule of cholinesterase hydrolyze in 1 s?

Table 18.2 **Turnover Numbers of Enzymes**

Enzyme	*Turnover Number (molecules of substrate per minute)*
Carbonic anhydrase	36,000,000
α-Glucosidase	1,000,000
Glutamate dehydrogenase	30,000
Phosphoglucomutase	1,240
Chymotrypsin	100
DNA polymerase	15

See **the Problem**

The problem asks us to find the number of acetylcholine molecules hydrolyzed by one molecule of cholinesterase in 1 s.

Think **It Through**

From the turnover number, we know the number of molecules hydrolyzed in 1 min. We also know that 1 min equals 60 s. From this information we can set up two conversion factors to solve the problem.

Execute **the Math**

$$\frac{1.5 \times 10^6 \text{ molecules}}{1 \cancel{\text{min}}} \times \frac{1 \cancel{\text{min}}}{60 \text{ s}} = 2.5 \times 10^4 \frac{\text{molecules}}{\text{s}}$$

Prepare **the Answer**

The value given for the turnover number has two significant figures, so our answer must also have two significant figures: 2.5×10^4 molecules/s.

Check Your Understanding

1. Define the following terms in your own words:

(a) apoenzyme **(h)** turnover number
(b) coenzyme **(i)** hydrolases
(c) holoenzyme **(j)** oxidoreductases
(d) cofactor **(k)** transferases
(e) proenzyme **(l)** lyases
(f) substrate **(m)** isomerases
(g) active site **(n)** ligases

2. Chymotrypsin catalyzes the hydrolysis of proteins in the small intestine. How many moles of protein can be hydrolyzed in 1 h by 100 molecules of chymotrypsin?

18.5 Specificity

One main difference between enzymes and inorganic catalysts is the specificity of enzymes. For example, platinum catalyzes several different types of reactions. Each enzyme, however, catalyzes only one type of reaction, and in some cases limits its activity to only one type of substrate.

Pancreatic lipase, for example, catalyzes the hydrolysis of the ester linkage between glycerol and fatty acids in lipids, but it has no effect on the hydrolysis of proteins or carbohydrates. Kidney phosphatase catalyzes the hydrolysis of esters of phosphoric acid, but at a different rate for each substrate. Urease is even more specialized; it catalyzes only the hydrolysis of urea. An extreme example of the specificity of enzyme action is given by the enzyme aspartase, which catalyzes only the following reversible reaction.

$$\underset{\text{fumaric acid}}{(H)(HOOC)C{=}C(COOH)(H)} + NH_3 \underset{}{\overset{\text{aspartase}}{\rightleftharpoons}} \underset{\text{L-aspartic acid}}{H_2N{-}C(COOH)(H){-}CH_2COOH}$$

This enzyme does not catalyze the addition of ammonia to any other unsaturated acid—not even maleic acid, which is the *cis* isomer of fumaric acid.

$$\underset{\text{maleic acid}}{(HOOC)(H)C{=}C(COOH)(H)} \qquad \underset{\text{D-aspartic acid}}{H{-}C(COOH)(NH_2){-}CH_2COOH}$$

Moreover, aspartase does not catalyze the removal of ammonia from D-aspartic acid.

Lock-and-Key Theory

Explain the lock-and-key theory of enzyme action.

How can we explain the specificity of enzymes? On the surface of the enzyme molecule is a particular area, called the **active site,** where the substrate attaches during a reaction. The shape, or configuration, of the amino acid R groups around the active site is especially designed for a specific substrate. This specificity results from the fact that enzymes are formed from L-amino acids; therefore, the active sites are asymmetrical. If the left hand makes an impression in clay, the right hand won't fit in it. That is why, for example, only polypeptides formed from L-amino acids and not D-amino acids fit in the active site and are hydrolyzed by the digestive enzyme trypsin. Because the configuration of the active site is determined by the amino acid sequence of the enzyme, the native configuration of the entire enzyme molecule must be intact for the active site to have the correct shape.

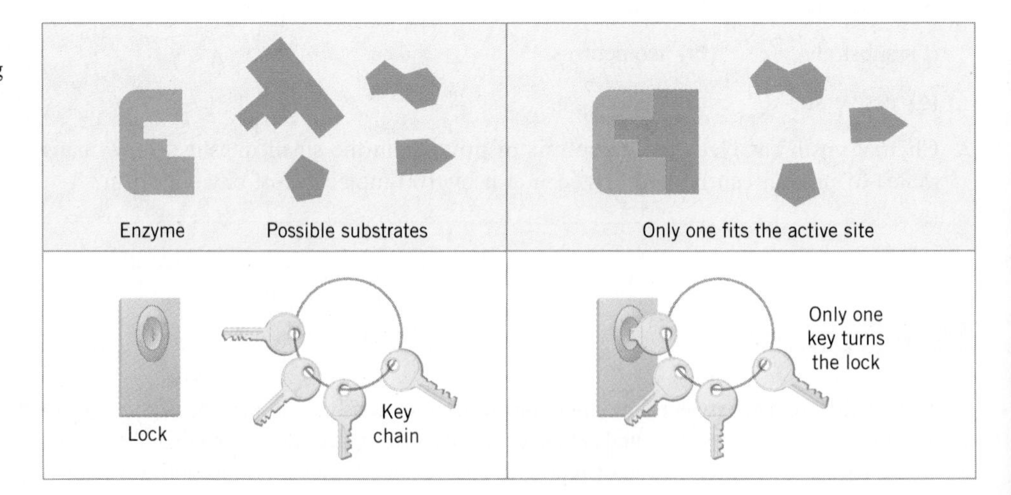

Figure 18.2 Many enzymes are very specific, often limiting their action to only one substrate (just as a lock is specific for only one key).

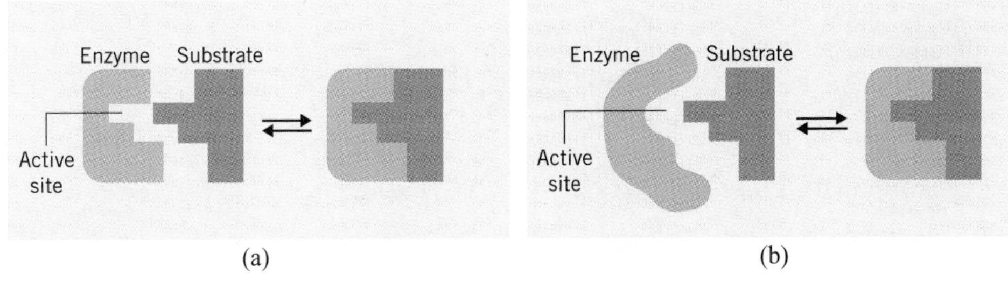

Figure 18.3 Two theories of enzyme action: (a) In the lock-and-key theory the active site conforms exactly to the substrate molecule; (b) In the induced-fit theory, the substrate induces the active site to take on a shape complementary to the shape of the substrate molecule.

When this occurs, the substrate then fits into the active site of the enzyme in much the same way as a key fits into a lock. The shape of the lock is specific for only one key; no other keys turn the lock (Figs. 18.2 and 18.3a).

As an example of the importance of the fit between an enzyme and the active site, researchers at Northwestern University and the Scripps Research Institute have discovered how a mutation causes amyotrophic lateral sclerosis (ALS). Often called Lou Gehrig's disease, ALS is the crippling condition suffered by cosmologist Stephen Hawking. The mutation affects the enzyme superoxide dismutase, which rids cells of damaging chemicals that build up during metabolism. In ALS, the amino acid sequence of this enzyme has changed. As a result the enzyme does not fold properly to form the active site and so is unable to catalyze the reactions that rid cells of the toxic chemicals.

Not only is the geometry of the active site important but so also is the arrangement of charged R groups around the active site. An electric attraction exists between the substrate and the enzyme, which pulls them together. The reaction products that are formed, however, have a different distribution of charges. The repulsion between the products and the active site causes the products to be "kicked off" the enzyme.

Explain the induced-fit theory of enzyme action.

Induced-Fit Theory

The model of a lock-and-key fit, which helps to explain the action of some enzymes, must be slightly changed to explain other enzymes. In these cases a better comparison a hand slipping into a glove—which then causes, or induces, a fit. Enzyme molecules are flexible. Although the active site of some enzymes may not initially match the substrate, the substrate itself, as it is drawn to the enzyme, may induce the enzyme to take on a shape that matches the substrate (Figs. 18.3b and 18.4). Even in this case the active site is still specific to the substrate, just as a left-hand glove cannot normally fit a right hand.

Figure 18.4 These two space-filling models of (a) the yeast enzyme hexokinase and (b) the complex of the enzyme with glucose (purple) illustrate the induced-fit theory.

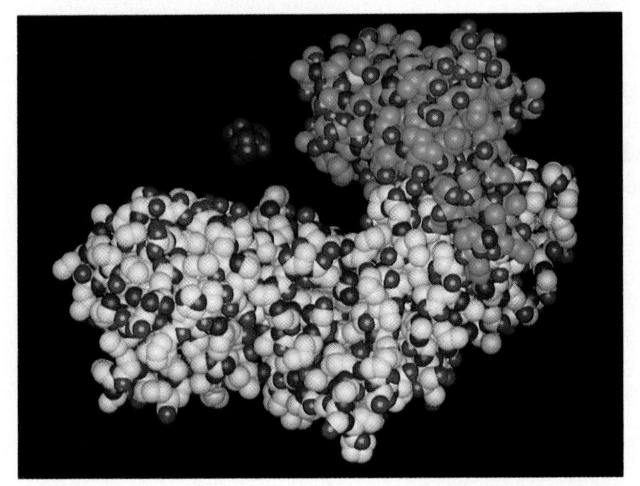

(a)

(b)

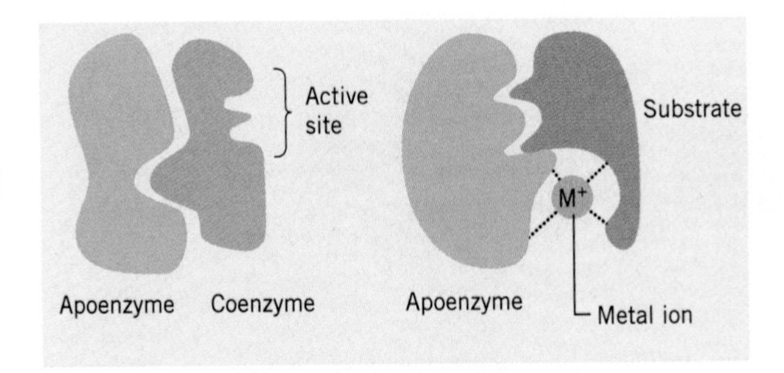

Figure 18.5 Cofactors may contribute to the activity of the enzyme (a) by providing the active site or (b) by forming a bridge between the enzyme and the substrate.

Enzymes that are secreted within the body as proenzymes have their active sites blocked. To activate the enzyme, these sites must be unblocked by the hydrolysis of part of the molecule. In other enzymes, cofactors must be present for the enzyme to be active because the cofactors provide the arrangement of molecules necessary for the active site to form a bridge between the substrate and the enzyme (Fig. 18.5).

18.6 Factors Affecting Enzyme Activity

Explain how changes in pH and temperature can affect enzyme activity.

We have seen that factors such as pH, temperature, solvents, and salt concentrations can change the structure of a protein (see Section 17.11). Such factors, therefore, have an effect on the activity levels of enzymes.

pH

Changes in the pH of the surrounding medium can change the secondary or tertiary structure of an enzyme. This may alter the geometry of the active site or the surrounding charge distribution. Each enzyme has a certain pH at which it is most active. For example, pepsin (which is active in the stomach) has an optimum pH of 1.5, whereas trypsin (which is active in the small intestine) has its maximum activity at pH 8 (Fig. 18.6).

Some enzymes can tolerate fairly large pH changes, whereas others are so sensitive that even a slight change in pH greatly decreases their activity. Extreme changes in pH denature all enzymes, however, which is why body fluids contain buffer systems to protect against such changes.

Temperature

Temperature also affects the rate at which enzyme-catalyzed reactions occur. Most body enzymes have their highest activity at temperatures from 35° to 45°C. Above this range the enzyme begins to denature, and the reaction rate decreases. When you run a high fever, for example, you feel ill partly because your high body temperature slows enzyme activity.

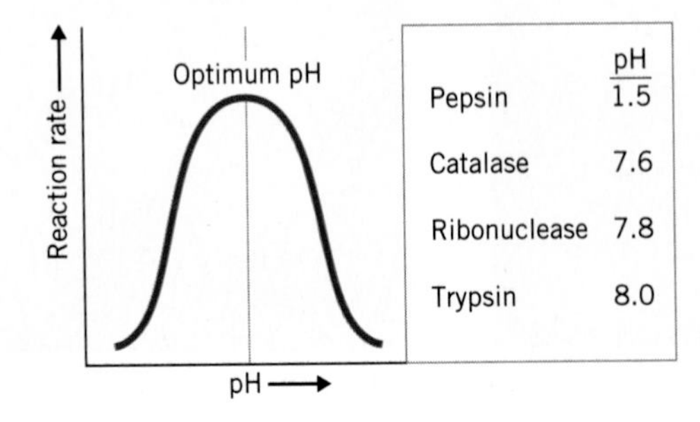

Figure 18.6 Each enzyme has its highest activity at a specific pH.

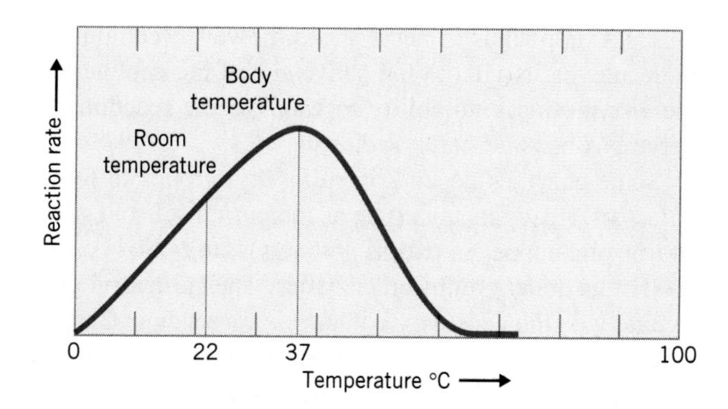

Figure 18.7 Temperature affects enzyme activity.

Above 80°C enzymes become permanently denatured. Below 35°C the reaction rate slowly decreases until it essentially stops as the enzymes become inactive (Fig. 18.7). Enzymes, however, are not denatured at low temperatures and resume activity if the temperature is again raised. It is this fact that allows researchers to preserve cell cultures, human tissues for transplants, and organisms such as bacteria for further studies. The preservation of food by refrigeration is based on the fact that the enzyme-catalyzed reactions that cause spoilage are slowed or stopped at low temperatures.

Check Your Understanding

1. At physiological pH (pH 7.4) and 37°C, it takes an hour or more for the following reaction to reach equilibrium:

$$CO_2 + H_2O \rightleftharpoons H_2CO_3$$

If the enzyme carbonic anhydrase is added to the system, equilibrium is reached in less than a minute.

(a) How does the enzyme increase the reaction rate?

(b) What effect (if any) does raising the pH to 9 have on the rate of the reaction? Why?

2. Pepsin and trypsin are both digestive enzymes that catalyze the hydrolysis of proteins. Do you expect these enzymes to hydrolyze all peptide bonds? Why or why not?

18.7 Inhibition of Enzyme Activity

Irreversible Inhibition

Describe three ways an enzyme can be inhibited, and give examples of each.

The activity of enzymes can be inhibited in several ways. Research into the processes of enzyme inhibition has produced a great deal of knowledge about the specificity of enzymes and the nature of their active sites. The action of many poisons and drugs is due to their ability to inhibit specific enzymes.

Irreversible inhibition of enzyme activity occurs when a functional group in the active site or a cofactor required for the activity of the enzyme is destroyed or modified. For example, cholinesterase is an enzyme that catalyzes a reaction taking place at the juncture of nerve cells. This enzyme is necessary for normal transmission of nerve impulses. Compounds in nerve gases, however, combine with the —OH group on a serine molecule that is vital to the active site of the cholinesterase enzyme. When this happens, the enzyme loses its ability to catalyze the reaction. This is why animals poisoned by nerve gas become paralyzed.

Some enzymes depend on sulfhydryl groups to form tight covalent bonds with metal cofactors. Heavy metals, such as mercury, lead, and silver (and even such essential metals as iron and copper in excess amounts), are toxic because they bind irreversibly with free —SH functional groups on enzymes. The sulfhydryl groups, then, are no longer available to bind with the necessary cofactor. As we saw in Chapter 17, lead inactivates the enzyme that catalyzes the insertion of iron into the heme molecule, thereby affecting the synthesis of hemoglobin and causing anemia. Lead also inhibits essential enzymes and proteins in the muscles and nervous system, resulting in muscle weakness and lack of coordination. Lead can be removed from the blood with chemicals called chelating agents that bind tightly to the lead. The complex formed between the lead and the chelating agent is water-soluble and can be excreted by the kidneys.

Reversible Inhibition

Reversible inhibition of enzyme activity occurs in two ways: noncompetitive inhibition and competitive inhibition (Fig. 18.8). **Noncompetitive inhibition** occurs when the inhibitor combines reversibly with some portion of the enzyme (other than the active site) that is essential to enzyme function. In the muscles, for example, iodoacetic acid inhibits the conversion of glucose to lactic acid by attaching reversibly to the sulfhydryl groups found on the enzyme glyceraldehyde phosphate dehydrogenase.

$$\underset{\text{active enzyme}}{E{-}CH_2{-}SH} + \underset{\text{iodoacetic acid}}{ICH_2COOH} \rightleftharpoons \underset{\text{inactive enzyme}}{E{-}CH_2{-}S{-}CH_2COOH} + HI$$

Competitive inhibition occurs when a compound with a structure very similar to the substrate competes with the substrate for the active site on the enzyme. When such inhibitors become bound to active sites, fewer enzyme molecules are available to the substrate. Enzyme activity therefore decreases.

$$\{\underset{\text{enzyme}}{E} + \underset{\text{inhibitor}}{I} \rightleftharpoons EI\} \text{ competes with } \{\underset{\text{enzyme}}{E} + \underset{\text{substrate}}{S} \rightleftharpoons ES\}$$

For example, succinic acid and malonic acid have very similar structures. Succinate dehydrogenase catalyzes the removal of hydrogen from succinic acid, and malonic acid inhibits this reaction.

$$\underset{\text{succinic acid}}{\begin{matrix} COOH \\ | \\ CH_2 \\ | \\ CH_2 \\ | \\ COOH \end{matrix}} + \underset{\uparrow \text{ succinate dehydrogenase}}{E{-}FAD} \longrightarrow \underset{\text{fumaric acid}}{\begin{matrix} COOH \\ | \\ HC \\ \| \\ CH \\ | \\ COOH \end{matrix}} + E{-}FADH_2$$

$$\underset{\text{competitive inhibitor: malonic acid}}{\begin{matrix} COOH \\ | \\ CH_2 \\ | \\ COOH \end{matrix}}$$

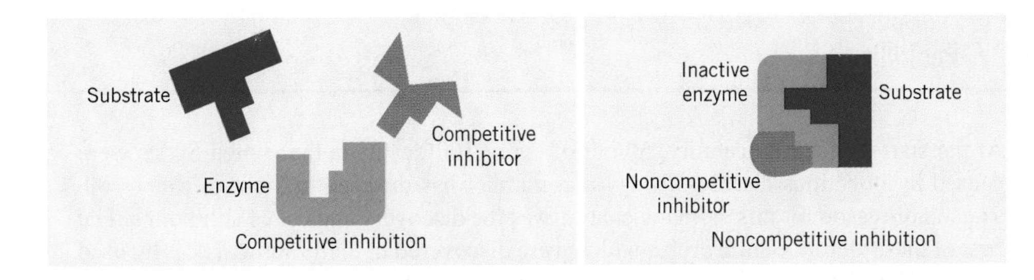

Figure 18.8 In competitive inhibition, the inhibitor and the substrate compete for the active site on the enzyme. In noncompetitive inhibition, the inhibitor combines with some other portion of the enzyme molecule that is essential for the activity of the enzyme.

Antibiotics

Explain how antibiotics can inhibit the growth of bacteria.

Antibiotics are chemicals, extracted from organisms (such as molds, bacteria, and yeasts), that inhibit growth or destroy other microorganisms. They are used to treat many diseases in humans and animals, and in small amounts they are used to speed the growth of poultry and livestock. Antibiotics can function in many ways to prevent the growth or to destroy a disease-causing organism. For example, the sulfa drug sulfanilamide prevents bacterial growth by competitive inhibition. Bacteria use *p*-aminobenzoic acid in the synthesis of the vitamin folic acid.

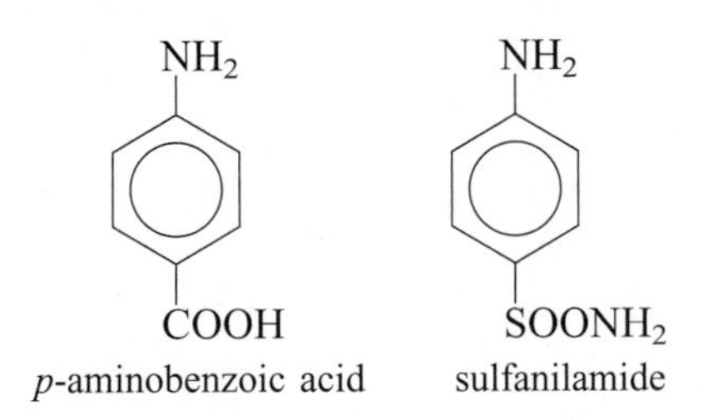

The structure of sulfanilamide closely resembles that of *p*-aminobenzoic acid, and sulfanilamide molecules compete for the active site on the bacterial enzyme. This, then, inhibits folic acid synthesis. Chemicals that inhibit enzyme action are called **antimetabolites.**

The most widely used antibiotics, the penicillins and cephalosporins, belong to the class of compounds called β-lactams. These compounds work by interfering with the process used by bacteria to construct their cell walls. The structure of the β-lactams is very similar to that of the compounds normally used by the bacteria in building the framework of the cell wall. The β-lactam antibiotics successfully compete for the active site on an enzyme necessary to this process, critically weakening the cell wall and destroying the bacteria. Because animal cells have cell membranes but lack cell walls, they are unaffected by these antibiotics.

Chemotherapy

As we learned at the beginning of this chapter, antimetabolites are also used in cancer therapy. Methotrexate, an antimetabolite whose structure resembles the vitamin folic acid, inhibits the enzyme dihydrofolate reductase. By competing with folic acid for the active site on the enzyme and then binding tightly to the enzyme, methotrexate ties up a pathway for the synthesis of DNA in rapidly dividing cancer cells.

Most antitumor agents have been discovered by a chance observation or the random screening of chemicals. However, Charles Heidelberger of the University of Wisconsin purposefully designed and synthesized the antimetabolite 5-fluorouracil (5-FU) after determining that malignant cells used the base uracil more efficiently than normal cells. 5-Fluorouracil is used to treat many solid tumors, including tumors of the breast and colon. The fluorine–carbon bond in 5-FU is much stronger than the hydrogen–carbon

Historical Perspective / Penicillin

At the start of the 20th century, 60 out of every 100 deaths in the United States were caused by infectious diseases. Today that number has dropped to 3 out of every 100. The major reason for this dramatic change was the discovery and use of antibiotics. The first of these were the sulfa drugs, which were discovered in the 1930s and heavily used during World War II to prevent infections in wounds. Another antibiotic, penicillin, was first tried in 1941. Since World War II it has been the most widely used antibiotic drug.

Alexander Fleming discovered penicillin in 1929. While conducting studies on infectious bacteria, he noticed that one of his cultures had become contaminated with a blue mold. He was curious because no bacterial colonies seemed to be growing near the mold. It appeared that the mold was producing a substance that interfered with the growth of the bacteria. Fleming obtained from the mold (*Penicillium notatum*) a crude extract of the active substance, which he named penicillin.

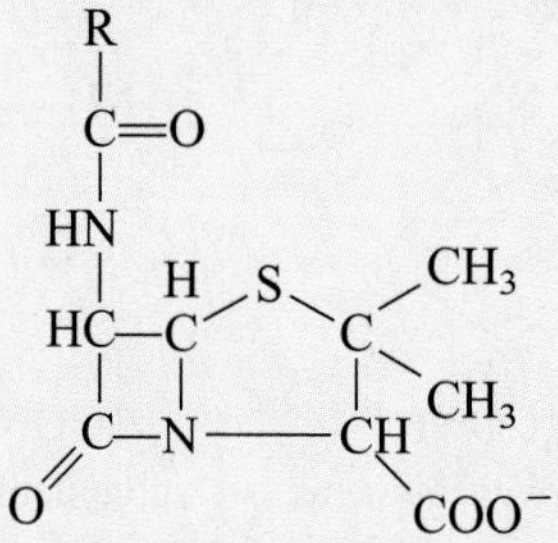

penicillin

Fleming's original penicillin extract contained a mixture of compounds with different R groups. By adding various compounds to the nutrient solution on which the *Penicillium* mold was grown, chemists have produced many different derivatives of penicillin. This is especially important because some organisms have developed a

bond in uracil and does not undergo the substitution reaction necessary for the formation of thymine (an important part of the DNA molecule to be discussed in Chapter 20).

uracil 5-fluorouracil thymine

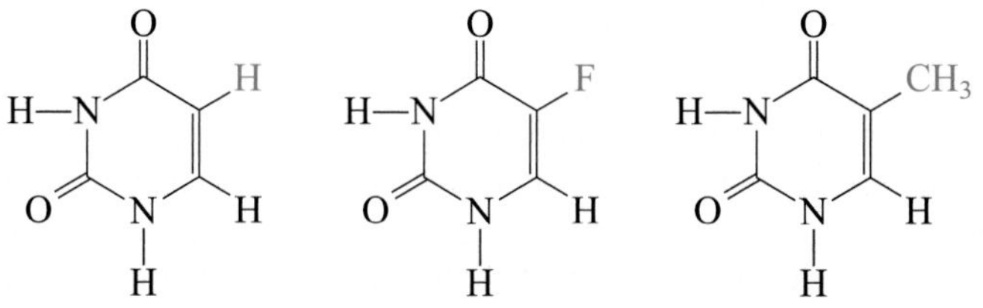

Cancer cells use both 5-FU and uracil in their synthetic pathways, but the compounds formed from 5-FU inhibit the enzyme thymidylate synthase and disrupt DNA synthesis, thus destroying the cell. These 5-FU compounds are also incorporated into the structure of RNA (another key molecule to be discussed in Chapter 20) thus disrupting the function of RNA in the cell.

Alexander Fleming (1881–1955)

resistance to the original forms of the drug. Bacteria develop resistance by producing an enzyme, penicillinase, which can destroy penicillin. To counter this resistance, researchers have synthesized new derivatives of penicillin that are resistant to penicillinase. Two of these penicillin derivatives are amoxicillin and ampicillin. With the overuse of these drugs, however, new bacterial strains are appearing that are resistant to the penicillin derivatives.

One new drug, clavulanic acid, may be the answer to penicillin-resistant strains of bacteria. Although not itself an antibacterial drug, clavulanic acid irreversibly binds to the active site of the penicillinase produced by the bacteria, rendering it useless. When given with penicillin, clavulanic acid destroys any resistance by the bacteria to the action of penicillin.

Check Your Understanding

Sodium arsenite ($NaAsO_2$) has a strong attraction for sulfhydryl groups.

(a) What effect might this attraction have on cellular enzymes?

(b) What type of enzymes are most affected?

18.8 Regulatory Enzymes

Enzymes give living systems the ability to function. But if all enzymes were equally active in the cell all the time, the cell would probably ''burn itself out'' and die. Living systems, therefore, can not only function but can also control their functioning. For example, enzymes in the bloodstream catalyze the formation of blood clots when we bleed, but they are not active and do not form clots under normal conditions. Enzymes catalyze the contraction of muscle fibers when we walk but are inactive when we sit or rest. The

Health Perspective / The Use and Abuse of Antibiotics

Nothing worked. For nine months Dr. Cynthia Gilbert, an infectious disease specialist, had tried one antibiotic after another on a patient suffering from blood poisoning. The man's blood was full of enterococcus bacteria, which were slowly poisoning his red blood cells. The infection seemed to clear up, but then came back even stronger as the resistant bacteria multiplied. After nine months there was nothing else to try, and the patient died of a massive bacterial infection of the blood and heart.

Antibiotics, the miracle drug of the 20th century, are being outsmarted by one of the most primitive organisms on earth: bacteria. How has this happened? There was a time, before antibiotics, when such infectious diseases as tuberculosis and pneumonia killed thousands of people. And now, just when medicine had declared that such infectious diseases have been eliminated or controlled, these diseases are again on the rise. The problem is that bacteria can slowly adapt to any antibiotic, and when antibiotics are misused, the process of adaptation is greatly increased. Many patients nowadays have come to expect a prescription for antibiotics no matter what their disease, and doctors have complied by writing prescriptions without waiting to determine if the infection is viral or bacterial. (Antibiotics do *not* affect virus infections.) And even if the antibiotic is properly prescribed, there is no guarantee it will be used correctly. Often a patient does not take the full course of antibiotic but stops when she is feeling better. When the medication is stopped early, however the bacteria have time to develop a supply of enzymes to counter the antibiotic. Or the bacteria might be able to change to a form that is resistant to the antibiotic, and then share that resistance with other bacteria in the gut. In addition to quitting the treatment early, patients often save the unused pills to take later (without consulting a doctor) or to give to friends. Taking antibiotics for infections against which they are not effective just increases the risk of creating resistant strains of bacteria.

control mechanisms of the living system involve the control of enzyme activity and the control of enzyme concentrations.

As we shall see, most biological chemical reactions occur in a sequence of reactions that eventually produces a specific metabolic end product. Because a separate enzyme catalyzes each reaction in such a sequence, these sequences of reaction are called **multienzyme systems.** For example, a multienzyme system is involved in the conversion of glucose to lactic acid in muscle cells, and other multienzyme systems are involved in the synthesis of different amino acids.

Define a multienzyme system, and explain how it is regulated.

In most multienzyme systems, the enzyme that catalyzes the first reaction of the series is the **regulatory,** or **allosteric, enzyme.** This enzyme controls the rate of the entire process. Regulatory enzymes are usually complex, high-molecular-weight molecules containing several polypeptide chains and cofactors These enzymes usually have more than one site for the attachment of molecules—one or more for the substrate, and one or more for regulatory molecules. A site for a regulatory molecule is called a **regulatory,** or **allosteric** (meaning other space or location), **site.** The **regulator molecule** itself can either inhibit or increase the activity of the enzyme. Such an allosteric enzyme is a flexible molecule, and the regulatory molecule causes a slight change in the shape of the enzyme. This, in turn, changes the shape of the active site, making it either more or less receptive to the substrate.

Many regulatory enzymes are inhibited by the end product of the multienzyme system—a process called **feedback inhibition.**

But the group most guilty of overusing antibiotics is the American farmer. Farm animals receive 30 times more antibiotics than do people. Antibiotics can increase an animal's weight gain, but resistant strains of bacteria emerge in animals just as they do in humans. These newly resistant organisms remain in the animal's flesh and are spread to humans through raw or undercooked meat. About 6.5 million people fall ill each year from eating contaminated meat and poultry. Milk could pose an even greater threat. Milk is allowed to contain a certain concentration of 80 different antibiotics. Recent tests show that the allowed level of antibiotics in milk can produce resistant bacteria strains in milk drinkers.

People with healthy immune systems can usually repel most bacteria, even if the bacteria are resistant to antibiotics. But when the immune system is weakened, such as during surgery, illness, or old age, these resistant bacteria are very difficult to control. This accounts for many deaths among patients in nursing homes and hospitals. For example, *Staphylococcus aureus,* which infects surgical wounds and can cause pneumonia, became resistant to penicillin and tetracycline in the 1950s and 1960s. The antibiotic methicillin controlled these infections for a decade, but now methicillin-resistant staphylococci are common in hospitals and nursing homes worldwide. Of new worldwide concern are the emerging strains of drug-resistant tuberculosis, malaria, syphilis, cholera, and dysentery.

There is hope that researchers and scientists can develop new and more powerful antibiotics. For the present, it is critical that doctors and patients use antibiotics with care and stop the abuse of antibiotics that are still effective.

Your Perspective: In some parts of the world, antibiotics are available without prescription as over-the-counter drugs. Imagine you are a member of the World Health Organization in Bangkok, Thailand, and you want to convince local officials to make antibiotics available by prescription only. Write a short speech presenting both sides of this issue, but be as persuasive as you can in explaining why the arguments for regulating antibiotics outweigh those against it.

$$\mathrm{A} \xrightarrow{E_1} \mathrm{B} \xrightarrow{E_2} \mathrm{C} \xrightarrow{E_3} \mathrm{D} \xrightarrow{E_4} \mathrm{F} \xrightarrow{E_5} \mathrm{G}$$

In this example, E_1 is the regulatory enzyme for the process, and it is inhibited by high concentrations of product G (Fig. 18.9).

Isocitrate dehydrogenase is a regulatory enzyme in the citric acid cycle, a series of reactions in which the compound AMP (adenosine monophosphate) is converted to the energy-rich molecule ATP (adenosine triphosphate). Specifically, this enzyme catalyzes

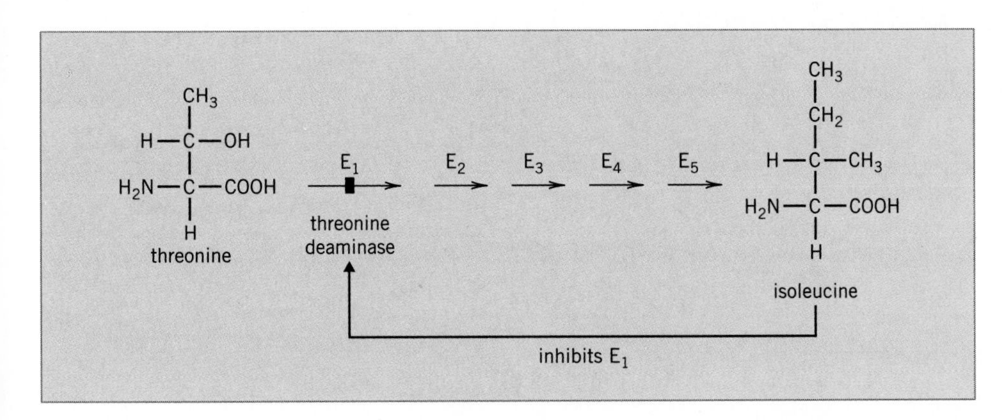

Figure 18.9 In the synthesis of the amino acid isoleucine from the amino acid threonine, the first enzyme (threonine deaminase) is the regulatory enzyme. It is inhibited by the end product of the multienzyme system (isoleucine).

the conversion of isocitric acid (in the form of the negative ion isocitrate) to α-ketoglutaric acid (in the form of the negative ion α-ketoglutarate).

$$\underset{\text{isocitrate}}{\begin{array}{c} COO^- \\ | \\ H-C-OH \\ | \\ {}^-OOC-C-H \\ | \\ CH_2 \\ | \\ COO^- \end{array}} + NAD^+ \xrightarrow[\text{dehydrogenase}]{\text{isocitrate}} \underset{\alpha\text{-ketoglutarate}}{\begin{array}{c} COO^- \\ | \\ O=C \\ | \\ CH_2 \\ | \\ CH_2 \\ | \\ COO^- \end{array}} + CO_2 + NADH + H^+$$

The activity of isocitrate dehydrogenase is controlled by five regulatory molecules. High concentrations of citric acid, NAD^+, or AMP increase the activity of the enzyme, whereas high concentrations of NADH or ATP inhibit the enzyme.

Metabolic processes are controlled not only by the level of enzyme activity but also by the concentration of each enzyme. The enzyme concentration is determined by the rate of synthesis and the rate of breakdown of the enzyme. As we shall see in Chapter 20, the function of genes is to direct the synthesis of protein molecules; therefore, one way in which a cell can control its metabolic activity is to switch specific genes on and off when a particular enzyme is needed.

18.9 Hormones

The body has other control systems in addition to those within individual cells; these systems coordinate the actions between cells in multicellular systems. This higher level of

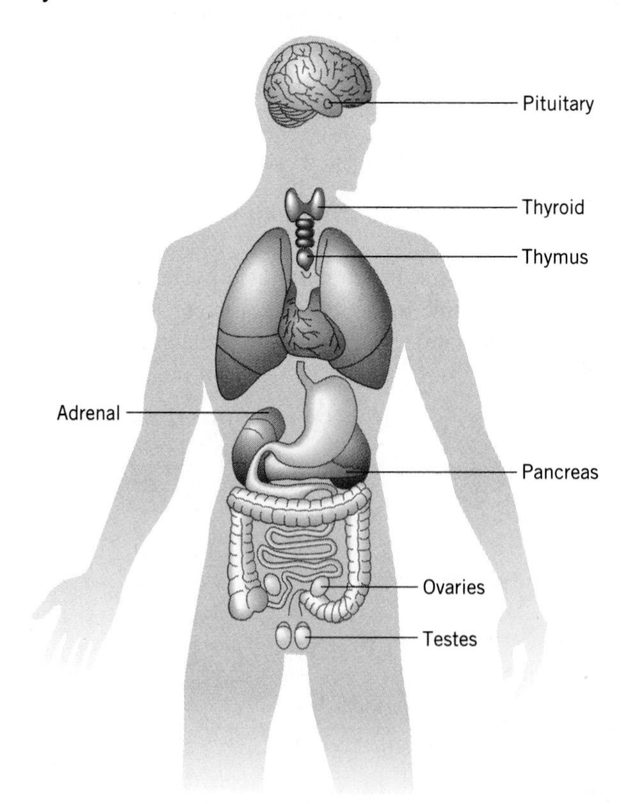

Figure 18.10 The endocrine glands.

communication between cells, tissues, and organs involves the nervous and endocrine systems. The endocrine system is a group of glands that produces and secretes chemical messengers called **hormones** into the body fluids, particularly the blood (Fig. 18.10 and Table 18.3). Each hormone has target organs or cells that are influenced by its presence. The secretion of a particular hormone by an endocrine gland may be triggered by the nervous system, as in the release of adrenalin triggered by hearing a nearby explosion, or by the concentration of a specific chemical compound, as in the release of insulin triggered by an increase in blood glucose levels.

Define hormone, and describe the two ways in which a hormone can regulate cellular processes.

Hormones regulate cellular processes in two ways. Steroid hormones, such as estrogen and testosterone, are nonpolar molecules that can pass through the cell membrane. They

Table 18.3 **Hormones of the Principal Endocrine Glands**

Gland or Tissue	*Hormone*	*Major Function of the Hormone*
Thyroid	1. Throxine	1. Stimulates rate of oxidative metabolism and regulates general growth and development
	2. Thyrocalcitonin	2. Lowers level of calcium in the blood
Parathyroid	Parathyroid hormone	Regulates the levels of calcium and phosphorus in the blood
Pancreas (islets of Langerhans)	1. Insulin 2. Glucagon	1. Decreases blood glucose level 2. Elevates blood glucose level
Adrenal medulla	Epinephrine	Various ''emergency'' effects on blood, muscle, temperature
Adrenal cortex	Cortisone and related hormones	Control carbohydrate, protein, mineral, salt, and water metabolism
Anterior pituitary	1. Thyrotropin	1. Stimulates thyroid gland functions
	2. Adrenocorticotropin (ACTH)	2. Stimulates development and secretion of adrenal cortex
	3. Growth hormone	3. Stimulates body weight and rate of growth of skeleton
	4. Gonadotropin (two hormones)	4. Stimulate gonads
	5. Prolactin	5. Stimulates lactation
Posterior pituitary	1. Oxytocin	1. Causes contraction of some smooth muscles
	2. Vasopressin	2. Inhibits excretion of water from the body by way of urine
Ovary (follicle)	Estrogen	Influences development of sex organs and female characteristics
Ovary (corpus luteum)	Progesterone	Influences menstrual cycle; prepares uterus for pregnancy; maintains pregnancy
Uterus (placenta)	Estrogen and progesterone	Function in maintenance of pregnancy
Testis	Androgens (testosterone)	Responsible for development and maintenance of sex organs and secondary male characteristics
Digestive system	Several gastrointestinal hormones	Integration of digestive processes

enter the cell and bind to a receptor site on a large protein molecule. This steroid–receptor group is then transferred to the nucleus of the cell, where it activates specific genes. On the other hand, peptide hormones such as insulin, human growth hormone, and hormones that regulate gastric and kidney secretions are polar, water-soluble molecules that cannot pass through the cell membrane. These hormones attach to specific receptor sites on the surface of the target cell membrane, triggering changes within the cell. The number of receptor sites on a target cell determines the level of response by that cell, and the number may increase or decrease depending on the concentration of hormones in the blood.

Cyclic Adenosine Monophosphate (Cyclic AMP)

Cyclic AMP is a molecule that acts within the cell as a secondary messenger, or intracellular hormone. Its formation is triggered by the attachment of a hormone molecule on the surface of a target cell. The system works as follows: A hormone, the primary messenger, is released into the bloodstream. It travels to a target cell, where it attaches to a specific receptor site on the outside of the cell membrane. This attachment causes a conversion of the enzyme adenylate cyclase from an inactive form to an active form on the inside of the cell membrane. This enzyme then catalyzes the formation of cyclic AMP from ATP.

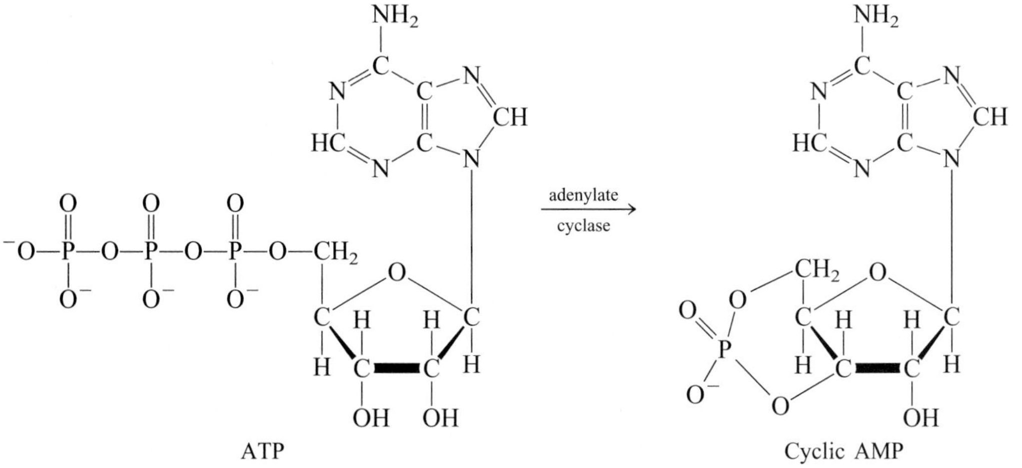

ATP $\xrightarrow{\text{adenylate cyclase}}$ Cyclic AMP

Cyclic AMP spreads through the cell as a secondary messenger, instructing the cell to respond to the hormone in the particular manner characteristic of that cell (Fig. 18.11). For example, the pituitary gland secretes the hormone thyrotropin, which travels to target cells in the thyroid gland. There it triggers the formation of cyclic AMP, to which the thyroid cells respond by secreting more thyroxine.

Prostaglandins

Prostaglandins belong to a class of compounds called eicosanoids (see Section 16.4) that resemble hormones in their effects, but are chemically quite different. They are 20-carbon fatty acids that are synthesized in the cell membrane from the polyunsaturated fatty acid arachidonic acid (Fig. 18.12). As with cyclic AMP, prostaglandins are present in almost all cells and tissues, and their synthesis in the cell is stimulated by the action of hormones, nerves, and muscles. They are among the most potent of all biological agents—as little as one-billionth of a gram is biologically active—yet a very small change in structure can completely alter their biological function. Prostaglandins can raise or lower blood pressure and can regulate gastric secretions. They play a role in both the process that prevents blood platelets from clumping together in healthy blood vessels and the process that

Describe five effects prostaglandins can have on the living organism.

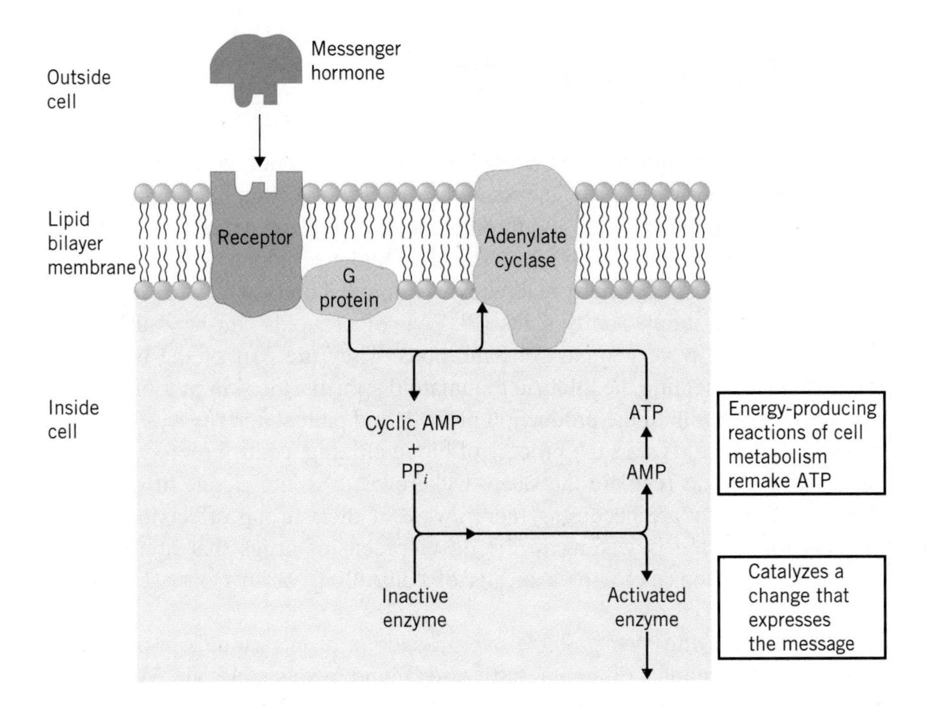

Figure 18.11 A hormone attaches to the cell membrane, triggering the formation of cyclic AMP. The cyclic AMP sets off a series of reactions that activates an enzyme in the cell.

makes them clump together (to form a blood clot) when a vessel is damaged. Prostaglandins produce fevers and inflammatory reactions; aspirin and steroid antiinflammatory drugs such as hydrocortisone work by inhibiting prostaglandin synthesis.

Knowledge of the functions of prostaglandins has led to effective treatment for many medical conditions. Prostaglandins cause uterine contractions and thus are used to induce labor. Drugs that block prostaglandin formation are now prescribed to treat severe menstrual cramps. Prostaglandins are used to inhibit the secretion of stomach acid in people suffering from peptic ulcers and to treat ulcers on the hands and feet of people suffering from Raynaud's disease, diabetes, or atherosclerosis. Other prostaglandins relax the smooth muscles and are used to relieve asthma and to treat high blood pressure. ''Blue'' babies are treated with prostaglandins to keep a fetal duct open, thereby increasing the oxygen content of their blood until they are strong enough to undergo corrective surgery.

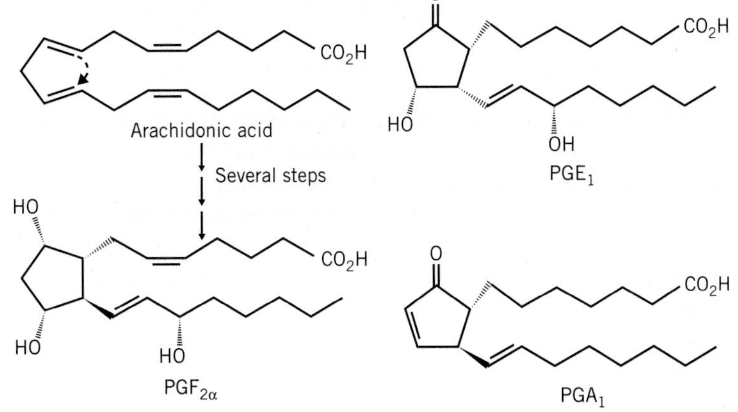

Arachidonic acid

Several steps

$PGF_{2\alpha}$

PGE_1

PGA_1

Figure 18.12 Important prostaglandins. The prostaglandins are synthesized from arachidonic acid.

Health Perspective / Eicosanoids and Polyunsaturated Fatty Acids

Many disorders of human health are now being linked to overproduction of hormone-like compounds called eicosanoids. The eicosanoids include such compounds as prostaglandins, leukotrienes, prostacyclins, and thromboxanes. They are synthesized in all mammalian cells (except red blood cells) from the fatty acid arachidonic acid. Arachidonic acid (20:4*n*-6) is a 20-carbon fatty acid with four double bonds, which belongs to the *n*-6 family of polyunsaturated fatty acids (PUFA). Eicosanoids can have drastic effects on the body even in very small concentrations. They are part of the body's inflammatory response, affecting the joints in rheumatoid arthritis, the skin in psoriasis, and the eyes. They play a role in the production of fever and pain and in the regulation of blood pressure. Eicosanoids start the process of blood clotting, control reproductive functions such as labor, and regulate the sleep–wake cycle. As knowledge of eiconsanoids increases, scientists are becoming more aware of their strong effects on human health. This knowledge is leading to the development of drugs that mimic or prevent the actions of eiconsanoids as a means of controlling various diseases and their symptoms.

Special concern is growing over possible overproduction of eicosanoids in people who consume excessive amounts of the *n*-6 fatty acids found in vegetable oils. Vegetable oils, such as safflower, soybean, and corn oils, all contain large amounts of linoleic acid (18:2*n*-6), which the liver uses to make arachidonic acid. We also absorb significant amounts of arachidonic acid from meat in our diets. Large amounts of arachidonic acid in the tissues may lead to blood clots, psoriasis, asthma, or arthritis, as well as accelerated tumor growth and metastasis.

Recent studies indicate that the harmful effects of excess arachidonic acid can be moderated by eating fish oils containing *n*-3 polyunsaturated fatty acids. The body produces a series of eicosanoids from the *n*-3 PUFAs that inhibit the production and moderate the effects of the eicosanoids produced from the *n*-6 PUFAs. For example, the *n*-3 PUFAs reduce the formation of blood clots by inhibiting the formation of thromboxane, a prostaglandin synthesized from arachidonic acid. Thromboxane causes blood platelets to bunch together to form blood clots. Several forms of cancer, especially breast cancers, are stimulated by the prostaglandin PGE_2, a compound that promotes tumor growth. Various *n*-3 PUFAs in the diet reduce the production of PGE_2 from arachidonic acid and slow the growth of certain tumors in animals.

The *n*-6 PUFAs, such as linoleic acid, are essential to health. A high consumption of vegetable oils containing these fatty acids, however, may lead to an overproduction of eicosanoids that promote certain diseases. To guard against this, you should reduce your *total* fat intake and include fish regularly in your diet.

Your Perspective: Public health clinics commonly publish brochures explaining general health issues to city residents. Suppose the clinic has asked you to write a brochure explaining some of the things that people should know about including fatty acids in their family's diet. Using the information from the preceding perspective and from the perspectives that discuss fish oils and *trans* fatty acids, write an explanation in language that can be understood by a local population having only basic skills in reading English.

Check Your Understanding

1. Construct a table showing the similarities and differences in functions, sites of synthesis, and sites of action among hormones, prostaglandins, and cyclic AMP.

2. The synthesis of cholesterol from acetyl-CoA requires many enzyme-catalyzed steps. In one of the middle steps, the enzyme HMG-CoA reductase converts 3-hydroxy-3-methylglutaryl-CoA (HMG-CoA) to mevalonate. Cholesterol inhibits this enzyme.

   (a) The sequence of reactions that produces cholesterol is an example of a ____________ system.

   (b) The process by which an end product inhibits one of the enzymes in a sequence of reactions is called ____________ inhibition.

   (c) HMG-CoA reductase is a ____________ enzyme.

   (d) HMG-CoA binds to the ____________ site on the enzyme molecule, and cholesterol binds to the ____________ site.

18.10 What Are Vitamins?

People have recognized the need for vitamins for over 200 years. Long ago, British sailors were give the slang name ''limeys'' from the lime juice they drank to prevent scurvy. The symptoms of scurvy—swollen gums, painful joints, hemorrhages under the skin, loss of weight, and muscular weakness—are prevented by vitamin C, which was first isolated in 1930.

But exactly what are vitamins? **Vitamins** are small organic molecules that are essential to maintain health; however, they either cannot be synthesized by the body or are synthesized in insufficient amounts. Vitamins must therefore be present in the diet (in trace amounts) for proper cellular function, growth, and reproduction. Some vitamins or derivatives of vitamins serve as coenzymes in cellular reactions, but the exact cellular function of other vitamins has yet to be determined. When vitamins are lacking in the diet, specific deficiency diseases result (Table 18.4). The National Research Council of the National Academy of Sciences has developed recommended dietary allowances (abbreviated RDA) for vitamins, set at levels that provide adequate nutrition for healthy individuals. Most healthy people who eat a balanced diet obtain the required amount of vitamins without taking vitamin supplements. Vitamins are classified into two main groups based on their solubility: they are either soluble in water (polar) or soluble in fat (nonpolar).

◆ **Explain what makes a substance a vitamin.**

18.11 Water-Soluble Vitamins

Because the vitamins in this group are water-soluble, they are easily eliminated from the body in the urine. They must therefore be present in the diet each day. Fruits and vegetables contain many water-soluble vitamins. Such foods should be eaten raw or should be cooked quickly in a very small amount of water. This will prevent both the loss of the vitamins in the water and the destruction of the vitamin's activity by the heat.

Table 18.4 Vitamins Required in Human Nutrition

Name	Recommended Dietary Allowances (RDA)[a]	Dietary Sources	Function	Symptoms of Deficiency
Water-Soluble				
Vitamin C (ascorbic acid)	60 mg	Citrus fruits, leafy vegetables, tomatoes, potatoes, cabbage	Synthesis of collagen, amino acid metabolism	Scurvy, bleeding gums, loosened teeth, swollen joints
Vitamin B_1 (thiamine)	1.5 mg (1.1 mg)	Whole grains, organ meats, legumes, nuts, pork	Coenzyme in carbohydrate metabolism	Beriberi, heart failure, mental disturbances
Vitamin B_2 (riboflavin)	1.7 mg (1.3 mg)	Milk, eggs, liver, leafy vegetables; synthesized by bacteria in the gut	Coenzyme in oxidation reactions	Fissures of the skin, visual disturbances, anemia
Niacin (nicotinic acid)	19 mg (15 mg)	Yeast, liver, lean meat, fish, whole grains, eggs, peanuts	NAD, NADP; coenzymes in redox reactions	Pellagra, skin lesions, diarrhea, dementia
Vitamin B_6 (pyridoxine)	2.0 mg (1.6 mg)	Whole grains, glandular meats, pork, milk, eggs	Coenzyme for amino acid and fatty acid metabolism	Convulsions in infants, skin disorders in adults
Vitamin B_{12} (cyanocobalamin)	2.0 μg	Liver, brain, kidney; synthesized by bacteria in the gut	Coenzymes in nucleic acid, amino acid, and fatty acid metabolism	Pernicious anemia, retarded growth
Pantothenic acid	4–7 mg[b]	Yeast, meats, whole grains, legumes, milk, vegetables, fruits	Forms part of coenzyme A (CoA)	Neuromotor, digestive, and cardiovascular disorders
Folacin[c] (folic acid)	200 μg (180 μg)	Yeast, leafy vegetables, liver, fruit, wheat germ	Coenzymes in nucleic acid and amino acid metabolism	Anemia, inhibition of cell division, digestive disorders
Biotin	30–100 μg[b]	Liver, egg yolk, legumes; synthesized by bacteria in the gut	Part of enzymes important in carbohydrate and fat metabolism	Skin disorders, anorexia, mental depression
Fat-Soluble				
Vitamin A	1000 μg (800 μg) of retinol	Green and yellow vegetables and fruits, fish oils, eggs, dairy products	Formation of visual pigments, maintenance of mucous membranes, transport of nutrients across cell membranes	Night blindness, skin lesions, eye disease (Excess: hyperirritability, skin lesions, bone decalcification, increased pressure on the brain)
Vitamin D	10 μg of cholecalciferol	Fish oils, liver; provitamins in skin activated by sunlight	Regulates calcium and phosphate metabolism	Rickets (Excess: retarded mental and physical growth in children)
Vitamin E (tocopherol)	10 mg (8 mg) of α-tocopherol	Green leafy vegetables, vegetable oils, wheat germ	Maintenance of cell membrane	Increased fragility of red blood cells
Vitamin K	70 μg (60 μg)	Green leafy vegetables; synthesized by bacteria in the gut	Synthesis of prothrombin and other blood-clotting factors in the liver	Failure of coagulation of blood (Excess: hemolytic anemia and liver damage)

[a] For men and women (in parentheses), 19 to 24 years, as set by the Food and Nutrition Board, National Academy of Sciences–National Research Council, 1989.

[b] Estimated safe and adequate dietary intake.

[c] Folacin is a generic term used for compounds having nutritional properties and a chemical structure similar to that of folic acid.

Health Perspective / Vitamin C and the Common Cold

Vitamin C is one of the best known of the essential nutrients we call vitamins, and it is also one of the most controversial. Linus Pauling, winner of two Nobel Prizes (chemistry, 1954; peace, 1962), started this controversy when he suggested taking very large daily doses of vitamin C as prevention against the common cold. Pauling also argued that taking large doses of vitamin C lessens the likelihood of colon cancer. The American Medical Association continues to disagree with these claims, and the federal government continues to recommend a dietary allowance of only 60 mg of vitamin C per day.

Linus Pauling (1901–1994)

Pauling wasn't alone in his belief that vitamin C is required in much larger amounts than the government suggests. He and others argued that human beings are one of only five species that do not synthesize vitamin C. These five species evolved in tropical regions where vitamin C was abundant in their food. Because this vitamin was so readily available, their bodies could concentrate on producing other vital compounds. This gave these species a metabolic advantage that ensured their survival.

Those species that evolved in areas where vitamin C was scarce had to produce it internally. Pauling calculated that if human beings were to produce their own vitamin C at a level equivalent to that produced by other species, it would equal several thousand milligrams per day. His critics argue that Pauling was exaggerating the need for vitamin C. Repeated studies have shown that massive doses of vitamin C do not decrease the number of colds, although it does reduce their severity. These studies showed that the human body could use only a limited amount of vitamin C, about 60 mg/day; the rest of the vitamin C is excreted. Pauling responded by pointing out that the amount of vitamin C needed in the body varies. Smokers, for example, need twice as much as nonsmokers. Studies done with schizophrenics showed that they needed huge amounts of the vitamin.

It is too soon to say just who is correct in this controversy. It may be years before enough evidence has been gathered to show whether Pauling was right or wrong. One thing both sides do agree on is that inexpensive synthetic vitamin C is chemically identical to natural vitamin C. The claim made by some people that natural vitamin C is better than the synthetic variety is just not true. Using fresh fruit as a source of vitamin C, however, does provide additional important ingredients essential to our health.

Vitamin C

Vitamin C (or ascorbic acid) is the most unstable of all the vitamins. It is easily destroyed by oxidation, a process that is speeded up by the presence of heat or a base.

◆ **List two water-soluble vitamins, and describe their functions.**

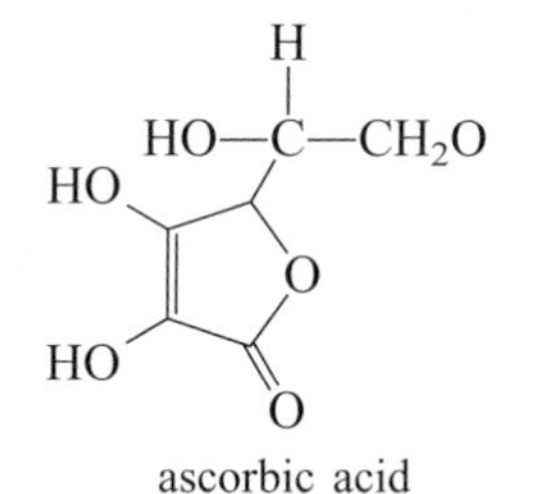

ascorbic acid

Health Perspective / Pellagra

Although pellagra was widespread in the rural South in the early 20th century, there was no sign of this nutritional disease among the Indians living in Mexico. This fact puzzled scientists because both populations used corn as the main food in their diets. The symptoms of pellagra can be eliminated if adequate amounts of niacin or nicotinamide are present in the diet or are synthesized by the body from the amino acid tryptophan. The protein in corn is too low in tryptophan to prevent pellagra. So where were the Indians getting the needed niacin or nicotinamide? It turns out that corn does have adequate amounts of nicotinamide to prevent pellagra but in a form that cannot be absorbed by the intestines until it is treated with base. What the scientists discovered was that a traditional food preparation practice was protecting the Mexican Indians from pellagra. The Indians soak the corn meal in lime water [a dilute $Ca(OH)_2$ solution] before using it to make tortillas. This practice allows the nicotanimide to be absorbed in the intestines, preventing the vitamin deficiency that causes pellagra.

Vitamin C is necessary for the enzymatic activity of prolyl hydroxylase. This enzyme is critical for the formation of collagen, the ''cement'' that holds tissues together. Collagen fibers are a major component of bones, teeth, cartilage, tendons, ligaments, skin, and blood vessels. A deficiency in vitamin C results in defective collagen, causing the symptoms of scurvy. Vitamin C also acts as a reducing agent in many metabolic processes, including the synthesis of enzymes and hormones.

The B Vitamins

The vitamins in the B family are thiamine, riboflavin, niacin, vitamin B_6 (pyridoxine), vitamin B_{12} (cyanocobalamin), pantothenic acid, biotin, and folacin (folic acid). They are all grouped together under the same letter because at the beginning of the 1900s the vitamins in this group were thought to be a single vitamin. Many B vitamins act as coenzymes in reactions essential to life. For example, the body converts thiamine (vitamin B_1) to thiamine pyrophosphate, a diphosphate ester. Thiamine pyrophosphate is an essential coenzyme for several reactions of carbohydrate and amino acid metabolism. A lack of thiamine results in beriberi, a disease that affects nerves and causes paralysis of the legs, enlargement of the heart, slowing of the heartbeat, and a reduction in appetite. No single food source is rich enough in vitamin B_1 to provide the complete daily requirement, so it must be obtained from a combination of many foods.

In Chapter 19 we will study three important coenzymes formed from the B vitamins. They serve as hydrogen carriers in oxidation–reduction reactions that produce the energy for all cellular processes. Two are derivatives of niacin (or nicotinic acid): NAD^+ (nicotinamide adenine dinucleotide) and $NADP^+$ (nicotinamide adenine dinucleotide phosphate). Note that both these molecules carry a positive charge. The other coenzyme, a derivative of riboflavin (vitamin B_2), is FAD (flavin adenine dinucleotide). We can show how these coenzymes act as hydrogen carriers by writing the following equation.

$$NAD^+ + XH_2 \xrightarrow{\text{enzyme}_1} NAD{-}H + H^+ + X$$

In this equation NAD^+ represents the positively charged NAD molecule, and XH_2 is a general way of representing some hydrogen-containing reactant molecule. The NADH

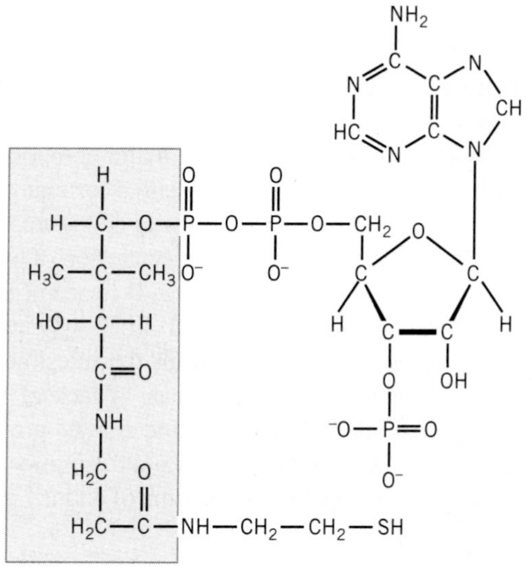

Figure 18.13 Structure of coenzyme A. The shaded portion of the molecule is formed from pantothenic acid, a B vitamin.

produced by this reaction can then serve as a hydrogen donor in a reaction requiring hydrogen. In this way, one molecule of NAD^+ can be used over and over again in the cell.

$$NADH + H^+ + Y \xrightarrow{\text{enzyme}_2} NAD^+ + YH_2$$

Meat and vegetables contain both riboflavin and niacin. Bacteria that are always present in the intestine also produce riboflavin. A deficiency of riboflavin in humans is therefore rare. A deficiency in niacin or nicotinamide results in pellagra, a disease with symptoms of dermatitis, diarrhea, and dementia.

Pantothenic acid is a necessary starting material for the synthesis by the cell of coenzyme A, one of the most important molecules in the metabolism of all nutrients (Fig. 18.13). Pantothenic acid is so widely distributed in food that there is little evidence of pantothenic acid deficiency in humans.

18.12 Fat-Soluble Vitamins

The fat-soluble vitamins (vitamins A, D, E, and K) are fairly stable, are not destroyed by heat, and are not soluble in water. We have seen that excess water-soluble vitamins are excreted by the kidneys. Excess fat-soluble vitamins, in contrast, are stored in the liver for release when needed by the body. As a result, too much of a fat-soluble vitamin can be just as toxic as too little. For example, vitamin A is required for normal growth, for vision in dim light, and for maintaining healthy skin and the mucous linings of the body. It is, however, a very toxic compound. As little as 7.5 mg each day over a period of 25 days can cause such toxic effects as increased cerebral spinal fluid pressure, headaches, irritability, patchy loss of hair, and dry skin with ulcerations. Toxic levels of vitamin A can result from taking vitamin supplements or eating vitamin-enriched foods in the place of foods with naturally occurring vitamin A.

Name the fat-soluble vitamins.

Explain why a dietary excess of vitamin A, but not vitamin C, can be harmful to your health.

Vitamin K is essential for the synthesis of compounds such as prothrombin that are part of the body's blood-clotting mechanism. These clotting proteins have a special "claw" to hold the Ca^{2+} ion in the right position for the clotting reactions to occur. Vitamin K participates in the reaction that adds the claw portion to the protein. Vitamin E acts as an antioxidant for unsaturated lipids. It protects cellular membranes from damage by reacting with free radicals, thereby preventing the free radicals from oxidizing the lipids in membranes.

Calcium ions (Ca^{2+}) are the major component of the minerals that make up bones and teeth. Calcium ions are also essential for many biological functions, including contraction of muscles, maintenance of body temperature, clotting of blood, and transmission of nerve impulses. Regulation of the concentration of calcium ions in the blood is therefore critical to normal body function. Vitamin D is one of three compounds that control the concentration of calcium ions in the blood. Vitamin D increases serum Ca^{2+} levels by increasing the absorption of calcium ions from the intestines and stimulating reabsorption of calcium ions from the bones and kidneys. A lack of vitamin D produces rickets, a disease that causes weak, deformed bones and stunted growth in children. However, excessive intake of vitamin D (2.5 mg for adults and 0.1 mg for children) over long periods is toxic and can cause loss of appetite, formation of kidney stones, and calcification of joints and soft tissues.

Check Your Understanding

1. Why should the dietary intake of fat-soluble vitamins be carefully controlled?

2. Egg whites contain the compound avidin, which combines with biotin and prevents it from being absorbed in the intestines.

(a) To which class of compounds does biotin belong?

(b) What effect might a diet of raw eggs have on a person's healthy? Why?

KEY CONCEPTS

Most enzymes are protein catalysts in biological systems. Enzymes may be simple proteins or conjugated proteins consisting of an apoenzyme and one or more nonprotein cofactors. A cofactor may be a metal ion or another organic molecule (called a coenzyme.) An enzyme is often secreted in an inactive form called a proenzyme, or zymogen, and then converted into the active form when needed.

For an enzyme to catalyze a reaction, the substrate must attach to the surface of the enzyme in a region called the active site. This activates the substrate, causing the formation of products that are then released from the surface of the enzyme. Enzymes catalyze only very specific reactions. Only the correct substrate fits into the active site, similar to a key turning a specific lock. The activity of an enzyme is affected by all the factors that can alter the structure of a protein. Each enzyme has an optimal pH at which it is the most active.

Enzyme activity can be inhibited in several ways. Inhibition is irreversible when a functional group or cofactor required for enzyme activity is permanently destroyed or modified. Enzyme activity can be inhibited reversibly when another molecule competes for the active site or when some portion of the enzyme required for its action is modified reversibly.

Metabolic processes are the result of sequences of enzyme-catalyzed reactions called multienzyme systems. Such sequences are controlled by a regulatory, or allosteric, enzyme—often the enzyme that catalyzed the first step in the sequence. In addition to the active site, such enzymes contain allosteric sites on other parts of the molecule to which regulatory molecules can attach, thereby inhibiting or increasing the activity of the enzyme.

Enzyme activity in an organism is regulated at many levels, both inside and outside the cell. Hormones, produced by the endocrine system, are one of the body's means of intercellular control. Hormones work in one of two ways: They can attach to a special area on the membrane of a target cell (activating the formation of a substance within the cell such as cyclic AMP or a prostaglandin), or they can themselves enter the cell and activate specific genes.

Vitamins are organic molecules that the body cannot synthesize but that are required in small amounts for normal cellular function. They are classified as water-soluble or fat-soluble. Insufficient amounts of a vitamin in the diet cause a specific deficiency disease. Water-soluble vitamins often function as coenzymes or important cellular enzymes.

Review Problems

Section 18.1

1. Define metabolism.

2. Define each of the following terms:
(a) catabolic reaction
(b) anabolic reaction

Section 18.2

3. Define the term enzyme. Why are enzymes needed in living organisms?

4. Why does only a small amount of a particular enzyme need to be present in a living organism?

Section 18.3

5. What type of reaction does each of the following enzymes catalyze?
(a) oxidoreductase
(b) transaminase
(c) hydrolase
(d) isomerase
(e) reductase
(f) oxidase
(g) transmethylase
(h) dehydrogenase

6. For each of the following enzymes, give the name of the substrate or class of substrates on which the enzyme acts:
(a) lipase
(b) sucrase
(c) cellulase
(d) peptidase
(e) lactase
(f) glucosidase
(g) protease
(h) esterase

7. Name the classes of enzymes that catalyze the following reactions:

(a) $CH_3CH_2CH_2OH \longrightarrow CH_3CH_2\overset{\displaystyle O}{\overset{\|}{C}}{-}H + H_2$

(b) $CH_3\overset{\displaystyle O}{\overset{\|}{C}}{-}O{-}CH_3 + H_2O \longrightarrow CH_3\overset{\displaystyle O}{\overset{\|}{C}}{-}OH + CH_3OH$

(c) $CH_3CH_2OH \longrightarrow CH_2{=}CH_2 + H_2O$

8. Name the classes of enzymes that catalyze the following reactions:

(a) $CH_3CH_2OH + ATP \longrightarrow CH_3CH_2O{-}\overset{\displaystyle O}{\overset{\|}{\underset{\underset{\displaystyle O^-}{|}}{P}}}{-}O^- + ADP$

(b) $CH_3{-}\overset{\displaystyle O}{\overset{\|}{C}}{-}\overset{\displaystyle OH}{\overset{|}{\underset{\underset{\displaystyle H}{|}}{C}}}{-}H \rightleftharpoons CH_3{-}\overset{\displaystyle OH}{\overset{|}{\underset{\underset{\displaystyle H}{|}}{C}}}{-}\overset{\displaystyle O}{\overset{\|}{C}}{-}H$

(c) $\underset{\text{(maltose)}}{C_{12}H_{22}O_{11}} + H_2O \longrightarrow \underset{\text{(glucose)}}{2C_6H_{12}O_6}$

Section 18.4

9. Phosphoglucomutase catalyzes the conversion of glucose-1-phosphate to glucose-6-phosphate.

$$\text{glucose-1-phosphate} \xrightleftharpoons{\text{phosphoglucomutase}} \text{glucose-6-phosphate}$$

   (a) In general terms, describe the four steps by which phosphoglucomutase accomplishes the conversion.

   (b) Phosphoglucomutase has a turnover number of 1×10^3. How many molecules of glucose-1-phosphate are converted in an hour by one molecule of phosphoglucomutase? How many moles?

10. How many molecules of carbon dioxide are converted to bicarbonate in 5 s by 10 molecules of carbonic anhydrase?

Section 18.5

11. Explain how the lock-and-key theory describes the method of enzyme action.

12. How does the induced-fit theory change the lock-and-key theory?

Section 18.6

13. Why doesn't pepsin, the enzyme released in the stomach to hydrolyze proteins, continue to function in the small intestine?

14. Why is boiling water an effective sterilizing agent?

Section 18.7

15. A noncompetitive inhibitor is often more toxic than a competitive inhibitor. Explain why this is usually the case.

16. Public health officials estimate that more than 200,000 children become ill from lead poisoning each year, many from eating lead-based paint chips. What chemical action of lead causes this illness?
17. What is an antimetabolite? Give two examples of compounds that act as antimetabolites.
18. What property of the antimetabolite methotrexate makes it useful in cancer therapy?

Section 18.8

19. **(a)** What is an allosteric enzyme?
    (b) Why are allosteric enzymes necessary for the normal operation of the cell?
    (c) What effect does the regulatory molecule have on the allosteric enzyme?
20. Explain, in your own words, the effect of an increase in isoleucine concentration on the rate of conversion of threonine to α-ketobutyric acid (see Fig. 18.9).

Section 18.9

21. Describe two ways in which hormones regulate cellular processes.
22. Prostaglandins are among the most potent biological agents.
    (a) Describe some of their various physiological effects.
    (b) In what way do they function like cyclic AMP?

Section 18.10

23. Define vitamin.
24. Vitamins are generally divided into two classes. What is the basis for this classification?

Section 18.11

25. Name the three coenzymes that are formed from the B vitamins and serve as hydrogen carriers.
26. What important coenzyme is synthesized from pantothenic acid?
27. Name the vitamin deficiency that results in each of the following diseases:
    (a) scurvy **(c)** pernicious anemia
    (b) pellagra **(d)** beriberi
28. Why should fruits and vegetables containing the water-soluble vitamins be eaten raw or cooked only briefly in a small amount of water?

Section 18.12

29. A deficiency of which vitamin causes rickets?
30. Excess vitamin A and D is stored in the body, but excess vitamin C and B_1 is readily excreted.
    (a) What property allows vitamins A and D to be stored in the body, whereas vitamins C and B_1 are excreted?
    (b) Explain why an excess of vitamin D in the diet can be as dangerous as a deficiency.

31. More and more foods are being fortified with vitamins. Breakfast foods, such as cereal, bread, margarine, and milk, may all be fortified with vitamin D in addition to other vitamins.

(a) Explain why this might pose a health hazard to young children who eat a daily breakfast of cereal with milk and toast with butter.

(b) Why should fortification of foods with vitamins be carefully controlled?

32. Excessive bleeding might result from a deficiency of what vitamin?

Applied Problems

33. At temperatures above 45°C, most body enzymes begin to denature. Which part of the enzyme—the coenzyme or the apoenzyme—is most likely to be affected by denaturation? Why?

34. In the late 1960s, detergents and presoaks containing enzymes appeared on the market. Concern over their safety eventually led to their being withdrawn from use. What type of enzymes were most likely present in these detergents?

35. Apples picked before completely ripening may be stored at low temperatures under an atmosphere of nitrogen to keep them fresh for many months. Explain why these storage conditions slow the ripening process.

36. The enzymes necessary for the formation of blood clots are always present in the blood plasma. Why don't blood clots form constantly in a normal blood vessel?

37. Under normal conditions a human sperm cell can live 24 to 36 h. However, doctors successfully perform artificial insemination using human sperm that have been frozen at −196.5°C for up to 6 months. Explain how it is possible for sperm cells to survive for 6 months and then resume normal activity.

38. Broccoli is a vegetable that contains high amounts of both vitamin A and vitamin C. In fact, half a cup of broccoli provides 40% of the RDA for vitamin A and 117% of the RDA for vitamin C. That is why this vegetable is often used in school lunch programs. If the broccoli is cooked in boiling water and then kept warm on a steam table, how are the original amounts of the two vitamins affected?

39. Many people believe that taking massive doses of vitamin C prevents or greatly reduces the symptoms of the common cold. Synthetic vitamin C is relatively inexpensive, whereas the ''natural vitamin C'' sold in health food stores costs much more. What is the difference between these two forms of vitamin C?

40. Sulfa drugs are effective because they resemble *p*-aminobenzoic acid, a compound vital to the formation of the coenzyme folic acid. Bacteria need folic acid to produce amino acids and nucleotides. When bacteria use the sulfa drug instead of the *p*-aminobenzoic acid in the production of folic acid, the resulting molecules won't work, and the bacteria die. Why don't sulfa drugs affect humans in the same way?

41. Although most stomach ulcers can be treated through careful eating habits and avoidance of substances like alcohol and coffee, surgical removal of part of the stomach is sometimes necessary. After such surgery, the patient is advised to regularly take vitamin B_{12} supplements. Why is this necessary?

42. Severe vitamin A deficiency can lead to dry-eye disease, a leading cause of blindness among children in developing countries. This deficiency also seems to lead to fatal infections. Recent trials in Indonesia have shown that high-dose vitamin A capsules taken twice a year can eliminate the blindness and greatly reduce the death rate in children. Explain why only two doses of vitamin A a year can be so effective.

43. Aspartate carbamoyltransferase (ATCase) catalyzes the formation of *N*-carbamoyl aspartate from carbamoyl phosphate and aspartate. This is the first step in the six-step synthesis of pyrimidines, important components of nucleic acids. Among the final products of the six enzyme-catalyzed reactions is cytidine triphosphate (CTP). CTP inhibits ATCase.
   (a) What is the general name given an enzyme such as ATCase?
   (b) When a cell is dividing, it synthesizes nucleic acids and rapidly uses up the supply of CTP. Explain how the level of CTP in the cell controls its own synthesis.

INTEGRATED PROBLEMS

1. Certain individuals cannot tolerate milk. For them, drinking milk results in gas and diarrhea.
   (a) What sugar is found only in milk? Which monosaccharides make up a molecule of this sugar?
   (b) Individuals with an intolerance for milk have low levels of a certain enzyme in their intestines. What is the name of this enzyme?
   (c) Stores now sell special milk that not only contains this enzyme but also contains 70% less milk sugar than regular low-fat milk. What sugars are found in this special milk?

2. Maltose is obtained by the partial hydrolysis of starch. For example, the malting process in making beer involves the conversion of the polysaccharide starch into the disaccharide maltose. Why does the storage of a given number of glucose molecules as starch rather than maltose greatly reduce the intracellular osmotic pressure and prevent plant cells from bursting?

3. Chitin is the main component of the exoskeletons of invertebrates, such as insects and crustaceans. It is almost as abundant in nature as cellulose. The structure of chitin differs from that of cellulose in that the —OH on carbon 2 is replaced by a acetoamino group, CH_3CONH—. Draw the structure of a four-monomer section of a chitin molecule.

4. As we saw in Chapter 16, a high concentration of low-density lipoproteins, or LDL cholesterol, is linked to the development of atherosclerosis. Cholesterol deposits form on the lining of the blood vessels. Cholesterol is transported in the blood as an ester in the LDL particle. Write the equation for the synthesis of the ester formed between cholesterol and stearic acid.

5. An enterprising student wants to make soap from the excess lard she has trimmed and saved from various pork products. If the average molecular weight of lard is 850 g, how many pounds of lye (sodium hydroxide) is required to convert 5 lb of this lard to soap?

6. Aspartame is an artificial sweetener (160 times sweeter than sugar) that is widely used in low-calorie products. It is a dipeptide and is digested as a protein. The two amino acids that form aspartame are aspartic acid and phenylalanine, with the methyl ester of phenylalanine forming the C-terminal end. Draw the structural formula of aspartame.

7. The role a protein has in the body as well as the shape it takes on are dependent on the primary structure of the protein. Sickle cell disease is one example of an abnormal protein primary structure. Over 400 other forms of abnormal hemoglobin have been identified. To get an idea of the relationship between primary structure and several properties of a protein, let's look at a simplified case. Write the structure for the two polypeptides I and II described in Section 17.7.

(a) Which polypeptide, I or II, is more soluble in water? Explain why.

(b) In which polypeptide, I or II, are hydrophobic interactions more likely to occur? Explain why.

(c) In which polypeptide, I or II, are salt bridges more likely to occur? Explain why.

8. Somatostatin is a polypeptide hormone secreted by the hypothalamus in the brain. One of its functions is to inhibit the release of insulin and glucagon from the pancreas. A somatostatin molecule contains one disulfide bridge and the following sequence of amino acids:

Ala-Gly-Cys-Lys-Asn-Phe-Phe-Trp-Lys-Thr-Phe-Thr-Ser-Cys

(a) Draw the structural formula for a molecule of somatostatin.

(b) Suggest a possible way in which somatostatin works on the cells of the pancreas to inhibit the release of insulin.

9. Bile is a fluid that helps in the digestion of lipids and the absorption of fat-soluble nutrients. Patients whose bile ducts are blocked, preventing bile from entering the small intestine, have blood that does not clot as readily as that of healthy individuals.

(a) What vitamin is important in the clotting process?

(b) Why might a lack of bile in the intestines result in a failure of blood clotting?

10. Beriberi used to be very common in the Far East where the main food of the diet was polished rice (rice with its outer coating removed). The introduction of enriched rice has mostly eliminated the disease.

(a) What causes beriberi?

(b) Why was beriberi occurring among this population?

(c) With what was the rice enriched?

11. Both ethanol and methanol are oxidized in the body: ethanol to acetic acid and methanol to formic acid. It is formic acid that causes the acidosis of methanol poisoning. The first step in the oxidation of both ethanol and methanol involves the enzyme alcohol dehydrogenase. One therapy for methanol poisoning is the use of a nearly intoxicating dose of ethanol. Use your knowledge of enzyme inhibition to explain why this therapy is effective.

Chapter 19
Pathways of Metabolism

Pete was hot and tired, but also excited as he left his job at the Philomath Sawmill. He had just been given a raise and was off to Squirrel's Tavern to celebrate with the other guys on his crew. Pete did have one nagging concern: it was six months since he had completed the alcohol and drug treatment program from his drunk driving arrest, and he had avoided going out with the guys since then. But Pete figured that one beer wouldn't hurt, and he would go right home after that. Much later that night Pete was standing on the highway beside his truck, trying to walk along a straight line between two police officers.

Pete spent 48 hours in jail; this was his third arrest for drunken driving. He was disgusted with himself. He really wanted to gain control over his life and his drinking problem. Hearing this, the judge recommended that Pete be considered for treatment with the drug disulfiram (also called Antabuse) as part of a six-month outpatient program.

At the county's drug and alcohol treatment center Pete learned that disulfiram was a liquid drug he would take three times a week under the supervision of a pharmacist. Disulfiram interferes with the metabolism of alcohol, causing a person to become very sick if he or she drinks alcohol. Such use of disulfiram is not a long-term treatment, nor will it cure alcoholism. It could, however, help Pete get through the first stages of treatment without relapsing, giving him time to discover releases other than alcohol for stress or tension.

Pete began taking 250 mg of disulfiram three times a week. In addition he attended a counseling session and two meetings of Alcoholics Anonymous each week. Pete joined the company's softball team

and bowling league, and was pleased to discover he could enjoy the company of his friends in places other than a tavern.

One hot day in late August, however, the foreman announced that the mill might have to lay off several workers. Pete spent the rest of the day worrying, and when he left work he simply had to stop by Squirrel's for a drink. Before he had finished his first beer Pete began to feel hot; then he felt his head begin to throb. He left the tavern quickly as nausea swept over him. Pete had to stop repeatedly on the way home to vomit out the door of his truck. He continued to vomit when he got home, feeling weaker and dizzier each moment. When the vomiting finally stopped, Pete fell into bed exhausted. Antabuse had done its job in keeping Pete from going on a drinking binge.

Why did Pete get so violently ill after only one glass of beer? Disulfiram (Antabuse) is a relatively nontoxic substance.

$$(H_5C_2)_2N-\underset{\underset{S}{\|}}{C}-S-S-\underset{\underset{S}{\|}}{C}-N(C_2H_5)_2$$

disulfiram (Antabuse)

Most people can take the drug for six months without any harmful effects unless alcohol (ethanol) is consumed. When ethanol enters the body, it normally is oxidized to acetaldehyde by an enzyme (alcohol dehydrogenase) in the liver. This is not a problem because another enzyme, aldehyde dehydrogenase, immediately oxidizes the acetaldehyde to acetic acid. Disulfiram, however, reacts with crucial sulfhydryl groups on the aldehyde dehydrogenase molecule to permanently inactivate the enzyme.

When Pete drank his beer, the ethanol he consumed was oxidized to acetaldehyde. The nonfunctioning aldehyde dehydrogenase enzyme allowed the acetaldehyde to accumulate in his blood and tissues to as much as 10 times the normal level. Such an accumulation causes very unpleasant symptoms, known as the acetaldehyde syndrome. Besides vomiting, the symptoms include respiratory difficulties, sweating, thirst, chest pain, low blood pressure, weakness, blurred vision, and confusion.

The oxidation of ethanol to acetaldehyde, and then from acetaldehyde to acetic acid is the body's normal pathway for the metabolism of ethanol. There are many metabolic pathways in the human body for the breakdown of other substances, as well as for the production of energy and the synthesis of compounds necessary for life. Any substance that inhibits the enzymes that catalyze these processes has drastic effects on the organism. In this chapter we study the major pathways for the metabolism of carbohydrates, lipids, and proteins.

19.1 Cellular Energy Requirements

The body has many energy requirements: energy is needed for synthesis reactions within cells, for muscle movement, and for the production of heat to maintain body temperature. Such energy is produced by the controlled combustion of compounds such as glucose within the cell. The cell, however, does not operate like a steam engine—converting chemical energy to heat energy and then using the heat energy to do work. Instead, the cell operates much more efficiently by using chemical energy directly to do such work as

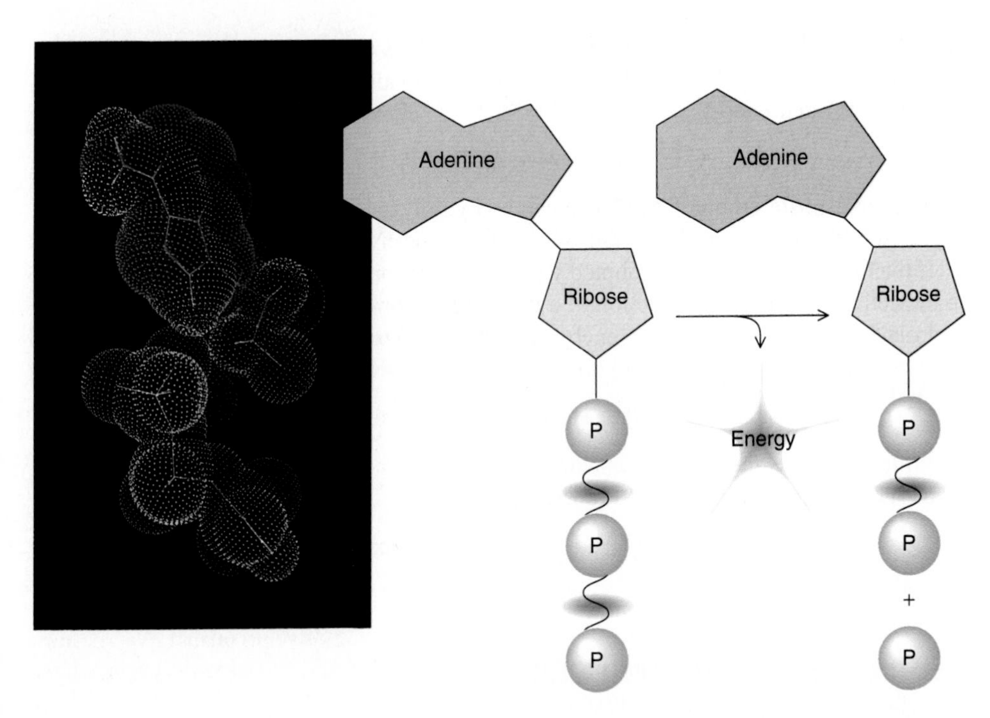

Figure 19.1 An ATP molecule when hydrolyzed releases ADP, inorganic phosphate (P), and energy. Left: a computer-generated model of ATP.

flexing muscles, transmitting nerve impulses, and synthesizing cellular components. For these tasks the cell requires **high-energy, or energy-rich,** compounds such as adenosine triphosphate (ATP) (Fig. 19.1). A molecule of ATP contains two oxygen-to-phosphorus bonds. These are called **high-energy phosphate bonds** and are often represented by a wavy line.

Define high-energy bond.

$$^{-}O-\overset{\overset{O}{\|}}{\underset{\underset{O^{-}}{|}}{P}}\sim O-\overset{\overset{O}{\|}}{\underset{\underset{O^{-}}{|}}{P}}\sim O-\overset{\overset{O}{\|}}{\underset{\underset{O^{-}}{|}}{P}}-O-CH_2-\text{(ribose, OH OH)}-\text{adenine }(NH_2)$$

AMP (adenosine monophosphate)

ADP (adenosine diphosphate)

ATP (adenosine triphosphate)

These same bonds are found in similar molecules, such as guanosine triphosphate (GTP) and uridine triphosphate (UTP). Hydrolysis of the high-energy phosphate bonds in compounds such as ATP is a highly exothermic reaction. It produces about twice as much energy (7 kcal/mol) as the hydrolysis of compounds containing low-energy phosphate bonds (2–5 kcal/mol). The large amount of chemical energy stored in the high-energy

Give examples of compounds that contain high-energy bonds.

phosphate bonds results from the specific arrangement of atoms in the molecule. When hydrolyzed, ATP forms ADP (adenosine diphosphate) and an inorganic phosphate (represented by P_i) and energy.

$$\text{ATP} + \text{H}_2\text{O} \longrightarrow \text{ADP} + \text{P}_i + 7.3 \text{ kcal}$$

The end phosphate group of ADP also can be hydrolyzed, yielding adenosine monophosphate (AMP) and the same amount of energy as the hydrolysis of ATP.

If the hydrolysis of ATP is coupled with an endothermic (or energy-requiring) reaction in the cell—such as the contraction of a muscle fiber or the synthesis of a molecule—the hydrolysis of ATP supplies the energy necessary for the other reaction to occur. Any excess energy provided by the hydrolysis of ATP is released as heat.

$$\text{relaxed muscle} + \text{ATP} + \text{H}_2\text{O} \longrightarrow \text{contracted muscle} + \text{ADP} + \text{P}_i$$

ATP also provides the energy for reactions in which the phosphate group is transferred from ATP to another molecule. For example,

$$\text{Glucose} + \text{ATP} \longrightarrow \text{glucose-6-phosphate} + \text{ADP}$$

19.2 Digestion and Absorption

Before a cell can produce ATP from nutrients contained in our diets, the nutrients must be digested and absorbed. **Digestion** is the process by which complex foods are broken down into simple molecules. Table 19.1 summarizes the digestion of carbohydrates, lipids, and proteins.

Describe the digestion of the potato, margarine, and steak consumed at a barbecue.

Carbohydrates

The digestion of carbohydrates begins in the mouth, where teeth break large pieces of food into smaller ones. Saliva, which contains an enzyme that begins the hydrolysis of starch to dextrins and maltose, mixes with the food in the mouth. When the food is swallowed it passes into the stomach, where the salivary enzymes become inactivated by the low pH. The churning action of the stomach further reduces the size of the food particles. As the

Table 19.1 **Summary of Human Digestion**

Site of Digestion	*Source of Enzymes*	*Nutrients Digested ⟶ Products Formed*
Mouth	Salivary glands	Starch ⟶ dextrins, maltose
Stomach	Stomach	Proteins ⟶ peptides, amino acids Emulsified fats ⟶ monoacylglycerols, fatty acids, glycerol
Small intestine	Pancreas	Starch ⟶ maltose Fats ⟶ monoacylglycerols, fatty acids, glycerol Proteins ⟶ peptides Peptides ⟶ amino acids
	Intestine	Peptides ⟶ amino acids Sucrose ⟶ glucose, fructose Maltose ⟶ glucose Lactose ⟶ glucose, galactose

food passes into the small intestine, the pancreas and cells in the intestinal wall secrete juices containing enzymes that completely hydrolyze the polysaccharides and disaccharides in the food. Glucose, fructose, and galactose—the monosaccharides formed by this process—pass directly through the intestinal wall into the bloodstream and are carried to the liver. There, liver enzymes convert the galactose to glucose. The fructose may either be converted to glucose or may enter into other metabolic reactions. The glucose may be used to meet immediate cellular energy needs, or it may be stored as glycogen in the liver and the muscles.

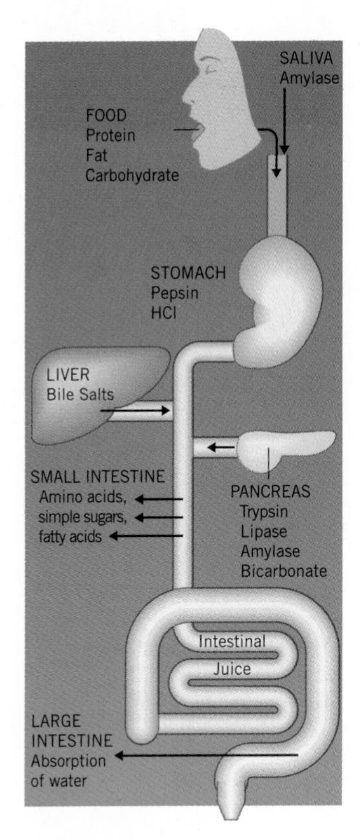

The digestive system

Lipids

Although some digestion of already emulsified food fats such as milk or eggs may occur in the stomach, the main digestion of lipids takes place in the small intestine. There, lipids mix with bile salts, which emulsify the fats. This provides the maximum surface area for the action of the enzymes in the pancreatic and intestinal juices. These enzymes hydrolyze the fat into glycerol, fatty acids, and monoacylglycerols and diacylglycerols.

Bile is a fluid that is continuously manufactured in the liver and stored in the gallbladder. It is released from the gallbladder by a muscle contraction that is triggered by a hormone produced when food enters the small intestine. Bile contains bile salts, bile pigments, phospholipids, and cholesterol. Bile salts are synthesized from cholesterol in the liver; they aid in the digestion of fats and in the absorption of fatty acids, fat-soluble vitamins, and other products of fat digestion. Bile salts are very efficiently reabsorbed from the intestine and then extracted from the blood by the liver to be used again. Bile pigments, mainly bilirubin, do not play a role in fat digestion. They are the waste products of the breakdown of hemoglobin in the liver and are the substances that give color to the feces.

If the bile duct becomes blocked, or if the breakdown of hemoglobin occurs at a very fast rate, bilirubin builds up in the blood. This will make the skin appear more yellow, which is one of the symptoms of jaundice. In some people the mucous membranes of the gallbladder absorb water, concentrating the bile. Under these conditions the cholesterol in bile, which is not very water-soluble, crystallizes out of solution with bile salts and bile pigments to form gallstones. These stones can cause infection and pain and can obstruct the flow of bile, which also results in jaundice.

Intestinal cells absorb the glycerol, fatty acids, and monoacylglycerols and diacylglycerols formed from the hydrolysis of fats. In these cells they are reformed into triacylglycerols and packaged as tiny droplets of lipoprotein called chylomicrons. The lymph system transports chylomicrons to points where they can enter the blood. After a meal rich in fat, the chylomicrons cause the blood to have a milky, opalescent appearance.

Proteins

The digestion of proteins involves yet another process. Proteins in the native state are not easily digested because their peptide bonds cannot be reached by digestive enzymes. Dietary proteins must therefore first be denatured by heat (in cooking), by chewing, and by the acids in the stomach. Protein digestion itself begins in the stomach. The protein-digesting enzymes are secreted by the cells of the stomach lining as proenzymes (inactive enzymes) and are activated by the HCl present in stomach acid. This process prevents the enzymes from digesting the proteins making up the cells of the stomach. It also allows time for the secretion of mucous, which protects the cells lining the stomach and intestines from the enzymes. Protein digestion continues in the small intestine, where enzymes in the pancreas and intestinal juices catalyze hydrolysis of the proteins. The amino acids and some very small peptides that are produced are absorbed through the intestinal wall and enter the bloodstream.

19.3 Carbohydrate Metabolism

We have said that carbohydrates, specifically glucose, are the major source of energy for cells. Most tissues can also use fatty acids to supply their energy needs, but the cells of the brain and nervous system can use only glucose. The brain, which contains no storehouse of glucose, depends completely on the glucose in the blood for its energy supplies. Therefore, the blood glucose level is critical to normal brain function. The normal level of glucose in the blood is 60 to 100 mg per 100 mL of blood. Any significant variation from normal, either too high or too low, can affect the brain and nervous system. Of course, we do not supply our bodies continuously with glucose. Instead, we eat large meals and then have long periods of fasting in between. To ensure a continuous supply of glucose to meet the energy needs of brain cells, the body has developed complicated control mechanisms for keeping the blood sugar level fairly constant.

Write the general equations for glycogenesis and glycogenolysis.

Right after a meal, the blood sugar level rises. To deal with this condition, the beta cells of the pancreas produce the hormone insulin. Insulin aids the passage of glucose into liver, muscle, and fatty tissue cells, and it encourages the storage of glucose in the liver as glycogen. The process of converting glucose to glycogen is called **glycogenesis.** This biosynthetic process involves the series of reactions shown in Figure 19.2, requiring several enzymes and two high-energy molecules: ATP and UTP. As time passes and the cells of the body absorb and metabolize the glucose in the blood, the blood sugar level falls. This causes the pancreas to stop producing insulin and to start producing another hormone, glucagon. Glucagon has an effect opposite that of insulin; it stimulates the liver to hydrolyze glycogen to form glucose. The series of reactions producing glucose from glycogen is called **glycogenolysis.** The glucose produced is released into the blood to maintain the blood sugar at normal levels (see Fig. 19.2).

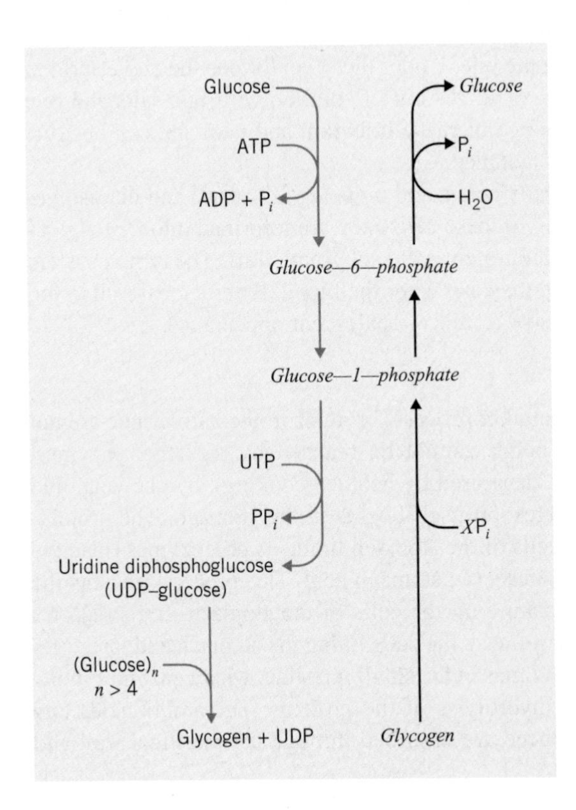

Figure 19.2 The reactions of glycogenesis and glycogenolysis. (P_i is phosphate and PP_i is pyrophosphate.)

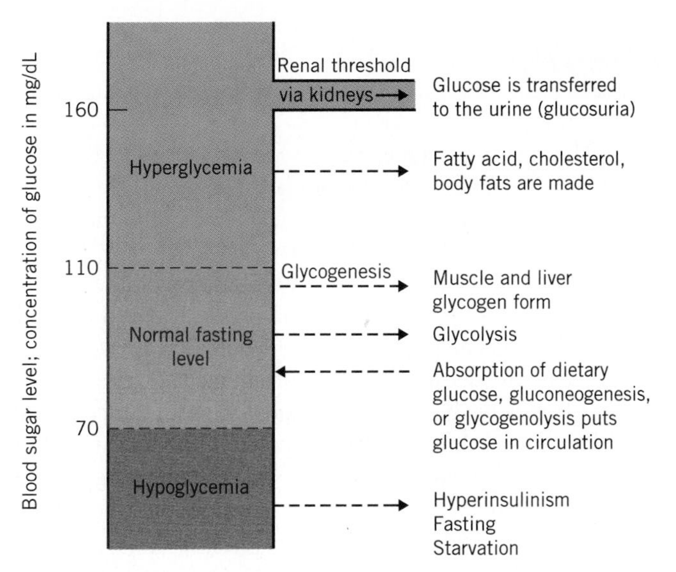

Figure 19.3 Blood sugar levels and the conditions of hypoglycemia and hyperglycemia.

Describe hyperglycemia and some conditions that can cause it.

Hyperglycemia is the condition that results when blood sugar levels are too high. Mild hyperglycemia occurs after meals. In severe hyperglycemia, the blood sugar level can rise so high that the glucose level exceeds the amount tolerated by the kidneys (called the renal threshold, 160 mg/100 mL). Glucose is then excreted in the urine (Fig. 19.3). Glycosuria, sugar in the urine, can result from conditions such as diabetes mellitus, emotional stress, kidney failure, or the administration of certain drugs.

Describe hypoglycemia and some conditions that can cause it.

Hypoglycemia is the condition that results when blood sugar levels are below normal. Under such conditions the brain becomes starved for glucose. Mild hypoglycemia produces irritability, dizziness, lethargy, grogginess, and fainting. Severe hypoglycemia can produce convulsions, shock, and coma. Such severe hypoglycemia can be produced by the overproduction or overinjection of insulin, a condition called hyperinsulinism. In hyperinsulinism the blood sugar level drops extremely rapidly, causing the person to go into convulsions and coma called *insulin shock*. For most people, the body's regulatory mechanisms keep blood sugar levels within normal ranges even after eating a meal rich in refined carbohydrates, such as the sugars in candy and cake. It is preferable, however, to eat complex carbohydrates, such as whole-grain cereals and pastas, which release glucose into the blood much more slowly.

19.4 Cellular Respiration: Oxidation of Carbohydrates

The glucose in the blood also diffuses into tissue cells. There it can be used in biosynthetic reactions or can be further metabolized to yield useful cellular energy through a process called **cellular respiration.** In cellular respiration, glucose is oxidized in a series of chemical reactions to form carbon dioxide, water, and ATP. The cellular processes of respiration are quite efficient; about 44% of the energy released from the oxidation of glucose is trapped in the high-energy bonds of ATP molecules and can be used to do work. For each molecule of glucose that is oxidized, 36 ATPs are produced. But what, then, happens to the other 56% of the energy that is released? It is given off as heat in quantities more than enough to maintain body temperature at 37°C. Any excess heat must be eliminated from the body through cooling mechanisms such as the evaporation of sweat.

Health Perspective / Epinephrine

A sudden squeal of brakes close by or a frightening movie can make your heart pound, your body perspire, and your stomach feel as though it were climbing into your throat. Each of these body reactions is caused by the hormone epinephrine (adrenalin). Epinephrine is produced by the adrenal glands, a pair of hat-shaped organs located on top of the kidneys. Under normal conditions, small amounts of epinephrine are released into the blood to help control blood pressure and to maintain the level of sugar in the blood. But this compound is also the body's way of meeting emergencies and dealing with such stresses as emotional excitement, exercise, extreme temperature change, severe hemorrhaging, or the administration of certain anesthetics.

When released, epinephrine acts like glucagon. It stimulates the liver to hydrolyze its supply of glycogen to glucose. The resulting surge of glucose into the bloodstream causes hyperglycemia, with blood sugar levels often exceeding the renal threshold. Such a large increase in available glucose delays fatigue in skeletal muscles and, together with epinephrine, increases the strength of their contractions. This enables ordinary people to do extraordinary things in emergency situations. In one case, a young mother saved the life of her child by lifting a car off the infant. Later, when trying to show others how she had done it, she could not budge the car.

Epinephrine also increases the rate and strength of the heart beat, which increases the flow of blood from the heart. It raises the blood pressure by causing constriction of blood vessels in all parts of the body, except the blood vessels in such vital organs as the skeletal muscles, heart, brain, and liver. It relaxes the smooth muscles in the lungs, which increases the rate and depth of breathing and enables more oxygen to enter the lungs. It slows the action of the digestive tract, and accelerates blood clotting. Each of these changes in normal body function caused by epinephrine enables the body to meet the initial challenge of the emergency or stress. The effects last only a short time because the liver inactivates the epinephrine in the bloodstream in about three minutes.

Figure 19.4 summarizes the reactions of glucose metabolism. The concentration of ATP and AMP in the cell controls the rates of each of the multienzyme systems. Cellular respiration (System 1 in Fig. 19.4) is a very complex process, involving many steps and requiring many enzymes and coenzymes. A specific description of each step must be left to other biochemistry courses, but the following is a general overview of the process.

◆ **Describe the two stages in the oxidation of glucose, indicating where they occur in the cell and in which step the most energy is produced.**

Anaerobic Stage: Glycolysis

The oxidation of glucose can be divided into two stages. The first, an anaerobic stage, requires no oxygen and occurs in the cytoplasm of a eukaryotic cell. This stage is called **glycolysis,** and it involves the conversion of glucose to form two molecules of lactic acid (in the form of the negative ion lactate; Fig. 19.5). The net energy production from this process is two ATPs, which can be summarized by the following overall reaction.

◆ **Write the general equation for glycolysis.**

$$\underset{\text{glucose}}{C_6H_{12}O_6} + 2ADP + 2P_i \longrightarrow 2\ \underset{\text{lactate}}{CH_3{-}\overset{\overset{\large OH}{|}}{C}H{-}\overset{\overset{\large O}{\|}}{C}{-}O^-} + 2ATP + 2H_2O + 2H^+$$

Glycolysis is the emergency energy-producing process (or pathway) for the cells. When not enough oxygen reaches the tissues, ATP levels can be maintained for a short

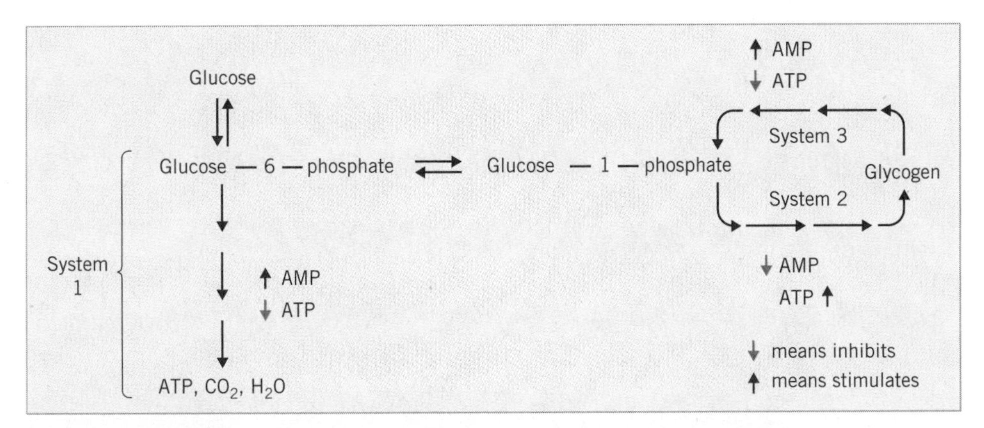

Figure 19.4 ATP and AMP have opposite regulatory effects on the three multienzyme systems controlling the level of ATP in the cell. System 1 involves the oxidation of glucose, System 2 involves glycogenesis, and System 3 glycogenolysis.

time through glycolysis. During childbirth, for example, the amount of oxygen available to the infant (and, therefore, present in the blood) is severely reduced. To ensure an adequate supply of oxygen to the brain, the circulation of the infant's blood decreases in all tissues except for the brain. The other tissues, then, depend on glycolysis for their energy until the circulation returns to normal after the delivery is over.

Aerobic Stage: Citric Acid Cycle and the Electron Transport Chain

The second part of glucose oxidation, the aerobic stage, requires oxygen. This stage in the oxidation of glucose produces most of the energy. These reactions also serve as the final stage of oxidation for other molecules used by the cell for energy production. The aerobic

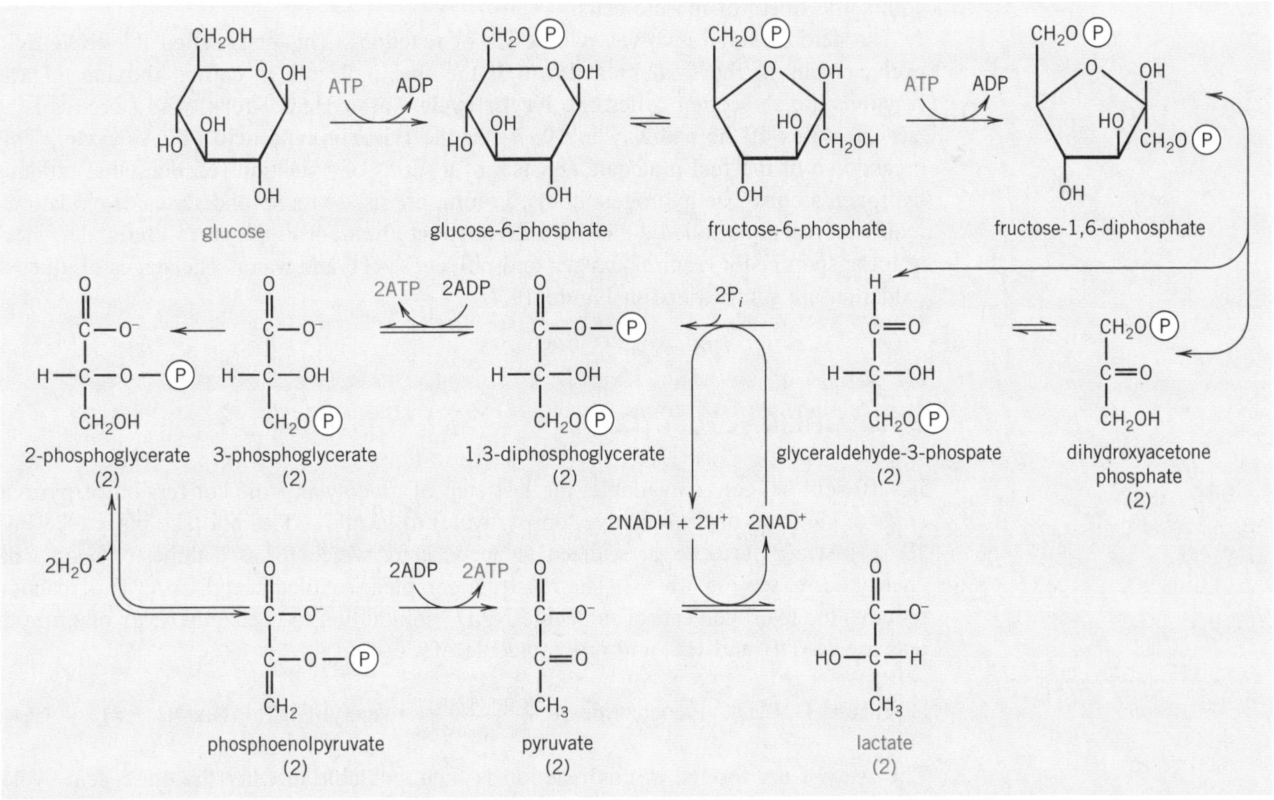

Figure 19.5 The reactions of glycolysis, the anaerobic stage of the oxidation of glucose, in which glucose is converted to lactic acid (in the form of the negative lactate ion). Enzymes are required for each reaction, but they are not listed in this diagram. (P) is phosphate.

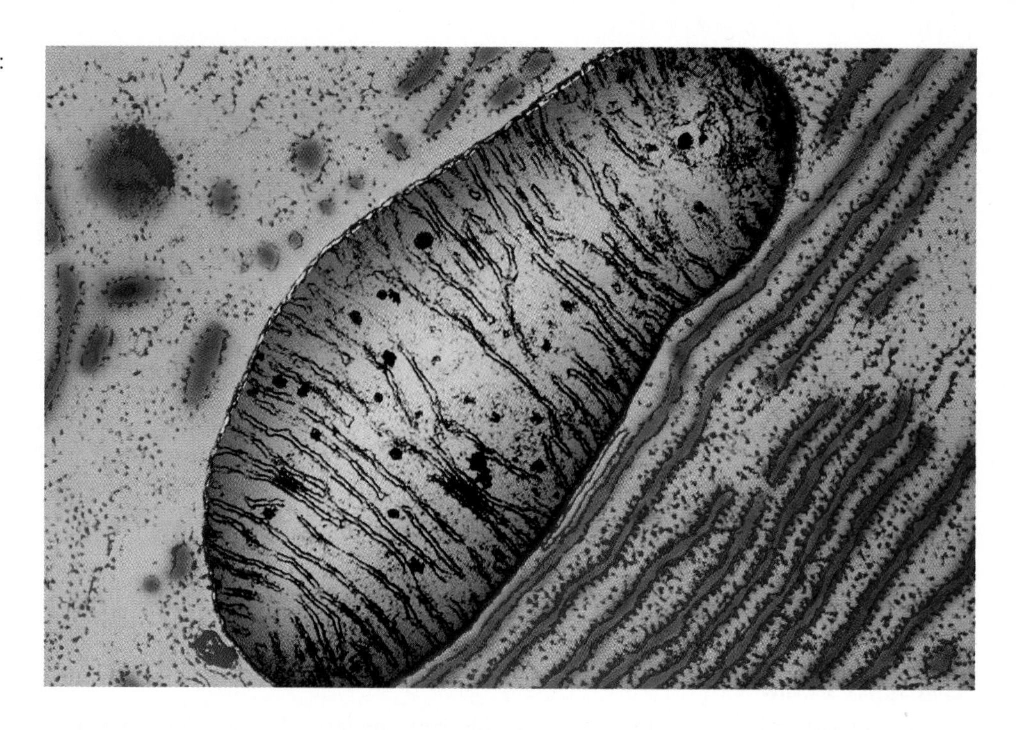

Figure 19.6 A mitochondrion: the site of cellular respiration.

stage takes place in the mitochondria, often called the ''power plants'' of the cell (Fig. 19.6). These structures are located near the parts of the cell that require energy, such as the contractile fibers of muscle cells.

The aerobic stage involves two series of reactions. The first, called the **citric acid cycle,** results in the final breakdown of the fuel molecule to carbon dioxide. [These reactions are also often called the **Krebs cycle**—after Hans Krebs, who suggested the basic features of the pathway in 1937—or the **tricarboxylic acid (TCA) cycle.**] This breakdown of the fuel molecule consists of a series of oxidation reactions that produce hydrogen atoms. The hydrogen atoms, in turn, are used in a second series of oxidation–reduction reactions called the **electron transport chain,** or **respiratory chain.** The electron transport chain requires oxygen and produces ATP and water. The stages of glucose oxidation are summarized in Figure 19.7.

19.5 Citric Acid Cycle

If sufficient oxygen is available, the last step of glycolysis—the conversion of pyruvic acid (in the form of the negative ion pyruvate) to lactate—does not take place (see Fig. 19.5). Instead, pyruvate is oxidized to acetic acid, which forms a thioester bond with coenzyme A (see Fig. 18.13). The resulting complex is called acetyl-CoA. This reaction requires the hydrogen carrier molecule NAD^+ in addition to coenzymeA; its other products are NADH and H^+, and one molecule of CO_2.

$$\text{Pyruvate} + NAD^+ + \text{coenzyme A} \xrightarrow[\text{and coenzymes}]{\text{other enzymes}} \text{acetyl-CoA} + NADH + H^+ + CO_2$$

The acetyl-CoA formed in this reaction is then available to enter the citric acid cycle.

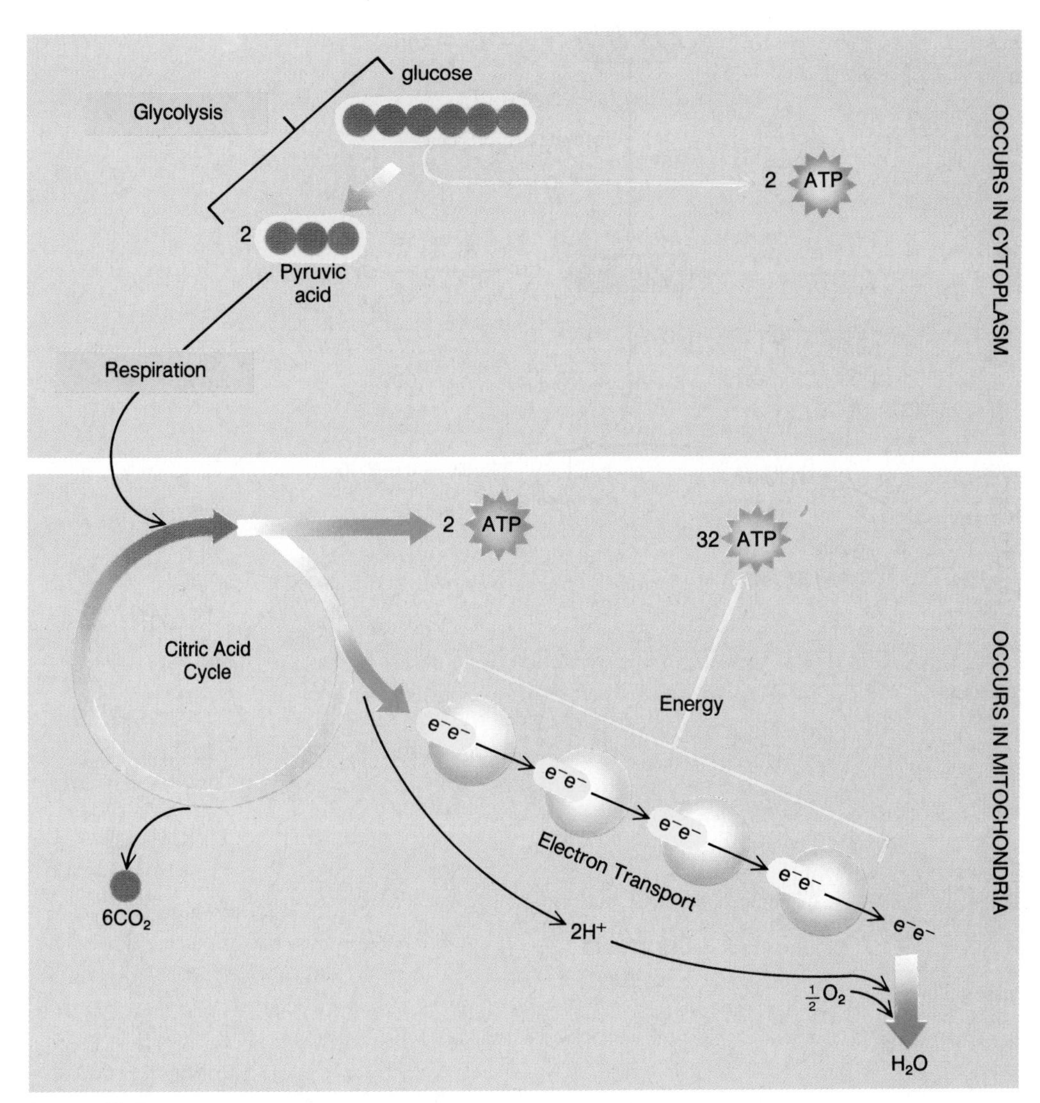

Figure 19.7 Summary of glucose oxidation.

The citric acid cycle, shown in Figure 19.8, oxidizes the two-carbon acetyl group to carbon dioxide. This process produces four pairs of hydrogen atoms and one high-energy phosphate bond in GTP (which transfers a phosphate to ADP, thus forming ATP). Three of the hydrogen atom pairs are captured and carried by the coenzyme NAD^+ and one by the coenzyme FAD. Cells, however, do not have an unlimited supply of NAD^+ and FAD. This means that the citric acid cycle cannot occur independently of the electron transport chain (in which these coenzymes donate their hydrogens to other hydrogen carriers and become available again as hydrogen acceptors in the citric acid cycle). It is important that you notice the cyclic (or circular) nature of the reactions in the citric acid cycle. In the first reaction, acetyl-CoA is joined with oxaloacetate to form citrate (remember that all of the acids found in the citric acid cycle exist as negative ions at cellular pH). The last reaction again produces oxaloacetate, which is then available to combine with another acetyl-CoA.

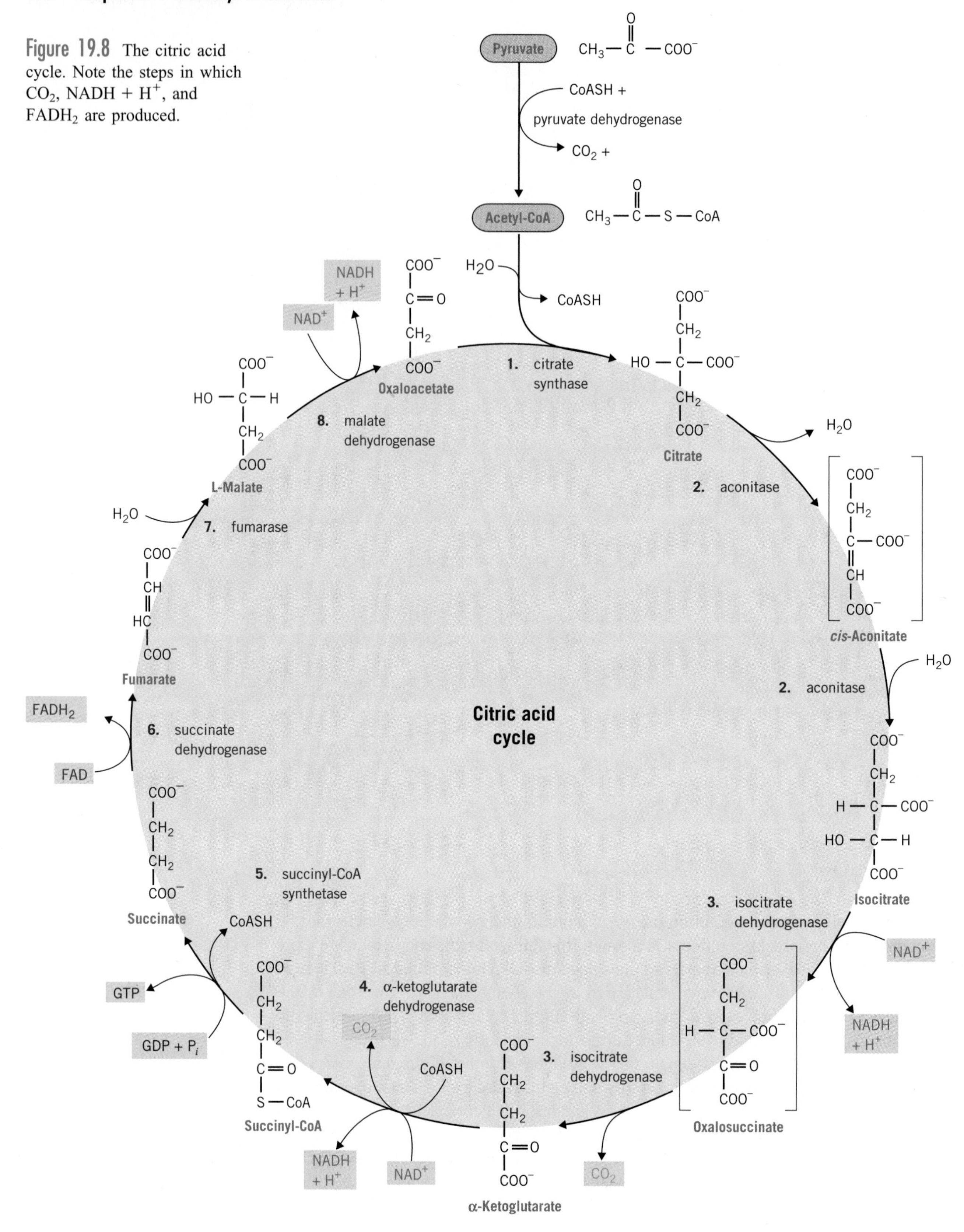

Figure 19.8 The citric acid cycle. Note the steps in which CO_2, NADH + H^+, and $FADH_2$ are produced.

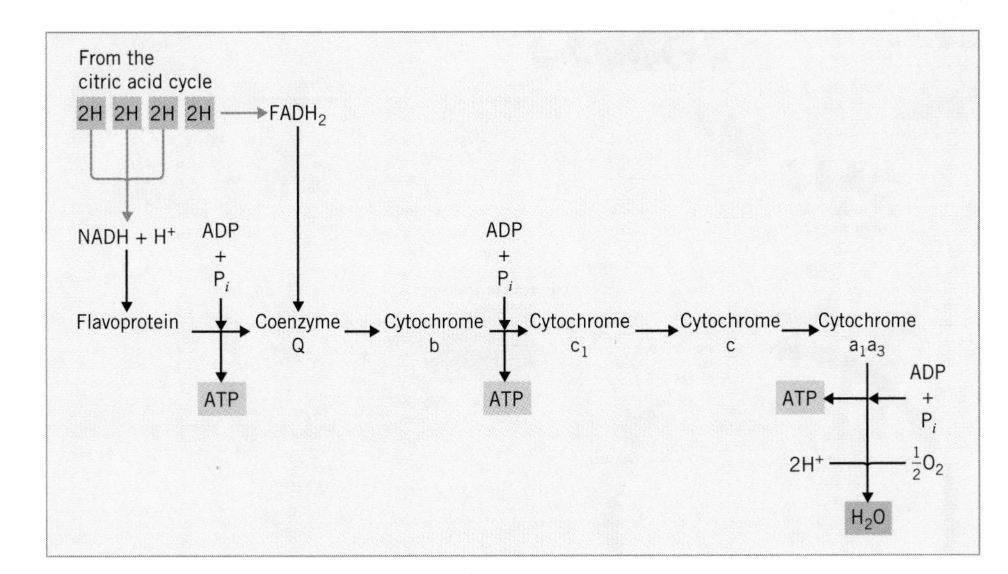

Figure 19.9 The electron transport chain. The hydrogens produced in the citric acid cycle are passed through a series of compounds until they combine with oxygen to form water. Most of the ATP formed in the oxidation of glucose is produced in this chain.

19.6 Electron Transport Chain and ATP Formation

The final hydrogen acceptor, or carrier, in aerobic oxidation is the oxygen we breathe. To arrive at this point, the hydrogens carried by NAD^+ and FAD from the citric acid cycle enter a series of coupled reactions called the **electron transport chain,** or **respiratory chain** (Fig. 19.9). In this sequence of redox reactions, the hydrogens (and later just their electrons) pass between a series of compounds until they combine with oxygen to form water. This process produces most of the ATP molecules formed in the oxidation of glucose by the cell. Each transfer is part of an oxidation–reduction reaction in which some energy is released. As the electrons flow through the sequence of reactions they can do work, just as the electrons flowing through a copper wire can do work. The kind of work done by the electrons in the cell, however, is the production of ATP from ADP and inorganic phosphate.

The exact way in which this process works is not well understood, but it is known at what points in the transport chain ATP is produced. From Figure 19.9 we see that each molecule of NADH produces three ATPs, and each molecule of $FADH_2$ produces two ATPs. As it passes through the citric acid cycle, one molecule of acetyl-CoA therefore generates a total of 12 ATPs: 1 from GTP produced in the citric acid cycle and 11 from NADH and $FADH_2$ in the electron transport chain (three NADHs produce nine ATPs, and one $FADH_2$ produces two ATPs).

To summarize, the glucose molecule absorbed by the cell is oxidized in three steps (Fig. 19.10):

1. Glycolysis: Glucose is oxidized to two pyruvate molecules in a reaction that produces two ATP and two NADH.
2. Citric Acid Cycle: The pyruvate molecule joins with acetyl-CoA (producing one NADH and one CO_2) and enters the citric acid cycle. Each turn of the citric acid cycle produces two CO_2, three NADH, one $FADH_2$, and one ATP. The two molecules of pyruvate generate a total of six CO_2, eight NADH, two FADH, and two ATP.

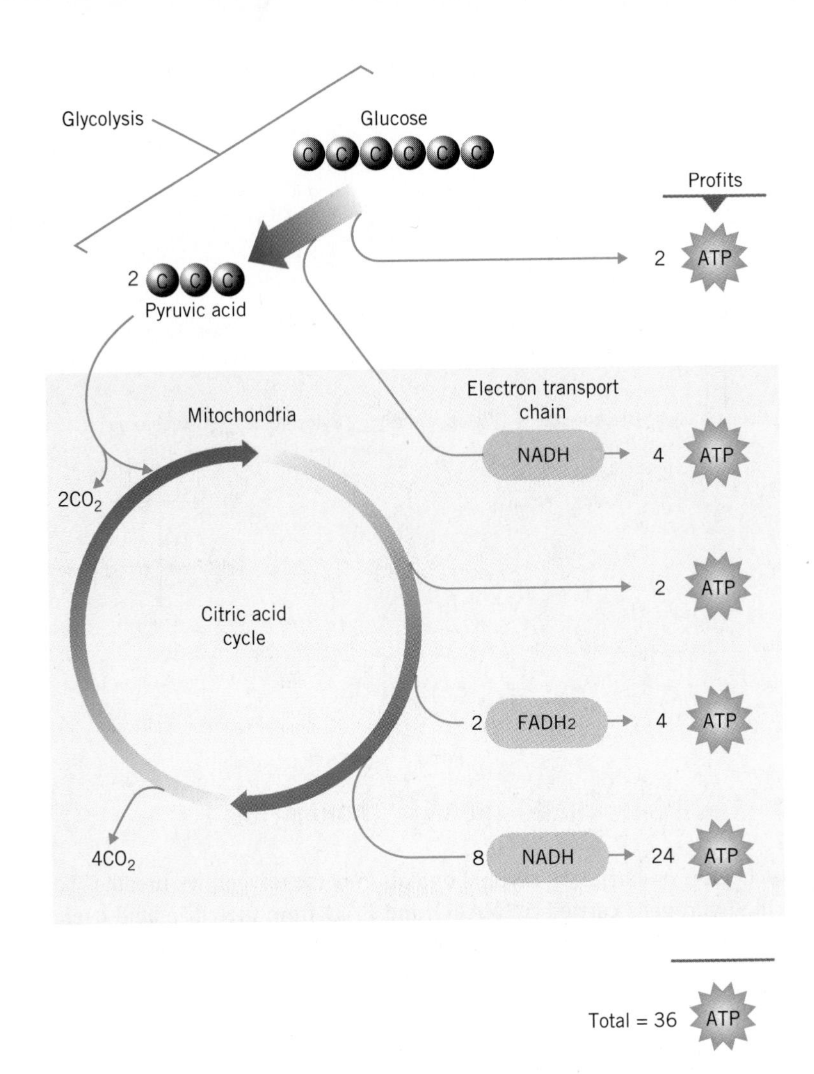

Figure 19.10 The energy produced from the aerobic oxidation of one molecule of glucose.

3. Electron Transport Chain: In eukaryotic cells, the processing of the NADH and $FADH_2$ from Steps 1 and 2 results in the production of 32 ATP and 12 H_2O. (The two NADH produced during glycolysis yield only four ATP in eukaryotic cells. In prokaryotic cells and in liver and heart muscle cells, however, the two NADH produced during glycolysis yield six ATP. So for those cells, the total ATP produced in the electron transport chain is 34.)

The oxidation of one molecule of glucose yields a total of 36 ATP as follows:

Glycolysis	2 ATP
Citric acid cycle	2 ATP
Electron transport chain	32 ATP
Total	36 ATP

19.7 Lactic Acid Cycle

A superbly trained athlete waits at the starting blocks for the gun, sprints 400 m at top speed, breaks the tape in record time, and then collapses on the track. Her legs feel like

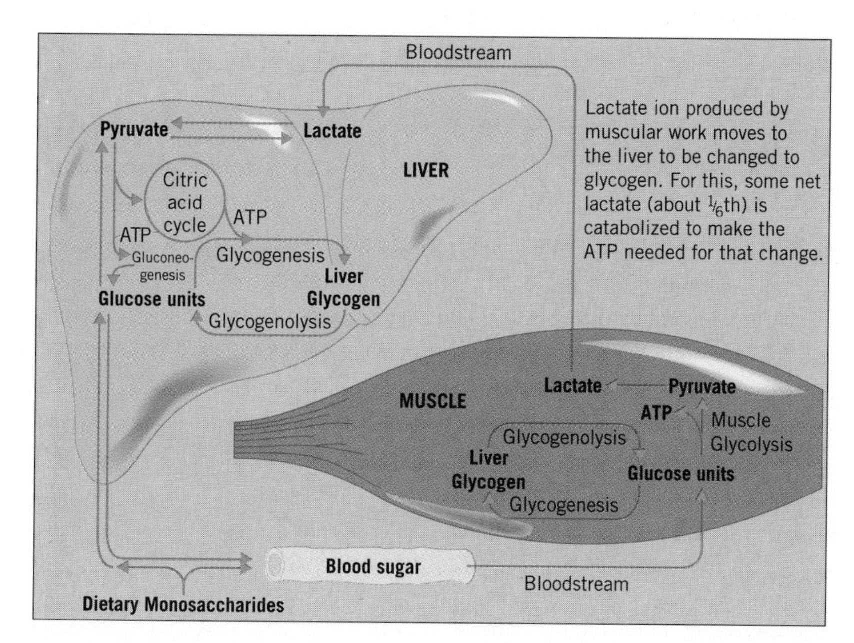

Figure 19.11 The lactic acid cycle.

rubber, she painfully gasps for breath, and she feels nauseated. If the athlete is well trained, why does it require a 90-min rest to completely recover from the exhaustion of this extreme effort?

When undergoing only moderate exercise, muscle cells have enough oxygen to carry out the respiration process aerobically. During strenuous exercise, however, the blood cannot supply oxygen to the muscles fast enough, and the muscle cells must rely on a backup system—the production of energy by glycolysis, or lactic acid fermentation. The lactic acid produced by glycolysis builds up in the muscle cells to the point where it hampers muscle performance, causing muscle fatigue and exhaustion. Such lactic acid buildup produces mild acidosis, which causes nausea, headache, lack of appetite, and impairment of oxygen transport, resulting in the difficult, painful gulping of air experienced by athletes after extreme physical efforts. Muscle cells are slow to recover from this condition. The lactic acid must be removed either by conversion to pyruvic acid in the muscle cells or by movement from these cells to the liver. The liver converts about 75% of the lactic acid to glycogen and the rest to pyruvate, which is oxidized in the citric acid cycle. The heavy breathing that occurs after strenuous exercise helps supply the oxygen necessary to oxidize the lactic acid (Fig. 19.11). A major difference between a well-conditioned athlete and a nonathlete is that the athlete can supply her muscle cells with the oxygen necessary to maintain aerobic respiration for a longer period than can the nonathlete. This means that her muscle cells use glycolysis to supply ATP only at much higher levels of exertion.

19.8 Fermentation

Fermentation is an anaerobic process that occurs in many microorganisms as well as in the muscle cells. The end products of fermentation vary depending on the organism, but in each case electrons are transferred from NADH to pyruvic acid or to a compound formed from pyruvic acid, resulting in the regeneration of NAD^+.

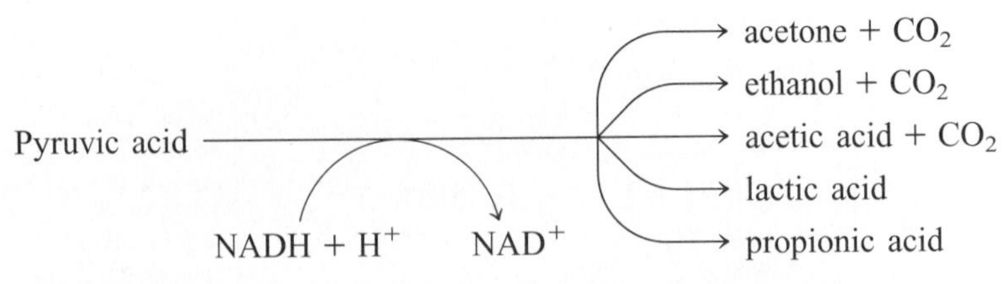

Write the general equation for alcoholic fermentation.

The most familiar type of fermentation, **alcoholic fermentation,** occurs in yeast cells. In alcoholic fermentation, yeast cells convert pyruvic acid to ethanol and carbon dioxide. Alcoholic beverages differ in the raw materials that supply the yeast with the sugars for fermentation: grapes are the raw material for wine, and malted cereals such as barley are the raw material for beers.

19.9 Lipid Metabolism

Carbohydrates, fats, and proteins that we consume in excess of our energy requirements are oxidized to acetyl-CoA, which is readily converted to fatty acids. These fatty acids (with other fatty acids from the diet) are stored as triacylglycerols in the fat cells of adipose tissue under the skin and around major organs. The synthesis of lipids is called **lipogenesis.** Storage fat serves several functions: it is an energy reserve, a support and a shock absorber for inner organs, and heat insulation for the body. Although the glycogen reserves in the liver and muscles are sufficient to last only a few hours, a 70-kg male has enough stored fat to sustain him for 70 days if water is available.

The lipids in adipose tissue are in dynamic equilibrium with the lipids in the blood. That is, stored fatty acids are constantly being exchanged with food fatty acids. The particular composition of the storage fat differs for different organisms, but in animals it can be altered by controlling the diet. Experiments are currently being carried out to determine if special diets for cattle will result in their muscle tissue having a higher concentration of unsaturated fats relative to saturated fats.

19.10 Oxidation of Fatty Acids

When oxidized, fats produce much more energy than carbohydrates or proteins: fats yield 9 kcal/g, whereas glycogen, starch, and proteins yield about 4 kcal/g. In vertebrates, oxidation of fatty acids provides at least half the energy needed by the liver, kidneys, heart, and the skeletal muscles at rest. In hibernating animals and migrating birds, fat is the only energy source.

The process of fat oxidation starts when fats stored in the adipose tissues are hydrolyzed to fatty acids and glycerol. These compounds are then carried to energy-requiring tissues where they are oxidized. The glycerol enters the glycolysis pathway, as discussed in Section 19.4. The fatty acids are oxidized, mainly in the mitochondria of the liver, heart, and skeletal muscles through a repeating series of reactions called the **fatty acid cycle,** or **β-oxidation.** This series of reactions is shown in Figure 19.12. We can briefly describe each step in this process as follows:

Describe beta oxidation of a fatty acid.

1. The first step involves joining the fatty acid with coenzyme A (CoA). This requires one ATP and produces AMP and a diphosphate (or pyrophosphate, denoted PP_i).
2. The second step is a dehydrogenation reaction that produces a double bond between the alpha and beta carbon. The hydrogens that are removed in this reaction are attached to the hydrogen carrier molecule FAD, producing $FADH_2$.

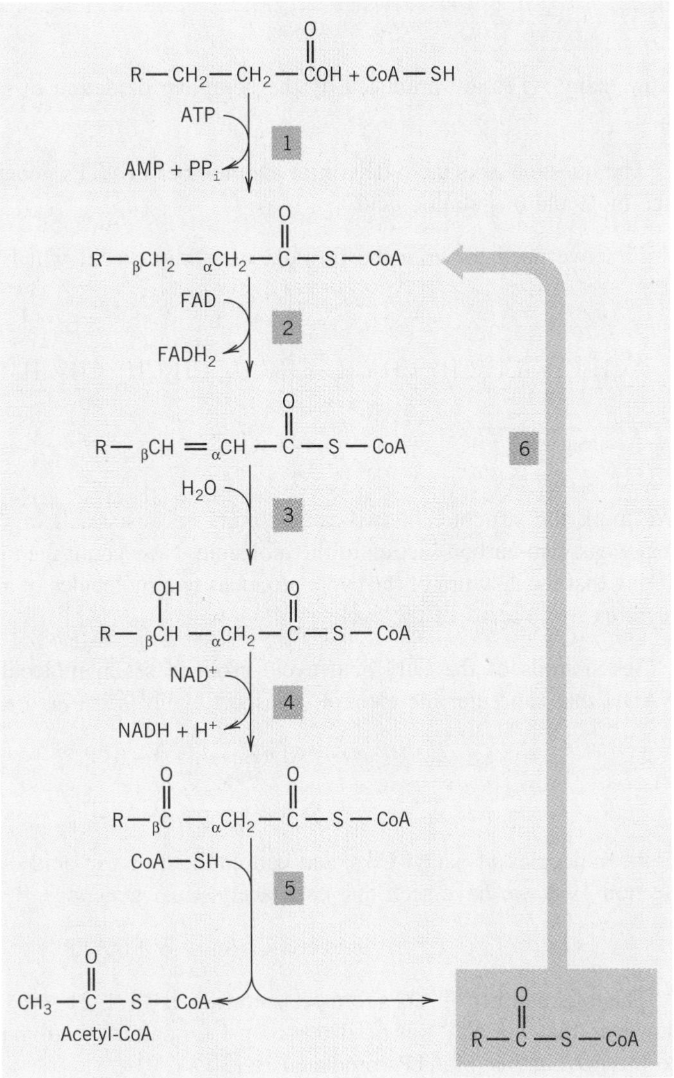

$$R-CH_2-CH_2-\overset{O}{\overset{\|}{C}}OH + CoA-SH$$

ATP → AMP + PP_i [1]

$$R-{}_{\beta}CH_2-{}_{\alpha}CH_2-\overset{O}{\overset{\|}{C}}-S-CoA$$

FAD → $FADH_2$ [2]

$$R-{}_{\beta}CH={}_{\alpha}CH-\overset{O}{\overset{\|}{C}}-S-CoA$$

H_2O [3]

$$R-{}_{\beta}\overset{OH}{\overset{|}{C}}H-{}_{\alpha}CH_2-\overset{O}{\overset{\|}{C}}-S-CoA$$

NAD^+ → NADH + H^+ [4]

$$R-{}_{\beta}\overset{O}{\overset{\|}{C}}-{}_{\alpha}CH_2-\overset{O}{\overset{\|}{C}}-S-CoA$$

CoA—SH [5]

$$CH_3-\overset{O}{\overset{\|}{C}}-S-CoA \qquad R-\overset{O}{\overset{\|}{C}}-S-CoA$$

Acetyl-CoA

[6]

Figure 19.12 The reactions of the fatty acid cycle.

3. In the next step water is added to the double bond, forming an alcohol group on the beta carbon.
4. The alcohol group on the beta carbon is then oxidized (from which comes the name *β oxidation*), producing NADH and H^+.
5. In the last step, coenzyme A breaks the bond between the alpha and beta carbons. This produces one acetyl-CoA and a new fatty acid (having two fewer carbon atoms) joined with coenzyme A.
6. The new fatty acid again enters the fatty acid cycle at Step 2.

The cycle continues to remove two-carbon units from the fatty acid until it has been completely oxidized. Each turn of the cycle produces one $FADH_2$ and one NADH that can enter the respiratory chain, along with one molecule of acetyl-CoA that can be used in the citric acid cycle for the production of ATP. Overall, about 50% of the energy released in the complete oxidation of a fatty acid is trapped in molecules of ATP.

Example 19-1

How many ATPs are produced by the complete oxidation of one molecule of palmitic acid?

See the Question

The question asks us to determine the number of ATPs generated by the oxidation of one molecule of palmitic acid.

Think It Through

First we need to write the formula of palmitic acid which is given in Table 16.2.

$$\underbrace{\underbrace{CH_3CH_2}\,\underbrace{CH_2CH_2}}_{7}\,\underbrace{CH_2CH_2}_{6}\,\underbrace{CH_2CH_2}_{5}\,\underbrace{CH_2CH_2}_{4}\,\underbrace{CH_2CH_2}_{3}\,\underbrace{CH_2CH_2}_{2}\,\underbrace{CH_2\overset{\overset{O}{\|}}{C}OH}_{1}$$

We mark the structure in two-carbon units because each turn of the fatty acid cycle removes a two-carbon section of the molecule. Now count the turns of the cycle, remembering that the last turn of the cycle produces two molecules of acetyl-CoA. Palmitic acid requires seven turns of the cycle.

Execute the Math

Seven turns of the fatty acid cycle produce seven molecules each of $FADH_2$ and NADH that can enter the electron transport chain (see Fig. 19.9).

$$7\ FADH_2 \longrightarrow 14\ ATP$$

$$7\ NADH \longrightarrow 21\ ATP$$

Eight molecules of acetyl-CoA that can enter the citric acid cycle are produced. From Section 19.6, we have seen that each acetyl-CoA generates 12 ATP. Therefore,

$$8\ \text{acetyl-CoA} \longrightarrow 96\ ATP$$

Prepare the Answer

The total number of ATPs formed, therefore, is $14 + 21 + 96 = 131$ ATP. Remember, however, that one ATP was needed at Step 1 for palmitic acid to enter the fatty acid cycle, so the net number of ATPs produced is 130.

Check Your Understanding

How many ATPs are produced in the complete oxidation of one molecule of myristic acid?

19.11 Metabolic Uses of Acetyl Coenzyme A

List six metabolic uses of acetyl-CoA.

Acetyl-CoA plays a central role in the metabolism of nutrients in the cell. Besides entering the citric acid cycle, acetyl-CoA can be used in the biosynthesis of compounds required by the cell (such as cholesterol, steroids, and long-chain fatty acids) or, when present in excess, can be used to synthesize ketone bodies. Figure 19.13 summarizes the key role played by acetyl-CoA in metabolism.

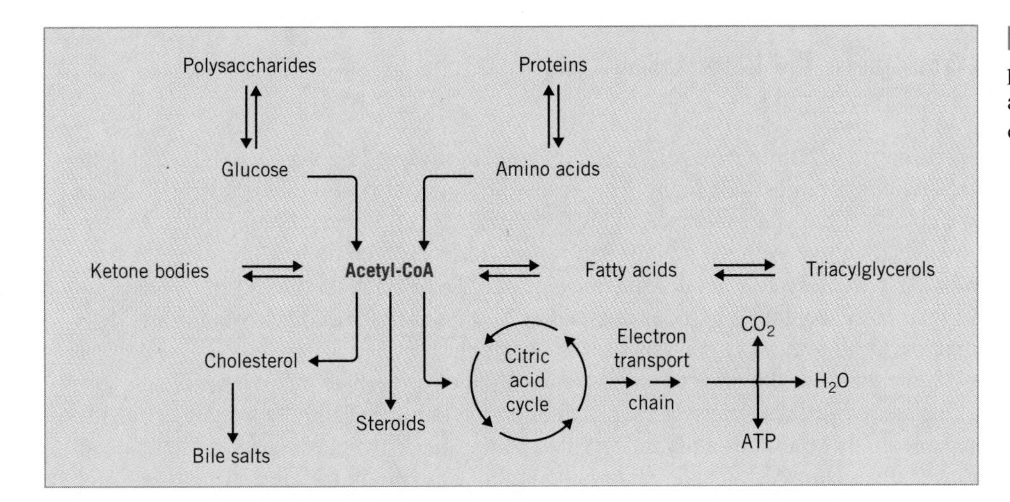

Figure 19.13 Acetyl CoA plays a central role in the anabolic and catabolic reactions of cellular metabolism.

19.12 Ketone Bodies

◆ **Describe the conditions that result in ketosis.**

Excess acetyl-CoA, produced by the oxidation of fatty acids, can be converted by the liver and the kidneys to acetoacetic acid (in the form of the negative ion, acetoacetate), acetone, and β-hydroxybutyric acid (in the form of its negative ion, β-hydroxybutyrate).

Fatty acids → (ATP)

Ketone bodies

$$\underset{\text{acetyl-CoA}}{CH_3\overset{\overset{O}{\|}}{C}-S-CoA} \rightarrow\rightarrow\rightarrow \underset{\text{acetoacetate}}{CH_3-\overset{\overset{O}{\|}}{C}-CH_2-\overset{\overset{O}{\|}}{C}-O^-}$$

$$\underset{\text{acetone}}{CH_3-\overset{\overset{O}{\|}}{C}-CH_3} \qquad \underset{\beta\text{-hydroxybutyrate}}{CH_3-\overset{\overset{OH}{|}}{CH}-CH_2-\overset{\overset{O}{\|}}{C}-O^-}$$

These compounds are called **ketone bodies** (even though β-hydroxybutyrate is not a ketone). Ketone bodies normally circulate in the blood in low concentrations, about 1 mg/100 mL. They are used by skeletal and heart muscles to produce ATP.

Any disruption of normal metabolism (such as liver damage, diabetes mellitus, starvation, or extreme dieting) that causes a restriction or decrease in glucose metabolism increases the rate of fat metabolism. This, in turn, increases the production of ketone bodies. If the level of ketone bodies in the blood exceeds the amount that can be used by the tissues, the ketone bodies accumulate in the blood and are excreted in the urine. This condition is called **ketosis.** Because two of the ketone bodies are acids, in ketosis the pH of the blood drops and acidosis occurs.

In diabetes, the glucose in the blood is not absorbed into cells but is instead excreted in the urine. To supply the body with energy, therefore, the untreated diabetic metabolizes large amounts of fats, resulting in a high production of ketone bodies. Acetone can be smelled on the breath, and acetoacetic acid and β-hydroxybutyric acid are found in high concentrations in the blood and urine. These conditions result in severe acidosis, producing nausea, dehydration, depression of the central nervous system, and, in extreme cases, coma and death.

Health Perspective / Crash Diets: The Never-Ending Cycle

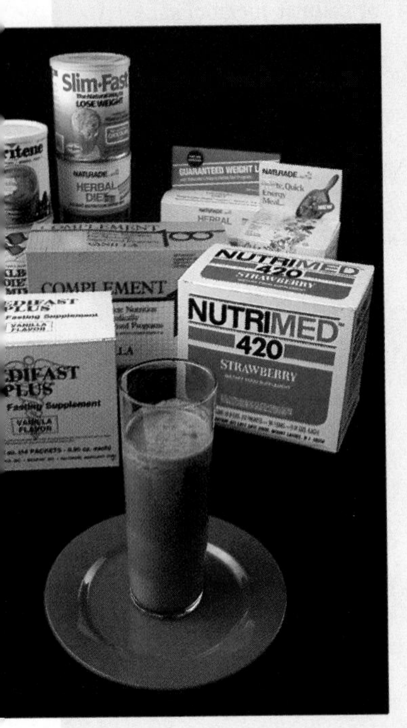

Mark Twain is said to have told a friend that quitting smoking was easy—he had done it hundreds of times! For many Americans the same thing could be said about losing weight. Over the years a person might lose hundreds of pounds through quick weight-loss diets, only to gain the pounds back again. In fact, many times a person gains back more than she lost. It is not unusual for repeated use of crash diets to result in a net gain in total body weight. Why does this happen? What is it about quick weight-loss diets that works to prevent a permanent loss of weight?

Some diets use diuretics to increase the output of urine. The initial large weight loss associated with these diets is simply due to dehydration. This lost weight is quickly regained when the body rehydrates. Most of the quick weight-loss diets eliminate or drastically reduce the intake of carbohydrates. Carbohydrates are the body's major source of glucose, which is required constantly by body cells as their source of "fuel." If the diet does not contain an adequate supply of glucose, the body must turn to its glycogen reserves as its source of glucose. Within 24 hours the glycogen supply in the liver is almost exhausted. Because glycogen molecules associate with many water molecules through hydrogen bonding, loss of the body's glycogen is accompanied by an even greater loss of water. For every pound of glycogen that is lost, 3 lb of water also eliminated. This is why, on the first day or two of a low-carbohydrate diet, you always lose 3 to 4 lb. After the body's glycogen supply is depleted, cells then turn to fatty acids from the body's stored fat to supply their energy needs. Brain cells, however, cannot use fatty acids as a source of fuel; they still need glucose. To supply glucose to the brain cells, the body must break down protein tissue. Any diet that greatly restricts carbohydrate intake is going to result in the loss of muscle tissues as well as stored fat.

It is important that dieters exercise to maintain and build their muscle tissue. Otherwise, any weight they regain will occur in the form of fat. This means they can end up with a higher percentage of body fat than when they began the diet. Because fatty tissue has a lower metabolic activity than muscle tissue, it requires less glucose to supply its energy needs. These unsuccessful dieters, therefore, have an overall reduction in their metabolic rate. If they return to their old eating habits, their lowered metabolic rate causes them to gain rather than maintain their weight. There is only one truly effective way to lose weight and keep it off: reduce total calories while keeping a balance of carbohydrate, protein, and fat, and exercise regularly. This results in a net gain in muscle tissue and a corresponding increase in metabolism. People with more muscle tissue tend to burn more body fat. So a regular exercise program actually allows them to eat more than they once did without gaining weight.

Your Perspective: Imagine that you are a nutritional consultant, and a local health club has asked you to design a program for their members that will allow them to lose 8 to 10 pounds over the course of six weeks. Describe how to run such a program, explaining the nutritional reasons behind your recommendations.

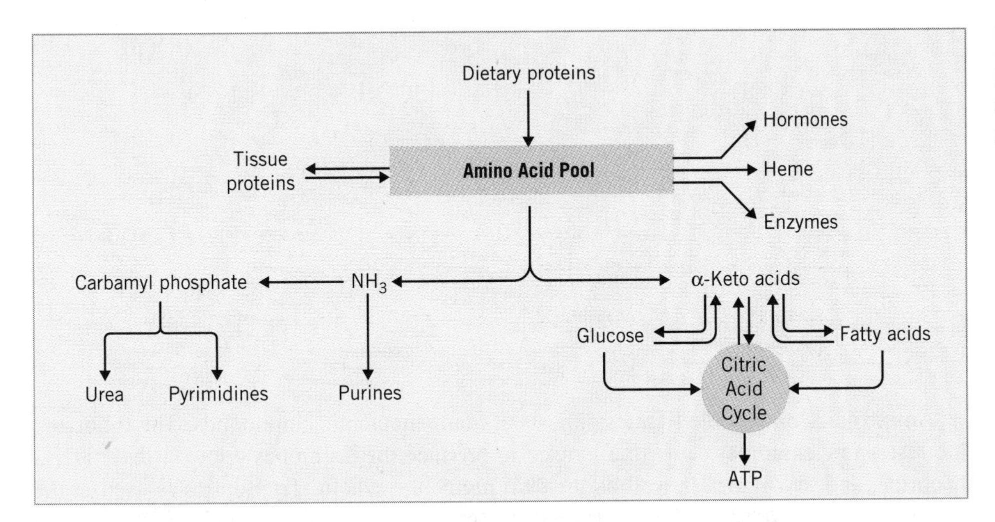

Figure 19.14 The amino acid pool in the human body is constantly changing as proteins are used for various functions.

19.13 Protein Metabolism

List four metabolic uses of amino acids.

The amino acids absorbed into the blood from the intestines have many uses in the body. They are our main dietary source of nitrogen, an element essential to life. Cells use amino acids to synthesize tissue protein in the formation of new cells or in the repair of old cells. The largest amounts of protein are required during periods of rapid growth (such as during infancy, adolescence, pregnancy, or when a mother is nursing) and during periods of extensive repair (such as after surgery, burns, hemorrhage, or infections). Certain amino acids are also used in the synthesis of nonessential amino acids, enzymes, hormones, antibodies, and nonprotein nitrogen-containing compounds, such as nucleic acids or heme groups (Fig. 19.14).

There is no storage form of amino acids in the body, as there is for carbohydrates (glycogen) and lipids (fat). Instead, the body maintains a constantly changing pool of amino acids as tissue protein is continually broken down and resynthesized. The turnover of amino acids in the liver and blood is relatively rapid—one-half is replaced every 6 days. This turnover is much slower in muscles and supportive tissue, where one-half of the amino acids are replaced every 180 days in muscles and every 1000 days in the collagen of supportive tissues.

The amino acids not used in such synthesis can be catabolized, or broken down, for energy. When this occurs, the amino group is removed from the amino acid through a process called oxidative deamination and enters the urea cycle (which we will discuss shortly). The remainder of the molecule—called the carbon skeleton of the amino acid—can enter the citric acid cycle to supply needed cellular energy, or it can be converted to body fat. This means that it is possible to become overweight by eating too much protein, just as from eating too much lipid or carbohydrate.

Transamination

The major way in which amino groups are removed from amino acids is through a process called **transamination.** Transamination involves the transfer of an amino group from an amino acid to an α-keto acid, thereby producing a new amino acid. (An α-keto acid is an acid containing a ketone functional group on the alpha carbon, the carbon next the carboxyl group.)

Write the general equation for the process of transamination.

$$\underset{\substack{\text{alanine}\\\text{(amino acid)}}}{H_2N-\overset{\displaystyle COOH}{\underset{\displaystyle CH_3}{C}}-H} + \underset{\substack{\alpha\text{-ketoglutaric acid}\\(\alpha\text{-keto acid})}}{\begin{matrix}COOH\\|\\C=O\\|\\CH_2\\|\\CH_2\\|\\COOH\end{matrix}} \rightleftharpoons \underset{\text{pyruvic acid}}{\begin{matrix}COOH\\|\\C=O\\|\\CH_3\end{matrix}} + \underset{\substack{\text{glutamic acid}\\\text{(new amino acid)}}}{\begin{matrix}COOH\\|\\H_2N-C-H\\|\\CH_2\\|\\CH_2\\|\\COOH\end{matrix}}$$

Amino acids play a role in the synthesis of many metabolic compounds. The following are just a few examples. Tyrosine is used to produce the hormones epinephrine, norepinephrine, and thyroxine, as well as the skin pigment melanin. Tryptophan is used in the synthesis of the neurotransmitter serotonin and the coenzymes NAD^+ and $NADP^+$. Serine is converted to ethanolamine, which is found in lipids, and cysteine, which is used in the synthesis of bile salts.

Oxidative Deamination

Amino acids (mainly glutamic acid produced through transamination) can be converted in the liver to ammonia, carbon dioxide, water, and energy. **Oxidative deamination** is the process by which the amino group is removed, in the form of ammonia, from an amino acid.

Write the general equation for oxidative deamination.

$$HO\overset{\displaystyle O}{\overset{\|}{C}}-\overset{\displaystyle R}{\underset{\displaystyle H}{C}}-\underset{\displaystyle H}{N}-H \xrightarrow[NAD^+ \;\; NADH + H^+]{} HO\overset{\displaystyle O}{\overset{\|}{C}}-\overset{\displaystyle R}{C}=NH \xrightarrow{H_2O} \underset{\alpha\text{-keto acid}}{HO\overset{\displaystyle O}{\overset{\|}{C}}-\overset{\displaystyle R}{C}=O} + \underset{\text{ammonia}}{NH_3}$$

The α-keto acids that are formed in oxidative deamination can be used in several ways. They may be oxidized in the citric acid cycle to produce energy, or they may be converted to other amino acids through transamination. The α-keto acids also may be used in the synthesis of carbohydrates and fats (see Fig. 19.14).

19.14 Urea Cycle

The ammonia that is formed in oxidative deamination is toxic to cells; a blood concentration of 5 mg/100 mL is toxic to humans. The liver disposes of the ammonia by converting it to the less toxic compound urea in a cyclic reaction called the **urea cycle,** or **Krebs ornithine cycle.** Ammonia enters the cycle as carbamyl phosphate and is joined to the amino acid ornithine. Each turn of the cycle produces one molecule of urea and one molecule of ornithine to begin the next turn of the cycle (Fig. 19.15). The net equation for the reaction is

Write the net equation for the reactions of the urea cycle.

$$2NH_3 + CO_2 \longrightarrow \underset{\text{urea}}{H_2N-\overset{\displaystyle O}{\overset{\|}{C}}-NH_2} + H_2O$$

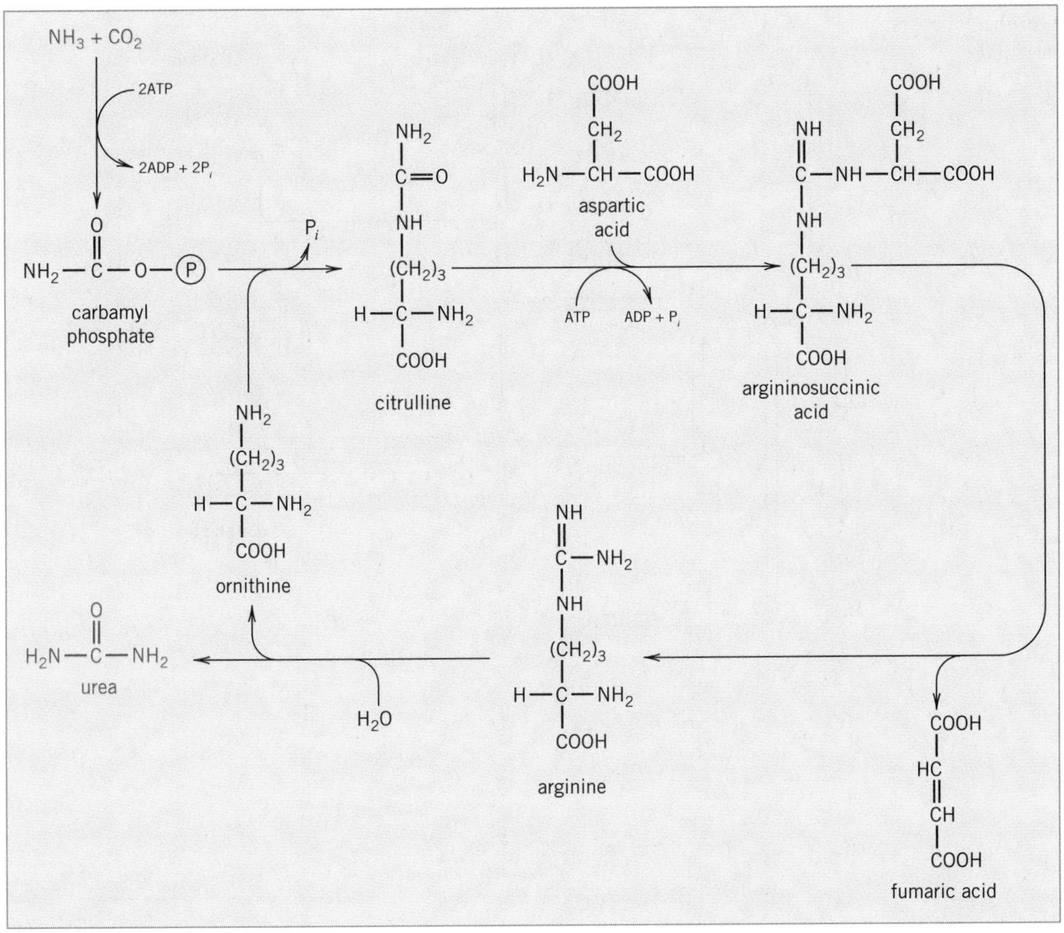

Figure 19.15 The urea cycle.

The urea produced in the liver can be safely transported by the blood to the kidneys, where it is eliminated in the urine. Any condition that impairs the elimination of urea by the kidneys can lead to uremia, a buildup of urea and other nitrogen wastes in the blood, which can be fatal. A person suffering from uremia feels nauseated, irritable, and drowsy. His blood pressure is elevated, and he may be anemic. He may experience hallucinations and in serious cases may lapse into convulsions and coma. To reverse this condition, either the cause of kidney failure must be removed or the patient must undergo hemodialysis to remove the nitrogen wastes from the blood.

19.15 Pathways of Metabolism

Describe how fasting affects carbohydrate, lipid, and protein metabolism.

We have discussed the metabolism of carbohydrates, lipids, and proteins separately in this chapter, but these pathways are all intricately interrelated, as shown in Figure 19.16. To illustrate the relationships among these pathways, let's consider the case of a college freshman who, when she entered college, was 5 ft 2 in. tall and weighed 167 lb. After

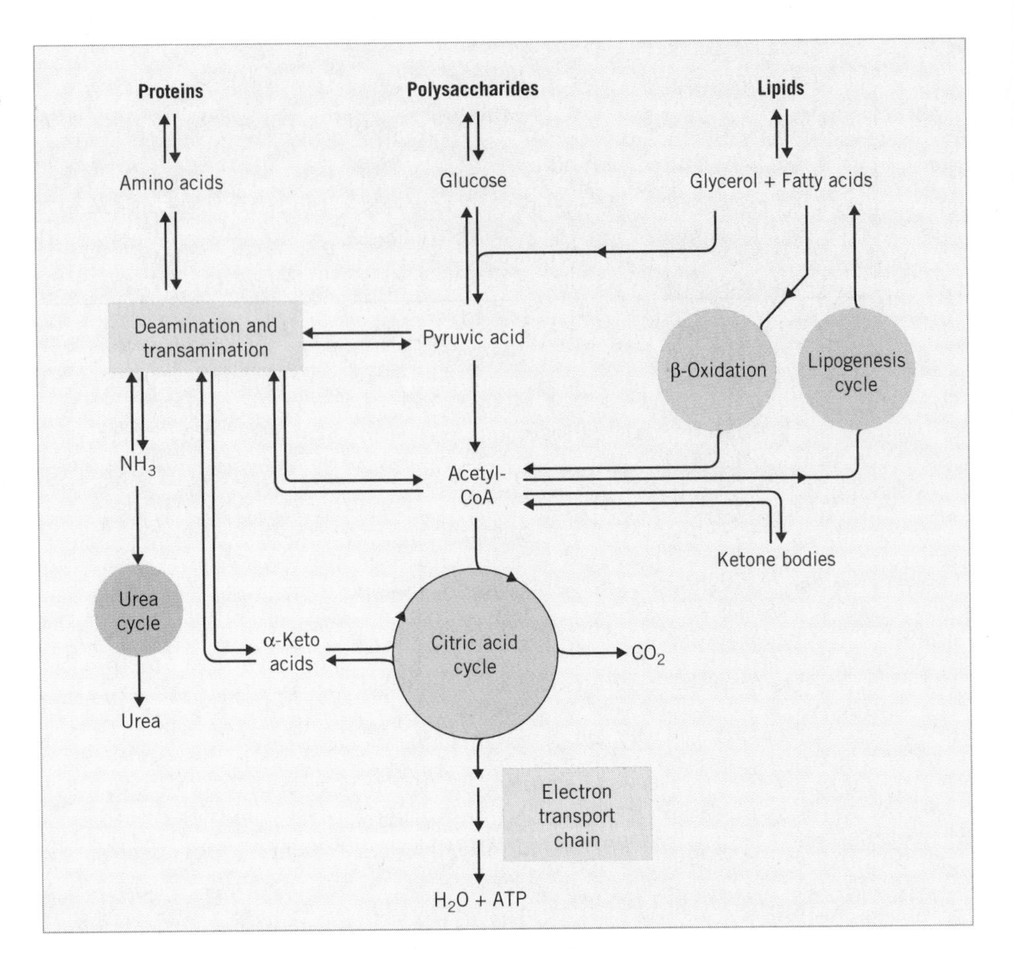

Figure 19.16 Summary of the pathways of carbohydrate, lipid, and protein metabolism.

about a month of dormitory living, she decided she would have to lose weight. To do it as quickly as possible, she began a fasting diet of water, black coffee, and two multiple vitamin pills daily. After several days on this diet, she became listless and stopped going to class. On the evening of the eighth day she collapsed while taking a shower and had to be rushed to the hospital. An examination at the hospital revealed the odor of acetone on her breath, a urinalysis that was positive for acetone but not glucose, and a urine pH of 5.5. Her blood glucose was 60 mg/100 mL. What was the cause of this sudden metabolic crisis?

Under normal conditions, the body tissues of a 70-kg person contain energy reserves as follows: 80 kcal of glucose in the blood, 280 kcal of liver glycogen, 480 kcal of muscle glycogen, 14,000 kcal of adipose fat, and 24,000 kcal of muscle protein. These body reserves become the energy source for a fasting individual. For a short while, the body can maintain a metabolic balance by using the circulating glucose. As the blood glucose reaches fasting level, the pancreas stops secreting insulin and begins to secrete glucagon. This stimulates glycogenolysis, which releases glucose from the glycogen reserves in the liver and muscles. When these glycogen reserves become depleted, the body must then find other sources of energy and blood glucose (remember, the brain requires glucose for normal function and must have 110 to 145 g of glucose a day). The body next begins to break down muscle and tissue proteins. The amino acids produced in this tissue breakdown are deaminated in the liver and are then used to synthesize the needed glucose, to

Health Perspective / Anorexia and Bulimia

From the time we first watch television or look at magazines we are flooded with pictures of the ''most desirable'' body type: slender and lithe. But only a few women or men are genetically endowed with such bodies. Unfortunately, the desire to achieve this type of body is driving more and more young women (and men) to crash diets or to the more serious conditions of anorexia and bulimia.

People with anorexia develop a deep fear of becoming fat, and develop an obsession with dieting. For these people, the image of their own body becomes distorted. Even though a patient may weigh as little as 80 lb, she still sees herself as fat and continues dieting (or even fasting). Anorexia sufferers also try to lose weight through excessive exercise, self-induced vomiting, or abuse of laxatives, enemas, and diuretics. Bulimia is a related condition that is marked by a tendency to binge eating (often as many as 15,000 kcal in a few hours), followed by purging (emptying the gastrointestinal tract with vomiting and diarrhea). The exact causes of bulimia and anorexia remain unclear, but most clinicians believe these conditions have social, psychological, and metabolic causes.

Both anorexia and bulimia, with their combination of starving and purging, can produce severe medical complications. Because so many cases involve adolescents, growth and development can be disturbed. Menstrual cycles disappear. The starving can cause severe metabolic imbalances, including acidosis and fatty degeneration of the liver. Purging causes a loss of potassium and chloride ions. The resulting low levels of ions in the blood can affect the heart and can be life-threatening in some instances. Continual vomiting of stomach acid can damage the teeth and can cause enlargement of the salivary glands.

Treatment of these conditions is a real problem. The patients tend to be secretive about their behavior, reluctant to change, and not persuaded by information and logic. Relapses are common. Successful treatment usually involves a long period of hospitalization and the use of powerful psychiatric medications in addition to counseling.

Your Perspective: For many people, some common foods can take on special meanings. There may be some foods that are eaten only at certain times or with certain people. Or there may be some foods that are avoided for certain personal reasons having nothing to do with the food's taste or appearance. Describe a food that has some special positive or negative meaning for you, and explain the reason why.

produce ATP, and to synthesize other compounds needed for the maintenance of the citric acid cycle. The large amounts of ammonia produced by this type of deamination are excreted in the urine in the form of the ammonium ion, NH_4^+ rather than as urea.

When insufficient glucose is available, the body uses fatty acids to provide the tissues with energy. This results in such a high level of fatty acid oxidation that the citric acid cycle cannot handle the large amounts of acetyl-CoA produced. The liver cells, therefore, convert the excess acetyl-CoA to ketone bodies. Two ketone bodies, acetoacetic acid and β-hydroxybutyric acid, are acids. A large increase of these ketone bodies places a strain on the buffering capacity of the blood, finally exceeding it and creating acidosis. After several days of fasting, the production of ketone bodies becomes so high that they are excreted in the urine producing further electrolyte imbalances.

You've probably gotten the idea by now that fasting is not a particularly effective means of weight reduction! The chemical imbalances produced by fasting slow down the basal metabolism rate, disrupting a person's concentration and causing physical movements to become sluggish and uncoordinated. Fasting puts a great strain on the liver (from increased fatty acid metabolism) and the kidneys (from the excretion of the waste products of amino acid and fatty acid metabolism). The acidosis and electrolyte imbalances from prolonged fasting can produce coma and death. It is important even when dieting, therefore, to eat sufficient amounts of complex carbohydrates to prevent the breakdown of tissue proteins and the formation of ketone bodies.

KEY CONCEPTS

The digestion of food begins in the mouth, where food is chewed into smaller pieces and mixed with saliva. The enzymes in saliva begin the digestion of starch. The food then passes to the stomach, where it is further reduced in size and where protein digestion begins. Pancreatic and intestinal juices (containing digestive enzymes) and bile are secreted as the food enters the small intestine. Fats are emulsified by the bile and hydrolyzed by enzymes to fatty acids, glycerol, and monoacylglycerols and diacylglycerols. These compounds are absorbed, reformed, and carried by the lymph system to the bloodstream. Also in the small intestines, proteins are hydrolyzed to amino acids, and carbohydrates to monosaccharides. These are absorbed directly into the bloodstream and carried to the liver. The liver regulates the concentration of these components in the blood.

The absorbed glucose may be converted by the liver to glycogen (through glycogenesis), may be used by cells to synthesize other compounds, or may be oxidized by the cell to produce energy through the series of reactions called cellular respiration. Cellular respiration is a two-step process, with an anaerobic stage called glycolysis, in which glucose is converted to lactic acid, and an aerobic stage, which involves the reactions of the citric acid cycle and the electron transport chain. This second stage produces most of the energy-rich molecules (ATP) produced from the oxidation of glucose. When there is not enough oxygen, cells use glycolysis to produce ATP. The lactic acid synthesized in glycolysis, however, builds up in the tissues to produce mild acidosis. Microorganisms such as yeast oxidize glucose under anaerobic conditions, but the products of this fermentation process are ethanol and carbon dioxide, rather than lactic acid.

Fats are the main energy reserve of the body, producing 9 kcal/g when oxidized. Lipids that are eaten can be stored in the fat cells of adipose tissue or can be hydrolyzed to glycerol (which enters the glycolysis pathway) and fatty acids (which are oxidized to acetyl-CoA in the fatty acid cycle). The acetyl-CoA that is produced may enter the citric acid cycle or may be used in biosynthesis. The liver converts excess acetyl-CoA to ketone bodies. In high concentrations ketone bodies cause acidosis and metabolic disruption.

The amino acids absorbed from the small intestine can be used to synthesize tissue proteins, nonessential amino acids, enzymes, hormones, antibodies, and nucleic acids. Amino acids are synthesized in the cell from other amino acids and from α-keto acids by transamination. Amino acids are catabolized by first removing the amino group in a process called oxidative deamination. The urea cycle converts the ammonia produced by oxidative deamination into urea. The α-keto acids that are also produced may enter the citric acid cycle or go to other biosynthetic pathways.

Maintenance of blood glucose levels is critical to the normal functioning of tissues, especially the brain. Diet and the hormones insulin, glucagon, and epinephrine all affect blood glucose concentrations. Hyperglycemia, or too much glucose in the blood, results from diabetes mellitus, emotional stress, kidney failure, or the administration of certain drugs. Hypoglycemia, or too little glucose in the blood, can occur with an overdose or overproduction of insulin or with starvation or fasting. When the blood glucose level is critically low, the body compensates by metabolizing proteins to synthesize glucose and by oxidizing fatty acids to produce energy. High levels of fatty acid oxidation lead to the formation of ketone bodies, which can lower blood pH and cause electrolyte imbalances.

Important Equations

Glycogenesis

$$n\text{Glucose} + n\text{UTP} + n\text{ATP} \longrightarrow \text{Glycogen} + n\text{UDP} + n\text{ADP} + 2n\text{P}_i$$

Glycogenolysis

$$\text{Glycogen} + (n-1)\text{H}_2\text{O} \longrightarrow n\text{Glucose}$$

Glycolysis

$$\underset{\text{glucose}}{\text{C}_6\text{H}_{12}\text{O}_6} + 2\text{ADP} + 2\text{P}_i \longrightarrow \underset{\text{lactic acid}}{2\text{C}_3\text{H}_6\text{O}_3} + 2\text{ATP} + 2\text{H}_2\text{O}$$

Alcoholic Fermentation

$$\underset{\text{glucose}}{\text{C}_6\text{H}_{12}\text{O}_6} + 2\text{ADP} + 2\text{P}_i \longrightarrow \underset{\text{ethanol}}{2\text{CH}_3\text{CH}_2\text{OH}} + 2\text{CO}_2 + 2\text{ATP}$$

Beta Oxidation

$$\underset{\text{fatty acid}}{\text{RCH}_2\text{CH}_2\text{—}\overset{\overset{\text{O}}{\|}}{\text{C}}\text{—OH}} + 2\text{CoA—SH} + \text{FAD} + \text{NAD}^+ \longrightarrow$$

$$\text{R—}\overset{\overset{\text{O}}{\|}}{\text{C}}\text{—S—CoA} + \underset{\text{acetyl-CoA}}{\text{CH}_3\overset{\overset{\text{O}}{\|}}{\text{C}}\text{—S—CoA}} + \text{FADH}_2 + \text{NADH} + \text{H}^+$$

Transamination

$$\underset{\text{amino acid}}{\text{H}_2\text{N—C(COOH)(R)—H}} + \underset{\alpha\text{-keto acid}}{\text{HOOC—C(=O)—CH}_2\text{—CH}_2\text{—COOH}} \longrightarrow \underset{\alpha\text{-keto acid}}{\text{HOOC—C(=O)—R}} + \underset{\substack{\text{new amino acid}\\ \text{(glutamic acid)}}}{\text{H}_2\text{N—CH(COOH)—CH}_2\text{—CH}_2\text{—COOH}}$$

Oxidative Deamination

$$\underset{\text{amino acid}}{H_2N-\overset{\overset{\large COOH}{|}}{\underset{\underset{\large R}{|}}{C}}-H} + NAD^+ + H_2O \longrightarrow \underset{\alpha\text{-keto acid}}{\overset{\overset{\large COOH}{|}}{\underset{\underset{\large R}{|}}{C}}=O} + NH_3 + NADH + H^+$$

Urea Cycle

$$2NH_3 + CO_2 \longrightarrow \underset{\text{urea}}{H_2N-\overset{\overset{\large O}{\|}}{C}-NH_2} + H_2O$$

Review Problems

Section 19.1

1. List three processes for which the body needs energy.
2. Draw the structure of AMP, ADP, and ATP.
3. **(a)** What is meant by a *high-energy* phosphate bond?
   (b) Circle the high-energy phosphate bonds in the structure of ATP.
4. Why are molecules such as ATP so important to the survival of a cell?

Section 19.2

5. What monosaccharides are produced by the digestion of each of the following:
   (a) starch **(b)** lactose **(c)** sucrose
6. What are the components of bile?
7. What role do the bile salts play in the digestion of lipids?
8. What classes of compounds are formed by the digestion of lipids?
9. What classes of compounds are formed by the digestion of proteins?
10. Describe the process by which the carbohydrates, proteins, and fats in a barbecue dinner of steak and a baked potato with margarine are digested and absorbed.

Section 19.3

11. Your brain requires a constant supply of glucose. Explain how your body maintains a fairly constant blood sugar level even though you eat (and thus obtain glucose) only a few times a day.
12. Suggest a diet for a patient who has hypoglycemia.

Section 19.4

13. Examine the sequence of reactions shown in Figure 19.5.
    (a) Write equations showing the two reactions in glycolysis that consume ATP.
    (b) Write equations showing the two reactions in glycolysis that produce ATP.
    (c) Write equations showing the three reactions in glycolysis that involve isomerization.

14. The oxidation of glucose by a cell can be divided into two separate stages: an anaerobic and an aerobic stage.
    (a) What is the name given to the anaerobic stage?
    (b) Write the equation for the overall reaction that occurs in the anaerobic stage.
    (c) Where in the cell does each stage occur?
    (d) Which stage produces more energy?
    (e) Name the two series of reactions that occur in the aerobic stage, and give a general description of the reactions that occur in each.

Section 19.5

15. Under what circumstances does the conversion of pyruvic acid to lactate not occur?
16. What are the products of the citric acid cycle?
17. Write a general summary equation for the citric acid cycle.
18. Examine the sequence of reactions in Figure 19.8. Identify the step or steps that involve each of the following types of reactions:
    (a) dehydration
    (b) oxidation
    (c) hydration
    (d) decarboxylation (loss of CO_2)

Section 19.6

19. In what way is the citric acid chain dependent on the respiratory chain?
20. What is the fate of the hydrogens carried to the respiratory chain by NAD^+ and FAD?
21. Construct a chart that summarizes the steps in the oxidation of glucose by the cell. For each step give the product(s) and the number of NADH, $FADH_2$ and ATP produced.
22. A 150-lb person walking at a fast clip burns approximately 300 kcal/h.
    (a) If all this energy comes from the hydrolysis of ATP (7.3 kcal/mol), how many moles of ATP are needed each hour?
    (b) If the ATP comes from the oxidation of glucose, how many grams of glucose must be burned each hour to provide this energy?

Section 19.7

23. What is the fate of the lactic acid produced in the muscles during anaerobic glycolysis?
24. What symptoms accompany the buildup of lactic acid in muscle cells?

Section 19.8

25. What products are produced during glycolysis in yeast cells?
26. How does glycolysis differ in yeast cells and animal cells? Use equations to support your answer.

Section 19.9

27. What is the fate of carbohydrates, proteins, and fats that we consume in excess of our energy requirements?
28. Stored fat serves several functions in the body; list three of these uses.

Section 19.10

29. Describe the steps involved in the oxidation of fats in the body.

30. How is a fatty acid activated before it can be metabolized in the mitochondria?

31. **(a)** What happens to the acetyl-CoA that is formed in the oxidation of a fatty acid?

(b) Why is this process referred to as β-oxidation?

32. How many ATPs are produced in the complete oxidation of one molecule of each of the following? Explain your answers.

(a) lauric acid **(b)** stearic acid **(c)** arachidic acid

33. Why is the fatty acid cycle shown in Figure 19.12 often referred to as the "fatty acid spiral"?

34. The glycerol formed during the hydrolysis of a fat enters the glycolysis pathway as glyceraldehyde-3-phosphate. Write the series of reactions that occur as it is converted to pyruvate.

Section 19.12

35. Give three examples of the importance of acetyl-CoA in the anabolic and catabolic reactions of metabolism.

36. **(a)** What is ketosis?

(b) What metabolic changes cause ketosis?

(c) What are the effects of ketosis on the body?

Section 19.13

37. Describe four possible metabolic uses of the amino acids contained in a hamburger.

38. Write the equation for the transamination reaction between threonine and pyruvic acid.

39. Write the equation for the deamination of alanine.

40. What is the fate of the α-keto acids produced in oxidative deamination of amino acids?

Section 19.14

41. Describe in general terms the metabolic fate of the ammonia produced in problem 39.

42. What is the name given to the condition in which urea and other nitrogen wastes build up in the blood? What are the symptoms of this condition?

Section 19.15

43. In the metabolism of lipids, proteins, and carbohydrates shown in Figure 19.16, what coenzyme is common to all?

44. Which body reserves are used as an energy source first for a fasting individual?

45. What reserves does the body use to provide tissues with energy when insufficient glucose is available?

46. What is the fate of the excess acetyl-CoA produced during high levels of fatty acid oxidation?

Applied Problems

47. **(a)** While muscles are at rest, what is their major source of energy?
(b) What is the major source of energy for muscles that are active?
(c) After an all-day bike ride, you wake up the next morning with very sore muscles in your legs. Why?

48. When comparing fats and carbohydrates as energy sources, we find that fats produce much more energy: 9 kcal/g compound with 4 kcal/g. Why are fats so much higher in energy?

49. A rare inherited disease called *maple syrup urine disease* results when a person lacks the ability to metabolize the α-keto acids that result from the transamination of valine, leucine, and isoleucine.
(a) Write the structures of the α-keto acids produced in the transamination of valine, leucine, and isoleucine.
(b) Suggest a possible reason for the name of this disease.

50. A 70-kg person can survive without food for about 70 days, but not without water. What processes that are important in the metabolism of a fasting individual require water?

51. **(a)** Why must insulin be given to diabetics intravenously rather than in the form of a pill?
(b) What is *insulin shock?*
(c) How does it occur?
(d) Explain why severe hyperinsulinism is more dangerous to a patient than the lack of insulin over a short period.

52. The glucose tolerance test is useful for diagnosing diabetes mellitus. In this test, a person who has fasted for 8 hours is given a drink containing 50 to 100 g of glucose in solution. The patient's blood sugar level is measured over the next several hours. The data for two different patients are shown here. Which patient may be suffering from diabetes mellitus? State the reason for your choice.

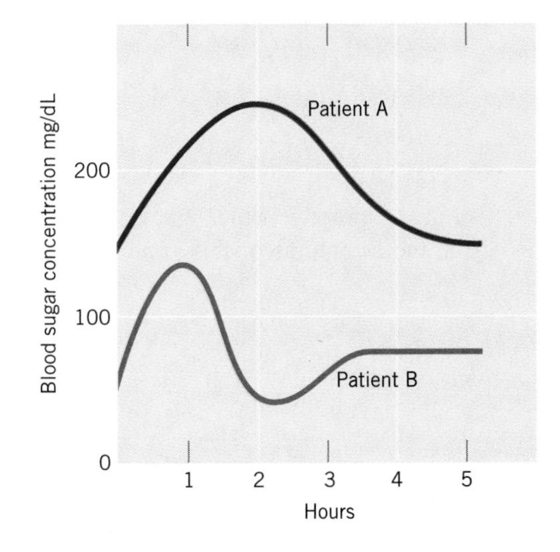

Chapter 20

Nucleic Acids

Virginia Smith watched her son George walk slowly down the street to the school bus stop. She was puzzled and concerned. He'd been complaining on and off for a year about pains in his joints and abdomen. Twice he'd run a low fever and suffered from vomiting, and she had kept him home from school. He was the youngest of her four children, and the only sickly one among them.

Early that afternoon Mrs. Smith received a call from the school nurse, who had just taken George to the hospital. George had collapsed during a hard game of soccer, screaming about a pain in his stomach. When Mrs. Smith reached the hospital, she found George under sedation and receiving liquids intravenously. The doctor first thought that George might have appendicitis, but after learning of George's symptoms over the last year, he decided to run several blood tests. As he had suspected, the doctor found that George had sickle cell anemia and was suffering from a sickle cell crisis. The pain in George's abdomen and the swelling in his joints slowly went away, and a week later he was able to return to school.

A few weeks later George and his mother went for a consultation with doctors at a large clinic specializing in the diagnosis and treatment of sickle cell anemia. There they learned that sickle cell anemia is an inherited disease that affects red blood cells. They were told that long ago a mutation had occurred in the evolution of the African black population, causing an error in the gene for hemoglobin (the protein molecule in red blood cells that carries oxygen to the body). Today, 1 out of every 2 African blacks carries the defective gene, and 1 in 10 American blacks carries the trait. Children who inherit two defective genes—one from each parent—suffer from sickle cell anemia, and almost all their hemoglobin is defective. They experience periodic sickle cell crises, resulting from the clogging of their capillaries by abnormal red blood cells. This causes severe pain, fever, swelling of the joints, and organ damage. Such crises can be set off by an infection, cold weather, trauma, strenuous exercise, or emotional stress. Children having sickle cell anemia are anemic, often jaundiced, susceptible to disease, and often die in childhood.

People with only one defective gene carry the sickle cell trait. About 40% of their hemoglobin is defective, but under normal conditions they show no clinical signs of the disease. Unusual stress, however, can bring on a crisis in these individuals also.

Sickle cell anemia was first described clinically in 1910, but it was not until 1949 that Linus Pauling demonstrated the molecular basis of the disease. He showed that sickle cell hemoglobin (abbreviated HbS) has a different rate of movement in an electric field than does normal hemoglobin (HbA). Hemoglobin (which has a molecular weight of 64,458) is a molecule consisting of a protein part, called globin, and a nonprotein part, called heme. The globin part contains four polypeptide chains: two alpha chains, each containing 141 amino acid units, and two beta chains, each containing 146 amino acid units. Each chain is wrapped around a heme group containing an Fe^{2+} atom; this is the group that actually carries the oxygen (see Fig. 17.8). Each hemoglobin molecule can therefore carry four molecules of oxygen.

In 1956, Vernon Ingram showed that the entire difference between HbS and HbA was in one amino acid on the beta chain. In HbS, a valine molecule is erroneously substituted for a glutamic acid molecule in position 6 on the beta chain.

$$H_3N^+—CH(COO^-)—CH(CH_3)—CH_3$$

valine

$$H_3N^+—CH(COO^-)—CH_2—CH_2—COO^-$$

glutamic acid

At pH 7, the glutamic acid has a negative charge, whereas the valine has no charge. Thus, a molecule of HbA has two more negative charges than a molecule of HbS, accounting for the difference in mobility of the two hemoglobin molecules in an electric field.

At low oxygen concentrations, red blood cells containing HbS take on abnormal shapes, some resembling sickles. In deoxygenated HbS, the nonpolar valine that replaced the polar glutamic acid undergoes hydrophobic bonding with a valine on a neighboring molecule. This causes the HbS molecules to lock together and polymerize into long fibers that distort the cell. The resulting abnormal cell shape usually corrects itself when the blood becomes oxygenated again, but some cells remain irreversibly sickled and are removed from the blood earlier than usual by the liver.

Red blood cells must be flexible, for they have to squeeze through capillaries smaller than they are. Although red blood cells containing HbS sickle when they give up their oxygen, there is a short delay between the moment HbS releases its oxygen to the tissues and the time the cell sickles. Under normal conditions, this delay is long enough for the cell to squeeze through the capillary and enter the vein. Under stress, however, the HbS molecule releases its oxygen sooner than usual. This can cause the rigid, sickled cells to wedge in the capillaries and block the blood flow, depriving the

tissues of further oxygen. Such blockage brings about the clinical symptoms of a sickle cell crisis.

Although people suffering from sickle cell anemia (and carriers of the trait) can be identified through blood screening tests, treatment is limited to supportive care during a crisis. Researchers are currently experimenting with drugs that inhibit the polymerization of HbS and that decrease the concentration of deoxygenated HbS by increasing the affinity of HbS for oxygen. A promising treatment uses the oral anticancer drug hydroxyurea or the food additive butyrate either alone or in combination with other drugs to induce the production of fetal hemoglobin (HbF). HbF, produced during embryonic development, has an enhanced affinity for oxygen. The fetal hemoglobin gene usually slacks off within the first six months of life, and individuals switch over to making the adult form of hemoglobin. HbF produced by adults with sickle cell anemia does not sickle and enhances their ability to transport oxygen. Many treated patients have produced enough fetal hemoglobin to alleviate their symptoms. These treatments are not without side effects, however, and clinical trials are continuing.

20.1 Molecular Basis of Heredity

Your hair color, height, blood type, and ability to metabolize nutrients are all inherited characteristics. These characteristics are the result of certain genes that you inherited from your parents. Genes are specific segments of molecules called DNA. In 1958, Francis Crick proposed a relationship between DNA, RNA, and proteins that he called the ''Central Dogma of Molecular Biology.'' It states that DNA directs the making of its own copy (replication) and the transfer of genetic information from DNA to RNA (transcription). RNA directs the transfer of this genetic information to the amino acid chain of a protein (translation). This dogma holds true in all cells, but recent research has uncovered exceptions to the dogma. In some viruses, for example, RNA can direct the synthesis of DNA and of RNA. To understand how the information carried by a gene directs the formation of a polypeptide chain, we must first become familiar with the molecular nature of the genetic material DNA and RNA.

20.2 Nucleotides

Deoxyribonucleic acid (DNA) and ribonucleic acid (RNA) belong to a class of compounds called **nucleic acids.** Nucleic acids are polymers of monomer units called **nucleotides.** Unlike the monomer units we have studied before, however, nucleotides can be further hydrolyzed to produce three components: a nitrogen-containing base, a five-carbon sugar, and phosphoric acid.

◆ **List the three components of nucleotides.**

The Nitrogen-Containing Bases

Name the sugars and bases found in DNA and RNA.

Two classes of nitrogen-containing bases are found in nucleotides: **pyrimidines** and **purines.** The bases derived from pyrimidine are cytosine (C), thymine (T), and uracil (U). Those derived from purine are adenine (A) and guanine (G). The base uracil is found only in nucleotides of RNA, and the base thymine is found only in nucleotides of DNA.

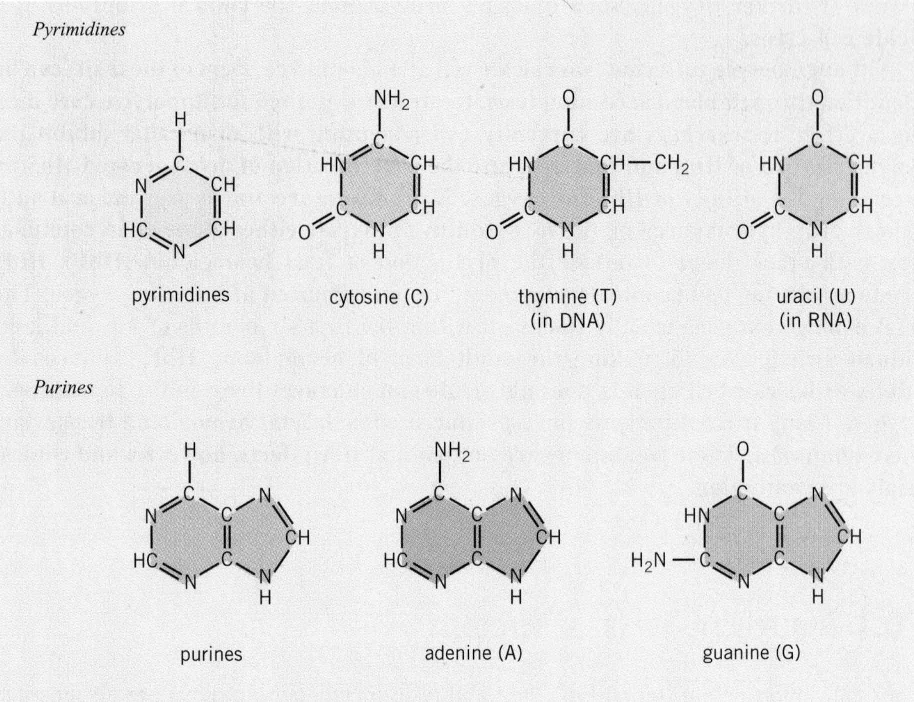

The Pentose Sugars

The second component of nucleotides is the pentose sugar. RNA contains the sugar ribose, and DNA contains a derivative of ribose, 2-deoxyribose.

$$\text{ribose} \qquad \text{2-deoxyribose}$$

(ribose: CH_2OH at C-5; ring O; C-1 with OH and H; C-2 with H and OH; C-3 with H and OH; C-4 with H. 2-deoxyribose: same, with C-2 bearing H and H.)

ribose　　2-deoxyribose

The Structure of a Nucleotide

Draw the general structure of a nucleotide.

The three components of the nucleotide are joined in the following manner:

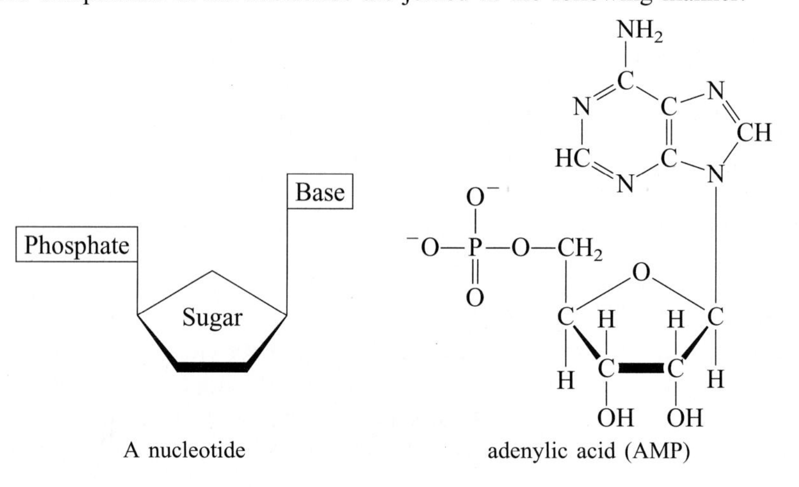

A nucleotide　　adenylic acid (AMP)

This example shows the nucleotide adenylic acid, or adenosine monophosphate (AMP), which contains adenine, ribose, and one phosphate. Other ribonucleotides have similar structures but different bases. The deoxyribonucleotides contain the sugar deoxyribose instead of ribose.

Cells contain large numbers of free nucleotides that perform many functions. Besides being the basic structural unit of nucleic acids, they also participate in biosynthetic reactions, serve as coenzymes, and are important in the transport of energy from energy-releasing reactions to energy-requiring reactions. ATP (adenosine triphosphate), GTP (guanosine triphosphate), and UTP (uridine triphosphate) are all energy-carrying mononucleotides.

Check Your Understanding

1. What are the three components that make up the nucleotides of DNA? Of RNA?
2. Draw the structure of the nucleotide guanosine monophosphate, which contains the sugar ribose.
3. What nucleotide is an essential component of the coenzyme FAD?

20.3 The Structure of DNA

Each human cell contains about 2 m of the nucleic acid DNA, packed into a set of 46 chromosomes having a total length of only about 200 μm. This reduction in length is possible because the DNA molecule wraps and folds itself around proteins, called histones, that are tightly bound to the DNA. DNA was first isolated in 1869 by the Swiss physician Johann Friedrich Miescher from white blood cells. Its structure was not determined until 1953, however, when James Watson and Francis Crick proposed a model to explain the physical and chemical properties of DNA. The Watson and Crick model has two helical polynucleotide chains coiled around the same axis, forming a double helix (Fig. 20.1). The hydrophilic sugar and negatively charged phosphate groups on the nucleotides are on the outside of the helix, and the hydrophobic bases are on the inside. In the years since 1953, researchers have discovered that DNA can assume several different structures, but the Watson and Crick double helix, called B-DNA, is the conformation (or shape) of DNA in its native state.

Describe the structure of a DNA molecule.

The nucleotides making up each strand of DNA are connected by ester bonds between the phosphate group and the deoxyribose sugar. This forms the ''backbone'' of each DNA strand, from which the bases extend (see Fig. 20.1). The bases of one strand of DNA pair with bases on the other strand through hydrogen bonding. This hydrogen bonding is very specific: the structure of adenine permits it to hydrogen bond only with thymine, and guanine bonds only with cytosine (Fig. 20.2). As a result, the two strands of DNA are not identical but rather are complementary—where thymine appears on one strand, adenine appears on the other. The hydrogen bonding is possible because the two strands of DNA run in opposite directions. Note in Figure 20.2 that the sugars on one side are drawn ''upside down'' to indicate that the DNA strands are running in opposite directions.

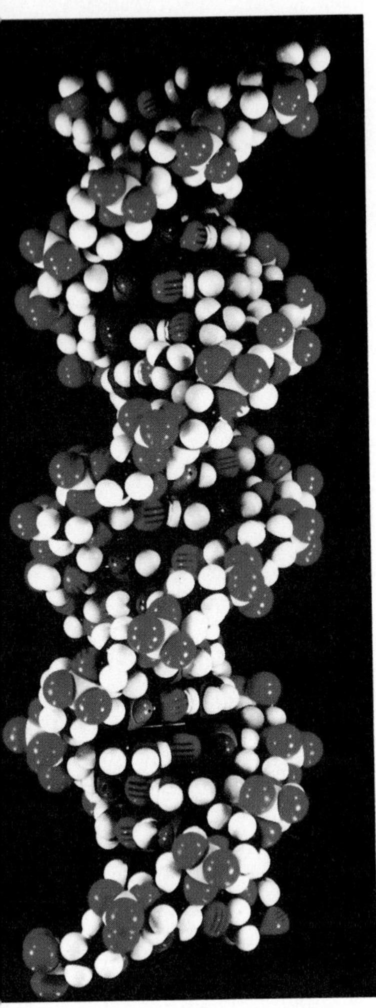

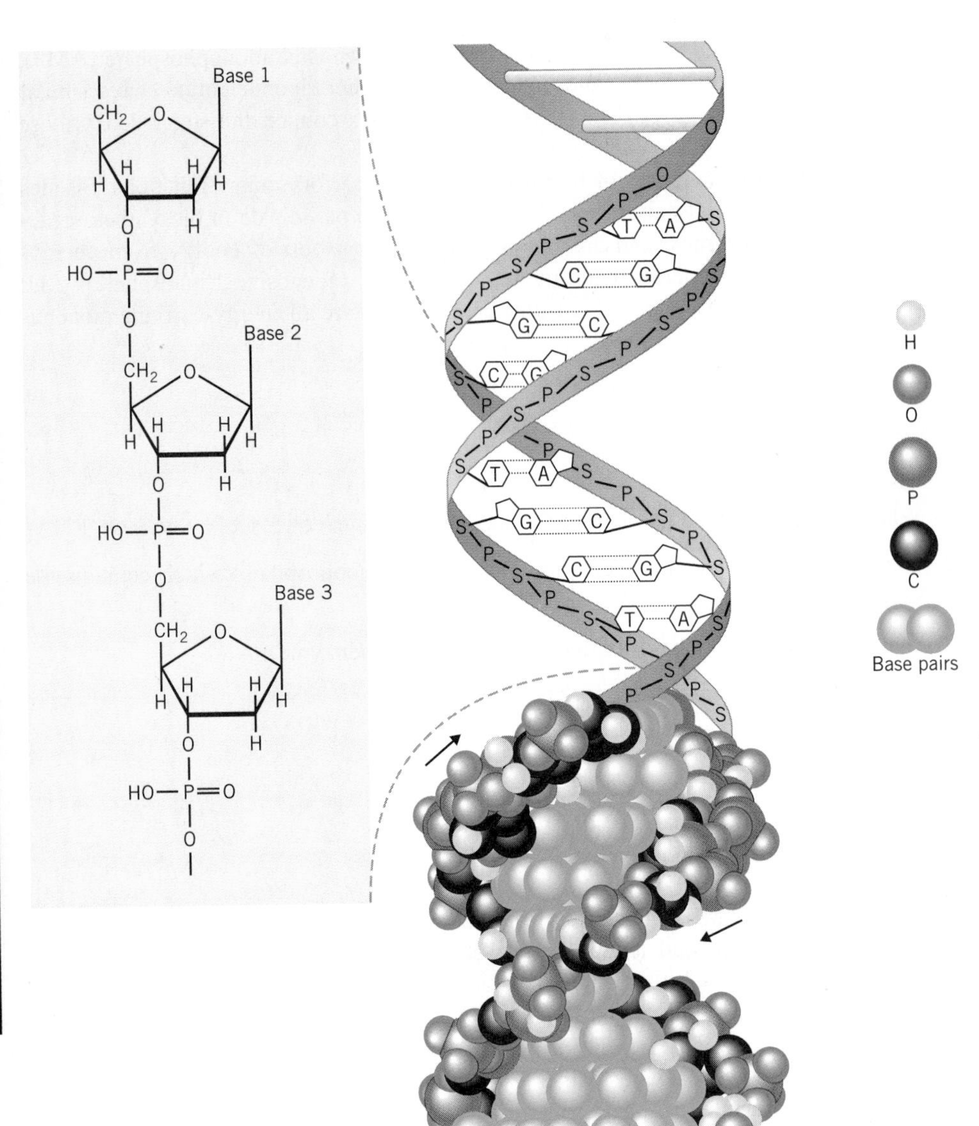

Figure 20.1 The double helix of the DNA molecule. The backbone of the DNA molecule consists of the sugar and phosphate groups of the nucleotides.

20.4 Replication of DNA

Describe how a molecule of DNA replicates.

DNA is the hereditary molecule in all cells. It must direct its own replication during cell division as well as direct the transfer of the genetic information it contains to RNA. Before a cell divides, the DNA molecules **replicate** (that is, make exact copies of themselves) so that each daughter cell has DNA identical to the parent cell. The replication of DNA is a complex process requiring a large number of enzymes. In this process, enzymes catalyze the progressive unwinding of small segments of the DNA helix. Each strand serves as a template or pattern for the synthesis of a new complementary strand of DNA by enzymes called DNA polymerases (Fig. 20.3). Replication is continuous on one strand, called the leading strand. It is discontinuous (carried out in separate pieces that are later attached together) on the other, or lagging strand. Each of the two daughter helices formed contain one original DNA strand and one newly made strand.

sugar
hydrogen bonds
sugar
Thymine
Adenine

sugar
sugar
Cytosine
Guanine

Figure 20.2 Hydrogen bonding in DNA. The bases on the nucleotides of DNA extend toward the inside of the helix, and the two strands of the helix are held together by hydrogen bonding between the bases.

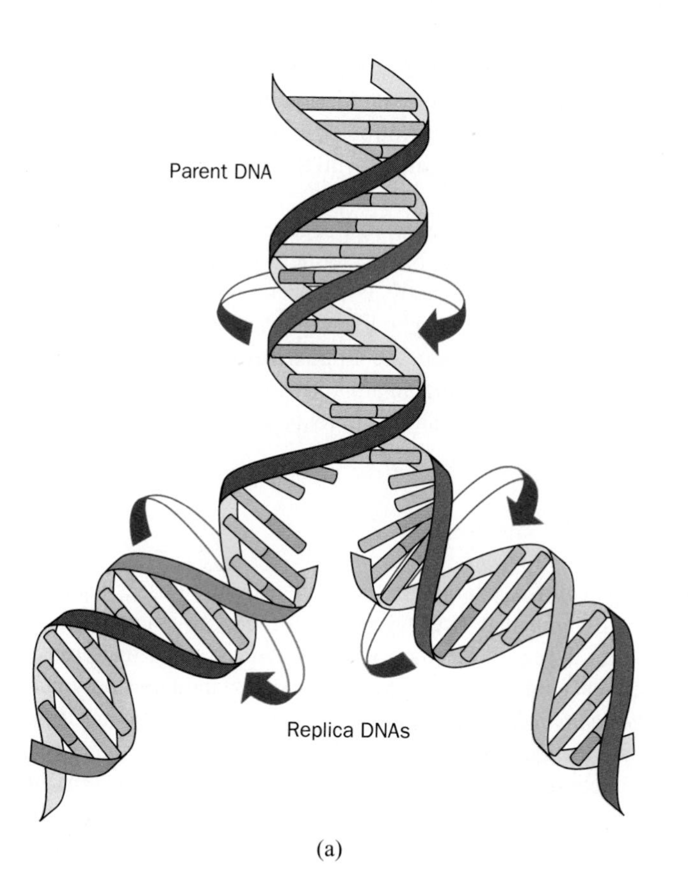

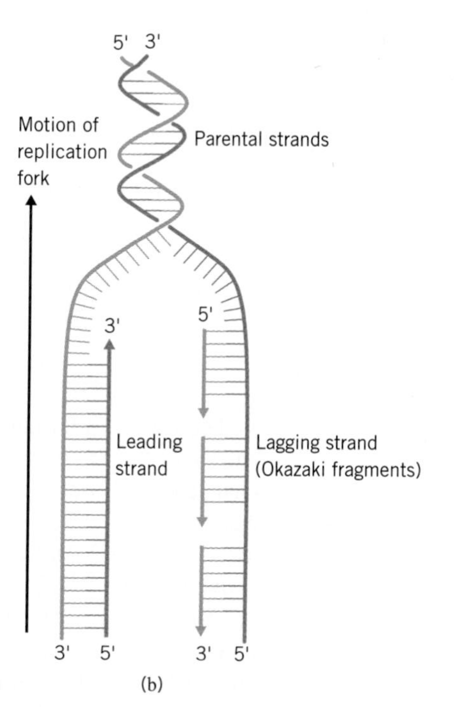

Figure 20.3 Replication of DNA. (a) Enzymes separate the DNA strands, and two new daughter strands are produced. (b) The leading strand is synthesized continuously, and the lagging strand is synthesized discontinuously. When complete, each daughter DNA helix contains one new strand and one old strand.

Health Perspective / Viruses

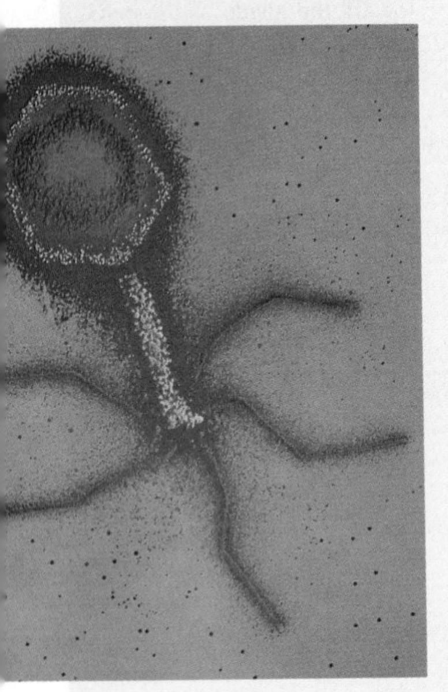

T4 Bacteriophage virus

A virus is so small that is is visible only with the aid of an electron microscope. Well-known diseases caused by viruses include smallpox, chicken pox, measles, German measles, influenza, and certain types of pneumonia and hepatitis. A virus is composed of a thin protein coat that encloses a single strand of DNA or RNA. The virus cannot make its own proteins, nor can it grow or multiply on its own. To reproduce, a virus must invade a living cell and use that cell's functions to meet its needs.

On entering a host cell, a virus begins to reproduce by forming DNA. The raw materials used to form the DNA are taken from the host cell, as is the energy needed for such reproduction. The newly formed viral DNA multiplies many times. Eventually, when the nutrition provided by the host cell has been nearly exhausted, the viral DNA uses the ribosomes of the host cell to synthesize protein molecules. These protein molecules surround the newly formed viral DNA, creating new virus particles. Eventually, the virus particles break out of the host cell and invade neighboring cells. In most cases, the emergence of the virus destroys the host cell.

Because the virus acts as a parasite and grows inside the host cell, it is highly resistant to drugs such as antibiotics that inhibit cellular growth and division. Our bodies are not defenseless against invasion by viruses, however. We have an immune system that gradually stops the spread of the invading virus. It even remembers and protects against any later infection by the same virus. The immune response is a complex series of reactions that involves different types of white blood cells. The job of the white blood cells is to kill the virus and virus-infected host cells. They do this by recognizing the presence of foreign molecules (such as the protein coat of the virus) and producing specialized proteins called antibodies. The foreign molecule is called an *antigen* because it induces the *anti*body *gen*eration. Antibodies are created specifically for each invading antigen. They are designed to combine with the antigen and render it harmless.

Although the body is slow to produce antibodies in response to the initial invasion by an antigen, a later infection by the same antigen brings about a very quick response. Because of this acquired immune response by the body, it has been possible to develop vaccines for most viral diseases. Such vaccines contain a weakened form of the live virus or a portion of the protein coat that brings about the immune response. The body's immune system responds to the vaccine by creating an antibody to it. Then, should the real virus invade the body, the immune system is able to quickly produce the antibody and prevent an infection.

The genetic information for the cell is contained in the sequence of the bases A, T, C, and G in the DNA molecule. Anything that alters the order of the sequence causes a change, or mutation, in the genes of the cell. Cells have many enzyme systems that protect against alteration of the base sequence on DNA. For example, the enzyme DNA polymerase I from the bacterium *Escherichia coli* (*E. coli*) is a complex molecule having three active sites. One is for adding nucleotides to the growing DNA chain, and the other two serve special "proofreading" functions. DNA polymerase I reads the DNA bases on the newly synthesized chain, removes wrongly placed bases, and replaces them with the correct ones. It also can cut out and replace DNA that has been altered by ultraviolet light or chemical mutagens.

Example 20-1

What is the sequence of bases on a strand of DNA that is complementary to a strand having the following sequence of bases?

A A T C G T A G G C A C

This question asks us to write the sequence of bases on a DNA strand that is complementary to the given sequence.

See the Question

In a DNA molecule, adenine (A) pairs only with thymine (T), and cytosine (C) only with guanine (G). So we can determine the sequence of bases on the complementary strand as follows.

Think It Through (Execute the Math)

Original strand: A A T C G T A G G C A C

Complementary strand: T T A G C A T C C G T G

Prepare the Answer

Check Your Understanding

1. Write the sequence of bases on the strand of DNA that is complementary to the following DNA strand:

   T A C T G A A A A C C G A T T

2. Using the two DNA strands from problem 1, draw the DNA segment that would be found in each daughter cell formed by the division of the cell containing the original DNA.

20.5 Ribonucleic Acid

Molecules of RNA make up 5% to 10% of the total weight of the cell and have more varied biological functions than DNA. The nucleotides of RNA contain ribose instead of deoxyribose and the base uracil in place of thymine. Unlike DNA, RNA is not always double-stranded; it usually consists of a single strand of nucleic acid. The additional hydroxyl group on the ribose is very important in forming hydrogen bonds that stabilize the tertiary structure of the nonhelical regions of the RNA molecule. Three of the types of RNA found in the cell are **messenger RNA (mRNA), ribosomal RNA (rRNA),** and **transfer RNA (tRNA).**

List three types of RNA, and describe their functions.

Messenger RNA

The first crucial step in the transmission of genetic information from a gene to a polypeptide chain comes in the **transcription** of this information from a segment of DNA to the complementary molecule of mRNA. Our first knowledge of the structure and function of mRNA came from *E. coli,* a common bacterium found in the intestines. In *E. coli,* mRNA is synthesized on one strand of the DNA by the enzyme RNA polymerase (Fig. 20.4). It therefore has a sequence of bases that is complementary to the sequence of the DNA strand (remember, though, that in RNA the base uracil is substituted for the base thymine). The cell synthesizes mRNA whenever it is needed. After being synthesized, mRNA migrates to the ribosomes, where it serves as a template, or pattern, for sequencing amino acids in the synthesis of proteins.

Health Perspective / Genetic Disease and DNA Repair

Jaime and Sherry Harrison will never play outside in the sunshine. They live in a house that has all its windows covered and is lit by candlelight. The two sisters can go out to play only at night after the sun has set. Jaime and Sherry suffer from the rare genetic disease called xeroderma pigmentosum (XP). People suffering from XP develop dry skin, excessive freckling, and skin and eye tumors from even brief exposure to sunlight or any other ultraviolet radiation.

When it penetrates the tissues, ultraviolet light reacts with DNA in the nucleus of cells. It causes the formation of cyclobutyl rings between adjacent thymine bases on the same strand of DNA. The joined thymines, called dimers, prevent the replication of DNA. Several enzyme systems in normal cells recognize these dimers and either reverse the dimerization or cut out and replace the altered section of DNA. People suffering from XP lack any one of the nine enzymes necessary for repairing this damage to the DNA. Unless they are totally protected from sunlight, individuals such as Jaime and Sherry often develop skin and eye cancers at a very early age.

Ribosomal RNA

In cells with a nucleus, rRNA is synthesized from a DNA template in small dark-staining nuclear bodies called nucleoli. Here rRNA joins with proteins to form the subunits of the ribosomes. **Ribosomes** are the sites of protein synthesis, and they are often located on the endoplasmic reticulum in the cytoplasm. The ribosomes contain two subunits that combine with mRNA and various enzymes to form the ''factory'' for the production of proteins.

Transfer RNA

Transfer RNA is the smallest of the RNA molecules; it is water-soluble and moves easily within the cell. The tRNA molecules are synthesized in the nucleus of the cell, and each is specifically designed for a particular amino acid. A tRNA molecule becomes *charged* when a specific amino acid is joined to the terminal adenine nucleotide present on each

Figure 20.4 Transcription of DNA to mRNA in *E. coli* occurs on one strand of DNA. The sequence of bases on the mRNA is complementary to the sequence on the DNA.

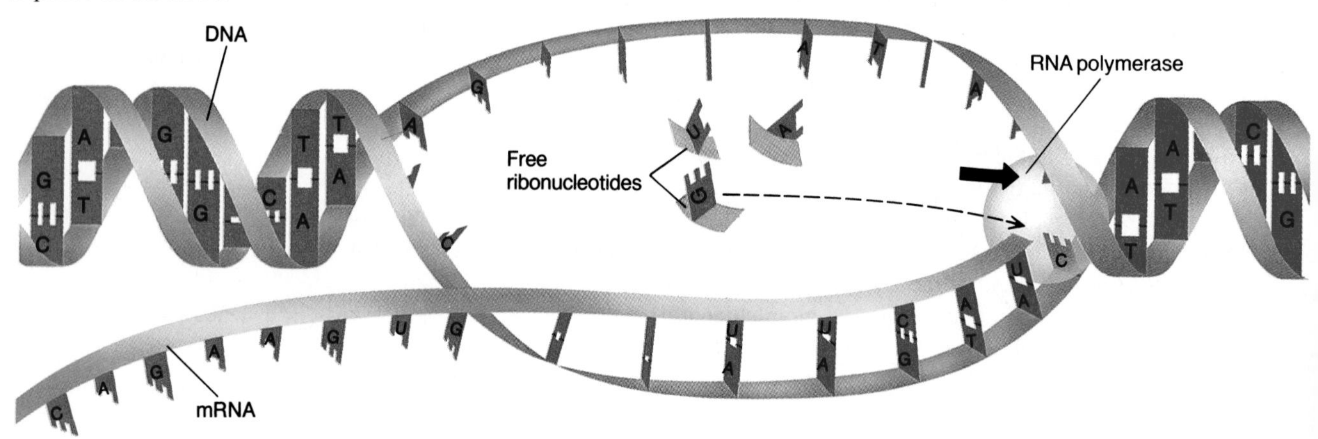

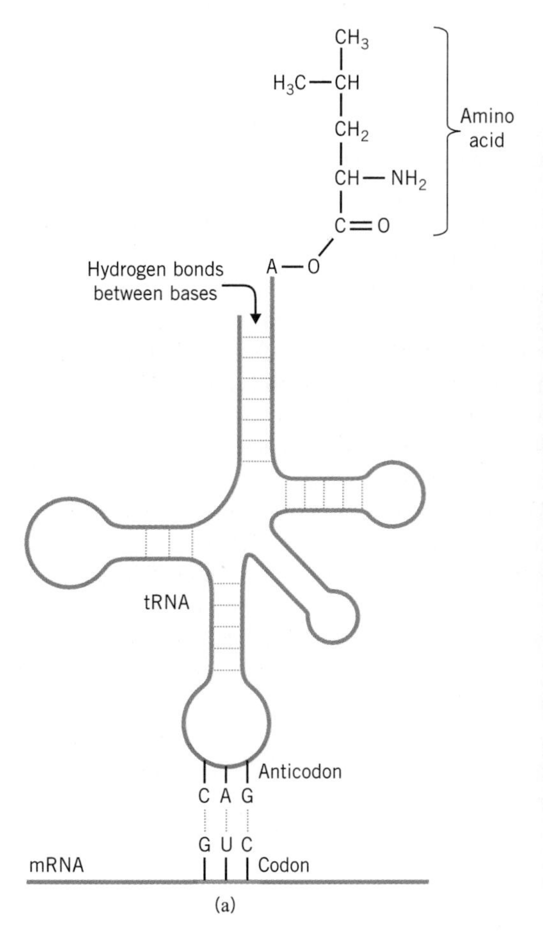

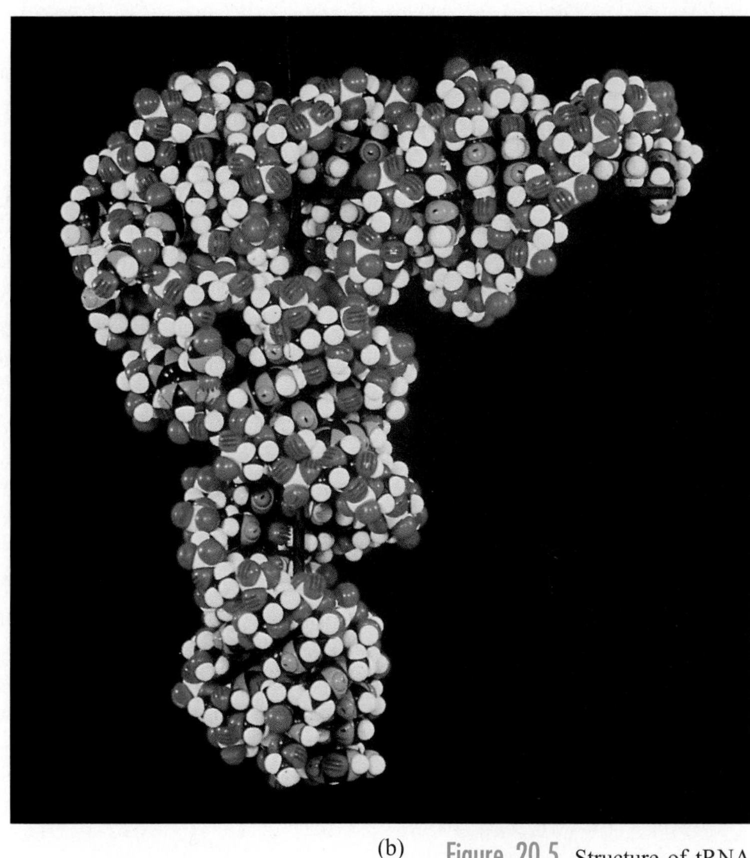

(b)

Figure 20.5 Structure of tRNA. (a) Schematic drawing of a leucine tRNA from *E. coli.* The three bases in the anticodon region are complementary to a codon on mRNA. (b) Tertiary structure of a tRNA.

tRNA polynucleotide chain (Fig. 20.5). The tRNA molecule then carries this amino acid to the ribosomes, where the amino acid is used in protein synthesis.

Example 20-2

What sequence of bases on the mRNA molecule is synthesized on the following strand of DNA?

DNA: T A T C T A C C T G G A

See the Question

This question asks us to write the sequence of bases on the mRNA that is synthesized on a given strand of DNA.

Think It Through (Execute the Math)

In the synthesis of RNA, the base cytosine (C) pairs only with guanine (G). Adenine (A) pairs with the thymine (T) on the DNA, but uracil (U) replaces the thymine base on the RNA strand.

Prepare the Answer

DNA: T A T C T A C C T G G A

mRNA: A U A G A U G G A C C U

Health Perspective / AIDS

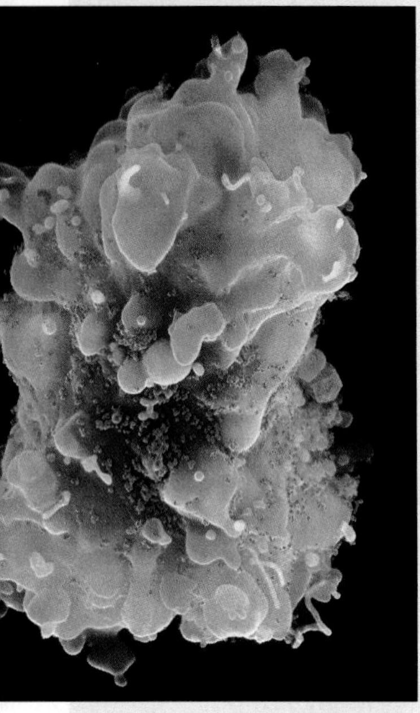

HIV virus (blue) coating the surface of a white blood cell.

As Sarah closed her textbook, the flood of tears came again. What was the point of studying? Her future seemed absolutely devoid of hope. Two months earlier, suffering from a mild rash, fever, and sore throat, and thinking that she might have mononucleosis, Sarah had gone to the college health service. When the mono tests came back negative, the nurse had tested her for human immunodeficiency virus (HIV). Although that test also was negative, the nurse suggested that Sarah return for a retest in six weeks because positive HIV results are often delayed. Just to be safe, Sarah decided to follow the nurse's advice. To her horror, the second test came back positive for HIV. Sarah now realized that her birth control pills had done nothing to protect her from the virus that causes the acquired immunodeficiency syndrome (AIDS), and she had seen no reason to use a condom since Jeff was her steady boyfriend.

The human immunodeficiency virus belongs to a class of viruses called retroviruses. Retroviruses contain a special enzyme called RNA-directed DNA polymerase (more commonly known as reverse transcriptase). This enzyme uses the biochemical machinery of the host cell to synthesize DNA from the viral RNA, forming an RNA–DNA hybrid. The enzyme then breaks apart the RNA strand and uses the DNA of the hybrid to synthesize a second strand of DNA. This double-stranded DNA contains all the information necessary to make the virus, and it directs the host cell to synthesize virus particles.

HIV invades the body's T4 lymphocytes, or ''helper T cells'' (also known as CD-4 cells). The T4 lymphocytes are the white blood cells critical to the body's entire immune system and are one of the body's major defenses against invading organisms.

Check Your Understanding

What sequence of bases on the mRNA is synthesized on the following strand of DNA?

DNA: T A C T G A A A A C C G A T T

20.6 The Genetic Code

A **gene** is a region on a DNA molecule composed of a sequence of bases that determines a hereditary trait. In the nucleus of a human cell, among the three billion bases of DNA, about 100,000 genes are distributed on 23 different chromosomes. Rapid developments in research techniques in the 1980s and 1990s have allowed an international group of scientists to embark on a project to map all 23 chromosomes, the entire human genome. The Human Genome Project was scheduled for completion in 2005, but a French researcher has almost finished a complete mapping. The next milestone will be determining the function of each gene. This knowledge might allow for future prenatal detection and correction of defective genes and chromosomes, curing genetic diseases after birth, and further modifying domestic animals to produce specific drug molecules that can be isolated and used for therapeutic purposes.

Even though a person may test positive for HIV, the immune system can keep the rapidly reproducing virus under control for as long as 8 to 15 years after the initial infection. It does this in part by replacing T cells almost as fast as the infection destroys them. When the virus finally gains the upper hand, the T cells are destroyed, leaving the patient vulnerable to diseases normally kept in check by the immune system. Therefore, people with AIDS eventually die from secondary infections or rare cancers that don't affect patients with normal immune systems.

The AIDS virus is transmitted through unprotected sexual intercourse whether oral, vaginal, or anal. Although donated blood is carefully screened by blood banks, HIV can also be transmitted through transfusions that contain infected blood, as well as through use of drug needles that contain traces of infected blood. The disease is also transmitted from an infected mother directly to her newborn infant. There is no known cure for AIDS, and development of a vaccine for AIDS continues to be hindered by the fact that the virus mutates so rapidly.

Anyone, no matter their age, sex, ethnicity, or sexual orientation, can contract AIDS. As many as 20% of the new AIDS cases diagnosed in 1994 were women. AIDS is not only one of the leading causes of death in men 25 to 44 years of age but is also the sixth leading cause of death for women in the same age group. The incidence of AIDS in the college population in 1990 was 1 in every 500 students. Current treatment for infection with HIV includes drugs that limit replication of the virus, such as zidovudine (AZT) often in combination with drugs such as bactrim to prevent secondary infections. These drug treatments do not hold much promise, however, and have toxic side effects. The best hope for the HIV-infected patient is to prolong life by beginning treatment against the opportunistic infections before they appear. Sarah will forever regret her ignorance of the hazards of AIDS as she looks toward a very uncertain future.

Describe the genetic code and its relationship to amino acids and polypeptide chains.

The conversion of the hereditary information carried on the bases of DNA requires a language, or **genetic code.** In the early 1960s the language of the genetic code was discovered through extensive studies using laboratory-synthesized mRNA molecules. It was discovered that a three-base sequence is necessary to code for each amino acid. (With the four-letter genetic alphabet—A, G, C, and U—64 three-letter combinations are possible, which are more than enough to code for the 20 amino acids.) The three-base sequence on mRNA that specifies a distinct amino acid is called a **codon.** There is more than one codon for most amino acids, as shown in Table 20.1. For example: GGU, GGC, GGA, and GGG are all codons for the amino acid glycine; UUU and UUC are the codons for the amino acid phenylalanine. Notice from Table 20.1 that the first two bases in the codon for a given amino acid are the same. The third base can vary and is less important in determining the amino acid being specified. The genetic code is universal for all organisms: the codon UUU specifies phenylalanine in a yeast, mushroom, or human cell. The genetic code on the mRNA is not punctuated; that is, the code is read one triplet after another, without a break from one end of the mRNA molecule to the other.

mRNA: G G U C A G U G C U C C . . .

amino acid: Gly - Gln - Cys - Ser - . . .

The codon AUG codes for the amino acid methionine and also serves as a "start" codon to signal the beginning of the amino acid sequence. Three codons (UAA, UAG,

Table 20.1 **Codons for the Amino Acids: The Genetic Code**

First Position	*Second Position*								*Third Position*
	U		*C*		*A*		*G*		
U	UUU	Phe	UCU	Ser	UAU	Tyr	UGU	Cys	U
	UUC		UCC		UAC		UGC		C
	UUA	Leu	UCA		UAA	Stop	UGA	Stop	A
	UUG		UCG		UAG	Stop	UGG	Trp	G
C	CUU	Leu	CCU	Pro	CAU	His	CGU	Arg	U
	CUC		CCC		CAC		CGC		C
	CUA		CCA		CAA	Gln	CGA		A
	CUG		CCG		CAG		CGG		G
A	AUU	Ile	ACU	Thr	AAU	Asn	AGU	Ser	U
	AUC		ACC		AAC		AGC		C
	AUA		ACA		AAA	Lys	AGA	Arg	A
	AUG	Met	ACG		AAG		AGG		G
G	GUU	Val	GCU	Ala	GAU	Asp	GGU	Gly	U
	GUC		GCC		GAC		GGC		C
	GUA		GCA		GAA	Glu	GGA		A
	GUG		GCG		GAG		GGG		G

UGA) do not code for any amino acid, but serve as ''stop,'' or terminal, codons. They cause the completed protein to be released from the ribosome. Each tRNA molecule contains a special three-base sequence, called the **anticodon,** that is complementary to one of the codons on an mRNA molecule (see Fig. 20.5).

20.7 Introns and Exons

Bacteria, such as *E. coli,* consist of a prokaryotic cell having neither a nucleus nor other organelles. Their DNA is found throughout the cell, with genes located in one continuous stretch along the DNA. Until recently, it was thought that the steps involved in the transcription of DNA to mRNA (that is, the passage of genetic information from DNA to mRNA) and then the translation of mRNA to protein (that is, the expression of the genetic information in the amino acid sequence of the protein)—all of which had been determined from research on *E. coli*—occurred in an identical fashion in all living cells. Several characteristics of eukaryotic cells (cells with nuclei), however, bothered scientists working in this field. Eukaryotic cells appear to have much too much DNA—about nine times more than a human cell needs to code for the proteins necessary for life. They also seem to have too much RNA. The nuclei of such cells have been found to contain huge strands of RNA, a far greater amount of RNA than that which actually exits the nucleus as mRNA. This nuclear RNA is called **hnRNA** (for **heterogeneous nuclear RNA**).

The solution to this mystery was found in 1977: the genes on the DNA of eukaryotic cells are ''interrupted'' by segments of DNA that do not code for amino acids. These intervening sequences of DNA have been given the name **introns;** the sequences of DNA

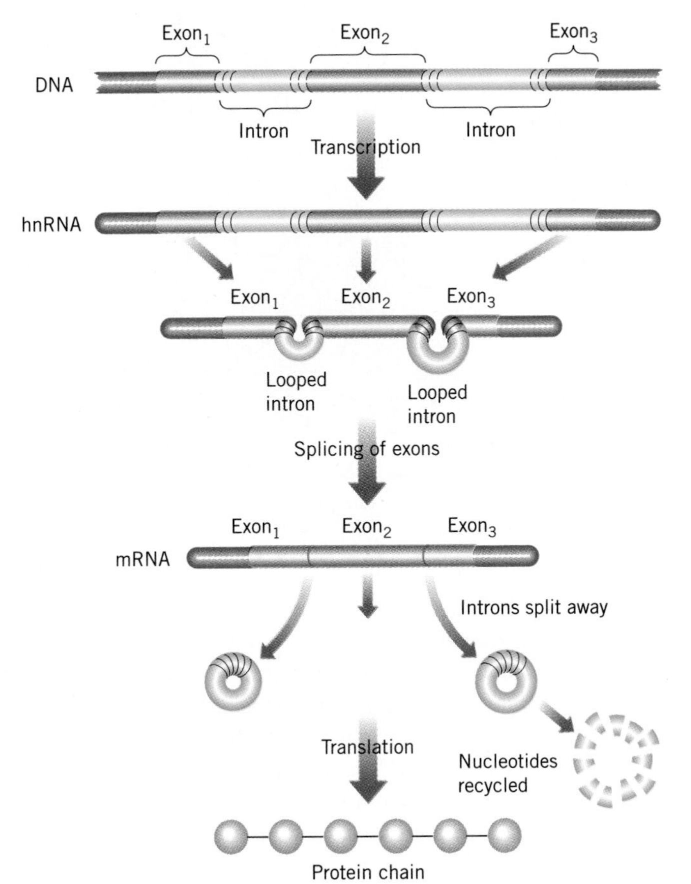

Figure 20.6 In a eukaryotic cell, the introns found on the hnRNA are removed by special enzymes through a series of steps in the nucleus that produce mRNA.

that do code for amino acids are called **exons.** The unexpectedly large amount of DNA in eukaryotic cells is explained by the finding that the exons are much shorter in length than the introns (exons are 100 to 300 bases in length, and introns are about 1000 bases long). The gene region that codes for a protein (in all but two of the proteins studied so far) is made up of alternating regions of exons and introns. Recent research has shown that genes for different proteins contain some of the same exons. Thus, the function of the introns may be to permit the "shuffling" of exons between different genes.

The hnRNA found in the nucleus of eukaryotic cells is a complete transcription of the entire DNA base sequence, including all introns and exons. Special enzyme complexes of RNA and protein then cut and splice the hnRNA in a series of reactions that produces mRNA (Fig. 20.6). The mRNA produced through this process contains only the sequence of bases that codes for the amino acids in the protein.

20.8 The Steps in Protein Synthesis

The second step in the transmission of genetic information from a gene to a polypeptide chain comes in the **translation** of this information from mRNA to the amino acid sequence of a protein. As stated earlier, protein synthesis takes place in the ribosomes. Each ribosome contains two subunits. The large subunit contains a short and long strand of rRNA and 32 different polypeptides. It is the place where the lengthening of the polypeptide chain occurs. The small subunit contains one strand of rRNA and 21 different poly-

Figure 20.7 Schematic drawing of translation, the synthesis of a polypeptide on mRNA.

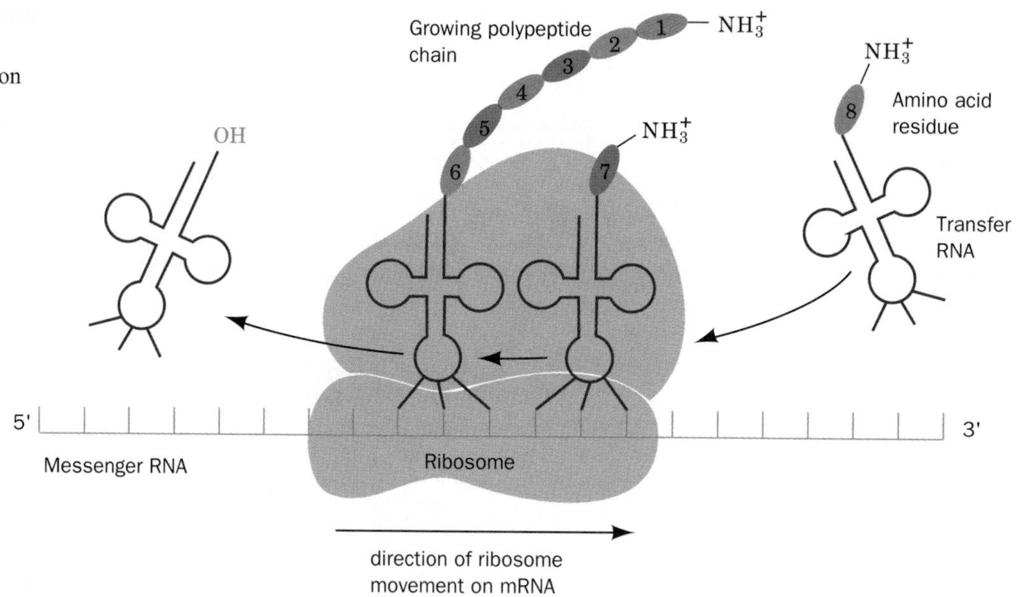

peptides. Its function is the correct matching of the charged tRNAs with the codon on the mRNA. Many ribosomes can be producing proteins from the same strand of mRNA. When viewed under a powerful electron microscope, these groups of ribosomes (called polyribosomes, or **polysomes**) appear as a series of dots.

Before protein synthesis can begin, mRNA must be synthesized as described in the previous sections. Then tRNAs attach themselves to amino acids, catalyzed by enzymes that are specific for each amino acid and each tRNA. The attachment of the amino acid to the tRNA (called the activation or charging of the tRNA) is a two-step process requiring the presence of ATP.

◆ **Describe the steps in protein synthesis.**

The specific process of protein synthesis, as it has been determined in *E. coli,* occurs as follows. First, the smaller subunit of the ribosome combines with three proteins (called initiation factors) and with the energy-rich molecule GTP, an mRNA, and a tRNA carrying a special "starting" amino acid. This complex then joins with the larger ribosomal subunit. The growth or elongation of the polypeptide chain takes place within the ribosome through a series of repeated steps in which amino acids brought by tRNAs (whose anticodon region matches the codon on the mRNA) form peptide bonds with the last amino acid in the growing chain (Fig. 20.7). These steps require the enzyme peptidyltransferase, other nonribosomal proteins called elongation factors, and two molecules of GTP. The elongation of the chain stops when a terminal codon is reached on the mRNA, at which time the ribosome separates, releasing the newly formed protein and the mRNA. It is through a similar synthesis of all the proteins required by a cell that the genes of a DNA molecule ultimately control the appearance and activities of that cell. Figure 20.8 summarizes the process of cellular protein synthesis.

Example 20-3

Write the amino acid sequence of the polypeptide chain that is coded for by the following segment of mRNA. Then, draw the structure of the polypeptide.

mRNA: U U C A C G G U C G U A G G G U A C

See the Question

The question asks us to determine the order of amino acids coded for by the bases on a segment of mRNA.

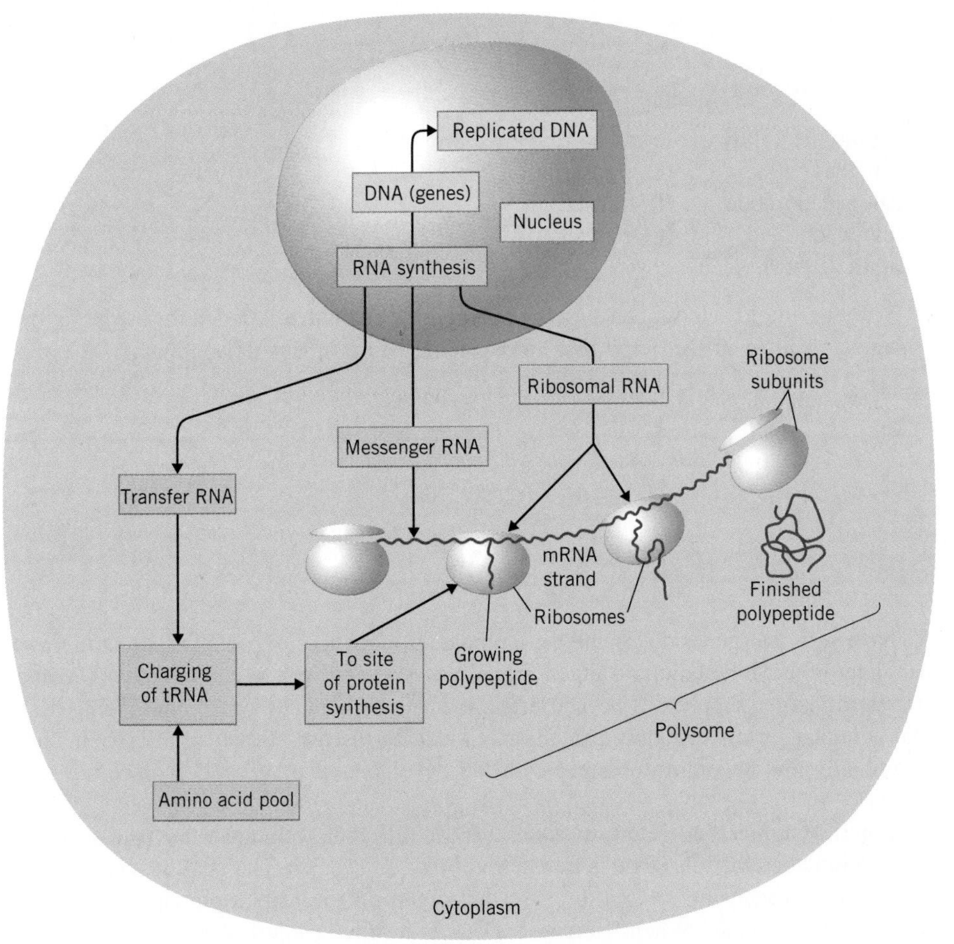

Figure 20.8 Relationships among DNA, the three types of RNA, and protein synthesis.

Think It Through (Execute the Math)

To determine the answer we must divide the sequence of bases on the mRNA into groups of three. We then use Table 20.1 to determine the amino acids corresponding to each three-base codon. Finally, we use Table 17.1 to determine the structure of the corresponding amino acids. Beginning with the first amino acid, we draw the structure of each amino acid forming a peptide bond between the carboxyl group on the first amino acid and the amino group on the next.

Prepare the Answer

mRNA:	U U C	A C G	G U C	G U A	G G G	U A C
Amino acids:	Phe -	Thr -	Val -	Val -	Gly -	Tyr

Polypeptide Structure

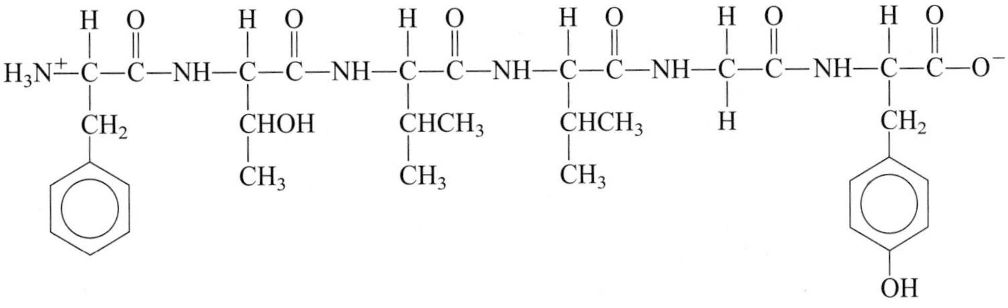

$$H_3\overset{+}{N}{-}\underset{\underset{C_6H_5}{|}}{\overset{H}{C}}H_2{-}\overset{O}{\overset{\|}{C}}{-}NH{-}\underset{CHOH(CH_3)}{\overset{H}{C}}{-}\overset{O}{\overset{\|}{C}}{-}NH{-}\underset{CHCH_3(CH_3)}{\overset{H}{C}}{-}\overset{O}{\overset{\|}{C}}{-}NH{-}\underset{CHCH_3(CH_3)}{\overset{H}{C}}{-}\overset{O}{\overset{\|}{C}}{-}NH{-}\underset{H}{\overset{H}{C}}{-}\overset{O}{\overset{\|}{C}}{-}NH{-}\underset{CH_2C_6H_4OH}{\overset{H}{C}}{-}\overset{O}{\overset{\|}{C}}{-}O^-$$

Check Your Understanding

1. Define the following terms in your own words:
   (a) gene
   (b) genetic code
   (c) codon
   (d) anticodon
   (e) transcription
   (f) translation
   (g) polysome
2. Write the amino acid sequence of the polypeptide chain that is coded for by the following segment of mRNA, and then draw the structure of this polypeptide.

   A U G A C U U U U G G C U A A

20.9 Genetic Engineering

We have seen that genes determine the amino acid sequences of proteins, and in that way also determine the hereditary traits of organisms. Over the years scientists have dreamed of creating larger supplies of needed proteins (such as human insulin) and adding desirable hereditary traits to plants and animals (such as disease resistance in crops). These dreams are now becoming possible with the development of genetic engineering techniques.

Explain how genetic engineering is used to synthesize drugs and biochemical compounds.

Genetic engineering refers to a number of different techniques by which genetic information contained in DNA is transferred between species. The goal of inserting new DNA into an organism is often to cause that organism to create a protein that it would otherwise never make. When ''foreign'' DNA is inserted as part of the organism's own DNA—that is, when the two types of DNA are *recombined*—the result is called **recombinant DNA.** The products of such recombinant DNA are revolutionizing the world of biochemistry.

Recombinant DNA is produced by ''splicing'' a specific segment of DNA (such as the gene that codes for human insulin) into a so-called cloning vector. One common type of cloning vector, called a **plasmid,** is a small circular DNA molecule found in the bacteria *E. coli.* The new genetically engineered plasmid—now containing the desired gene—is inserted into other *E. coli,* where it is replicated in the organism's normal reproductive cycle (Fig. 20.9). Because a single *E. coli* can produce up to millions of copies of the transplanted gene in a day, large quantities of the desired gene can be acquired in a relatively short time. This process of making large quantities of a desired gene is called **gene cloning.**

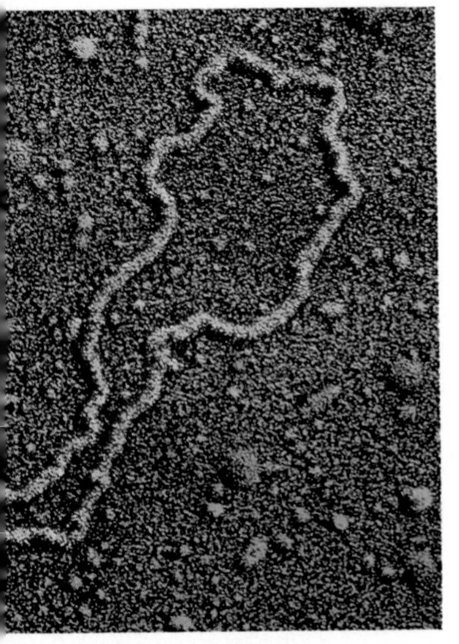

Bacterial plasmid

When they are not dividing, the bacteria produce proteins as directed by their genes. If their recombined DNA now contains the gene that codes for human insulin, the bacteria produce insulin that is identical to that found in humans. In addition to human insulin, such genetic engineering techniques have been used to produce human growth hormone and the brain hormone somatostatin. Using similar methods, scientists have produced sheep and goats that secrete human proteins in their milk (such as the clot-busting agent tPA) and mice that produce human antibodies.

Genetic engineering techniques are also used to generate tailor-made organisms having new properties. Researchers have already succeeded in introducing desired traits (such as resistance to frost, insect pests, or herbicides) into plant species. The first genetically

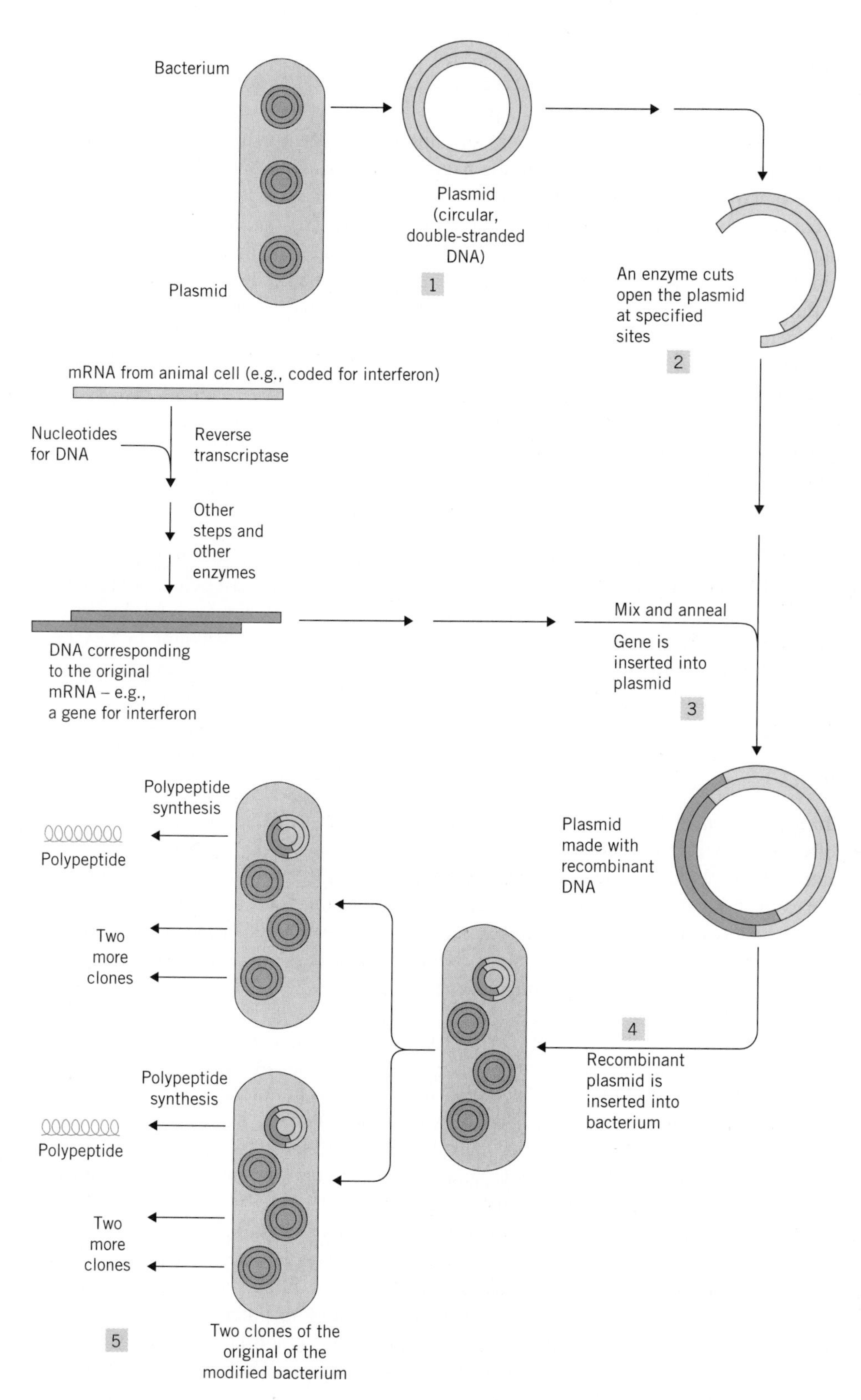

Figure 20.9 Steps in the recombinant DNA procedure. (1) The outer membranes of *E. coli* are dissolved, and small pieces of DNA called plasmids are isolated. (2) The plasmids are broken at specific positions by special enzymes. (3) The cleaved plasmids are mixed with pieces of foreign DNA that are then attached to the plasmids by the enzyme DNA ligase. (4) The recombined plasmids with their foreign DNA are inserted into new *E. coli.* (5) The segments of foreign DNA and the protein for which they code are reproduced in the normal life cycle of the *E. coli.*

Health Perspective / Clot Busters

Chicago Bears' coach Mike Ditka was still upset about the loss his football team had suffered the day before. He ignored the nagging chest pain and feeling of nausea that had swept over him during the day and was relieved when the discomfort went away by evening. Two days later, as he dressed after his morning workout, he began experiencing a crushing pain in his chest. Two assistant coaches noticed his distress and forced him to go to the hospital. In the emergency room, a doctor confirmed that Ditka was suffering a heart attack. She quickly administered a clot buster—a fibrolytic drug that breaks down blood clots. Within several hours of receiving the drug, Ditka's pain and discomfort went away. Amazingly, he was back on the Bears' sidelines just 11 days after suffering the heart attack.

A heart attack occurs when a blood clot blocks a coronary artery (an artery that supplies blood to the heart muscle), thereby stopping the blood supply to a portion of the heart. Until recently, doctors were able to treat only the symptoms of the heart attack, not the clot itself. Extensive tissue death often occurred in the heart muscle, requiring a recovery period of weeks or months.

Now, several clot-busting agents are available. If given within 6 hours of a heart attach, they break up the clot and restore the blood supply to the heart before permanent tissue damage can occur. Two of these clot busters are streptokinase (a protein produced by certain streptococci bacteria) and tissue plasminogen activator or tPA (a human glycoprotein produced by recombinant DNA technology). Because it is a human protein, tPA has fewer side effects than streptokinase. Regardless of side effects, however, the quick administration of one of these agents is saving lives, preventing heart muscle damage, and allowing patients to return quickly to normal, productive lives.

Your Perspective: Studies comparing streptokinase and tPA have shown that both are effective at dissolving blood clots. Patients appear to have a slightly higher survival rate with tPA, but this drug also leads to more deaths caused by hemorrhaging. Streptokinase costs about $200 per dose, whereas tPA (produced by genetic engineering) costs over $2000 per dose. Suppose the director of a hospital emergency room asked you for advice on which drug to use. Write a memo describing the various factors she should take into account in deciding which drug to use.

engineered whole food to receive approval from the U.S. Food and Drug Administration appeared on grocery shelves in 1994. The MacGregor Flavr Savr tomato was engineered to slow the rate at which it softened. This has allowed these tomatoes to remain on the vines a few extra days, ripening and gaining flavor without becoming too mushy to ship.

Another area of genetic engineering is human gene therapy. The first human testing of this technique began in 1990 with two young Ohio girls suffering from a genetic disease called ADA deficiency. People suffering this disease lack the enzyme adenosine deaminase. Without this enzyme, critical disease-fighting cells called T cells die off, crippling the immune system and leaving these people vulnerable to potentially fatal infections. Blood from each girl was withdrawn so that the T cells could be equipped with a new ADA gene. Four days later, the treated cells were infused back into their veins. The procedure was repeated seven times in the next year, and three times after that in an

Chemical Perspective / DNA Fingerprinting

Forensic DNA analysis, or genetic fingerprinting, is a powerful new way of identifying people. This technique analyzes different sections of cellular DNA to determine if two samples are likely to be from the same person. The chance that two people have the same DNA, unless they are identical twins, is almost zero. Although current laboratory procedures examine only a small part of the DNA sequence, the odds against two people having the same pattern can be millions or even billions to one. DNA fingerprinting is now being used in court as evidence in paternity suits, immigration cases, rape cases, and criminal investigations. It is even being used to help win the release of prisoners years after their conviction. For example, a man convicted of the rape and murder of a young child was released after nine years in prison when a recent test revealed that his DNA did not match semen found at the crime scene.

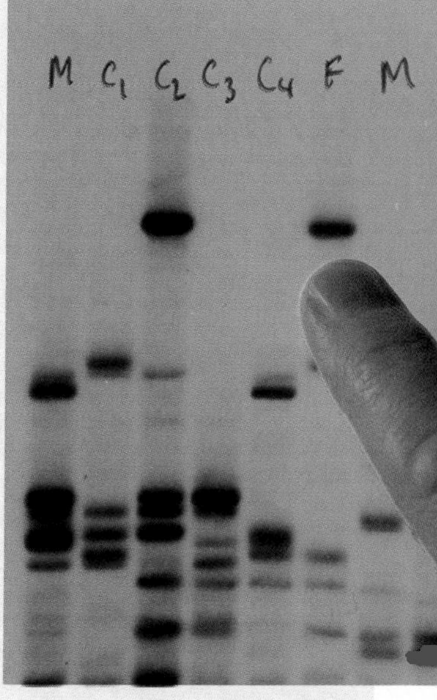

DNA fingerprints showing DNA of a father (F) matching the DNA of a child (C_2)

The most commonly used method to study DNA for court evidence is called restriction fragment length polymorphism (RFLP) analysis. Special enzymes are used to ''cut up'' samples of DNA extracted from semen, saliva, skin, hair roots, bone, or any other tissue that has cells with nuclei. These enzymes recognize specific sequences of DNA and break the strand into pieces called restriction fragments. Using gel electrophoresis, these fragments are sorted by size. Because the DNA break points are somewhat different for every individual (except for identical twins), the electrophoresis produces distinct patterns. The gel, which contains radioactive tags, is then exposed to a photographic film. This causes a ''fingerprint'' resembling a supermarket bar code to appear on the developed film. Comparison of DNA fingerprints from different individuals can lead to identification and conviction of a person accused of a crime.

If samples are too small to extract enough DNA for gel electrophoresis, forensic scientists can use a technique called polymerase chain reaction (PCR). This method uses a repetitive process that copies the DNA over and over again, amplifying DNA from as few as 30 to 60 cells in a tiny speck of blood or a single human hair. This is the test that allowed investigators to solve the World Trade Center bombing case by matching DNA extracted from a suspect's saliva with DNA extracted from dried saliva from a licked envelope.

Your Perspective: You are an expert witness in a rape trial and are asked to explain how DNA fingerprinting works. Describe the process of producing a DNA fingerprint, and explain why the fingerprint ''proves'' that the defendant is guilty. Write out your statement to the jury, using language that an ordinary citizen can understand.

attempt to fix as many T cells as possible. The results were dramatic: The girls no longer suffered multiple infections, their energy returned, and they were able to maintain their weight and resume normal growth.

20.10 Mutations

A **mutation** is a change in the sequence of bases on the DNA molecule. Some mutations may be beneficial to an organism, but most are detrimental to varying degrees. The sequence of bases on the DNA molecule is critical and very specific. Adding a base,

Define a mutation.

Table 20.2 **Diseases of Fat Metabolism That Are Hereditary**[a]

Disease	*Symptoms*
Fabry's disease	Reddish purple skin rash, kidney failure, and pain in legs
Gaucher's disease	Spleen and liver enlargement, mental retardation in infantile form, and erosion of long bones and pelvis
Generalized gangliosidosis	Mental retardation, liver enlargement, skeletal deformities, and red spot in retina in about 50% of cases; death by age 2
Niemann–Pick disease	Mental retardation, liver and spleen enlargement, and red spot in retina in about 30% of cases
Tay–Sachs disease	Mental retardation, red spot in retina, blindness, and muscular weakness; death by age 3

[a]In these diseases, sphingolipids accumulate in tissues because the enzymes that normally catalyze the cleavage of these lipids are defective.

deleting a base, or replacing a base in the sequence can throw off the code and result either in no protein or an altered or defective protein. Such mutations can occur spontaneously or may be caused by radiation, chemical agents, or viruses.

The majority of such changes in the DNA are repaired by special enzyme systems that function constantly in the nucleus. Any breakdown in these enzyme systems or any change in the DNA that is not successfully repaired, however, can cause a mutation.

Explain several ways, both positive and negative, that mutations can affect the normal functions of an organism.

The altered proteins produced through a mutation might improve the organism's chances of survival by providing new chemical pathways, or they might have no biological activity at all—resulting in the death of the cell. In other cases, the defective gene may cause abnormalities or disease. At the beginning of this chapter we saw that a mutation in the gene for hemoglobin occurred among the African black population, resulting in an error in one amino acid on the beta hemoglobin chain. This causes the production of the abnormal hemoglobin found in sickle cell anemia. To date, more than 3500 separate genetic disorders have been identified. Among these conditions and diseases resulting from mutations in genes are cystic fibrosis, albinism, hemophilia, thalassemia, galactosemia, color blindness, and phenylketonuria (Table 20.2).

20.11 Phenylketonuria

We end this chapter with the story of Billy. In 1958, Billy was brought home from the hospital as a happy, healthy baby. People who came to visit the family commented on how fair his skin and hair looked compared with the rest of the family. As Billy entered his fourth month, his mother started noticing that he no longer watched his mobile as it turned above the crib, and he rarely returned her smiles. Billy was slow in learning to sit up by himself. But then again, his brother had also been late in doing such things, and he had turned out to be a very active young child. As Billy grew older, however, his parents became increasingly concerned about his development and behavior. He had become irritable and would have temper tantrums for no reason at all. Although his parents worked very hard to teach Billy to talk, he was able to learn only a few words. Furthermore, his mother noticed a strange musty odor about him when she changed his diapers, and his skin

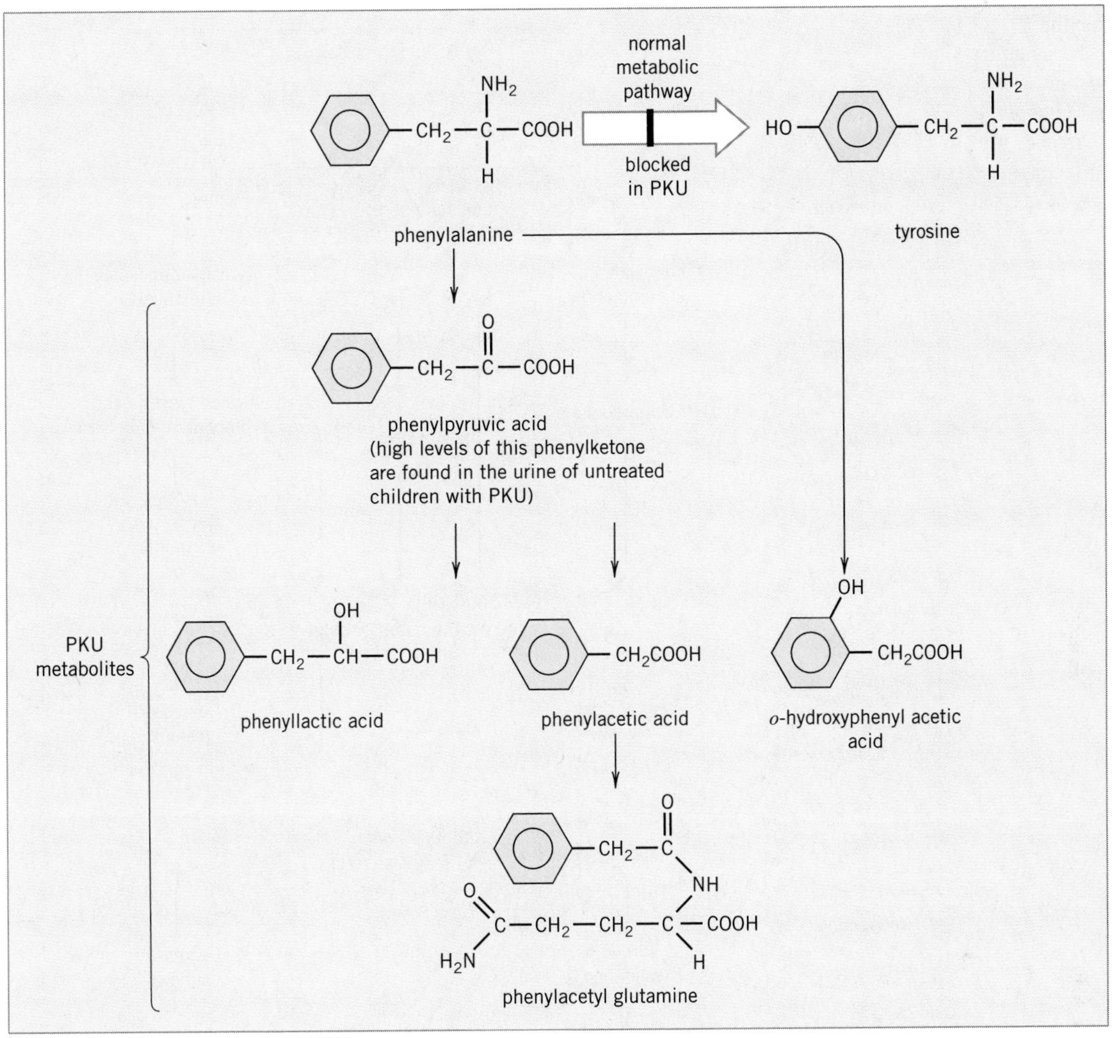

Figure 20.10 The blockage of a metabolic pathway in PKU results in the production of PKU metabolites as the body tries to metabolize phenylalanine by alternative chemical routes.

was often inflamed and flaky. As Billy neared the age of three, his parents began to admit that he was retarded. He still didn't walk or talk, and he was becoming uncontrollable.

Finally, a friend convinced them that the best thing for Billy and for themselves would be to take him to a clinic for diagnosis. At the clinic Billy was given a set of tests, which indicated that he had a disorder called phenylketonuria, or PKU for short. This disease results from a mutation in the gene that codes for the liver enzyme phenylalanine hydroxylase. The defective gene is carried by 2% of the population. A child must inherit the defective gene from both parents to suffer from PKU. In this disease the defective phenylalanine hydroxylase produced in the liver cannot catalyze the conversion of phenylalanine to tyrosine (Fig. 20.10). This results in low tyrosine concentrations and the buildup in the body of phenylalanine and PKU metabolites—substances produced by the metabolism of the excess phenylalanine.

Melanin, the dark pigment found in hair and skin, is formed from tyrosine, and the low level of tyrosine in untreated PKU children leads to the light hair and skin observed on Billy. The hormones epinephrine, norepinephrine, and thyroxine are also synthesized from tyrosine. A high level of phenylalanine and PKU metabolites (especially phenylacetic

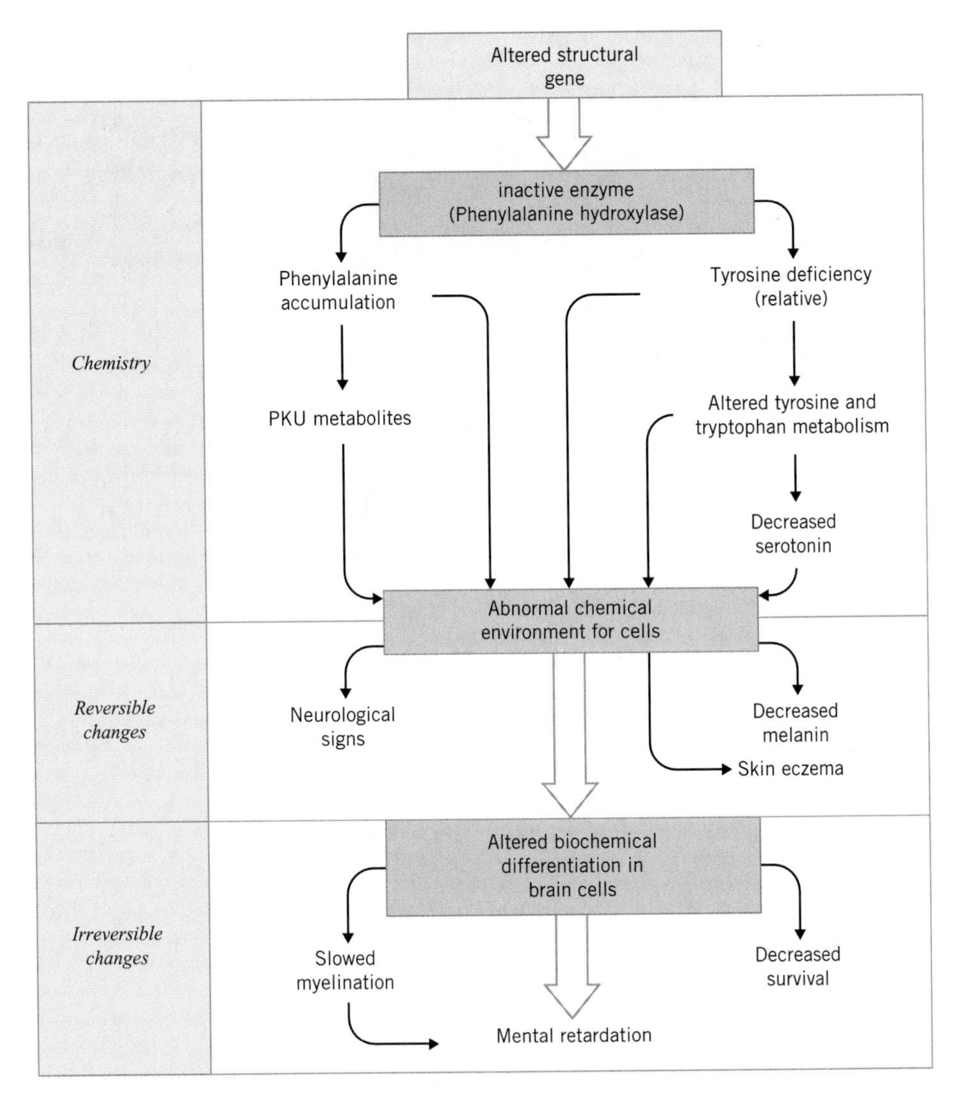

Figure 20.11 Metabolic changes caused by the mutation of PKU. (Adapted from Stanbury et al. *The Metabolic Basis of Inherited Disease.* Copyright ©1972, McGraw-Hill, Inc., New York, NY. Used with permission.)

acid) blocks energy-releasing reactions in the brain of an infant, preventing cells from obtaining the amount of energy necessary for normal functioning. This high level also seems to delay the formation of the myelin sheath, the protective coating around the nerves. In addition, large amounts of phenylalanine in the fluid of the brain prevent the normal uptake of nutrients by the brain cells, so these cells do not have the normal mix of chemicals from which to build their essential and permanent parts. The brain of an infant at birth is only 25% of its mature weight, and it grows rapidly, reaching 89% of its mature weight by 6 years of age. During these years of rapid growth, it is extremely important that the brain be surrounded by the correct chemical environment. The abnormal chemical environment of the brain of an untreated PKU child results in the formation of defective brain cells, which explains the mental retardation of such children (Fig. 20.11). Placing Billy on a special diet containing no phenylalanine improved Billy's behavior and skin condition but could do nothing to reverse the brain damage that had occurred.

A blood test can now detect PKU at birth. Children suffering from PKU eat special diets that control the level of phenylalanine in their bodies and prevent any retardation. These children grow up healthy and active (Fig. 20.12). As they reach maturity and their

Figure 20.12 Both of these children suffer from phenylketonuria. The 11-year-old boy whose condition was not treated when he was an infant, is severely retarded. His three-year-old sister, whose phenylketonuria was treated from birth, is normal.

brains complete their development, their diets can become more varied. As we learn more about genes and their products, more genetic diseases are being identified and tests developed for their identification.

KEY CONCEPTS

DNA is a nucleic acid molecule that carries all of the genetic information necessary for the development of a living organism. Human DNA combines with proteins to form 46 chromosomes in the nucleus of a human tissue cell. DNA is structured as a double helix, consisting of two helical polynucleotide chains coiled around each other. The two chains are held together by hydrogen bonding between the bases of the nucleotides. A nucleotide contains a sugar molecule, a nitrogen base, and a phosphate group or groups.

Genetic information is carried in the sequence of bases on the DNA molecule. A gene identifies a specific region of bases on a chromosome. DNA molecules can replicate themselves with the help of special enzymes, allowing the genetic information to be passed from a parent cell to daughter cells.

There are three major types of RNA in a cell: mRNA, rRNA, and tRNA. In cells with nuclei, a long RNA strand (called hnRNA) is synthesized on the DNA, then cut and spliced together by special enzymes to form shorter messenger RNA (mRNA) molecules. The mRNA molecules migrate to the cytoplasm where they join with ribosomal RNA (rRNA) and proteins to form ribosomes, the location of protein synthesis in the cell. The function of transfer RNA (tRNA) is to join with specific amino acids in the cytoplasm and bring them to the ribosome. There the anticodon region on the tRNA briefly pairs with the codon on the mRNA, allowing the tRNA to discharge its amino acid. This process results in the synthesis of a polypeptide having a specific amino acid sequence.

A mutation is any change in the order of bases on the DNA molecule. This change can result in the synthesis of no protein or an insufficient amount of protein, or the synthesis of a protein with an altered amino acid sequence. A mutation can produce a change that improves an organism's chances of survival, that results in the death of the organism, or that produces a specific genetic disease such as phenylketonuria (PKU).

Review Problems

Section 20.1

1. What are genes?
2. State, in your own words, the "Central Dogma of Molecular Biology."

Section 20.2

3. List three functions of nucleotides in the cell.
4. What are the three components making up a nucleotide?
   What is the difference between a nucleotide and a nucleic acid?
5. Which nitrogen bases found in DNA are pyrimidines? Which are purines?
6. Write the structure of the nucleotides containing:
   (a) the sugar deoxyribose, the base cytosine, and one phosphate
   (b) the sugar ribose, the base guanine, and three phosphates
7. From the structure of deoxyribose and ribose, explain the meaning of the prefix deoxy-.
8. Draw the structures of the nucleotides adenosine diphosphate (ADP) and deoxyadenosine diphosphate (dADP).

Section 20.3

9. How are the two polynucleotide chains held together in the double helix of DNA?
10. The complementary base pairs in DNA are always adenine and thymine or cytosine and guanine. What is it that keeps guanine and thymine from forming complementary base pairs in DNA?

Section 20.4

11. "The two strands of DNA are not identical but rather are complementary." Explain this statement.
12. In general terms, describe the replication of a DNA molecule.

Section 20.5

13. Name three types of RNA found in the cell. Describe the function of each type.
14. What are the three main structural differences between a molecule of DNA and a molecule of RNA?

Section 20.6

15. **(a)** How is genetic information carried on the DNA molecule?
    (b) Referring to Table 20.1, how many amino acids are coded for with the three-base sequences?
    (c) Which of the amino acids in Table 20.1 has only one codon? What is this codon?
16. During transcription, a portion of mRNA with the following base sequence is formed:

AUG-CCA-CAU-GUA-UUG-AAC-CCC-AUU-CUG-UGA

(a) What is the sequence of bases on the DNA from which this portion of mRNA was synthesized?

(b) What is the sequence of amino acids in the protein coded for by this section of mRNA?

Section 20.7

17. Define the terms *exon* and *intron.*

18. (a) What is the relationship between mRNA and hnRNA?

(b) Explain how genetic information is transferred from DNA to mRNA in eukaryotic cells.

Section 20.8

19. (a) Where does protein synthesis occur in a cell?

(b) Describe the steps that must occur for the synthesis of the following polypeptide.

Asp-Trp-Val-Arg-Asn-Ser-Phe-Cys-Gln-Gly-Pro-Tyr-Met

20. How does the sequence of steps described in your answer to problem 19(b) differ for the synthesis of the following polypeptide?

Met-Ile-Tyr-Trp-Val-Ser-Ser-Arg-Ala-Cys-Cys-Gly-Glu

21. Write the amino acid sequence on the polypeptide chain that is coded for by the following sequence on DNA.

TAC-TGT-TTT-CCT-CCC-CGT-ATA-TGA-GGG-ACT

22. Write the amino acid sequence on the polypeptide chain that is coded for by the following mRNA sequence.

AUG-ACG-AAA-AGA-AGG-GGA-GCC-GCU-UCC-UAA

Section 20.9

23. What is a plasmid?

24. Cortisone, a steroid hormone, is often used in the treatment of Addison's disease. Describe how recombinant DNA could be used to produce cortisone.

Section 20.10

25. (a) What is a mutation?

(b) Describe, on the molecular level, several ways in which a mutation can occur.

(c) How does the cell prevent mutations from occurring?

26. Describe three diseases that result from mutations in genes.

Section 20.11

27. (a) Explain, on a molecular level, what causes the disease PKU.

(b) How does this disease affect its victims?

28. (a) What can be done to minimize the effects of PKU?

(b) Could genetic engineering eventually be used to cure PKU? Explain.

Applied Problems

29. The amino acid sequence of bovine insulin is shown in Figure 17.3. The B-chain of bovine insulin has 30 amino acids in a single polypeptide chain.

(a) What is the minimum number of bases on the DNA necessary to code for the B-chain?

(b) Give an mRNA sequence that codes for its synthesis.

30. The following is a sequence of bases that might be found on the gene that codes for the human pituitary hormone, oxytocin:

DNA strand: TACACAATGTAAGTTTTGACGGGGGACCCTATC

(a) What is the sequence of bases on the complementary strand of DNA that forms a double helix with this strand?

(b) What sequence of bases is found on the mRNA molecule synthesized on the original DNA strand?

(c) Mark off the codons of the mRNA strand from part (b), and list the bases found on the anticodons of the tRNA molecules that join with the first three codons on the mRNA.

(d) What is the sequence of amino acids in a molecule of oxytocin? (The initiating methionine is not part of the oxytocin molecule.)

31. Methionine enkephalin is discussed in the *Check Your Understanding* of Section 17.7. List the sequence of bases on a molecule of DNA that codes for this compound.

32. Many people who have used bovine insulin to treat diabetes develop allergies to the injections. Scientists have now been able to produce commercial amounts of human insulin using recombinant DNA. Why is the insulin produced through recombinant DNA preferable for use by diabetics?

33. Sickle cell anemia resulted from the mutation of a specific region of DNA that codes for the polypeptide chain of a hemoglobin molecule.

(a) Suggest three possible causes of this mutation.

(b) The mRNA for the beta chain of normal hemoglobin has the base triplet GAA or GAG in the sixth position. What change in this base sequence results in the production of the beta chain of sickle cell hemoglobin?

(c) In general terms describe the steps involved in the production of one normal beta chain.

(d) Explain the difference in mobility of normal and sickle cell hemoglobin in an electric field.

(e) Explain how the substitution of one amino acid in the beta chain of hemoglobin can result in the symptoms of sickle cell anemia.

34. Aplastic anemia is a type of anemia in which the ability of the bone marrow to produce all types of blood cells is reduced. This type of anemia can be caused by prolonged exposure to the solvent benzene, which is a mutagen. Mutations in DNA may lead to a number of genetic diseases; however, mutations in RNA molecules rarely affect an organism. Explain.

35. Human insulin produced through genetic engineering is now available on the market. Recombinant DNA experiments are carried out using only specially weakened strains of *E. coli.* Why is this important, and what might happen if this precaution were not taken?

Chapter 21

Body Fluids

When Judy Bering finally returned to work, her secretary was delighted. Judy had been out for 10 days following an auto accident, and orders had certainly piled up while she was gone. Happily, Judy now felt fine and wasted no time getting started. Three days later, however, her secretary noticed a sudden change in Judy's work habits—she was away from her desk much more often than usual and seemed to be worried about something. Judy finally confided that all day long she had needed to go to the bathroom about every half-hour and yet had felt terribly thirsty the entire day. This intense thirst and frequent urination continued for two more days before Judy called her doctor for an appointment.

On hearing these symptoms, the doctor requested that the laboratory test the specific gravity of Judy's urine over a period of time. The initial specific gravity of her urine was measured at 1.003 (the normal range is 1.005 to 1.035), indicating that Judy's urine was less dense (or more dilute) than it should be. Over the course of the test Judy was not allowed to drink any liquids, yet the specific gravity of her urine did not increase as it normally should.

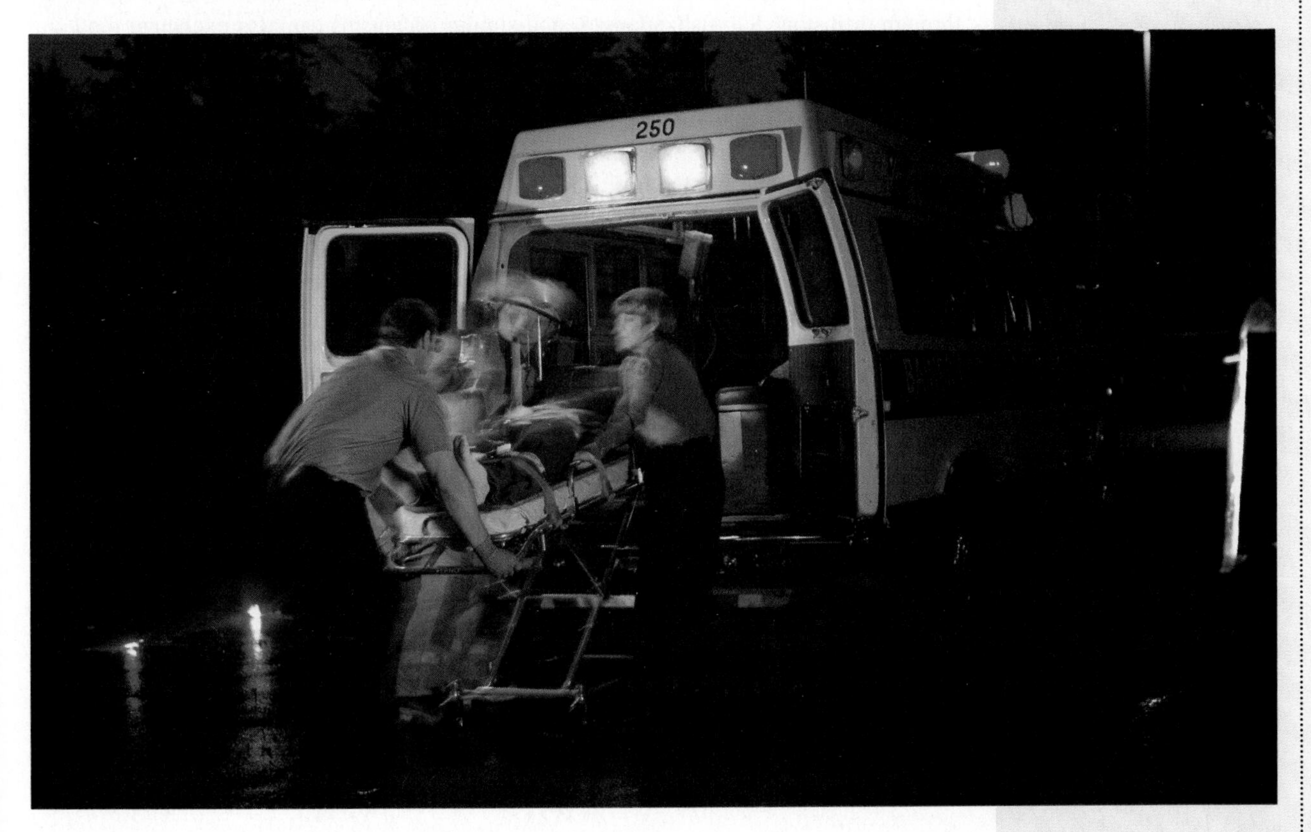

This indicated to the doctor that Judy might be suffering from a disease called diabetes insipidus. Although Judy had suffered these symptoms for only a few days, her diabetes insipidus actually resulted from the concussion she had received in the auto accident. That blow to her head had damaged a small organ called the pituitary gland, located just under the brain. The damaged pituitary had eventually stopped producing the hormone vasopressin (also called antidiuretic hormone, ADH), which causes the kidneys to reabsorb water from the urine. Without this hormone in her blood, Judy's kidneys were not functioning properly and were producing 15 to 20 L of urine each day (compared with the normal 1–2 L/day).

A person suffering from diabetes insipidus who does not constantly replace the water lost through urination becomes dehydrated. Fortunately, this condition is treatable, and Judy was given a nasal spray that replaces the vasopressin not supplied by the pituitary gland.

Laboratory tests such as those given to Judy require many different types of measurements to analyze accurately the various body fluids. This chapter examines the components that make up blood and urine and discusses several blood and urine abnormalities.

21.1 Body Fluids

You have probably observed simple single-celled organisms such as amoeba under a microscope. Such organisms exchange nutrients and wastes with their surroundings through diffusion. Jellyfish, sponges, and similar organisms can also exchange food and wastes by diffusion with the surrounding seawater. However, in the case of all vertebrates and most invertebrates, a different method of exchange is necessary. For such creatures diffusion is not efficient enough because their interior cells are too far from the surface. These more complex organisms must have a circulatory system that can bring nutrients to the cells and remove wastes from them.

Some organisms, such as the mollusks (snails, for example) and the arthropods (insects, for example), have open circulatory systems. Fluid is pumped from the heart through a series of tubes into an open cavity in the body tissues. Nutrients and wastes can then diffuse into and out of the cells making up the organism's tissues. The fluid is then pumped from the body cavity back to the heart, and the process starts all over again.

The circulatory system of vertebrates is a closed system (Fig. 21.1). In this type of system, blood is pumped from the heart into blood vessels (arteries) that have relatively large diameters. Fluid moves rapidly through these arteries into smaller branching arteries and then to the microscopic capillaries. As the blood slows down through the capillaries, nutrients and wastes are exchanged between the blood and the cells. The blood is then rapidly carried back to the heart through the veins.

Even in organisms with closed circulatory systems, not all of the body fluids are found within the circulatory system. The intracellular fluids are the fluids found within the cells. Fluids that are not in the cells are called the extracellular fluids. Blood is only one of the many extracellular fluids. Others include lymph, digestive juices, the synovial fluid that lubricates the joints, and the interstitial fluid—the fluid that fills the spaces (interstices) between the cells.

We study just two of the extracellular fluids in this chapter: blood and urine. We have seen that nutrients and other important compounds, such as vitamins, minerals, hormones,

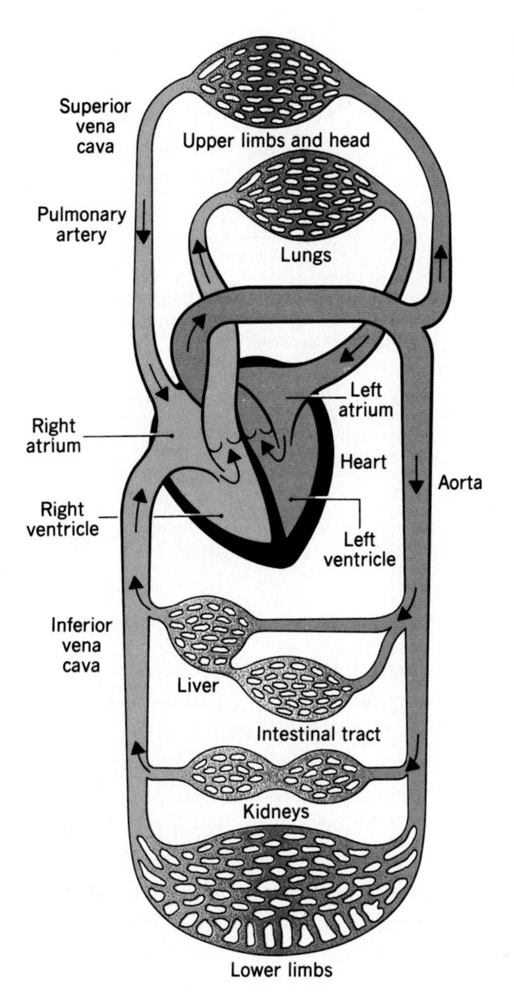

Figure 21.1 Human circulatory system.

and oxygen, must be brought to the cells. Similarly, wastes, such as carbon dioxide and urea, must be transported away from the cells. Blood and urine play an important role in this process.

21.2 Blood Components

Approximately one-twelfth of the human body is blood. This amounts to about 30 mL/lb of body weight for males, and slightly less for females. You can think of blood as both a solution of chemical compounds dissolved in water and as a suspension. As we saw in Chapter 9, true solutions do not separate but suspensions do. If fresh blood is centrifuged, the suspended material separates to the bottom of the tube. This suspended material makes up about 45% of the blood and is usually referred to as the **formed elements.** The exact percentage of formed elements is called the **hematocrit.** The solution remaining after the formed elements are removed is called the **blood plasma.** Plasma is about 92% water and about 7% protein (the remaining 1% is made up of salts, amino acids, creatinine, and other

List the major components of blood.

Describe the difference between blood plasma and blood serum.

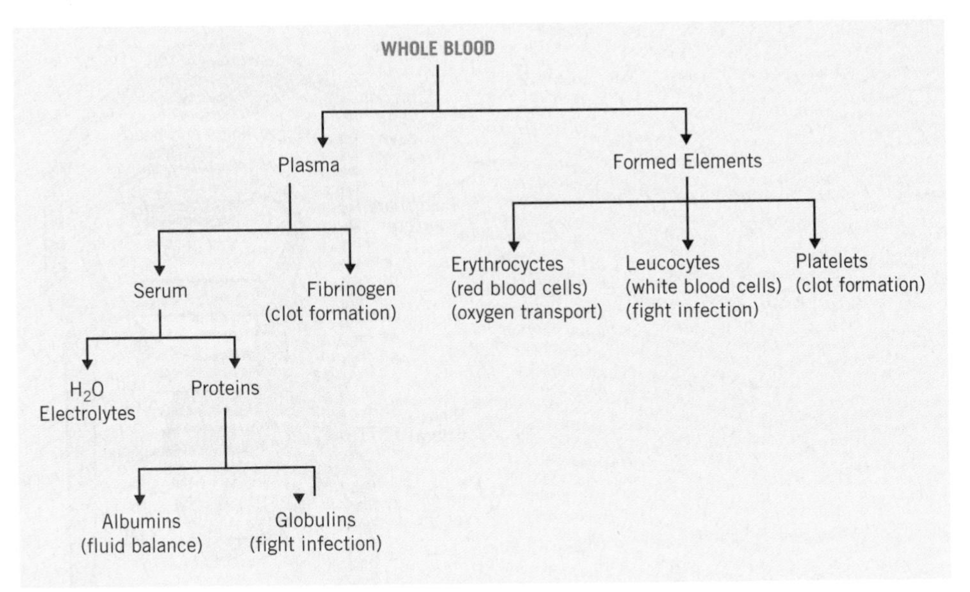

Figure 21.2 Blood components.

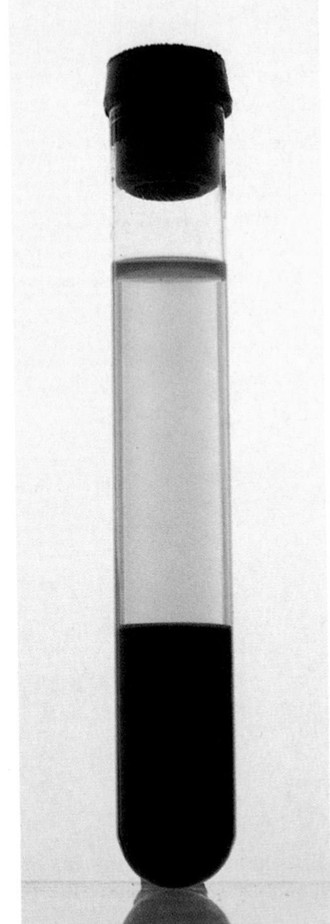

When blood is centrifuged, the formed elements are separated from the blood serum.

compounds). Blood allowed to stand in the open clots. The solution that remains after the clot is removed is called **blood serum** (Fig. 21.2).

The proteins that make up the blood plasma include the albumins, globulins, and fibrinogen. Approximately 54% to 58% of the total proteins are albumins, 40% to 44% are globulins, and the remaining 3% to 5% is fibrinogen. The albumins help maintain osmotic pressure. The globulins are involved in metal ion transport and the transport of oxygen, while the gamma globulins play an important role in fighting infectious diseases. Fibrinogen participates in the formation of blood clots. Because blood serum is the solution remaining after the formation of blood clots, blood plasma contains fibrinogen but serum does not. Finally, the formed elements consist of red blood cells (erythrocytes), white blood cells (leucocytes), and platelets.

21.3 Functions of Blood

Blood plays many roles in the body (Table 21.1). These include

- **The transport of oxygen and carbon dioxide between the lungs and the tissues.** The cells require oxygen as a reactant in the respiratory chain as part of the aerobic stage of glycolysis (see Section 19.4). Oxygen is carried by hemoglobin in red blood cells. Carbon dioxide is a byproduct produced in cells as part of the citric acid cycle.
- **Maintenance of fluid balance.** Organisms must be able to maintain a balanced flow of fluids into and out of their bodies as well as across cell membranes. Imagine what would happen if fluid balance were not regulated and maintained. Either the cells and the organism itself would absorb too much fluid and become bloated or would lose too much fluid and shrivel.
- **Maintenance of the proper pH in the body.** Many of the equilibrium reactions occurring in living organisms involve the gain or loss of hydrogen ions (H^+). As we saw in Chapter 8, equilibria shift when the concentration of any reactant or product is altered.

Table 21.1 Functions of the Blood

Functions of the Blood
Transport of oxygen and carbon dioxide between the lungs and tissue
Maintenance of fluid balance
Maintenance of proper pH
Formation of clots
Transport of food and wastes
Regulation of body temperature
Fighting invading microorganisms

Disruption of the body's pH balance can have serious effects on such vital processes as respiration.

List five important functions of blood.

- **Formation of clots.** Because we have a closed circulatory system, it is important that the system not develop leaks. If we experience a cut or other rupture of our circulatory system, a method is needed for resealing the broken blood vessel.
- **The transport of food and wastes.** In a multicellular organism, nutrients must be carried to the cells and waste products must be carried away. In our bodies, reactions in the digestive system hydrolyze carbohydrates, proteins, and lipids. These compounds are carried to the cells where they take part in the metabolic processes discussed in Chapter 19. The waste products of metabolism must then be carried to the kidneys and lungs.
- **Regulation of body temperature.** Because of its high specific heat (see Chapter 2), water acts as a moderating influence to resist rapid temperature changes. This helps us maintain a relatively constant body temperature with only small fluctuations.
- **Fighting invading microorganisms.** Our cells are constantly being invaded by viruses, bacteria, and other foreign substances. The white blood cells, or leukocytes, are responsible for defending our bodies against such invaders.

All of these important functions are performed by the blood in our bodies. In this chapter we focus on the transport of blood gases between the lungs and the tissues, and the way in which blood helps maintain the body's fluid balance, regulates pH, and forms clots. Before continuing with this chapter, you should review chemical equilibrium (Sections 8.4–8.6), buffers (Section 11.7), and osmosis (Section 10.8).

21.4 Blood Gas Transport

Write the equations that describe the transport of oxygen and carbon dioxide in blood. Use Le Chatelier's principle to explain how inhaled air affects these equilibria.

As we have seen, the blood transports oxygen from the lungs to the tissues, and carbon dioxide from the tissues to the lungs (Figs. 21.3 and 21.4). Because the solubility of oxygen in blood is very low, most of the oxygen is transported as oxyhemoglobin. About 10% of the carbon dioxide transported by the blood is in the form of the dissolved gas, 20% is transported as a complex with hemoglobin, and the other 70% is in the form of bicarbonate.

$$\text{Equilibrium I:}\quad \underset{\text{hemoglobin}}{HHb} + O_2 \rightleftharpoons H^+ + \underset{\text{oxyhemoglobin}}{HbO_2}$$

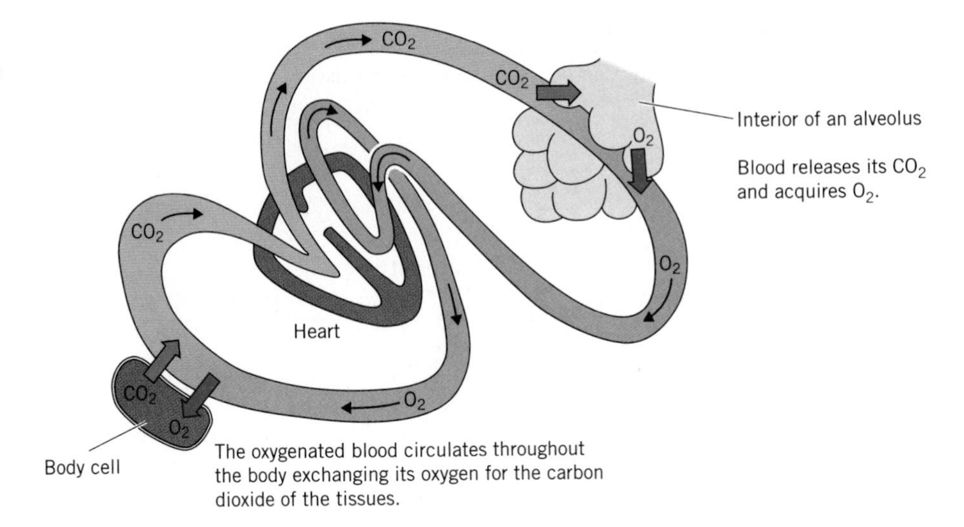

Figure 21.3 Oxygen and carbon dioxide exchange in the circulatory system.

Equilibrium I shows how hemoglobin (HHb) and oxygen combine to form the oxygen–hemoglobin complex (HbO_2). Note that a proton is part of this equilibrium. Now, from Le Chatelier's principle we know that a system at equilibrium, when subjected to stress, shifts to relieve that stress. The addition or removal of a component of the equilibrium system is a form of stress. In the lungs, the partial pressure of oxygen in the inhaled air is high. This increases the solubility of oxygen in the blood that is in contact with the inhaled air. The increased oxygen in the blood shifts Equilibrium I to the right, forming more of the oxyhemoglobin complex. This shift also increases the concentration of protons in the blood.

$$\text{Equilibrium I:} \quad \text{HHb} + O_2 \rightleftharpoons H^+ + HbO_2$$

Equilibrium II shows how bicarbonate and acid combine to give carbonic acid, which dissociates into carbon dioxide and water. The reverse of this reaction occurs when carbon dioxide is dissolved in water. Note the role of the proton in this equilibrium also.

$$\text{Equilibrium II:} \quad H^+ + HCO_3^- \rightleftharpoons H_2CO_3 \rightleftharpoons H_2O + CO_2(g)$$

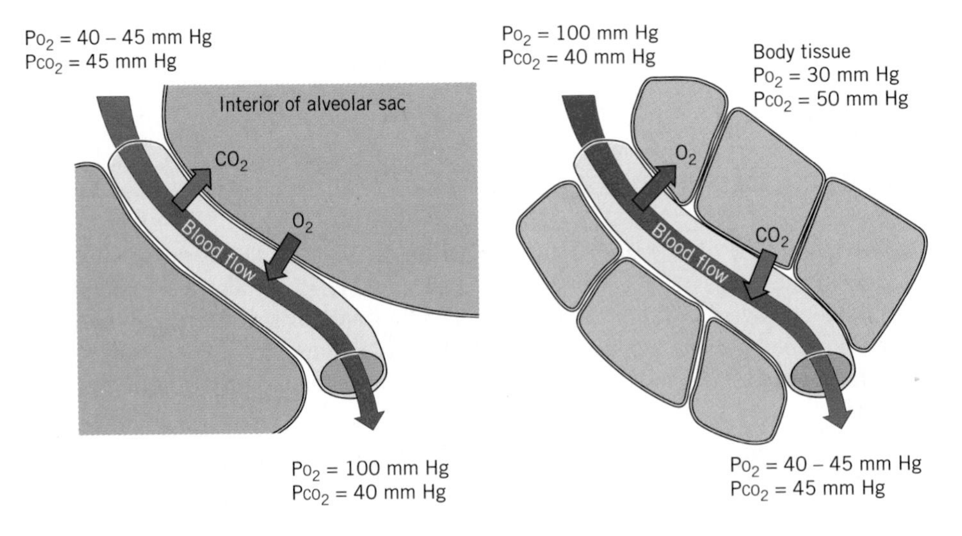

Figure 21.4 The effect of partial pressures on the flow of oxygen and carbon dioxide in the blood.

In the Lungs

We saw that increased oxygen in the lungs shifts Equilibrium I to the right. The resulting increased proton concentration also shifts Equilibrium II to the right, releasing carbon dioxide from the blood.

$$\text{Equilibrium II:} \quad H^+ + HCO_3^- \rightleftharpoons H_2CO_3 \rightleftharpoons H_2O + CO_2(g)$$

Because the partial pressure of carbon dioxide in inhaled air is low, its solubility is low, and it escapes from the blood into the lungs where it is exhaled.

Equilibrium III is the formation of the carbamino complex ($HbNHCO_2H$) between hemoglobin and carbon dioxide. To emphasize that carbon dioxide bonds to N-terminal amino acid groups ($—NH_2$) in the protein part of hemoglobin, the structures of hemoglobin and the complex with carbon dioxide are drawn somewhat differently than in Equilibrium I.

$$\text{Equilibrium III:} \quad HbNH_2 + CO_2 \rightleftharpoons HbNHCO_2H$$

We can see that a low concentration of carbon dioxide causes this equilibrium to shift to the left. The carbon dioxide that is then released from the blood is exhaled through the lungs.

$$\text{Equilibrium III:} \quad HbNH_2 + CO_2(g) \rightleftharpoons HbNHCO_2H$$

In the Tissues

In the tissues, where oxygen is consumed and carbon dioxide is released in the oxidative processes of metabolism, the partial pressure of carbon dioxide is high and that of oxygen is low. The high partial pressure of carbon dioxide increases its solubility in blood, causing Equilibrium II to shift to the left and Equilibrium III to shift to the right. More bicarbonate and more of the carbamino complex with hemoglobin are formed.

$$\text{Equilibrium II:} \quad H^+ + HCO_3^- \rightleftharpoons H_2CO_3 \rightleftharpoons H_2O + CO_2$$

$$\text{Equilibrium III:} \quad HbNH_2 + CO_2 \rightleftharpoons HbNHCO_2H$$

The shift to the left of Equilibrium II releases protons that in turn cause Equilibrium I to shift to the left. This releases oxygen from its complex with hemoglobin and makes the oxygen available to the tissues (Fig. 21.5).

$$\text{Equilibrium I:} \quad HHb + O_2 \rightleftharpoons H^+ + HbO_2$$

Example 21-1

In what direction does the equilibrium for the formation of oxyhemoglobin shift when the partial pressure of oxygen increases?

See the Question

The question asks us to determine which direction Equilibrium I shifts if the concentration of oxygen increases.

Think It Through (Execute the Math)

We need to apply Le Chatelier's principle to Equilibrium I. The equilibrium equation is

$$\text{Equilibrium I:} \quad \underset{\text{hemoglobin}}{HHb} + O_2 \rightleftharpoons H^+ + \underset{\text{oxyhemoglobin}}{HbO_2}$$

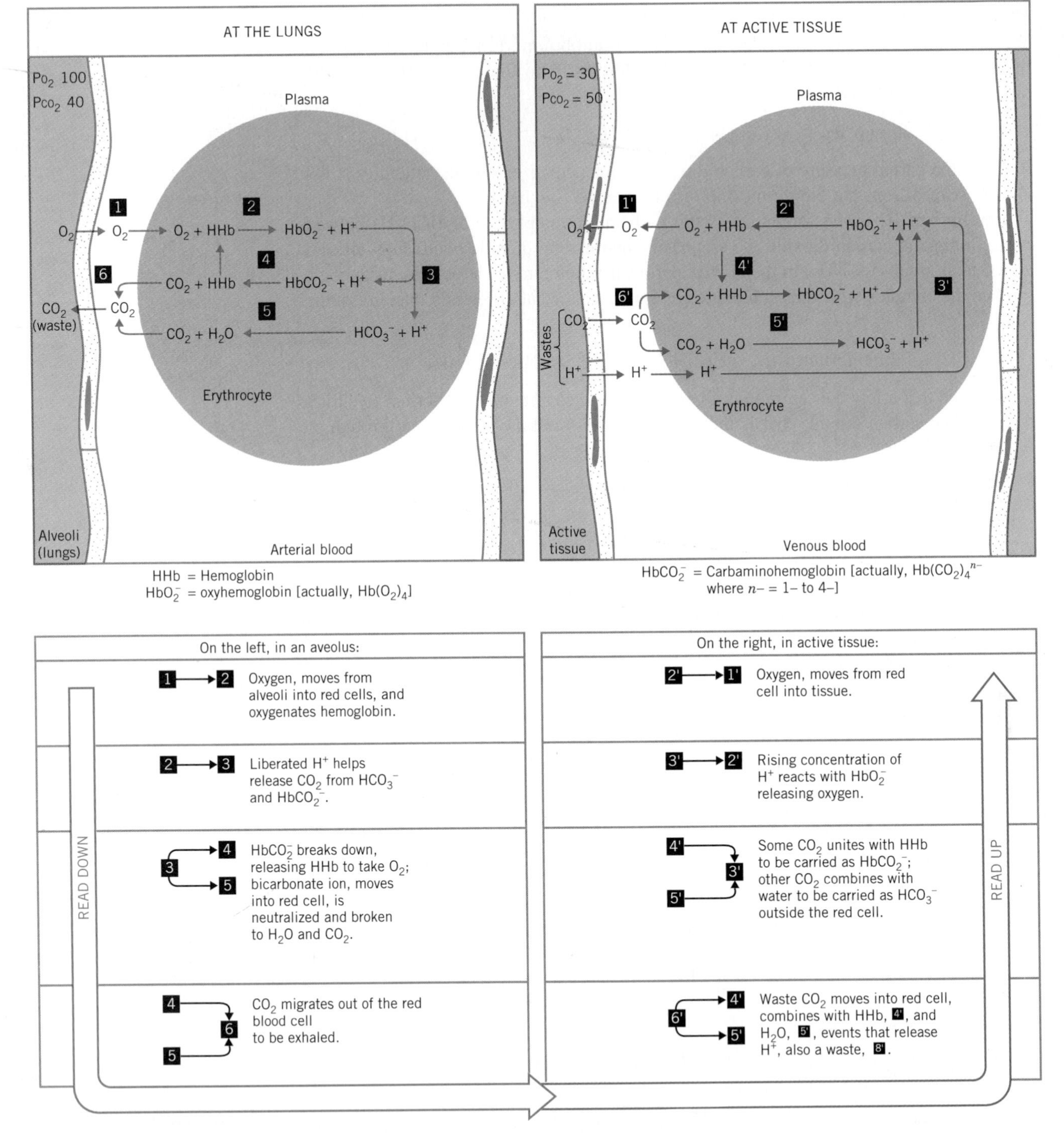

Figure 21.5 Oxygen–carbon dioxide equilibrium in the lungs and tissues.

Le Chatelier's principle states that when a system at equilibrium is subjected to stress, it shifts to relieve that stress. Increasing the oxygen partial pressure, which is on the left side of the equation, shifts the equilibrium to the right.

$$HHb + O_2 \rightleftharpoons H^+ + HbO_2$$

Prepare the Answer

Check Your Understanding

1. In what direction does the equilibrium for the formation of oxyhemoglobin shift if the pH increases?
2. In what direction does the equilibrium for the formation of carbonic acid (H_2CO_3) shift when the partial pressure of carbon dioxide increases?

21.5 Fluid Balance

Osmotic pressure (see Section 10.8) results from concentration differences across differentially permeable membranes. Remember that osmotic pressure depends on the number of dissolved particles, not on the type of substance those particles happen to be. When solutions of different concentration are separated by a differentially permeable membrane, an osmotic pressure difference occurs across that membrane. There is a net flow of solvent across the membrane from the side with lower osmotic pressure toward the side with higher osmotic pressure. The regulation of osmotic pressure is vitally important to living organisms if they are to maintain fluid balance. Although this is true for all living organisms, it's especially important for animals that live in water. Freshwater fish live in an environment that is at lower osmotic pressure than their blood. That is, freshwater is hypotonic compared with blood. If osmosis were not controlled, these fish would take in excess fluid. On the other hand, marine fish live in an environment with a salinity (salt content) of about 3.5%. That means that their environment is hypertonic relative to their blood. If osmosis were not controlled, these fish would lose fluids and suffer dehydration. One genus of fish (*Cyprinodon,* also known as pup fish) lives in water that can be twice as salty as the ocean. This fish would rapidly dehydrate if it weren't able to overcome the force of osmotic pressure.

To maintain osmotic pressure across a membrane, the membrane must prevent the passage of ions. But the walls of the blood capillaries are only partially able to prevent electrolytes from passing through. This means that an increase in the salt content of the blood results in only a temporary concentration difference across the membrane, and thus only a temporary osmotic pressure. In a few hours, the salts diffuse through the membrane and the osmotic pressure difference disappears.

Describe how fluid balance is maintained between the blood and tissues.

So how is fluid balance maintained? The plasma proteins are larger than the electrolytes and, more importantly, they are larger than the pores in the walls of the blood capillaries. It is the plasma proteins that maintain the concentration difference across the capillary walls and, therefore, the osmotic pressure difference. The albumins make up more than half of the total plasma proteins, so they have the greatest effect on blood osmotic pressure. Let's see how this works using a rather simplified example (Fig. 21.6).

In the example, the osmotic pressure is maintained constant at 30 units. (The actual units are not important here; what *is* important are the different magnitudes of the units.)

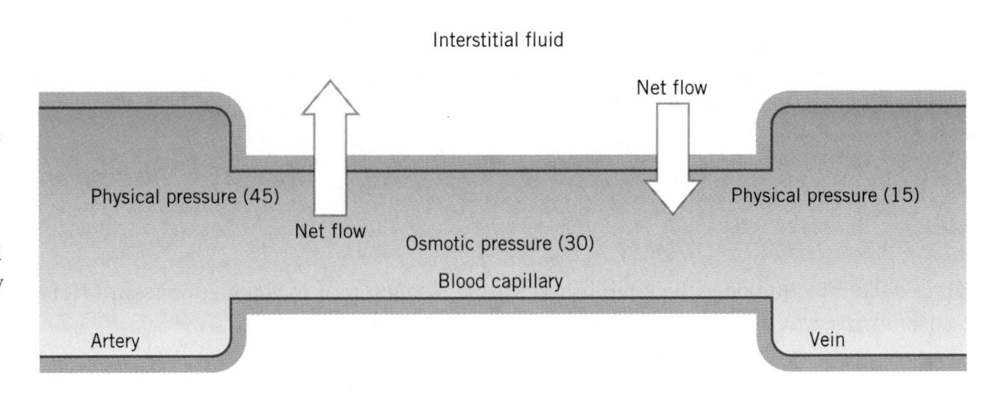

Figure 21.6 Normal fluid balance in a capillary. At the arterial end of the capillary blood pressure exceeds osmotic pressure and the net flow of fluid is into the tissues. At the venous end of the capillary osmotic pressure exceeds blood pressure and there is a net flow of fluid into the blood vessel. There is a balance between fluid flow into the tissues and fluid flow out of the tissues.

But the physical pressure of the blood is greater at the arterial end of the capillary than it is at the venous end. In this case, the blood pressure drops from 45 units at the arterial end to 15 units at the venous end. At the arterial end of the capillary, the physical pressure of the blood is greater than the osmotic pressure. This causes fluid to pass through the pores of the capillaries into the tissues. At the venous end of the capillary, the osmotic pressure is greater than the physical pressure of the blood. This causes fluid to flow out of the tissues into the blood vessel. In this example, therefore, a balance is maintained between fluid flowing into the tissues and fluid flowing out of the tissues.

Explain how changes in blood pressure, such as during cardiac failure, affect fluid balance.

When a person is suffering from cardiac failure, the heart is not able to pump blood from the veins efficiently. This causes a buildup of pressure in the veins. Let's look at our example again, but this time with a higher blood pressure in the vein and a lower pressure in the artery (Fig. 21.7). This small change in blood pressure has upset the delicate balance between the flow of fluids in the capillary and the interstitial spaces. Fluid builds up in the tissues in a condition known as **edema.** Fluid retention, particularly in the extremities, is a common problem for persons suffering from damaged hearts. Changes in the concentration of albumin in the blood also affect osmotic pressure and fluid retention.

Example 21-2

How does an increase in albumin concentration in the blood affect fluid retention in the tissues?

See the Question

We are asked to determine how a change in concentration of a component of the blood affects the osmotic pressure of blood and the exchange of fluids between the blood and tissues.

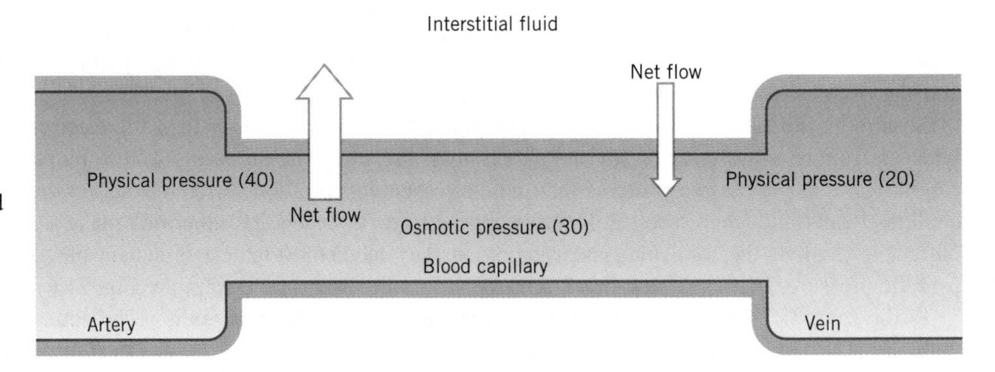

Figure 21.7 Fluid flow in edema. Higher blood pressure in the vein has upset fluid balance. The net flow of fluid into the tissues continues at the arterial end of the capillary. However, the flow of fluid out of the tissues into the blood vessel is reduced because there is less difference between blood pressure and osmotic pressure at the venous end of the capillary. This imbalance causes fluid to build up in the tissues.

Health Perspective / Congestive Heart Failure and Edema

Heart failure is a condition in which the heart becomes unable to pump blood efficiently. This may be caused by damage to the heart muscle from a heart attack or from other disease or may be caused by a failure of the heart valves to close properly.

Because the heart's pumping action is weakened, heart failure causes blood to accumulate in the veins. This accumulation results in "back pressure" within the veins. In Section 21.5 we saw that when blood pressure is sufficiently high at the venous end of the capillaries, osmotic pressure cannot draw fluids back into the blood vessels. That means that fluids build up in the tissues. If the right side of the heart is not working efficiently, fluid accumulates in all of the tissues except the lungs. The resulting edema is usually most obvious in the lower legs and ankles. If the problem is with the left side of the heart, the high venous back pressure causes fluid to accumulate in the alveoli of the lungs. Patients with such pulmonary edema suffer shortness of breath. Inadequate movement of blood also reduces the ability of the kidneys to remove water from the body. This also contributes to the accumulation of fluids throughout the body, increasing the level of edema.

Your Perspective: Doctors often advise a person with edema to take diuretics and to eat a low-salt diet. Suppose you are a nurse practitioner who is asked to explain these recommendations to an elderly female patient suffering from edema. Write down the words that you would say, just as if you were talking directly to her.

Think It Through (Execute the Math)

1. An increase in the blood albumin concentration causes an increase in blood osmotic pressure.
2. An increase in blood osmotic pressure at the arterial end of the capillary reduces the difference between osmotic pressure and physical blood pressure, which in turn reduces the passage of fluid into the tissues.
3. At the venous end of the capillary, the increased osmotic pressure increases the flow of fluid from the tissues into the blood vessel.

Prepare the Answer

The net result of these changes is a loss of fluid from the tissues, resulting in dehydration.

Check Your Understanding

When a person suffers severe trauma, such as a major injury or burn, albumin and other plasma proteins are lost from the blood. What effect does such a decrease in blood albumin concentration have on fluid balance?

21.6 Regulation of Blood pH

Give an example of the way in which the blood plasma is protected against large changes in pH.

In the human body the blood plasma has a normal pH of 7.4. A drop in pH below 7.0 or a rise above 7.8 is usually fatal. In Section 11.7, we saw that the major buffer system in the blood is the carbonic acid–bicarbonate system.

$$H_2CO_3 \rightleftharpoons HCO_3^- + H^+$$

The buffer systems in the blood are very effective in protecting this fluid from large changes in pH. For example, if 1 mL of a 10.0 *M* HCl solution is added to 1 L of unbuffered physiological saline (0.15 *M* NaCl) at a pH of 7, the pH falls to 2. If, however, 1 mL of a 10.0 *M* HCl solution is added to 1 L of blood plasma at a pH 7.4, the pH drops only to 7.2.

Various factors can cause an abnormal increase in acid levels in the blood. High acid levels can result from hypoventilation caused by emphysema, congestive heart failure or bronchopneumonia; an increase in the production of metabolic acids, caused by diabetes mellitus or some low-carbohydrate/high-fat diets; ingestion of excess acids; excess loss of bicarbonate in severe diarrhea; or decreased excretion of hydrogen ions through kidney failure. Each of these conditions causes an increase in the hydrogen ion level in the blood and a decrease in the concentration of basic components (such as bicarbonate), known as the alkaline reserve. Under such conditions the pH of the blood can drop to 7.1 or 7.2, resulting in a condition known as **acidosis** (called respiratory acidosis if its origin is in the respiratory system, and metabolic acidosis if the origin is other than respiratory) (Table 21.2). The body has ways to restore the blood pH to normal. First, it can expel the excess carbon dioxide formed from the carbonic acid, through an increase in the rate of breathing. Second, it can increase the excretion of H^+ and the retention of HCO_3^- by the kidneys, resulting in acidic urine (pH about 4).

◆ **Define acidosis and alkalosis, and describe conditions in the body that cause each of these.**

Table 21.2 **Acidosis and Alkalosis**

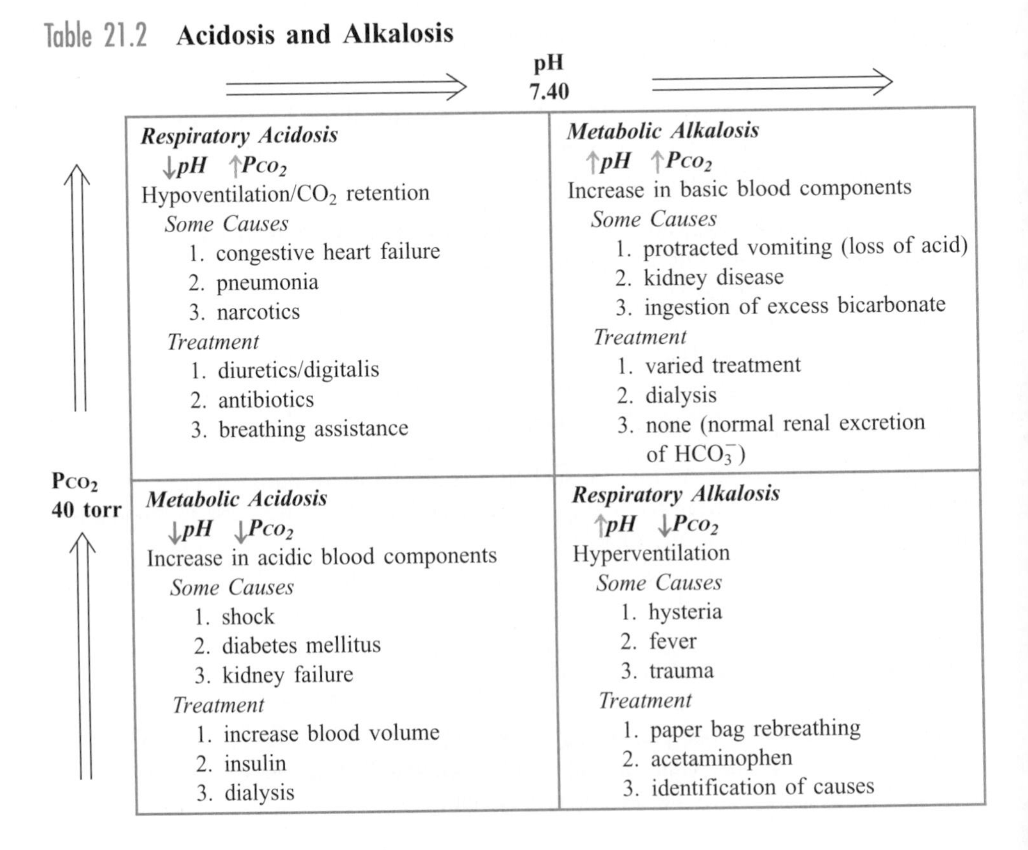

pH 7.40 (arrows →)

P_{CO_2} 40 torr (arrows ↑)

Respiratory Acidosis ***↓pH ↑P_{CO_2}*** Hypoventilation/CO_2 retention *Some Causes* 1. congestive heart failure 2. pneumonia 3. narcotics *Treatment* 1. diuretics/digitalis 2. antibiotics 3. breathing assistance	***Metabolic Alkalosis*** ***↑pH ↑P_{CO_2}*** Increase in basic blood components *Some Causes* 1. protracted vomiting (loss of acid) 2. kidney disease 3. ingestion of excess bicarbonate *Treatment* 1. varied treatment 2. dialysis 3. none (normal renal excretion of HCO_3^-)
Metabolic Acidosis ***↓pH ↓P_{CO_2}*** Increase in acidic blood components *Some Causes* 1. shock 2. diabetes mellitus 3. kidney failure *Treatment* 1. increase blood volume 2. insulin 3. dialysis	***Respiratory Alkalosis*** ***↑pH ↓P_{CO_2}*** Hyperventilation *Some Causes* 1. hysteria 2. fever 3. trauma *Treatment* 1. paper bag rebreathing 2. acetaminophen 3. identification of causes

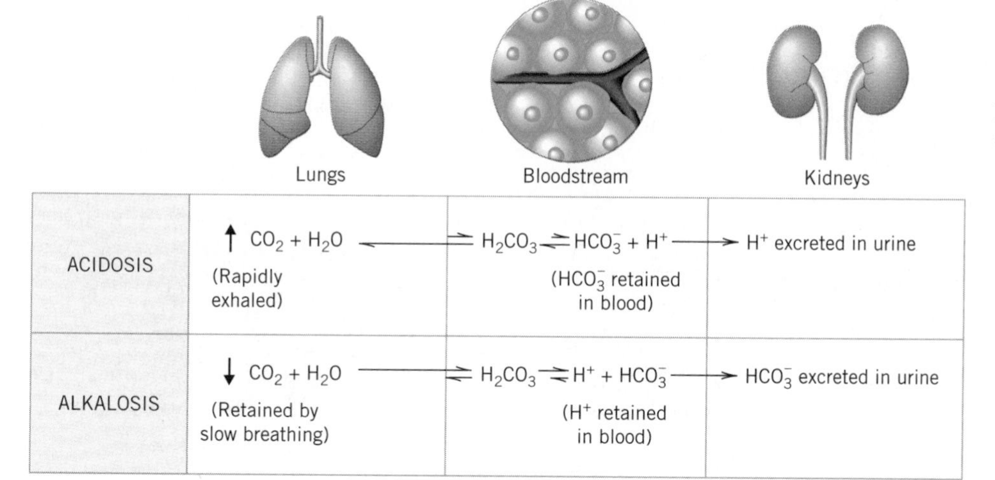

Lungs
Bloodstream
Kidneys
ACIDOSIS
↑ $CO_2 + H_2O$
(Rapidly exhaled)
$H_2CO_3 \rightleftharpoons HCO_3^- + H^+$
(HCO_3^- retained in blood)
H^+ excreted in urine
ALKALOSIS
↓ $CO_2 + H_2O$
(Retained by slow breathing)
$H_2CO_3 \rightleftharpoons H^+ + HCO_3^-$
(H^+ retained in blood)
HCO_3^- excreted in urine

Figure 21.8 The carbonic acid–bicarbonate buffer system acts in the blood to prevent both acidosis and alkalosis.

The bicarbonate buffer system also protects against an addition of strong base to the system. A base reacts with the hydrogen ions to produce water, decreasing the concentration of hydrogen ions in the system. This drives the reaction to the right.

$$H_2CO_3 \rightleftharpoons HCO_3^- + H^+$$

Such an increase in base in the blood can occur in cases of hyperventilation during extreme fevers or hysteria, from excessive ingestion of basic substances such as antacids, and in severe vomiting. The pH of the blood can increase to pH 7.5, resulting in a condition known as **alkalosis** (see Table 21.2). Alkalosis is not as common as acidosis. The body's means for returning the pH to normal are a decrease in expulsion of carbon dioxide by the lungs and an increase in excretion of HCO_3^- by the kidneys, resulting in an alkaline urine (pH greater than 7) (Fig. 21.8).

Drawing arterial blood gases (ABG) gives an accurate measure of the pH, $P\text{CO}_2$, and $P\text{O}_2$ of the blood that is available to the cells in the tissues. The pH reading tells whether a patient is suffering from acidosis or alkalosis, and the $P\text{CO}_2$ value determines whether the cause is respiratory or metabolic. The $P\text{O}_2$ value is a vitally important clinical reading because the brain stores no oxygen and requires a constant supply. The concentration of blood oxygen affects the acid–base balance only indirectly, however.

Another buffer system, active mainly within the cells, is the phosphate buffer system, which has a maximum buffering action at a pH of 7.2.

$$H_2PO_4^- \rightleftharpoons HPO_4^{2-} + H^+$$

Adding strong acid to this system drives the reaction to the left, increasing the concentration of $H_2PO_4^-$, which is only weakly acidic. Large amounts of $H_2PO_4^-$ result in acidosis, but the body eliminates the excess in the urine. Adding strong base to the system drives the reaction to the right, as the hydrogen ions react with the base to form water. Large amounts of HPO_4^{2-} are found in alkalosis, but under normal kidney function the HPO_4^{2-} is also excreted in the urine (Fig. 21.9).

Normal metabolic reactions in the body result in the continuous production of acids. Body cells produce an average of about 10 to 20 mol of carbonic acid each day, which is equivalent to 1 or 2 L of concentrated HCl. This acid must be removed from the cells and carried to the organs of excretion without disrupting the pH of the blood. It is through the

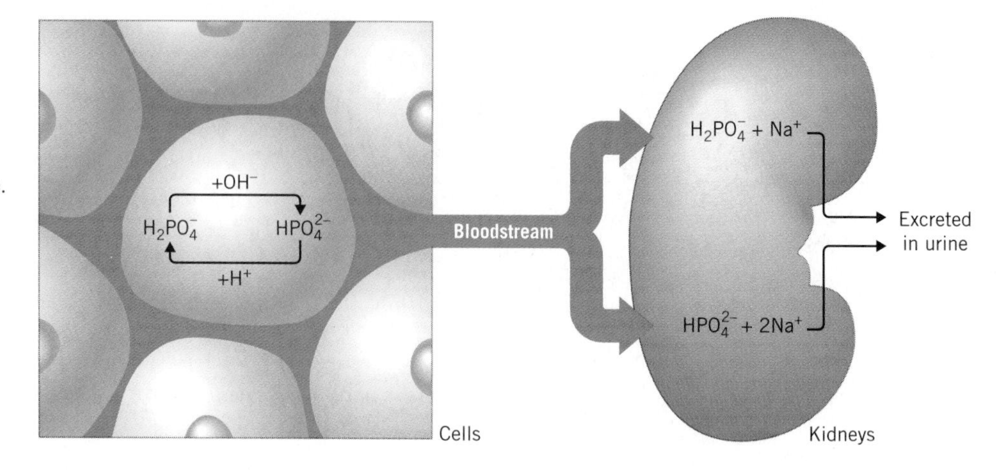

Figure 21.9 The phosphate buffer system acts in the cells to prevent changes in pH. The kidneys remove any excess HPO_4^{2-} or $H_2PO_4^-$ from the blood for excretion in the urine.

action of the buffer systems in the cells and extracellular fluid that our bodies are protected from deadly changes in pH that would otherwise be caused by these acids.

21.7 Formation of Blood Clots

The clotting of blood is a very complicated process involving several different chemical pathways. In addition to an adequate supply of blood platelets, it requires a variety of proteins, enzymes, and metal ions. Anything that impairs the clotting process can result in hemorrhage. For this reason, careful monitoring of blood clotting times can be critical to such medical decisions as drug dosages and treatment strategies.

When blood vessels are injured, blood cells called platelets immediately begin to adhere to one another and ''plug'' leaks in the blood vessels. Drugs, disease, surgery, or blood transfusions can cause the blood to have too few platelets or can impair platelet function. Taking just one aspirin a day can decrease platelet function. For this reason, aspirin is being prescribed for patients at high risk of heart attack or stroke, conditions caused by dangerous clot formation. Similarly, aspirin can cause bleeding problems in routine surgery if the patient does not report use of this medication. Another fairly common cause of easy bleeding is ITP, a condition in which a patient has low platelet counts for no apparent reason.

The buildup of blood platelets around the injured tissue begins the process of blood clotting or coagulation, which is the body's major defense against blood loss. Blood clotting requires the participation of 20 different substances, most of which are glycoproteins produced in the liver. Scientists refer to the clotting factors by their common names and by roman numerals. For example, fibrinogen is factor I, and prothrombin is factor II. These factors react in a cascading fashion through two pathways (called the intrinsic and the extrinsic pathways) that eventually lead into a common pathway that results in the formation of a blood clot. This process is extremely complex and confusing, but by understanding where problems occur in these pathways, physicians are better able to diagnose and treat bleeding disorders.

◆

Describe the processes that lead to the formation of blood clots.

Let's take a somewhat simplified look at the chemistry of the clotting process. When the blood platelets come in contact with tissue—such as when the skin is cut—they disintegrate. This disintegration causes the release of a platelet factor that interacts with a number of compounds in the plasma to produce thromboplastin.

Health Perspective / Strokes and Anticlotting Drugs

"Stroke" is the name given to the conditions that occur when the blood supply to part of the brain is reduced or completely cut off. The symptoms of stroke can range from a temporary deterioration in speech or vision to a complete loss of consciousness. The medical term for stroke is cerebrovascular accident.

The most common form of stroke is caused by an insufficient supply of blood through an artery. This commonly occurs when lipids deposit on the arterial walls, calcium salts deposit on top of the lipids, and then a fibrous net forms over all of this. When this whole layer (called a plaque) sticks out into the blood vessel, it triggers the formation of a clot. As the plaque and the clot grow, the entire artery can become blocked, shutting off the blood supply to the tissues. If this blockage occurs in an artery supplying blood to the brain, a stroke may occur. Or blood flow to the brain might become obstructed by a clot that formed elsewhere in the body and was carried through the bloodstream to the brain. The nerve tissues that rely on the blocked artery for their blood supply then become unable to function and may even die.

Although a stroke cannot be cured once it occurs, the likelihood of future strokes can be reduced. Patients who have suffered a stroke, or who may be at high risk to have a stroke, can be treated with compounds called *blood thinners.* This name is misleading because these compounds don't actually make the blood any thinner or less viscous. A better term for such compounds is anticoagulants, because they act to reduce the likelihood of clot formation.

Heparin is the most commonly used anticoagulant, and it is also used before and after surgery. Another use for heparin is in preventing clotting in capillary tubes used to collect small samples of blood. Warfarin (Coumadin) is often used for the long-term prevention of clots. Similar compounds are used in some rodent poisons because they cause rodents who eat them to die of internal bleeding. Aspirin is particularly effective in preventing strokes resulting from the accumulation of platelets in the bloodstream. Drugs like streptokinase are powerful clot busters that actually dissolve clots.

These anticoagulants and clot busters can't be used to treat another, less common type of stroke known as a cerebral hemorrhage. In this type of stroke an artery in the brain leaks. The accumulation of blood in the brain compresses the brain tissue, causing it to die. Use of anticoagulants for this type of stroke would only worsen the leakage and thus increase the damage to the brain.

Your Perspective: Heart attacks and strokes are the leading causes of death among males. The American Heart Association recommends that *all* men over 50 take an aspirin each day. Write a 30-second public-service radio announcement explaining the reasons for this recommendation and urging that listeners start doing it.

$$\text{Platelet factors} \longrightarrow \text{thromboplastin}$$

Thromboplastin catalyzes the conversion of prothrombin to thrombin.

$$\text{Thromboplastin} + Ca^{2+} + \text{prothrombin} \longrightarrow \text{thrombin}$$

Prothrombin is the inactive form of the enzyme thrombin. It is produced in the liver and depends for its synthesis on vitamin K intake. Low levels of vitamin K interfere with efficient formation of blood clots.

Health Perspective / Drawing Blood

Blood is drawn for many different laboratory tests. Because the collection tubes may differ for certain tests, they are commonly color-coded. For example, collection tubes that have red tops are used for tests on blood serum, whereas green tops indicate tubes for plasma studies.

For tests in which whole blood or blood plasma is needed, an anticoagulating agent is added to the tube. This anticoagulant might be EDTA or heparin. If blood serum is needed, no anticoagulating agent is added. If the test involves a study of glucose, an agent that inhibits glycolysis is added to the collection tube.

When collecting blood samples, care must be taken to prevent infection. Technicians collecting blood should wash their hands and put on protective-weight latex gloves to prevent cross-contamination from other samples being collected, as well as to prevent possible self-infection. Before drawing blood, the technician should thoroughly clean the puncture site with alcohol or a povidone-iodine sponge to prevent infection by organisms on the surrounding skin.

Blood samples for most tests are taken from a vein. The technician most commonly punctures a vein in the forearm but may instead use a vein in the wrist. A tourniquet is used above the puncture site. It should be tight enough to partially stop the flow of blood from the veins to the heart while still allowing flow in the arteries. This causes the veins in the arm to dilate so they are easier to find and to puncture.

Blood samples are used for a wide variety of tests. The most common blood test, the complete blood count (CBC), is performed to determine the number of each type of blood cell in a given volume of blood. A test for lipids is called for when the physician suspects possible coronary artery disease. This test is used to determine the levels of lipoproteins, triglycerides, and cholesterol in the blood. Other tests determine the concentration of particular electrolytes in the blood, such as potassium. Tests for enzymes can be used to determine if a change has occurred in the functioning of certain organs, such as the liver or heart.

For certain tests blood is collected from an artery rather than a vein. This process is known as arterial blood gas (ABG) analysis. An ABG analysis determines the partial pressures of oxygen and carbon dioxide and the pH of the blood. The partial pressure of oxygen in arterial blood is an indicator of the ability of the lungs to transfer this gas to the blood, whereas the partial pressure of carbon dioxide indicates the ability of the lungs to remove this gas. The determination of blood pH provides information about the acid–base balance of the blood.

The most common problems that can occur during the collection of blood samples are hematoma and infection. A hematoma causes tenderness and a bruise around the puncture site. It results from a leakage of blood from the puncture into the surrounding tissue. An infection may be caused by inadequate cleansing of the skin around the collection site.

Because the Ca^{2+} ion is required for the synthesis of thrombin, low calcium levels in the blood also slow down clot formation. When physicians need to prevent blood from clotting, they add to the blood sample compounds that ''tie up'' (or complex) calcium. The citrate ion and EDTA are used for this purpose.

Thrombin is the enzyme that catalyzes the conversion of fibrinogen to fibrin.

$$\text{Fibrinogen} \xrightarrow{\text{thrombin}} \text{fibrin}$$

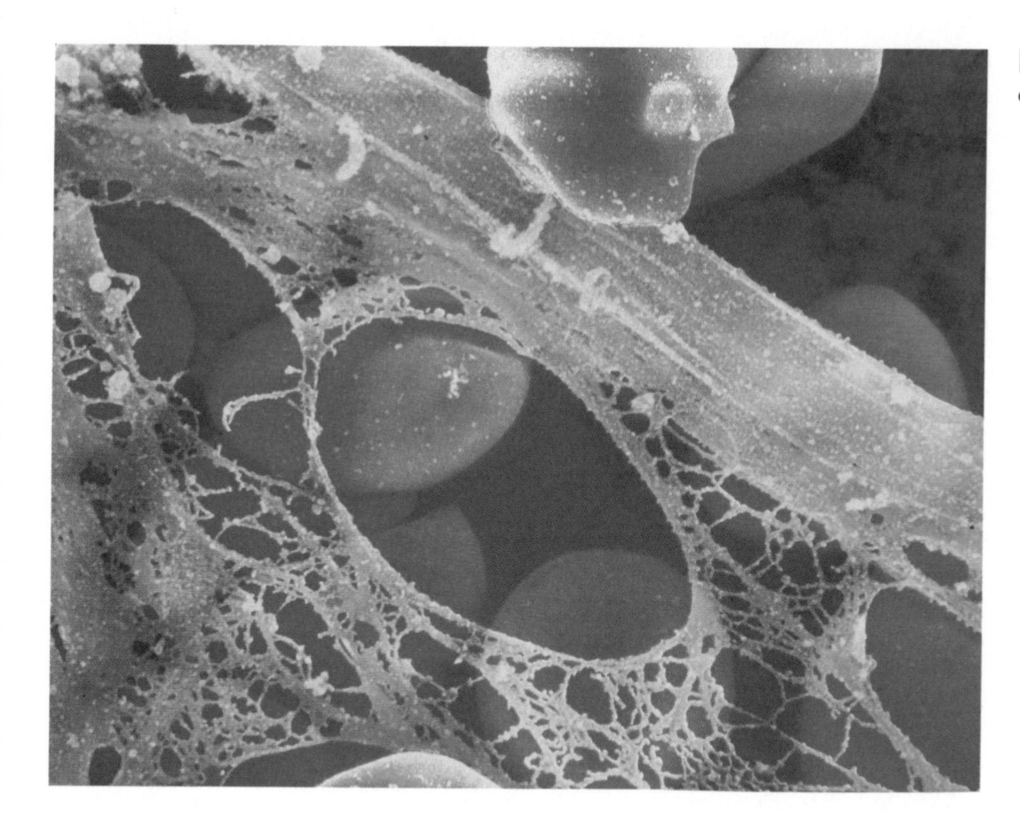

Figure 21.10 SEM of a blood clot showing fibrin network.

Fibrinogen is a soluble globular plasma protein, whereas fibrin makes up the insoluble clot (Fig. 21.10). You can think of the clotting process as a series of reactions that convert fibrinogen to fibrin, which thereby stop the bleeding.

21.8 Urine Specific Gravity

Each day an adult excretes between 800 mL and 2000 mL of urine. This amount varies depending on various conditions that can influence fluid loss and fluid intake. For example, the loss of fluid through heavy sweating in hot weather reduces urine amounts. The minimum volume of urine needed to remove body wastes from an adult is about 500 mL/day.

The specific gravity of urine can be an indicator of disease or disruption in normal metabolism. As you recall from our earlier discussions, specific gravity measures the concentration of dissolved solids in a liquid without indicating what those solids are. In the past, hydrometers were used to measure the specific gravity of urine. Today urine specific gravity can be quickly determined using a specially coated dipstick or a small refractometer. Normal urine has a specific gravity of about 1.010, with a range of 1.005 to 1.035. (Remember that pure water has a specific gravity of 1.000.) The greater the concentration of dissolved solids in the urine, the higher the specific gravity. Also, the lower the volume of excreted urine, the greater the concentration of dissolved solids—and, therefore, the higher the specific gravity.

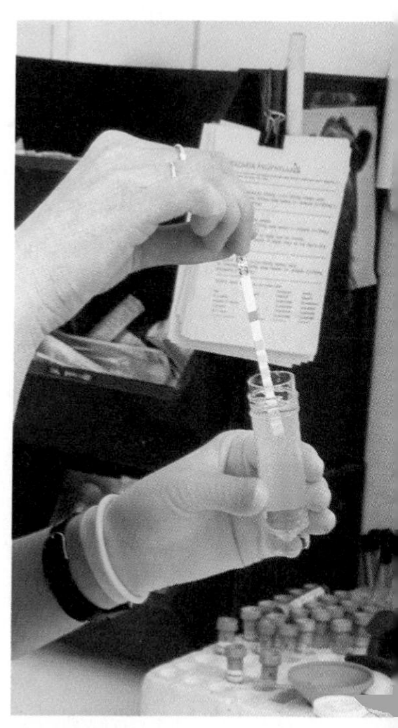

Testing for urine specific gravity with a specially coated dip-stick.

If a person's urine has a specific gravity reading below this normal range, it may indicate that the flow of water from the body into the urine is occurring at an abnormally

Define polyuria and oliguria, and list conditions under which each of these might occur.

high rate. Production of large amounts of urine, known as **polyuria,** may be due to high fluid intake or result from several other causes, including diabetes insipidus. This disease, characterized by an extremely high urine flow, is caused by a lack of antidiuretics. Antidiuretics are compounds produced in the brain that regulate urine flow. (You may have heard of diuretic drugs—compounds that increase urine flow—which are often prescribed for treatment of high blood pressure.) Polyuria also might indicate that the kidney is losing its ability to concentrate the urine.

Diabetes mellitus also causes polyuria, although the symptoms are less extreme than in diabetes insipidus. The names of these two conditions are derived from Greek and Latin words. *Diabetes* means "to pass through"; *mellitus* means "sweet," and *insipidus* means "without taste." In diabetes mellitus, glucose is excreted in the urine, making it taste sweet—hence the term "mellitus." The polyuria of diabetes mellitus is accompanied by high specific gravity because of the glucose and other compounds dissolved in the urine. The polyuria of diabetes insipidus is accompanied by low specific gravity, and because there is a low concentration of dissolved solids the urine is "without taste."

Oliguria is the term used to describe low urine volume. Low urine volume occurs whenever the body loses an excessive amount of water through sweating, fever, vomiting, bleeding, or diarrhea. This leads to high concentrations of urinary constituents like glucose and proteins, which causes high specific gravity. Some underlying causes of high urine specific gravity may be diabetes mellitus, dehydration, adrenal insufficiency, hepatitis, edema, or congestive cardiac failure. The fluid loss (dehydration) that accompanies diarrhea is a major cause of infant mortality in some parts of the world.

21.9 Urine pH

The pH of urine ranges between 4.8 and 7.5, with an average value of around 6.0. This slight acidity of urine is important because it helps to keep metals ions in solution. Metals typically precipitate from basic solutions. Note that in the case of blood, whose pH is slightly on the basic side (7.35–7.45), metals are prevented from precipitating by the formation of complexes with proteins.

Low urine pH is caused by excess acid in the system. Such excess acid can result from a high-protein diet or can be caused by diabetes mellitus. In this disease, a high level of ketone bodies are found in the blood. Because two of the ketone bodies are carboxylic acids, their presence lowers the pH of the blood. Low blood pH leads to low urine pH as the body tries to maintain its acid–base balance. The presence of ketone bodies in the blood is known as ketonemia, and their presence in the urine is called ketonuria.

High urine pH may occur after meals and is sometimes called the "alkaline tide." When metabolized, most fruits and vegetables produce the basic compound bicarbonate. Some urinary tract infections also cause a high urine pH. For this reason, foods that lower urine pH (such as cranberry juice) are sometimes recommended for the treatment of urinary tract infections.

Although it is not equal to the pH of the blood, the pH of the urine mirrors blood pH. If the pH of the blood falls, that of the urine also falls. As blood pH increases, the pH of the urine increases. This is one of the ways the body is able to eliminate or retain protons, thereby maintaining its pH balance.

21.10 Normal Components of Urine

Inorganic Components

List the major inorganic components of urine and the sources of these substances.

The major cations in the urine are sodium (Na^+) and potassium (K^+). The amount of sodium chloride excreted in the urine is about 10 to 15 g each day. As you might expect, the exact amount of these ions in the urine varies with the amount in the diet (Table 21.3).

The ammonium ion (NH_4^+) is produced by the deamination of amino acids in the kidneys. Ammonia is an important compound in the regulation of pH because it accepts a proton to form the ammonium ion, which can then be excreted in the urine. This process gives our bodies a way to remove protons from the blood. High levels of the ammonium ion in the blood can be an indication of acidosis, such as found in diabetes. It can also result from a high-protein diet.

Phosphate is found in the urine as both HPO_4^{2-} and $H_2PO_4^-$. As with the cations, the total amount of phosphate present in the urine reflects the amount of phosphates in the diet. The ratio of the two phosphate salts is determined by the acid–base balance of the body. Consider the following equilibrium:

$$H_2PO_4^- \rightleftharpoons HPO_4^{2-} + H^+$$

If excess protons cause the blood pH to fall, this equilibrium shifts to the left in order to remove protons. But when the pH is high, protons need to be retained by the body, so the equilibrium shifts to the right. The ratio $HPO_4^{2-}/H_2PO_4^-$ in blood is 1.6 : 1. A lower than normal ratio in the urine is an indication of lower than normal blood pH.

Sulfur is found in the urine primarily as the sulfate ion (SO_4^{2-}). Some is also present in the form of sulfur-containing organic compounds. Because proteins are our main source of sulfur in foods, the amount of sulfur in our urine reflects the amount of protein in our diet.

Table 21.3 Components of Normal Urine

Component		*Dietary Source*
Inorganic		
ammonium	NH_4^+	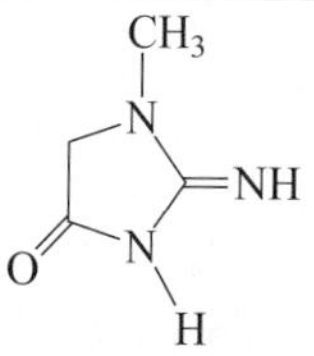 proteins
calcium ions	Ca^{2+}	milk, some vegetables
chloride	Cl^-	salt
phosphate	HPO_4^{2-}, $H_2PO_4^-$	phosphoproteins, phospholipids 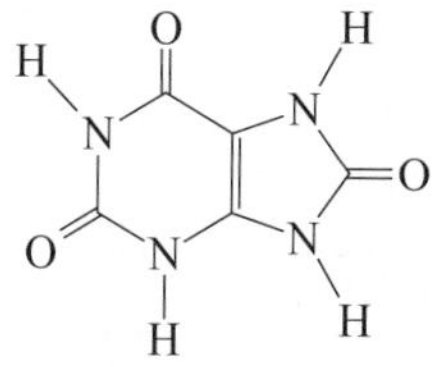
sodium ions	Na^+	salt
sulfate	SO_4^{2-}	proteins
Organic		
creatinine	(structure: ring with N–CH_3, $=NH$, $C=O$, N–H)	proteins
urea	$H_2N-C(=O)-NH_2$	proteins
uric acid	(structure: purine ring with three $C=O$ and four N–H)	purines (legumes and organ meats)

List the major organic components of urine and the sources of these substances.

Organic Components

Urea is the major organic component of urine. About 25 g of urea is produced each day in the urea cycle (see Fig. 19.14). Excreting urea through the urine is the chief method for removing nitrogen from the body.

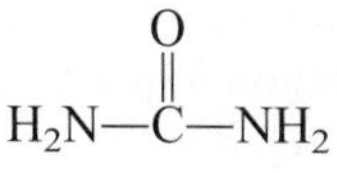

$$H_2N-\overset{\overset{\displaystyle O}{\|}}{C}-NH_2$$

urea

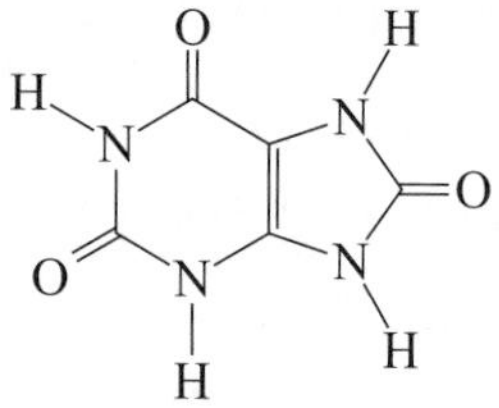

uric acid

Uric acid is also excreted in the urine, but less than 1 g is excreted daily. Excess uric acid can cause a condition called gout. Uric acid is produced by the metabolism of the purines—guanine and adenine—which are components of nucleic acids (see Chapter 20).

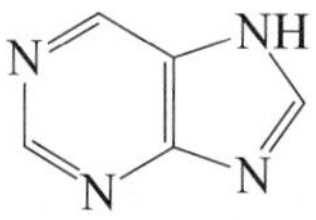

purine 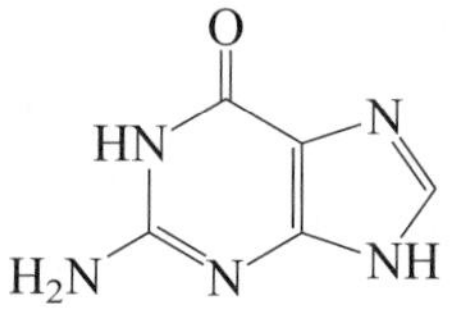   guanine 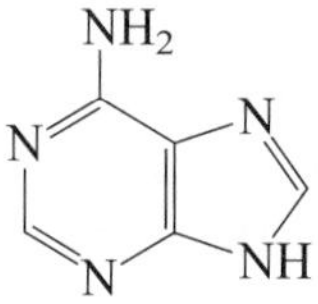   adenine

Creatine is found in muscle tissue as phosphocreatine.

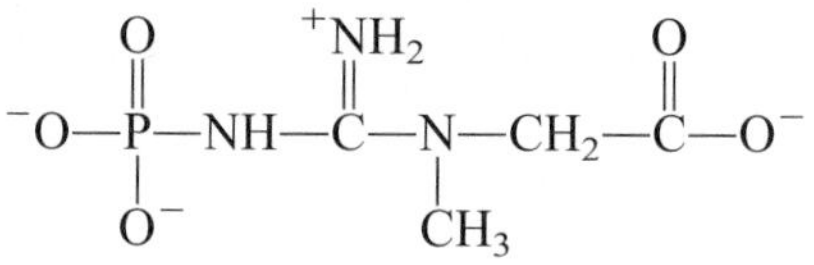

$$^{-}O-\overset{\overset{\displaystyle O}{\|}}{\underset{\underset{\displaystyle O^{-}}{|}}{P}}-NH-\overset{\overset{\displaystyle ^{+}NH_2}{\|}}{C}-\underset{\underset{\displaystyle CH_3}{|}}{N}-CH_2-\overset{\overset{\displaystyle O}{\|}}{C}-O^{-}$$

phosphocreatine

Phosphocreatine acts as a high-energy compound and is used in the conversion of ADP to ATP.

$$\text{Phosphocreatine} + \text{ADP} \rightleftharpoons \text{creatine} + \text{ATP}$$

Creatinine, which is excreted in the urine, is a waste product produced from creatine. The amount of creatinine in the urine is fairly constant for any given person. Normally, the urine contains about 0.8 to 2.0 g of creatinine per day. The amount of creatinine per day per kilogram of body weight is known as the creatinine coefficient (Table 21.4).

$$\text{Creatinine coefficient} = \text{mg creatinine/day/kg body weight}$$

Table 21.4 Creatine Concentrations in Body Fluids

	Men	*Women*
Normal values	*Creatine Concentration*	
Plasma	0.9–1.5 mg/100 mL	0.8–1.2 mg/100 mL
Urine	1.0–2.0 g/day	0.8–1.8 g/day
Creatinine coefficient[a]	20–26	14–22

[a]Creatinine coefficient = mg of creatinine excreted per day per kg of body weight.

Health Perspective / Uric Acid and Gout

In the old comic strip ''Maggie and Jiggs,'' Jiggs suffered from a malady called ''the gout.'' He often used this ailment as an excuse to avoid doing chores that Maggie wanted him to do. Despite the humor in the comic strip, gout is no laughing matter. The disease causes extremely painful swelling of the joints, especially in the foot and big toe. The culprit causing gout is uric acid.

Most of the nitrogen produced by the body in the metabolism of nitrogen-containing compounds is eliminated as urea. Humans, however, also produce a small amount of uric acid in the breakdown of purines formed from the hydrolysis of nucleic acids. Most other mammals convert uric acid to the water-soluble compound allantoin. The enzyme responsible for this conversion is called urate oxidase. Unfortunately, humans do not synthesize this enzyme and cannot convert uric acid to a water-soluble compound. In the human body, therefore, the insoluble uric acid must be carried by the blood to the kidneys for elimination. The normal blood concentration of uric acid is 4 mg/dL. Gout results from hyperuricemia, which is an abnormally high amount of uric acid in the blood. When the blood concentration of uric acid rises above 7 mg/dL, the uric acid precipitates in tissues, especially in joints and the tendons surrounding them. Although gout mostly affects the joints of the foot and the big toe, uric acid crystals can also form in the joints of the fingers. Wherever they form, these needlelike crystals cause extremely painful inflammation and swelling of the joints. In advanced stages of the disease, cartilage can be destroyed, and the joints may become permanently deformed and stiff. Treatment of gout involves elevation of the affected areas, a change of diet, and administration of drugs to prevent synthesis of uric acid and promote its excretion.

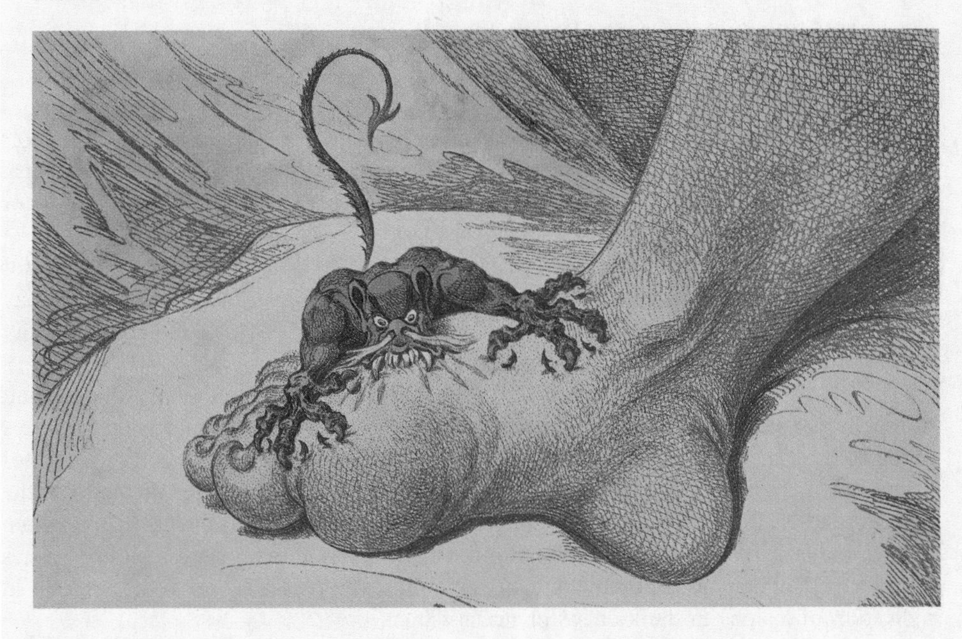

Gout, a cartoon by James Gilroy (Yale University Historical Library)

The creatinine coefficient is actually an indication of a person's percentage of muscle tissue. The higher the percentage of muscle tissue the higher the creatinine coefficient. An increase in the creatinine coefficient is an indication of an increase in muscle tissue, whereas a decrease in muscle mass (such as in a wasting disease) is accompanied by a decrease in the creatinine coefficient.

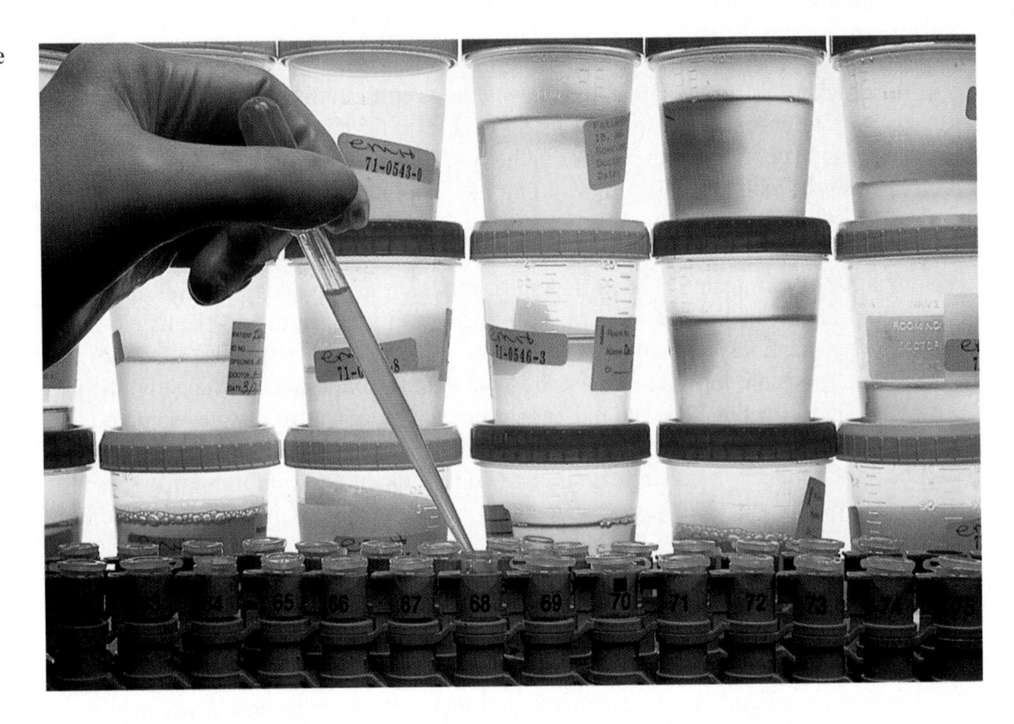

The color and clarity of urine samples may indicate the presence of abnormal components.

21.11 Abnormal Components of Urine

Urine should be clear when eliminated from the body. After a while it slowly becomes cloudy as the pH level rises (causing metals to precipitate from the urine) and as it cools (causing other solutes to precipitate). This increase in pH is due to the breakdown of urea. Cloudiness in a fresh sample of urine is an abnormality that may be due to pus, blood, or bacteria.

The odor of urine may be influenced by what you eat. You've probably noticed, for example, that eating asparagus produces a distinctive urine odor. An ammonia smell may be due to the breakdown of urea, or it could be an indication of bacterial infection. A fruity smell could be due to the presence of ketone bodies.

The yellow color associated with urine is due to the compound urochrome. The amount of urochrome produced by the body each day is almost constant. This means that changes in the color of the urine must be due to changes in the volume of urine. As the volume of urine decreases, the concentration of urochrome increases, which causes the color of the urine to darken from pale yellow to almost brown. Urine may also pick up colors from dyes in foods or medicines. Fresh blood in the urine colors it red, whereas blood that has been in the bladder for some time is dark brown or even black. Blood in the urine is an indication of lesions in the kidneys or urinary tract.

Glucose

Glucose should not be present in the urine. If glucose is present it should be found in such a low concentration that it is not detectable by the usual tests. The detectable presence of glucose in the urine is called **glycosuria.** There are numerous possible causes of glycosuria, including

- Diabetes mellitus
- Renal diabetes, which is a condition characterized by a low renal threshold (see Fig. 19.3). The renal threshold is the concentration of glucose in the blood which, if exceeded, causes glucose to be excreted by the kidneys. A person with a low renal threshold is more likely to exceed this limit than persons with a normal renal threshold.
- Consumption of large amounts of carbohydrates over a short period of time, which may lead to a blood glucose concentration that exceeds the renal threshold
- Emotional stress, which causes the release of epinephrine and an ensuing increase in blood sugar level that may exceed the renal threshold
- Failure of the liver to remove glucose from the bloodstream due to moderate damage to the liver

The tests for glucose in the urine, such as Benedict's test (see page 388), are based on the ease of oxidation of aldehydes. Because such tests are positive for all reducing sugars, substances in the urine other than glucose can also give positive tests. Such a positive test for reducing sugars other than glucose is called false glycosuria. For example, lactose is a reducing sugar that is present in the urine during the late stages of pregnancy and lactation. A fermentation test distinguishes between glucose and lactose, because lactose does not ferment but glucose does. Also, eating a large quantity of fruit increases the level of pentose sugars in the blood and perhaps in the urine. As we have learned, all monosaccharides are reducing sugars. Pentoses therefore give a positive test.

Plasma Proteins

The plasma proteins from the blood should not pass through the membranes in the kidneys; therefore, they should not be found in the urine. Damage to the glomerular membrane of the kidneys, however, may permit the proteins to pass into the urine. Because the albumins are the smallest of the plasma proteins, they are the most likely to show up in the urine. This condition can be detected by heating the urine or by adding nitric acid. Heating causes the albumin to coagulate, just like heating eggs causes egg albumin to coagulate. If concentrated nitric acid is carefully poured down the side of a test tube containing urine, a white ring appears at the interface between the acid and the urine if albumin is present. Albumin in the urine can result from kidney disease, severe heart disease, and certain infectious diseases. The globulins are larger than the albumins, so their appearance in the urine is an indication of even more serious damage to the kidneys.

Ketone Bodies

The presence of ketone bodies in the urine is known as **ketonuria.** The ketone bodies are produced by the metabolism of fats. Normally, only very small amounts of these compounds are present in the urine. However, when carbohydrates are not being efficiently metabolized, the level of ketone bodies in the blood and then in the urine increases. The causes of ketonuria include

◆ **Define ketonuria, and describe how this condition arises.**

- **Starvation.** The body is able to store only small amounts of carbohydrate in the form of glycogen. During starvation, this storage is used up very quickly. The body then turns to using fats, resulting in the production of ketone bodies.
- **Diabetes mellitus.** If not treated by the use of insulin or careful control of the diet (or both), this condition leads to the buildup of ketone bodies.

Health Perspective / Dialysis

Cellular membranes are not osmotic membranes. For a cell to live, the membrane must allow the passage of not only water molecules but also ions, nutrients, and waste products. Membranes that allow crystalloids to pass through them, but not large molecules or colloids, are called **dialyzing membranes. Dialysis** is the movement of ions and small molecules through dialyzing membranes. Most animal membranes are dialyzing membranes.

Dialysis can be a useful laboratory tool for obtaining purified colloids from solutions of cell contents that may contain both crystalloids and colloids. For example, biochemists often use dialysis to separate protein molecules from aqueous ions. By controlling the nature of the dialyzing membrane, they also can separate large protein molecules from smaller protein molecules.

The function of the kidneys is to remove waste products from the blood. If the kidneys are damaged and fail to function, waste products of crystalloid size continue to build up in the blood, and the patient dies. A device called the artificial kidney is now saving the lives of patients with kidney failure. This machine passes blood from an artery in the arm or leg through long, coiled cellophane tubes that act as dialyzing membranes. Surrounding these tubes is a solution, called the dialysate, that is isotonic for all the components that are to remain in the blood. These solutes therefore pass in and out of the blood at an equal rate. Because the dialysate contains no waste products, however, the wastes pass out of the blood at a rate greater than they return. In this way the patient's blood is cleansed. Each week, dialysis patients require two or three treatments that can last 5 to 7 hours to remove a sufficient amount of waste products from their blood.

Your Perspective: Imagine you are a doctor with a 12-year-old patient who is suffering from kidney failure. The boy is waiting for a kidney transplant and is frightened by the thought of going through weeks of kidney dialysis. Write a letter explaining to him why dialysis is necessary and how the procedure will make him feel better and keep him alive.

A patient undergoing hemodialysis

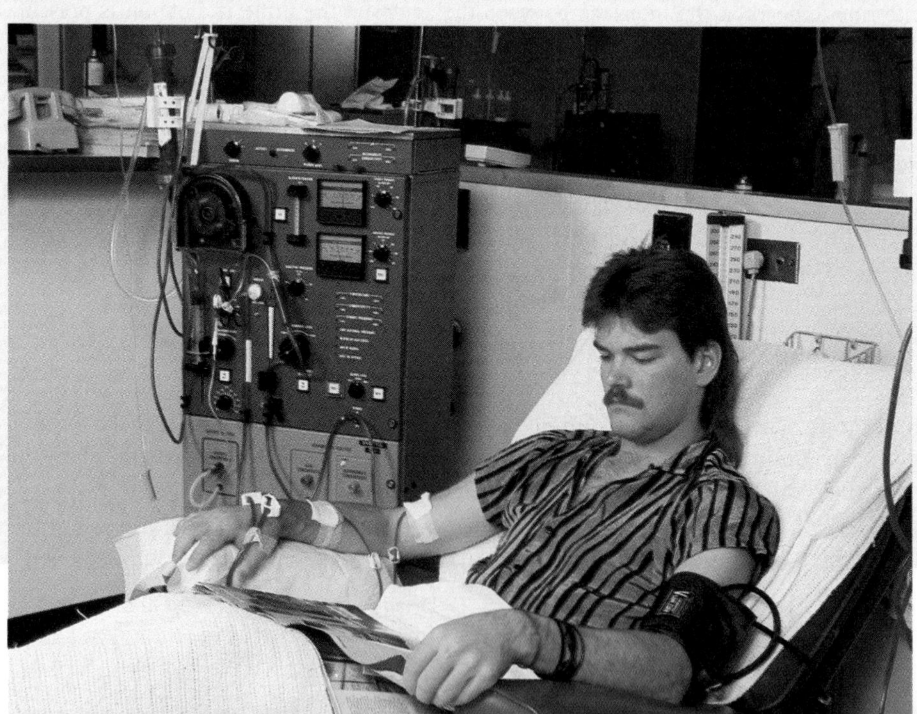

- **Severe liver damage.** The main storage organ for glycogen is the liver. If the liver is severely damaged, the body's ability to store glycogen is reduced. Again, this causes the body to turn to the metabolism of fats, which increases the production of ketone bodies.
- **Diet.** Sometimes special diets are prescribed for the treatment of urinary tract infections. These diets, known as ketogenic diets, are designed to lower urine pH. These diets also cause ketonuria.

KEY CONCEPTS

Intracellular fluids exist within the cells, and extracellular fluids are found outside the cells. Blood and urine are two of the extracellular fluids found in the body.

The suspended material that separates when blood is centrifuged is called the formed elements. The formed elements consist of the blood cells and platelets.

Blood plasma is the fluid that remains after removal of the formed elements. It is mostly water but also contains the plasma proteins, salts, amino acids, and numerous other compounds.

Oxygen and carbon dioxide are transported between the lungs and the tissues by the blood. The transport of these blood gases can be explained by a series of chemical equilibria.

Fluid balance between the blood and the tissues is maintained by a delicate interplay between osmotic pressure and blood pressure differences. If proper fluid balance is not maintained, either edema or dehydration occurs.

Blood clots are made up of the insoluble protein fibrin. The formation of clots involves a complex series of reactions triggered by the rupture or cut of a blood vessel.

The specific gravity of urine is a measure of the concentration of dissolved solids. The specific gravity of normal urine is about 1.010. This value changes with the volume of urine produced and with such conditions as diabetes mellitus and diabetes insipidus.

Urine normally contains many dissolved inorganic and organic substances. Among the inorganic components are sodium, potassium, and ammonium cations, as well as phosphate and sulfate anions. The organic components include urea, creatinine, and uric acid, among other substances.

The presence of certain substances in the urine is abnormal. Glucose, blood, and ketone bodies are examples of substances whose presence in the urine indicates some abnormality.

REVIEW PROBLEMS

Section 21.1

1. By what method do single-celled organisms exchange nutrients and wastes with their surroundings?
2. Why is the method named in problem 1 not efficient for vertebrates like humans?

3. What is meant by a closed circulatory system?
4. Name three extracellular fluids.

Section 21.2

5. Define the following:
   (a) blood plasma (c) formed elements
   (b) hematocrit (d) blood serum
6. Name three plasma proteins.

Section 21.3

7. List seven functions that the blood performs in our bodies.
8. Why must an organism with a closed circulatory system have a method of clot formation?
9. What property of water enables it to help maintain relatively constant body temperature?
10. What component of blood acts to defend against bacteria?

Section 21.4

11. Why is most of the oxygen in the blood transported as oxyhemoglobin?
12. In what form is most of the carbon dioxide in the blood?
13. State Le Chatelier's principle.
14. Apply Le Chatelier's principle to Equilibrium I in Section 21.4 to explain the effect of increased blood pH on the ability of blood to transport oxygen.
15. What is the effect of a high carbon dioxide partial pressure in the tissues on Equilibria II and III in Section 21.4?
16. What effect does the shift in Equilibrium II you predicted in problem 15 have on Equilibrium I?

Section 21.5

17. Why must living organisms regulate osmotic pressure?
18. Define hypertonic and hypotonic.
19. Why do the albumins have the greatest effect on the osmotic pressure of the blood?
20. When two liquids of different osmotic pressures are separated by a differentially permeable membrane, the net flow of solvent is in which direction?
21. Why does an increase in the concentration of salts in the blood only temporarily affect osmotic pressure?
22. Describe how cardiac failure can lead to edema.

Section 21.6

23. Define acidosis and alkalosis, and list three possible causes of each.
24. Write equations to show the effect of adding acid and of adding base to the phosphate buffer system.

Section 21.7

25. What is the relationship between prothrombin and thrombin?
26. Why do low calcium levels in the blood slow clot formation?
27. What is the first step in the formation of a blood clot?
28. What is the final step in the formation of a blood clot?

Section 21.8

29. Define (a) polyuria and (b) oliguria.
30. What is the relationship between urine specific gravity and the concentration of dissolved solids in the urine?
31. List three conditions that cause polyuria.
32. List three conditions that cause oliguria.
33. What is the normal relationship between changes in urine volume and changes in urine specific gravity?
34. Why is the polyuria of diabetes mellitus accompanied by high urine specific gravity?

Section 21.9

35. What are two circumstances in which the pH of urine is low?
36. What are two circumstances in which the pH of urine is high?

Section 21.10

37. What are the two major cations in the urine?
38. What is the source of the ammonium ion in the urine?
39. Why does acidosis result in an increase in the level of ammonium ion in the urine?
40. What effect does high blood pH have on the ratio $HPO_4^{2-}/H_2PO_4^-$ in the urine?
41. Why does consumption of a high protein diet lead to increased amounts of sulfur in the urine?
42. List three organic components of urine. Which of these is the main nitrogen-containing waste product of the body?

Section 21.11

43. Why might a urine sample, which is initially clear, turn cloudy after standing?
44. Why do samples of urine collected when a person is experiencing polyuria usually have a much lighter yellow color than during oliguria?
45. What is glycosuria? List five possible causes of this condition.
46. Why should proteins not be present in the urine?
47. What is ketonuria? List four possible causes of this condition.

Applied Problems

48. The atmosphere is composed of about 20% oxygen. It has been reported that a rock star, in an apparently misguided attempt to retard normal aging processes, has spent

periods of time in a chamber with a much higher oxygen partial pressure than is found in the atmosphere. What effect do you expect this to have on blood pH?

49. A student entered a chemistry class to take a final examination. After seeing the questions on the exam he had a panic attack. He started to breathe very rapidly and felt he could not get enough air. Soon thereafter he started to feel faint and he got a tingling sensation in his neck. What effect does his hyperventilation have on the pH of his blood? Can you suggest a possible quick remedy for this condition?

50. A 200-lb male patient was found to have 2.25 g of creatinine in a 24-h urine sample. Do you think this person engages in regular exercise or is he a couch potato? Explain your answer.

51. Benedict's test has been the classic method for the determination of glucose in urine. Other substances may also give a positive Benedict's test, which could lead to misdiagnosis. Newer methods based on the use of enzymes specific for glucose have been developed. Can you name other compounds that might be in urine that give a positive Benedict's test?

52. More than 20 substances that may be found in urine reduce Benedict's reagent. Some of these are abnormal, and others are temporary results of diet; for example, fructose may be present in the urine after eating honey. Another reducing sugar that is a normal component of urine under certain circumstances is lactose. Under what circumstances is lactose likely to be present in urine? How can you distinguish between lactose and glucose?

53. Triamterene and hydrochlorothiazide combined in one tablet are sometimes prescribed for the treatment of hypertension (high blood pressure). These drugs work by increasing the amount of salt and water eliminated from the body and by relaxing the walls of smaller blood vessels. How does the action of these drugs reduce blood pressure? What effect do you think these drugs have on fluid balance?

Integrated Problems

1. A disruption of the citric acid cycle occurs in the cells of patients suffering from pellagra. Explain why this disruption occurs.

2. Glucose in the urine does not always indicate diabetes mellitus. Shortly after a meal rich in carbohydrates (especially refined carbohydrates), the renal threshold for glucose may be exceeded. The same thing can occur when a person is frightened, causing high levels of epinephrine to be released into the blood. How does epinephrine cause glycosuria?

3. HMG-CoA reductase is the enzyme that controls the rate of the multistep synthesis of cholesterol. This enzyme is regulated by low-density lipoprotein (LDL).

(a) What is the general name given to an enzyme that controls the rate of a chemical synthesis?

(b) The activity of HMG-CoA reductase is suppressed by as little as 5.0 μg/mL of LDL. What is this concentration of LDL in moles per liter (M)? (*Note:* The mass of one LDL particle is 3.0×10^6 daltons.)

(c) People suffering from familial hypercholesterolemia (FH) have mutations in their LDL receptor gene. Suggest three ways in which a mutation can affect the LDL receptor and the transport of LDL into the cell.

(d) Because of their defective LDL receptors, the cells of FH patients do not take up LDL at the normal rate. Explain how this can affect the synthesis of cholesterol within the cell and the development of atherosclerosis.

(e) Because liver cells have the most LDL receptors in the body, they are very important in controlling the level of circulating LDL. Explain how each of the following experimental treatments might be effective in treating heterozygous FH patients (those having one normal gene and one abnormal gene for LDL receptors).

(1) preventing reabsorption of bile acids from the intestines

(2) using fungal antimetabolites that inhibit HMG-CoA reductase

(f) Why are the treatments suggested in part (e) not effective in treating homozygous FH patients (those having two abnormal genes for LDL receptors)?

4. In 1967, Akira Endo isolated from penicillin mold a substance called compactin. The side chain of the compactin molecule closely resembles the structure of the natural substrate of HMG-CoA reductase. Explain why compactin is being tested as a drug to lower the blood LDL level in patients suffering from atherosclerosis.

5. When DNA is subjected to high temperatures, the secondary structure is disrupted, causing the DNA to unwind into random single strands. The higher the proportion of A–T basepairs, the lower the temperature required to carry out this denaturation of a DNA molecule. Conversely, the higher the proportion of G–C basepairs, the higher the temperature required for denaturation of the DNA. Explain this difference.

6. In one form of the genetic disease β-thalassemia, the normal sequence AAG at codon 17 on the mRNA for the beta chain of hemoglobin becomes UAG. What effect does this have on protein synthesis and on the resulting beta chain? (*Note:* The beta chain of normal hemoglobin contains 146 amino acids.)

7. Kidney stones are a rather common and very painful disorder. By age 70 about 1 in 10 men and 1 in 20 women have had at least one kidney stone. Of the many types of kidney stones, one type contains calcium oxalate. The oxalate ion is the conjugate base of oxalic acid (HOOCCOOH). Persons suffering from this condition are advised to avoid foods containing oxalates, including tea and chocolates. Can you name a food that can be toxic because of its high oxalate content? Vitamin C supplements should also be avoided under these circumstances. Vitamin C, ascorbic acid, is necessary in our diet to prevent scurvy and is an antioxidant, which means that it reduces compounds that oxidize other compounds. Why should large doses of vitamin C be avoided when suffering from oxalate-containing kidney stones?

Appendix 1

Numbers in Exponential Form

When working with very large or very small numbers, it is convenient to write them in **exponential form** (also called scientific notation). Numbers written in exponential form are expressed as a number between 1 and 10, called the **coefficient,** multiplied by 10 raised to some power. The **exponent** is the power to which the number 10 is raised and is written as a superscript next to the number 10.

$$4.75 \times 10^3 \quad \text{exponent}$$

A positive exponent indicates how many times the coefficient is to be multiplied by the number 10. For example,

$$10^3 = 1 \times 10^3 = 1 \times 10 \times 10 \times 10 = 1000$$

$$4.5 \times 10^6 = 4.5 \times 10 \times 10 \times 10 \times 10 \times 10 \times 10 = 4{,}500{,}000$$

A negative exponent tells us how many times the coefficient is to be divided by the number 10. For example,

$$10^{-3} = 1 \times 10^{-3} = \frac{1}{10 \times 10 \times 10} = 0.001$$

$$4.5 \times 10^{-6} = \frac{4.5}{10 \times 10 \times 10 \times 10 \times 10 \times 10} = 0.0000045$$

The table shows how various numbers look in exponential form.

Number	*Exponential Form*	*Number*	*Exponential Form*
10	1×10^1	45	4.5×10^1
100	1×10^2	356	3.56×10^2
1000	1×10^3	8400	8.4×10^3
10,000	1×10^4	24,500	2.45×10^4
100,000	1×10^5	680,000	6.8×10^5
1,000,000	1×10^6	7,450,000	7.45×10^6
0.1	1×10^{-1}	0.5	5×10^{-1}
0.01	1×10^{-2}	0.037	3.7×10^{-2}
0.001	1×10^{-3}	0.004	4×10^{-3}
0.0001	1×10^{-4}	0.00056	5.6×10^{-4}
0.00001	1×10^{-5}	0.000082	8.2×10^{-5}
0.000001	1×10^{-6}	0.0000091	9.1×10^{-6}
0.0000001	1×10^{-7}	0.0000002	2×10^{-7}

Multiplying Numbers in Exponential Form

To multiply two numbers expressed in exponential form,

1. *Multiply* the two coefficients.
2. Then, *add* the two exponents to determine the new power of 10 to use in the product.

For example,

(a) $(1 \times 10^4) \times (1 \times 10^6) = 1 \times 10^{(4+6)} = 1 \times 10^{10}$
(b) $(4 \times 10^2) \times (6 \times 10^5) = (4 \times 6) \times 10^{(2+5)} = 24 \times 10^7 = 2.4 \times 10^8$
(c) $(2 \times 10^4) \times (3 \times 10^{-6}) = (2 \times 3) \times 10^{[4+(-6)]} = 6 \times 10^{-2}$

Example (b) illustrates that in exponential form we always rewrite the coefficient so that it represents a number between 1 and 10.

Dividing Numbers in Exponential Form

To divide two numbers expressed in exponential form,

1. *Divide* the two coefficients.
2. Then, *subtract* the exponent in the denominator from the exponent in the numerator.

For example,

(a) $\dfrac{1 \times 10^6}{1 \times 10^4} = \dfrac{1}{1} \times 10^{(6-4)} = 1 \times 10^2$

(b) $\dfrac{8 \times 10^7}{2 \times 10^5} = \dfrac{8}{2} \times 10^{(7-5)} = 4 \times 10^2$

(c) $\dfrac{8 \times 10^4}{3 \times 10^{-2}} = \dfrac{8}{3} \times 10^{[4-(-2)]} = 2.67 \times 10^6$

(d) $\dfrac{4 \times 10^{-3}}{8 \times 10^2} = \dfrac{4}{8} \times 10^{(-3-2)} = 0.5 \times 10^{-5} = 5 \times 10^{-6}$

Adding and Subtracting Numbers in Exponential Form

To add or subtract numbers expressed in exponential form,

1. Select the number with the largest exponent.
2. Rewrite the other numbers so their exponents are the same as the exponent of the number selected.
3. Add or subtract the coefficients.
4. The answer is the sum or the difference times the power of 10 selected.

For example,

(a) Add the following numbers:
6.02×10^5
5.82×10^6
8.35×10^4

The value with the largest exponent is 6.02×10^6. The other numbers are rewritten so their exponents are also 6. If we raise the exponent by 1, we are making the number 10 times larger. The coefficient must therefore be decreased by a power of 10. To decrease the coefficient by a power of 10, move the decimal one place to the left. (If we want to increase the coefficient by a power of 10, we move the decimal one place to the right.) If we raise the exponent by 2, we must decrease the coefficient by 100. To do so we move the decimal two places to the left.

Changing the exponents, we have,

$$\begin{aligned} 6.02 \times 10^5 &= 0.602 \times 10^6 \\ 5.82 \times 10^6 &= 5.82 \times 10^6 \\ 8.35 \times 10^4 &= \underline{0.00835 \times 10^6} \\ & 6.43035 \times 10^6 \end{aligned}$$

(*Note:* This sum is not the answer to the problem. The answer must be written with the correct number of significant figures. See Appendix 2.)

(b) Complete the following subtraction:

$$\begin{array}{r} 5.854 \times 10^7 \\ \underline{-4.327 \times 10^6} \end{array}$$

This problem is rewritten by converting both exponents to the seventh power.

$$\begin{aligned} 5.854 \times 10^7 &= 5.854 \times 10^7 \\ \underline{-4.327 \times 10^6} &= \underline{-0.4327 \times 10^7} \\ & 5.4213 \times 10^7 \end{aligned}$$

Again, this difference must be written in the correct number of significant figures to be the final answer. See Appendix 2.

Check Your Understanding

1. Write the following numbers in exponential form:

Number	*Answer*
(a) 56	5.6×10^1
(b) 476.54	4.7654×10^2
(c) 0.00046	4.6×10^{-4}
(d) 75,340,000	7.534×10^7
(e) 1278	1.278×10^3
(f) 0.03	3×10^{-2}
(g) 0.6	6×10^{-1}
(h) 890,000	8.9×10^5
(i) 0.00009	9×10^{-5}
(j) 0.0000000000012	1.2×10^{-12}

2. Perform the following operations:

Problem	*Answer*
(a) $\dfrac{(3 \times 10^3)(8 \times 10^{10})}{(6 \times 10^4)(1 \times 10^6)}$	4×10^3
(b) $\dfrac{(1.5 \times 10^2)(4.0 \times 10^6)}{(5.0 \times 10^{10})(2.5 \times 10^5)}$	4.8×10^{-8}
(c) $\dfrac{(7.5 \times 10^{-3})(9.0 \times 10^6)}{(1.5 \times 10^2)(2.5 \times 10^{-8})}$	1.8×10^{10}
(d) $\dfrac{(2.0 \times 10^{-6})(4.2 \times 10^{-2})}{(1.4 \times 10^{-11})(1.0 \times 10^5)}$	6.0×10^{-2}

3. Convert the following numbers into standard exponential form (in which the coefficient is between 1 and 10):

Number	*Answer*
(a) 160×10^5	1.60×10^7
(b) 0.0068×10^8	6.8×10^5
(c) 10.76×10^{-3}	1.076×10^{-2}
(d) 0.075×10^{-5}	7.5×10^{-7}

4. Perform the following additions:

Problem	*Answer*
(a) $1.207 \times 10^3 + 1.421 \times 10^4 + 4.228 \times 10^2$	1.58398×10^4
(b) $4.86 \times 10^{10} + 3.85 \times 10^{10} + 4.91 \times 10^9$	9.201×10^{10}
(c) $5.86 \times 10^{-2} + 4.99 \times 10^{-3} + 6.65 \times 10^{-1}$	7.2859×10^{-1}
(d) $2.27 \times 10^{-9} + 3.65 \times 10^{-7} + 4.28 \times 10^{-8}$	4.1007×10^{-7}

5. Perform the following subtractions:

Problem	*Answer*
(a) $6.825 \times 10^5 - 4.307 \times 10^4$	6.3943×10^5
(b) $5.928 \times 10^{-2} - 7.66 \times 10^{-3}$	5.162×10^{-2}
(c) $1.876 \times 10^3 - 9.885 \times 10^2$	8.875×10^2
(d) $8.755 \times 10^{-8} - 9.027 \times 10^{-9}$	7.8523×10^{-8}

Appendix 2
Using Significant Figures

In Section 1.8, we introduced the concept of **significant figures,** or **significant digits.** Significant figures are used to indicate the precision with which a measurement is made. When solving a problem, we are often required to perform mathematical operations on experimental data. In such cases it is important to maintain the correct number of significant figures. The underlying principle of significant figures is that we must never report our results with more precision than the least precise piece of experimental data.

Determining the Number of Significant Figures in a Measurement

The following guidelines can be used to determine the number of significant figures in a given measurement.

1. The digits 1 through 9 are all significant; therefore, the number 27, for example, has two significant figures and 3.584 has four significant figures.
2. The particular placement of the digit zero in a number determines whether or not the zero is significant.

   (a) The zero is significant if it is located between two nonzero digits. For example, 1003 has four significant figures and 1.03 has three significant figures.

   (b) The zero is significant if it is the final digit to the *right* of the decimal point. For example, 39.0, 3.90, and 0.390 all have three significant figures.

   (c) The zero is not significant when it is used to mark the position of the decimal point in a number less than 1. For example, both 0.178 and 0.00178 have three significant figures.

   (d) A zero used to mark the decimal place in a number greater than 1 is usually not significant. Specifically, if such zeros are used simply as place markers for the decimal point rather than being part of the actual measurement, they are not significant. There are several ways that we can write numbers to eliminate confusion about whether the zero is significant. Writing the number in exponential form can show clearly whether or not the zeros are significant. For example, the number 4800 can be written as

4.8×10^3	2 significant figures
4.80×10^3	3 significant figures
4.800×10^3	4 significant figures

Check Your Understanding

Identify the number of significant figures in each of the following numbers:

Number	*Answer*
(a) 458.7	4
(b) 0.004	1
(c) 1.704	4
(d) 325	3
(e) 63.0	3
(f) 0.27650	5
(g) 3×10^3	1
(h) 1.0003	5
(i) 9.00×10^4	3
(j) 0.0056	2
(k) 45.67	4

Note: The zeros in questions (b) and (j) are not significant figures. They are used to mark the position of the decimal point in these numbers that are less than 1.

The number in (f) is also less than 1. The first zero is used to mark the decimal point and is therefore not a significant figure. The zero at the end of this number is a significant figure because it is used to show precision not to mark the decimal point.

Exact Numbers

Not all numbers used in chemical calculations come from measurements. Numbers that are given in definitions (1 meter equals 100 millimeters) or that result from counting objects (1 dozen eggs equals 12 eggs) are called **exact numbers.** When using exact numbers in calculations, we think of them as having an infinite number of significant figures. We do not therefore have to take exact numbers into account when determining the number of significant figures in our answer.

Electronic Calculators

A calculator can be of real help in performing mathematical computations, but its use presents a special kind of problem in calculations based on experimental data: The calculator usually shows answers with too many significant digits. Including all of these digits erroneously indicates that our data are more precise than they actually are.

Rules for Determining the Correct Number of Significant Figures in a Calculated Answer

Rule 1. *Addition and Subtraction.* When numbers are added or subtracted, the number of decimal places in the answer equals the *smallest number of decimal places* among all of the numbers added or subtracted. In this way, the answer indicates the first decimal place in which uncertainty exists in any of the measurements.

Rule 2. *Multiplication and Division.* When numbers are multiplied and divided, the number of significant figures in the answer can be no more than the *smallest number of significant figures* among all of the numbers multiplied or divided.

Rule 3. *Rounding Off.* To obtain the correct number of significant figures in a calculated result, the nonsignificant digits in the calculator's answer are rounded off.

(a) If the nonsignificant digit is less than 5, the digit is dropped. For example, 32.233 to four significant figures is 32.23.

(b) If the nonsignificant digit is greater than 5, the digit is dropped and the last significant digit is increased by one. For example, 32.236 to four significant figures is 32.24.

(c) If the nonsignificant digit is 5 (or 5 followed by zeros), the 5 is dropped and the last significant digit is increased by one if it is odd and is not changed if it is even. For example, 32.235 to four significant figures is 32.24 but 32.225 to four significant figures is 32.22.

Example 1

1. Add the following numbers: 25, 1.278, 127.1, and 5.45.

$$\begin{array}{r} 25 \\ 1.278 \\ 127.1 \\ 5.45 \\ \hline 158.828 \end{array} \quad \text{calculator answer}$$

Uncertainty in the answer begins in the ones column (as indicated by the number 25); therefore, the answer should be rounded off to the ones place, or to three significant figures. Rounded off, the answer with the correct number of significant figures is 159.

2. Subtract 1.286 from 19.57.

$$\begin{array}{r} 19.57 \\ -1.286 \\ \hline 18.284 \end{array} \quad \text{calculator answer}$$

Uncertainty begins in the hundredths place, so the answer should be rounded to four significant figures: 18.28.

3. Multiply 13.6 by 0.004.

$$13.6 \times 0.004 = 0.0544 \quad \text{calculator answer}$$

The number 13.6 has three significant figures and 0.004 has one significant figure. Therefore, our answer must have one significant figure: 0.0544 rounded to one significant figure is 0.05.

4. Divide 67.0 by 563.

$$\frac{67.0}{563} = 0.1190053 \quad \text{calculator answer}$$

Both the dividend and the divisor have three significant figures. Therefore, we must round our answer to three significant figures: 0.1190053 rounded to three significant figures is 0.119.

•••••••••••••••••••••••••••

Check Your Understanding

1. Round each of the following numbers to two significant figures:

Number	*Answer*
(a) 1.598	1.6
(b) 7.35	7.4
(c) 26.3	26
(d) 386	390 or 3.9×10^2
(e) 4.250	4.2
(f) 0.03457	0.035
(g) 0.9246	0.92
(h) 0.1486	0.15

2. Perform the following operations and round your answer to the correct number of significant figures:

Problem	*Answer*	*Calculator Answer*
(a) $43.67 + 27.4 + 0.0265$	71.1	71.0965
(b) $156 + 32.7 + 4.38$	193	193.08
(c) $1.4651 - 0.53$	0.94	0.9351
(d) $256 - 139.48$	117	116.52
(e) $1.48 \times 39.1 \times 0.312$	18.1	18.054816
(f) $67.84 \div 4.6$	15	14.747826
(g) $\frac{9.50 \times 784}{1465}$	5.08	5.083959
(h) $\frac{0.036 \times 25.78}{1.4865 \times 169}$	0.0037	0.0036943

Appendix 3

Selected Answers to Text Problems

Answers to Check Your Understanding

Section 1.9

1. **(a)** $\dfrac{1 \text{ kilogram}}{2.2 \text{ pounds}}$ $\dfrac{2.2 \text{ pounds}}{1 \text{ kilogram}}$

(b) $\dfrac{1 \text{ mile}}{5280 \text{ feet}}$ $\dfrac{5280 \text{ feet}}{1 \text{ mile}}$

(c) $\dfrac{1 \text{ liter}}{1000 \text{ milliliters}}$ $\dfrac{1000 \text{ milliliters}}{1 \text{ liter}}$

(d) $\dfrac{1\ \mu\text{m}}{10^{-6} \text{ meter}}$ $\dfrac{10^{-6} \text{ meter}}{1\ \mu\text{m}}$

2. **(a)** $\dfrac{1 \text{ kilogram}}{2.2 \text{ pounds}}$

(b) $\dfrac{5280 \text{ feet}}{1 \text{ mile}}$

(c) $\dfrac{1 \text{ liter}}{1000 \text{ milliliters}}$

(d) $\dfrac{1\ \mu\text{m}}{10^{-6} \text{ meter}}$

Section 1.11

(a) 42.2 km
(b) 4.92 minutes per mile
(c) 3.05 minutes per kilometer

Section 1.12

1. **(a)** 0.253 mg **(d)** 12 oz
(b) 3200 g **(e)** 1.50 lb
(c) 5000 mg **(f)** 849 g

2. 1.83 lb

Section 1.13

1. **(a)** 2500 mL **(d)** 1250 L
   (b) 0.345 L **(e)** 5.6 qt
   (c) 25 cc **(f)** 2.00 pt
2. 36 oz = 1066 mL. The three 12-oz bottles are a better buy.

Section 1.14

1. **(a)** −65°C **(d)** 89.6°F
   (b) 105°F **(e)** −131°C
   (c) 55.6°C **(f)** 310
2. 106°F
3. 327°C, 600 K

Section 1.15

1. 11.3 g/cm^3
2. 3.0 cm^3
3. 373 g
4. **(a)** 1.11 g/mL
   (b) 1.11
   (c) ethylene glycol

Section 2.4

1. 9200 kcal
2. 16.5 kcal

Section 3.3

0.868 atm, 660 torr

Section 3.4

1. P_2 = 600 torr 2. V_2 = 884 mL

Section 3.5

1. T_2 = 35°C 2. 70 mL

Section 3.7

1. P_{O_2} = 731 torr
2. **(a)** Charles's law **(b)** Boyle's law **(c)** Henry's law **(d)** Charles's law

Section 4.2

1. **(a)** $p = 88$, $e^- = 88$, $n = 134$
   (b) $p = 24$, $e^- = 24$, $n = 27$
   (c) $p = 80$, $e^- = 80$, $n = 123$

2. Z = 15, M = 31, symbol = P

Section 4.4

atomic weight = 28.1

Section 4.6

(a) Sodium, $1s^2 2s^2 2p^6 3s^1$

$1s$	$2s$	$2p_x$	$2p_y$	$2p_z$	$3s$
↑↓	↑↓	↑↓	↑↓	↑↓	↑

(b) Phosphorus, $1s^2 2s^2 2p^6 3s^2 3p^3$

$1s$	$2s$	$2p_x$	$2p_y$	$2p_z$	$3s$	$3p_x$	$3p_y$	$3p_z$
↑↓	↑↓	↑↓	↑↓	↑↓	↑↓	↑	↑	↑

(c) Chlorine, $1s^2 2s^2 2p^6 3s^2 3p^5$

$1s$	$2s$	$2p_x$	$2p_y$	$2p_z$	$3s$	$3p_x$	$3p_y$	$3p_z$
↑↓	↑↓	↑↓	↑↓	↑↓	↑↓	↑↓	↑↓	↑

Section 5.3

1. (a) $\dot{\text{Rb}}$ **(b)** $\cdot\ddot{\text{Si}}\cdot$ **(c)** $:\ddot{\text{I}}:$

2. (a) $\text{K}\cdot + \cdot\ddot{\text{I}}: \longrightarrow \text{K}^+ :\ddot{\text{I}}:^-$

(b) $:\ddot{\text{I}}\cdot + \text{Mg} + \cdot\ddot{\text{I}}: \longrightarrow \text{Mg}^{2+} + 2[:\ddot{\text{I}}:]^-$

Section 5.5

(a) $\text{H}:\ddot{\text{Cl}}:$ H—Cl

(b) $\text{H}:\ddot{\text{N}}:\text{H}$ with H below N H—N—H with H below N

(c) $\ddot{\text{S}}::\text{C}::\ddot{\text{S}}$ S=C=S

Section 5.11

1. (a) magnesium iodide
(b) iron(II) oxide or ferrous oxide
(c) sulfur dioxide
(d) hydrogen sulfide
(e) sodium hydrogen sulfate or sodium bisulfate
(f) potassium dichromate

2. (a) MgO **(b)** $NiCl_2$ **(c)** K_2S **(d)** SO_3 **(e)** SiF_4 **(f)** N_2O_5
(g) $SnCl_4$ **(h)** $Cu(HSO_4)_2$ **(i)** $Ca_3(PO_4)_2$

Section 6.2

(a) $4P + 5O_2 \rightarrow P_4O_{10}$
(b) $2NOCl \rightarrow 2NO + Cl_2$
(c) $CH_4 + 2O_2 \rightarrow CO_2 + 2H_2O$
(d) $Ca(OH)_2 + 2HCl \rightarrow CaCl_2 + 2H_2O$
(e) $2Mg + O_2 \rightarrow 2MgO$
(f) $2PbS + 3O_2 \rightarrow 2PbO + 2SO_2$
(g) $Na_2CO_3 + Mg(NO_3)_2 \rightarrow MgCO_3 + 2NaNO_3$

Section 6.4

1. **(a)** 9.85 g **(b)** 131 g **(c)** 3.21 g **(d)** 1.01×10^{-2} g
2. **(a)** 0.300 mol Ag **(c)** 1.00×10^{-3} mol Ne
(b) 34.9 mol Si **(d)** 19.9 mol U

Section 6.5

1. **(a)** 254 **(b)** 103 **(c)** 239 **(d)** 123 **(e)** 342 **(f)** 58.1
2. **(a)** 103 g **(b)** 123 g **(c)** 342 g
3. 254 g

Section 6.6

42.3 g

Section 6.7

1. **(a)** 127 g **(b)** 674 g **(c)** 4.30 g **(d)** 1370 g
2. **(a)** 25.0 mol **(b)** 0.299 mol **(c)** 0.0107 mol
3. 1.50×10^{22} molecules

Section 6.8

Page 144

1. 1 molecule of methane reacts with 2 molecules of oxygen gas to produce 1 molecule of carbon dioxide plus 2 molecules of water.
2. 1 mole of methane reacts with 2 moles of oxygen gas to produce 1 mole of carbon dioxide plus 2 moles of water.
3. 16 grams of methane reacts with 64 grams of oxygen gas to produce 44 grams of carbon dioxide plus 36 grams of water.

Page 148

1. 177 g MgO
2. 25.6 g HCl, 0.146 g HCl

Section 7.1

1. **(a)** $^{226}_{88}\text{Ra} \longrightarrow {}^{222}_{86}\text{Rn} + {}^{4}_{2}\text{He}$
   (b) $^{28}_{13}\text{Al} \longrightarrow {}^{28}_{14}\text{Si} + {}^{0}_{-1}e$
   (c) $^{227}_{89}\text{Ac} \longrightarrow {}^{223}_{87}\text{Fr} + {}^{4}_{2}\text{He}$
2. **(a)** $^{45}_{20}\text{Ca} \longrightarrow {}^{45}_{21}\text{Sc} + {}^{0}_{-1}e$
   (b) $^{14}_{6}\text{C} \longrightarrow {}^{14}_{7}\text{N} + {}^{0}_{-1}e$
   (c) $^{149}_{62}\text{Sm} \longrightarrow {}^{145}_{60}\text{Nd} + {}^{4}_{2}\text{He}$
3. $^{137}_{55}\text{Cs} \longrightarrow {}^{137}_{56}\text{Ba} + {}^{0}_{-1}e + \gamma$

Section 7.2

1. 7.5×10^{-3} mg
2. 79.8 hours, or after 3 days

Section 8.2

1. **(a)** endothermic **(b)** absorbed
   (c) $C(s) + H_2O(g) + 31.4\ \text{kcal} \rightarrow CO(g) + H_2(g)$
2.

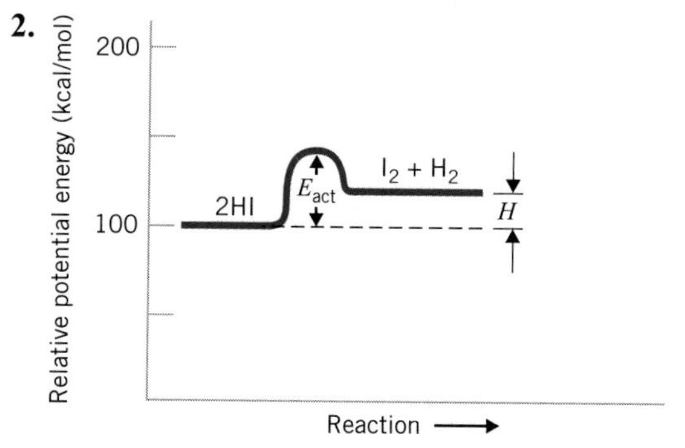

Section 8.6

1. The equilibrium concentration of O_2 decreases.
2. The equilibrium concentration of O_2 decreases.
3. The equilibrium concentration of O_2 decreases.
4. The equilibrium concentration of O_2 increases.
5. No change occurs in the equilibrium concentration of O_2.

Section 9.9

1. **(a)** Yes **(b)** No **(c)** Yes **(d)** Yes
2. **(a)** $Mg^{2+}(aq) + 2OH^-(aq) \rightarrow Mg(OH)_2(s)$
   (c) $2Cl^-(aq) + Pb^{2+}(aq) \rightarrow PbCl_2(s)$
   (d) $Ba^{2+}(aq) + SO_4^{2-}(aq) \rightarrow BaSO_4(s)$
3. $Ca^{2+}(aq) + CO_3^{2-}(aq) \rightarrow CaCO_3(s)$

Section 10.2

1. **(a)** Dissolve 5.30 g Na_2CO_3 in enough water to make 250 mL of solution.
   (b) Dissolve 110 g H_3PO_4 in enough water to make 1.5 L of solution.
   (c) Dissolve 14.2 g $KMnO_4$ in enough water to make 150 mL of solution.
2. 625 mL
3. Yes, 139 mmol Na^+/L

Section 10.3

1. **(a)** Start with 16.1 g of NaCl and add enough water to make 350 mL of solution.
   (b) Start with 0.121 g K_2CO_3 and add enough water to make 55.0 mL of solution.
   (c) Start with 20 mg of glucose and add enough water to make 25 mL of solution.
2. 80 mL blood
3. 33.0 g glucose

Section 10.4

CCl_4	Yes, 1.6 ppb	Selenium	No, 0.0052 ppm
Lead	No, 0.044 ppm	Mercury	No, 0.84 ppb

Section 10.5

1. **(a)** 39.1 g **(b)** 35.5 g **(c)** 48.0 g **(d)** 48.0 g
2. 6.72 g
3. 94 mmol/L

Section 10.6

1. **(a)** Dissolve 265 g NaCl in a small amount of water and then add enough water to make 1.65 L of solution.
   (b) Start with 1.10 L of 4.12 *M* NaCl and add enough water to make 1.65 L of solution.
2. Take 1 volume of 2.74 *M* NaCl and add 5 equal volumes of water; 0.458 *M* NaCl

Section 11.3

$$NaOH + HNO_3 \longrightarrow H_2O + NaNO_3$$

$$H^+(aq) + OH^-(aq) \longrightarrow H_2O$$

Section 11.5

(a) $[H^+] = 1 \times 10^{-11}\ M$ **(d)** $[H^+] = 2.5 \times 10^{-11}\ M$
(b) $[H^+] = 2 \times 10^{-9}\ M$ **(e)** $[H^+] = 1 \times 10^{-12}\ M$
(c) $[H^+] = 5 \times 10^{-12}\ M$ **(f)** $[H^+] = 6.25 \times 10^{-8}\ M$

Section 11.6

1. **(a)** basic $[H^+] = 1 \times 10^{-11}\ M$ $[OH^-] = 1 \times 10^{-3}\ M$
   (b) acidic $[H^+] = 1 \times 10^{-2}\ M$ $[OH^-] = 1 \times 10^{-12}\ M$

(c) acidic $[H^+] = 1 \times 10^{-5}\,M$ $[OH^-] = 1 \times 10^{-9}\,M$

(d) basic $[H^+] = 1 \times 10^{-9}\,M$ $[OH^-] = 1 \times 10^{-5}\,M$

2. pH = 7

Section 12.3

1. (a)

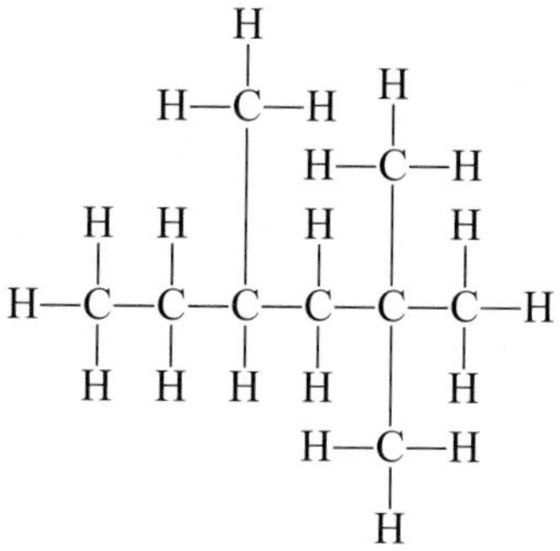

$$
\begin{array}{ccccccccccccc}
 & & & & H & & & & & & & & \\
 & & & & | & & & & & & & & \\
 & & & H- & C & -H & & & H & & & & \\
 & & & & | & & & & | & & & & \\
 & & & & | & & & H- & C & -H & & & \\
 & H & & H & | & & H & & | & & H & & \\
 & | & & | & | & & | & & | & & | & & \\
H- & C & - & C & - C - & & C & - & C & - & C & -H & \\
 & | & & | & | & & | & & | & & | & & \\
 & H & & H & H & & H & & | & & H & & \\
 & & & & & & & H- & C & -H & & & \\
 & & & & & & & & | & & & & \\
 & & & & & & & & H & & & &
\end{array}
$$

(b)

$$
\begin{array}{ccccc}
 & & H & & \\
 & & | & & \\
 & H- & C & -H & \\
 & H & | & H & \\
 & | & | & | & \\
H- & C - & C & - C & -H \\
 & | & | & | & \\
 & H & | & H & \\
 & H- & C & -H & \\
 & & | & & \\
 & & H & &
\end{array}
$$

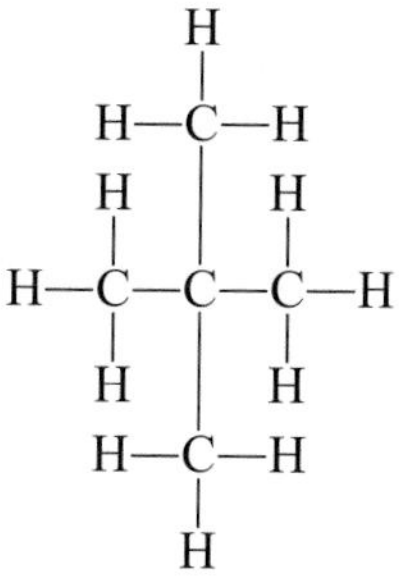

2. (a) $CH_3CH_2CH_2CH_2CH_2CH_3$ **(b)** $CH_3CH_2CH(CH_3)CH(CH_3)_2$

Section 12.6

1. 2,4-dimethylpentane
2. 3-ethyl-2-methylhexane
3. 3-ethyloctane

Section 12.7

$$2CH_3{-}(CH_2)_4{-}CH_3 + 19O_2 \longrightarrow 12CO_2 + 14H_2O$$

$$CH_3{-}CH_3 + Cl_2 \longrightarrow CH_3{-}CH_2{-}Cl + HCl$$

Section 12.9

1. (a) 4-methyl-2-heptene

(b) 1,1-dichloropropene

2. (a)

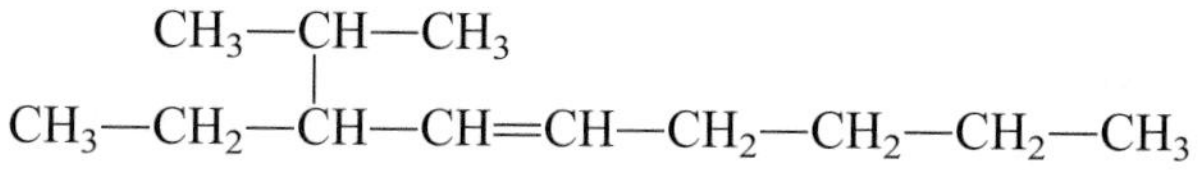

$$
\begin{array}{l}
\qquad\qquad CH_3{-}CH{-}CH_3 \\
\qquad\qquad\qquad\;\; | \\
CH_3{-}CH_2{-}CH{-}CH{=}CH{-}CH_2{-}CH_2{-}CH_2{-}CH_3
\end{array}
$$

(b)

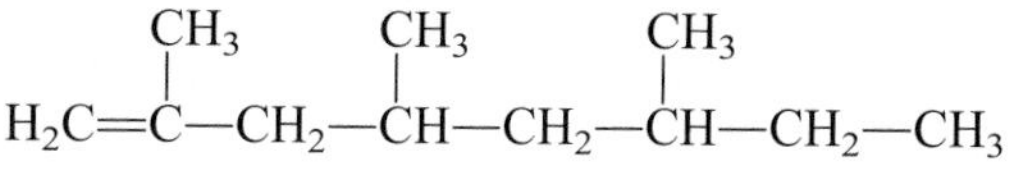

$$
\begin{array}{l}
\qquad\;\; CH_3 \qquad\qquad\quad CH_3 \qquad\qquad\quad CH_3 \\
\qquad\quad | \qquad\qquad\qquad\; | \qquad\qquad\qquad\; | \\
H_2C{=}C{-}CH_2{-}CH{-}CH_2{-}CH{-}CH_2{-}CH_3
\end{array}
$$

Section 12.11

$$
\begin{array}{ccc}
H & & H \\
 & \diagdown\; C{=}C \;\diagup & \\
CH_3 \diagup & & \diagdown CH_2CH_3
\end{array}
\qquad\qquad
\begin{array}{ccc}
H & & CH_2CH_3 \\
 & \diagdown\; C{=}C \;\diagup & \\
CH_3 \diagup & & \diagdown H
\end{array}
$$

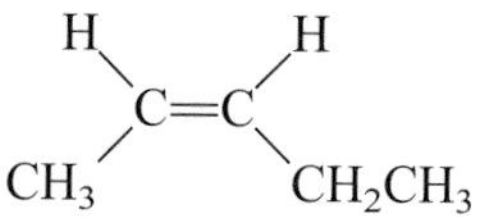

cis-2-pentene 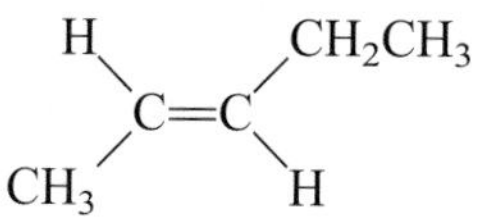 *trans*-2-pentene

Section 12.12

1. $CH_3{-}(CH_2)_2{-}CH{=}CH_2 + Br_2 \longrightarrow CH_3{-}(CH_2)_2{-}\underset{}{\overset{Br}{\overset{|}{C}}}H{-}CH_2{-}Br$

1-pentene → 1,2-dibromopentane

2. $\text{cis-}CH_3CH{=}CHCH_3 + H_2 \longrightarrow CH_3{-}(CH_2)_2{-}CH_3$

cis-2-butene → butane

3. $\text{cis-}CH_3CH{=}CHCH_3 + H_2O \longrightarrow CH_3{-}CH_2{-}\overset{OH}{\overset{|}{C}}H{-}CH_3$

Section 12.14

1. **(a)** chlorocyclohexane
   (b) *trans*-1,3-dimethylcyclopentane
   (c) 1,1-dibromocyclobutane

2. **(a)** 1,2-dichlorocyclohexane (Cl, Cl)
   (b) cis-1,3-dibromocyclopentane (Br, Br; H, H)
   (c) 1,1-dimethylcyclohexane (CH_3, CH_3)

Section 13.2

1. **(a)** ethyl propyl ether
   (b) ethyl isopropyl ether

2. **(a)** cyclohexyl$-O-CH_3$
   (b) $CH_3CH_2CH_2CH_2OCH_3$

Section 13.3

1. **(a)** 2-methyl-2-hexanol, tertiary
   (b) 2,2-dimethyl-1-butanol, primary
   (c) cyclohexanol, secondary

2. **(a)** $CH_3{-}C(CH_3)_2{-}CH(OH){-}CH_3$
   (b) $CH_3{-}CH(OH){-}CH_2{-}OH$
   (c) $CH_3{-}CH(CH_3){-}CH_2{-}CH(CH(CH_3){-}CH_3){-}CH_2{-}CH_2{-}OH$

Section 13.6

1. $CH_3{-}CH{=}CH_2$

2. 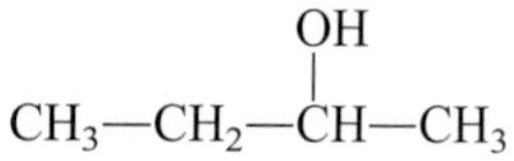 $$\begin{array}{c} \phantom{CH_3{-}CH_2{-}}OH \\ \phantom{CH_3{-}CH_2{-}}| \\ CH_3{-}CH_2{-}CH{-}CH_3 \end{array}$$

3. $H_2C{=}O$

4. 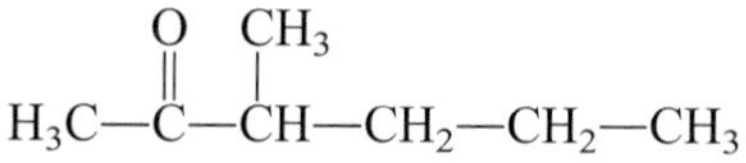 $$H_3C{-}\overset{\overset{\displaystyle O}{\|}}{C}{-}\overset{\overset{\displaystyle CH_3}{|}}{C}H{-}CH_2{-}CH_2{-}CH_3$$

Section 13.7

1. **(a)** 2-methylpropanal **(b)** 3-pentanone **(c)** 4-methyl-2-hexanone

2. **(a)** 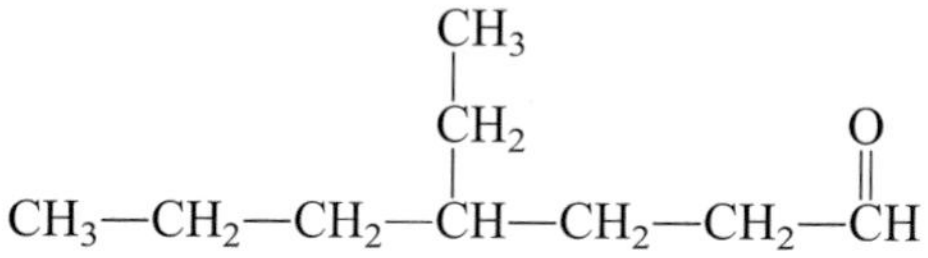 $$CH_3{-}CH_2{-}CH_2{-}\overset{\overset{\overset{\displaystyle CH_3}{|}}{\displaystyle CH_2}}{\underset{|}{C}}H{-}CH_2{-}CH_2{-}\overset{\overset{\displaystyle O}{\|}}{C}H$$

(b) $$CH_3{-}CH_2{-}CH_2{-}CH_2{-}\overset{\overset{\overset{\displaystyle CH_3}{|}}{\displaystyle CH_2}}{\underset{|}{C}}H{-}\overset{\overset{\displaystyle O}{\|}}{C}{-}\overset{\overset{\displaystyle CH_3}{|}}{C}H{-}CH_3$$

(c) 2-methylcyclohexanone: cyclohexane ring with C=O and adjacent CH_3

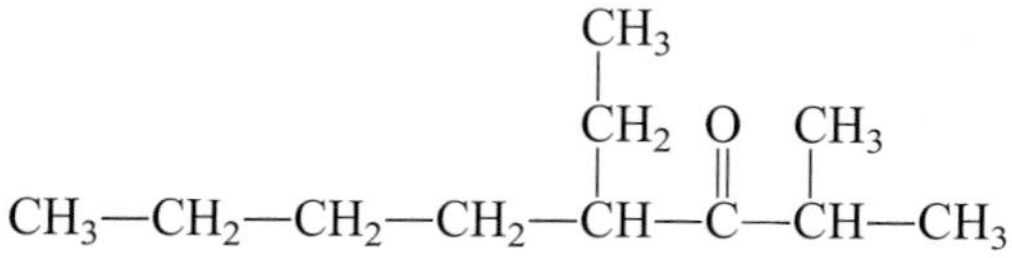

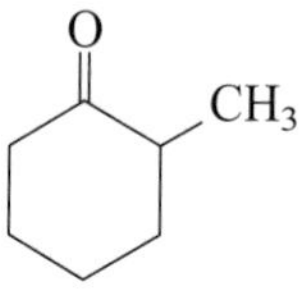

Section 13.10

1. $CH_3{-}CH_2{-}CH_2{-}CH_2{-}CH_2{-}OH$

2. 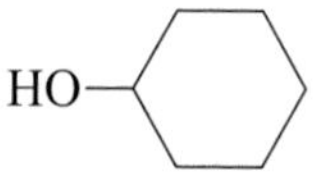 HO–(cyclohexane ring)

3. 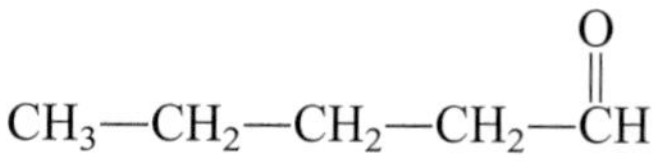 $$CH_3{-}CH_2{-}CH_2{-}CH_2{-}\overset{\overset{\displaystyle O}{\|}}{C}H$$

4. 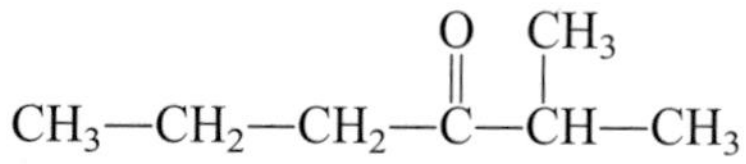 $$CH_3{-}CH_2{-}CH_2{-}\overset{\overset{\displaystyle O}{\|}}{C}{-}\overset{\overset{\displaystyle CH_3}{|}}{C}H{-}CH_3$$

5. 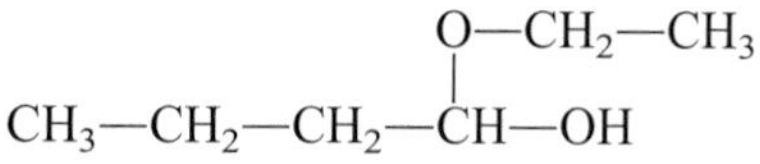 $$CH_3{-}CH_2{-}CH_2{-}\overset{\overset{\displaystyle O{-}CH_2{-}CH_3}{|}}{C}H{-}OH$$

6. 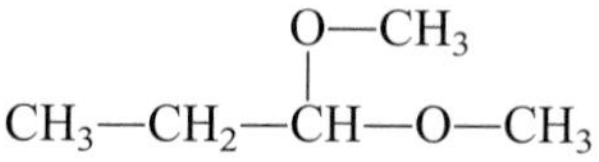 $$CH_3{-}CH_2{-}\overset{\overset{\displaystyle O{-}CH_3}{|}}{C}H{-}O{-}CH_3$$

7. 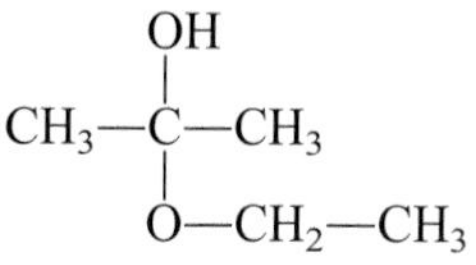 $$CH_3{-}\overset{\overset{\displaystyle OH}{|}}{\underset{\underset{\displaystyle O{-}CH_2{-}CH_3}{|}}{C}}{-}CH_3$$

Section 13.11

1. **(a)** 3-phenylpropanoic acid
   (b) 4,7-dimethyloctanoic acid

2. **(a)** 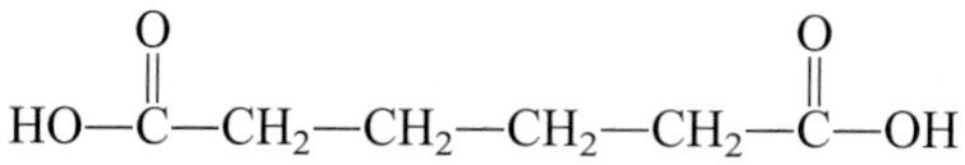 $$HO{-}\overset{\overset{\displaystyle O}{\|}}{C}{-}CH_2{-}CH_2{-}CH_2{-}CH_2{-}\overset{\overset{\displaystyle O}{\|}}{C}{-}OH$$

(b) 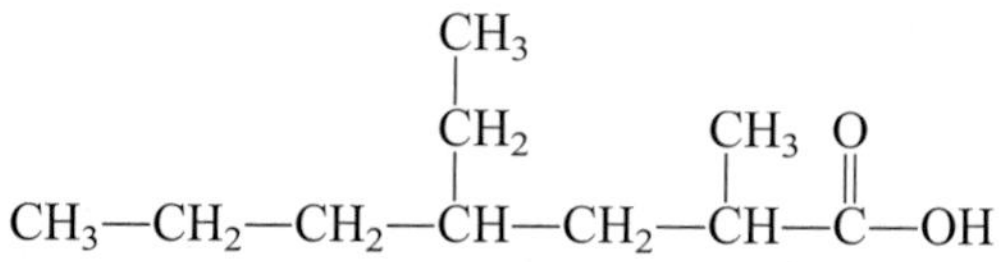 $$CH_3{-}CH_2{-}CH_2{-}\overset{\overset{\overset{\displaystyle CH_3}{|}}{\displaystyle CH_2}}{\underset{|}{C}}H{-}CH_2{-}\overset{\overset{\displaystyle CH_3}{|}}{C}H{-}\overset{\overset{\displaystyle O}{\|}}{C}{-}OH$$

Section 13.15

1. 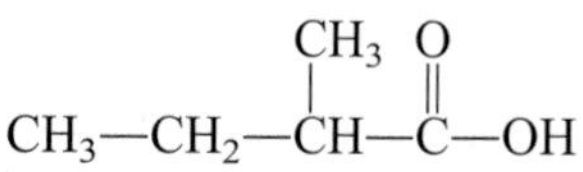 $$CH_3{-}CH_2{-}\overset{\overset{\displaystyle CH_3}{|}}{C}H{-}\overset{\overset{\displaystyle O}{\|}}{C}{-}OH$$

2.

$$
\begin{array}{c}
\qquad\qquad CH_3 \qquad\qquad\quad O \qquad\qquad\qquad\qquad\qquad \\
\qquad\qquad | \qquad\qquad\qquad \| \qquad\qquad\qquad\qquad\qquad \\
CH_3—CH_2—CH—CH_2—C—O—CH_2—CH_2—CH_3
\end{array}
$$

Section 14.2

(a) 1-aminooctane, primary

(b) triethylamine, tertiary

(c) 3-amino-1-propanol, primary

(d) 3-amino-2-methylbutanal, primary

(e) 3-N-methylaminopropanoic acid, secondary

2. (a)

$$
\begin{array}{c}
CH_3 \\
| \\
CH_2 \\
| \\
CH_3—CH_2—N—CH_2—CH—CH_3 \\
| \\
CH_3
\end{array}
$$

(b)

$$
\begin{array}{c}
CH_2—CH_2 \\
H_2C \qquad\qquad CH—NH—CH_3 \\
CH_2—CH_2
\end{array}
$$

(c)

$$
\begin{array}{c}
NH_2 \\
| \\
CH_3—CH_2—CH—CH_2—CH_2—CH_2—NH_2
\end{array}
$$

(d)

$$
\begin{array}{c}
CH_3—N—C_6H_5 \\
| \\
CH_2 \\
| \\
CH_3
\end{array}
$$

Section 14.5

1. **(a)** ethanamide
   (b) *N*-propylbenzamide
   (c) *N*-propylpentanamide
   (d) *N,N*-diethylethanamide

2. (a)

$$
\begin{array}{c}
O \\
\| \\
CH_3—CH_2—CH_2—CH_2—C—NH_2
\end{array}
$$

(b)

$$
\begin{array}{c}
O \qquad CH_3 \\
\| \qquad\quad | \\
CH_3—CH_2—CH_2—C—NH—CH \\
| \\
CH_3
\end{array}
$$

(c)

$$
\begin{array}{c}
CH_3 \\
| \\
O \quad CH_2 \\
\| \qquad | \\
CH_3—CH_2—CH_2—CH_2—C—N—CH_2—CH_3
\end{array}
$$

(d) $C_6H_5-C(=O)-N(CH_2CH_3)(CH_2CH_2CH_2CH_3)$

Section 15.7

1. ribose, aldopentose
2. fructose, ketohexose
3. threose, aldotetrose
4. galactose, aldohexose
5. mannose, aldohexose
6. ribulose, ketopentose

Section 15.10

Disaccharide	*Source*	*Monosaccharide Subunits*	*Reducing Sugar?*	*Type of Linkage*
Sucrose	Fruit, honey, vegetables	Glucose, fructose	No	α, 1:2
Lactose	Milk	Glucose, galactose	Yes	β, 1:4
Maltose	Starch, glycogen	Glucose	Yes	α, 1:4

Section 15.15

1. glycogen
2. dextrins
3. starch
4. cellulose
5. glucose

Section 16.5

1. **(a)** Simple lipids yield only fatty acids and an alcohol on hydrolysis, whereas compound lipids yield fatty acids, an alcohol, and some other compounds on hydrolysis.

   (b) Essential fatty acids are required for good health but are either not produced or not produced in adequate amounts by our bodies.

   (c) Triacylglycerols, or triglycerides, are simple fats that are esters of glycerol and three fatty acids.

   (d) Waxes are esters of long-chain fatty acids and long-chain monohydric alcohols.

   (e) Saturated fatty acids have only carbon-to-carbon single bonds and are waxy solids at room temperatures. Unsaturated fatty acids have one or more carbon-to-carbon double bonds and are liquids at room temperatures.

2. Since the fatty acids are all different, this is a mixed triacylglycerol.

$$\begin{array}{l} CH_2OC(=O)(CH_2)_{14}CH_3 \\ | \\ CHOC(=O)(CH_2)_{16}CH_3 \\ | \\ CH_2OC(=O)(CH_2)_7CH{=}CH(CH_2)_7CH_3 \end{array}$$

3. $CH_3(CH_2)_{24}CH_2OC(=O)(CH)_{24}CH_3$

Section 16.10

1. Low iodine number. This compound is likely to be a solid at room temperature.

2.

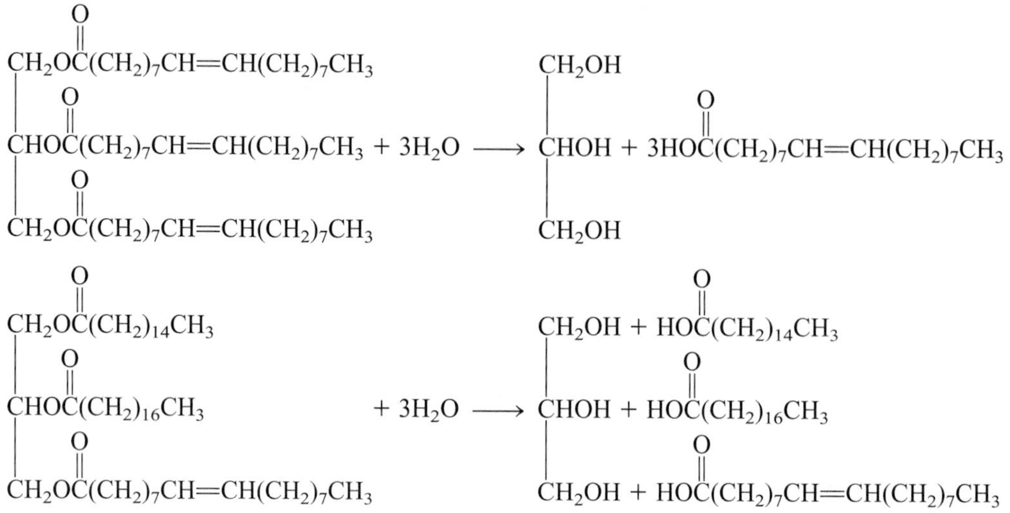

$$\begin{array}{l} CH_2O\overset{\overset{O}{\|}}{C}(CH_2)_7CH{=}CH(CH_2)_7CH_3 \\ | \\ CHO\overset{\overset{O}{\|}}{C}(CH_2)_7CH{=}CH(CH_2)_7CH_3 \\ | \\ CH_2O\overset{\overset{O}{\|}}{C}(CH_2)_7CH{=}CH(CH_2)_7CH_3 \end{array} + 3H_2O \longrightarrow \begin{array}{l} CH_2OH \\ | \\ CHOH \\ | \\ CH_2OH \end{array} + 3HO\overset{\overset{O}{\|}}{C}(CH_2)_7CH{=}CH(CH_2)_7CH_3$$

$$\begin{array}{l} CH_2O\overset{\overset{O}{\|}}{C}(CH_2)_{14}CH_3 \\ | \\ CHO\overset{\overset{O}{\|}}{C}(CH_2)_{16}CH_3 \\ | \\ CH_2O\overset{\overset{O}{\|}}{C}(CH_2)_7CH{=}CH(CH_2)_7CH_3 \end{array} + 3H_2O \longrightarrow \begin{array}{l} CH_2OH + HO\overset{\overset{O}{\|}}{C}(CH_2)_{14}CH_3 \\ | \\ CHOH + HO\overset{\overset{O}{\|}}{C}(CH_2)_{16}CH_3 \\ | \\ CH_2OH + HO\overset{\overset{O}{\|}}{C}(CH_2)_7CH{=}CH(CH_2)_7CH_3 \end{array}$$

3.

$$\begin{array}{l} CH_2O\overset{\overset{O}{\|}}{C}(CH_2)_7CH{=}CH(CH_2)_7CH_3 \\ | \\ CHO\overset{\overset{O}{\|}}{C}(CH_2)_7CH{=}CH(CH_2)_7CH_3 \\ | \\ CH_2O\overset{\overset{O}{\|}}{C}(CH_2)_7CH{=}CH(CH_2)_7CH_3 \end{array} + 3NaOH \longrightarrow \begin{array}{l} CH_2OH \\ | \\ CHOH \\ | \\ CH_2OH \end{array} + 3CH_3(CH_2)_7CH{=}CH(CH_2)_7\overset{\overset{O}{\|}}{C}O^-Na^+$$

$$\begin{array}{l} CH_2O\overset{\overset{O}{\|}}{C}(CH_2)_{14}CH_3 \\ | \\ CHO\overset{\overset{O}{\|}}{C}(CH_2)_{16}CH_3 \\ | \\ CH_2O\overset{\overset{O}{\|}}{C}(CH_2)_7CH{=}CH(CH_2)_7CH_3 \end{array} + 3NaOH \longrightarrow \begin{array}{l} CH_2OH + CH_3(CH_2)_{14}\overset{\overset{O}{\|}}{C}O^-Na^+ \\ | \\ CHOH + CH_3(CH_2)_{16}\overset{\overset{O}{\|}}{C}O^-Na^+ \\ | \\ CH_2OH + CH_3(CH_2)_7CH{=}CH(CH_2)_7\overset{\overset{O}{\|}}{C}O^-Na^+ \end{array}$$

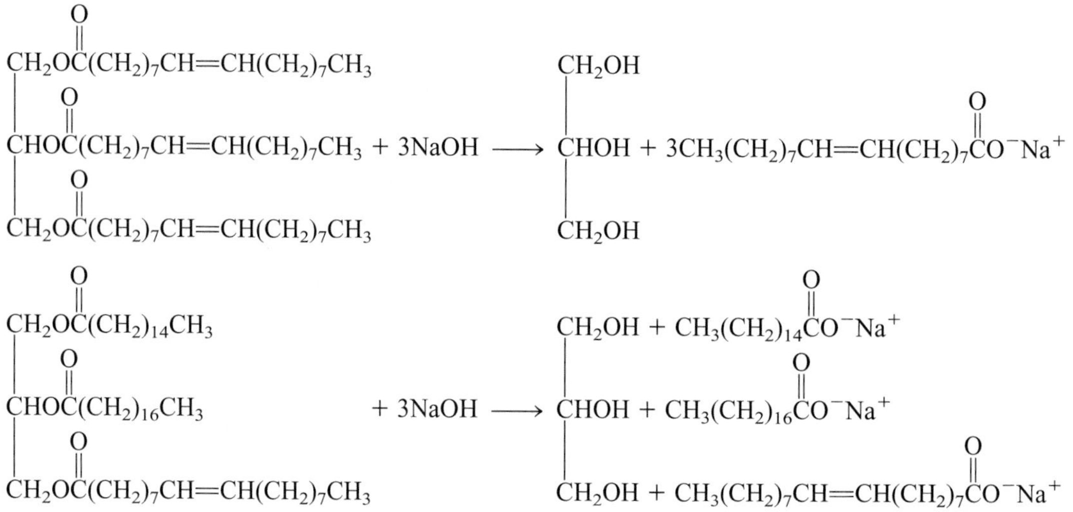

Section 16.13

1. cholesterol, cortisone, vitamin D, testosterone, or progesterone

2. phosphoglycerides, sphingolipids, and cholesterol are major components. Membranes also contain proteins and carbohydrates in the form of glycolipids and glycoproteins.

3. (a)

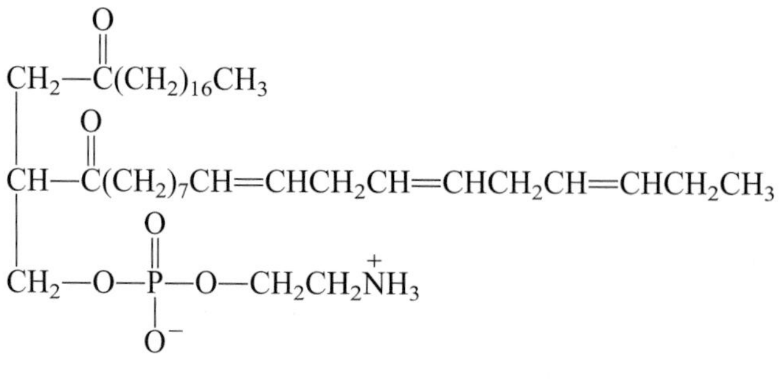

$$\begin{array}{l} CH_2{-}\overset{\overset{O}{\|}}{C}(CH_2)_{16}CH_3 \\ | \\ CH{-}\overset{\overset{O}{\|}}{C}(CH_2)_7CH{=}CHCH_2CH{=}CHCH_2CH{=}CHCH_2CH_3 \\ | \\ CH_2{-}O{-}\underset{\underset{O^-}{|}}{\overset{\overset{O}{\|}}{P}}{-}O{-}CH_2CH_2\overset{+}{N}H_3 \end{array}$$

(b) 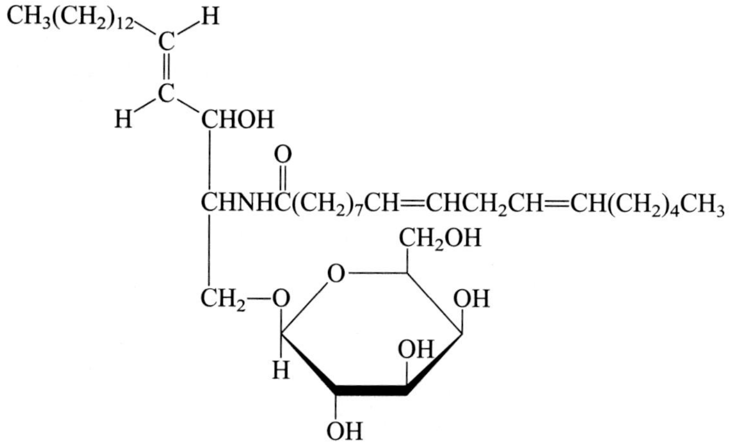 $CH_3(CH_2)_{12}CH{=}CH{-}CHOH{-}CH(NHC(=O)(CH_2)_7CH{=}CHCH_2CH{=}CH(CH_2)_4CH_3){-}CH_2{-}O{-}$(galactose ring: CH_2OH, OH, OH, OH, H)

Section 17.5

1. (a) Amino acids can act as buffers because they act as either acids or bases in aqueous solution.

(b)

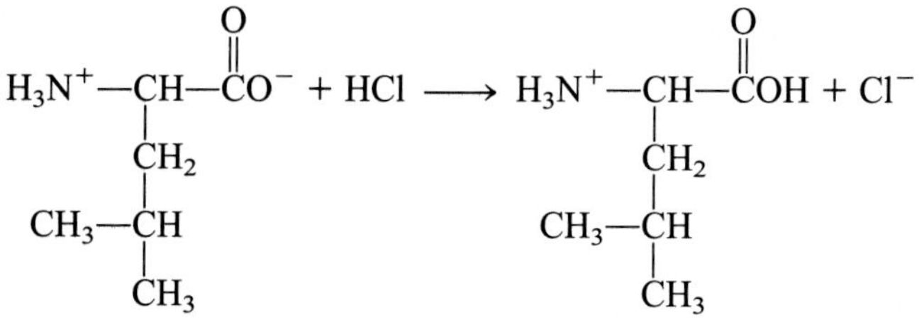

$$H_3N^+{-}\underset{\underset{\underset{CH_3}{|}}{CH_3{-}CH}}{\underset{|}{\underset{CH_2}{|}}}{CH}{-}\overset{O}{\overset{\|}{C}}O^- + HCl \longrightarrow H_3N^+{-}CH(CH_2CH(CH_3)CH_3){-}\overset{O}{\overset{\|}{C}}OH + Cl^-$$

(c)

$$H_3N^+{-}CH(CH_2C_6H_5){-}\overset{O}{\overset{\|}{C}}O^- + NaOH \longrightarrow H_2N{-}CH(CH_2C_6H_5){-}\overset{O}{\overset{\|}{C}}O^- + Na^+ + H_2O$$

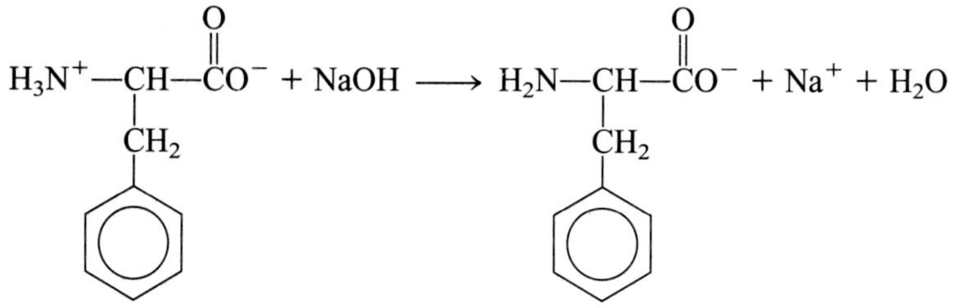

2. (a) The p*I* of phenylalanine is 5.5.

(b) The pH of 7 is more basic than pH 5.5; therefore, the phenylalanine has a net negative charge and moves toward the positive electrode at pH 7.

Section 17.7

1. serine-glutamine-arginine

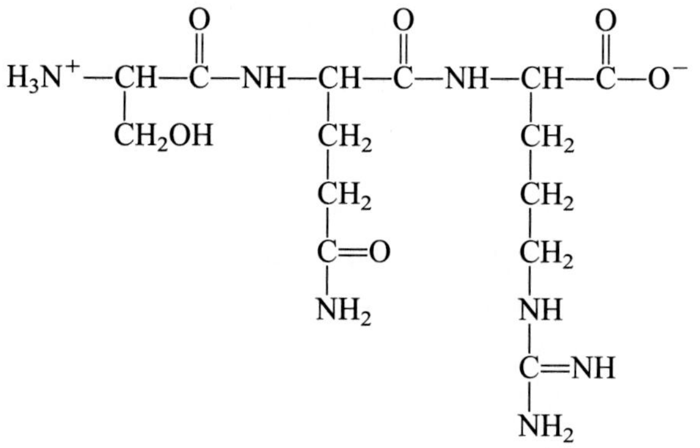

$$H_3N^+{-}CH(CH_2OH){-}\overset{O}{\overset{\|}{C}}{-}NH{-}CH(CH_2CH_2C(=O)NH_2){-}\overset{O}{\overset{\|}{C}}{-}NH{-}CH(CH_2CH_2CH_2NHC(=NH)NH_2){-}\overset{O}{\overset{\|}{C}}{-}O^-$$

```
         H   O          H   O          H   O          H   O          H  O
         |   ||         |   ||         |   ||         |   ||         |  ||
2. H3N+—C———C—N————————C———C—N————————C———C—N————————C———C—N————————C——C—O−
         |      |       |      |       |      |       |      |       |
        CH2     H       H      H       H      H      CH2     H      CH2
         |                                            |              |
     (benzene ring)                             (benzene ring)      CH2
         |                                                           |
         OH                                                          S
                                                                     |
                                                                    CH3
```

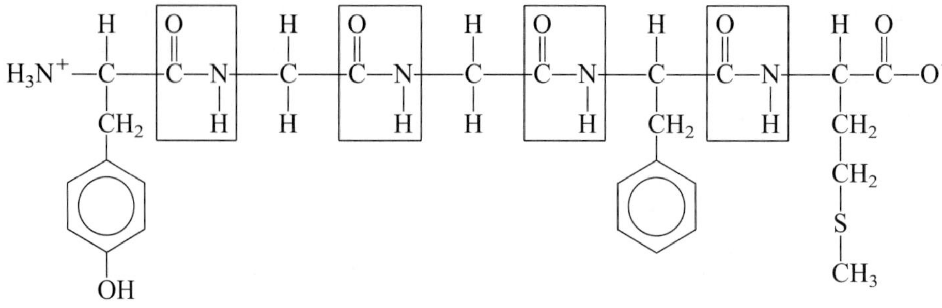

Section 17.11

1. **(a)** peptide bonds
   (b) hydrogen bonds
   (c) hydrogen bonds, disulfide bridges, salt bridges, and hydrophobic interactions
   (d) hydrogen bonds, hydrophobic interactions, and salt bridges
2. The native state of a protein is the most stable structural arrangement for that protein.
3. **(a)** hydrogen bonds and salt bridges
   (b) hydrogen bonds and salt bridges
   (c) salt bridges

Section 18.4

1. **(a)** Apoenzyme is the term for the protein part of the enzyme molecule.
   (b) A coenzyme is a complex organic molecule other than a protein that is necessary for the activity of an enzyme.
   (c) Holoenzyme is the term for the active enzyme that is made up of an apoenzyme and a cofactor.
   (d) Cofactors are the additional chemical groups necessary for the activity of an enzyme.
   (e) A proenzyme is the inactive form of an enzyme.
   (f) The substrate is the chemical substance or substances on which the enzyme acts.
   (g) The active site is the specific area of the enzyme to which the substrate attaches during the reaction.
   (h) The turnover number of an enzyme is the number of substrate molecules transformed per minute by one molecule of the enzyme under optimal conditions of temperature and pH.
   (i) Hydrolases are enzymes that catalyze hydrolysis reactions.
   (j) Oxidoreductases are enzymes that catalyze oxidation–reduction reactions.
   (k) Transferases are enzymes that catalyze reactions involved in the transfer of functional groups.
   (l) Lyases are enzymes that catalyze the addition to double bonds.
   (m) Isomerases are enzymes that catalyze the interconversion of isomers.
   (n) Ligases are enzymes that, in conjunction with ATP, catalyze the formation of new bonds.
2. 1×10^{-18} mol/h

Section 18.6

1. **(a)** Carbonic anhydrase acts as a catalyst, lowering the activation energy of the reaction.

   (b) The change in pH slows the reaction because increasing the pH causes some denaturation of the enzyme.

2. Because of the specificity necessary between enzyme and substrate, pepsin and trypsin are not able to hydrolyze all peptide bonds in proteins.

Section 18.9

1.

Function	*Site of Synthesis*	*Site of Action*
Hormones		
Coordinate actions between cells, tissues, and organs	Endocrine glands	Circulate in the blood to target cells or organs
Prostaglandins		
Large variety of diverse functions	Cell membrane	Within the cell
Cyclic AMP		
Regulates cell's response to hormones	Cell membrane	Within the cell

2. **(a)** multienzyme **(b)** feedback **(c)** allosteric or regulatory **(d)** active, allosteric, or regulatory

Section 18.12

1. Fat-soluble vitamins are not excreted from the body but are stored in the liver and can be toxic at high concentrations.
2. **(a)** Biotin is a vitamin.

   (b) A vitamin deficiency can result from a diet high in raw eggs because biotin is not absorbed in the intestine as a result of the eggs.

Section 18.13

(a) Sodium arsenite can bind with the sulfhydryl groups of an enzyme and thereby inhibit the enzyme.

(b) Enzymes for which the sulfhydryl groups are important to the activity of the enzyme are most affected.

Section 19.10

The complete oxidation of one molecule of myristic acid produces 113 molecules of ATP.

Section 20.2

1. The nucleotides of DNA are composed of 2-deoxyribose, phosphoric acid, and one of the nitrogen-containing bases: cytosine, thymine, adenine, or guanine. The nucleotides of RNA are composed of ribose, phosphoric acid, and one of the nitrogen-containing bases: cytosine, uracil, adenine, or guanine.

2.

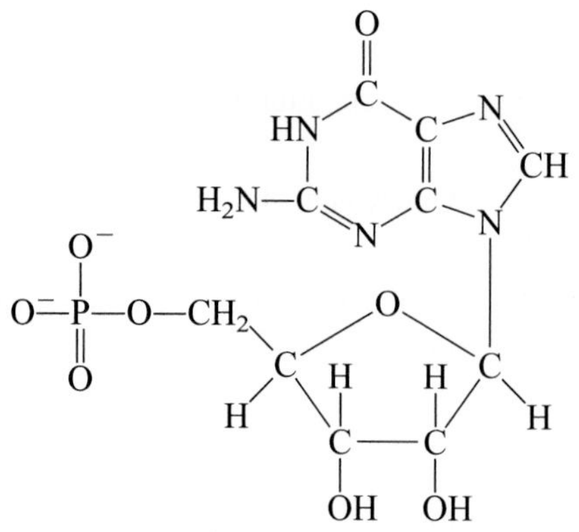

```
                    O
                    ‖
                 HN   C   N
                   \ / \ / \
                   |    C    CH
                   |    ‖   /
           H2N—C     C—N
                  \\ /    |
     O−             N     |
     |                    |
O=—P—O—CH2        O       |
     ‖        \  /   \    |
     O         C        C
              / H     H  \
             H   |    |   H
                 C————C
                 |    |
                 OH   OH
```

3. adenosine diphosphate

Section 20.4

1. A T G A C T T T T G G C T A A
2. T A C T G A A A A C C G A T T
   A T G A C T T T T G G C T A A

Section 20.5

A U G A C U U U U G G C U A A

Section 20.8

1. **(a)** A gene is a sequence of bases on a DNA molecule that translates into the specific sequence of amino acids making up a protein.
   (b) The genetic code is the general term used to describe the sequence of bases found in the gene.
   (c) A codon is the three-base sequence on mRNA that is necessary to code for an amino acid.
   (d) An anticodon is the three-base sequence on a tRNA molecule that is complementary to the codon on a mRNA molecule.
   (e) Transcription is the passage of genetic information from DNA to mRNA.
   (f) Translation is the formation of a protein from the genetic information in the mRNA.
   (g) Polysomes are groups of ribosomes producing proteins from the same strand of mRNA.
2. methionine-threonine-phenylalanine-glycine
3. $H_3N^+—CH(CH_2CH_2SCH_3)—C(=O)—NH—CH(CHOHCH_3)—C(=O)—NH—CH(CH_2C_6H_5)—C(=O)—NH—CH(H)—C(=O)—O^-$

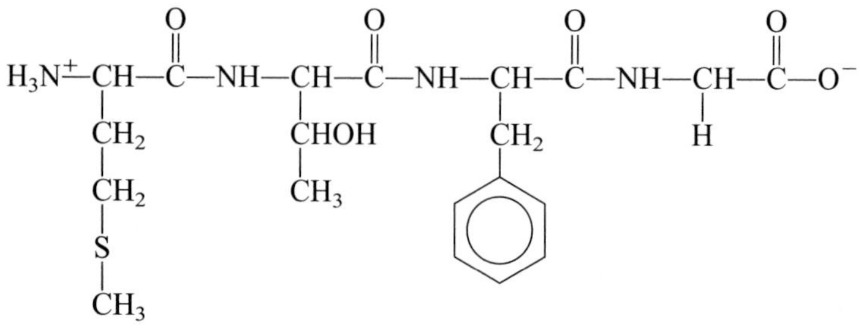

Section 21.4

1. It shifts to the right
2. It shifts to the left

Section 21.5

A decrease in blood albumin concentration reduces the osmotic pressure of the blood. Fluids are retained in the tissues.

ANSWERS TO ODD-NUMBERED PROBLEMS

Chapter 1

1. The mass of an object is a measure of how hard it is to start the object moving or how hard it is to stop its speed or direction if it is moving. The weight of an object is a measure of the attraction of gravity on the object.

3. An atom is the smallest unit of an element having the properties of that element. A molecule is a chemical unit that contains two or more atoms.

5. A compound always consists of the same elements in the same proportion by weight. Mixtures are made up of two or more substances mixed in any proportion.

7. **(a)** chemical change **(b)** chemical change **(c)** physical change **(d)** physical change **(e)** physical change **(f)** physical change

9. A hypothesis is a suggested explanation of why matter behaves as it does. If the hypothesis continues to be supported by additional data, it becomes a scientific law.

11. The analytical balance is more precise because the two values are closer together than those from the platform balance. We cannot determine which balance is more accurate without knowing the true weight of the sample.

13. **(a)** 3 **(b)** 2 **(c)** 5 **(d)** 3 **(e)** 6 **(f)** 3 **(g)** 1 **(h)** 3 **(i)** 7

15. **(a)** $\dfrac{1000 \text{ grams}}{1 \text{ kilogram}}$ $\dfrac{1 \text{ kilogram}}{1000 \text{ grams}}$

    (b) $\dfrac{946 \text{ milliliters}}{1 \text{ quart}}$ $\dfrac{1 \text{ quart}}{946 \text{ milliliters}}$

    (c) $\dfrac{1 \text{ ton}}{2000 \text{ pounds}}$ $\dfrac{2000 \text{ pounds}}{1 \text{ ton}}$

17. **(a)** $1.524 \text{ kg} \times \dfrac{1000 \text{ g}}{1 \text{ kg}}$ **(d)** $150 \text{ g} \times \dfrac{1 \text{ kg}}{1000 \text{ g}}$

    (b) $2.55 \text{ qt} \times \dfrac{946 \text{ mL}}{1 \text{ qt}}$ **(e)** $550 \text{ mL} \times \dfrac{1 \text{ qt}}{946 \text{ mL}}$

    (c) $1525 \text{ lb} \times \dfrac{1 \text{ ton}}{2000 \text{ lb}}$ **(f)** $2.1 \text{ tons} \times \dfrac{2000 \text{ lb}}{1 \text{ ton}}$

19. Calculator answer is in parentheses.

(a) 18 ft (17.712)
(b) 3480 ft (3476.8)
(c) 30.0 in.
(d) 53 mi (52.785)
(e) 5000 mL
(f) 16.5 qt (16.536)
(g) 7.6 L (7.58)
(h) 170 g (169.8)
(i) 0.231 lb (0.2312775)
(j) 109 yd (109.33333)
(k) 410 g (410.35)
(l) 3.5 kg (3.4958)
(m) 76 cm (76.2)
(n) 5900 ft (5904)
(o) 5.00×10^{-4} lb
(p) 0.464 gal (0.4643799)
(q) 50.0 in.
(r) 250 g
(s) 2.4 L (2.3584905)
(t) 10 pt (9.964)
(u) 0.583 pt

21. **(a)** 22°C **(b)** 194°F **(c)** −122°C **(d)** 12°F **(e)** −45°C **(f)** 423 K **(g)** −5°C

23. **(a)** **(1)** 0.79 g/mL **(2)** 0.79 **(3)** 198 g **(4)** 1.3 mL
(b) **(1)** 1.1 g/mL **(2)** 1.1 **(3)** 275 g **(4)** 0.91 mL
(c) **(1)** 0.98 g/mL **(2)** 0.98 **(3)** 245 g **(4)** 1.0 mL

25. 456 mL, 0.456 L

27. 663 km

29. 0.004 in.

31. 0.750 L, 0.795 qt

33. **(a)** 23 yards
(b) 250,000 kg

35. **(a)** 9000 g, 20 lb **(b)** 1500 m, 0.93 mi

37. Paris: 72°F Rome: 90°F

39. 104°F, 115°F

41. 0.969 g/cm^3

43. $5\frac{1}{2}$ tablets/dose

45. Better buy: 1 L for $0.50
The two 16-oz bottles cost $0.62 for 0.947 L

47. 13.0 kg/month

49. 1.3 lb, 80 lb

Chapter 2

1. Energy is the capacity to do work.

3. **(a)** 100 g of steam
(b) a car moving at 50 mph
(c) the football linebacker

5. **(a)** snow in the mountains
(b) a pendulum at the top of its swing
(c) a drawn bow and arrow

7. **(a)** 3500 cal **(d)** 250 cal
   (b) 125,000 cal **(e)** 34.9 cal
   (c) 10 cal **(f)** 32,300 cal
9. 11 kcal, 11 food Cal
11. **(a)** endothermic **(d)** endothermic
    (b) exothermic **(e)** exothermic
    (c) endothermic
13. yellow light
15. The rest of the energy is lost to the surroundings as heat energy.
17. **(a)** spilled milk
    (b) unassembled model airplane
    (c) virgin forest
    (d) wood chips
19. 90 kcal/1 h
21. **(a)** 4.6 miles **(b)** 8.9 lb
23. 620 kcal (600 kcal to one significant figure)
25. 252 cal
27. 0.98 h
29. Skiing in the Rocky Mountains. There is less atmosphere to screen out the ultraviolet rays at high elevations.
31. Entropy increases as the sugar molecules leave the ordered state of the solid sugar crystal and enter the disordered, random state of sugar molecules dissolved in water.

Chapter 3

1. The particles of an amorphous solid are arranged in a completely random pattern, whereas those in a crystalline solid are arranged in a regular, repeating pattern. Glass is an amorphous solid, and quartz is a crystalline solid.
3. **(a)** motor oil **(b)** honey **(c)** water at 10°C
5. 27 kcal
7. Strongest: sugar Weakest: ammonia
9. **(a)** $5.0\,\cancel{\text{atm}} \times \frac{760\ \text{mmHg}}{1\,\cancel{\text{atm}}} =$

   (b) $63\,\cancel{\text{torr}} \times \frac{133.3\ \text{Pa}}{1\,\cancel{\text{torr}}} =$
11. 240 mL
13. 2 L
15. 598 mL
17. 3.3 L
19. The container in which the pressure is 2.5 atm. According to Henry's law, the higher the pressure on a gas, the greater its solubility in water.
21. 684 mm Hg

23. The partial pressure of carbon dioxide in the lungs is lower than its partial pressure in the cells. Carbon dioxide is produced in the cells.

25. A gas consists of very small particles randomly moving at very high speeds. Gas particles move in straight lines until they collide elastically with each other or the sides of the container. Gas particles have negligible volume, and essentially no interaction takes place between the particles of an ideal gas.

27. The water on the surface of the grapes freezes first, releasing energy (equal to the heat of fusion) to the grapes and surrounding air. This helps keep the grapes from freezing.

29. On a hot humid day, the air around us quickly becomes saturated with water vapor, slowing evaporation of sweat. A fan moves the saturated air away from us and replaces it with air that can accept more water. Evaporation of sweat increases, and we feel cooler.

31. **(a)** The atmospheric pressure is greater than the pressure inside the balloon, and when released, the pressure on the exterior of the balloon forces the air out of the balloon, propelling the balloon around the room.

(b) When you suck on a straw, you lower the pressure inside the straw below that of atmospheric pressure. The atmospheric pressure pushes down on the surface of the liquid, forcing it up into the straw.

(c) The Heimlich maneuver rapidly decreases the volume inside the chest, increasing the pressure inside the lungs which expels the obstruction out of the windpipe.

33. The particles in warm air have more kinetic energy and are moving more rapidly than the particles of cold air. Warm air particles occupy a greater volume at the same atmospheric pressure than the same number of cold air particles and are less dense.

35. 47 torr

37. 50.4 torr

Chapter 4

1.

Subatomic Particle	*Location*	*Charge*	*Relative Mass*
(1) proton	nucleus	1+	1
(2) neutron	nucleus	0	1
(3) electron	around nucleus	1−	1/1837

3.

	Atomic Number	*Symbol*	*Mass Number*	*p*	e^-	*n*
(a)	3	Li	7	3	3	4
(b)	16	S	32	16	16	16
(c)	11	Na	23	11	11	12
(d)	26	Fe	56	26	26	30
(e)	35	Br	80	35	35	45
(f)	38	Sr	88	38	38	50
(g)	50	Sn	116	50	50	66
(h)	80	Hg	200	80	80	120
(i)	88	Ra	226	88	88	138

5. **(a)** $p = 25$, $n = 30$, $e^- = 25$
(b) $p = 51$, $n = 71$, $e^- = 51$
(c) $p = 19$, $n = 20$, $e^- = 19$

7. boron-10: $p = 5$, $n = 5$, $e^- = 5$
boron-11: $p = 5$, $n = 6$, $e^- = 5$

9. The isotopes of zinc differ in the number of neutrons in their nuclei.

11. Br, 80.0 Zn, 65.5

13. The quantum mechanical model of the atom has the protons and neutrons in a small, dense nucleus with the electrons surrounding the nucleus in probability regions called orbitals. Each orbital can hold two electrons spinning in different directions. The farther from the nucleus, the larger the number of orbitals and the greater the energy of the electrons in those orbitals.

15.

	Symbol	*Atomic Number*	*Atomic Weight*	*Electron Configuration*
(a) lithium	Li	3	6.94	$1s^2 2s^1$
(b) nitrogen	N	7	14.01	$1s^2 2s^2 2p^3$
(c) neon	Ne	10	20.18	$1s^2 2s^2 2p^6$
(d) magnesium	Mg	12	24.30	$1s^2 2s^2 2p^6 3s^2$
(e) aluminum	Al	13	26.98	$1s^2 2s^2 2p^6 3s^2 3p^1$
(f) chlorine	Cl	17	35.45	$1s^2 2s^2 2p^6 3s^2 3p^5$

17. **(a)** sodium **(c)** vanadium **(e)** arsenic
(b) fluorine **(d)** iron

19. **(a)** cation **(b)** anion **(c)** anion **(d)** cation

21. Periodicity is the repeating nature of certain properties of the elements.

23. **(a)** 1 **(b)** 3 **(c)** 5 **(d)** 8 **(e)** 2 **(f)** 4 **(g)** 7 **(h)** 6

25. **(a)** Sr **(b)** Ti **(c)** F **(d)** Be
(a) and **(d)** are in the same group, so they have similar chemical properties.

27. **(a)** metal **(e)** metal
(b) metal **(f)** metalloid
(c) nonmetal **(g)** nonmetal
(d) nonmetal **(h)** nonmetal

29. **(a)** B, Al, Ga **(c)** Si, Ga, Cd **(e)** F, O, Cl
(b) Te, Sb, Sn **(d)** Cl, P, As

31. **(a)** **(1)** Na **(2)** Cl **(e)** **(1)** Xe **(2)** Ne
(b) **(1)** C **(2)** O **(f)** **(1)** Sb **(2)** N
(c) **(1)** Rb **(2)** Li **(g)** **(1)** Sr **(2)** Si
(d) **(1)** As **(2)** F **(h)** **(1)** Fe **(2)** Br

33. hydrogen, carbon, nitrogen, and oxygen

35. 1.107×10^{30} electrons

37. **(a)** lithium-7 **(1)** 3 **(2)** 7 **(3)** 3 **(4)** 4
lithium-6 **(1)** 3 **(2)** 6 **(3)** 3 **(4)** 3

(b) 6.93

39. atomic number 120
configuration (2 8 18 32 32 18 8 2)

41. The electrons in the *s* orbital have less energy than the 3*d* orbitals, so the *s* orbital is filled first.

Chapter 5

1. The elements of group VIIIA are referred to as ''noble'' because they are unreactive.

3. Ionic bonds are formed between metals and nonmetals.

5. **(a)** Ċs **(b)** ·Ge· **(c)** Ca· **(d)** :Ne: **(e)** ·As· **(f)** Al·

(g) ·S: **(h)** ·I:

7. Covalent bonds are formed between nonmetals.

9. A single bond is a covalent bond in which two electrons are shared between two atoms. For example, H : H. A double bond is a covalent bond in which four electrons are shared between two atoms. For example, O : : C : : O. A triple bond is a covalent bond in which six electrons are shared between two atoms. For example, N : : : N.

11. **(a)** H:I: **(d)** H:C:::N: **(g)** H:C:I: (with H above C and :I: below C)

(b) :F:F: **(e)** :O:F: (with :F: below O) **(h)** :O::S:O:

(c) H:C:Cl: (with H above and H below C) **(f)** :Cl:P:Cl: (with :Cl: below P) **(i)** :Br:C:Br: (with H above C and :Br: below C)

13. **(a)** H, C, Br, Cl **(b)** B, P, O, F **(c)** K, Al, C, N

15. **(a)** Ionic bonds form between elements that have large differences in electronegativity.

(b) Covalent bonds form between elements with the same or very similar electronegativities.

17. Yes. A compound can be nonpolar if its polar bonds are arranged symmetrically in space so that the center of positive charge for the molecule coincides with the center of negative charge.

19. **(a)** polar **(d)** nonpolar

(b) nonpolar **(e)** polar

(c) polar **(f)** polar

21. Yes. The hydrogen is attached to the highly electronegative fluorine. Therefore, the partially positive hydrogen region on one hydrogen fluoride molecule will form a hydrogen bond with the partially negative fluorine on another molecule.

23. **(a)** K_2CO_3
(b) Na_2S
(c) $Ca(NO_3)_2$
(d) SrS
(e) $CrCl_3$
(f) $Fe_2(HPO_4)_3$
(g) $Cu(C_2H_3O_2)_2$
(h) $BaSO_4$
(i) $Al_2(SO_3)_3$
(j) $Sn(NO_3)_4$
(k) BeF_2
(l) $CsBr$

25. **(a)** 2Li, 2C, 3H, 1O
(b) 1Mg, 2H, 2C, 6O
(c) 2Al, 3Cr, 12O
(d) 3Sr, 2P
(e) 2Co, 6C, 12O
(f) 3Sn, 2P, 8O
(g) 2Ag, 1S, 3O
(h) 2Mn, 3O
(i) 1Pb, 8H, 4P, 16O
(j) 3N, 12H, 1N

27. **(a)** potassium carbonate
(b) sodium sulfide
(c) calcium nitrate
(d) strontium sulfide
(e) chromium(III) chloride
(f) iron(III) monohydrogen phosphate
(g) copper(II) acetate
(h) barium sulfate
(i) aluminum sulfite
(j) tin(IV) nitrate
(k) beryllium fluoride
(l) cesium bromide

29. **(a)** ammonium iodide
(b) phosphorus trichloride
(c) calcium hydroxide
(d) carbon tetrabromide
(e) iron(II) chloride
(f) ammonium dichromate
(g) dioxygen difluoride
(h) zinc nitrite
(i) diphosphorus pentoxide
(j) sodium hydride
(k) hydrogen iodide
(l) sodium permanganate
(m) lithium sulfide
(n) potassium sulfite
(o) calcium carbide
(p) aluminum sulfate

31. Because the compound XY_2 has polar bonds but is not polar, the compound must be linear.

33. $CaCl_2$. Ionic. A crystal of this substance consists of Ca^{2+} and Cl^- ions.

35. H : C ⋮⋮ N :   : C ⋮⋮ N :

37. Potassium loses an electron and achieves an octet in its valence shell. Bromine gains an electron and also achieves a valence shell octet. The electron configuration for the K^+ ion is $1s^22s^22p^63s^23p^6$, which is the same as that of Ar. The electron configuration of the Br^- ion is $1s^22s^22p^63s^23p^64s^23d^{10}4p^6$, which is the same as that for Kr.

Chapter 6

1. **(a)** $H_2 + Br_2 \longrightarrow 2HBr$
(b) $Ca(HCO_3)_2 \longrightarrow CaCO_3 + H_2O + CO_2$
(c) $2AgNO_3 + Cu \longrightarrow Cu(NO_3)_2 + 2Ag$
(d) $3H_2 + N_2 \longrightarrow 2NH_3$
(e) $CH_4 + 4Cl_2 \longrightarrow CCl_4 + 4HCl$

3. **(a)** $6HCl + 2Cr \longrightarrow 2CrCl_3 + 3H_2$

(b) $2FeCl_3 + 3Na_2CO_3 \longrightarrow Fe_2(CO_3)_3 + 6NaCl$

(c) $PbS + 4H_2O_2 \longrightarrow PbSO_4 + 4H_2O$

(d) $2C_4H_{10} + 13O_2 \longrightarrow 8CO_2 + 10H_2O$

(e) $2CrO_4^{2-} + 2H^+ \longrightarrow Cr_2O_7^{2-} + H_2O$

5. **(a)** Ca-oxidized Cl-reduced

(b) Mg-oxidized O-reduced

7. **(a)** 28.8 g **(c)** 15.8 g

(b) 130 g **(d)** 4.92 g

9.

	Group I	*Group II*	*Group III*
(a)	40.0 amu	74.6 amu	74.1 amu
(b)	111 amu	98.1 amu	132 amu
(c)	64.1 amu	263 amu	16.0 amu
(d)	164 amu	44.0 amu	36.5 amu

11. 22.4 L

13. V = 115 L

15.

	Group I	*Group II*	*Group III*
(a)	6.0 g	313 g	70 g
(b)	278 g	147 g	607 g
(c)	51 g	3.9 g	200 g
(d)	82 g	242 g	0.91 g
(e)	839 g	27 g	53 g
(f)	1.6 g	11.8 g	85 g
(g)	8.48 g	6.3 g	403 g
(h)	0.018 g	9.6 g	0.01 g

17. **(a)** 0.60 mol **(b)** 18 g

19. **(a)** 1.8 mol

(b) 205 g

(c) 28.76 g

21. **(a)** 0.225 mol

(b) 0.110 mol

(c) 3.3 mol

23. **(a)** limiting reactant: Mg

$$\text{Mg: } 25.0\cancel{\text{g}} \times \frac{1 \text{ mol}}{24.3\cancel{\text{g}}} = 1.03 \text{ mol}$$

$$\text{S: } 34.0\cancel{\text{g}} \times \frac{1 \text{ mol}}{32.1\cancel{\text{g}}} = 1.06 \text{ mol}$$

(b) limiting reactant: SO_3

$$SO_3\text{: } 94.4\,\text{g} \times \frac{1\text{ mol}}{80.1\,\text{g}} = 1.18\text{ mol}$$

$$HNO_3\text{: } 250\,\text{g} \times \frac{1\text{ mol } HNO_3}{63.0\,\text{g}} \times \frac{1\text{ mol } SO_3}{2\text{ mol } HNO_3} = 1.98\text{ mol } SO_3$$

(c) limiting reactant: Fe

$$\text{Fe: } 16.8\,\text{g} \times \frac{1\text{ mol}}{55.8\,\text{g}} = 0.301\text{ mol}$$

$$H_2O\text{: } 12.0\,\text{g} \times \frac{1\text{ mol } H_2O}{18.0\,\text{g}} \times \frac{3\text{ mol Fe}}{4\text{ mol } H_2O} = 0.500\text{ mol Fe}$$

25. 2.56 kg ethyl alcohol

27. 0.390 g O_2

29. **(a)** $2\,NH_3 + CO_2 \longrightarrow (NH_2)_2CO + H_2O$

(b) 10.6 metric tons urea

31. 7.1 tons lauryl alcohol

Integrated Problems

1. **(a)** 610 mg/kg

(b) 33 g, 25 g

3. 4×10^{16} miles

5. 81 km/kg

7. **(a)** 722 ft

(b) 125°F

9. **(a)** $4.98 \times 10^{-12}\ \mu g$

(b) 22×10^{-6} mm

11. **(a)** $2SO_2 + O_2 \longrightarrow 2SO_3$

(b) 800 g SO_2

13. 622°F, 2.01×10^5 mm Hg

Chapter 7

1. A substance is radioactive when its unstable nucleus emits particles or gives off gamma radiation to become more stable.

3. ${}^{222}_{86}\text{Rn} \longrightarrow {}^{218}_{84}\text{Po} + {}^{4}_{2}\text{He}$

5. ${}^{230}_{90}\text{Th} \longrightarrow {}^{226}_{88}\text{Ra} + {}^{4}_{2}\text{He} + \gamma$

7. **(a)** ${}^{8}_{4}\text{Be} \longrightarrow {}^{4}_{2}\text{He} + {}^{4}_{2}\text{He}$

(b) ${}^{135}_{53}\text{I} \longrightarrow {}^{135}_{54}\text{Xe} + {}^{0}_{-1}e$

(c) ${}^{20}_{8}\text{O} \longrightarrow {}^{20}_{9}\text{F} + {}^{0}_{-1}e$

(d) ${}^{245}_{97}\text{Bk} \longrightarrow {}^{241}_{95}\text{Am} + {}^{4}_{2}\text{He}$

9. 600 years

11. The third hour. One hour can be thought of as the "half-life" of the money you have to gamble; after each hour, one-half of your money has "decayed." In radioactive decay, the radioactive element decays to a daughter nucleus, which remains. When you gamble and lose, you no longer have that money.

13. neutrons

15. Mill tailings, low-level wastes, transuranic wastes, high-level wastes. Transuranic and high-level wastes pose the most threat to human health.

17. Atomic fission is the breaking apart of a large, unstable nucleus into smaller, more stable nuclei and is the process presently used in power plants. Atomic fusion is the combining of small nuclei to form larger, more stable nuclei.

19. Ionizing radiation is radiation that can produce highly reactive charged particles when it hits living tissue. Alpha, beta, and gamma radiation are all forms of ionizing radiation.

21. Alpha particles are the least penetrating; they can be stopped by a piece of paper. Beta particles are more penetrating than alpha particles; they can pass through a piece of paper but are stopped by a piece of wood. Gamma radiation easily passes through paper and wood, but they are reduced to background levels by lead blocks or thick concrete.

23. 9.4×10^{13} disintegrations

25. These tests have doubled the patient's annual background radiation compared with the average in the United States.

27. The radiation 3 ft from the source is one-ninth of that 1 ft from the source.

29. Images produced by ordinary X rays are two-dimensional, and often details are masked by overlying tissues. A CAT scan produces cross-sectional images that can be enhanced by color to differentiate specific tissue or other body materials.

31. **(a)** The energy of the gamma rays can be easily detected by cameras now in use.
    (b) No alpha or beta radiation is given off.
    (c) The half-life allows time for the isotope to localize in the body and allows administration in doses that limit radiation exposure to that comparable to an X-ray procedure.

33. Iridium-192 is better for this treatment. Strontium-90 is taken up by the bones in place of calcium and is therefore retained in the body.

35. After 3 half-lives, 2.50 g remain.

37. Twenty-five percent remains after two half-lives, so the sample was 11,460 years old.

39. 2×10^{15} kJ

41. Radium is in the same group on the periodic table as calcium and has similar chemical properties. The body confuses radium with calcium and deposits radium in the bones. Radium is an alpha and gamma emitter. When concentrated in the bones, it does most of its damage to bone tissue through direct action.

43. This treatment is very specific. Only the tumor tissue that has absorbed the boron compound is irradiated by the alpha particles. None of the normal body tissues are affected. In other less specific treatments, normal tissue is killed as well as the cancerous tissue.

Chapter 8

1. A minimum energy is required to break the glass. When the glass remains intact, this energy requirement has not been met. On the occasion that the glass breaks, the minimum energy requirement has been met.

3.

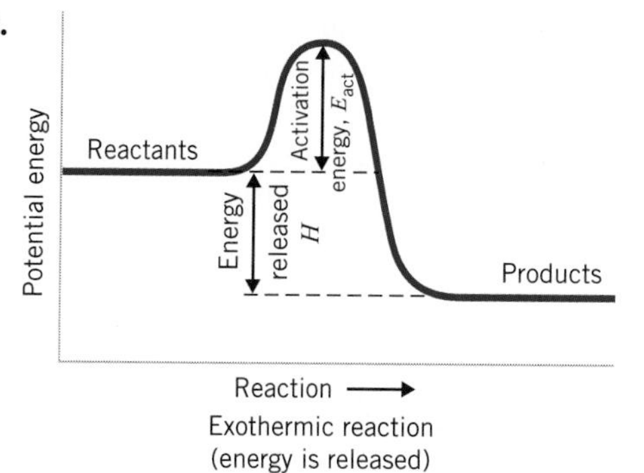

Exothermic reaction
(energy is released)

Potential energy
Reactants
Activation energy, E_{act}
Energy absorbed H
Products
Reaction

Endothermic reaction
(energy is absorbed)

5. **(a)** exothermic **(b)** 54 kcal **(c)** 32 kcal **(d)** 86 kcal

7. Add a catalyst, add more reactants, grind up a solid reactant.

9. A catalyst is a substance that increases the rate of a chemical reaction without being consumed in the reaction.

11. **(a)** The addition of platinum increases the rate of reaction because the lower activation energy increases the chances for the molecules to collide with enough energy to react.

(b)

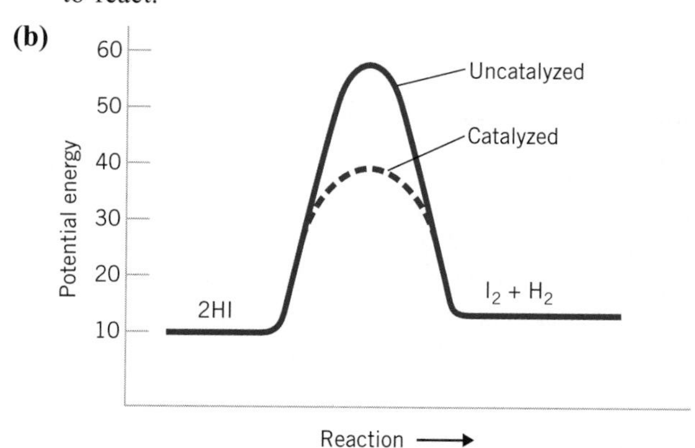

(c) uncatalyzed: $E_{act} = 40.8$ kcal catalyzed: $E_{act} = 26$ kcal

(d) $\Delta H = 3$ kcal $\Delta H = 3$ kcal

13. An equilibrium is dynamic because the system is constantly changing, but the forward and reverse reactions occur at equal rates.

15. changes in concentration or temperature

17. When the lid is off a can of ether, the system is not closed, the ether continues to evaporate, and no equilibrium is established. When the lid is on, the system is closed and the ether that evaporates cannot escape from the can. The rate of the reverse condensing reaction soon equals the rate of evaporation, and an equilibrium is established.

19. statement of Le Chatelier's principle

21. The equilibrium shifts to the right; the forward reaction occurs at a greater rate than the reverse reaction.

23. **(a)** decrease **(b)** decrease **(c)** increase **(d)** decrease

25. Covering the peaches decreases exposure to the air and lowers the concentration of the oxygen that causes the browning. Placing the peaches in the refrigerator decreases the rate of the browning reaction by lowering the temperature.

27. Lowering body temperature slows the rates of reactions in all body tissues, slowing respiration, heart rate, cognitive reactions, and leads to drowsiness and sleep.

29. (a)

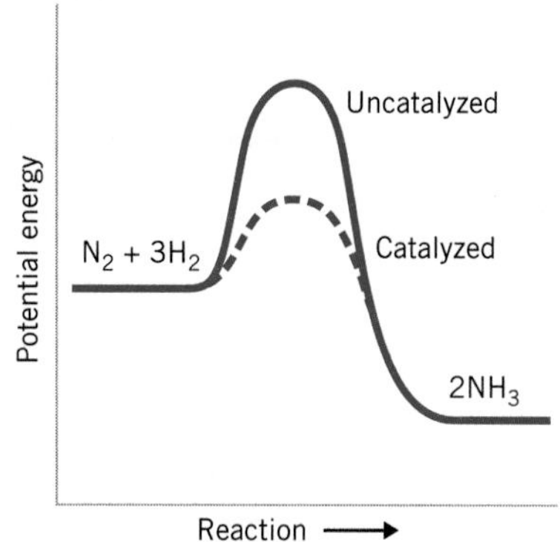

(b) Increase the concentration of N_2, increase the concentration of H_2, and increase the pressure.

31. (a) increase **(b)** decrease **(c)** decrease **(d)** decrease **(e)** no change

Integrated Problems

1. (a) 0.03 g, 30 mg

(b) 0.12 cent

3. Strontium is in the same chemical family as calcium, and the body treats strontium in the same way it does calcium. The strontium is therefore distributed throughout the body, especially in the bones and teeth. Substances that are beta emitters do not pose much risk to health outside of the body, but within the cells and tissues, the beta radiation can damage molecules critical to the life of the cells.

5. (a) ${}^{130}_{52}\text{Te} + {}^{1}_{0}n \longrightarrow {}^{131}_{53}\text{I} + {}^{0}_{-1}e$

(b) ${}^{131}_{53}\text{I} \longrightarrow {}^{131}_{54}\text{Xe} + {}^{0}_{-1}e + \gamma$

(c) 2.50 g (24 days = 3 half-lives)

(d) Iodine-131 is a beta and gamma emitter. The beta radiation does not pass through the patient's body, but the gamma radiation does. This gamma radiation, however, is reduced to background levels by lead shielding. The nurse must wear a film badge to keep track of any radiation exposure.

(e) Iodine-131 is present in the patient's blood and urine. If any blood or urine were to get on the nurse's skin, the ${}^{131}\text{I}$ could be absorbed through the skin into the body.

(f) Iodine-131 is a polar ion and is eliminated from the body in the urine. The rate of clearance from the body is quite rapid, and after 4 days very little remains in the patient's body.

7. As carbon dioxide is produced in the tissues, the second equilibrium is shifted toward the formation of carbonic acid and then bicarbonate, which can be carried by the blood to the lungs. Since the carbon dioxide concentration is low in the lungs, the equilibrium shifts to the reverse reaction in the lungs. Carbonic acid forms carbon dioxide, which is exhaled. Since the carbon dioxide is constantly being removed, the reaction is constantly driven in the reverse direction in the lungs. The abundance of oxygen in the lungs causes the first equilibrium to shift to the production of oxyhemoglobin, thus allowing the blood to transport oxygen to the tissues. In the tissues, the high concentration of hydrogen ions and the low oxygen concentration drive this equilibrium toward the production of oxygen, which diffuses into the tissues.

Chapter 9

1. Water is called the universal solvent because it dissolves so many different substances, both ionic and polar covalent.

3. 10,930 cal

$$15.2\,\text{g} \times \frac{80\text{ cal}}{\text{g}} = 1220\text{ cal}$$

$$15.2\,\text{g} \times \frac{1\text{ cal}}{\text{g}\,^\circ\text{C}} \times 100^\circ\text{C} = 1520\text{ cal}$$

$$15.2\,\text{g} \times \frac{539\text{ cal}}{\text{g}} = 8190\text{ cal}$$

5. 4.2×10^2 cal (to two significant figures)

$$34\,\text{g} \times \frac{1\text{ cal}}{\text{g}\,^\circ\text{C}} \times 45^\circ\text{C} + 34\,\text{g} \times \frac{80\text{ cal}}{\text{g}} = 4250\text{ cal}$$

7. 32°C

9. solutions, colloids, and suspensions

11. **(a)** Use the Tyndall effect to distinguish between a solution and a colloidal dispersion.

(b) Let the two mixtures stand. The suspension settles out, but the colloidal dispersion does not.

13. **(a)** liquid in solid **(d)** solid in liquid
(b) solid in solid **(e)** solid in liquid
(c) liquid in gas **(f)** gas in liquid

15. description of the steps that occur when NaCl is dissolved in water

17. **(a)** no, nonelectrolyte **(c)** yes, electrolyte
(b) yes, electrolyte

19. **(a)** anion **(b)** cation **(c)** anion **(d)** cation

21. The solubility of a gas in a liquid decreases with an increase in temperature. Hence, more carbon dioxide is dissolved in the soft drink when it is cold than when it is at room temperature.

23. **(a)** no **(b)** yes **(c)** yes **(d)** no **(e)** yes **(f)** no

25. $H^+(aq) + OH^-(aq) \longrightarrow H_2O(\ell)$

27. Final temperature (T_f): 40°C

$$\left(250\,\text{g} \times \frac{80\text{ cal}}{\text{g}}\right) + \left[250\,\text{g} \times \frac{1\text{ cal}}{\text{g}\,^\circ\text{C}} \times (T_f - 0^\circ\text{C})\right] = \left[3000\,\text{g} \times \frac{1\text{ cal}}{\text{g}\,^\circ\text{C}} \times (50^\circ\text{C} - T_f)\right]$$

$$T_f = 40^\circ\text{C}$$

29. $Ag^+(aq) + Br^-(aq) \longrightarrow AgBr(s)$

31. Water is necessary to remove urea from the body as urine. After a high-protein meal, therefore, a lot of water is used in removing the large amount of urea formed. You feel thirsty from the need to replace this water.

Chapter 10

1. There is no standard definition of dilute and saturated solutions; however, solutions that have only a few solute particles in a given volume are dilute, and those with many solute particles are concentrated.
3. The solubility decreases and, usually, crystals appear.
5. **(a)** Dissolve 0.40 g of NaOH in enough water to make 50 mL of solution.
   (b) Dissolve 3.9 g of Na_2SO_4 in enough water to make 250 mL of solution.
   (c) Dissolve 2.7 g of NH_4Cl in enough water to make 0.10 L of solution.
7. 0.20 M H_2SO_4
9. 4.44 g hexachlorophene
11. Dissolve 0.62 mg in enough water to make 25 mL of solution.
13. **(a)** Dissolve 49.5 g of lactic acid in enough water to make 225 mL of solution.
    (b) Dissolve 0.29 g of KCl in enough water to make 35 mL of solution.
15. 0.154 *M* NaCl
17. 1.7 ppb
19. **(a)** 31.7 g **(b)** 79.9 g **(c)** 68.7 g **(d)** 30.0 g
21. **(a)** 3.15 mEq
    (b) 16.7 mEq
    (c) 9.11 mEq
23. **(a)** 1.2 *M* H_2SO_4
    (b) 0.4 ppm Cd^{2+}
25. **(a)** 200 mL of stock solution plus enough water to make 250 mL of solution
    (b) 45 mL of stock solution plus enough water to make 150 mL of solution
    (c) 75 mL of stock solution plus enough water to make 750 mL of solution
27. Salt lowers the freezing point of water, thus melting the ice.
29. **(a)** side 1 **(b)** side 2 **(c)** side 1
31. When celery has gone limp, it has lost water from its cells. Water is hypotonic to the cells of celery. When the celery is placed in water, water molecules move into the cells of the celery, restoring its crispness.
33. Osmotic pressure is a colligative property, a property that depends only on the number of particles in solution.
35. When potassium sulfate dissolves in water, it dissociates to form two potassium ions and one sulfate ion. The alcohol molecules remain intact when they dissolve in water. Therefore, a 1 *M* potassium sulfate solution contains 3 mol of ions and has three times the osmotic pressure of a 1 *M* alcohol solution.
37. **(a)** crenation **(b)** no change **(c)** hemolysis
39. **(a)** Add 13 g of procaine hydrochloride to enough water to make 250 mL of solution.
    (b) 0.18 *M*
41. Yes, the blood concentration of Li^+ is 2.0 mEq/L.

43. Na^+: 122 mmol/L

K^+: 3.2 mmol/L

Cl^-: 83 mmol/L

45. The saltwater is hypertonic to the cellular fluid of the freshwater fish, and the fish cells lose fluid. The salts also diffuse into the cells, increasing the concentration of salt in the cell and disrupting normal cellular function.

47. **(a)** As the salt builds up in the soil, the water–salt solution in the soil becomes hypertonic to the cells of the plant. Water moves from the cells to the soil, dehydrating and eventually killing the plant.

(b) Plants that live in saltwater marshes have cells that have adapted to the higher osmotic pressure of the water by absorbing salt to balance the concentration of the dissolved substances within their tissues to that of the soil–water environment outside the cells.

Chapter 11

1. **(a)** $\underset{\text{base}}{H_2O} + \underset{\text{acid}}{HCl} \rightleftharpoons \underset{\text{acid}}{H_3O^+} + \underset{\text{base}}{Cl^-}$

(b) $\underset{\text{acid}}{H_2O} + \underset{\text{base}}{CH_3NH_2} \rightleftharpoons \underset{\text{acid}}{CH_3NH_3^+} + \underset{\text{base}}{OH^-}$

3. These solutions are sour, they turn blue litmus paper red, and they react with certain metals to produce hydrogen gas.

5. **(a)** HCO_3^- **(d)** $H_2PO_4^-$

(b) NH_4^+ **(e)** NH_3

(c) H_2SO_3 **(f)** H_2O

7. $H_2SO_4 + H_2O \rightleftharpoons HSO_4^- + H_3O^+$

$HSO_4^- + H_2O \rightleftharpoons SO_4^{2-} + H_3O^+$

9. **(a)** OH^- **(b)** PO_4^{3-} **(c)** NH_3 **(d)** CO_3^{2-}

11. **(a)** $NH_4Cl + NaOH \longrightarrow H_2O + NH_3 + NaCl$

(b) $NH_4^+ + OH^- \longrightarrow H_2O + NH_3$

(c) The odor of ammonia is greater after the addition of sodium chloride because ammonia is a product of the reaction between ammonium chloride and sodium hydroxide.

13. **(a)** $1 \times 10^{-6}\,M$ **(c)** $3.3 \times 10^{-12}\,M$

(b) $1.5 \times 10^{-12}\,M$ **(d)** $3.8 \times 10^{-12}\,M$

15.

	$[H^+]$	$[OH^-]$
(a)	1×10^{-1}	1×10^{-13}
(b)	1×10^{-6}	1×10^{-8}
(c)	1×10^{-12}	1×10^{-2}

17. **(a)** 2 **(b)** 8 **(c)** 10

19. **(a)** pH 3.0 $[H^+] = 5 \times 10^{-6}$ mol/5 mL
$[OH^-] = 5 \times 10^{-14}$ mol/5 mL

(b) pH 4.0 $[H^+] = 5 \times 10^{-7}$ mol/5 mL
$[OH^-] = 5 \times 10^{-13}$ mol/5 mL

21. The following equilibrium exists in the buffer system:

$$CH_3COOH + H_2O \rightleftharpoons CH_3COO^- + H_3O^+$$

(a) When an acid is added to the system, the equilibrium shifts to the left, thus lowering the H^+ concentration.

(b) When a base is added to the system, the equilibrium shifts to the right, to replace the H^+ that reacts with the base.

23. When acid is added, the equilibrium shifts to the left, forming more carbonic acid.

$$H_2CO_3 \rightleftharpoons HCO_3^- + H^+$$

25. $CO_3^{2-} + 2H^+ \longrightarrow H_2CO_3$

27. $HCO_3^- + H^+ \longrightarrow H_2CO_3 \longrightarrow CO_2(g) + H_2O$

29. The tea contains a compound that acts as an acid–base indicator and changes color when the pH of the tea changes as the acidic lemon juice is added.

31. The pH of the blood does not change with the addition of lactic acid because the blood has several buffer systems that protect it against large changes in pH.

Integrated Problems

1. The water on the surface of the bag evaporates, lowering the average kinetic energy of the water, thus cooling the water in the bag.

3. **(a)** 0.0986% aspartame

(b) $3.35 \times 10^{-3}\ M$

5. By actively transporting sodium and chloride ions into the interstitial fluid, the gallbladder cells cause the interstitial fluid to become hypertonic to the contents of the cells. Water moves from the gallbladder cells to the interstitial fluid concentrating the bile in the gallbladder.

7. **(a)** $BaCl_2(aq) + Na_2SO_4(aq) \longrightarrow BaSO_4(s) + 2NaCl(aq)$

$$Ba^{2+}(aq) + SO_4^{2-}(aq) \longrightarrow BaSO_4(s)$$

(b) 0.005 mol $BaSO_4$

9. **(a)** Shift the equilibrium to the left.

(b) Shift the equilibrium to the right.

11. Sodium ions are found mainly in the extracellular fluids. An increase in the concentration of sodium ions in these fluids increases the osmotic pressure of the fluid and results in the movement of water from the blood into the tissues, causing edema. This retention of fluids causes an increase in the blood pressure.

13. 0.1 mg% corresponds to 1 ppm. Metacresol is purple at $pH > 9$. Add acid and halogen disinfectant.

Chapter 12

1. the study of the compounds of carbon

3. There are so many more organic compounds than inorganic compounds because of the unique bonding ability of carbon. It can form single, double, and triple bonds; rings; chains; and branched chains; and it can bond to many other elements.

5. Hybrid orbitals are formed when atomic orbitals mix together. These hybrid orbitals allow for greater overlap and the formation of stronger bonds.

7. The molecular formula of a compound indicates only the number and type of atoms in a molecule. The structural formula shows the geometric arrangement of these atoms.

9. (a)

```
              H
              |
          H—C—H
     H        |        H
     |        |        |
  H—C————C————C—H
     |        |        |
     H        |        H
          H—C—H
     H        |
     |        |
  H—C————C—H
     |        |
     H        |
          H—C—H
              |
              |
          H—C—H
              |
              H
```

(b)

```
                        H
                        |
                    H—C—H
            H           |         H
            |           |         |
        H—C—H       |     H—C—H
   H        |           |         |         H
   |        |           |         |         |
H——C————C————C————C————C——H
   |        |           |         |         |
   H        H           |         H         H
                    H—C—H
                        |
                        H
```

(c)

```
            H                H
            |                |
        H—C—H        H—C—H
    H       |        H       |        H
    |       |        |       |        |
H——C————C————C————C————C——H
    |       |        |       |        |
    H       H        |       H        H
                 H—C—H
                     |
                     H
```

11. All of the carbon-to-carbon bonds are single bonds in alkanes.

13. (a) the same
(b) constitutional isomers
(c) unrelated
(d) the same
(e) the same
(f) constitutional isomers
(g) constitutional isomers
(h) constitutional isomers
(i) the same
(j) unrelated
(k) unrelated
(l) constitutional isomers
(m) unrelated
(n) unrelated
(o) unrelated
(p) the same

15. (a) propane
(b) nonane
(c) methylbutane
(d) 2,4-dimethylpentane
(e) methylpropane
(f) 4-ethyl-3-methyl-heptane
(g) 3-isopropylhexane
(h) 4-methylheptane
(i) 4-isopropyl-4-propyloctane

17. (a) 2-nitropropane
(b) 3-ethyl-3-methylpentane

(c) 2-iodo-3-methylbutane

(d) 3-chloro-3-ethyl-2,2-dimethylpentane

19. Only the carbon backbone is shown.

(a) C—C—C—C—C—C—C heptane

(b) 2-methylhexane

```
C—C—C—C—C—C
  |
  C
```

(c) 3-methylhexane

```
C—C—C—C—C—C
    |
    C
```

(d) 2,3-dimethylpentane

```
C—C—C—C—C
  | |
  C C
```

(e) 2,4-dimethylpentane

```
C—C—C—C—C
  |   |
  C   C
```

(f) 3-ethylpentane

```
C—C—C—C—C
    |
  C—C
```

(g) 2,2-dimethylpentane

```
  C
  |
C—C—C—C—C
  |
  C
```

(h) 3,3-dimethylpentane

```
    C
    |
C—C—C—C—C
    |
    C
```

(i) 2,2,3-trimethylbutane

```
  C C
  | |
C—C—C—C
  |
  C
```

21. (a)

```
      C
      |
C—C—C—C—C
```

2-methylpentane: The carbons should be numbered so that the substituted group has the lowest possible number.

(b)

```
C—C C
  | |
C—C—C—C—C—C
```

3,4-dimethylheptane: The longest carbon chain has seven carbons.

(c)

```
      C
      |
C—C—C—C—C—C
      |
      C
```

3,3-dimethylhexane: The carbons should be numbered so that the substituted groups have the lowest possible number.

(d)

```
C—C
  |
C—C—C—C—C
  |
  C—C
```

3-ethyl-3-methylhexane: The longest carbon chain has six carbons.

(e)

```
C—C—C—C—C—C—C
  |
C—C—C—C
```

3,4-dimethylnonane: The longest carbon has nine carbons.

23. (a) $C_3H_8 + 5O_2 \longrightarrow 3CO_2 + 4H_2O$ + energy

(b) $2C_8H_{18} + 25O_2 \longrightarrow 16CO_2 + 18H_2O$ + energy

(c) $2C_6H_{14} + 19O_2 \longrightarrow 12CO_2 + 14H_2O$ + energy

25. Two electrons are shared by carbons in a single bond, four electrons in a double bond, and six electrons in a triple bond.

27. (a)

```
CH3CHCH=CHCH2CH2CH3
    |
    CH3
```

(b)

```
     CH3  CH3
     |    |
CH3—C====C—CH3
```

(c) $CH_2{=}CH{-}CH{=}CH{-}CH{=}CHCH_2CH_3$

(d)

```
          Br   Br
          |    |
CH2=CH—C———C—CH2—CH3
          |    |
          Br   Br
```

(e)

```
               CH3
               |
CH3CH=CHCH2CCH3
               |
               CH3
```

(f)

```
      CH3   CH2CH2CH3
      |     |
CH2=C—CHCHCHCH2CH3
         |   |
         CH3 CH3
```

29. (a) one sigma and one pi bond

(b) one sigma and two pi bonds

31. (a)

```
CH3       CH2CH3     CH3        H
   \     /              \      /
    C=C                  C=C
   /     \              /      \
 H        H           H        CH2CH3
```

(b)

```
Cl       Cl     Cl       H
  \     /         \     /
   C=C             C=C
  /     \         /     \
H        H      H        Cl
```

(c)

```
 (cyclopentane ring)   (cyclopentane ring)
  H    H                 Cl   H
  |    |                 |    |
  C————C                 C————C
  |    |                 |    |
  Cl   Cl                H    Cl
```

33. (a) $C_4H_8 + 6O_2 \longrightarrow 4CO_2 + 4H_2O$

(b) $CH_3CH{=}CHCH_2CH_3 + H_2 \longrightarrow CH_3CH_2CH_2CH_2CH_3$

35. (a)

```
          CH3
          |
          C===CH
         /      \
CH3—CH          CH2
       \        /
         CH2
```

(c)

```
     HC===CH          CH3
    /       \         |
H2C          CH—CHCH3
    \       /
      CH2
```

(b)

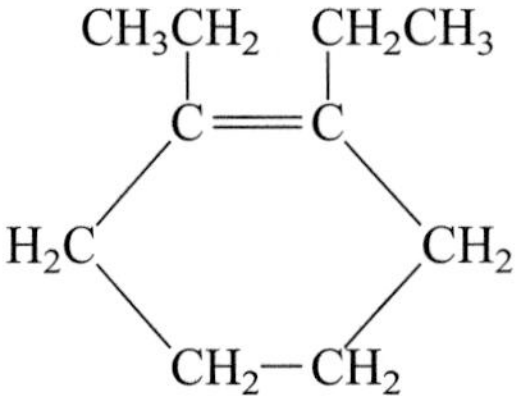

```
  CH3CH2  CH2CH3
      |    |
      C == C
     /      \
  H2C        CH2
     \      /
     CH2—CH2
```

37. (a) H_2 + cyclopentene ⟶ cyclopentane

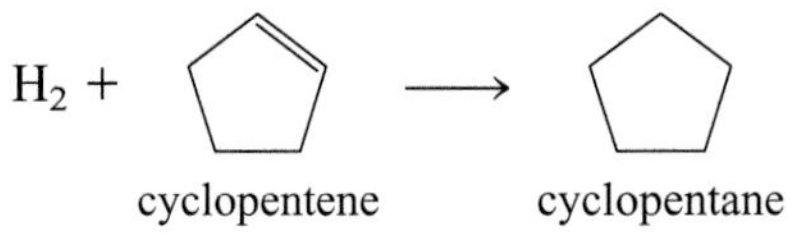

(b) cyclohexane + $9O_2 \longrightarrow 6CO_2 + 6H_2O$

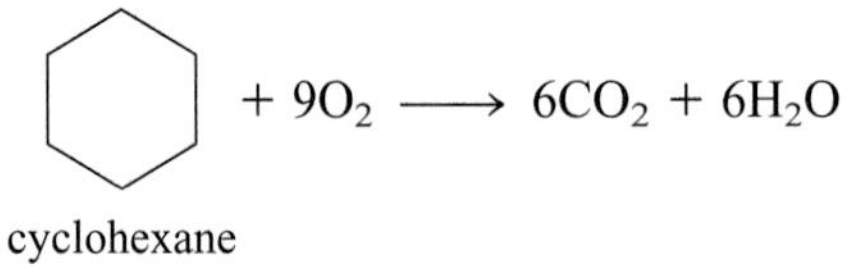

39. (b) and (c) are aromatic.

41. (a) alkane, saturated
(b) aromatic, unsaturated
(c) alkane, saturated
(d) alkyne, unsaturated
(e) alkene, unsaturated
(f) alkane, saturated
(g) alkane, saturated
(h) alkene, unsaturated

43. $CH_3CH_2CH_2Cl$ 1-chloropropane
$CH_3CHClCH_3$ 2-chloropropane

45. dichloromethane, CH_2Br_2

47.

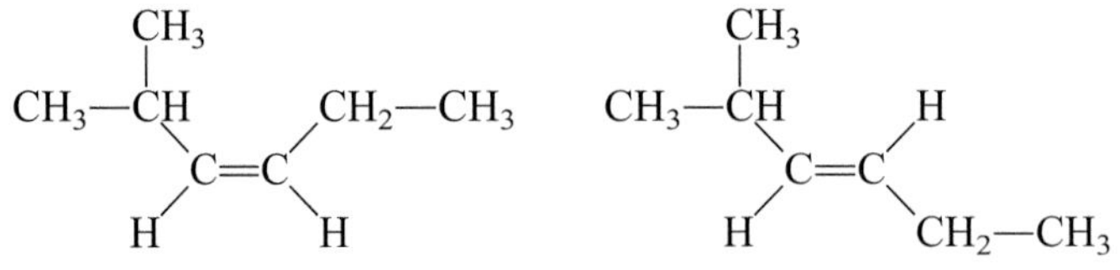

```
        CH3                               CH3
        |                                 |
  CH3—CH         CH2—CH3        CH3—CH          H
       \        /                    \        /
        C == C                        C == C
       /      \                      /      \
      H        H                    H        CH2—CH3
          cis                             trans
```

In the *cis* isomer, the large isopropyl and ethyl groups are very close together in space and tend to repel one another, making the molecules less stable than the more spread-out *trans* isomer.

49. Test each liquid with a bromine–water solution. If the reddish brown color of bromine disappears, the liquid is cyclohexene.

Chapter 13

1. Functional groups are reactive areas within molecules that give the molecule certain specific chemical properties.

3. (a) $CH_3CH_2CH_2OCH_2CH_2CH_3$

(b)

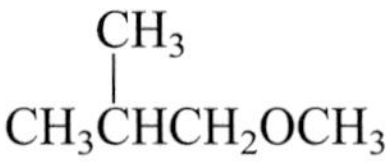

```
     CH3
     |
  CH3CHCH2OCH3
```

(c) C_6H_5—O—CH_2CH_3

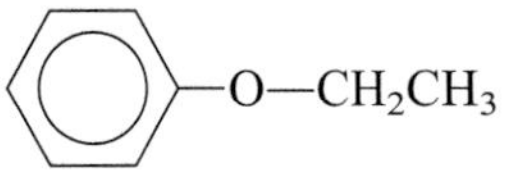

5. **(a)** 2-butanol, secondary
(b) benzyl alcohol, primary
(c) 2-methyl-2-butanol, tertiary
(d) 1-propanol, primary
(e) 3,3-dimethyl-2-butanol, secondary
(f) 2-phenyl-2-propanol, tertiary

7. **(a)** $CH_2ClCH_2CH_2OH$

(b)
```
      OH    CH3
      |     |
CH3CHCH2CCH3
            |
            CH3
```

(c) $HOCH_2CH_2CHOHCH_2CH_3$

(d)
```
        OH
        |
        CH—CH—CH3
       /      \
   H2C          CH2
       \      /
        CH2—CH2
```

9. The hydroxyl group creates a polar spot on the otherwise nonpolar hydrocarbon chain and can form a hydrogen bond with a water molecule. Increasing the number of hydroxyl groups on a molecule, therefore, increases its solubility in water.

11. Only the carbon chain and alcohol group are shown.

—C—C—C—C—C—OH, primary
pentanol

```
      OH
      |
—C—C—C—C—C—, secondary
```
2-pentanol

```
        OH
        |
—C—C—C—C—C—, secondary
```
3-pentanol

```
      C
      |
—C—C—C—C—OH, primary
```
3-methyl-1-butanol

```
        C
        |
—C—C—C—C—OH, primary
```
2-methyl-1-butanol

```
      C
      |
—C—C—C—OH, primary
      |
      C
```
2,2-dimethyl-1-propanol

```
        OH
        |
—C—C—C—C—, secondary
      |
      C
```
3-methyl-2-butanol

```
        OH
        |
—C—C—C—C, tertiary
        |
        C
```
2-methyl-2-butanol

13. Methanol is added to ethanol to make it undrinkable.

15. **(a)**
```
     CH3 CH3
     |   |
CH3—CH—C=CH2
```

(b) $CH_3—CH=CH_2$

(c)
```
     CH—CH2
    //     \
  HC        CH—CH3
    \      /
     CH2-CH2
```

17. **(a)** 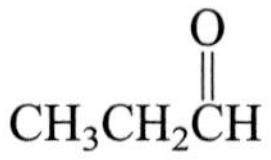 $CH_3CH_2\overset{O}{\overset{\|}{C}}H$

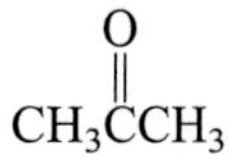

(b) $CH_3\overset{O}{\overset{\|}{C}}CH_3$

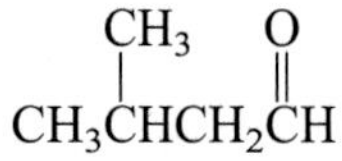

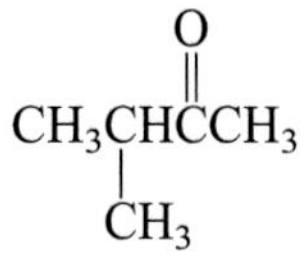

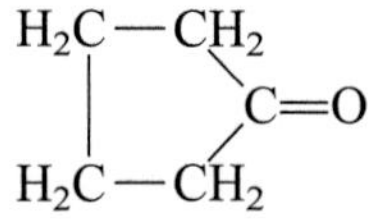

(c) $CH_3\underset{}{\overset{CH_3}{\overset{|}{C}}}HCH_2\overset{O}{\overset{\|}{C}}H$

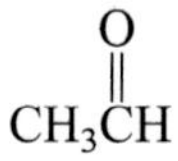

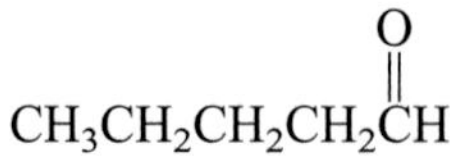

(d) $CH_3\underset{\underset{CH_3}{|}}{C}H\overset{O}{\overset{\|}{C}}CH_3$

(e) cyclopentanone ring: $H_2C—CH_2$, $H_2C—CH_2$, $C{=}O$

(f) $CH_3\overset{O}{\overset{\|}{C}}H$

(g) $CH_3CH_2CH_2CH_2\overset{O}{\overset{\|}{C}}H$

(h) $CH_3CH_2CH_2CH_2\overset{O}{\overset{\|}{C}}CH_3$

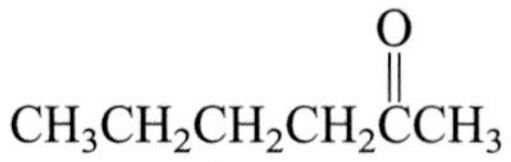

19. **(a)** 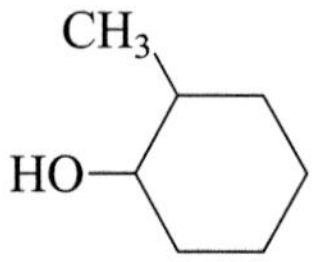 HO— cyclohexane ring with adjacent CH_3

(b) 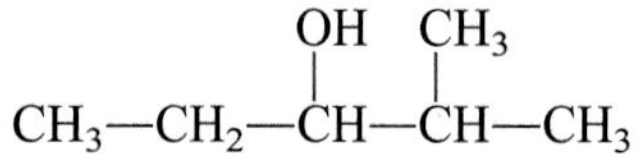 $CH_3—CH_2—\overset{OH}{\overset{|}{C}}H—\overset{CH_3}{\overset{|}{C}}H—CH_3$

(c) $CH_3—CH_2—CH_2—\overset{CH_3}{\overset{|}{C}}H—CH_2—OH$

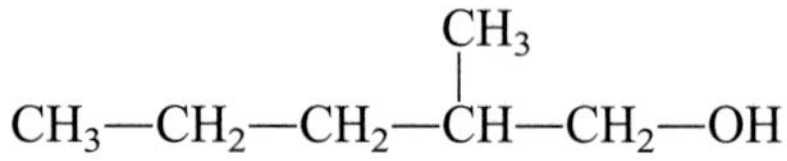

(d) 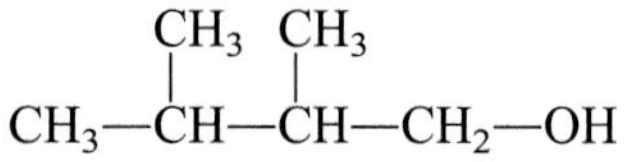 $CH_3—\overset{CH_3}{\overset{|}{C}}H—\overset{CH_3}{\overset{|}{C}}H—CH_2—OH$

21. **(a)** 2-methylcyclohexanone **(b)** 2-methyl-3-pentanone **(c)** 2-methylpentanal **(d)** 2,3-dimethylbutanal

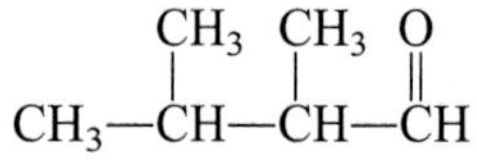

23. 167 g

25. **(a)** 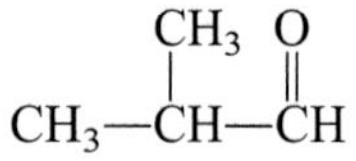 $CH_3—\overset{CH_3}{\overset{|}{C}}H—\overset{O}{\overset{\|}{C}}H$

(b) $CH_3—\overset{CH_3}{\overset{|}{C}}H—\overset{CH_3}{\overset{|}{C}}H—\overset{O}{\overset{\|}{C}}H$

(c) cyclohexane ring $—CH_2—\overset{O}{\overset{\|}{C}}H$

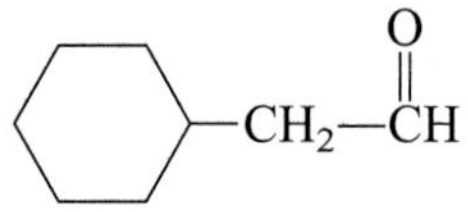

27. **(a)** 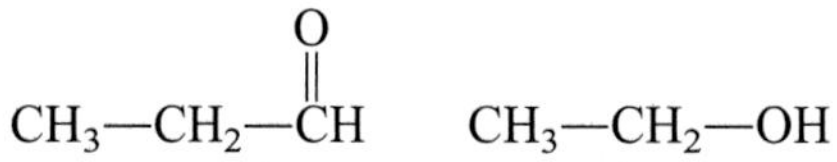 $CH_3—CH_2—\overset{O}{\overset{\|}{C}}H$ $CH_3—CH_2—OH$

propanal ethanol

(b) 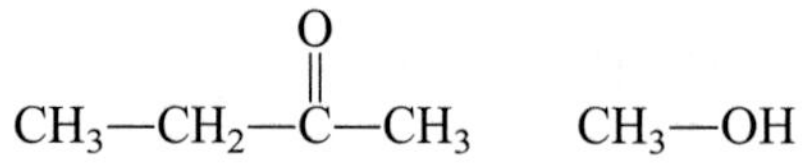 $CH_3—CH_2—\overset{O}{\overset{\|}{C}}—CH_3$ $CH_3—OH$

2-butanone methanol

(c) 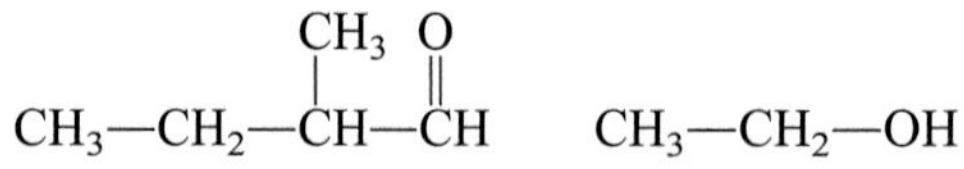 $CH_3—CH_2—\overset{CH_3}{\overset{|}{C}}H—\overset{O}{\overset{\|}{C}}H$ $CH_3—CH_2—OH$

2-methylbutanal ethanol

(d) 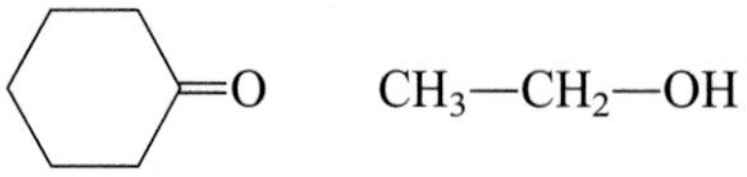 cyclohexane ring $=O$ $CH_3—CH_2—OH$

cyclohexanone ethanol

29. **(a)** heptanoic acid
(b) 3-isopropylhexanoic acid
(c) 3-ethylpentanoic acid
(d) 2,3-dichlorohexanoic acid

31. lactic acid

33. **(a)** $CH_3-CH_2-CH_2-CH_2-CH_2-OH \qquad CH_3-CH_2-CH_2-CH_2-\overset{O}{\overset{\|}{C}}H$

(b) $C_6H_5-CH_2-OH \qquad C_6H_5-CH{=}O$

(c) $Cl-CH_2-\overset{Cl}{\overset{|}{C}}H-CH_2-OH \qquad Cl-CH_2-\overset{Cl}{\overset{|}{C}}H-\overset{O}{\overset{\|}{C}}H$

(d) $CH_3-CH_2-\underset{CH_3}{\underset{|}{\overset{CH_3}{\overset{|}{C}}}}-CH_2-OH \qquad CH_3-CH_2-\underset{CH_3}{\underset{|}{\overset{CH_3}{\overset{|}{C}}}}-\overset{O}{\overset{\|}{C}}H$

35. Hydrochloric acid is a much stronger acid than benzoic acid.

37. **(a)**

$$CH_3CH_2CH_2\overset{O}{\overset{\|}{C}}OH + HOCH_3 \longrightarrow CH_3CH_2CH_2\overset{O}{\overset{\|}{C}}OCH_3 + H_2O$$

(b)

$$C_6H_5-\overset{O}{\overset{\|}{C}}OH + HOCH_2CH_3 \longrightarrow C_6H_5-\overset{O}{\overset{\|}{C}}OCH_2CH_3 + H_2O$$

(c)

$$H\overset{O}{\overset{\|}{C}}OH + HOCH_2CH_2CH_3 \longrightarrow H\overset{O}{\overset{\|}{C}}OCH_2CH_2CH_3 + H_2O$$

(d)

$$CH_3\overset{O}{\overset{\|}{C}}OH + CH_3\overset{OH}{\overset{|}{C}}HCH_2CH_3 \longrightarrow CH_3\overset{O}{\overset{\|}{C}}O\overset{CH_3}{\overset{|}{C}}HCH_2CH_3 + H_2O$$

(e)

$$o\text{-}HO{-}C_6H_4{-}\overset{O}{\overset{\|}{C}}OH + HOCH_2CH_3 \longrightarrow o\text{-}HO{-}C_6H_4{-}\overset{O}{\overset{\|}{C}}OCH_2CH_3 + H_2O$$

(f)

$$o\text{-}HO{-}C_6H_4{-}\overset{O}{\overset{\|}{C}}OH + HO\overset{O}{\overset{\|}{C}}CH_2CH_3 \longrightarrow o\text{-}(CH_3CH_2\overset{O}{\overset{\|}{C}}O){-}C_6H_4{-}\overset{O}{\overset{\|}{C}}OH + H_2O$$

39. **(a)** 2-methylpropanoic acid
(b) ethyl propyl ether
(c) ethyl butanoate
(d) 2-methylpropanal
(e) 3-hexanone
(f) 3-hydroxybutanoic acid
(g) diphenyl ether
(h) 3-ethyl-3-hexanol
(i) pentanal
(j) 2,4-dimethyl-3-pentanone
(k) phenyl pentanoate
(l) 5-isopropyl-6-methyl-4-octanol

41. 1.8×10^{-3}

43. **(a)**

$$CH_3O\overset{O}{\overset{\|}{C}}CH_3 + NaOH \longrightarrow CH_3OH + NaO\overset{O}{\overset{\|}{C}}CH_3$$

(b)

$$CH_3CH_2O\overset{O}{\overset{\|}{C}}CH_2CH_3 + NaOH \longrightarrow CH_3CH_2OH + NaO\overset{O}{\overset{\|}{C}}CH_2CH_3$$

(c)

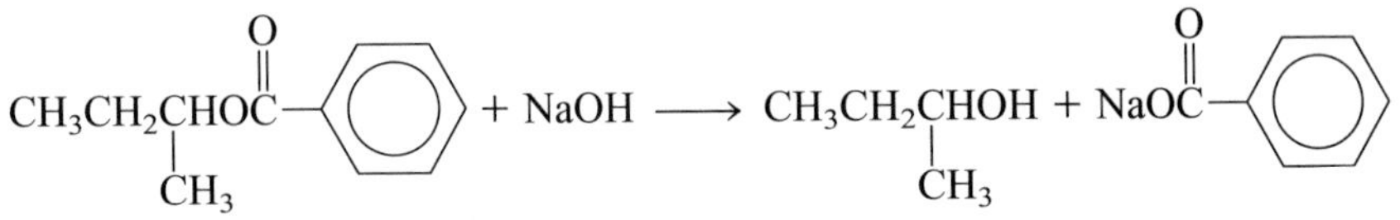

$$CH_3CH_2\underset{CH_3}{\underset{|}{C}}HO\overset{O}{\overset{\|}{C}}-C_6H_5 + NaOH \longrightarrow CH_3CH_2\underset{CH_3}{\underset{|}{C}}HOH + NaO\overset{O}{\overset{\|}{C}}-C_6H_5$$

(d)

$$CH_3CH_2O\overset{O}{\overset{\|}{C}}-\overset{O}{\overset{\|}{C}}OCH_2CH_3 + 2NaOH \longrightarrow 2CH_3CH_2OH + NaO\overset{O}{\overset{\|}{C}}-\overset{O}{\overset{\|}{C}}ONa$$

45.

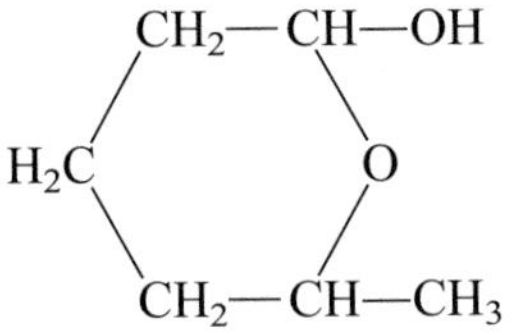

Six-membered ring: $CH_2-CH-OH$ (top), H_2C (left), O (right), $CH_2-CH-CH_3$ (bottom)

47. A condensation reaction between salicylic acid and methanol produces oil of wintergreen.

49. To convert estradiol into estrone, oxidize the secondary alcohol group. The phenol group on the left ring is not oxidized.

Chapter 14

1. three covalent bonds

3. **(a)** primary
(b) tertiary
(c) primary
(d) secondary
(e) tertiary
(f) primary

5. **(a)**

$$CH_3-CH_2-\overset{H}{\overset{|}{N}}H^+-CH_2-CH_3 + OH^-$$

(b)

$$CH_3-CH_2-\overset{O}{\overset{\|}{C}}-O^- + CH_3-\overset{H}{\overset{|}{N}}H^+-CH_3$$

(c)

$$CH_3-CH_2-CH_2-\overset{+}{N}H_3 + CH_3-\overset{O}{\overset{\|}{C}}-O^-$$

7. **(a)**

$$\underset{\text{}}{CH_3\overset{\overset{\displaystyle CH_3}{|}}{C}HCH_2NH_2} + HCl \longrightarrow \underset{\text{isobutylammonium chloride}}{CH_3\overset{\overset{\displaystyle CH_3}{|}}{C}HCH_2NH_3Cl}$$

(b)

$$(CH_3CH_2)_3N + HCl \longrightarrow \underset{\text{triethylammonium chloride}}{(CH_3CH_2)_3NHCl}$$

(c)

$$(CH_3CH_2)_3N + CH_3CH_2Br \longrightarrow \underset{\text{tetraethylammonium bromide}}{(CH_3CH_2)_4NBr}$$

9. **(a)** pentamide
   (b) *N,N*-dimethylacetamide
   (c) *N*-phenylpropanamide
   (d) *N*-ethyl-*N*-methylhexanamide

11. alkene, two amine groups, aromatic ring, and an amide group

13.

$$CH_3{-}CH_2{-}CH_2{-}CH_2{-}\overset{\overset{\displaystyle O}{\|}}{C}{-}NH_2 + H_2O \longrightarrow CH_3{-}CH_2{-}CH_2{-}CH_2{-}\overset{\overset{\displaystyle O}{\|}}{C}{-}OH + NH_3$$

$$CH_3{-}\overset{\overset{\displaystyle O}{\|}}{C}{-}\overset{\overset{\displaystyle CH_3}{|}}{N}{-}CH_3 + H_2O \longrightarrow CH_3{-}\overset{\overset{\displaystyle O}{\|}}{C}{-}OH + CH_3{-}NH{-}CH_3$$

$$CH_3{-}CH_2{-}\overset{\overset{\displaystyle O}{\|}}{C}{-}NH{-}C_6H_5 + H_2O \longrightarrow CH_3{-}CH_2{-}\overset{\overset{\displaystyle O}{\|}}{C}{-}OH + H_2N{-}C_6H_5$$

$$CH_3{-}CH_2{-}CH_2{-}CH_2{-}CH_2{-}\overset{\overset{\displaystyle O}{\|}}{C}{-}\overset{\overset{\overset{\displaystyle CH_3}{|}}{\overset{\displaystyle CH_2}{|}}}{N}{-}CH_3 + H_2O \longrightarrow$$

$$CH_3{-}CH_2{-}CH_2{-}CH_2{-}CH_2{-}\overset{\overset{\displaystyle O}{\|}}{C}{-}OH + CH_3{-}NH{-}CH_2{-}CH_3$$

15. Any two structures from the table of heterocycles in the text are correct.

17. **(a)** epinephrine
    (b) codeine
    (c) nicotine
    (d) epinephrine

19. Trimethylamine does not have a hydrogen on the nitrogen and therefore cannot form hydrogen bonds. This causes the compound to have a lower boiling point.

21. Repeated use of morphine can cause both an addiction and tolerance to the drug.

23. **(a)** Nicotine is a stimulant to the central nervous system, causing both irregular heartbeat and increased blood pressure, and inducing vomiting and diarrhea.
    (b) After repeated exposure, the body of a smoker develops chemical pathways to protect itself against the toxic effects of nicotine. As a result, a smoker can absorb more nicotine than can a nonsmoker before feeling any ill effects.

25. The hydrogen on the hydroxyl group on the benzene ring must be replaced by a methyl group, thus converting the group into an ether.

Integrated Problems

1. **(a)** A compound that is carcinogenic is capable of producing cancer in living organisms.

   (b) The answer to this problem is a complicated one and will be the subject of debate for many years. In general, a carcinogenic compound should probably not be commercially available, but the specific use of the compound must also be taken into account before it is totally banned.

3. **(a)**

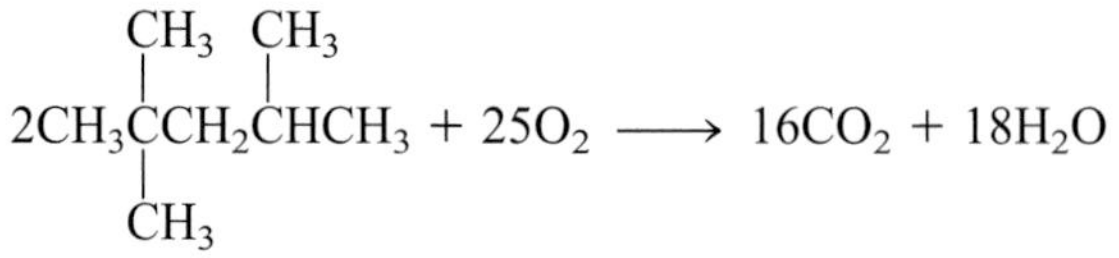

$$2CH_3\overset{\overset{CH_3}{|}}{\underset{\underset{CH_3}{|}}{C}}CH_2\overset{\overset{CH_3}{|}}{C}HCH_3 + 25O_2 \longrightarrow 16CO_2 + 18H_2O$$

(C_8H_{18})

   (b) $1000\,\cancel{g} \times \dfrac{1\,\cancel{mol}\;C_8H_{18}}{114\,\cancel{g}} \times \dfrac{16\text{ mol }CO_2}{2\,\cancel{mol}\;C_8H_{18}} = 70.2\text{ mol }CO_2$

$70.2\,\cancel{mol}\;CO_2 \times \dfrac{22.4\text{ L}}{1\,\cancel{mol}\text{ gas}} = 1590\text{ L}$

5. **(a)** $HOCH_2CH_2OH$

   (b) 1,2-ethanediol, or ethylene glycol, is a colorless sweet liquid that is as toxic as methanol when consumed by humans. Wine containing ethylene glycol would pose a serious health problem.

7. b, a, e, c, d

9. **(a)**

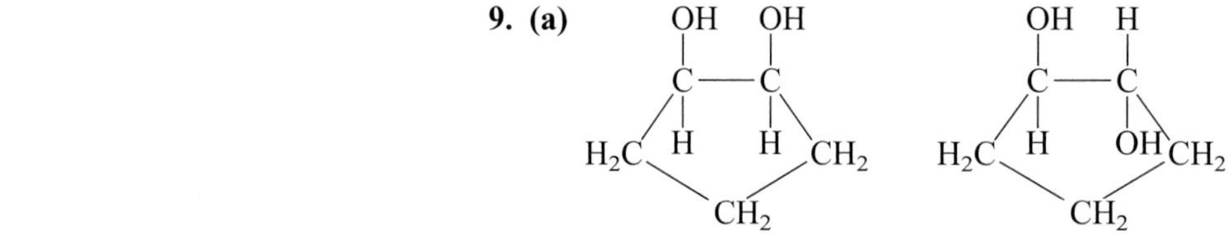

cis: cyclopentane ring (H_2C, CH_2, CH_2) with two adjacent C atoms each bearing OH (up) and H (down).

trans: cyclopentane ring (H_2C, CH_2, CH_2) with adjacent C atoms; one bearing OH (up) and H (down), the other H (up) and OH (down).

cis   *trans*

   (b) Intramolecular hydrogen bonding in the *cis* isomer prevents the formation of intermolecular hydrogen bonding that forms between the *trans* isomers and raises the boiling point of the *trans* isomer.

11. **(a)** 0.15%

$7.5\,\cancel{mL} \times \dfrac{0.020\text{ g}}{100\,\cancel{mL}} = 0.0015\text{ g}$

   (b) yes

$\dfrac{0.0015}{1\text{ mL}} = \dfrac{0.15}{100\text{ mL}} = 15\%$

Chapter 15

1. Monosaccharides are carbohydrates that cannot be broken down into smaller units on hydrolysis. Disaccharide molecules produce two monosaccharide molecules on hydrolysis, and polysaccharides produce three or more monosaccharide molecules on hydrolysis.

3. (a)

```
      H
      |
      C=O
      |
  H—C—OH
      |
 HO—C—H
      |
  H—C—OH
      |
  H—C—OH
      |
      CH2OH
```

(b)

```
        CH2OH
        |
        CH—O
  H   /      \   H
  |  /        \  |
  C   OH   H   C
  | \ |    |  / |
 HO   C————C    OH
      |    |
      H    OH
```

α-glucose

(c)

```
        CH2OH
        |
        CH—O
  H   /      \   OH
  |  /        \  |
  C   OH   H   C
  | \ |    |  / |
 HO   C————C    H
      |    |
      H    OH
```

β-glucose

5. (a)

```
      H
      |
      C=O
      |
  H—C—OH
      |
 HO—C—H
      |
      CH2OH
```

(b)

```
      H
      |
      C=O
      |
  H—C—OH
      |
 HO—C—H
      |
 HO—C—H
      |
      CH2OH
```

(c)

```
      H
      |
      C=O
      |
 HO—C—H
      |
 HO—C—H
      |
  H—C—OH
      |
 HO—C—H
      |
      CH2OH
```

7.

```
 CH2OH
 |       O        OH
 |    /     \     |
 C   H   HO   C
 | \ |    |  / |
 H   C————C    CH2OH
     |    |
     HO   H
```

9.

```
     CH2OH
     |————O
 HO  H         H
  |             |
 H   OH   H     OH
     |    |
     H    OH
```

α-galactose

⇌

```
      CHO
      |
  H—C—OH
      |
 HO—C—H
      |
 HO—C—H
      |
  H—C—OH
      |
      CH2OH
```

D-galactose

⇌

```
     CH2OH
     |————O
 HO  H         OH
  |             |
 H   OH   H     H
     |    |
     H    OH
```

β-galactose

11. (a) pentose
(b) alpha
(c)

```
      H
      |
      C=O
      |
  H—C—OH
      |
  H—C—OH
      |
  H—C—OH
      |
      CH2OH
```

(d) ribose

13. Fructose, ribose, lactose and maltose all give a positive Benedict's test because they all have a free, or potentially free, aldehyde or ketone group.

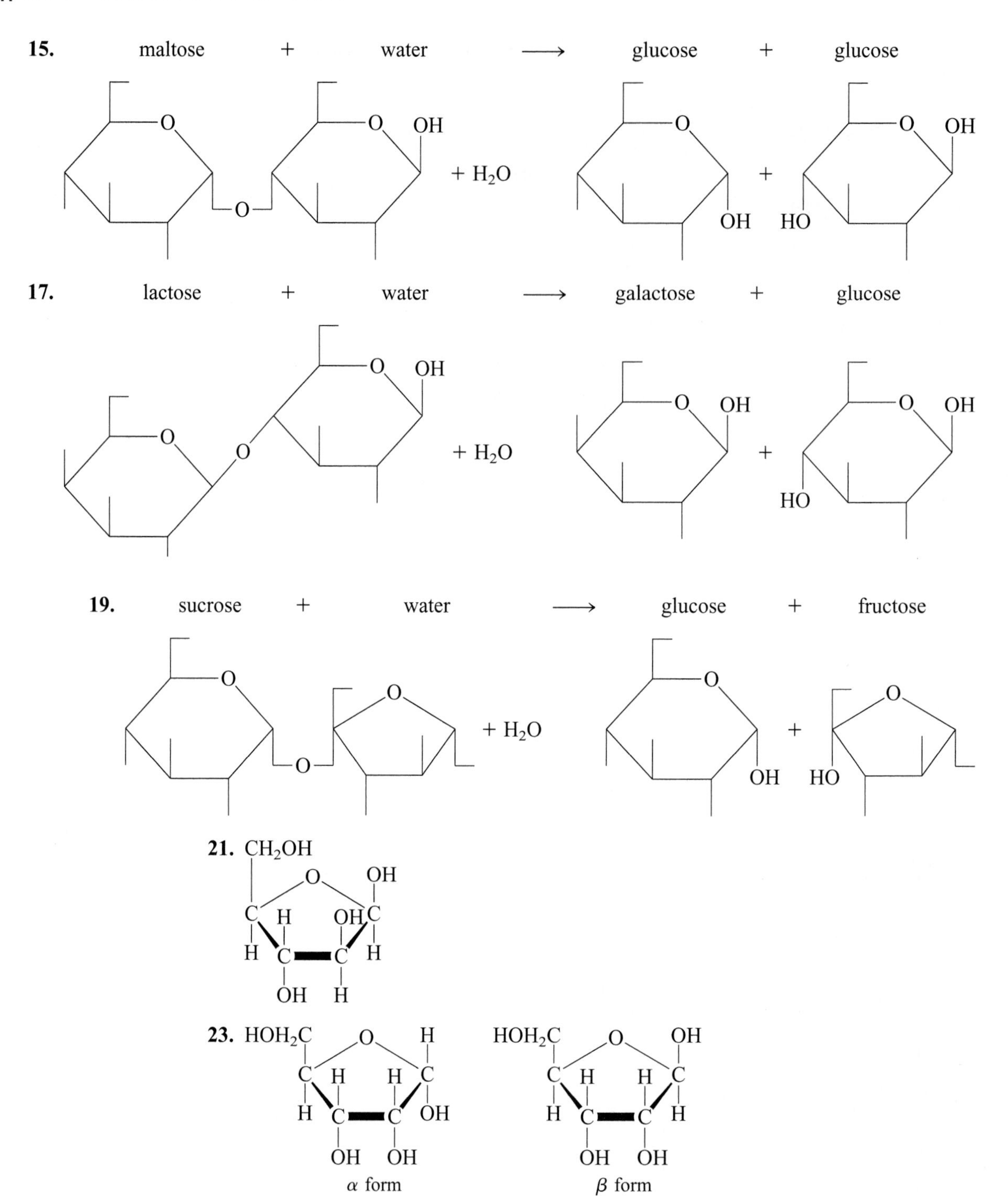

25. **(a)** amylose and amylopectin

(b) Amylose is a linear polymer of glucose, whereas amylopectin is a highly branched polymer of glucose.

27. Glycogen is the storage form of glucose in animals. Its structure is very similar to amylopectin.

29. Cellulose and amylose are both polymers of glucose, and they both have 1:4 glycosidic linkages between the glucose units. They differ in that the glycosidic linkages are alpha in amylose and beta in cellulose.

31. Place 15 drops of iodine test solution on a glass plate that is resting on a white sheet of paper. Add a small amount of starch solution to a flask that is in a warm water bath. Add several drops of hydrochloric acid solution to the starch to promote hydrolysis, and immediately remove a drop of this mixture and place it on the first drop of the iodine test solution. Take drops from the reaction mixture at fixed intervals, and add them to the iodine test solution to monitor the progress of the reaction.

33. **(a)** carbon dioxide, water, and chlorophyll

(b) chloroplasts of plant cells

35. No, the Clinitest tablet measures the presence of a reducing sugar in the urine, not specifically the presence of glucose in the urine. Other monosaccharides and disaccharides could be present in the urine and give a positive test.

37. Celery is composed of cellulose, which is not digested in the human digestive tract; therefore, the glucose it contains cannot be absorbed by the body. Potato chips are composed of starch, which can be digested in the digestive tract and the glucose formed is absorbed and used by the body.

39. **(a)** No, there is no free or potentially free aldehyde or ketone group.

(b) α,1:6 linkage and α,1:2 linkage

(c)

galactose glucose fructose

41. The glucosidic linkage forms between the alcohol groups on carbon 1 (in the hemiacetal form) in both molecules of glucose.

43. **(a)**

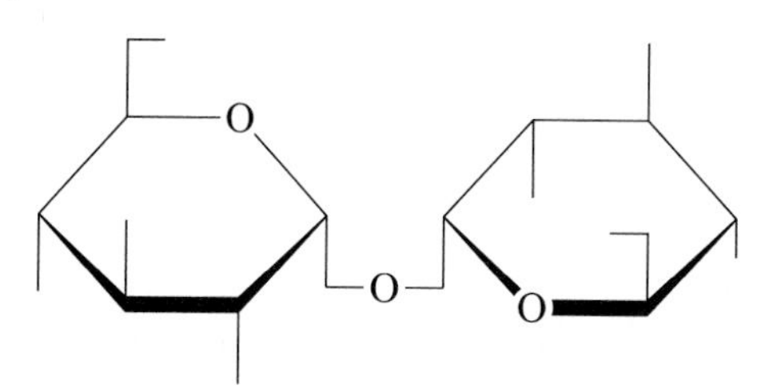

(b) no

Chapter 16

1. A saponifiable lipid is one that can be hydrolyzed by a base. A nonsaponifiable lipid is one that cannot be hydrolyzed by a base.

3. Simple triacylglycerols contain the same fatty acid on all three positions on the glycerol molecule, whereas mixed triacylglycerols contain two or more different fatty acids.

5. A saturated fatty acid contains all carbon-to-carbon single bonds. An unsaturated fatty acid contains a carbon-to-carbon double bond. A polyunsaturated fatty acid contains more than one carbon-to-carbon double bond.

7. **(a)** saturated: myristic, lauric
unsaturated: oleic, linolenic

(b) oleic

9. **(a)** $CH_3CH_2CH{=}CHCH_2CH{=}CH(CH_2)_{10}COOH$

(b) $CH_3(CH_2)_4CH{=}CHCH_2CH{=}CHCH_2CH{=}CH(CH_2)_6COOH$

11. An essential fatty acid is one that is not produced by the body in sufficient amounts to maintain good health and must be provided in the diet. A nonessential fatty acid can be synthesized by the body.

13. $CH_3{-}(CH_2)_{18}{-}\overset{\overset{\large O}{\|}}{C}{-}O{-}(CH_2)_{29}CH_3$

15. The iodine number of a lipid gives information on how many unsaturated bonds are in the lipid's fatty acids.

17. The unsaturated fatty acids in corn oil are partially hydrogenated. This produces a product that has a higher melting point and that is therefore solid at room temperature.

19. **(a)** The fats in butter undergo hydrolysis, producing the fatty acids that give butter a rancid taste and smell.

(b) Refrigeration retards the rate of hydrolysis.

(c) butyric acid

21. **(a)**

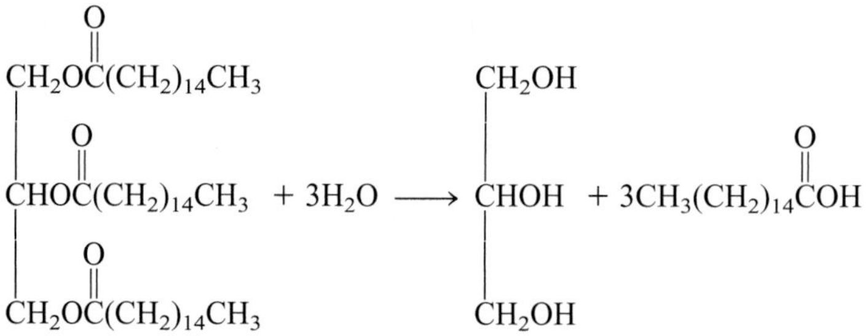

$$\begin{array}{l} CH_2O\overset{\overset{O}{\|}}{C}(CH_2)_{14}CH_3 \\ | \\ CHO\overset{\overset{O}{\|}}{C}(CH_2)_{14}CH_3 \\ | \\ CH_2O\overset{\overset{O}{\|}}{C}(CH_2)_{14}CH_3 \end{array} + 3H_2O \longrightarrow \begin{array}{l} CH_2OH \\ | \\ CHOH \\ | \\ CH_2OH \end{array} + 3CH_3(CH_2)_{14}\overset{\overset{O}{\|}}{C}OH$$

(b)

$$\begin{array}{l} CH_2O\overset{\overset{O}{\|}}{C}(CH_2)_{18}CH_3 \\ | \\ CHO\overset{\overset{O}{\|}}{C}(CH_2)_{16}CH_3 \\ | \\ CH_2O\overset{\overset{O}{\|}}{C}(CH_2)_7CH{=}CH(CH_2)_7CH_3 \end{array} + 3H_2O \longrightarrow \begin{array}{l} CH_2OH + HO\overset{\overset{O}{\|}}{C}(CH_2)_{18}CH_3 \\ | \\ CHOH + HO\overset{\overset{O}{\|}}{C}(CH_2)_{16}CH_3 \\ | \\ CH_2OH + HO\overset{\overset{O}{\|}}{C}(CH_2)_7CH{=}CH(CH_2)_7CH_3 \end{array}$$

(c)

$$\begin{array}{l} CH_2O\overset{\overset{O}{\|}}{C}(CH_2)_7CH{=}CH(CH_2)_7CH_3 \\ | \\ CHO\overset{\overset{O}{\|}}{C}(CH_2)_7CH{=}CH(CH_2)_7CH_3 \\ | \\ CH_2O\overset{\overset{O}{\|}}{C}(CH_2)_7CH{=}CH(CH_2)_7CH_3 \end{array} + 3H_2O \longrightarrow \begin{array}{l} CH_2OH \\ | \\ CHOH \\ | \\ CH_2OH \end{array} + 3HO\overset{\overset{O}{\|}}{C}(CH_2)_7CH{=}CH(CH_2)_7CH_3$$

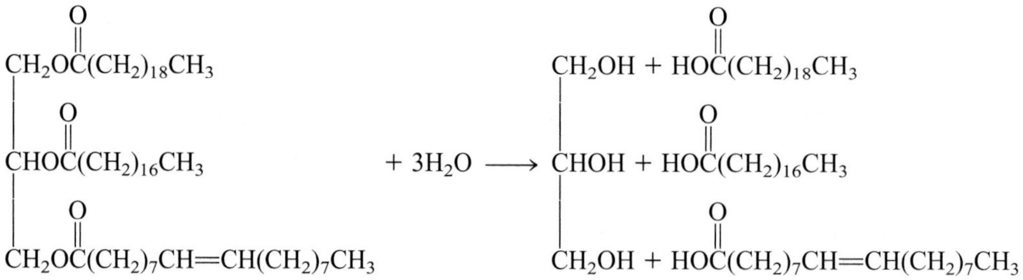

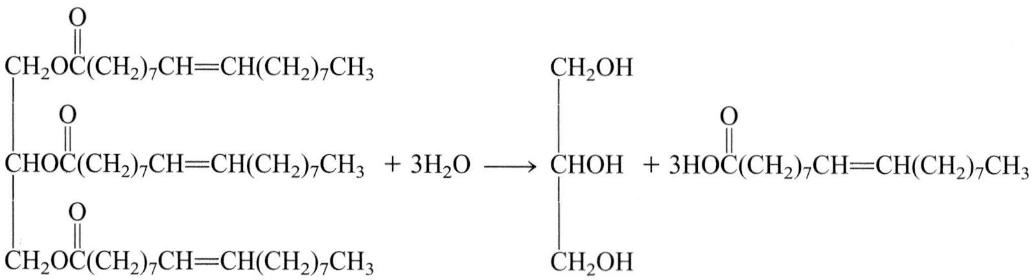

23. **(a)** The nonpolar tails of the soap molecules dissolve in the grease, whereas the polar heads remain dissolved in the water. This breaks the grease into colloidal droplets that can be washed away by the water.

(b) The soap molecules can disrupt and dissolve the bacterial cell membrane, thus killing the bacteria.

25. **(a)** A triacylglycerol is a simple lipid containing glycerol and three fatty acids. Phospholipids are compound lipids containing an alcohol, fatty acids, a phosphate group, and a nitrogen-containing compound.

(b) Phosphatidylcholine is a phosphatide containing choline, whereas sphingomyelin is a sphingolipid containing choline.

(c) The difference between glycolipids and phospholipids is that glycolipids contain a sugar group in place of the phosphate group in the phospholipid molecule.

27. **(a)** Phosphoglycerides form cell membranes.

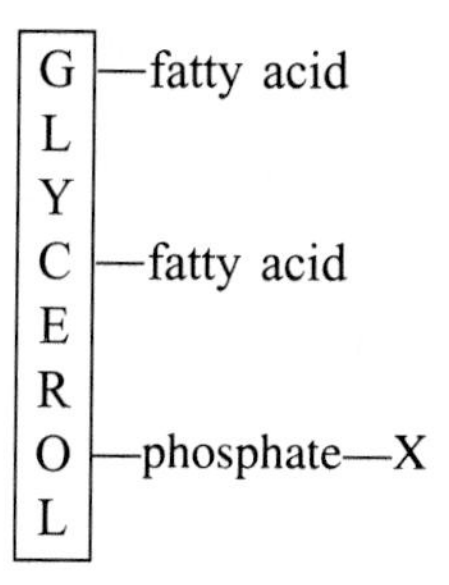

(b) Phosphatidylcholine is important in the transport and metabolism of fats and is a source of inorganic phosphate for tissue formation.

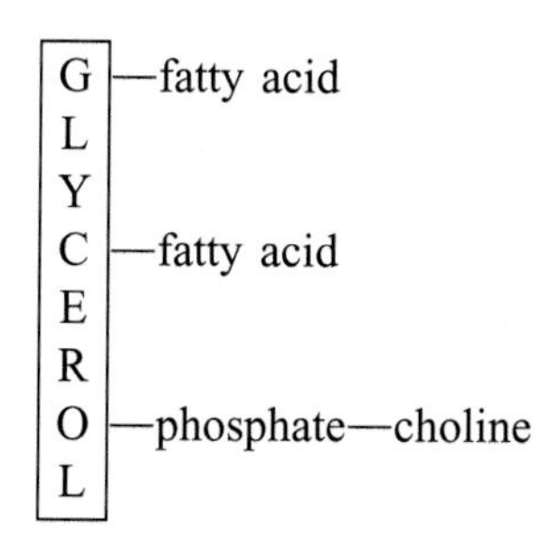

(c) Sphingolipids are found in the brain and nervous tissue and form part of the myelin sheath.

SPHINGOSINE
—fatty acid
—phosphate—nitrogen compound

(d) Sphingomyelins form part of the myelin sheath around nerves.

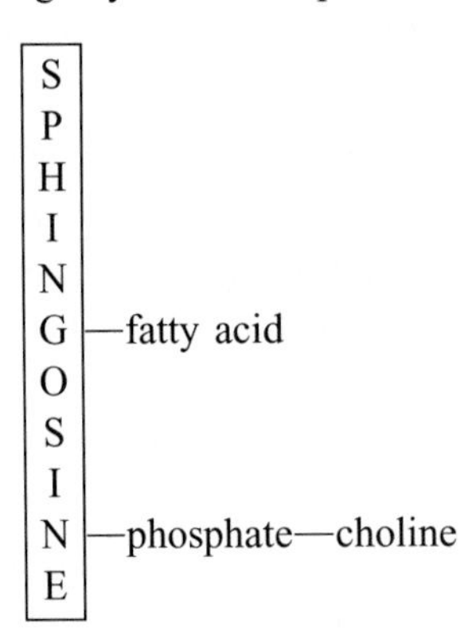

(e) Glycolipids are components of brain and nerve tissue.

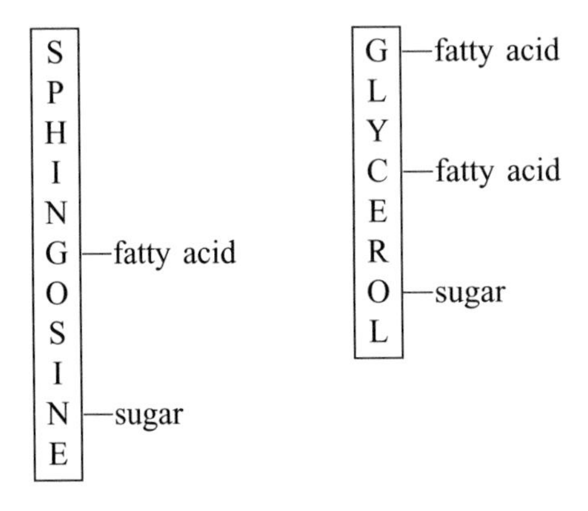

29. (a) steroids

(b) cholesterol

(c) They differ in one functional group: testosterone has an alcohol group where progesterone has a ketone group.

31. an alcohol and four alkene groups

33. (a) The cell membrane consists of a lipid bilayer—two rows of phospholipids, each with their polar hydrophilic heads toward the outside of the membrane and their nonpolar hydrophobic tails toward the water-free interior. The protein components may be on the surface of, imbedded in, or extending completely through the lipid bilayer.

(b) The polar head and nonpolar tail of the phospholipids allow them to form a bilayer.

35. (a)

$$\begin{matrix} H & & H \\ & C{=}C & \\ CH_3(CH_2)_6CH_2 & & CH_2(CH_2)_6COOH \end{matrix}$$

oleic acid

$$\begin{matrix} CH_3(CH_2)_6CH_2 & & H \\ & C{=}C & \\ H & & CH_2(CH_2)_6COOH \end{matrix}$$

elaidic acid

(b) *Trans* fatty acids do not normally exist in nature. These compounds may therefore not be recognized by the enzymes in the body and may be metabolized in entirely different pathways from *cis* fatty acids, possibly leading to the production of harmful products.

37. Double bonds add rigid bends to the molecules of fat, preventing them from getting close and attracting one another, thus lowering the melting point.

39. Safflower oil is more unsaturated than olive oil and, therefore, when heated it is more reactive and has a tendency to break down during cooking.

41.

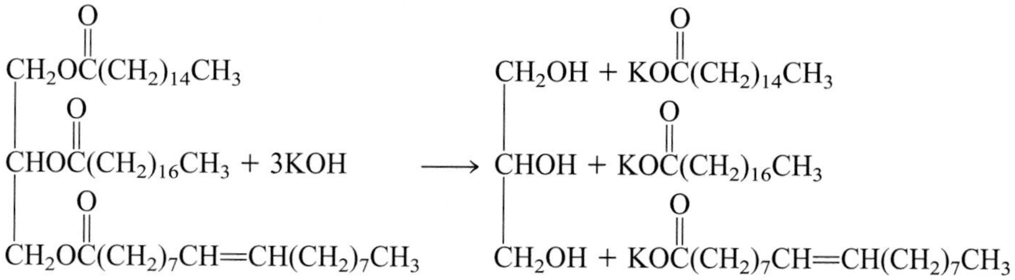

$$\begin{array}{l} CH_2OC(=O)(CH_2)_{14}CH_3 \\ | \\ CHOC(=O)(CH_2)_{16}CH_3 + 3KOH \\ | \\ CH_2OC(=O)(CH_2)_7CH{=}CH(CH_2)_7CH_3 \end{array} \longrightarrow \begin{array}{l} CH_2OH + KOC(=O)(CH_2)_{14}CH_3 \\ | \\ CHOH + KOC(=O)(CH_2)_{16}CH_3 \\ | \\ CH_2OH + KOC(=O)(CH_2)_7CH{=}CH(CH_2)_7CH_3 \end{array}$$

Chapter 17

1. Amino acids are the monomer units from which the protein polymers are formed.

3. (a)

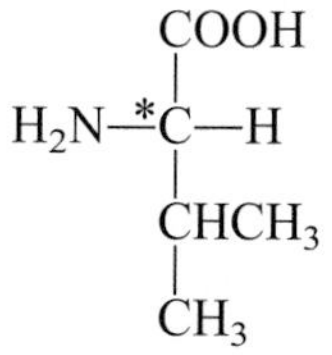

$$\begin{array}{c} COOH \\ | \\ H_2N{-}{}^{*}C{-}H \\ | \\ CHCH_3 \\ | \\ CH_3 \end{array}$$

(b)

$$\begin{array}{c} COOH \\ | \\ H{-}{}^{*}C{-}NH_2 \\ | \\ CH_2CH_2COOH \end{array}$$

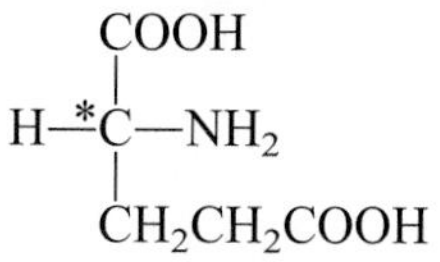

5. (a) Adequate protein is protein that contains all the essential amino acids.

(b) Most vegetable protein is inadequate protein. A vegetarian diet must contain a mixture of vegetable proteins to ensure that all the essential amino acids are found in the diet.

7. Amino acids exist as dipolar ions and, therefore, have a strong attraction for one another that results in high melting points.

9.

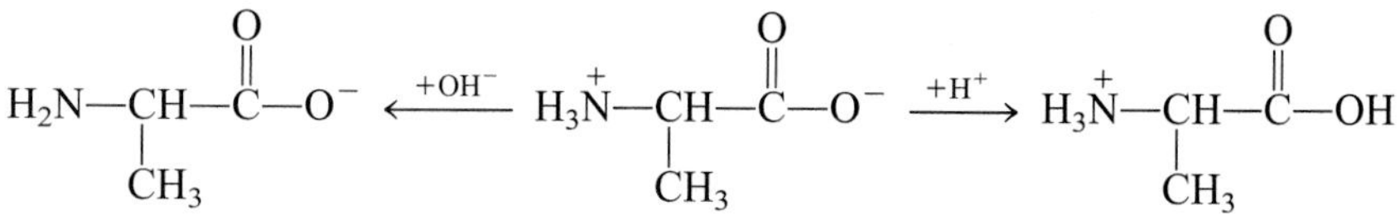

$$H_2N{-}CH(CH_3){-}C(=O){-}O^- \xleftarrow{+OH^-} H_3\overset{+}{N}{-}CH(CH_3){-}C(=O){-}O^- \xrightarrow{+H^+} H_3\overset{+}{N}{-}CH(CH_3){-}C(=O){-}OH$$

11. (a) Urease migrates toward the positive pole.

(b) Myoglobin does not migrate.

(c) Chymotrypsin migrates toward the negative pole.

13. **(a)** $H_3\overset{+}{N}-CH(CH_3C_6H_5)-\overset{O}{\overset{\|}{C}}-OH$ **(c)** $H_2N-CH(CH_3)-\overset{O}{\overset{\|}{C}}-O^-$

(b) $H_2N-CH(CH_3C_6H_5)-\overset{O}{\overset{\|}{C}}-O^-$ **(d)** $H_3\overset{+}{N}-CH(CH_3)-\overset{O}{\overset{\|}{C}}-OH$

15. **(a)** A simple protein yields only amino acids on hydrolysis, but a conjugated protein yields amino acids and other organic or inorganic substances on hydrolysis.

(b) A globular protein is soluble in water, fragile, and has a catalyzing or transporting function, whereas a fibrous protein is insoluble in water, physically tough, and has a structural or protective function.

(c) A glycoprotein is a conjugated protein in which the prosthetic group is a carbohydrate. A lipoprotein is a conjugated protein in which the prosthetic group is a lipid.

17. $H_3\overset{+}{N}-CH(CH_2C_6H_4OH)-\overset{O}{\overset{\|}{C}}-NH-CH(CH_2COOH)-\overset{O}{\overset{\|}{C}}-O^-$ $H_3\overset{+}{N}-CH(CH_2COOH)-\overset{O}{\overset{\|}{C}}-NH-CH(CH_2C_6H_4OH)-\overset{O}{\overset{\|}{C}}-O^-$

19. The N-terminal end of a protein molecule is the end of the polypeptide chain with the free amino group. The C-terminal end of a protein molecule is the end of the polypeptide chain with the free carboxylic acid group.

21.

$H_2N-CH(CH_2C_6H_5)-\overset{O}{\overset{\|}{C}}-OH$

phenylalanine

$H_2N-CH(CHCH_3(CH_3))-\overset{O}{\overset{\|}{C}}-OH$

valine

$H_2N-CH(CH_2-C{=}O(NH_2))-\overset{O}{\overset{\|}{C}}-OH$

asparagine

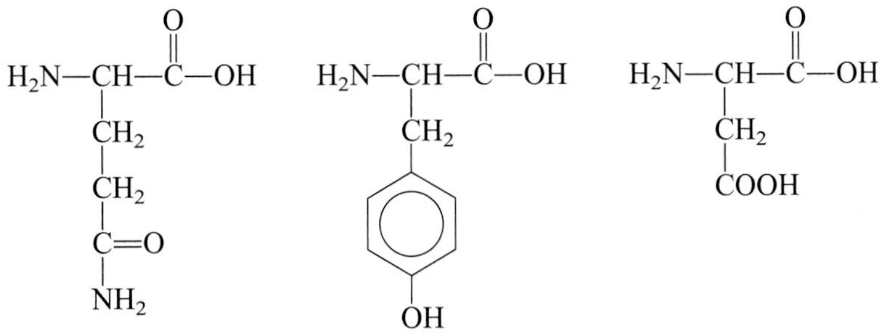

$$\underset{\text{glutamine}}{H_2N{-}\underset{\substack{|\\ CH_2\\ |\\ CH_2\\ |\\ C{=}O\\ |\\ NH_2}}{CH}{-}\overset{\substack{O\\ \|}}{C}{-}OH} \qquad \underset{\text{tyrosine}}{H_2N{-}\underset{\substack{|\\ CH_2\\ |\\ C_6H_4\\ |\\ OH}}{CH}{-}\overset{\substack{O\\ \|}}{C}{-}OH} \qquad \underset{\text{aspartic acid}}{H_2N{-}\underset{\substack{|\\ CH_2\\ |\\ COOH}}{CH}{-}\overset{\substack{O\\ \|}}{C}{-}OH}$$

23. (a) The primary structure of a protein is the sequence of amino acids in the molecule, and the secondary structure is the geometric arrangement formed by that sequence of amino acids.

(b) A dipeptide contains two amino acids, and a polypeptide contains three or more amino acids connected by peptide bonds.

(c) An alpha helix is a secondary configuration of a protein in which the amino acids form loops held together by hydrogen bonds. A beta configuration is a secondary structure of a protein in which several polypeptide chains are held together by hydrogen bonding in a zigzag fashion.

25. More sulfide bridges exist between the alpha-helices of the keratin in hooves than in the keratin of wool.

27. (a) The native shape of a protein is the shape that is most energetically stable for that protein.

(b) hydrogen bonds, disulfide bridges, hydrophobic interactions, and salt bridges

29. Tertiary structure is the three-dimensional structure of a globular protein. Quaternary structure refers to the way in which polypeptide chains and prosthetic groups fit together in proteins containing more than one polypeptide chain.

31. (a) The interactions that hold the protein in its native shape are broken causing the protein to unwind.

(b) In some cases, the process can be reversed.

33. Protein molecules are of colloidal size and normally do not pass through the differentially permeable membrane of the kidney. Proteins in the urine, therefore, may indicate damage to the kidney membrane.

35. (a)

```
                  O                O                 O
                  ‖                ‖                 ‖
   H2N—CH—C—NH—CH—C—NH—CH—C
        |           |          |    |
       CH2         CH2        CH2   |
        |           |          |    |
        S        (C6H4)      (C6H5) |
        |           |               |
        S          OH               |
        |                           |
  O    CH2     O           O        |
  ‖     |      ‖           ‖        |
  C—HC—HN—C—HC—HN—C—HC—HN
  |             |          |
  |            CH2        CH2
  |             |          |
  |            C=O        CH2
  |             |          |
  |            NH2        C=O
  |                        |
  |                       NH2
  |
    NH+   O          O                 O
H2C/  \   ‖          ‖                 ‖
 |     CH—C—NH—CH—C—NH—CH2—C—NH2
H2C\  /             |
    CH2             CH2CH2CH2NHC=NH
                                |
                               NH2
```

(b) disulfide

37. During hair straightening, the disulfide bridges between the protein molecules of hair are broken by reducing agents. They are then reformed while the hair is pulled straight.

39. Silver nitrate precipitates proteins, killing the bacteria.

41. (a) Egg whites are rich in proteins that bind with the lead ions while they are still in the stomach.

(b) The victim must be made to vomit to void the stomach of the poison before the lead ions are released again by the normal digestive processes.

Chapter 18

1. Metabolism is the term given to all the enzyme-catalyzed reactions of the body.

3. Enzymes are proteins that are biological catalysts. Without enzymes the reactions in our bodies occur at rates too slow to support life.

5. (a) oxidation–reduction reactions

(b) transfer of amino groups

(c) hydrolysis reactions

(d) interconversion of isomers

(e) reduction reaction

(f) oxidation reaction

(g) transfer of methyl groups

(h) removal of hydrogen

7. **(a)** dehydrogenase **(b)** esterase **(c)** lyase

9. **(a)** Step 1. The glucose-1-phosphate attaches to the active site of the phosphoglucomutase molecule.
Step 2. The glucose-1-phosphate becomes activated.
Step 3. Glucose-6-phosphate forms on the surface of the phosphoglucomutase molecule.
Step 4. The glucose-6-phosphate is released from the active site on the phosphoglucomutase molecule.

(b) 6×10^4 molecules/h: 1×10^{-19} moles/h

11. The lock-and-key analogy explains the specificity of enzyme action by stating that the conformation of the active site is specific to the geometry of the substrate molecule, just as the configuration of a lock is specific for only one key.

13. Pepsin has an optimum activity at pH 1.5, and its activity decreases as the pH increases. The small intestines have an alkaline pH and, as a result, the activity of pepsin is very low there.

15. A noncompetitive inhibitor combines reversibly with a portion of the enzyme molecule that is essential to enzyme function. It can affect all enzyme molecules. A competitive inhibitor competes with the substrate for the active site on the molecule. Sometimes it is effective in attaching to the active site, and sometimes it is not, so a competitive inhibitor affects only a portion of the enzyme molecules.

17. An antimetabolite is a chemical substance that inhibits enzyme activity. Sulfanilamide and penicillin are antimetabolites.

19. **(a)** An allosteric enzyme is an enzyme that controls the reaction rate of a multienzyme system.

(b) If all possible reactions occurred in a cell at a high rate, the cell would certainly die. Allosteric enzymes allow the cell to control the rate of reactions, producing only those products required by the cell.

(c) The regulatory molecule can either inhibit or increase the activity of the allosteric enzyme by attaching to the allosteric site on the enzyme molecule.

21. Hormones regulate cellular processes by attaching to the cell membrane triggering changes within the cell, or they enter the cell and migrate to the nucleus where they activate specific genes.

23. Vitamins are small organic molecules that essential to maintain health but that cannot be synthesized in the body or cannot be synthesized in sufficient quantity.

25. nicotinamide adenine dinucleotide
nicotinamide adenine dinucleotide phosphate
flavin adenine dinucleotide

27. **(a)** vitamin C **(c)** vitamin B_{12}
(b) niacin **(d)** vitamin B_1

29. vitamin D

31. **(a)** Vitamin D is a fat-soluble vitamin, which when eaten in excess of daily requirements, is stored in fatty tissue. In high concentrations it can be harmful to the health of a child. By eating so many foods that are fortified with vitamin D, a child could easily exceed his daily requirements.

(b) Fortification of foods with vitamins should be carefully controlled to protect the consumer from an excess of fat-soluble vitamins, which is potentially as harmful as a deficiency.

33. The apoenzyme is the protein portion of the enzyme molecule and is more sensitive to heat denaturation than the coenzyme, which is the organic, nonprotein part of the molecule.

35. Lowering the temperature of the apples slows the rate of the ripening reactions, and removing oxygen from the surrounding air lowers the concentration of a reactant, also slowing the ripening process.

37. Enzyme activity decreases with a decrease in temperature, but the enzymes are not denatured as happens with high temperatures. At very low temperatures cellular activity ceases. After being frozen for six months, the enzymes in the sperm again catalyze the reactions in the cell when the sperm are warmed to body temperature.

39. There is no chemical difference between these two types of vitamin C.

41. Vitamin B_{12} is absorbed through the stomach and with less surface area to the stomach less of the vitamin B_{12} is absorbed. Increasing the concentration of vitamin B_{12} in the stomach increases its rate of absorption.

43. **(a)** allosteric or regulatory enzyme

(b) A high level of CTP in the cell acts as an allosteric, or regulatory, molecule and inhibits ATCase. As CTP is used up, it no longer inhibits ATCase, and ATCase can then catalyze the reaction that leads to the formation of more CTP.

Integrated Problems

1. **(a)** lactose; glucose and galactose

(b) lactase

(c) lactose, glucose and galactose

3.

6, 5, 4, 3, 2, 1; O O O O; O O O O

CH_3CONH CH_3CONH CH_3CONH CH_3CONH n

5. 0.7 lb of NaOH

7.

(I) $H_2N{-}CH{-}C({=}O){-}NH{-}CH{-}C({=}O){-}NH{-}CH{-}C({=}O){-}NH{-}CH{-}C({=}O){-}NH{-}CH{-}C({=}O){-}OH$

Side chains: $CH_2CH_2CH_2CH_2NH_2$; CH_2COOH; CH_2OH; $CH_2C({=}O)NH_2$; CH_2CH_2COOH

(II) $H_2N{-}CH{-}C({=}O){-}NH{-}CH{-}C({=}O){-}NH{-}CH{-}C({=}O){-}NH{-}CH{-}C({=}O){-}NH{-}CH{-}C({=}O){-}OH$

Side chains: $CH_3{-}CH{-}CH_3$; CH_2–benzene ring; CH_3; CH_2–indole (HN); $CH_2{-}HC{-}CH_3$, CH_3

(a) Polypeptide I is more soluble because its structure contains more polar R groups.

(b) Hydrophobic interactions are more likely to occur between the nonpolar R groups in polypeptide II.

(c) Salt bridges are more likely to occur between the carboxyl and amino side chains on polypeptide I.

9. (a) vitamin K

   (b) Bile aids in the absorption of fat-soluble vitamins. Vitamin K is fat-soluble, and without enough bile in the intestines, vitamin K would not be absorbed in sufficient quantity to promote blood clotting.

11. The ethanol and methanol compete for the active site on the alcohol dehydrogenase, reducing the amount of toxic formic acid produced. This allows the body to eliminate the formic acid, and prevents acidosis from occurring.

Chapter 19

1. for synthesis reactions, for muscle movement, and for heat to maintain the body's temperature

3. (a) A high-energy phosphate bond is one that yields twice as much energy on hydrolysis as an ordinary phosphate bond.

   (b)

NH_2

N N N N

$^{-}O-\overset{O}{\overset{\|}{\underset{O^-}{\underset{|}{P}}}}\sim O-\overset{O}{\overset{\|}{\underset{O^-}{\underset{|}{P}}}}\sim O-\overset{O}{\overset{\|}{\underset{O^-}{\underset{|}{P}}}}-O-CH_2$

O

OH OH

5. (a) glucose

   (b) glucose and galactose

   (c) glucose and fructose

7. They emulsify lipids.

9. Amino acids and very small peptides.

11. The level of glucose in the blood is carefully controlled by the liver and by several hormones that can cause the liver to take up glucose or to produce it. After a meal, glucose enters the blood in large amounts. Insulin is produced by the pancreas to promote the uptake of glucose by the liver and muscle cells, where it is converted into glycogen. As the glucose in the blood is used by the tissues, the level of glucose in the blood drops. Glucagon, another hormone produced by the pancreas, then causes an increase in the rate of glycogenolysis in the liver and, therefore, an increase in the concentration of glucose in the blood. In this way the concentration of glucose in the blood is maintained at a fairly constant level throughout the day.

13. **(a)** glucose + ATP ⟶ glucose-6-phosphate + ADP
fructose-6-phosphate + ATP ⟶ fructose-1,6-phosphate

(b) 1,3-diphosphoglycerate + ADP ⟶ 3-phosphoglycerate + ATP
phosphoenolpyruvate + ADP ⟶ pyruvate + ATP

(c) glucose-6-phosphate ⟶ fructose-6-phosphate
glyceraldehyde-3-phosphate ⟶ dihydroxyacetone phosphate
3-phosphoglycerate ⟶ 2-phosphoglycerate

15. When there is sufficient oxygen present.

17. $CH_3CO \longrightarrow 2CO_2 + ATP + 3\ NADH + FADH_2$
(acetyl group)

19. The citric acid cycle cannot operate unless oxygen is available for the respiratory chain to function.

21.

Step	*Products*	*NADH/FAD/ATP*
glycolysis	pyruvic acid	2ATP + 2NADH
citric acid cycle	carbon dioxide	1ATP + 4NADH + 1$FADH_2$ (per turn)
electron transport chain	water	34ATP

23. It is converted to pyruvic acid in the muscle cells or in the liver.

25. ethanol, water, and ATP

27. They are converted to acetyl-CoA, which is then converted into fatty acids. These fatty acids are stored as triacylglycerols in fat cells of adipose tissue.

29. For fats to be oxidized, they must first be hydrolyzed to glycerol and fatty acids. The glycerol enters the glycolysis pathway. Fatty acids are oxidized by the process of beta oxidation to acetyl-CoA, which then enters the citric acid cycle.

31. **(a)** The acetyl-CoA can enter the citric acid cycle or can be used in the synthesis of compounds used in the cell.

(b) This process creates an alcohol on the beta carbon, which is then oxidized to a ketone.

33. The fatty acid cycle is cyclic, but each turn of the cycle begins with a fatty acid that is two carbons shorter than the last until two acetyl-CoAs remain.

35. Acetyl-CoA is important in the citric acid cycle; it can be used in the biosynthesis of compounds required by the cell, and, if present in excess, it can be used to synthesize ketone bodies.

37. Amino acids are used to synthesize tissue protein for formation or repair of cells and to synthesize other amino acids, enzymes, hormones, and antibodies. Amino acids are used to synthesize nonprotein nitrogen-containing compounds, such as nucleic acids or heme groups. Amino acids can also be broken down for energy.

39.

$$\mathrm{H{-}C(COOH)(CH_3){-}NH_2} \xrightarrow[H_2O]{\text{enzymes}} \mathrm{C(COOH)(CH_3){=}O} + NH_3$$

41. The ammonia produced by the oxidative deamination of proteins is converted, by cells in the liver, to urea, through the reactions of the urea cycle. The urea enters the blood, is carried to the kidneys, and is excreted.

43. coenzyme A

45. Glycogen reserves in the liver and muscles are used, and then muscle and tissue protein is broken down.

47. **(a)** aerobic respiration

(b) at first aerobic respiration and then as the level of activity increases, anaerobic respiration

(c) because lactic acid builds up in the muscle tissue during strenuous exercise and causes muscle fatigue and soreness

49.

$$\begin{array}{c}COOH\\|\\C{=}O\\|\\H{-}C{-}CH_3\\|\\CH_3\end{array}\qquad\begin{array}{c}COOH\\|\\C{=}O\\|\\CH_2\\|\\H{-}C{-}CH_3\\|\\CH_3\end{array}\qquad\begin{array}{c}COOH\\|\\C{=}O\\|\\H{-}C{-}CH_3\\|\\CH_2\\|\\CH_3\end{array}$$

51. **(a)** Insulin is a protein molecule and would be broken down into amino acids by enzymes in the digestive tract if it were taken in pill form.

(b) Insulin shock is the condition that results when the blood sugar drops drastically after an overdose of insulin.

(c) As a result, brain cells are starved for glucose and can die, causing convulsions and coma.

(d) The ill effects of a high blood glucose concentration caused by a lack of insulin are produced at a much slower rate than those produced by hyperinsulinism.

Chapter 20

1. Genes are segments in molecules of DNA that result in inherited characteristics.

3. Free nucleotides in the cell participate in biosynthetic reactions, serve as coenzymes, and are important in the transport of energy from energy-releasing reactions to energy-requiring reactions.

5. Thymine and cytosine are pyrimidines. Adenine and guanine are purines.

7. The structure of deoxyribose differs from that of ribose at carbon number 2. Deoxyribose does not have a hydroxyl group at that carbon.

9. The two polynucleotide chains are held together by hydrogen bonding between the bases.

11. The pairing of bases on the two polynucleotide chains of DNA is very specific: thymine pairs only with adenine, and cytosine and guanine; therefore, the order of bases on each chain is not identical. Where there is a thymine on one chain, there is always an adenine on the other. That is, the chains are complementary.

13. The three types of RNA are messenger RNA, transfer RNA, and ribosomal RNA. mRNA serves as the template for protein synthesis; tRNA transports the amino acids to the ribosomes for use in protein synthesis; and rRNA along with protein forms the ribosomes, the sites of protein synthesis.

15. **(a)** Genetic information is carried in the sequence of bases on the DNA molecule.

(b) 20 amino acids

(c) methionine −AUG and tryptophan − UGG

17. DNA in eukaryotic cells contains exon segments that code for amino acids and intron segments that do not code for amino acids.

19. **(a)** Protein synthesis occurs in the ribosomes.

(b) **(1)** mRNA is synthesized in the nucleus and migrates to the cytoplasm.
(2) The smaller subunit of the ribosome combines with the mRNA, three initiation factors, and a tRNA carrying the start amino acid.
(3) This complex joins with the larger half of the ribosome. The growth of the amino acid chain takes place through a series of repeated steps in which the anticodon region of a charged tRNA is matched with the codon on the mRNA and peptide bonds are formed between the amino acids on the tRNAs.
(4) The elongation of the chain stops when the terminal codon is reached on the mRNA.

21. (start) Met-Thr-Lys-Gly-Gly-Ala-Tyr-Thr-Pro (stop)

23. A plasmid is a small circular piece of DNA from specially developed strains of the bacteria *E. coli.*

25. **(a)** A mutation is any change in the base sequence on DNA that results in a change in the amino acids on a polypeptide chain.

(b) Mutations can occur when one base is incorrectly inserted into the DNA molecule, when segments of the DNA are deleted, or additional segments of DNA are added to the gene.

(c) The nucleus of a cell contains special enzymes that continually read the DNA and correct base errors and remove additional bases or connect broken pieces of DNA to prevent mutations.

27. **(a)** PKU is caused by an error in the gene that codes for phenylalanine hydroxylase.

(b) PKU disrupts the metabolism of the amino acid phenylalanine and the metabolites that accumulate in the blood can, over time, cause permanent brain damage.

29. **(a)** 90 bases

(b) UUU GUU AAU CAA CAU UUA UGU GGU AGU CAC UUG GUC GAA GCU CUU UAU CUC GUA UGC GGC GAG CGU GGA UUC UUC UAC ACU CCC AAA GCA
Other answers are correct since several triplets of bases code for one amino acid.

31. ATACCACCGAAATAC
There are other correct answers depending on the codon chosen for each amino acid.

33. **(a)** A mutation can result spontaneously, can be caused by chemicals, or can be caused by ionizing radiation.

(b) The base sequence is changed to one of four sequences that code for valine: GUA, GUG, GUU, or GUC.

(c) Messenger RNA is synthesized in the nucleus. It travels to the cytoplasm where it joins with the two segments of a ribosome. tRNA brings amino acids to the ribosome, and the beta chain is synthesized on the mRNA template. When a terminal codon is reached, the beta chain is completed and released from the ribosome.

(d) Normal hemoglobin contains glutamic acid in position 6 on the beta chain, and sickle cell hemoglobin contains a valine in that position. At a pH of 7, glutamic acid carries a negative charge, but valine is neutral, which accounts for the differences in mobility of the two hemoglobins.

(e) When the nonpolar valine is substituted for the polar glutamic acid on the beta chain, hydrophobic interactions can arise between HbS molecules at low oxygen concentrations, causing them to aggregate in long chains and resulting in the bizarre shapes of the sickled red blood cells. These sickled cells can block capillaries, depriving tissues of oxygen and causing the symptoms of sickle cell crisis.

35. If the *E. coli.* used in the recombinant DNA experiments were not a weakened strain and they got into the general population they could cause the overproduction of insulin (hyperinsulinism) in affected individuals.

Chapter 21

1. by diffusion
3. In a closed circulatory system blood is pumped from the heart into the arteries, then into the capillaries, where food and waste are exchanged with the cells and then into the veins, where it is carried back to the heart.
5. **(a)** Blood plasma is the solution remaining after blood is centrifuged.
   (b) Hematocrit is the exact percentage of formed elements in the blood.
   (c) The formed elements is the suspended material separated when blood is centrifuged. It consists of the red blood cells, white blood cells, and platelets.
   (d) Blood serum is the solution that remains after blood clots.
7. Transport oxygen and carbon dioxide, maintain fluid balance, maintain proper pH, form clots. Transport food and wastes, regulate body temperature, and fight invading microorganisms
9. its high specific heat
11. because of its low solubility in water
13. When a system at equilibrium is subjected to stress, it shifts to relieve that stress.
15. High carbon dioxide partial pressure causes equilibrium II to shift to the left and equilibrium III to shift to the right.
17. Living organisms must regulate osmotic pressure in order to maintain fluid balance.
19. because they constitute more than half of the plasma proteins
21. The walls of the blood capillaries are only partially able to prevent the passage of salts, and thus only a temporary concentration difference occurs.
23. When blood pH falls to 7.1 or 7.2 or lower the condition is known as acidosis. This condition may be due to hypoventilation, diabetes, ingestion of excess acids, or a decrease in the removal of hydrogen ions through kidney failure. When blood pH increases to as much as 7.5 or more, the condition is known as alkalosis. This condition may be due to hyperventilation, excessive intake of basic substances, or severe vomiting.
25. Prothrombin is the inactive form of the enzyme thrombin.
27. The first step in clot formation is the buildup of platelets around injured tissue.
29. **(a)** Polyuria is the production of an abnormally high volume of urine.
    (b) Oliguria is the production of an abnormally low volume of urine.
31. high fluid intake, diabetes insipidus, diabetes mellitus
33. As urine volume increases, specific gravity decreases, and vice versa.
35. Low urine pH results from diabetes mellitus or diets high in protein.
37. K^+ and Na^+
39. because the ammonium ion is one of the methods the body has for removing excess protons
41. because most of the sulfur in our diet comes from proteins
43. The breakdown of urea in the urine sample causes an increase in pH. Metals ions are less soluble at higher pH, and they precipitate out.

45. Glucosuria is the presence of glucose in the urine. Possible causes include diabetes mellitus, renal diabetes, emotional stress, moderate liver damage, and consumption of large amounts of carbohydrates over a short period of time.

47. Ketonuria is the presence of ketone bodies in the urine. Possible causes include starvation, diabetes mellitus, severe liver damage, and consumption of a ketogenic diet.

49. Hyperventilation causes the removal of too much carbon dioxide, resulting in alkalosis. Having the person breathe into a paper bag restores blood gas balance.

51. Any reducing sugar gives a positive Benedict's test.

53. By increasing the loss of fluids through urination and by relaxing the blood vessels, there is less blood to flow in a larger space, hence there is less pressure. Lower blood pressure shifts the fluid balance to the opposite of that shown in Figure 21.7. There is a net flow of fluid out of the tissue, thus reducing edema.

Integrated Problems

1. Pellagra is a disease resulting from a deficiency of niacin. This vitamin forms part of the structure of NAD^+, a hydrogen carrier important in the citric acid cycle.

3. **(a)** regulatory, or allosteric, enzyme

(b) 1.7×10^{-9} M

$$\frac{1\ \cancel{\text{molecule}}\ \text{LDL}}{3.0 \times 10^{6}\ \cancel{\text{amu}}} \times \frac{1\ \text{mole}}{6.03 \times 10^{23}\ \cancel{\text{molecules}}} \times \frac{1\ \cancel{\text{amu}}}{1.66 \times 10^{-24}\ \text{g}} = \frac{1\ \text{mole LDL}}{30 \times 10^{5}\ \text{g}}$$

$$\frac{5.0\ \cancel{\mu\text{g}}}{1\ \cancel{\text{mL}}} \times \frac{1000\ \cancel{\text{mL}}}{1\ \text{L}} \times \frac{1\ \text{mol}}{30 \times 10^{5}\ \cancel{\text{g}}} \times \frac{1 \times 10^{-6}\ \cancel{\text{g}}}{1\ \cancel{\mu\text{g}}} = 0.17 \times 10^{-8}\ \text{M}$$

(c) **(1)** No receptor is synthesized.

(2) The receptor is synthesized, but it does not bind to LDL.

(3) The receptor is synthesized, but it fails to cluster in the coated pits.

(d) It increases the synthesis of cholesterol by the cell, because cholesterol is not reaching the cell from the blood. It increases the rate of developing atherosclerosis because the level of circulating LDL is elevated.

(e) **(1)** Bile is synthesized in the liver from cholesterol. If bile acids weren't recycled, the liver would increase the absorption of cholesterol from the blood by increasing the number of LDL receptors and would also increase the synthesis of cholesterol within the liver cells.

(2) Inhibiting HMG-CoA reductase prevents the synthesis of cholesterol by the cells so that the cholesterol needed by the cell has to be absorbed from the blood by increasing the number of LDL receptors.

(f) People suffering from homozygous FH have two defective genes and do not produce normal LDL receptors at all. The treatment described in part (e) depends on the cell being able to produce some normal receptors.

5. The adenine–thymine pairs are only held together by two hydrogen bonds, which can be more easily disrupted by heat than the three hydrogen bonds that hold the guanine–cytosine pairs together.

7. Rhubarb contains a high concentration of oxalate. As an antioxidant, oxalate is itself oxidized. The products of this oxidation include ascorbic acid.

Glossary

The section numbers in brackets indicate where the term is first introduced.

A

Absolute zero 0 K or −273.15°C: the temperature at which all motion within a substance stops. [1.14]

Accuracy The closeness of a measured value to the true value. [1.7]

Acetal linkage The linkage formed by a condensation reaction between a hemiacetal and an alcohol. [15.8]

Acid (Brønsted–Lowry definition), a substance that donates hydrogen ions (protons); a proton donor. [11.1]

Acid–base indicator A chemical dye that changes color at a specific pH. [11.6]

Acidosis A condition that occurs when the blood pH falls below 7.3. [11.7]

Actinide series The 14 elements with atomic numbers from 90 to 103. [4.9]

Activated complex An unstable combination of particles that is an intermediate state between reactants and products in a chemical reaction. [8.1]

Activation energy The minimum amount of energy with which two particles must collide for a chemical reaction to occur. [8.1]

Active site The area on an enzyme molecule to which the substrate attaches. [18.2]

Addition reaction In organic chemistry, a reaction in which a reagent reacts with a double or triple bond, allowing other substances to be added to the molecule. [12.12]

Adenosine diphosphate (ADP) A high-energy diphosphate ester that is produced by the hydrolysis of ATP in the cell. [19.1]

Adenosine monophosphate (AMP) A low-energy monophosphate ester produced by the hydrolysis of ATP and ADP. This nucleotide is also used in the formation of the nucleic acids, DNA and RNA. [19.1]

Adenosine triphosphate (ATP) A very high-energy triphosphate ester that provides the energy required for the reactions of metabolism. [19.1]

Adequate protein A protein that contains all the essential amino acids required by humans. [17.3]

Adipose tissue Tissue that contains fat storage cells. [19.9]

ADP (see adenosine diphosphate).

Adrenal glands Endocrine glands located on top of the kidneys that produce epinephrine, norepinephrine, and the corticosteroids. [14.9]

Aerobic Requiring oxygen. [19.4]

Alcohol An organic compound containing a hydroxyl group (—OH) bonded to a carbon atom having four single bonds. [13.3]

Aldehyde An organic compound containing a terminal carbonyl group

$$(-\overset{\displaystyle O}{\overset{\|}{C}}-H). \quad [13.7]$$

Aldose General term for a monosaccharide containing an aldehyde group. [15.1]

Aliphatic hydrocarbons Those hydrocarbons not containing benzene rings in their structures. [12.15]

Alkali metals The elements in group IA on the periodic table. [4.9]

Alkaline Basic (not acidic); pH $>$ 7.

Alkaline earth metals The elements in group IIA on the periodic table. [4.9]

Alkaloids A large class of complex nitrogen-containing compounds, many of which are produced by plants as part of their defense system and that often have strong physiological effects on humans. [14.9]

Alkalosis A condition that occurs when the blood pH rises above 7.5. [11.7]

Alkane Any hydrocarbon containing only single bonds. [12.4]

Alkene Any hydrocarbon containing one or more carbon-to-carbon double bonds. [12.8]

Alkyl group A substituted group that is derived from an alkane. [12.6]

Alkyne Any hydrocarbon that contains one or more carbon-to-carbon triple bonds. [12.13]

Allosteric enzyme A regulatory enzyme that controls the rate of a series of reactions. [18.8]

Allosteric site The site on the allosteric enzyme (other than the active site) to which the regulatory molecule attaches to inhibit or increase the activity of the enzyme. [18.8]

Alpha helix (α helix) A secondary configuration of proteins in which the polypeptide chain is held in a twisted coil by hydrogen bonds. [17.8]

Alpha radiation Ionizing radiation consisting of streams of high-energy helium nuclei (symbol: ${}^{4}_{2}He$). [7.1]

Amide An organic compound that contains the amide group. [14.5]

Amide group The functional group containing a carbon double-bonded to an oxygen and single-bonded to a nitrogen

$$(-\overset{\overset{\displaystyle O}{\|}}{C}-\overset{|}{N}-).$$ [14.5]

Amide linkage The bond between the carbon and the nitrogen in the amide functional group. [14.5]

Amine An organic compound containing a nitrogen attached to one, two, or three carbons. [14.2]

Amino acid A monomer unit of a protein, containing both an amino group and a carboxylic acid group. (general formula:

$$H_2N-\underset{\underset{\displaystyle R}{|}}{\overset{\overset{\displaystyle H}{|}}{C}}-\overset{\overset{\displaystyle O}{\|}}{C}-OH).$$ [17.1]

Amino group The group of atoms $-NH_2$. [14.2]

Amorphous solid A noncrystalline solid whose particles are randomly arranged. [3.1]

AMP (see adenosine monophosphate).

Amphoteric Having both acidic and basic properties. [11.1]

Amylopectin The highly branched polymer of α-glucose found in starch. [15.11]

Amylose The linear polymer of α-glucose found in starch. [15.11]

Anabolic reactions Reactions in which the cell uses energy to produce molecules needed for growth and repair of the cell. [18.1]

Anaerobic Not requiring oxygen.

Analgesic A drug that reduces pain without causing the loss of consciousness.

Anemia A group of diseases that results in a low number of red blood cells. Examples are hemolytic anemia, iron deficiency anemia, and pernicious anemia.

Anesthetic Any drug that produces a partial or total loss of the sense of pain. [12.14]

Angina pectoris A sharp, intense pain in the chest caused by reduced blood flow to the heart.

Anion A negatively charged ion. [4.7]

Antibiotic A drug that is extracted from microorganisms and that acts as an antimetabolite. [18.7]

Anticodon The three-base sequence on tRNA that is complementary to the codon on mRNA. [20.6]

Antimetabolite Any compound that inhibits enzyme activity.

Antioxidant A substance added to food to prevent the oxidation of unsaturated fats or oils present in the food. [16.8]

Antipyretic A drug that reduces or prevents fever. [13.16]

Antiseptic A substance that inhibits the action of microorganisms and prevents infection. [13.4]

Apoenzyme The protein portion of an enzyme molecule. [18.2]

Aqueous solution A solution containing water as the solvent. [9.6]

Aromatic hydrocarbons The class of hydrocarbons containing benzene and its derivatives. [12.15]

Arteriosclerosis The disease of the arteries commonly known as hardening of the arteries.

Atherosclerosis The most common form of arteriosclerosis, in which the inner layer of the arterial wall becomes thickened by lipid deposits.

Atmosphere (atm) The unit of pressure equal to the force per unit area that supports a column of mercury 760 mm high. [3.3]

Atom The smallest unit of an element having the properties of that element. [1.3]

Atomic mass unit (amu) An arbitrary unit of measure established to allow comparison of the relative masses of the elements. [4.4]

Atomic number For each element, the number of protons in the nucleus of any atom of that element. [4.2]

Atomic weight The weighted average of the masses of the naturally occurring isotopes of an element, expressed in atomic mass units. [4.4]

ATP (see adenosine triphosphate).

Avogadro's law This law states that under identical conditions of temperature and pressure, equal volumes of different gases contain equal numbers of moles of gas. [6.6]

Avogadro's number The number of particles in one mole: 6.02×10^{23}. [6.4]

B

Background radiation Ionizing radiation that comes from natural sources. [7.8]

Basal metabolism rate (BMR) The minimum amount of energy required daily by the body to maintain the basic processes of life. [2.4]

Base (Brønsted–Lowry definition), a substance that accepts hydrogen ions (protons): a proton acceptor. [11.1]

Bends A painful disorder caused by the formation of nitrogen bubbles in the tissues of deep-sea divers who are brought to the ocean's surface too quickly.

Benedict's test The test widely used for the detection of a reducing sugar in the urine. [13.9]

Benign Not malignant.

Beriberi Deficiency disease resulting from a lack of thiamine, vitamin B_1. [18.11]

Beta configuration (β configuration) (see beta pleated sheet).

Beta-oxidation (see fatty acid cycle).

Beta pleated sheet (β pleated sheet) A secondary structure of a protein in which polypeptide chains lie next to one another, held together by hydrogen bonds and having the R groups extending above and below the sheet. [17.8]

Beta radiation Ionizing radiation consisting of streams of high-energy electrons (symbol: ${}_{-1}^{0}e$). [7.1]

Bile A fluid (containing bile salts, bile pigments, and cholesterol) that is produced in the liver, stored in the gallbladder, and released in the small intestine. [19.2]

Binary compound A compound containing only two elements.

Blood plasma Solution remaining after the formed elements are removed from the blood. [21.2]

Blood serum The solution that remains after blood is allowed to clot. [21.2]

Blood sugar Termed often used for glucose.

Blood sugar level The concentration of glucose in the blood (usually expressed in milligrams per 100 mL of blood). [19.3]

Boiling point The temperature at which a substance boils (that is, changes state from liquid to gas). The normal boiling point of a substance is the temperature at which it boils when the atmospheric pressure equals 760 mm Hg. [3.2]

Bond axis The imaginary line connecting two nuclei that are held together by a chemical bond.

Boyle's law This law states that the volume of a gas is inversely proportional to the pressure when the temperature and number of moles of gas remain constant. [3.4]

Brownian movement The erratic, random movement of particles in a colloid. [9.5]

Buffer Any substance that, when added to a solution, protects against sudden changes in the pH of that solution. [11.7]

C

Calorie (cal) The amount of energy necessary to raise the temperature of one gram of water exactly one degree Celsius. [2.4]

Calorimeter The instrument used to determine the caloric content of a substance. [2.4]

Cancer A disease resulting from uncontrolled cell division. [7.6]

Carbohydrate Any of a class of compounds with a general formula of $C_x(H_2O)_y$ whose function is energy storage or structural support in living organisms. [15.1]

Carbon cycle The cycling of carbon in different chemical forms, from the atmosphere, through organisms, to the earth and back to the atmosphere. [15.15]

Carbon-14 dating A method for fixing the age of once-living material using the ratio of carbon-14 to carbon-12 in the artifact. [7.2]

Carbonyl group The functional group containing a carbon atom double-bonded to an oxygen atom,

$$(\text{—}\overset{\overset{\displaystyle O}{\|}}{C}\text{—}). \quad [13.7]$$

Carboxyl group The functional group containing a carbon atom bonded to a hydroxyl group and double-bonded to an oxygen atom

$$(\text{—}\overset{\overset{\displaystyle O}{\|}}{C}\text{—OH}). \quad [13.11]$$

Carboxylic acid An organic acid containing the carboxyl group. [13.11]

Carcinogen Any chemical that can cause cancer in animals.

Catabolic reaction A cellular reaction in which large molecules are broken down to produce smaller molecules and cellular energy. [18.1]

Catalyst A substance that increases the rate of a chemical reaction without being consumed in the reaction. [8.3]

Cation A positively charged ion. [4.7]

Cellular respiration The series of reactions by which glucose is oxidized to form CO_2, H_2O, and ATP. [19.4]

Cellulose A linear polymer of β-glucose that is the main structural molecule in plants. [15.13]

Celsius (°C) The temperature scale in the metric system with 100 degrees between the freezing point of water (set at 0°C) and the boiling point of water (100°C). [1.14]

Chain reaction A self-sustaining process in which the product of one reaction causes one or more additional reactions to occur. [7.3]

Change in state Occurs when matter goes from one phase or state to another, such as when a solid melts or a liquid evaporates. [2.5]

Charles's law This law states that the volume of a gas is inversely proportional to the temperature (in kelvins) if the pressure and number of moles of gas remain constant. [3.5]

Chemical bond The force of attraction that holds atoms together in chemical compounds.

Chemical change A transformation that involves a change in the basic chemical composition of the reactants. [1.5]

Chemical energy Potential energy stored in substances, which results from the spatial arrangement and forces between atoms in the substance. [2.3]

Chemical equation A shorthand way of representing what occurs in a chemical reaction. [6.1]

Chemical equilibrium A dynamic state in which the rate of the forward reaction equals the rate of the reverse reaction. [8.4]

Chemical family (see group).

Chemical formula A shorthand way of representing the composition of a substance by using the chemical symbols of the elements and subscripts to indicate the number of atoms of each element in a reacting unit of the substance. [5.10]

Chemical symbol A one- or two-letter abbreviation representing one atom of an element. [1.4]

Chemistry The science that studies the composition and interactions of matter.

Chemotherapy The treatment of cancer and other diseases using drugs such as antimetabolites. [18.7]

Chiral center A carbon that has four different groups attached to it and that, therefore, is asymmetrical. [15.3]

Chiral molecule A molecule that contains a chiral center and cannot be superimposed on its mirror image. [15.3]

Chlorophyll The green pigment found in plant cells that is involved in the light reactions of photosynthesis. [15.15]

Chloroplast Site of the light reactions of photosynthesis in a plant cell. [15.15]

Cholesterol A sterol produced by all cells and found in cell membranes. It is used to synthesize bile and some hormones. [16.12]

***Cis* isomer** A stereoisomer with the specified atoms or groups of atoms on the same side of the double bond or the ring. [12.11]

***Cis–trans* isomers** Molecules that have the same molecular formula and have atoms connected in the same order, but that differ in the arrangement of those atoms in space. [12.11]

Citric acid cycle A cyclic series of reactions (occurring in the mitochondria of cells) that convert acetyl-CoA to two molecules of CO_2, one ATP, and eight hydrogens that enter into the respiratory chain. [19.5]

Cloning vector A DNA molecule to which the desired gene becomes linked. Used in recombinant DNA technology. [20.9]

Codon The three-base sequence on mRNA that codes for a specific amino acid. [20.6]

Coenzyme A cofactor that is a complex organic molecule other than a protein. [18.2]

Cofactor A metal ion or organic molecule required for an enzyme to function properly. [18.2]

Colligative properties Those properties of a solution that depend only on the number of solute particles in the solution. [10.7]

Colloid A mixture containing relatively large particles (from 1 to 1000 nm) that do not settle out. [9.5]

Competitive inhibition Inhibition of an enzyme that occurs when another molecule competes with the substrate for the active site. [18.7]

Complex carbohydrate A polysaccharide.

Compound A substance formed in a chemical change and composed of two or more elements that combine in a definite proportion by weight. [1.3]

Compound lipid A saponifiable lipid that when hydrolyzed yields fatty acids, an alcohol, and some other compound. [16.1]

Computerized axial tomography (CT) scanner A diagnostic X-ray machine that uses a computer to analyze the signals produced by many X-ray beams passing through the body and to produce detailed cross-sectional pictures. [7.10]

Concentration A numerical measure of the relative amount of solute in a solution. [10.2–10.5]

Condensation The conversion of a substance in the gaseous state to the liquid state. [3.2]

Condensation reaction In organic chemistry, a reaction in which a water molecule is removed from two reactant molecules, thereby forming one product molecule. [13.14]

Conjugate acid The substance formed when a base accepts a hydrogen ion. [11.1]

Conjugate base The substance formed when an acid donates a hydrogen ion. [11.1]

Conjugated double bonds An arrangement of double and single bonds alternating between carbon atoms in a hydrocarbon molecule.

Conjugated protein A protein that produces amino acids and other organic or inorganic groups on hydrolysis. [17.6]

Constitutional isomers Molecules that have the same molecular formula but the atoms are connected in a different order. [12.5]

Conversion factor A ratio or fraction that is formed from an equality and is equal to the number 1. [1.9]

Coordinate covalent bond A covalent bond in which one atom donates both of the electrons shared in the bond. [11.4]

Covalent bond The type of bond formed when electrons are shared between two atomic nuclei. [5.4]

Covalent compound A compound consisting of single, electrically neutral units called molecules. [5.4]

Crenation The shrinking of red blood cells when they are placed in a hypertonic solution. [10.9]

Critical mass The minimum amount of a fissionable isotope that must be present for a nuclear chain reaction to occur. [7.3]

Crystal lattice The repeating, three-dimensional arrangement of particles in a crystalline solid. [5.2]

Crystalline solid A solid whose particles are arranged in a regular repeating pattern. [3.1]

Crystalloid A substance containing particles that are small in size (less than 1 nm) and that form a true solution when placed in water. [9.6]

Curie (Ci) The unit of measure that describes the activity of a radioactive source (1 curie = 3.7×10^{10} transformations/s). [7.7]

Cyclic AMP A chemical messenger within cells, whose formation is triggered when a hormone attaches to a receptor site on the cell membrane. [18.9]

Cyclic compound A compound containing a ring structure. [12.14]

Cytoplasm The gel-like fluid in the interior of a cell.

D

Dalton (d) A unit of mass equal to 1 amu. [4.4]

Dalton's law This law states that the total pressure of a mixture of gases is equal to the sum of the partial pressures of each gas in the mixture. [3.7]

Decay series (see transformation series).

Deficiency disease A disease resulting from too little of an essential nutrient in the diet. [18.10]

Dehydration (a) a medical condition resulting from excessive water loss; (b) in organic chemistry, a reaction involving the removal of a molecule of water from another molecule. [13.6]

Denaturation A reversible or irreversible disruption of the normal arrangement of atoms (the native state) of a protein. [17.11]

Density The mass of one unit of volume of a substance (expressed in g/cm^3 for solids, g/mL for liquids, and g/L for gases). [1.15]

Deoxyribonucleic acid (DNA) The nucleic acid that is a polymer of deoxyribonucleotides, and whose base sequence carries genetic information. [20.3]

Diabetes mellitus A disease resulting from low levels or a total lack of the hormone insulin; commonly known as "diabetes."

Dialysis The movement of ions and small molecules (but not colloidal particles) through a membrane.

Diatomic Containing two atoms. [5.4]

Differentially permeable membrane A membrane that allows water, but not solute particles, to pass through. [10.8]

Diffusion The spontaneous mixing of particles from regions of higher to regions of lower concentration.

Digestion The process by which food is broken down into simple molecules that can be absorbed through the lining of the intestinal tract. [19.2]

Dipeptide A molecule that can be hydrolyzed to form two amino acids. [17.7]

Dipolar ion An ion, such as an amino acid, that has a positive and a negative region.

Disaccharide A compound composed of two monosaccharides. [15.1]

Disintegration series (see transformation series).

Distillation A procedure that separates the compounds in a mixture on the basis of their boiling points.

Disulfide bridge Covalent linkage formed when the sulfhydryl groups on cysteine molecules undergo oxidation. This linkage can form between amino acids on the same protein molecule or different protein molecules. [17.9]

DNA (see deoxyribonucleic acid).

Double bond A covalent bond in which two pairs of electrons (four electrons) are shared by two atomic nuclei. [5.5]

Double helix The structure of a DNA molecule—two helical polynucleotide chains coiled around the same axis. [20.3]

E

Edema A swelling of the tissues caused by an increase in the amount of water in the extracellular fluid. [10.9]

Effusion The movement of a gas through a tiny hole to a region of lower concentration.

Eicosanoids A class of compounds containing 20 carbons that have profound physiological effects at very low concentrations. [16.4]

Electrolyte Any substance that, in aqueous solution, conducts electricity. [9.7]

Electromagnetic energy Energy transmitted by wave-like fluctuations in the strength of electric and magnetic fields. [2.6]

Electromagnetic spectrum The entire range of radiant energy, which includes visible light, radiowaves, and infrared, ultraviolet, X-ray, and gamma radiation. [2.6]

Electron A subatomic particle that is found in certain regions (called orbitals) around the nucleus. It has a mass of 1/1837 amu and 1 unit of negative charge. [4.1]

Electron affinity The amount of energy released when an electron is added to a neutral gaseous atom of an element. [4.11]

Electron configuration The most stable arrangement of electrons in probability regions (orbitals) around the nucleus of an atom. [4.6]

Electron dot diagram A schematic way of representing the valence electrons in an atom of an element. [5.3]

Electronegativity A measure of the ability of an atom to attract toward itself the electrons that it shares in a covalent bond. [5.6]

Electron transport chain The series of oxidation–reduction reactions that is linked to the citric acid cycle and that produces the majority of ATP in the oxidation of molecules in the cell. [19.6]

Electrophoresis A laboratory technique that separates proteins and amino acids on the basis of their different mobilities in an electric field at a specific pH. [17.5]

Element A substance that contains atoms, all of which have the same number of protons in their nuclei. [1.3]

Elongation factors Proteins involved in adding amino acids to the polypeptide chain during protein synthesis in the ribosome. [20.8]

Emphysema A disease in which the lung tissue is so badly damaged that adequate levels of oxygen cannot be maintained in the blood, resulting in very labored breathing.

Enantiomers A chiral molecule and its mirror image. [15.3]

Endocrine gland Any of a specialized group of glands in the body that produce hormones and regulate their secretion into the blood. [18.9]

Endorphins Compounds produced by the brain that produce the same effects as opiates.

Endothermic reaction A reaction in which energy in the form of heat is required to keep the reaction occurring. [2.5]

End point The pH at which an acid–base indicator changes color in a titration. [11.6]

Energy The capacity to do work. [2.1]

Energy level An energy region around the nucleus occupied by electrons. [4.5]

Entropy A measure of the randomness, or disorder, in a system. [2.8]

Enzyme A protein molecule that functions as a biological catalyst. [18.2]

Epinephrine An adrenal hormone more commonly known as adrenalin. [14.9]

Equilibrium (see chemical equilibrium).

Equivalence point The pH at which all the hydrogen (or hydroxide) ions in a solution have been neutralized. [11.6]

Equivalent (Eq) (a) the amount of a substance containing one mole of charge (either + or −). [10.5]

Equivalent weight Weight of a substance, in grams, that contains one equivalent of ions. [10.5]

Erythrocyte A red blood cell.

Essential A term referring to 10 amino acids and two fatty acids that must be present in the diet for normal growth and development. [16.4, 17.3]

Ester Any organic compound formed in a condensation reaction between an alcohol and an organic acid, and containing the functional group

$$-\overset{\overset{\displaystyle O}{\|}}{C}-O-\underset{|}{\overset{|}{C}}-. \quad [13.15]$$

Esterification The condensation reaction between an alcohol and a carboxylic acid, producing an ester. [13.14]

Ether Any organic compound containing an oxygen bonded to two carbons

$$(-\underset{|}{\overset{|}{C}}-O-\underset{|}{\overset{|}{C}}-). \quad [13.2]$$

Eukaryotic cell A cell that contains a nucleus.

Evaporation The conversion of a substance in the liquid state to the gaseous state. [3.2]

Exact number A number that is given in a definition or that results from counting objects and, therefore, has an infinite number of significant digits.

Excited atom An atom having one or more electrons in an energy level higher than normal. [4.7]

Exons Sequences of bases on the DNA of eukaryotic cells that code for amino acids. [20.7]

Exothermic reaction A reaction in which energy in the form of heat is produced. [2.5]

Exponential form The expression of any number as a number between 1 and 10 times a power of 10.

Extracellular fluids Any fluids found in the body tissues but not contained inside the cells.

F

Fahrenheit (°F) Temperature scale in the English system with 180 degrees between the freezing point of water (set at 32°F) and the boiling point of water (212°F). [1.14]

Familial hypercholesterolemia (FH) A genetic disease that causes blood cholesterol levels to be 2 to 10 times higher than normal.

Family (see group).

Fat A triacylglycerol that is a solid at room temperature and that contains mainly saturated fatty acids. [16.2]

Fatty acid An organic compound containing one carboxylic acid group and usually having a long carbon chain. [16.3]

Fatty acid cycle A repeating series of reactions in which a fatty acid is oxidized in two-carbon units. [19.10]

Feedback inhibition Regulation of multienzyme systems wherein the end product of the system inhibits the allosteric, or regulatory, enzyme. [18.8]

Fermentation An anaerobic process that occurs in many microorganisms as well as muscle cells. In alcoholic fermentation in yeast, the end products are ethanol and carbon dioxide.

First law of thermodynamics This law states that energy can neither be created nor destroyed, but only changed in form. [2.7]

Fission The process by which a large unstable nucleus, when bombarded by neutrons, breaks apart to form two smaller nuclei, several neutrons, and a tremendous amount of energy. [7.3]

Formed elements Suspended material in the blood consisting of red blood cells, white blood cells, and platelets. [21.2]

Formula (see chemical formula).

Formula weight The sum of the atomic weights of all the atoms appearing in the chemical formula of a substance. [6.5]

Fractional distillation A process by which compounds in a mixture are separated by means of their boiling points.

Free radical A highly reactive uncharged particle. [7.6]

Frequency (ν) The number of wavelengths to pass a specific point per unit of time. [2.6]

Functional group A group of atoms that gives characteristic chemical properties to all molecules containing that group. [12.1]

Fusion The process by which several small nuclei combine to form a larger, more stable nucleus and tremendous amounts of energy. [7.5]

G

Gamma radiation A naturally occurring high-energy electromagnetic radiation, similar to X rays, with high penetrating power. [2.6]

Gas One of the three states of matter, in which the particles are very far apart and moving very rapidly in a chaotic, random fashion. [3.3]

Gene A sequence of bases on a DNA molecule that codes for a specific protein. [20.6]

Genetic code The sequence of bases on a DNA molecule. [20.6]

Genetic disease A disease that results from a defective gene that is inherited from one or both parents. [20.10]

Genetic engineering A number of different techniques by which genetic information on DNA is transferred between species. [20.9]

Geometric isomerism Stereoisomerism that results from restricted rotation around a double bond or in a cyclic molecule: *cis-trans* isomerism. [12.11]

Glucose An aldohexose that is the principal carbohydrate in the blood and a building block of many disaccharides and polysaccharides. Also called dextrose or blood sugar. [15.2]

Glucose tolerance test A series of tests for blood sugar level taken after the ingestion of a high dose of glucose; often used to diagnose diabetes mellitus.

Glycogen The highly branched polymer of α-glucose that is the glucose storage molecule in animals. [15.12]

Glycogenesis The series of reactions by which glucose molecules are joined together to form glycogen. [19.3]

Glycogenolysis The series of reactions by which glycogen molecules are hydrolyzed or broken down to form glucose molecules. [19.3]

Glycolipid A compound lipid that contains fatty acids, an alcohol, and a sugar group that is either galactose or glucose. [16.11]

Glycolysis The series of reactions by which glucose is converted to two molecules of lactic acid and two molecules of ATP; occurs in the cellular cytoplasm. [19.4]

Glycosidic linkage An acetal linkage formed between two monosaccharides. [15.8]

Glycosuria A medical condition in which sugar molecules are found in the urine. [21.11]

Gram (g) The metric unit of mass (454 grams = 1 pound). [1.12]

Ground state The term applied to an atom having all of its electrons in the lowest possible energy levels. [4.7]

Group A vertical column of elements on the periodic table. [4.9]

H

Half-life The length of time required for one-half of the atoms of a radionuclide in a given sample to undergo radioactive decay. [7.2]

Halogen Any of the elements in group VIIA on the periodic table. [4.9]

Halogenation A reaction in which a halogen substitutes for a hydrogen bonded to a carbon in a hydrocarbon. [12.7]

Heat of fusion The amount of energy (in calories) required to change one gram of a substance (at the melting point) from a solid to a liquid. [3.1]

Heat of reaction (ΔH) The amount of heat energy (in kilocalories per mole) absorbed or released in a chemical reaction. [8.2]

Heat of vaporization The amount of energy (in calories) required to change one gram of a substance (at the boiling point) from a liquid to a gas. [3.2]

Hematocrit The percentage of formed elements in a blood sample. [21.2]

Hemiacetal A compound formed in the reaction between an aldehyde and an alcohol. [13.9]

Hemiketal A compound formed in the reaction between a ketone and an alcohol. [13.9]

Hemodialysis The process of removing wastes from the blood by dialysis.

Hemoglobin The oxygen-carrying molecule in red blood cells. It consists of four polypeptide chains and four nonprotein heme groups, each of which can carry one molecule of oxygen. [17.7]

Hemolysis The rupturing of red blood cells that results when the cells are placed in a hypotonic solution (or from other causes). [10.9]

Henry's law This law states that the higher the pressure, the greater the solubility of gas in a liquid. [3.6]

Heterocyclic compound A compound having a ring structure that contains two or more different types of atoms making up the ring. [14.8]

Heterogeneous Nonuniform.

Heterogeneous nuclear RNA (hnRNA) RNA that is synthesized on the DNA of eukaryotic cells and that contains many more bases than are found on mRNA. [20.7]

Hexose A monosaccharide containing six carbon atoms. [15.1]

High-density lipoprotein (HDL) Particle that transports cholesterol from tissue cell surfaces to the liver; a cholesterol scavenger. [16.12]

High-energy phosphate bond The phosphorus-to-oxygen bond, found in molecules, such as ATP and UTP, which, on hydrolysis, releases large amounts of energy. [19.1]

Holoenzyme A conjugated protein containing the apoenzyme and cofactors that allow it to catalyze a reaction. [18.2]

Homeostasis The maintenance by the body of a nearly constant internal environment while the external environment may vary tremendously.

Homogeneous Uniform throughout.

Hormone Any of the chemical messengers produced by cells to have an effect on other targeted cells. [18.9]

Hybridization The process by which hybrid orbitals are formed. [12.2]

Hybrid orbitals Orbitals formed by a combination of *s*, *p*, or *d* atomic orbitals. [12.2]

Hydrated Surrounded by water molecules. [9.6]

Hydration reaction In organic chemistry, the addition of a water molecule to a double or triple bond. [12.12]

Hydrocarbon An organic compound containing only atoms of carbon and hydrogen. [12.1]

Hydrogenation In organic chemistry, the addition of a hydrogen molecule to a double or triple bond. [16.7]

Hydrogen bond A weak force of attraction between a partially positive hydrogen and a partially negative atom, such as oxygen, fluorine, or nitrogen, on another molecule or on another region of the same molecule. [5.8]

Hydrogen ion A proton. This term is also often used in acid–base chemistry to mean the hydronium ion (H_3O^+). [11.4]

Hydrolysis The addition of a water molecule to a reactant, thereby breaking the reactant into two product molecules. [13.17]

Hydrometer A bulb-shaped instrument used to measure the specific gravity of a liquid. [1.15]

Hydronium ion, H_3O^+ The ion formed when a hydrogen ion (a proton) joins to a water molecule. [11.4]

Hydrophilic Water-attracting; a term given to substances or groups of atoms that are generally very polar or ionic.

Hydrophobic Water-repelling; a term given to substances or groups of atoms that are nonpolar.

Hydroxide ion, OH^- The negative ion formed when a water molecule ionizes. [11.4]

Hydroxyl group The functional group containing an oxygen atom single-bonded to a hydrogen atom, —O—H. [13.3]

Hyper- Prefix used to indicate "higher than normal." For example, hypertension, hypercholesterolemia, and hypertonic.

Hyperglycemia A condition resulting from higher than normal blood glucose levels. [19.3]

Hypertension High blood pressure.

Hyperthyroid A condition in which the thyroid gland produces higher than normal amounts of the hormone thyroxine.

Hypertonic solution A solution with a higher solute concentration than the standard solution. [10.9]

Hypo- Prefix used to indicate "lower than normal." For example, hypoglycemia, hypotonic, and hypothermia.

Hypoglycemia A condition resulting from lower than normal blood glucose levels. [19.3]

Hypothermia A lowering of the body's internal temperature.

Hypothesis A suggested explanation of scientific data that serves as a basis for further investigation and experimentation. [1.7]

Hypotonic solution A solution with a lower solute concentration than the standard solution. [10.9]

I

Indicator A chemical dye that changes color at a specific hydrogen ion concentration. [11.6]

Induced-fit theory Theory of enzyme action stating that the active site of some enzymes is induced by the substrate to fit the shape of the substrate molecule. [18.5]

Infrared radiation Electromagnetic energy that cannot be seen but can be felt as heat. [2.6]

Inhibition The prevention of, or interference with, the action of an enzyme, thus lowering its activity. May be reversible or irreversible. [18.7]

Initiation factors Three proteins involved with starting the process of protein synthesis in the ribosome. [20.8]

Inner transition elements The metals having atomic numbers 58 to 71 and 90 to 103, found in the two long rows at the bottom of the periodic table. [4.9]

Insulin The hormone, produced by beta cells in the pancreas, that controls blood glucose levels by increasing the absorption of glucose from the blood and the rate of glycogenesis. [19.3]

Insulin shock Convulsion and coma resulting from an overproduction or overdose of insulin, which causes the blood glucose level to decrease very fast. [19.3]

Intracellular fluid Fluid found within cells.

Intravenous Administered into a vein.

Introns Intervening sequences of bases, found on the DNA of eukaryotic cells, that do not code for amino acids. [20.7]

Inverse square law This law states that the intensity of radiation on a given surface area decreases by the square of the distance from the source. [7.9]

Iodine number Indicates the amount of unsaturation in a compound: the higher the iodine number, the more unsaturated the compound. [16.6]

Ion A positively or negatively charged particle. [4.7]

Ionic bond The attraction between ions formed when one or more electrons are transferred from one atom to another. [5.2]

Ionic compound A compound consisting of an orderly arrangement of oppositely charged ions, which are combined in a ratio such that the compound is electrically neutral. [5.2]

Ionization energy The amount of energy that must be added to a gaseous atom to remove its outermost or most loosely held electron. [4.11]

Ionizing radiation Radiation, such as alpha, beta, or gamma radiation, that can produce unstable and highly reactive ions in living tissue. [7.6]

Ion product constant of water (K_w) $K_w = [H^+][OH^-] = 1 \times 10^{-14}$. [11.5]

Isoelectric point The pH at which an amino acid or protein is electrically neutral and does not migrate in an electric field. [17.5]

Isomers Compounds having the same molecular formula but different structures. [12.3]

Isotonic solution A solution with a solute concentration equal to the standard solution. [10.9]

Isotopes Atoms of the same element that differ in the number of neutrons in their nuclei. [4.3]

IUPAC Abbreviation for the International Union of Pure and Applied Chemistry. [12.1]

J

Jaundice A condition caused by a high level of bilirubin in the blood, which results from a blockage of the bile duct or a malfunction of the liver. [2.6]

Joule (J) A unit of energy in the SI system (4.184 J = 1 cal). [2.4]

K

Kelvin (K) The temperature scale in the SI system, with 100 kelvins between the freezing point of water (273.15 K) and the boiling point of water (373.15 K). [1.14]

Ketone An organic compound containing a carbonyl group bonded to two other carbons

$$(-\underset{|}{\overset{|}{C}}-\overset{\overset{O}{\|}}{C}-\underset{|}{\overset{|}{C}}-). \quad [13.7]$$

Ketone bodies Compounds produced from the metabolism of excess acetyl-CoA (which, in turn, is produced when large amounts of fats are oxidized to supply cellular energy). [19.12]

Ketonurea The presence of ketone bodies in the urine. [21.11]

Ketose The general term for a monosaccharide containing a ketone group. [15.1]

Ketosis A condition caused by higher than normal levels of ketone bodies in the blood. [19.12]

Kilocalorie (kcal) The amount of heat energy necessary to raise the temperature of 1000 grams of water one degree Celsius (1000 cal = 1 Calorie = 1 kcal). [2.4]

Kinetic energy Energy of motion. [2.2]

Kinetic theory The theory that explains the behavior of a gas in terms of the motion of its particles. [3.9]

Krebs cycle (see citric acid cycle).

Krebs ornithine cycle (see urea cycle).

L

Lactic acid fermentation The anaerobic process that occurs in muscle cells during strenuous exercise in which pyruvic acid is converted to lactic acid. [19.7]

Lanthanide series The 14 elements with atomic numbers from 58 to 71. [4.9]

Law of conservation of energy (see first law of thermodynamics).

Law of conservation of mass This law states that in a chemical reaction the mass of the products equals the mass of the reactants. [6.1]

Law of definite proportions This law states that a compound is composed of specific elements in a definite proportion by weight. [1.3]

Le Chatelier's principle This principle states that a system at equilibrium resists changes in temperature, pressure, or concentration of reactants or products. [8.6]

Lipid Any of a large class of nonpolar, organic compounds that have oily or waxy properties. [16.1]

Lipogenesis The synthesis of lipids in cells. [19.9]

Liquid One of the three states of matter, in which the particles are fairly close together but can slip over one another. [3.2]

Liter (litre) (L) A unit of volume equal to 1000 cubic centimeters (1 liter = 1.06 quarts). [1.13]

Lock-and-key theory The theory of enzyme action stating that only a specific substrate fits the active site of an enzyme, just as only a specific key turns a lock. [18.5]

Low-density lipoprotein (LDL) The primary cholesterol transport molecule in the blood. High levels are associated with heart disease. [16.12]

Luminescence The release of energy as visible light by an excited atom.

M

Macromineral One of seven elements (K, Mg, Na, Ca, P, S, Cl) that are required in small amounts for normal cell growth and development. [4.12]

Magnetic resonance imaging (MRI) A body scanning technique that uses low-level electromagnetic radiation produced within the body to create images of the body used for diagnosis.

Malignant Term describing cells that are growing and dividing in an uncontrolled fashion.

Mass A measure of the resistance of an object to a change in speed or direction. [1.1]

Mass number The sum of the number of protons and neutrons in the nucleus of an atom. [4.2]

Matter Anything that has mass and occupies space. [1.1]

Melanin A brown skin pigment produced by the body to protect against the effects of ultraviolet radiation.

Melting point The temperature at which a solid breaks down to form a liquid. [3.1]

Messenger RNA (mRNA) The RNA that carries the genetic code for a specific protein; mRNA is synthesized in the nucleus and then migrates to the cytoplasm where it attaches to ribosomes and serves as the template for protein synthesis. [20.5]

Metabolism All of the enzyme-catalyzed reactions in the body. [18.1]

Metal An element that is shiny, dense, and easily worked, that has a high melting point, and that conducts electricity. [4.10]

Metalloid An element that acts like a metal in some ways and like a nonmetal in other ways. These elements are found between the metals and the nonmetals on the periodic table. [4.10]

Metastable In an energy state higher than normal. [7.11]

Metastasis The spread of cancer cells from a tumor to other parts of the body. [7.11]

Meter (metre) Unit of length in the metric system (1 meter = 3.28 feet). [1.11]

Metric system A system of measure based on the decimal system. [1.10]

Microwave radiation Electromagnetic radiation used in cooking foods, navigation, engineering, and broadcast transmissions. [2.6]

Milliequivalent (mEq) Unit of measure used to express the concentration of ions in the blood (1000 mEq = 1 Eq). [10.5]

Millimeter of mercury (mm Hg) Unit of pressure equal to 1/760 atmosphere. [3.3]

Mitochondria Structures in the cell where the citric acid cycle and the respiratory chain occur. [19.4]

Mixture Two or more substances combined in any proportion. [1.3]

Molarity (*M*) Unit of solution concentration defined as the number of moles of solute per liter of solution. [10.2]

Mole (mol) The amount of a substance that has the same number of particles as there are atoms in 12 grams of carbon-12 (6.02×10^{23} atoms). [6.4]

Molecular cloning (see recombinant DNA technology).

Molecule An electrically neutral unit formed when two or more atoms are joined together by covalent bonds. [1.3]

Monomer A single unit that joins with many other similar units to form a polymer. [12.12]

Monosaccharide A carbohydrate that cannot be broken into smaller units by hydrolysis. [15.1]

Multienzyme system A sequence of enzyme-catalyzed reactions that produces a specific metabolic result. [18.8]

Mutagen Any chemical or physical agent that is capable of producing a mutation.

Mutant Containing altered DNA. [7.6]

Mutation A change in the sequence of the four bases that are the informational code on the DNA molecule. [20.10]

Myelin sheath The protective coating surrounding nerves. [16.11]

Myocardial infarction A heart attack.

N

Native state (native configuration) The shape of a protein that is energetically the most stable. [17.9]

Net ionic equation A chemical equation showing only the ions that take part in the reaction. [9.9]

Neurotransmitters Compounds synthesized by nerve cells that act as chemical messengers between nerve cells.

Neutral Term applied to a solution that has neither acidic nor basic properties, or to a particle that has no net electric charge.

Neutralization The process by which an acidic or basic solution is converted to a neutral solution. [11.3]

Neutron A subatomic nuclear particle with a mass of 1 amu and no charge. [4.1]

Noble gas Any of the elements in group VIIIA (helium, neon, argon, krypton, xenon, and radon), all of which have great chemical stability. [4.9]

Nonelectrolyte A substance that does not conduct electricity when placed in solution. [9.7]

Nonmetal An element that is brittle, has low density and a low melting point, and does not conduct electricity. [4.10]

Nonpolar The term applied to covalent bonds and covalent molecules when the centers of positive and negative charge coincide. [5.6]

Nonsaponifiable lipid Any lipid that cannot be hydrolyzed by an aqueous solution of base. [16.12]

Nuclear reactor A facility that uses controlled nuclear fission to produce energy. [7.3]

Nucleic acid A polymer of nucleotides: either DNA (deoxyribonucleic acid) or RNA (ribonucleic acid). [20.2]

Nucleotide The monomer unit of nucleic acids, whose structure contains a five-carbon sugar, a nitrogen-containing base, and a phosphate group. [20.2]

Nucleus (a) the dense center of an atom, containing protons and neutrons; [4.1] (b) a cellular organelle surrounded by a nuclear membrane and containing chromosomes and nucleoli.

O

Octet rule The tendency of elements in groups IA to VIIA to form bonds that result in eight valence electrons in the outer energy level of each atom (except in the first energy level, where the tendency is toward two electrons). [5.1]

Oil A triacylglycerol, extracted from vegetable seeds or fruits, that is a liquid at room temperature and that contains mainly unsaturated fatty acids. [16.2]

Oligurea Low urine volume. [21.8]

Opiates Alkaloids or derivatives of alkaloids extracted from the opium poppy, such as morphine, codeine, and heroin.

Optical isomers Compounds that are mirror images and that differ only in the way they interact with polarized light. [15.3]

Orbital A region around the nucleus of an atom in which there is a high probability of finding one or two electrons. The four kinds of orbitals are called *s*, *p*, *d*, and *f* orbitals. [4.5]

Orbital diagram A schematic way of showing the electron configuration of an atom. [4.6]

Organic chemistry The study of carbon compounds. [12.1]

Osmole (osmol) One mole of any combination of particles. [10.8]

Osmolarity A unit of concentration that describes the total number of particles in solution; expressed in osmols per liter. [10.8]

Osmosis The movement of water molecules through a differentially permeable membrane from a region of lower solute concentration to a region of higher solute concentration. [10.8]

Osmotic pressure The amount of pressure that must be applied to a solution to prevent osmosis if the solution were separated from pure water by a differentially permeable membrane. [10.8]

Oxidation (a) the loss of one or more electrons by an atom, ion, or molecule. [6.3] (b) In organic chemistry, the loss of hydrogen or the gain of oxygen by an organic molecule or ion. [12.7]

Oxidation–reduction reaction (see redox reaction).

Oxidative deamination The removal of an amino group from an amino acid, producing an α-keto acid and ammonia. [19.13]

Oxidizing agent A substance that causes the oxidation of a reactant molecule. [6.3]

P

Partial pressure The pressure exerted by a specified gas in a mixture of gases. [3.7]

Parts per billion (ppb) Unit of concentration: the number of micrograms of solute per liter of solution. [10.4]

Parts per million (ppm) Unit of concentration: the number of milligrams of solute per liter of solution. [10.4]

Pascal (Pa) The SI unit of pressure (133.3 Pa = 1 torr). [3.3]

Pellagra A deficiency disease resulting from a lack of the vitamin niacin. [20.10]

Pentose A monosaccharide containing five carbon atoms. [15.6]

Peptide bond An amide linkage formed by a condensation reaction between two amino acids. [17.7]

Percentage concentration [10.3]

Weight/volume (w/v) percent: The number of grams of solute per 100 milliliters of solution.

Milligram percent (mg%): The number of milligrams of solute per 100 milliliters of solution.

Period A horizontal row on the periodic table. [4.9]

Periodicity The repeating nature of chemical properties of the elements when they are arranged in order of atomic number. [4.8]

Periodic law This law states that many properties of the elements repeat periodically as the atomic number of the elements increases. [4.11]

Periodic table An arrangement of the elements in order of increasing atomic number that illustrates chemical similarities between groups of elements. [4.8]

pH A measure of the hydrogen ion concentration of an aqueous solution; $[H^+] = 1 \times 10^{-pH}$. [11.6]

Phenylalanine An amino acid that accumulates in the body of a child with PKU. [20.11]

Phenylketonuria (PKU) An inherited disease in which an enzyme responsible for the conversion of phenylalanine to tyrosine is defective. [20.11]

Phosphoglyceride A phospholipid containing the alcohol glycerol. [16.11]

Phospholipid A compound lipid whose structure contains an alcohol, fatty acids, and a phosphate group. [16.11]

Photosynthesis The process by which green plants use sunlight as the source of energy to produce glucose and oxygen from water and carbon dioxide. [15.15]

Physical change A transformation during which a substance changes form but keeps its chemical identity. [1.5]

Pi bond (π bond) A bond formed by the sideways overlap (above and below the bond axis) of two *p* orbitals. [12.10]

Plaque (a) in dentistry, the sticky substance produced by bacteria in the mouth that adheres to the teeth; (b) in cardiology, a deposit of smooth muscle cells, fats, and scar tissue on the interior of an arterial wall.

Plasmid A circular piece of DNA, often from *E. coli,* that is used as a cloning vector in recombinant DNA technology. [20.9]

Polar Term applied to covalent bonds and covalent molecules when the center of positive charge and the center of negative charge do not coincide, thus forming an electric dipole. [5.6]

Polyatomic ion An electrically charged group of covalently bonded atoms that stays together as a unit in most chemical reactions. [5.9]

Polycythemia The excessive formation of red blood cells; called *polycythemia vera* when the increase is caused by a tumor. [7.12]

Polyester A polymer of ester molecules, used to make fibers for fabrics. [15.16]

Polyhydric Containing more than one hydroxyl group.

Polymer A very large molecule made up of repeating units called monomers. [12.12]

Polymerization A chemical reaction in which single molecules called monomers react with each other to form large molecules called polymers. [12.12]

Polypeptide A polymer composed of amino acids connected by peptide bonds. [17.7]

Polyprotic acid An acid that can donate more than one hydrogen ion. [11.1]

Polysaccharide A polymer of three or more monosaccharide molecules. [15.1]

Polysome A group of ribosomes all synthesizing protein on the same molecule of mRNA. [20.8]

Polyunsaturated The term describing a triacylglycerol whose molecules have two or more double bonds. [12.8]

Polyurea The production of large amounts of urine. [21.8]

Positron emission tomography (PET) A scanning technique that uses positron-emitting radioactive tracers that concentrate in specific tissues to produce images that show metabolic changes. [7.10]

Potential energy Energy of position. [2.3]

Precipitate A solid that forms in a solution as a result of a chemical reaction. [9.9]

Precision The degree to which measurements are reproducible. [1.7]

Pressure A force exerted per unit of area. [3.3]

Primary structure The sequence of amino acids (connected by peptide bonds) in the polypeptide chain of a protein. [17.7]

Product A substance that results from a chemical reaction. [6.1]

Proenzyme The inactive form of an enzyme. [18.2]

Prokaryotic cell Cells that contain no nucleus or cellular organelles.

Prostaglandin Any of a class of 20-carbon fatty acids that are derived from prostanoic acid and that have a wide variety of potent physiological effects. [18.9]

Prosthetic group A cofactor that is tightly bound to an apoenzyme. [18.2]

Protein A polymer of amino acids. [17.1]

Proton A subatomic nuclear particle having a mass of 1 amu and one unit of positive charge. [4.1]

Ptomaine Any amine produced in the natural decay of living organisms. [14.2]

Pulmonary Having to do with the lungs.

Purines Heterocyclic amines whose derivatives (adenine and guanine) are essential parts of DNA and RNA molecules. [20.2]

Pyrimidines Heterocyclic amines whose derivatives (cytosine, thymine, and uracil) are essential parts of DNA and RNA molecules. [20.2]

Q

Quantum mechanical model A model of the atom in which the position of the electrons around the nucleus are described in probability regions called orbitals. [4.5]

Quaternary structure The overall structure of a protein that contains more than one polypeptide chain. [17.10]

R

Radioactive decay The process by which an unstable nucleus gives off nuclear particles or gamma radiation (or both) to become more stable. [7.1]

Radioactive tracer A chemical that contains radioactive atoms but has the same chemical nature and behavior as naturally occurring compounds; used to follow metabolic pathways. [7.10]

Radioactivity The giving off, or emission, of radiation from certain isotopes. [7.11]

Radionuclide A radioactive isotope. [7.1]

Rancid A term applied to foods containing fats and oils that have undergone hydrolysis or oxidation, forming substances that give the food a bad smell or taste. [16.8]

Reactant A starting substance in a chemical reaction. [6.1]

Recombinant DNA technology A technique that produces large amounts of the product of a gene that is inserted into a specially developed strain of *E. coli.* [20.9]

Recommended dietary allowance (RDA) The amount of a nutrient needed in the diet (as established by the Food and Nutrition Board of the National Academy of Sciences) to meet the daily requirements of a healthy individual. [18.10]

Redox reaction Abbreviation for *reduction–oxidation* reaction, a reaction in which electrons are transferred from one reactant to another. [6.3]

Reducing agent A substance that causes the reduction of a reactant molecule. [6.3]

Reducing sugar Any carbohydrate that can act as a reducing agent and produce a positive Benedict's test. [15.7]

Reduction The gaining of electrons by a reagent. [6.3] In organic chemistry, reduction occurs when an organic molecule or ion gains hydrogen atoms or loses oxygen atoms. [12.12]

Regulatory enzyme (see allosteric enzyme).

Regulatory molecule Molecule that either inhibits or increases the activity of an enzyme when it attaches to the allosteric site. [18.8]

Regulatory site (see allosteric site).

Rem The unit of absorbed dose of radiation that produces the same biological effect as 1 Rad of therapeutic X rays. [7.7]

Renal threshold The concentration of glucose in the blood above which glucose begins to appear in the urine. [19.3]

Replicate To make an exact copy; in cell division the DNA molecules replicate, producing two daughter cells with identical DNA. [20.4]

Representative element Any element in the A groups on the periodic table. [4.9]

Resonance The term used to describe the delocalization of electrons among atoms, as in benzene. [12.15]

Resonance hybrid The term used to describe the structure of the benzene molecule. [12.15]

Respiratory chain (see electron transport chain).

Ribonucleic acid (RNA) A group of nucleic acids, polymers of ribonucleotides, synthesized on the DNA strand and having different cellular functions. [20.5]

Ribosomal RNA (rRNA) The RNA that, with proteins, forms granules called ribosomes in the cytoplasm. [20.5]

Ribosomes Granules in the cytoplasm made up of two subunits formed by rRNA and protein that combine with mRNA to synthesize polypeptides. [20.5]

Rickets A disease that results from too little vitamin D in the diet and that causes bones to soften and bend out of shape. [18.12]

RNA (see ribonucleic acid).

S

Saline solution (a) a solution that contains salt; (b) a solution that is isotonic to blood plasma (normal or physiological saline). [10.9]

Salt A compound formed alone with water in a neutralization reaction. [11.3]

Salt bridge A force of attraction that occurs between the charged R groups on the polypeptide chains of a protein; similar to the attraction between ions in an ionic crystal. [17.9]

Saponifiable lipid A lipid that can be hydrolyzed in a basic aqueous solution. [16.1]

Saponification The hydrolysis of an ester in an aqueous solution of strong base. [13.17]

Saturated compound Any hydrocarbon or its derivative that contains only carbon-to-carbon single bonds. [12.4]

Saturated solution A solution that contains as many solute particles as can dissolve in the solvent at that temperature. [10.1]

Scientific method The study of nature through observation, followed by development of a hypothesis and testing of the hypothesis through further experimentation. [1.6]

Scurvy A deficiency disease caused by a lack of vitamin C (ascorbic acid). [18.11]

Secondary structure The localized spatial arrangement of atoms in the backbone of a polypeptide chain. [17.8]

Second law of thermodynamics This law states that the entropy, or disorder, of the universe is increasing. [2.8]

Semimetal (see metalloid).

SI units Abbreviation for the International System of Units, a system of weights and measures that is the successor to the metric system. [1.10]

Sickle cell anemia Anemia caused by an inherited defect in the hemoglobin molecule.

Sievert (Sv) The SI unit of absorbed dose (1 Sv = 100 rem).

Sigma bond (σ bond) A bond formed by the overlap of orbitals along the bond axis. [12.10]

Significant figures (significant digits) The number of digits in a measured value that are known, plus one digit that is uncertain. [1.8]

Simple lipid A saponifiable lipid that, when hydrolyzed, yields fatty acids and an alcohol. [16.1]

Simple protein A protein that yields only amino acids on hydrolysis. [17.6]

Single bond A chemical bond in which one pair of electrons (two electrons) are shared by two atomic nuclei. [5.5]

Soap A salt of a fatty acid, produced by the saponification of triacylglycerols. [16.10]

Solid One of the three states of matter, in which the particles are closely packed in a rigid structural arrangement. [3.1]

Solubility The amount of a solute that dissolves in a fixed volume or weight of a solvent. [9.8]

Solute The substance being dissolved in a solution. [9.6]

Solution A homogeneous mixture of two or more substances. [9.6]

Solvent The substance in which the solute is dissolved. [9.6]

Specific gravity A comparison of the mass of a liquid with the mass of the same volume of pure water. [1.15]

Specific heat (specific heat capacity) The amount of heat energy required to raise the temperature of one gram of a substance one degree Celsius. [2.4]

Sphingolipid A phospholipid containing the alcohol sphingosine. [16.11]

Standard atmospheric pressure The pressure that supports a column of mercury 760 mm high at a temperature of 0°C. [3.3]

Starch A mixture of amylose and amylopectin; the energy storage molecule in plants. [15.11]

States of matter Solid, liquid, and gas. [1.5]

Stereoisomerism (see *cis-trans* isomers).

Steroid Any of a large class of nonsaponifiable lipids, all of which contain a complicated four-ring framework. [16.12]

STEP method A four-step method for solving problems; See the question, Think it through, Execute the math, and Prepare the answer.

STP Standard temperature and pressure: 0°C and 1 atm. [3.3]

Structural formula A diagram that shows the arrangement of atoms in the molecule. [12.3]

Structural isomers (see constitutional isomers).

Sublimation The process of changing from the solid state directly to the gaseous state. [3.2]

Substituent Atoms or groups of atoms (other than hydrogen) attached to the carbon chain in a hydrocarbon. [12.6]

Substitution reaction A chemical reaction in which an atom or group of atoms is substituted for some atom on a reactant molecule. [12.7]

Substrate A reactant in an enzyme-catalyzed reaction that attaches to the surface of the enzyme. [18.2]

Sulfhydryl group The functional group, —S—H.

Supersaturated The term applied to a solution that has more solute particles dissolved in it than it can hold at equilibrium at that temperature. [10.1]

Surface tension The resistance of the particles on the surface of a liquid to the expansion of that liquid. [3.2]

Surfactant (surface active agent), a substance that acts to reduce the surface tension of water. [9.2]

Suspension A heterogeneous mixture having particles that, in time, settle out. [9.4]

Symbol A shorthand way of representing one atom of an element. For example, C for carbon. [1.4]

T

Temperature A measure of the average kinetic energy of the particles of a substance. [1.14]

Tertiary structure The three-dimensional structure of the entire polypeptide chain. [17.9]

Thiol group (see sulfhydryl group).

Thyroid The endocrine gland, located in the neck, that produces the hormones thyroxine and calcitonin.

Thyroxine An iodine-containing hormone produced by the thyroid.

Titration A laboratory procedure for measuring an unknown concentration of an acidic or basic solution. [11.6]

Torr A unit of pressure equal to 1 Mm Hg. [3.3]

Trace element An element required in minute amounts for normal cell growth and development. [4.12]

Transamination The transfer of an amino group from an amino acid to an α-keto acid, thus producing a new amino acid. [19.13]

Transcription The transfer of genetic information from DNA to mRNA; the synthesis of mRNA on a segment of DNA. [20.5]

Transfer RNA (tRNA) RNAs that each carry a specific amino acid to the ribosomes and place the amino acid, by pairing the bases in the anticodon region of tRNA with the codon of mRNA, in the proper sequence for the formation of the polypeptide chain. [20.5]

Transformation series A series of radioactive decays or disintegrations by which an unstable nucleus becomes a stable nucleus. [7.1]

***Trans* isomer** A stereoisomer with the specified atoms or groups of atoms on opposite sides of the double bond or the ring. [12.11]

Transition element Any of the B group metals on the periodic table. [4.9]

Transition state (see activated complex).

Translation The expression of genetic information in the amino acid sequence of a protein; the synthesis of protein molecules on one mRNA molecule. [20.8]

Triacylglycerol An ester of glycerol and three fatty acids; general term for fats and oils. [16.2]

Tricarboxylic acid (TCA) cycle (see citric acid cycle).

Triglyceride (see triacylglycerol).

Triple bond A chemical bond in which three pairs of electrons (six electrons) are shared by two atomic nuclei. [5.5]

Turnover number The number of substrate molecules transformed per minute by one molecule of enzyme under optimal conditions. [18.4]

Tyndall effect The scattering of light by the particles in a colloid. [9.5]

Tyrosine An amino acid that is lacking in an untreated child with PKU. [20.11]

U

Ultraviolet radiation High-energy electromagnetic radiation that causes sunburn. [2.6]

Unit factor (see conversion factor).

Unsaturated compound Any hydrocarbon or its derivative that contains one or more double or triple bonds. [12.8]

Unsaturated solution A solution in which more solute can be dissolved at that temperature. [10.1]

Urea cycle The cyclic series of reactions in the liver by which urea is produced from ammonia and carbon dioxide. [19.14]

Uremia A condition, resulting from a damaged kidney, in which urea builds up in the blood. [19.14]

V

Valence electron An electron in the outermost energy level of an atom. [4.9]

Vapor pressure The pressure of the gas above a liquid in a container. [3.7]

Vasopressin An antidiuretic hormone; that is, a substance that controls the release of water into the urine by the kidneys. [17.7]

Viscosity A measure of how easily a liquid flows. [3.2]

Vitamin An organic nutrient that the body cannot synthesize but which is necessary for normal body function. [18.10]

Vitamin deficiency disease A disease that results only from the lack of a specific vitamin in the diet. [18.10]

Volatile Term applied to a substance that evaporates quickly.

W

Wavelength (λ) The distance between crests in the wave-like fluctuations of electromagnetic radiation. [2.6]

Wax An ester of a long-chain fatty acid and a long-chain alcohol. [16.5]

Weight A measure of the attraction of gravity on an object. [1.1]

X

X ray High-energy ionizing radiation, similar to gamma rays, produced in X-ray tubes. [2.6]

Z

Zwitterion A dipolar ion; one that has a positively charged area and a negatively charged area. For example, an amino acid. [17.4]

Zymogen (see proenzyme).

Photo Credits

Chapter 1 *Opener:* Tom Stewart/The Stock Market. *Figure 1.1:* NASA/The Image Works. *Figure 1.4:* Michael Keller/FPG International. *Page 15:* Courtesy New York Public Library. *Figure 1.11:* Runk/Schoenberger/Grant Heilman Photography. *Page 27:* Yoav Levy/Phototake.

Chapter 2 *Opener:* Charles Gupton/Tony Stone Images/New York, Inc. *Figure 2.1:* John Taylor/FPG International. *Figure 2.3(a):* John Gerlach/Earth Scenes. *Figure 2.3(b):* Milan Chukovich. *Page 46:* Michael A. Keller/The Stock Market. *Figure 2.8:* Photo Researchers. *Page 49:* Bettmann Archive.

Chapter 3 *Opener:* Courtesy DeVilbiss Health Care, Inc. *Figure 3.2(a):* Charles Krebs/The Stock Market. *Figure 3.2(b):* Eric Grave/Science Source/Photo Researchers. *Figure 3.4:* Courtesy R. Stimson Wilcox/SUNY Binghamton. *Page 62:* Yoav Levy/Phototake. *Page 66:* Fred Bavendam/Peter Arnold, Inc.

Chapter 4 *Opener:* Carole Banger/Praeses Productions. *Page 79:* Courtesy AIP Neils Bohr Library. *Figure 4.1:* Bryan Yablonsky/Duomo Photography, Inc. *Page 91:* Courtesy Chemical Heritage Foundation. *Page 93:* Richard Megna Fundamental Photographs. *Page 94:* Andy Washnik.

Chapter 5 *Opener:* Michael Melford/The Image Bank. *Page 106:* Michael Watson. *Figure 5.1:* M. Claye/Iacana/Photo Researchers. *Figure 5.6:* Dr. J. Burgess/Photo Researchers. *Page 117:* Courtesy Slack Myers & Barshinger, Inc. and Nutrasweet Corporation.

Chapter 6 *Opener:* Bob Daemmrich/The Image Works. *Page 136:* Bettmann Archive/photo by Brady. *Figure 6.2:* Ken Karp. *Page 139:* Courtesy of NBS Archives/AIP Neils Bohr Library. *Page 143:* Jim Riffle/Astroworks, Corp.

Chapter 7 *Opener:* Stephen Frisch/Stock, Boston. *Page 158:* Biomedical Communications/Photo Researchers. *Page 163:* Courtesy AIP Emilio Segré Visual Archives. *Page 167:* Patrick Mesner/Gamma Liaison. *Page 169:* Otto Hahn: *A Scientific Autobiography New York:* Charles Scribner's Sons, 1966, Courtesy of AIP Emilio Segré Visual Archives, 1966. *Figure 7.7:* Yoav Levy/Phototake. *Figure 7.9(a):* Courtesy Dimensional Medicine, Inc. *Figure 7.9(b):* Courtesy Dr. David Vining, Bowman Gray School of Medicine. *Figure 7.8:* Simon Fraser/Science Photo Library/Photo Researchers. *Page 178 (top):* Scott Camazine/Photo Researchers. *Page 178 (bottom):* Hank Morgan/Rainbow. *Figure 7.10:* Courtesy Good Samaritan Hospital, Corvallis, Oregon. *Figure 7.11:* Erik Hilderbrandt/Photo Researchers.

Chapter 8 *Opener:* Terje Rakke/The Image Bank. *Page 192:* Jim Brandenburg/Minden Pictures, Inc. *Figure 8.5:* Science VU/Visuals Unlimited. *Page 195:* A. D'Arazien/The Image Bank. *Page 203:* Ken Karp/Omni-Photo Communications.

Chapter 9 *Opener:* Les Stone/Sygma. *Figure 9.3:* COMSTOCK, Inc. *Page 215 (margin):* Yoav Levy/Phototake. *Figure 9.4:* OPC, Inc. *Figure 9.5:* Whit Bronaugh. *Page 220:* Ken Karp. *Figure 9.8:* Michael Watson.

Chapter 10 *Opener:* Mug Shots/The Stock Market. *Figure 10.1:* Andy Washnik. *Figure 10.2:* Peter Lerman. *Page 232:* Andy Washnik. *Figure 10.4:* OPC, Inc. *Page 241:* Stephen Marks/The Image Bank. *Page 242:* Andy Washnik.

Chapter 11 *Opener:* Doug Goodman/Monkmeyer Press Photo. *Page 248:* Robert Capece. *Page 249:* Andy Washnik. *Page 252:* Martin Bond/Science Photo Library/Photo Researchers. *Page 255:* Arthur Beck/Photo Researchers. *Page 259:* Courtesy Fisher Scientific.

Chapter 12 *Opener:* Courtesy NASA. *Page 271:* Phillip Hayson/Photo Researchers. *Page 272:* Don Spiro/Tony Stone Images/New York, Inc. *Page 281:* Brett Froomer/The Image Bank. *Figure 12.9 (left):* David Austen/Tony Stone Images/New York, Inc. *Figure 12.9 (right):* Laurie Rubin/The Image Bank. *Page 297:* Scott Camazine/Photo Researchers.

Page 298 (top): Andy Washnik. *Page 298 (bottom):* Courtesy Institute of Scrap Recycling Industries. *Page 305:* National Cancer Institute/Science Photo Library/Photo Researchers.

Chapter 13 *Opener:* Ted Russell/The Image Bank. *Page 320:* Courtesy Massachusetts General Hospital. *Page 325:* Andy Washnik. *Page 326:* Courtesy New York Public Library. *Page 327:* Doug Plummer/Photo Researchers. *Page 334:* Andy Washnik. *Page 339:* Andy Washnik.

Chapter 14 *Opener:* Will & Deni McIntyre/Photo Researchers. *Page 366:* Andy Washnik. *Page 369:* A. Tannenbaum/Sygma.

Chapter 15 *Opener:* Joe McNally/Sygma. *Figure 15.1:* Harald Sund/The Image Bank. *Figure 15.3* and *Page 391:* Andy Washnik. *Figure 15.7:* J. Litway/Visuals Unlimited. *Page 396:* James Dennis/Phototake.

Chapter 16 *Opener:* Charles Krebs/Tony Stone Images/Seattle. *Figure 16.1:* Biophoto Associates/Photo Researchers. *Page 405:* E. Hossler/Visuals Unlimited. *Page 407:* Ken Karp. *Page 408:* Steve Allen/The Image Bank. *Figure 16.2:* Ken Karp. *Page 415:* Lennart Nilsson/Boehringer Ingelheim International. *Figure 16.5:* Francis Leroy/Science Photo Library/Photo Researchers. *Page 420:* Steven E. Sutton/Duomo Photography, Inc.

Chapter 17 *Opener:* Mugshots/The Stock Market. *Page 434:* Joe Munroe/Photo Researchers. *Page 438:* Stanley Flegler/Visuals Unlimited. *Page 440 (top):* CNRI/Science Photo Library/Photo Researchers. *Page 440 (center):* J. Bross, F.O. Schmitt, D.W. Fawcett/Visuals Unlimited. *Page 440 (bottom):* Hans Pfletschinger/Peter Arnold, Inc. *Page 442:* Jean-Jacques Grezet/Sygma. *Figure 17.8:* The Studio of Wood Ronsaville Harlin, Inc.

Chapter 18 *Opener:* Don and Pat Valenti/Tony Stone Images/Seattle. *Figure 18.1(b):* Oxford Molecular Biophysics Lab/Science Photo Library/Photo Researchers. *Page 458:* Paul Silverman/Fundamental Photographs. *Figure 18.4:* Courtesy Robert Stodola Fox Chase Research Center. *Page 467:* Bettmann Archive. *Page 477:* Courtesy Linus Pauling Institute of Science and Medicine.

Chapter 19 *Opener:* G. Steisler/FPG International. *Figure 19.6:* K.R. Porter/Photo Researchers. *Page 501:* David Madison/Duomo Photography, Inc. *Page 506:* L. Kolvoord/The Image Works.

Chapter 20 *Opener:* Bob Daemmrich/Stock, Boston. *Figure 20.1(b):* Courtesy Sung Hou Kim/University of California, Berkeley. *Page 526:* Lee D. Simon/Science Source/Photo Researchers. *Figure 20.5(b):* Courtesy S.H. Kim, Duke University Medical Center. *Page 530:* Lennart Nilsson/Boehringer Ingelheim International. *Page 536:* Stanley N. Cohen/Photo Researchers. *Page 539:* David Parker/Science Photo Library/Photo Researchers. *Figure 20.12:* Photo by Willard R. Centerwall, M.D., From *Phenylketonuria,* Frank L. Lyman Ed., Charles C. Thomas, 1963.

Chapter 21 *Opener:* Pete Saloutos/Tony Stone Images/New York, Inc. *Page 550:* Yoav Levy/Phototake. *Page 562:* Charles Gupton/Tony Stone Images/New York, Inc. *Figure 21.10:* Lennart Nilsson/Boehringer Ingleheim International. *Page 563 (bottom):* Hattie Young/Science Photo Library/Photo Researchers. *Page 568:* Alan Goldsmith/The Stock Market. *Page 570:* SIU/Photo Researchers.

Figure Credits

*R*eprinted by Permission of John Wiley & Sons, Inc.

Section 3 Opener Biology, Exploring Life, 2nd edition, Gil Brum, Larry McKane, Gerry Karp, Copyright © 1994 John Wiley & Sons, Inc.

Chapter 1 *Section 1.1:* The Extraordinary Chemistry of Ordinary Things, Carl Snyder, Copyright © 1992 John Wiley & Sons, Inc. *Figure 1.5:* Basic Concepts of Chemistry, Leo Malone, Copyright © 1992 John Wiley & Sons, Inc. *Section 1.8:* Chemistry, An Experimental Science, 2nd edition, George M. Bodner and Harry L Pardue, Copyright © 1995 John Wiley & Sons, Inc.

Chapter 2 *Figure 2.5:* Chemistry, An Experimental Science, 2nd edition, George M. Bodner and Harry L Pardue, Copyright © 1995 John Wiley & Sons, Inc. *Figure 2.6:* Chemistry, An Experimental Science, 2nd edition, George M. Bodner and Harry L Pardue, Copyright © 1995 John Wiley & Sons, Inc.

Chapter 3 *Figure 3.5:* Chemistry, An Experimental Science, 2nd edition, George M. Bodner and Harry L Pardue, Copyright © 1995 John Wiley & Sons, Inc. *Figure 3.6:* The Extraordinary Chemistry of Ordinary Things, Carl Snyder, Copyright © 1992 John Wiley & Sons, Inc. *Figure 3.8:* The Extraordinary Chemistry of Ordinary Things, Carl Snyder, Copyright © 1992 John Wiley & Sons, Inc. *Figure 3.10:* Fundamentals of General, Organic, and Biological Chemistry, 5th edition, John R. Holum, Copyright © 1994 John Wiley & Sons, Inc.

Chapter 4 *Figure 4.2:* Fundamentals of General, Organic, and Biological Chemistry, 5th edition, John R. Holum, Copyright © 1994 John Wiley & Sons, Inc. *Figure 4.3:* Fundamentals of General, Organic, and Biological Chemistry, 5th edition, John R. Holum, Copyright © 1994 John Wiley & Sons, Inc. *Figure 4.4:* Chemistry, An Experimental Science, 2nd edition, George M. Bodner and Harry L Pardue, Copyright © 1995 John Wiley & Sons, Inc.

Chapter 5 *Figure 5.7:* Fundamentals of General, Organic, and Biological Chemistry, 5th edition, John R. Holum, Copyright © 1994 John Wiley & Sons, Inc. *Figure 5.10:* Biochemistry, Donald Voet and Judith Voet, Copyright © 1990 John Wiley & Sons, Inc.

Chapter 6 *Section 6.1:* Chemistry, An Experimental Science, 2nd edition, George M. Bodner and Harry L Pardue, Copyright © 1995 John Wiley & Sons, Inc.

Chapter 7 *Figure 7.1:* The Extraordinary Chemistry of Ordinary Things, Carl Snyder, Copyright © 1992 John Wiley & Sons, Inc. *Section 7.3:* The Extraordinary Chemistry of Ordinary Things, Carl Snyder, Copyright © 1992 John Wiley & Sons, Inc. *Figure 7.4:* The Extraordinary Chemistry of Ordinary Things, Carl Snyder, Copyright © 1992 John Wiley & Sons, Inc. *Figure 7.5:* The Extraordinary Chemistry of Ordinary Things, Carl Snyder, Copyright © 1992 John Wiley & Sons, Inc.

Chapter 8 *Figure 8.6:* Chemistry, An Experimental Science, 2nd edition, George M. Bodner and Harry L Pardue, Copyright © 1995 John Wiley & Sons, Inc.

Chapter 9 *Figure 9.6:* Biology, Exploring Life, 2nd edition, Gil Brum, Larry McKane, Gerry Karp, Copyright © 1994 John Wiley & Sons, Inc.

Chapter 10 *Figure 10.5:* Fundamentals of General, Organic, and Biological Chemistry, 5th edition, John R. Holum, Copyright © 1994 John Wiley & Sons, Inc. *Figure 10.7:* Fundamentals of General, Organic, and Biological Chemistry, 5th edition, John R. Holum, Copyright © 1994 John Wiley & Sons, Inc.

Chapter 11 *Figure 11.1:* Fundamentals of General, Organic, and Biological Chemistry, 5th edition, John R. Holum, Copyright © 1994 John Wiley & Sons, Inc. *Section 11.4:* Fundamentals of General, Organic, and Biological Chemistry, 5th edition, John R. Holum, Copyright © 1994 John Wiley & Sons, Inc.

Chapter 12 *Figure 12.4:* The Extraordinary Chemistry of Ordinary Things, Carl Snyder, Copyright © 1992 John Wiley & Sons, Inc. *Figure 12.5:* Fundamentals of General, Organic, and Biological Chemistry, 5th edition, John R. Holum, Copyright © 1994 John Wiley & Sons, Inc. *Figure 12.11:* Chemistry, The Study of Matter and Its Changes, James E. Brady and John R. Holum, Copyright © 1993 John Wiley & Sons, Inc. *Figure 12.12:* Chemistry, The Study of Matter and Its Changes, James E. Brady and John R. Holum, Copyright © 1993 John Wiley & Sons, Inc. *Figure 12.13:* Chemistry, The Study of Matter and Its Changes, James E. Brady and John R. Holum, Copyright © 1993 John Wiley & Sons, Inc.

Chapter 13 *Figure 13.1:* Chemistry, An Experimental Science, 2nd edition, George M. Bodner and Harry L Pardue, Copyright © 1995 John Wiley & Sons, Inc.

Chapter 15 *Figure 15.2:* Fundamentals of General, Organic, and Biological Chemistry, 5th edition, John R. Holum, Copyright © 1994 John Wiley & Sons, Inc. *Section 15.3:* Fundamentals of General, Organic, and Biological Chemistry, 5th edition, John R. Holum, Copyright © 1994 John Wiley & Sons, Inc. *Figure 15.5:* Biochemistry, Donald Voet and Judith Voet, Copyright © 1990 John Wiley & Sons, Inc. *Figure 15.6:* Biochemistry, Donald Voet and Judith Voet, Copyright © 1990 John Wiley & Sons, Inc. *Figure 15.8:* Biology, Exploring Life, 2nd edition, Gil Brum, Larry McKane, Gerry Karp, Copyright © 1994 John Wiley & Sons, Inc.

Chapter 16 *Section 16.10:* The Extraordinary Chemistry of Ordinary Things, Carl Snyder, Copyright © 1992 John Wiley & Sons, Inc. *Figure 16.4:* Biochemistry, Donald Voet and Judith Voet, Copyright © 1990 John Wiley & Sons, Inc.

Chapter 17 *Figure 17.1:* Biochemistry, Donald Voet and Judith Voet, Copyright © 1990 John Wiley & Sons, Inc. *Figure 17.2:* Biology, Exploring Life, 2nd edition, Gil Brum, Larry McKane, Gerry Karp, Copyright © 1994 John Wiley & Sons, Inc. *Figure 17.4(a):* Copyright © Irving Geis. *Figure 17.4(b):* Fundamentals of General, Organic, and Biological Chemistry, 5th edition, John R. Holum, Copyright © 1994 John Wiley & Sons, Inc. *Figure 17.5:* Copyright © Irving Geis. *Figure 17.6:* Reprinted with permission from John Wiley & Sons, Inc. From Voet and Voet, *Biochemistry,* copyright © 1990. (After Kannan, K. K., Liljas, A., Waara, I., Bergsten, P.-C., Lovgren, S., Strandberg, B., Bengtsson, J., Carlbom, U., Friedborg, K., Jarup, L. and Petef, M., *Cold Spring Harbor Symp. Quant. Biol.* **36,** 221 (1971). *Figure 17.7:* Chemistry, An Experimental Science, 2nd edition, George M. Bodner and Harry L Pardue, Copyright © 1995 John Wiley & Sons, Inc.

Chapter 18 *Figure 18.1(b):* Chemistry, An Experimental Science, 2nd edition, George M. Bodner and Harry L Pardue, Copyright © 1995 John Wiley & Sons, Inc. *Section 18.4:* Fundamentals of General, Organic, and Biological Chemistry, 5th edition, John R. Holum, Copyright © 1994 John Wiley & Sons, Inc. *Figure 18.11:* Fundamentals of General, Organic, and Biological Chemistry, 5th edition, John R. Holum, Copyright © 1994 John Wiley & Sons, Inc.

Chapter 19 *Figure 19.1:* Biology, Exploring Life, 2nd edition, Gil Brum, Larry McKane, Gerry Karp, Copyright © 1994 John Wiley & Sons, Inc. *Section 19.2:* Biology, Exploring Life, 2nd edition, Gil Brum, Larry McKane, Gerry Karp, Copyright © 1994 John Wiley & Sons, Inc. *Figure 19.3:* Fundamentals of General, Organic, and Biological Chemistry, 5th edition, John R. Holum, Copyright © 1994 John Wiley & Sons, Inc. *Figure 19.7:* Biology, Exploring Life, 2nd edition, Gil Brum, Larry McKane, Gerry Karp, Copyright © 1994 John Wiley & Sons, Inc. *Figure 19.8:* Biochemistry, Donald Voet and Judith Voet, Copyright © 1990 John Wiley & Sons, Inc. *Figure 19.10:* Biology, Exploring Life, 2nd edition, Gil Brum, Larry McKane, Gerry Karp, Copyright © 1994 John Wiley & Sons, Inc. *Figure 19.11:* Fundamentals of General, Organic, and Biological Chemistry, 5th edition, John R. Holum, Copyright © 1994 John Wiley & Sons, Inc.

Chapter 20 *Figure 20.3(a, b):* Biochemistry, Donald Voet and Judith Voet, Copyright © 1990 John Wiley & Sons, Inc. *Figure 20.4(a):* Biology, Exploring Life, 2nd edition, Gil Brum, Larry McKane, Gerry Karp, Copyright © 1994 John Wiley & Sons, Inc. *Figure 20.6:* Biology, Exploring Life, 2nd edition, Gil Brum, Larry McKane, Gerry Karp, Copyright © 1994 John Wiley & Sons, Inc. *Figure 20.7:* Biochemistry, Donald Voet and Judith Voet, Copyright © 1990 John Wiley & Sons, Inc.

Figure 20.9: Fundamentals of General, Organic, and Biological Chemistry, 5th edition, John R. Holum, Copyright © 1994 John Wiley & Sons, Inc.

Chapter 21 *Figure 21.1:* Fundamentals of General, Organic, and Biological Chemistry, 5th edition, John R. Holum, Copyright © 1994 John Wiley & Sons, Inc. *Figure 21.3:* The Extraordinary Chemistry of Ordinary Things, Carl Snyder, Copyright © 1992 John Wiley & Sons, Inc. *Figure 21.4:* The Extraordinary Chemistry of Ordinary Things, Carl Snyder, Copyright © 1992 John Wiley & Sons, Inc. *Figure 21.5:* Fundamentals of General, Organic, and Biological Chemistry, 5th edition, John R. Holum, Copyright © 1994 John Wiley & Sons, Inc.

*I*ndex

Page numbers in italics indicate material in a Table. Page numbers in boldface indicate material in a Perspective.

	Family											
	Alkane	**Alkene**	**Alkyne**	**Aromatic**	**Alcohol**	**Ether**	**Amine**	**Aldehyde**	**Ketone**	**Carboxylic Acid**	**Ester**	**Amide**
Specific Example	CH_3CH_3	$CH_2{=}CH_2$	$HC{\equiv}CH$	⌬	CH_3CH_2OH	CH_3OCH_3	CH_3NH_2	$CH_3\overset{O}{\overset{\|}{C}}H$	$CH_3\overset{O}{\overset{\|}{C}}CH_3$	$CH_3\overset{O}{\overset{\|}{C}}OH$	$CH_3\overset{O}{\overset{\|}{C}}OCH_3$	$CH_3\overset{O}{\overset{\|}{C}}NH_2$
IUPAC Name	Ethane	Ethene or Ethylene	Ethyne or Acetylene	Benzene	Ethanol	Dimethyl ether	Methyl-amine	Ethanal	Propanone	Ethanoic Acid	Methyl ethanoate	Ethanamide
General Formula	RH	$RCH{=}CH_2$ $RCH{=}CHR$ $R_2C{=}CHR$ $R_2C{=}CR_2$	$RC{\equiv}CH$ $RC{\equiv}CR$	ArH	ROH	ROR	RNH_2 R_2NH R_3N	$R\overset{O}{\overset{\|}{C}}H$	$R\overset{O}{\overset{\|}{C}}R$	$R\overset{O}{\overset{\|}{C}}OH$	$R\overset{O}{\overset{\|}{C}}OR$	$R\overset{O}{\overset{\|}{C}}NH_2$ $R\overset{O}{\overset{\|}{C}}NHR$ $R\overset{O}{\overset{\|}{C}}NR_2$
Functional Group	C—H and C—C bonds	$>C{=}C<$	$-C{\equiv}C-$	⌬	$-\overset{\|}{\underset{\|}{C}}-OH$	$-\overset{\|}{\underset{\|}{C}}-O-\overset{\|}{\underset{\|}{C}}-$	$-\overset{\|}{\underset{\|}{C}}-\underset{\|}{N}-$	$-\overset{O}{\overset{\|}{C}}-H$	$-\overset{\|}{\underset{\|}{C}}-\overset{O}{\overset{\|}{C}}-\overset{\|}{\underset{\|}{C}}-$	$-\overset{O}{\overset{\|}{C}}-OH$	$-\overset{O}{\overset{\|}{C}}-O-\overset{\|}{\underset{\|}{C}}-$	$-\overset{O}{\overset{\|}{C}}-\underset{\|}{N}-$